Springer Texts in Statistics

Advisors:
George Casella Stephen Fienberg Ingram Olkin

Springer
New York
Berlin
Heidelberg
Hong Kong
London
Milan
Paris
Tokyo

Springer Texts in Statistics

Alfred: Elements of Statistics for the Life and Social Sciences
Berger: An Introduction to Probability and Stochastic Processes
Bilodeau and Brenner: Theory of Multivariate Statistics
Blom: Probability and Statistics: Theory and Applications
Brockwell and Davis: Introduction to Times Series and Forecasting, Second Edition
Chow and Teicher: Probability Theory: Independence, Interchangeability, Martingales, Third Edition
Christensen: Advanced Linear Modeling: Multivariate, Time Series, and Spatial Data; Nonparametric Regression and Response Surface Maximization, Second Edition
Christensen: Log-Linear Models and Logistic Regression, Second Edition
Christensen: Plane Answers to Complex Questions: The Theory of Linear Models, Third Edition
Creighton: A First Course in Probability Models and Statistical Inference
Davis: Statistical Methods for the Analysis of Repeated Measurements
Dean and Voss: Design and Analysis of Experiments
du Toit, Steyn, and Stumpf: Graphical Exploratory Data Analysis
Durrett: Essentials of Stochastic Processes
Edwards: Introduction to Graphical Modelling, Second Edition
Finkelstein and Levin: Statistics for Lawyers
Flury: A First Course in Multivariate Statistics
Jobson: Applied Multivariate Data Analysis, Volume I: Regression and Experimental Design
Jobson: Applied Multivariate Data Analysis, Volume II: Categorical and Multivariate Methods
Kalbfleisch: Probability and Statistical Inference, Volume I: Probability, Second Edition
Kalbfleisch: Probability and Statistical Inference, Volume II: Statistical Inference, Second Edition
Karr: Probability
Keyfitz: Applied Mathematical Demography, Second Edition
Kiefer: Introduction to Statistical Inference
Kokoska and Nevison: Statistical Tables and Formulae
Kulkarni: Modeling, Analysis, Design, and Control of Stochastic Systems
Lange: Applied Probability

E.L. Lehmann George Casella

Theory of Point Estimation

Second Edition

Springer

E.L. Lehmann
Department of Statistics
University of California, Berkeley
Berkeley, CA 94720
USA

George Casella
Department of Statistics
University of Florida
Gainesville, FL 32611
USA

Library of Congress Cataloging-in-Publication Data
Lehmann, E.L. (Erich Leo), 1917-
Theory of point estimation. — 2nd ed./ E.L. Lehmann, George Casella
p. cm. — (Springer texts in statistics)
Includes bibliographical references and index.

1. Fix-point estimation. I. Casella, George. II. Title.
III. Series
QA276.8 .L43 1998
519.5'44—dc21 98-16687

ISBN 978-1-4419-3130-6 Printed on acid-free paper.

9 8 7 6 5 SPIN 11013402

Springer-Verlag is a part of *Springer Science+Business Media*

springeronline.com

To our children

Stephen, Barbara, and Fia
ELL

Benjamin and Sarah
GC

Preface to the Second Edition

Since the publication in 1983 of *Theory of Point Estimation*, much new work has made it desirable to bring out a second edition. The inclusion of the new material has increased the length of the book from 500 to 600 pages; of the approximately 1000 references about 25% have appeared since 1983.

The greatest change has been the addition to the sparse treatment of Bayesian inference in the first edition. This includes the addition of new sections on Equivariant, Hierarchical, and Empirical Bayes, and on their comparisons. Other major additions deal with new developments concerning the information inequality and simultaneous and shrinkage estimation. The Notes at the end of each chapter now provide not only bibliographic and historical material but also introductions to recent development in point estimation and other related topics which, for space reasons, it was not possible to include in the main text. The problem sections also have been greatly expanded. On the other hand, to save space most of the discussion in the first edition on robust estimation (in particular L, M, and R estimators) has been deleted. This topic is the subject of two excellent books by Hampel et al (1986) and Staudte and Sheather (1990). Other than subject matter changes, there have been some minor modifications in the presentation. For example, all of the references are now collected together at the end of the text, examples are listed in a Table of Examples, and equations are references by section and number within a chapter and by chapter, section and number between chapters.

The level of presentation remains the same as that of TPE. Students with a thorough course in theoretical statistics (from texts such as Bickel and Doksum 1977 or Casella and Berger 1990) would be well prepared. The second edition of TPE is a companion volume to "Testing Statistical Hypotheses, Second Edition (TSH2)." Between them, they provide an account of classical statistics from a unified point of view.

Many people contributed to TPE2 with advice, suggestions, proofreading and problem-solving. We are grateful to the efforts of John Kimmel for overseeing this project; to Matt Briggs, Lynn Eberly, Rich Levine and Sam Wu for proofreading and problem solving, to Larry Brown, Anirban DasGupta, Persi Diaconis, Tom DiCiccio, Roger Farrell, Leslaw Gajek, Jim Hobert, Chuck McCulloch, Elias Moreno, Christian Robert, Andrew Rukhin, Bill Strawderman and

Larry Wasserman for discussions and advice on countless topics, and to June Meyermann for transcribing most of TPE to LaTeX. Lastly, we thank Andy Scherrer for repairing the near-fatal hard disk crash and Marty Wells for the almost infinite number of times he provided us with needed references.

E.L. Lehmann
Berkeley, California

George Casella
Ithaca, New York

March 1998

Preface to the First Edition

This book is concerned with point estimation in Euclidean sample spaces. The first four chapters deal with exact (small-sample) theory, and their approach and organization parallel those of the companion volume, *Testing Statistical Hypotheses* (TSH). Optimal estimators are derived according to criteria such as unbiasedness, equivariance, and minimaxity, and the material is organized around these criteria. The principal applications are to exponential and group families, and the systematic discussion of the rich body of (relatively simple) statistical problems that fall under these headings constitutes a second major theme of the book.

A theory of much wider applicability is obtained by adopting a large sample approach. The last two chapters are therefore devoted to large-sample theory, with Chapter 5 providing a fairly elementary introduction to asymptotic concepts and tools. Chapter 6 establishes the asymptotic efficiency, in sufficiently regular cases, of maximum likelihood and related estimators, and of Bayes estimators, and presents a brief introduction to the local asymptotic optimality theory of Hajek and LeCam. Even in these two chapters, however, attention is restricted to Euclidean sample spaces, so that estimation in sequential analysis, stochastic processes, and function spaces, in particular, is not covered.

The text is supplemented by numerous problems. These and references to the literature are collected at the end of each chapter. The literature, particularly when applications are included, is so enormous and spread over the journals of so many countries and so many specialties that complete coverage did not seem feasible. The result is a somewhat inconsistent coverage which, in part, reflects my personal interests and experience.

It is assumed throughout that the reader has a good knowledge of calculus and linear algebra. Most of the book can be read without more advanced mathematics (including the sketch of measure theory which is presented in Section 1.2 for the sake of completeness) if the following conventions are accepted.

1. A central concept is that of an integral such as $\int f\,dP$ or $\int f\,d\mu$. This covers both the discrete and continuous case. In the discrete case $\int f\,dP$ becomes $\Sigma f(x_i)P(x_i)$ where $P(x_i) = P(X = x_i)$ and $\int f\,d\mu$ becomes $\Sigma f(x_i)$. In the continuous case, $\int f\,dP$ and $\int f\,d\mu$ become, respectively, $\int f(x)p(x)\,dx$ and $\int f(x)\,dx$. Little is

lost (except a unified notation and some generality) by always making these substitutions.

2. When specifying a probability distribution P, it is necessary to specify not only the sample space $\mathcal{X}$, but also the class $\mathcal{A}$ of sets over which P is to be defined. In nearly all examples $\mathcal{X}$ will be a Euclidean space and $\mathcal{A}$ a large class of sets, the so-called Borel sets, which in particular includes all open and closed sets. The references to $\mathcal{A}$ can be ignored with practically no loss in the understanding of the statistical aspects.

A forerunner of this book appeared in 1950 in the form of mimeographed lecture notes taken by Colin Blyth during a course I taught at Berkeley; they subsequently provided a text for the course until the stencils gave out. Some sections were later updated by Michael Stuart and Fritz Scholz. Throughout the process of converting this material into a book, I greatly benefited from the support and advice of my wife, Juliet Shaffer. Parts of the manuscript were read by Rudy Beran, Peter Bickel, Colin Blyth, Larry Brown, Fritz Scholz, and Geoff Watson, all of whom suggested many improvements. Sections 6.7 and 6.8 are based on material provided by Peter Bickel and Chuck Stone, respectively. Very special thanks are due to Wei-Yin Loh, who carefully read the complete manuscript at its various stages and checked all the problems. His work led to the correction of innumerable errors and to many other improvements. Finally, I should like to thank Ruth Suzuki for her typing, which by now is legendary, and Sheila Gerber for her expert typing of many last-minute additions and corrections.

E.L. Lehmann
Berkeley, California,

March 1983

Contents

List of Tables

List of Figures

List of Examples

Chapter 1: Preparations

Chapter 2: Unbiasedness

Chapter 3: Equivariance

Chapter 4: Average Risk Optimality

Chapter 5: Minimaxity and Admissibility

Chapter 6: Asymptotic Optimality

Table of Notation

The following notation will be used throughout the book.
We present this list for easy reference.

Quantity	Notation	Comment
Random variable	$X, Y,$	uppercase
Sample space	$\mathcal{X}, \mathcal{Y}$	uppercase script Roman letters
Parameter	θ, λ	lowercase Greek letters
Parameter space	Θ, Ω	uppercase script Greek letters
Realized values (data)	x, y	lowercase
Distribution function (cdf)	$F(\mathbf{x}), F(\mathbf{x}\|\theta), P(\mathbf{x}\|\theta)$ $F_\theta(x), P_\theta, (x)$	continuous or discrete
Density function (pdf)	$f(\mathbf{x}), f(\mathbf{x}\|\theta), p(\mathbf{x}\|\theta)$ $f_\theta(x), P_\theta(x)$	notation is "generic", i.e., don't assume $f(x\|y) = f(x\|z)$
Prior distribution	$\Lambda(\gamma), \Lambda(\gamma\|\lambda)$	
Prior density	$\pi(\gamma), \pi(\gamma\|\lambda)$	may be improper
Probability triple	$(\mathcal{X}, \mathcal{P}, \mathcal{B})$	sample space, probability distribution, and sigma-algebra of sets

Quantity	Notation	Comment
Vector	$\mathbf{h} = (h_1, \ldots, h_n) = \{h_i\}$	boldface signifies vectors
Matrix	$H = \{h_{ij}\} = \|\|h_{ij}\|\|$	uppercase signifies matrices
Special matrices and vectors	I	Identity matrix
	$\mathbf{1}$	vector of ones
	$J = \mathbf{11}'$	matrix of ones
Dot notation	$h_{i\cdot} = \frac{1}{J}\sum_{j=1}^{J} h_{ij}$	average across the dotted subscript
Gradient	$\nabla h(\mathbf{x}) = \left(\frac{\partial}{\partial x_1}h(\mathbf{x}), \ldots, \frac{\partial}{\partial x_n}h(\mathbf{x})\right)$ $= \left\{\frac{\partial}{\partial x_i}h(\mathbf{x})\right\}$	vector of partial derivatives
Hessian	$\nabla\nabla h(\mathbf{x}) = \left\{\frac{\partial^2}{\partial x_i \partial x_j}h(\mathbf{x})\right\}$	matrix of partial second derivatives
Jacobian	$\left\{\frac{\partial}{\partial x_j}h_i(\mathbf{x})\right\}$	matrix of derivatives
Laplacian	$\sum_i \frac{\partial^2}{\partial x_i^2}h(\mathbf{x})$	sum of second derivatives
Euclidean norm	$\|\mathbf{x}\|$	$(\Sigma x_i^2)^{1/2}$
Indicator function	$I_A(x)$, $I(x \in A)$ or $I(x < a)$	equals 1 if $x \in A$, 0 otherwise
Big "Oh," little "oh"	$O(n)$, $o(n)$ or $O_p(n)$, $o_p(n)$	As $n \to \infty$ $\frac{O(n)}{n} \to$ constant, $\frac{o(n)}{n} \to 0$ subscript p denotes *in probability*

CHAPTER 1

Preparations

1 The Problem

Statistics is concerned with the collection of data and with their analysis and interpretation. We shall not consider the problem of data collection in this book but shall take the data as given and ask what they have to tell us. The answer depends not only on the data, on what is being observed, but also on background knowledge of the situation; the latter is formalized in the assumptions with which the analysis is entered. There have, typically, been three principal lines of approach:

Data analysis. Here, the data are analyzed on their own terms, essentially without extraneous assumptions. The principal aim is the organization and summarization of the data in ways that bring out their main features and clarify their underlying structure.

Classical inference and decision theory. The observations are now postulated to be the values taken on by random variables which are assumed to follow a joint probability distribution, P, belonging to some known class $\mathcal{P}$. Frequently, the distributions are indexed by a parameter, say θ (not necessarily real-valued), taking values in a set, Ω, so that

$$\mathcal{P} = \{P_\theta, \theta \in \Omega\}. \tag{1.1}$$

The aim of the analysis is then to specify a plausible value for θ (this is the problem of point estimation), or at least to determine a subset of Ω of which we can plausibly assert that it does, or does not, contain θ (estimation by confidence sets or hypothesis testing). Such a statement about θ can be viewed as a summary of the information provided by the data and may be used as a guide to action.

Bayesian analysis. In this approach, it is assumed in addition that θ is itself a random variable (though unobservable) with a *known* distribution. This prior distribution (specified according to the problem) is modified in light of the data to determine a posterior distribution (the conditional distribution of θ given the data), which summarizes what can be said about θ on the basis of the assumptions made *and the data.*

These three methods of approach permit increasingly strong conclusions, but they do so at the price of assumptions which are correspondingly more detailed and possibly less reliable. It is often desirable to use different formulations in conjunction; for example, by planning a study (e.g., determining sample size) under rather detailed assumptions but performing the analysis under a weaker set which appears more trustworthy. In practice, it is often useful to model a problem in a number of different ways. One may then be satisfied if there is reasonable agreement among the conclusions; in the contrary case, a closer examination of the different sets of assumptions will be indicated.

In this book, Chapters 2, 3, and 5 will be primarily concerned with the second formulation, Chapter 4 with the third. Chapter 6 considers a large-sample treatment of both. (A book-length treatment of the first formulation is Tukey's classic *Exploratory Data Analysis*, or the more recent book by Hoaglin, Mosteller, and Tukey 1985, which includes the interesting approach of Diaconis 1985.) Throughout the book we shall try to specify what is meant by a "best" statistical procedure for a given problem and to develop methods for determining such procedures. Ideally, this would involve a formal decision-theoretic evaluation of the problem resulting in an optimal procedure.

Unfortunately, there are difficulties with this approach, partially caused by the fact that there is no unique, convincing definition of optimality. Compounding this lack of consensus about optimality criteria is that there is also no consensus about the evaluation of such criteria. For example, even if it is agreed that squared error loss is a reasonable criterion, the method of evaluation, be it Bayesian, frequentist (the classical approach of averaging over repeated experiments), or conditional, must then be agreed upon.

Perhaps even more serious is the fact that the optimal procedure and its properties may depend very heavily on the precise nature of the assumed probability model (1.1), which often rests on rather flimsy foundations. It therefore becomes important to consider the *robustness* of the proposed solution under deviations from the model. Some aspects of robustness, from both Bayesian and frequentist perspectives, will be taken up in Chapters 4 and 5.

The discussion so far has been quite general; let us now specialize to point estimation. In terms of the model (1.1), suppose that g is a real-valued function defined over Ω and that we would like to know the value of $g(\theta)$ (which may, of course, be θ itself). Unfortunately, θ, and hence $g(\theta)$, is unknown. However, the data can be used to obtain an estimate of $g(\theta)$, a value that one hopes will be close to $g(\theta)$.

Point estimation is one of the most common forms of statistical inference. One measures a physical quantity in order to estimate its value; surveys are conducted to estimate the proportion of voters favoring a candidate or viewers watching a television program; agricultural experiments are carried out to estimate the effect of a new fertilizer, and clinical experiments to estimate the improved life expectancy or cure rate resulting from a medical treatment. As a prototype of such an estimation problem, consider the determination of an unknown quantity by measuring it.

Example 1.1 The measurement problem. A number of measurements are taken of some quantity, for example, a distance (or temperature), in order to obtain an estimate of the quantity θ being measured. If the n measured values are $x_1, \ldots, x_n$, a common recommendation is to estimate θ by their mean

$$\bar{x} = \frac{(x_1 + \cdots + x_n)}{n}.$$

The idea of averaging a number of observations to obtain a more precise value seems so commonplace today that it is difficult to realize it has not always been in use. It appears to have been introduced only toward the end of the seventeenth century (see Plackett, 1958). But why should the observations be combined in just this way? The following are two properties of the mean, which were used in early attempts to justify this procedure.

(i) An appealing approximation to the true value being measured is the value a, for which the sum of squared difference $\Sigma(x_i - a)^2$ is a minimum. That this *least squares* estimate of θ is $\bar{x}$ is seen from the identity

$$\Sigma(x_i - a)^2 = \Sigma(x_i - \bar{x})^2 + n(\bar{x} - a)^2, \tag{1.2}$$

since the first term on the right side does not involve a and the second term is minimized by $a = \bar{x}$. (For the history of least squares, see Eisenhart 1964, Plackett 1972, Harter 1974–1976, and Stigler 1981. Least squares estimation will be discussed in a more general setting in §3.4.)

(ii) The least squares estimate defined in (i) is the value minimizing the sum of the squared residuals, the residuals being the differences between the observations x_i and the estimated value. Another approach is to ask for the value a for which the sum of the residuals is zero, so that the positive and negative residuals are in balance. The condition on a is

$$\Sigma(x_i - a) = 0, \tag{1.3}$$

and this again immediately leads to $a = \bar{x}$. (That the two conditions lead to the same answer is, of course, obvious since (1.3) expresses that the derivative of (1.2) with respect to a is zero.)

These two principles clearly belong to the first (data analytic) level mentioned at the beginning of the section. They derive the mean as a reasonable descriptive measure of the center of the observations, but they cannot justify $\bar{x}$ as an estimate of the true value θ since no explicit assumption has been made connecting the observations x_i with θ. To establish such a connection, let us now assume that the x_i are the observed values of n independent random variables which have a common distribution depending on θ. Eisenhart (1964) attributes the crucial step of introducing such probability models for this purpose to Simpson (1755).

More specifically, we shall assume that $X_i = \theta + U_i$, where the measurement error U_i is distributed according to a distribution F symmetric about 0 so that the X_i are symmetrically distributed about θ with distribution

$$P(X_i \le x) = F(x - \theta). \tag{1.4}$$

In terms of this model, can we now justify the idea that the mean provides a more precise value than a single observation? The second of the approaches mentioned at the beginning of the section (classical inference) suggests the following kind of consideration.

If the X's are independent and have a finite variance σ^2, the variance of the mean $\bar{X}$ is σ^2/n; the expected squared difference between $\bar{X}$ and θ is therefore only $1/n$ of what it is for a single observation. However, if the X's have a Cauchy distribution, the distribution of $\bar{X}$ is the same as that of a single X_i (Problem 1.6), so that nothing is gained by taking several measurements and then averaging them. Whether $\bar{X}$ is a reasonable estimator of θ thus depends on the nature of the X_i. ||

This example suggests that the formalization of an estimation problem involves two basic ingredients:

(a) A real-valued function g defined over a parameter space Ω, whose value at θ is to be estimated; we shall call $g(\theta)$ the *estimand*. [In Example 1.1, $g(\theta) = \theta$.]

(b) A *random observable* X (typically vector-valued) taking on values in a sample space $\mathcal{X}$ according to a distribution P_θ, which is known to belong to a family $\mathcal{P}$ as stated in (1.1). [In Example 1.1, $X = (X_1, \ldots, X_n)$, where the X_i are independently, identically distributed (iid) and their distribution is given by (1.4). The observed value x of X constitutes the *data*.]

The problem is the determination of a suitable *estimator*.

Definition 1.2 An *estimator* is a real-valued function δ defined over the sample space. It is used to estimate an *estimand*, $g(\theta)$, a real-valued function of the parameter.

Of course, it is hoped that $\delta(X)$ will tend to be close to the unknown $g(\theta)$, but such a requirement is not part of the formal definition of an estimator. The value $\delta(x)$ taken on by $\delta(X)$ for the observed value x of X is the *estimate* of $g(\theta)$, which will be our "educated guess" for the unknown value.

One could adopt a slightly more restrictive definition than Definition 1.2. In applications, it is often desirable to restrict δ to possible values of $g(\theta)$, for example, to be positive when g takes on only positive values, to be integer-valued when g is, and so on. For the moment, however, it is more convenient not to impose this additional restriction.

The estimator δ is to be close to $g(\theta)$, and since $\delta(X)$ is a random variable, we shall interpret this to mean that it will be close on the average. To make this requirement precise, it is necessary to specify a measure of the average closeness of (or distance from) an estimator to $g(\theta)$. Examples of such measures are

$$P(|\delta(X) - g(\theta)| < c) \quad \text{for some} \quad c > 0 \tag{1.5}$$

and

$$E|\delta(X) - g(\theta)|^p \quad \text{for some} \quad p > 0. \tag{1.6}$$

(Of these, we want the first to be large and the second to be small.) If g and δ take on only positive values, one may be interested in

$$E\left|\frac{\delta(X)}{\delta(\theta)} - 1\right|^p,$$

which suggests generalizing (1.6) to

$$\kappa(\theta)E|\delta(X) - g(\theta)|^p. \tag{1.7}$$

Quite generally, suppose that the consequences of estimating $g(\theta)$ by a value d are measured by $L(\theta, d)$. Of the *loss function* L, we shall assume that

$$L(\theta, d) \geq 0 \quad \text{for all} \quad \theta, d \tag{1.8}$$

and

$$L[\theta, g(\theta)] = 0 \quad \text{for all} \quad \theta, \tag{1.9}$$

so that the loss is zero when the correct value is estimated. The accuracy, or rather inaccuracy, of an estimator δ is then measured by the *risk function*

$$R(\theta, \delta) = E_\theta\{L[\theta, \delta(X)]\}, \tag{1.10}$$

the long-term average loss resulting from the use of δ. One would like to find a δ which minimizes the risk for all values of θ.

As stated, this problem has no solution. For, by (1.9), it is possible to reduce the risk at any given point θ_0 to zero by making $\delta(x)$ equal to $g(\theta_0)$ for all x. There thus exists no *uniformly best* estimator, that is, no estimator which simultaneously minimizes the risk for all values of θ, except in the trivial case that $g(\theta)$ is constant.

One way of avoiding this difficulty is to restrict the class of estimators by ruling out estimators that too strongly favor one or more values of θ at the cost of neglecting other possible values. This can be achieved by requiring the estimator to satisfy some condition which enforces a certain degree of impartiality. One such condition requires that the *bias* $E_\theta[\delta(X)] - g(\theta)$, sometimes called the systematic error, of the estimator δ be zero, that is, that

$$E_\theta[\delta(X)] = g(\theta) \quad \text{for all} \quad \theta \in \Omega. \tag{1.11}$$

This condition of *unbiasedness* ensures that, in the long run, the amounts by which δ over- and underestimates $g(\theta)$ will balance, so that the estimated value will be correct "on the average." A somewhat similar condition is obtained by considering not the amount but only the frequency of over- and underestimation. This leads to the condition

$$P_\theta[\delta(X) < g(\theta)] = P_\theta[\delta(X) > g(\theta)] \tag{1.12}$$

or slightly more generally to the requirement that $g(\theta)$ be a median of $\delta(X)$ for all values of θ. To distinguish it from this condition of *median-unbiasedness*, (1.11) is called *mean-unbiasedness* if there is a possibility of confusion.

Mean-unbiased estimators, due to Gauss and perhaps the most classical of all frequentist constructions, are treated in Chapter 2. There, we will also consider performance assessments that naturally arise from unbiasedness considerations. [A more general unbiasedness concept, of which (1.11) and (1.12) are special cases, will be discussed in Section 3.1.]

A different impartiality condition can be formulated when symmetries are present in a problem. It is then natural to require a corresponding symmetry to hold for the estimator. The resulting condition of *equivariance* will be explored in Chapter 3 and will also play a role in the succeeding chapters.

In many important situations, unbiasedness and equivariance lead to estimators that are uniformly best among the estimators satisfying these restrictions. Nevertheless, the applicability of both conditions is limited. There is an alternative approach which is more generally applicable. Instead of seeking an estimator which minimizes the risk uniformly in θ, one can more modestly ask that the risk function be low only in some overall sense. Two natural global measures of the size of the risk are the average

$$(1.13)\qquad \int R(\theta, \delta)w(\theta)d\theta$$

for some weight function w and the maximum of the risk function

$$(1.14)\qquad \sup_{\Omega} R(\theta, \delta).$$

The estimator minimizing (1.13) (discussed in Chapter 4) formally coincides with the Bayes estimator when θ is assumed to be a random variable with probability density w. Minimizing (1.14) leads to the *minimax* estimator, which will be considered in Chapter 5.

The formulation of an estimation problem in a concrete situation along the lines described in this chapter requires specification of the probability model (1.1) and of a measure of inaccuracy $L(\theta, d)$. In the measurement problem of Example 1.1 and its generalizations to linear models, it is frequently reasonable to assume that the measurement errors are approximately normally distributed. In other situations, the assumptions underlying a binomial or Poisson distribution may be appropriate. Thus, knowledge of the circumstances and previous experience with similar situations will often suggest a particular parametric family $\mathcal{P}$ of distributions. If such information is not available, one may instead adopt a nonparametric model, which requires only very general assumptions such as independence or symmetry but does not lead to a particular parametric family of distributions. As a compromise between these two approaches, one may be willing to assume that the true distribution, though not exactly following a particular parametric form, lies within a stated distance of some parametric family. For a theory of such neighborhood models see, for example, Huber (1981) or TSH2, Section 9.3.

The choice of an appropriate model requires judgment and utilizes experience; it is also affected by considerations of convenience. Analogous considerations for choice of the loss function L appear to be much more difficult. The most common fate of a point estimate (for example, of the distance of a star or the success probability of an operation) is to wind up in a research report or paper. It is likely to be used on different occasions and in various settings for a variety of purposes which cannot be foreseen at the time the estimate is made. Under these circumstances, one wants the estimator to be accurate, but just what measure of accuracy should be used is fairly arbitrary.

This was recognized very clearly by Laplace (1820) and Gauss (1821), who compared the estimation of an unknown quantity, on the basis of observations with random errors, with a game of chance and the error in the estimated value with the loss resulting from such a game. Gauss proposed the square of the error as a measure of loss or inaccuracy. Should someone object to this specification as arbitrary, he writes, he is in complete agreement. He defends his choice by an appeal to mathematical simplicity and convenience. Among the infinite variety of possible functions for the purpose, the square is the simplest and is therefore preferable.

When estimates are used to make definite decisions (for example, to determine the amount of medication to be given a patient or the size of an order that a store should place for some goods), it is sometimes possible to specify the loss function by the consequences of various errors in the estimate. A general discussion of the distinction between inference and decision problems is given by Blyth (1970) and Barnett (1982).

Actually, it turns out that much of the general theory does not require a detailed specification of the loss function but applies to large classes of such functions, in particular to loss functions $L(\theta, d)$, which are convex in d. [For example, this includes (1.7) with $p \geq 1$ but not with $p < 1$. It does not include (1.5)]. We shall develop here the theory for suitably general classes of loss functions whenever the cost in complexity is not too high. However, in applications to specific examples — and these form a large part of the subject — the choice of squared error as loss has the twofold advantage of ease of computation and of leading to estimators that can be obtained explicitly. For these reasons, in the examples we shall typically take the loss to be squared error.

Theoretical statistics builds on many different branches of mathematics, from set theory and algebra to analysis and probability. In this chapter, we will present an overview of some of the most relevant topics needed for the statistical theory to follow.

2 Measure Theory and Integration

A convenient framework for theoretical statistics is measure theory in abstract spaces. The present section will sketch (without proofs) some of the principal concepts, results, and notational conventions of this theory. Such a sketch should provide sufficient background for a comfortable understanding of the ideas and results and the essentials of most of the proofs in this book. A fuller account of measure theory can be found in many standard books, for example, Halmos (1950), Rudin (1966), Dudley (1989), and Billingsley (1995).

The most natural example of a "measure" is that of the length, area, or volume of sets in one-, two-, or three-dimensional Euclidean space. As in these special cases, a measure assigns non-negative (not necessarily finite) values to sets in some space $\mathcal{X}$. A measure μ is thus a set function; the value it assigns to a set A will be denoted by $\mu(A)$.

In generalization of the properties of length, area, and volume, a measure will be required to be *additive*, that is, to satisfy

(2.1) $$\mu(A \cup B) = \mu(A) + \mu(B) \quad \text{when } A, B \text{ are disjoint,}$$

where $A \cup B$ denotes the union of A and B. From (2.1), it follows immediately by induction that additivity extends to any finite union of disjoint sets. The measures with which we shall be concerned will be required to satisfy the stronger condition of *sigma-additivity*, namely that

(2.2) $$\mu\left(\bigcup_{i=1}^{\infty} A_i\right) = \sum_{i=1}^{\infty} \mu(A_i)$$

for any countable collection of disjoint sets.

The domain over which a measure μ is defined is a class of subsets of $\mathcal{X}$. It would seem easiest to assume that this is the class of all subsets of $\mathcal{X}$. Unfortunately, it turns out that typically it is not possible to give a satisfactory definition of the measures of interest for all subsets of $\mathcal{X}$ in such a way that (2.2) holds. [Such a negative statement holds in particular for length, area, and volume (see, for example, Halmos (1950), p. 70) but not for the measure μ of Example 2.1 below.] It is therefore necessary to restrict the definition of μ to a suitable class of subsets of $\mathcal{X}$. This class should contain the whole space $\mathcal{X}$ as a member, and for any set A also its *complement* $\mathcal{X} - A$. In view of (2.2), it should also contain the union of any countable collection of sets of the class. A class of sets satisfying these conditions is called a *σ-field* or *σ-algebra*. It is easy to see that if $A_1, A_2, \ldots$ are members of a σ-field $\mathcal{A}$, then so are their union and intersection (Problem 2.1).

If $\mathcal{A}$ is a σ-field of subsets of a space $\mathcal{X}$, then $(\mathcal{X}, \mathcal{A})$ is said to be a *measurable space* and the sets A of $\mathcal{A}$ to be *measurable*. A *measure* μ is a nonnegative set function defined over a σ-field $\mathcal{A}$ and satisfying (2.2). If μ is a measure defined over a measurable space $(\mathcal{X}, \mathcal{A})$, the triple $(\mathcal{X}, \mathcal{A}, \mu)$ is called a *measure space.*

A measure is *σ-finite* if there exist sets A_i in $\mathcal{A}$ whose union is $\mathcal{X}$ and such that $\mu(A_i) < \infty$. All measures with which we shall be concerned in this book are σ-finite, and we shall therefore use the term *measure* to mean a σ-finite measure.

The following are two important examples of measure spaces.

Example 2.1 Counting measure. Let $\mathcal{X}$ be countable and $\mathcal{A}$ the class of all subsets of $\mathcal{X}$. For any A in $\mathcal{A}$, let $\mu(A)$ be the number of points of A if A is finite, and $\mu(A) = \infty$ otherwise. This measure μ is called *counting measure.* That μ is σ-finite is obvious. ||

Example 2.2 Lebesgue measure. Let $\mathcal{X}$ be n-dimensional Euclidean space E_n, and let $\mathcal{A}$ be the smallest σ-field containing all open rectangles

(2.3) $$A = \{(x_1, \ldots, x_n) : a_i < x_i < b_i\}, \quad -\infty < a_i < b_i < \infty.$$

We shall then say that $(\mathcal{X}, \mathcal{A})$ is *Euclidean.* The members of $\mathcal{A}$ are called *Borel sets.* This is a very large class which contains, among others, all open and all closed subsets of $\mathcal{X}$. There exists a (unique) measure μ, defined over $\mathcal{A}$, which assigns to (2.3) the measure

(2.4) $$\mu(A) = (b_1 - a_1) \cdots (b_n - a_n),$$

that is, its volume; μ is called *Lebesgue measure.* ||

The intuitive meaning of measure suggests that any subset of a set of measure zero should again have measure zero. If $(\mathcal{X}, \mathcal{A}, \mu)$ is a measure space, it may, however, happen that a subset of a set in $\mathcal{A}$ which has measure zero is not in $\mathcal{A}$ and hence not measurable. This difficulty can be remedied by the process of completion. Consider the class $\mathcal{B}$ of all sets $B = A \cup C$ where A is in $\mathcal{A}$ and C is a subset of a set in $\mathcal{A}$ having measure zero. Then, $\mathcal{B}$ is a σ-field (Problem 2.7). If μ' is defined over $\mathcal{B}$ by $\mu'(B) = \mu(A)$, μ' agrees with μ over $\mathcal{A}$, and $(\mathcal{X}, \mathcal{B}, \mu')$ is called the *completion* of the measure space $(\mathcal{X}, \mathcal{A}, \mu)$.

When the process of completion is applied to Example 2.1 so that $\mathcal{X}$ is Euclidean and $\mathcal{A}$ is the class of Borel sets, the resulting larger class $\mathcal{B}$ is the class of Lebesgue measurable sets. The measure μ' defined over $\mathcal{B}$, which agrees with Lebesgue measure over the Borel sets, is also called Lebesgue measure.

A third principal concept needed in addition to σ-field and measure is that of the integral of a real-valued function f with respect to a measure μ. However, before defining this integral, it is necessary to specify a suitable class of functions f. This will be done in three steps.

First, consider the class of real-valued functions s called *simple,* which take on only a finite number of values, say $a_1, \ldots, a_m$, and for which the sets

$$A_i = \{x : s(x) = a_i\} \tag{2.5}$$

belong to $\mathcal{A}$. An important special case of a simple function is the *indicator* I_A of a set A in $\mathcal{A}$, defined by

$$I_A(x) = I(x \in A) = \begin{cases} 1 & \text{if } x \in A \\ 0 & \text{if } x \notin A. \end{cases} \tag{2.6}$$

If the set A is an interval, for example $(a, b]$, the indicator function of the interval may be written in the alternate form $I(a < x \leq b)$.

Second, let $s_1, s_2, \ldots$ be a nondecreasing sequence of non-negative simple functions and let

$$f(x) = \lim_{n \to \infty} s_n(x). \tag{2.7}$$

Note that this limit exists since for every x, the sequence $s_1(x), s_2(x), \ldots$ is nondecreasing but that $f(x)$ may be infinite. A function with domain $\mathcal{X}$ and range $[0, \infty)$, that is, non-negative and finite valued, will be called $\mathcal{A}$-measurable or, for short, *measurable* if there exists a nondecreasing sequence of non-negative simple functions such that(2.7) holds for all $x \in \mathcal{X}$.

Third, for an arbitrary function f, define its *positive and negative part* by

$$f^+(x) = \max(f(x), 0), \quad f^-(x) = -\min(f(x), 0),$$

so that f^+ and f^- are both non-negative and

$$f = f^+ - f^-.$$

Then a function with domain $\mathcal{X}$ and range $(-\infty, \infty)$ will be called *measurable* if both its positive and its negative parts are measurable. The measurable functions constitute a very large class which has a simple alternative characterization.

It can be shown that a real-valued function f is $\mathcal{A}$-measurable if and only if, for every Borel set B on the real line, the set

$$\{x : f(x) \in B\}$$

is in $\mathcal{A}$. If follows from the definition of Borel sets that it is enough to check that $\{x : f(x) < b\}$ is in $\mathcal{A}$ for every b. This shows in particular that if $(\mathcal{X}, \mathcal{A})$ is Euclidean and f continuous, then f is measurable. As another important class, consider functions taking on a countable number of values. If f takes on distinct values $a_1, a_2, \ldots$ on sets $A_1, A_2, \ldots$, it is measurable if and only if $A_i \in \mathcal{A}$ for all i.

The integral can now be defined in three corresponding steps.

(i) For a non-negative simple function s taking on values a_i on the sets A_i, define

$$\int s\,d\mu = \Sigma a_i \mu(A_i), \tag{2.8}$$

where $a\mu(A)$ is to be taken as zero when $a = 0$ and $\mu(A) = \infty$.

(ii) For a non-negative measurable function f given by (2.7), define

$$\int f d\mu = \lim_{n\to\infty} \int s_n d\mu. \tag{2.9}$$

Here, the limit on the right side exists since the fact that the functions s_n are nondecreasing implies the same for the numbers $\int s_n d\mu$. The definition (2.9) is meaningful because it can be shown that if $\{s_n\}$ and $\{s_n'\}$ are two nondecreasing sequences with the same limit function, their integrals also will have the same limit. Thus, the value of $\int f d\mu$ is independent of the particular sequence used in (2.7).

The definitions (2.8) and (2.9) do not preclude the possibility that $\int s\,d\mu$ or $\int f d\mu$ is infinite. A non-negative measurable function is *integrable* (with respect to μ) if $\int f d\mu < \infty$.

(iii) An arbitrary measurable function f is said to be integrable if its positive and negative parts are integrable, and its integral is then defined by

$$\int f d\mu = \int f^+ d\mu - \int f^- d\mu. \tag{2.10}$$

Important special cases of this definition are obtained by taking, for μ, the measures defined in Examples 2.1 and 2.2.

Example 2.3 Continuation of Example 2.1. If $\mathcal{X} = \{x_1, x_2, \ldots\}$ and μ is counting measure, it is easily seen from (2.8) through (2.10) that

$$\int f d\mu = \Sigma f(x_i).$$ ||

Example 2.4 Continuation of Example 2.2. If μ is Lebesgue measure, then $\int f d\mu$ exists whenever the Riemann integral (the integral taught in calculus courses) exists and the two agree. However, the integral defined in (2.8) through (2.10) exists for many functions for which the Riemann integral is not defined. A simple example is the function f for which $f(x) = 1$ or 0, as x is rational or irrational. It follows from (2.22) below that the integral of f with respect to Lebesgue measure is zero; on the other hand, f is not Riemann integrable (Problem 2.11). ||

In analogy with the customary notation for the Riemann integral, it will frequently be convenient to write the integral (2.10) as $\int f(x)d\mu(x)$. This is especially true when f is given by an explicit formula.

The integral defined above has the properties one would expect of it. In particular, for any real numbers $c_1, \ldots, c_m$ and any integrable functions $f_1, \ldots, f_m$, $\Sigma c_i f_i$ is also integrable and

$$\int (\Sigma c_i f_i)\, d\mu = \Sigma c_i \int f_i d\mu. \tag{2.11}$$

Also, if f is measurable and g integrable and if $0 \leq f \leq g$, then f is also integrable, and

$$\int f\, d\mu \leq \int g\, d\mu. \tag{2.12}$$

We shall often be dealing with statements that hold except on a set of measure zero. If a statement holds for all x in $\mathcal{X} - N$ where $\mu(N) = 0$, the statement is said to hold a.e. (almost everywhere) μ (or a.e. if the measure μ is clear from the context).

It is sometimes required to know when $f(x) = \lim f_n(x)$ or more generally when

$$f(x) = \lim f_n(x) \quad (\text{a.e. } \mu) \tag{2.13}$$

implies that

$$\int f d\mu = \lim \int f_n d\mu. \tag{2.14}$$

Here is a sufficient condition.

Theorem 2.5 (Dominated Convergence) *If the f_n are measurable and satisfy (2.13), and if there exists an integrable function g such that*

$$|f_n(x)| \leq g(x) \quad \textit{for all} \quad x, \tag{2.15}$$

then the f_n and f are integrable and (2.14) holds.

The following is another useful result concerning integrals of sequences of functions.

Lemma 2.6 (Fatou) *If $\{f_n\}$ is a sequence of non-negative measurable functions, then*

$$\int \left(\liminf_{n\to\infty} f_n\right) d\mu \leq \liminf_{n\to\infty} \int f_n d\mu \tag{2.16}$$

with the reverse inequality holding for limsup.

Recall that the *liminf* and *limsup* of a sequence of numbers are, respectively, the smallest and largest limit points that can be obtained through subsequences. See Problems 2.5 and 2.6.

As a last extension of the concept of integral, define

$$\int_A f d\mu = \int I_A f d\mu \tag{2.17}$$

when the integral on the right exists. It follows in particular from (2.8) and (2.17) that

$$\int_A d\mu = \mu(A). \tag{2.18}$$

Obviously such properties as (2.11) and (2.12) continue to hold when $\int$ is replaced by $\int_A$.

It is often useful to know under what conditions an integrable function f satisfies

$$\int_A f\,d\mu = 0. \tag{2.19}$$

This will clearly be the case when either

$$f = 0 \quad \text{on} \quad A \tag{2.20}$$

or

$$\mu(A) = 0. \tag{2.21}$$

More generally, it will be the case whenever

$$f = 0 \quad \text{a.e. on } A, \tag{2.22}$$

that is, f is zero except on a subset of A having measure zero.

Conversely, if f is a.e. non-negative on A,

$$\int_A f\,d\mu = 0 \Rightarrow f = 0 \quad \text{a.e. on } A, \tag{2.23}$$

and if f is a.e. positive on A, then

$$\int_A f\,d\mu = 0 \Rightarrow \mu(A) = 0. \tag{2.24}$$

Note that, as a special case of (2.22), if f and g are integrable functions differing only on a set of measure zero, that is, if $f = g$ (a.e. μ), then

$$\int f\,d\mu = \int g\,d\mu.$$

It is a consequence that functions can never be determined by their integrals uniquely but at most up to sets of measure zero.

For a non-negative integrable function f, let us now consider

$$\nu(A) = \int_A f\,d\mu \tag{2.25}$$

as a set function defined over $\mathcal{A}$. Then ν is non-negative, σ-finite, and σ-additive and hence a measure over $(\mathcal{X}, \mathcal{A})$.

If μ and ν are two measures defined over the same measurable space $(\mathcal{X}, \mathcal{A})$, it is a question of central importance whether there exists a function f such that (2.25) holds for all $A \in \mathcal{A}$. By (2.21), a necessary condition for such a representation is clearly that

$$\mu(A) = 0 \Rightarrow \nu(A) = 0. \tag{2.26}$$

When (2.26) holds, ν is said to be *absolutely continuous* with respect to μ. It is a surprising and basic fact known as the *Radon-Nikodym* theorem that (2.26) is not only necessary but also sufficient for the existence of a function f satisfying (2.25) for all $A \in \mathcal{A}$. The resulting function f is called the *Radon-Nikodym derivative* of ν with respect to μ. This f is not unique because it can be changed on a set of μ-measure zero without affecting the integrals (2.25). However, it is *unique a.e.* μ

in the sense that if g is any other integrable function satisfying (2.25), then $f = g$ (a.e. μ). It is a useful consequence of this result that

$$\int_A f\,d\mu = 0 \quad \text{for all} \quad A \in \mathcal{A}$$

implies that $f = 0$ (a.e. μ).

The last theorem on integration we require is a form of Fubini's theorem which essentially states that in a repeated integral of a non-negative function, the order of integration is immaterial. To make this statement precise, define the Cartesian product $A \times B$ of any two sets A, B as the set of all ordered pairs (a, b) with $a \in A, b \in B$. Let $(\mathcal{X}, \mathcal{A}, \mu)$ and $(\mathcal{Y}, \mathcal{B}, \nu)$ be two measure spaces, and define $\mathcal{A} \times \mathcal{B}$ to be the smallest σ-field containing all sets $A \times B$ with $A \in \mathcal{A}$ and $B \in \mathcal{B}$. Then there exists a unique measure λ over $\mathcal{A} \times \mathcal{B}$ which to any product set $A \times B$ assigns the measure $\mu(A) \cdot \nu(B)$. The measure λ is called the *product measure* of μ and ν and is denoted by $\mu \times \nu$.

Example 2.7 Borel sets. If $\mathcal{X}$ and $\mathcal{Y}$ are Euclidean spaces E_m and E_n and $\mathcal{A}$ and $\mathcal{B}$ the σ-fields of Borel sets of $\mathcal{X}$ and $\mathcal{Y}$ respectively, then $\mathcal{X} \times \mathcal{Y}$ is Euclidean space E_{m+n}, and $\mathcal{A} \times \mathcal{B}$ is the class of Borel sets of $\mathcal{X} \times \mathcal{Y}$. If, in addition, μ and ν are Lebesgue measure on $(\mathcal{X}, \mathcal{A})$ and $(\mathcal{Y}, \mathcal{B})$, then $\mu \times \nu$ is Lebesgue measure on $(\mathcal{X} \times \mathcal{Y}, \mathcal{A} \times \mathcal{B})$. ‖

An integral with respect to a product measure generalizes the concept of a double integral. The following theorem, which is one version of Fubini's theorem, states conditions under which a double integral is equal to a repeated integral and under which it is permitted to change the order of integration in a repeated integral.

Theorem 2.8 (Fubini) Let $(\mathcal{X}, \mathcal{A}, \mu)$ and $(\mathcal{Y}, \mathcal{B}, \nu)$ be measure spaces and let f be a non-negative $\mathcal{A} \times \mathcal{B}$-measurable function defined on $\mathcal{X} \times \mathcal{Y}$.

Then

$$\text{(2.27)} \qquad \int_{\mathcal{X}} \left[\int_{\mathcal{Y}} f(x, y) d\nu(y) \right] d\mu(x) = \int_{\mathcal{Y}} \left[\int_{\mathcal{X}} f(x, y) d\mu(x) \right] d\nu(y) = \int_{\mathcal{X}\times\mathcal{Y}} f\,d(\mu \times \nu).$$

Here, the first term is the repeated integral in which f is first integrated for fixed x with respect to ν, and then the result with respect to μ. The inner integrals of the first two terms in (2.27) are, of course, not defined unless $f(x, y)$, for fixed values of either variable, is a measurable function of the other. Fortunately, under the assumptions of the theorem, this is always the case. Similarly, existence of the outer integrals requires the inner integrals to be measurable functions of the variable that has not been integrated. This condition is also satisfied.

3 Probability Theory

For work in statistics, the most important application of measure theory is its specialization to probability theory. A measure P defined over a measure space

$(\mathcal{X}, \mathcal{A})$ satisfying

$$P(\mathcal{X}) = 1 \tag{3.1}$$

is a *probability measure* (or *probability distribution*), and the value $P(A)$ it assigns to A is the *probability* of A. If P is absolutely continuous with respect to a measure μ with Radon-Nikodym derivative, p, so that

$$P(A) = \int_A p\, d\mu, \tag{3.2}$$

p is called the *probability density* of P with respect to μ. Such densities are, of course, determined only up to sets of μ-measure zero.

We shall be concerned only with situations in which $\mathcal{X}$ is Euclidean, and typically the distributions will either be discrete (in which case μ can be taken to be counting measure) or absolutely continuous with respect to Lebesgue measure.

Statistical problems are concerned not with single probability distributions but with families of such distributions

$$\mathcal{P} = \{P_\theta, \theta \in \Omega\} \tag{3.3}$$

defined over a common measurable space $(\mathcal{X}, \mathcal{A})$. When all the distributions of $\mathcal{P}$ are absolutely continuous with respect to a common measure μ, as will usually be the case, the family $\mathcal{P}$ is said to be *dominated* (by μ).

Most of the examples with which we shall deal belong to one or the other of the following two cases.

(i) *The discrete case.* Here, $\mathcal{X}$ is a countable set, $\mathcal{A}$ is the class of subsets of $\mathcal{X}$, and the distributions of $\mathcal{P}$ are dominated by counting measure.

(ii) *The absolutely continuous case.* Here, $\mathcal{X}$ is a Borel subset of a Euclidean space, $\mathcal{A}$ is the class of Borel subsets of $\mathcal{X}$, and the distributions of $\mathcal{P}$ are dominated by Lebesgue measure over $(\mathcal{X}, \mathcal{A})$.

It is one of the advantages of the general approach of this section that it includes both these cases, as well as mixed situations such as those arising with censored data (see Problem 3.8).

When dealing with a family $\mathcal{P}$ of distributions, the most relevant null-set concept is that of a $\mathcal{P}$-*null set*, that is, of a set N satisfying

$$P(N) = 0 \quad \text{for all} \quad P \in \mathcal{P}. \tag{3.4}$$

If a statement holds except on a set N satisfying (3.4), we shall say that the statement holds (a.e. $\mathcal{P}$). If $\mathcal{P}$ is dominated by μ, then

$$\mu(N) = 0 \tag{3.5}$$

implies (3.4). When the converse also holds, μ and $\mathcal{P}$ are said to be *equivalent*.

To bring the customary probabilistic framework and terminology into consonance with that of measure theory, it is necessary to define the concepts of random variable and random vector. A random variable is the mathematical representation of some real-valued aspect of an experiment with uncertain outcome. The experiment may be represented by a space $\mathcal{E}$, and the full details of its possible outcomes by the points e of $\mathcal{E}$. The frequencies with which outcomes can be expected to fall into different subsets E of $\mathcal{E}$ (assumed to form a σ-field $\mathcal{B}$) are given by a probability distribution over $(\mathcal{E}, \mathcal{B})$. A *random variable* is then a real-valued function X defined over $\mathcal{E}$. Since we wish the probabilities of the events $X \leq a$ to be defined, the function X must be measurable and the probability

$$(3.6) \qquad F_X(a) = P(X \leq a)$$

is simply the probability of the set $\{e : X(e) \leq a\}$. The function F_X defined through (3.6) is the *cumulative distribution function (cdf)* of X.

It is convenient to digress here briefly in order to define another concept of absolute continuity. A real-valued function f on $(-\infty, \infty)$ is said to be *absolutely continuous* if given any $\varepsilon > 0$, there exits $\delta > 0$ such that for each finite collection of disjoint bounded open intervals (a_i, b_i),

$$(3.7) \qquad \Sigma(b_i - a_i) < \delta \quad \text{implies} \quad \Sigma|f(b_i) - f(a_i)| < \varepsilon.$$

A connection with the earlier concept of absolute continuity of one measure with respect to another is established by the fact that a cdf F on the real line is absolutely continuous if and only if the probability measure it generates is absolutely continuous with respect to Lebesgue measure. Any absolutely continuous function is continuous (Problem 3.2), but the converse does not hold. In particular, there exist continuous cumulative distribution functions which are not absolutely continuous and therefore do not have a probability density with respect to Lebesgue measure. Such distributions are rather pathological and play little role in statistics.

If not just one but n real-valued aspects of an experiment are of interest, these are represented by a measurable vector-valued function $(X_1, \ldots, X_n)$ defined over $\mathcal{E}$, with the joint cdf

$$(3.8) \qquad F_X(a_1, \ldots, a_n) = P[X_1 \leq a_1, \ldots, X_n \leq a_n]$$

being the probability of the event

$$(3.9) \qquad \{e : X_1(e) \leq a_1, \ldots, X_n(e) \leq a_n\}.$$

The cdf (3.8) determines the probabilities of $(X_1, \ldots X_n)$ falling into any Borel set A, and these agree with the probabilities of the events

$$\{e : [X_1(e), \ldots, X_n(e)] \in A\}.$$

From this description of the mathematical model, one might expect the starting point for modeling a specific situation to be the measurable space $(\mathcal{E}, \mathcal{B})$ and a family $\mathcal{P}$ of probability distributions defined over it. However, the statistical analysis of an experiment is typically not based on a full description of the experimental outcome (which would, for example, include the smallest details concerning all experimental subjects) represented by the points e of $\mathcal{E}$. More often, the starting point is a set of observations, represented by a random vector $X = (X_1, \ldots, X_n)$, with all other aspects of the experiment being ignored. The specification of the model will therefore begin with X, the *data*; the measurable space $(\mathcal{X}, \mathcal{A})$ in which X takes on its values, the *sample space*; and a family $\mathcal{P}$ of probability distributions to which the distribution of X is known to belong. Real-valued or vector-valued measurable functions $T = (T_1, \ldots, T_k)$ of X are called *statistics*; in particular, estimators are statistics.

The change of starting point from $(\mathcal{E}, \mathcal{B})$ to $(\mathcal{X}, \mathcal{A})$ requires clarification of two definitions: (1) In order to avoid reference to $(\mathcal{E}, \mathcal{B})$, it is convenient to require T to be a measurable function over $(\mathcal{X}, \mathcal{A})$ rather than over $(\mathcal{E}, \mathcal{B})$. Measurability with respect to the original $(\mathcal{E}, \mathcal{B})$ is then an automatic consequence (Problem 3.3). (2) Analogously, the expectation of a real-valued integrable T is originally defined as

$$\int T[X(e)]dP(e).$$

However, it is legitimate to calculate it instead from the formula

$$E(T) = \int T(x)dP_X(X)$$

where P_X denotes the probability distribution of X.

As a last concept, we mention the *support* of a distribution P on $(\mathcal{X}, \mathcal{A})$. It is the set of all points x for which $P(A) > 0$ for all open rectangles A [defined by (2.3)] which contain x.

Example 3.1 Support. Let X be a random variable with distribution P and cdf F, and suppose the support of P is a finite interval I with end points a and b. Then, I must be the closed interval $[a, b]$ and F is strictly increasing on $[a, b]$ (Problem 3.4). ‖

If P and Q are two probability measures on $(\mathcal{X}, \mathcal{A})$ and are equivalent (i.e., each is absolutely continuous with respect to the other), then they have the same support; however, the converse need not be true (Problems 3.6 and 3.7).

Having outlined the mathematical foundation on which the statistical developments of the later chapters are based, we shall from now on ignore it as far as possible and instead concentrate on the statistical issues. In particular, we shall pay little or no attention to two technical difficulties that occur throughout.

(i) The estimators that will be derived are statistics and hence need to be measurable. However, we shall not check that this requirement is satisfied. In specific examples, it is usually obvious. In more general constructions, it will be tacitly understood that the conclusion holds only if the estimator in question is measurable. In practice, the sets and functions in these constructions usually turn out to be measurable although verification of their measurability can be quite difficult.

(ii) Typically, the estimators are also required to be integrable. This condition will not be as universally satisfied in our examples as measurability and will therefore be checked when it seems important to do so. In other cases, it will again be tacitly assumed.

4 Group Families

The two principal families of models with which we shall be concerned in this book are *exponential families* and *group families*. Between them, these families cover many of the more common statistical models. In this and the next section, we shall discuss these families and some of their properties, together with some of the more important special cases. More details about these and other special distributions can be found in the four-volume reference work on statistical distributions by Johnson and Kotz (1969-1972), and the later editions by Johnson, Kotz, and Kemp (1992) and Johnson, Kotz, and Balakrishnan (1994,1995).

One of the main reasons for the central role played by these two families in statistics is that in each of them, it is possible to effect a great simplification of the data. In an exponential family, there exists a fixed (usually rather small) number of statistics to which the data can be reduced without loss of information, regardless of the sample size. In a group family, the simplification stems from the fact that the different distributions of the family play a highly symmetric role. This symmetry in the basic structure again leads essentially to a reduction of the dimension of the data since it is then natural to impose a corresponding symmetry requirement on the estimator.

A *group family* of distributions is a family obtained by subjecting a random variable with a fixed distribution to a suitable family of transformations.

Example 4.1 Location-scale families. Let U be a random variable with a fixed distribution F. If a constant a is added to U, the resulting variable

$$X = U + a \tag{4.1}$$

has distribution

$$P(X \leq x) = F(x - a). \tag{4.2}$$

The totality of distributions (4.2), for fixed F and as a varies from $-\infty$ to ∞, is said to constitute a *location family*.

Analogously, a *scale family* is generated by the transformations

$$X = bU, \quad b > 0, \tag{4.3}$$

and has the form

$$P(X \leq x) = F(x/b). \tag{4.4}$$

Combining these two types of transformations into

$$X = a + bU, \quad b > 0, \tag{4.5}$$

one obtains the *location-scale* family

$$P(X \leq x) = F\left(\frac{x - a}{b}\right). \tag{4.6}$$

In applications of these families, F usually has a density f with respect to Lebesgue measure. The density of (4.6) is then given by

$$\frac{1}{b} f\left(\frac{x - a}{b}\right). \tag{4.7}$$

Table 4.1 exhibits several such densities, which will be used in the sequel. ||

In each of (4.1), (4.3), and (4.5), the class of transformations has the following two properties.

Table 4.1. *Location-Scale Families* $-\infty < a < \infty, b > 0$

Density	Support	Name	Notation
$\frac{1}{\sqrt{2\pi}b} e^{-(x-a)^2/2b^2}$	$-\infty < x < \infty$	Normal	$N(a, b^2)$
$\frac{1}{2b} e^{-\|x-a\|/b}$	$-\infty < x < \infty$	Double exponential	$DE(a, b)$
$\frac{b}{\pi} \frac{1}{b^2+(x-a)^2}$	$-\infty < x < \infty$	Cauchy	$C(a, b)$
$\frac{1}{b} \frac{e^{-(x-a)/b}}{[1+e^{-(x-a)/b}]^2}$	$-\infty < x < \infty$	Logistic	$L(a, b)$
$\frac{1}{b} e^{-(x-a)/b} I_{[a,\infty)}(x)$	$a < x < \infty$	Exponential	$E(a, b)$
$\frac{1}{b} I_{[a-b/2, a+b/2]}(x)$	$a - \frac{b}{2} < x < a + \frac{b}{2}$	Uniform	$U\left(a - \frac{b}{2}, a + \frac{b}{2}\right)$

(i) *Closure under composition.* Application of a 1:1 transformation g_1 from $\mathcal{X}$ to $\mathcal{X}$ followed by another, g_2, results in a new such transformation called the composition of g_1 with g_2 and denoted by $g_2 \cdot g_1$. For the transformation (4.1), addition first of a_1 and then of a_2 results in the addition of $a_1 + a_2$. For (4.3), multiplication by b_1 and then by b_2 is equivalent to multiplication by $b_2 \cdot b_1$. The composition rule (4.5) is slightly more complicated. First transforming u to $x = a_1 + b_1 u$ and then the result to $y = a_2 + b_2 x$ results in the transformation

$$y = a_2 + b_2(a_1 + b_1 u) = (a_2 + b_2 a_1) + b_2 b_1 u. \tag{4.8}$$

A class $\mathcal{J}$ of transformations is said to be *closed* under composition if $g_1 \in \mathcal{J}$, $g_2 \in \mathcal{J}$ implies that $g_2 \cdot g_1 \in \mathcal{J}$. We have just shown that the three classes of transformations,

$$\begin{aligned} &(4.1) \quad \text{with } -\infty < a < \infty, \\ &(4.3) \quad \text{with } 0 < b, \\ &(4.5) \quad \text{with } -\infty < a < \infty, 0 < b, \end{aligned} \tag{4.9}$$

are all closed with respect to composition. On the other hand, the class (4.1) with $|a| < 1$ is not, since $U + 1/2$ and $U + 2/3$ are both members of the class but their composition is not.

(ii) *Closure under inversion.* Given any $1:1$ transformation $x' = gx$, let g^{-1}, the *inverse* of g, denote the transformation which undoes what g did, that is, takes x' back to x so that $x = g^{-1}x'$. For the transformation which adds a, the inverse subtracts a; the inverse in (4.3) of multiplication by b is division by b; and the inverse of $a + bu$ is $(x - a)/b$. A class $\mathcal{J}$ is said to be closed under inversion if $g \in \mathcal{J}$ implies $g^{-1} \in \mathcal{J}$. The three classes listed in (4.9) are all closed under inversion. On the other hand, (4.1) with $0 \le a$ is not.

The structure of the class of transformations possessing these properties is a special case of a more general mathematical object, simply called a *group*.

Definition 4.2 A set G of elements is called a *group* if it satisfies the following four conditions.

(i) There is defined an operation, group multiplication, which with any two elements $a, b \in G$ associates an element c of G. The element c is called the product of a and b and is denoted ab.

(ii) Group multiplication obeys the associative law

$$(ab)c = a(bc).$$

(iii) There exists an element $e \in G$, called the *identity*, such that

$$ae = ea = a \quad \text{for all} \quad a \in G.$$

(iv) For each element $a \in G$, there exists an element a^{-1}, its *inverse*, such that

$$aa^{-1} = a^{-1}a = e.$$

Both the identity element and the inverse a^{-1} of any element a can be shown to be unique.

The groups of primary interest in statistics are *transformation groups*.

Definition 4.3 A class G of transformations is called a *transformation group* if it is closed under both composition and inversion.

It is straightforward to verify (Problem 4.4) that a transformation group is, in fact, a group. In particular, note that the *identity transformation* $x \equiv x$ is a member of any transformation group G since $g \in G$ implies $g^{-1} \in G$ and hence $g^{-1}g \in G$, and by definition, $g^{-1}g$ is the identity. Note also that the inverse $(g^{-1})^{-1}$ of g^{-1} is g, so that gg^{-1} is also the identity.

A transformation group G which satisfies

$$g_2 \cdot g_1 = g_1 \cdot g_2$$

for all $g_1, g_2 \in G$ is called *commutative*. The first two groups of transformations of (4.9) are commutative, but the third is not.

Example 4.4 Continuation of Example 4.1. The group families (4.2), (4.4), and (4.6) generalize easily to the case that $\mathbf{U}$ is a vector $\mathbf{U} = (U_1, \ldots, U_n)$, if one defines

(4.10) $\quad \mathbf{U} + a = (U_1 + a, \ldots, U_n + a) \quad \text{and} \quad b\mathbf{U} = (bU_1, \ldots, bU_n).$

This covers in particular the case that $X_1, \ldots, X_n$ are iid according to one of the previous families, for example, one of the densities of Table 4.1. Larger group families are obtained in the same way by letting

(4.11) $\quad \mathbf{U} + \mathbf{a} = (U_1 + a_1, \ldots, U_n + a_n) \text{ and } \mathbf{b}\mathbf{U} = (b_1U_1, \ldots, b_nU_n).$

‖

Example 4.5 Multivariate normal distribution. As a more special but very important example, suppose next that $\mathbf{U} = (U_1, \ldots, U_p)$ where the U_i are independently distributed as $N(0, 1)$ and let

$$\begin{pmatrix} X_1 \\ \vdots \\ X_p \end{pmatrix} = \begin{pmatrix} a_1 \\ \vdots \\ a_p \end{pmatrix} + B \begin{pmatrix} U_1 \\ \vdots \\ U_p \end{pmatrix} \tag{4.12}$$

where B is nonsingular $p \times p$ matrix. The resulting family of distributions in p-space is the family of nonsingular *p-variate normal distributions*. If the three columns of (4.12) are denoted by $\mathbf{X}$, $\mathbf{a}$, and $\mathbf{U}$, respectively,[1] (4.12) can be written as

$$\mathbf{X} = \mathbf{a} + B\mathbf{U}. \tag{4.13}$$

From this equation, it is seen that the covariance matrix Σ of $\mathbf{X}$ are given by

$$E(\mathbf{X}) = \mathbf{a} \quad \text{and} \quad \Sigma = E[(\mathbf{X} - \mathbf{a})(\mathbf{X} - \mathbf{a})'] = BB'. \tag{4.14}$$

To obtain the density of $\mathbf{X}$, write the density of $\mathbf{U}$ as

$$\frac{1}{(\sqrt{2\pi})^p} e^{-(1/2)\mathbf{u}'\mathbf{u}}.$$

Now $\mathbf{U} = B^{-1}(\mathbf{X} - \mathbf{a})$ and the Jacobian of the linear transformation (4.13) is just the determinant $|B|$ of B. Thus, by the usual formula for transforming densities, the density of $\mathbf{X}$ is seen to be

$$\frac{|B|^{-1}}{(\sqrt{2\pi})^p} e^{-(\mathbf{x}-\mathbf{a})'\Sigma^{-1}(\mathbf{x}-\mathbf{a})/2}. \tag{4.15}$$

For the case $p = 2$, this reduces to (Problem 4.6)

$$\frac{1}{2\pi\sigma\tau\sqrt{1-\rho^2}} e^{-[(x-\xi)^2/\sigma^2 - 2\rho(x-\xi)(y-\eta)/\sigma\tau + (y-\eta)^2/\tau^2]/2(1-\rho^2)} \tag{4.16}$$

where we write (x, y) for (x_1, x_2) and (ξ, η) for (a_1, a_2), and where $\sigma^2 = \text{var}(X)$, $\tau^2 = \text{var}(Y)$, and $\rho\sigma\tau = \text{cov}(X, Y)$. ||

There is a difference between the transformation groups (4.1), (4.3), and (4.5), on the one hand, and (4.13), on the other. In the first three cases, different transformations of the group lead to different distributions. This is not true of (4.13) since the distributions of

$$\mathbf{a}_1 + B_1\mathbf{U} \quad \text{and} \quad \mathbf{a}_2 + B_2\mathbf{U}$$

coincide provided $\mathbf{a}_1 = \mathbf{a}_2$ and $B_1B_1' = B_2B_2'$. This occurs when $\mathbf{a}_1 = \mathbf{a}_2$ and $(B_2^{-1}B_1)(B_2^{-1}B_1)'$ is the identity matrix, that is, when $B_2^{-1}B_1$ is orthogonal. The same family of distributions can therefore be generated by restricting the matrices B in (4.13) to belong to a smaller group. In particular, it is enough to let G be the group of lower triangular matrices, in which all elements above the main diagonal are zero (Problems 4.7 - 4.9).

[1] When it is not likely to cause confusion, we shall use $\mathbf{U}$ and so on to denote both the vector and the column with elements U_i.

Example 4.6 The linear model. Let us next consider a different generalization of a location-scale family. As before, let $\mathbf{U} = (U_1, \ldots, U_n)$ have a fixed joint distribution and consider the transformations

$$X_i = a_i + bU_i, \quad i = 1, \ldots, n, \tag{4.17}$$

where the translation vector $\mathbf{a} = (a_1, \ldots, a_n)$ is restricted to be in some s-dimensional linear subspace Ω of n-space, that is, to satisfy a set of linear equations

$$a_i = \sum_{j=1}^{s} d_{ij}\beta_j \quad (i = 1, \ldots, n). \tag{4.18}$$

Here, the d_{ij} are fixed (without loss of generality the matrix $D = (d_{ij})$ is assumed to be of rank s) and the β_j are arbitrary.

The most important case of this model is that in which the U's are iid as $N(0, 1)$. The joint distribution of the X's is then given by

$$\frac{1}{(\sqrt{2\pi}b)^n} \exp\left[-\frac{1}{2b^2}\Sigma(x_i - a_i)^2\right] \tag{4.19}$$

with $\mathbf{a}$ ranging over Ω. ||

We shall next consider a number of models in which the groups (and hence the resulting families of distributions) are much larger than in the situations discussed so far.

Example 4.7 A nonparametric iid family. Let $U_1, \ldots, U_n$ be n independent random variables with a fixed continuous common distribution, say $N(0, 1)$, whose support is the whole real line, and let G be the class of all transformations

$$X_i = g(U_i) \tag{4.20}$$

where g is any continuous, strictly increasing function satisfying

$$\lim_{u \to -\infty} g(u) = -\infty, \quad \lim_{u \to \infty} g(u) = \infty. \tag{4.21}$$

This class constitutes a group. The X_i are again iid with common distribution, say F_g. The class $\{F_g : g \in G\}$ is the class of all continuous distributions whose support is $(-\infty, \infty)$, that is, the class of all distributions whose cdf is continuous and strictly increasing on $(-\infty, \infty)$.

In this example, one may wish to impose on g the additional restriction of differentiability for all μ. The resulting family of distributions will be as before but restricted to have probability density with respect to Lebesgue measure. ||

Many variations of this basic example are of interest, we shall mention only a few.

Example 4.8 Symmetric distributions. Consider the situation of Example 4.7 but with g restricted to be odd, that is, to satisfy $g(-u) = -g(u)$ for all u. This leads to the class of all distributions whose support is the whole real line and which are symmetric with respect to the origin. If instead we let $X_i = g(u_i) + a$, $-\infty < a < \infty$, the resulting class is that of all distributions whose support is the real line and which are symmetric with the point a of symmetry being specified. ||

Example 4.9 Continuation of Example 4.7. In Example 4.7, replace $N(0, 1)$ as the initial distribution of the U_i with the uniform distribution on $(0, 1)$, and let G be the class of all strictly increasing continuous functions g on $(0, 1)$ satisfying $g(0) = 0, g(1) = 1$. If, then, $X_i = a + bg(U_i)$ with $-\infty < a < \infty, 0 < b$, the resulting group family is that of all continuous distributions whose support is an interval. ||

The examples of group families considered so far are of two types. In Examples 4.1 - 4.6, the distributions within a family were naturally indexed by a relatively small number of parameters (a and b in Example 4.1; the elements of the matrix B and the vector $\mathbf{a}$ in Example 4.4; the quantities b and $\beta_1, \ldots, \beta_s$ in Example 4.6). On the other hand, in Examples 4.7 - 4.9, the distribution of the X_i was fairly unrestricted, subject only to conditions such as independence, identity of distribution, nature of support, continuity, and symmetry. The next example is the prototype of a third kind of model arising in survey sampling.

Example 4.10 Sampling from a finite population. To motivate this model, consider a finite population of N elements (or subjects) to each of which is attached a real number (for example, the age or income of the subject) and an identifying label. A random sample of n elements drawn from this population constitutes the observations. Let the observed values and labels be $(X_1, J_1), \ldots, (X_n, J_n)$. The following group family provides a possible model for this situation.

Let $v_1, \ldots, v_N$ be any fixed N real numbers, and let the pairs $(U_1, J_1), \ldots,$ (U_n, J_n) be n of the pairs $(v_1, 1), \ldots, (v_N, N)$ selected at random, that is, in such a way that all

$$\binom{N}{n} \quad \text{possible choices of } n \text{ pairs are equally likely.}$$

Finally, let G be the group of transformations

$$X_1 = U_1 + a_{J_1}, \ldots, X_n = U_n + a_{J_n} \tag{4.22}$$

where the N-tuple $(a_1, \ldots, a_N)$ ranges over all possible N-tuples $-\infty < a_1,$ $a_2, \ldots, a_N < \infty$. If we put $y_i = v_i + a_i$, then the pairs $(X_1, J_1), \ldots, (X_n, J_n)$ are a random sample from the population $(y_1, 1), \ldots, (y_N, N)$, the y values being arbitrary.

This example can be extended in a number of ways. In particular, the sampling method, reflecting some knowledge concerning the population of y values, may be more complex. In *stratified sampling*, for instance, the population of N is divided into, say, s subpopulations of $N_1, \ldots, N_s$ members ($\Sigma N_i = N$) and a sample of n_i is drawn at random from the ith subpopulation (Problem 4.12). This and some other sampling schemes will be considered in Section 3.7. A different modification places some restrictions on the y's such as $0 < y_i < \infty$, or $0 < y_i < 1$ (Problem 4.11). ||

It was stated at the beginning of the section that in a group family, the different members of the family play a highly symmetric role. However, the general construction of such a family $\mathcal{P}$ as the distributions of gU, where U has a fixed distribution P_0 and g ranges over a group G of transformations, appears to single out the distribution P_0 of U (which is a member of $\mathcal{P}$ since the identity transformation is a member of G) as the starting point of the construction. This asymmetry is only apparent. Let P_1 be any distribution of $\mathcal{P}$ other than P_0 and consider the family $\mathcal{P}'$ of distributions of gV as g ranges over G, where V has distribution P_1. Since P_1 is an element of $\mathcal{P}$, there exists an element g_0 of G for which g_0U is distributed according to P_1. Thus, g_0U can play the role of V, and $\mathcal{P}'$ is the family of distributions of gg_0U as g ranges over G. However, as g ranges over G, so does gg_0 (Problem 4.5), so that the family of distributions of gg_0U, $g \in G$, is the same as the family of $\mathcal{P}$ of gU, $g \in G$. A group family is thus independent of which of its members is taken as starting distribution.

If one cannot find a group generating a given family $\mathcal{P}$ of distributions, the question arises whether such a group exists, that is, whether $\mathcal{P}$ is a group family. In principle, the answer is easy. For the sake of simplicity, suppose that $\mathcal{P}$ is a family of univariate distributions with continuous and strictly increasing cumulative distribution functions. Let F_0 and F be two such cdf's and suppose that U is distributed according to F_0. Then, if g is strictly increasing, $g(U)$ is distributed according to F if and only if $g = F^{-1}(F_0)$ (Problem 4.14). Thus, the transformations generating the family must be the transformations

$$\{F^{-1}(F_0),\ F \in \mathcal{P}\}. \tag{4.23}$$

The family $\mathcal{P}$ will be a group family if and only if the transformations (4.23) form a group, that is, are closed under composition and inversion. In specific situations, the calculations needed to check this requirement may not be easy. For an important class of problems, the question has been settled by Borges and Pfanzagl (1965).

5 Exponential Families

A family $\{P_\theta\}$ of distributions is said to form an s-dimensional exponential family if the distributions P_θ have densities of the form

$$p_\theta(x) = \exp\left[\sum_{i=1}^{s} \eta_i(\theta)T_i(x) - B(\theta)\right] h(x) \tag{5.1}$$

with respect to some common measure μ. Here, the η_i and B are real-valued functions of the parameters and the T_i are real-valued statistics, and x is a point in the sample space $\mathcal{X}$, the *support* of the density. Frequently, it is more convenient to use the η_i as the parameters and write the density in the *canonical form*

$$p(x|\eta) = \exp\left[\sum_{i=1}^{s} \eta_i T_i(x) - A(\eta)\right] h(x). \tag{5.2}$$

It should be noted that the form (5.2) is not unique. We can, for example, multiply η_i by a constant c if, at the same time, T_i is replaced by T_i/c. More generally, we can make linear transformations of the η's and T's.

Both (5.1) and (5.2) are redundant in that the factor $h(x)$ could be absorbed into μ. The reason for not doing so is that it is then usually possible to take μ to be either Lebesgue measure or counting measure rather than having to define a more elaborate measure.

The function p given by (5.2) is non-negative and is therefore a probability density with respect to the given μ, provided its integral with respect to μ equals 1. A constant $A(\eta)$ for which this is the case exists if and only if

$$\int e^{\Sigma \eta_i T_i(x)} h(x) d\mu(x) < \infty. \tag{5.3}$$

The set Ξ of points $\eta = (\eta_1, \ldots, \eta_s)$ for which (5.3) holds is called the *natural parameter space* of the family (5.2) and η is called the *natural parameter*. It is not difficult to see that Ξ is convex (TSH2, Section 2.7, Lemma 7). In most applications, it turns out to be open, but this need not be the case (Problem 5.1). In the parametrization (5.1), the natural parameter space is the set of θ values for which $[\eta_1(\theta), \ldots, \eta_s(\theta)]$ is in Ξ.

Example 5.1 Normal family. If X has the $N(\xi, \sigma^2)$ distribution, then $\theta = (\xi, \sigma^2)$ and the density with respect to Lebesgue measure is

$$p_\theta(x) = \exp\left[\frac{\xi}{\sigma^2}x - \frac{1}{2\sigma^2}x^2 - \frac{\xi^2}{2\sigma^2}\right]\frac{1}{\sqrt{2\pi}\sigma},$$

a two-parameter exponential family with natural parameters $(\eta_1, \eta_2) = (\xi/\sigma^2, -1/2\sigma^2)$ and natural parameter space $\Re \times (-\infty, 0)$. ‖

Some other examples of one- and two-parameter exponential families are shown in Table 5.1.

If the statistics $T_1, \ldots, T_s$ satisfy linear constraints, the number s of terms in the exponent of (5.1) can be reduced. Unless this is done, the parameters η_i are statistically meaningless; they are *unidentifiable* (see Problem 5.2).

Definition 5.2 If X is distributed according to p_θ, then θ is said to be *unidentifiable on the basis of* X if there exist $\theta_1 \neq \theta_2$ for which $P_{\theta_1} = P_{\theta_2}$.

A reduction is also possible when the η's satisfy a linear constraint. In the latter case, the natural parameter space will be a convex set which lies in a linear subspace of dimension less than s. If the representation (5.2) is minimal in the sense that neither the T's nor the η's satisfy a linear constraint, the natural parameter space will then be a convex set in E_s containing an open s-dimensional rectangle. If (5.2) is minimal and the parameter space contains an s-dimensional rectangle, the family (5.2) is said to be of *full rank*.

Example 5.3 Multinomial. In n independent trials with $s+1$ possible outcomes, let the probability of the ith outcome be p_i in each trial. If X_i denotes the number of trials resulting in outcome i ($i = 0, 1, \ldots, s$), then the joint distribution of the X's is the *multinomial distribution* $M(p_0, \ldots, p_s; n)$

$$P(X_0 = x_0, \ldots, X_s = x_s) = \frac{n!}{x_0! \cdots x_s!} p_0^{x_0} \cdots p_s^{x_s}, \tag{5.4}$$

which can be rewritten as

$$\exp(x_0 \log p_0 + \cdots + x_s \log p_s) h(x).$$

Table 5.1. *Some One- and Two-Parameter Exponential Families*

Density*	Name	Notation	Support
$\frac{1}{\Gamma(a)b^a}x^{a-1}e^{-x/b}$	Gamma(a, b)	$\Gamma(a, b)$	$0 < x < \infty$
$\frac{1}{\Gamma(f/2)2^{f/2}}x^{f/2-1}e^{-x/2}$	Chi-squared(f)	χ_f^2	$0 < x < \infty$
$\frac{\Gamma(a+b)}{\Gamma(a)\Gamma(b)}x^{a-1}(1-x)^{b-1}$	Beta(a, b)	$B(a, b)$	$0 < x < 1$
$p^x(1-p)^{n-x}$	Bernoulli(p)	$b(p)$	$x = 0, 1$
$\binom{n}{x}p^x(1-p)^{n-x}$	Binomial(p, n)	$b(p, n)$	$x = 0, 1, \ldots, n$
$\frac{1}{x!}\lambda^x e^{-\lambda}$	Poisson(λ)	$P(\lambda)$	$x = 0, 1, \ldots$
$\binom{m+x-1}{m-1}p^m q^x$	Negative binomial(p, m)	$Nb(p, m)$	$x = 0, 1, \ldots$

*The density of the first three distributions is with respect to Lebesgue measure, and that of the last four with respect to counting measure.

Since the x_i add up to n, this can be reduced to

$$\exp[n \log p_0 + x_1 \log(p_1/p_0) + \cdots + x_s \log(p_s/p_0)]h(x). \tag{5.5}$$

This is an s-dimensional exponential family with

$$\eta_i = \log(p_i/p_0), \quad A(\eta) = -n \log p_0 = n \log\left[1 + \sum_{i=1}^{s} e^{\eta_i}\right]. \tag{5.6}$$

The natural parameter space is the set of all $(\eta_1, \ldots, \eta_s)$ with $-\infty < \eta_i < \infty$. ‖

In the normal family of Example 5.1, it might be the case that the mean and the variance are related. [Such a model can be useful in data analysis, where the variance may be modeled as a power of the mean (see, for example, Snedecor and Cochran 1989, Section 15.10).] In such cases, when the natural parameters of the distribution are related in a nonlinear way, we say that (5.1) or (5.2) forms a *curved exponential family* (see Note 10.6).

Example 5.4 Curved normal family. For the normal family of Example 5.1, assume that $\xi = \sigma$, so that

$$p_\theta(x) = \exp\left[\frac{1}{\xi}x - \frac{1}{2\xi^2}x^2 - \frac{1}{2}\right]\frac{1}{\sqrt{2\pi}\xi}, \quad \xi > 0. \tag{5.7}$$

Although this is formally a two-parameter exponential family with natural parameter $(\frac{1}{\xi}, -\frac{1}{2\xi^2})$, this parameter is, in fact, generated by the single parameter ξ. The two-dimensional parameter $(\frac{1}{\xi}, -\frac{1}{2\xi^2})$ lies on a curve in $\Re^2$, making (5.7) a curved exponential family. ||

The underlying parameter in a curved exponential family need not be one dimensional. The following is an example in which it is two dimensional.

Example 5.5 Logit model. Let X_i be independent $b(p_i, n_i)$, $i = 1, \ldots, m$, so that their joint distribution is

$$P(X_1 = x_1, \ldots, X_m = x_m) = \prod_{i=1}^{m} \binom{n_i}{x_i} p_i^{x_i}(1 - p_i)^{n_i - x_i}.$$

This can be written as

$$\text{(5.8)} \qquad \exp\left\{\sum_{i=1}^{m} x_i \log \frac{p_i}{1 - p_i}\right\} \prod_{i=1}^{m} \binom{n_i}{x_i} (1 - p_i)^{n_i},$$

an m-dimensional exponential family with natural parameters $\eta_i = \log[(p_i/(1 - p_i)]$, $i = 1, \ldots, m$. The quantity $\log[p/(1 - p)]$ is known as the *logit* of p.

If the η's satisfy

$$\text{(5.9)} \qquad \eta_i = \alpha + \beta z_i, \quad i = 1, \ldots, m,$$

for known covariates z_i, the model only contains the two parameters α and β and (5.8) becomes a curved exponential family (see Note 10.6). ||

Note that the parameter space of an s-dimensional curved exponential family cannot contain an s-dimensional rectangle, so a curved exponential family is not of full rank. Nevertheless, as long as the T's are not rank deficient, a curved exponential family shares many of the following properties of a full rank family. (An exception is completeness of the sufficient statistic, discussed in the next section.) A more detailed treatment can be found in Brown (1986a) or Barndorff-Nielsen and Cox (1994).

Let X and Y be independently distributed according to s-dimensional exponential families (not necessarily full rank) with densities

$$\text{(5.10)} \qquad \exp[\Sigma\eta_i T_i(x) - A(\eta)]\, h(x) \quad \text{and} \quad \exp[\Sigma\eta_i U_i(y) - C(\eta)]\, k(y)$$

with respect to measures μ and ν over $(\mathcal{X}, \mathcal{A})$ and $(\mathcal{Y}, \mathcal{B})$, respectively. Then, the joint distribution of X, Y is again an exponential family, and by induction, the result extends to the joint distribution of more than two factors. The most important special case is that of iid random variables X_i, each distributed according to (5.1): The exponential structure is preserved under random sampling. The joint density of $X = (X_1, \ldots, X_n)$ is

$$\text{(5.11)} \qquad \exp\left[\Sigma\eta_i(\theta)T_i'(x) - \eta B(\theta)\right] h(x_1) \cdots h(x_n)$$

with $T_i'(x) = \Sigma_{j=1}^{n} T_i(x_j)$.

Example 5.6 Normal sample. Let X_i $(i = 1, \ldots, n)$ be iid according to $N(\xi, \sigma^2)$. Then, the joint density of $X_1, \ldots, X_n$ with respect to Lebesgue measure in E_n is

$$\exp\left(\frac{\xi}{\sigma^2}\Sigma x_i - \frac{1}{2\sigma^2}\Sigma x_i^2 - \frac{n}{2\sigma^2}\xi^2\right) \cdot \frac{1}{(\sqrt{2\pi}\,\sigma)^n}. \tag{5.12}$$

As in the case $n = 1$ (Example 5.1), this constitutes a two-parameter exponential family with natural parameters $(\xi/\sigma^2, -1/2\sigma^2)$. ||

Example 5.7 Bivariate normal. Suppose that (X_i, Y_i), $i = 1, \ldots, n$, is a sample from the bivariate normal density (4.16). Then, it is seen that the joint density of the n pairs is a five-parameter exponential density with statistics

$$T_1 = \Sigma X_i, \quad T_2 = \Sigma X_i^2, \quad T_3 = \Sigma X_i Y_i, \quad T_4 = \Sigma Y_i, \quad T_5 = \Sigma Y_i^2.$$

This example easily generalizes to the p-variate case (Problem 5.3). ||

A useful property of exponential families is given by the following theorem, which is proved, for example, in TSH2 (Chapter 2, Theorem 9) and in Barndorff-Nielsen (1978, Section 7.1).

Theorem 5.8 *For any integrable function f and any η in the interior of Ξ, the integral*

$$\int f(x) \exp[\Sigma \eta_i T_i(x)] h(x)\, d\mu(x) \tag{5.13}$$

is continuous and has derivatives of all orders with respect to the η's, and these can be obtained by differentiating under the integral sign.

As an application, differentiate the identity

$$\int \exp[\Sigma \eta_i T_i(x) - A(\eta)] h(x)\, d\mu(x) = 1$$

with respect to η_j to find

$$E_\eta(T_j) = \frac{\partial}{\partial \eta_j} A(\eta). \tag{5.14}$$

Differentiating (5.14), in turn, with respect to η_k leads to

$$\text{cov}(T_j, T_k) = \frac{\partial^2}{\partial \eta_j \partial \eta_k} A(\eta). \tag{5.15}$$

(For the corresponding formulas in terms of (5.1), see Problem 5.6.)

Example 5.9 Continuation of Example 5.3. From (5.6), (5.14), and (5.15), one easily finds for the multinomial variables of Example 5.3 that (Problem 5.15)

$$E(X_i) = np_i, \quad \text{cov}(X_j, X_k) = \begin{cases} np_j(1 - p_j) & \text{if } k = j \\ -np_j p_k & \text{if } k \neq j. \end{cases} \tag{5.16}$$

||

As will be discussed in the next section, in an exponential family the statistics $T = (T_1, \ldots, T_s)$ carry all the information about η or θ contained in the data, so that all statistical inferences concerning these parameters will be based on the T's.

For this reason, we shall frequently be interested in calculating not only the first two moments of the T's given by (5.14) and (5.15) but also some of the higher moments

$$\alpha_{r_1,\ldots,r_s} = E(T_1^{r_1} \cdots T_s^{r_s}) \tag{5.17}$$

and central moments

$$\mu_{r_1,\ldots,r_s} = E\{[T_1 - E(T_1)]^{r_1} \cdots [T_s - E(T_s)]^{r_s}\}. \tag{5.18}$$

A tool that often facilitates such calculations is the *moment generating function*

$$M_T(u_1, \ldots, u_s) = E(e^{u_1 T_1 + \cdots + u_s T_s}). \tag{5.19}$$

If M_T exists in some neighborhood $\Sigma u_i^2 < \delta$ of the origin, then all moments $\alpha_{r_1,\ldots,r_s}$ exist and are the coefficients in the expansion of M_T as a power series

$$M_T(u_1, \ldots, u_s) = \sum_{(r_1,\ldots,r_s)} \alpha_{r_1,\ldots,r_s} u_1^{r_1} \cdots u_s^{r_s} / r_1! \cdots r_s! \tag{5.20}$$

As an alternative, it is sometimes more convenient to calculate, instead, the *cumulants* $\kappa_{r_1,\ldots,r_s}$, defined as the coefficients in the expansion of the *cumulant generating function*

$$\begin{aligned} K_T(u_1, \ldots, u_s) &= \log M_T(u_1, \ldots, u_s) \\ &= \sum_{(r_1,\ldots,r_s)} \kappa_{r_1,\ldots,r_s} u_1^{r_1} \cdots u_s^{r_s} / r_1! \cdots r_s! \end{aligned} \tag{5.21}$$

From the cumulants, the moments can be determined by formal comparison of the two power series (see, for example, Cramér 1946a, p. 186, or Stuart and Ord 1987, Chapter 3.). For $s = 1$, one finds, for example (Problem 5.7),

$$\begin{aligned} \alpha_1 = \kappa_1, \quad \alpha_2 = \kappa_2 + \kappa_1^2, \quad \alpha_3 &= \kappa_3 + 3\kappa_1\kappa_2 + \kappa_1^3, \\ \alpha_4 &= \kappa_4 + 3\kappa_2^2 + 4\kappa_1\kappa_3 + 6\kappa_1^2\kappa_2 + \kappa_1^4. \end{aligned} \tag{5.22}$$

For exponential families, the moment and cumulant generating functions can be expressed rather simply as follows.

Theorem 5.10 *If X is distributed with density (5.2), then for any η in the interior of Ξ, the moment and cumulant generating functions $M_T(u)$ and $K_T(u)$ of the T's exist in some neighborhood of the origin and are given by*

$$K_T(u) = A(\eta + u) - A(\eta) \tag{5.23}$$

and

$$M_T(u) = e^{A(\eta+u)} / e^{A(\eta)} \tag{5.24}$$

respectively.

Frequently, the calculation of moments becomes particularly easy when they can be represented as the sum of independent terms. We shall illustrate two examples for the case $s = 1$.

(a) Suppose $X = X_1 + \cdots + X_n$, where the X_i are independent with moment and cumulant generating functions $M_{X_i}(u)$ and $K_{X_i}(u)$, respectively. Then

$$M_X(u) = E[e^{(x_1 + \cdots + x_n)}] = M_{X_1}(u) \cdots M_{X_n}(u)$$

and therefore

$$K_X(u) = \sum_{i=1}^{n} K_{X_i}(u).$$

From the definition of cumulants, it then follows that

$$\kappa_r = \sum_{i=1}^{n} \kappa_{ir} \tag{5.25}$$

where κ_{ir} is the rth cumulant of X_i.

(b) The situation is also very simple for low central moments. If $\xi_i = E(X_i)$, $\sigma_i^2 = \text{var}(X_i)$ and the X_i are independent, one easily finds (Problem 5.7)

$$\begin{aligned} &\text{var}(\Sigma X_i) = \Sigma\sigma_i^2, \quad E[\Sigma(X_i - \xi_i)]^3 = \Sigma E(X_i - \xi_i)^3, \\ &E[\Sigma(X_i - \xi_i)]^4 = \Sigma E(X_i - \xi_i)^4 + 6\sum_{i<j}\sigma_i^2\sigma_j^2. \end{aligned} \tag{5.26}$$

For the case of identical components with $\xi_i = \xi$, $\sigma_i^2 = \sigma^2$, this reduces to

$$\begin{aligned} &\text{var}(\Sigma X_i) = n\sigma^2, \quad E[\Sigma(X_i - \xi)]^3 = nE(X_1 - \xi)^3, \\ &E[\Sigma(X_i - \xi)]^4 = nE(X_1 - \xi)^4 + 3n(n-1)\sigma^4. \end{aligned} \tag{5.27}$$

The following are a few of the many important special cases of exponential families and some of their moments. Additional examples are given in the problems; see also Johanson (1979), Brown (1986a), or Hoffmann-Jorgensen (1994, Chapter 12).

Example 5.11 Binomial moments. Let X have the binomial distribution $b(p, n)$ so that for $x = 0, 1, \ldots, n$,

$$P(X = x) = \binom{n}{x} p^x q^{n-x} \quad (0 < p < 1;\ q = 1 - p). \tag{5.28}$$

This is the special case of the multinomial distribution (5.4) with $s = 1$. The probability (5.28) can be rewritten as

$$\binom{n}{x} e^{x \log(p/q) + n \log q},$$

which defines an exponential family, with μ being counting measure over the points $x = 0, 1, \ldots, n$ and with

$$\eta = \log(p/q), \quad A(\eta) = n \log(1 + e^\eta). \tag{5.29}$$

From (5.24) and (5.29), one finds that (Problem 5.8)

$$M_X(u) = (q + pe^u)^n. \tag{5.30}$$

An easy way to obtain the expectation and the first three central moments of X is to use the fact that X arises as the number of successes in n Bernoulli trials with success probability p, and hence that $X = \Sigma X_i$, where X_i is 1 or 0, as the ith trial is or is not a success. From (5.27) and the moments of X_i, one then finds (Problem 5.8)

(5.31)
$$E(X) = np, \quad E(X-np)^3 = npq(q-p),$$
$$\operatorname{var}(X) = npq, \quad E(X-np)^4 = 3(npq)^2 + npq(1-6pq). \quad \|$$

Example 5.12 Poisson moments. A random variable X has the Poisson distribution $P(\lambda)$ if

(5.32)
$$P(X = x) = \frac{\lambda^x}{x!}e^{-\lambda}, \quad x = 0, 1, \ldots; \lambda > 0.$$

Writing this as an exponential family in canonical form, we find

(5.33)
$$\eta = \log \lambda, \quad A(\eta) = \lambda = e^{\eta}$$

and hence

(5.34)
$$K_X(u) = \lambda(e^u - 1), \quad M_X(u) = e^{\lambda(e^u - 1)},$$

so that, in particular, $\kappa_r = \lambda$ for all r. The expectation and first three central moments are given by (Problem 5.9)

(5.35)
$$E(X) = \lambda, \quad E(X-\lambda)^3 = \lambda,$$
$$\operatorname{var}(X) = \lambda, \quad E(X-\lambda)^4 = \lambda + 3\lambda^2. \quad \|$$

Example 5.13 Normal moments. Let X have the normal distribution $N(\xi, \sigma^2)$ with density

(5.36)
$$\frac{1}{\sqrt{2\pi}\sigma}e^{-(x-\xi)^2/2\sigma^2}$$

with respect to Lebesgue measure. For fixed σ, this is a one-parameter exponential family with

(5.37)
$$\eta = \xi/\sigma^2 \quad \text{and} \quad A(\eta) = \eta^2\sigma^2/2 + \text{constant}.$$

It is thus seen that

(5.38)
$$M_X(u) = e^{\xi u + (1/2)\sigma^2 u^2},$$

and hence in particular that

(5.39)
$$E(X) = \xi.$$

Since the distribution of $X - \xi$ is $N(0, \sigma^2)$, the central moments μ_r of X are simply the moments α_r of $N(0, \sigma^2)$, which are obtained from the moment generating function

$$M_0(u) = e^{\sigma^2 u^2/2}$$

to be

(5.40)
$$\mu_{2r+1} = 0, \quad \mu_{2r} = 1 \cdot 3 \cdots (2r-1)\sigma^{2r}, \quad r = 1, 2, \ldots.$$

$\|$

Example 5.14 Gamma moments. A random variable X has the *gamma distribution* $\Gamma(\alpha, b)$ if its density is

(5.41)
$$\frac{1}{\Gamma(\alpha)b^{\alpha}}x^{\alpha-1}e^{-x/b}, \quad x > 0, \alpha > 0, b > 0,$$

with respect to Lebesgue measure on $(0, \infty)$. Here, b is a scale parameter, whereas α is called the *shape parameter* of the distribution. For $\alpha = f/2$ (f an integer), $b = 2$, this is the χ^2-distribution χ_f^2 with f degrees of freedom. For fixed-shape parameter α, (5.41) is a one-parameter exponential family with $\eta = -1/b$ and

$$A(\eta) = \alpha \log b = -\alpha \log(-\eta).$$

Thus, the moment and cumulant generating functions are seen to be

$$(5.42) \quad M_X(u) = (1 - bu)^{-\alpha} \text{ and } K_X(u) = -\alpha \log(1 - bu), u < 1/b.$$

From the first of these formulas, one finds

$$(5.43) \quad E(X^r) = \alpha(\alpha + 1) \cdots (\alpha + r - 1)b^r = \frac{\Gamma(\alpha + r)}{\Gamma(\alpha)} b^r$$

and hence (Problem 5.17)

$$(5.44) \quad \begin{aligned} E(X) &= \alpha b, \quad E(X - \alpha b)^3 = 2\alpha b^3, \\ \text{var}(X) &= \alpha b^2, \quad E(X - \alpha b)^4 = (3\alpha^2 + 6\alpha)b^4. \end{aligned}$$

‖

Another approach to moment calculations is to use an identity of Charles Stein, which was given a thorough treatment by Hudson (1978). Stein's identity is primarily used to establish minimaxity of estimators, but it is also useful in moment calculations.

Lemma 5.15 (Stein's identity) *If X is distributed with density (5.2) and g is any differentiable function such that $E|g'(X)| < \infty$, then*

$$(5.45) \quad E\left\{\left[\frac{h'(X)}{h(X)} + \sum_{i=1}^{s} \eta_i T_i'(X)\right] g(X)\right\} = -Eg'(X),$$

provided the support of X is $(-\infty, \infty)$. If the support of X is the bounded interval (a, b), then (5.45) holds if $\exp\{\sum \eta_i T_i(x)\}h(x) \to 0$ *as $x \to a$ or b.*

The proof of the lemma is quite straightforward and is based on integration by parts (Problem 5.18). We illustrate its use in the normal case.

Example 5.16 Stein's identity for the normal. If $X \sim N(\mu, \sigma^2)$, then (5.45) becomes

$$E\{g(X)(X - \mu)\} = \sigma^2 E g'(X).$$

This immediately shows that $E(X) = \mu$ (take $g(x) = 1$) and $E(X^2) = \sigma^2 + \mu^2$ (take $g(x) = x$). Higher-order moments are equally easy to calculate (Problem 5.18). ‖

Not only are the moments of the statistics T_i appearing in (5.1) and (5.2) of interest but also the family of distributions of the T's. This turns out again to be an exponential family.

Theorem 5.17 *If X is distributed according to an exponential family with density (5.1) with respect to a measure μ over $(\mathcal{X}, \mathcal{A})$, then $T = (T_1, \ldots, T_s)$ is distributed according to an exponential family with density*

$$(5.46) \quad \exp[\Sigma \eta_i t_i - A(\eta)] k(t)$$

with respect to a measure ν *over* E_s.

For a proof, see, for example, TSH2, Section 2.7, Lemma 8.

Let us now apply this theorem to the case of two independent exponential families with densities (5.10). Then it follows from Theorem 5.17 that $(T_1+U_1, \ldots, T_s+U_s)$ is also distributed according to an s-dimensional exponential family, and by induction, this result extends to the sum of more than two independent terms. In particular, let $X_1, \ldots, X_n$ be independently distributed, each according to a one-parameter exponential family with density

$$\exp[\eta T_i(x_i) - A_i(\eta)]\, h_i(x_i). \tag{5.47}$$

Then, the sum $\sum_{i=1}^n T_i(X_i)$ is again distributed according to a one-parameter exponential family. In fact, the sum of independent Poisson or normal variables again has a distribution of the same type, and the same is true for a sum of independent binomial variables with common p, or a sum of independent gamma variables $\Gamma(\alpha_i, b)$ with common b.

The normal distributions $N(\xi, \sigma^2)$ for fixed σ constitute both a one-parameter exponential family (Example 5.12) and a location family (Table 4.1). It is natural to ask whether there are any other families that enjoy this double advantage. Another example is obtained by putting $X = \log Y$ where Y has the gamma distribution $\Gamma(\alpha, b)$ given by (5.41), and where the location parameter θ is $\theta = \log b$. Since multiplication of a random variable by a constant $c \neq 0$ preserves both the exponential and location structure, a more general example is provided by the random variable $c \log Y$ for any $c \neq 0$. It was shown by Dynkin (1951) and Ferguson (1962) that the cases in which X is normal or is equal to $c \log Y$, with Y being gamma, provide the only examples of exponential location families.

The $\Gamma(\alpha, b)$ distribution, with known parameter α, constitutes an example of an exponential scale family. Another example of an exponential scale family is provided by the inverse Gaussian distribution (see Problem 5.22), which has been extensively studied by Tweedie (1957). For a general treatment of these and other results relating exponential and group families, see Barndorff-Nielsen et al. (1992) or Barndorff-Nielsen (1988).

6 Sufficient Statistics

The starting point of a statistical analysis, as formulated in the preceding sections, is a random observable X taking on values in a sample space $\mathcal{X}$, and a family of possible distributions of X. It often turns out that some part of the data carries no information about the unknown distribution and that X can therefore be replaced by some statistic $T = T(X)$ (not necessarily real-valued) without loss of information. A statistic T is said to be *sufficient* for X, or for the family $\mathcal{P} = \{P_\theta, \theta \in \Omega\}$ of possible distributions of X, or for θ, if the conditional distribution of X given $T = t$ is independent of θ for all t.

This definition is not quite precise and we shall return to it later in this section. However, consider first in what sense a sufficient statistic T contains all the information about θ contained in X. For that purpose, suppose that an investigator reports the value of T, but on being asked for the full data, admits that they have been discarded. In an effort at reconstruction, one can use a random mechanism (such as a pseudo-random number generator) to obtain a random quantity X' distributed according to the conditional distribution of X given t. (This would not be possible, of course, if the conditional distribution depended on the unknown θ.) Then the unconditional distribution of X' is the same as that of X, that is,

$$P_\theta(X' \in A) = P_\theta(X \in A) \quad \text{for all } A,$$

regardless of the value of θ. Hence, from a knowledge of T alone, it is possible to construct a quantity X' which is completely equivalent to the original X. Since X and X' have the same distribution for all θ, they provide exactly the same information about θ (for example, the estimators $\delta(X)$ and $\delta(X')$ have identical distributions for any θ).

In this sense, a sufficient statistic provides a reduction of the data without loss of information. This property holds, of course, only as long as attention is restricted to the model $\mathcal{P}$ and no distributions outside $\mathcal{P}$ are admitted as possibilities. Thus, in particular, restriction to T is not appropriate when testing the validity of $\mathcal{P}$.

The construction of X' is, in general, effected with the help of an independent random mechanism. An estimator $\delta(X')$ depends, therefore, not only on T but also on this mechanism. It is thus not an estimator as defined in Section 1, but a randomized estimator. Quite generally, if X is the basic random observable, a *randomized estimator* of $g(\theta)$ is a rule which assigns to each possible outcome x of X a random variable $Y(x)$ with a known distribution. When $X = x$, an observation of $Y(x)$ will be taken and will constitute the estimate of $g(\theta)$. The risk, defined by (1.10), of the resulting estimator is then

$$\int_{\mathcal{X}} \left[\int_{\mathcal{Y}} L(\theta, y) dP_{Y|X=x}(y) \right] dP_{X|\theta}(x),$$

where the probability measure in the inside integral does not depend on θ. With this representation, the operational significance of sufficiency can be formally stated as follows.

Theorem 6.1 *Let X be distributed according to $P_\theta \in \mathcal{P}$ and let T be sufficient for $\mathcal{P}$. Then, for any estimator $\delta(X)$ of $g(\theta)$, there exists a (possibly randomized) estimator based on T which has the same risk function as $\delta(X)$.*

Proof. Let X' be constructed as above so that $\delta'(X)$ is an (possibly randomized) estimator depending on the data only through T. Since $\delta(X)$ and $\delta'(X)$ have the same distribution, they also have the same risk function. □

Example 6.2 Poisson sufficient statistic. Let X_1, X_2 be independent Poisson variables with common expectation λ, so that their joint distribution is

$$P(X_1 = x_1, X_2 = x_2) = \frac{\lambda^{x_1+x_2}}{x_1!x_2!} e^{-2\lambda}.$$

Then, the conditional distribution of X_1 given $X_1 + X_2 = t$ is given by

$$P(X_1 = x_1 | X_1 + X_2 = t) = \frac{\lambda^t e^{-2\lambda}/x_1!(t - x_1)!}{\sum_{y=0}^t \lambda^t e^{-2\lambda}/y!(t - y)!}$$

$$= \frac{1}{x_1!(t-x_1)!}\left(\frac{1}{\sum_{y=0}^{t} 1/y!(t-y)!}\right)^{-1}.$$

Since this is independent of λ, so is the conditional distribution given t of $(X_1, X_2 = t - X_1)$, and hence $T = X_1 + X_2$ is a sufficient statistic for λ. To see how to reconstruct (X_1, X_2) from T, note that

$$\Sigma \frac{1}{y!(t-y)!} = \frac{1}{t!}2^t$$

so that

$$P(X_1 = x_1 | X_1 + X_2 = t) = \binom{t}{x_1}\left(\frac{1}{2}\right)^{x_1}\left(\frac{1}{2}\right)^{t-x_1},$$

that is, the conditional distribution of X_1 given t is the binomial distribution $b(1/2, t)$ corresponding to t trials with success probability 1/2. Let X'_1 and $X'_2 = t - X'_1$ be respectively the number of heads and the number of tails in t tosses with a fair coin. Then, the joint conditional distribution of (X'_1, X'_2) given t is the same as that of (X_1, X_2) given t. ||

Example 6.3 Sufficient statistic for a uniform distribution. Let $X_1, \ldots, X_n$ be independently distributed according to the uniform distribution $U(0, \theta)$. Let T be the largest of the n X's, and consider the conditional distribution of the remaining $n-1$ X's given t. Thinking of the n variables as n points on the real line, it is intuitively obvious and not difficult to see formally (Problem 6.2) that the remaining $n-1$ points (after the largest is fixed at t) behave like $n-1$ points selected at random from the interval $(0, t)$. Since this conditional distribution is independent of θ, T is sufficient. Given only $T = t$, it is obvious how to reconstruct the original sample: Select $n-1$ points at random on $(0, t)$. ||

Example 6.4 Sufficient statistic for a symmetric distribution. Suppose that X is normally distributed with mean zero and unknown variance σ^2 (or more generally that X is symmetrically distributed about zero). Then, given that $|X| = t$, the only two possible values of X are $\pm t$, and by symmetry, the conditional probability of each is 1/2. The conditional distribution of X given t is thus independent of σ and $T = |X|$ is sufficient. In fact, a random variable X' with the same distribution as X can be obtained from T by tossing a fair coin and letting $X' = T$ or $-T$ as the coin falls heads or tails. ||

The definition of sufficiency given at the beginning of the section depends on the concept of conditional probability, and this, unfortunately, is not capable of a treatment which is both general and elementary. Difficulties arise when $P_\theta(T = t) = 0$, so that the conditioning event has probability zero. The definition of conditional probability can then be changed at one or more values of t (in fact, at any set of t values which has probability zero) without affecting the distribution of X, which is the result of combining the distribution of T with the conditional distribution of X given T.

In elementary treatments of probability theory, the conditional probability $P(X \in A|t)$ is considered for fixed t as defining the conditional distribution of X given

$T = t$. A more general approach can be obtained by a change of viewpoint, namely by considering $P(X \in A|t)$ for fixed A as a function of t, defined in such a way that in combination with the distribution of T, it leads back to the distribution of X. (See TSH2, Chapter 2, Section 4 for details.) This provides a justification, for instance, of the assignment of conditional probabilities in Example 6.4 and Example 6.10.

In the same way, the conditional expectation $\eta(t) = E[\delta(X)|t]$ can be defined in such a way that

$$E\eta(T) = E\delta(X), \tag{6.1}$$

that is, so that the expected value of the conditional expectation is equal to the unconditional expectation.

Conditional expectation essentially satisfies the usual laws of expectation. However, since it is only determined up to sets of probability zero, these laws can only hold a.e. More specifically, we have with probability 1

$$E[af(X) + bg(X)|t] = aE[f(X)|t] + bE[g(X)|t]$$

and

$$E[b(T)f(X)|t] = b(t)E[f(X)|t]. \tag{6.2}$$

As just discussed, the functions $P(A|t)$ are not uniquely defined, and the question arises whether determinations exist which, for each fixed t, define a conditional probability. It turns out that this is not always possible. [See Romano and Siegel (1986), who give an example due to Ash (1972). A more detailed treatment is Blackwell and Ryll-Nardzewsky (1963).] It is possible when the sample space is Euclidean, as will be the case throughout most of this book (see TSH2, Chapter 2, Section 5). When this is the case, a statistic T can be defined to be sufficient if there exists a determination of the conditional distribution functions of X given t which is independent of θ.

The determination of sufficient statistics by means of the definition is inconvenient since it requires, first, guessing a statistic T that might be sufficient and, then, checking whether the conditional distributions of X given t is independent of θ. However, for dominated families, that is, when the distributions have densities with respect to a common measure, there is a simple criterion for sufficiency.

Theorem 6.5 (Factorization Criterion) *A necessary and sufficient condition for a statistic T to be sufficient for a family $\mathcal{P} = \{P_\theta, \theta \in \Omega\}$ of distributions of X dominated by a σ-finite measure μ is that there exist non-negative functions g_θ and h such that the densities p_θ of P_θ satisfy*

$$p_\theta(x) = g_\theta[T(x)]h(x) \quad (a.e.\ \mu). \tag{6.3}$$

Proof. See TSH2, Section 2.6, Theorem 8 and Corollary 1. □

Example 6.6 Continuation of Example 6.2. Suppose that $X_1, \ldots, X_n$ are iid according to a Poisson distribution with expectation λ. Then

$$P_\lambda(X_1 = x_1, \ldots, X_n = x_n) = \lambda^{\Sigma x_i} e^{-n\lambda} / \Pi(x_i!).$$

This satisfies (6.3) with $T = \Sigma X_i$, which is therefore sufficient. ||

Example 6.7 Normal sufficient statistic. Let $X_1, \ldots, X_n$ be iid as $N(\xi, \sigma^2)$ so that their joint density is

$$(6.4)\qquad p_{\xi,\sigma}(x) = \frac{1}{(\sqrt{2\pi}\sigma)^n} \exp\left[-\frac{1}{2\sigma^2}\Sigma x_i^2 + \frac{\xi}{\sigma^2}\Sigma x_i - \frac{n}{2\sigma^2}\xi^2\right].$$

Then it follows from the factorization criterion that $T = (\Sigma X_i^2, \Sigma X_i)$ is sufficient for $\theta = (\xi, \sigma^2)$. Sometimes it is more convenient to replace T by the equivalent statistic $T' = (\bar{X}, S^2)$ where $\bar{X} = \Sigma X_i/n$ and $S^2 = \Sigma(X_i - \bar{X})^2 = \Sigma X_i^2 - n\bar{X}^2$. The two representations are equivalent in that they identify the same points of the sample space, that is, $T(x) = T(y)$ if and only if $T'(x) = T'(y)$. ||

Example 6.8 Continuation of Example 6.3. The joint density of a sample $X_1, \ldots, X_n$ from $U(0, \theta)$ is

$$(6.5)\qquad p_\theta(\mathbf{x}) = \frac{1}{\theta^n} \prod_{i=1}^{n} I(0 < x_i) I(x_i < \theta)$$

where the indicator function, $I(\cdot)$ is defined in (2.6). Now

$$\prod_{i=1}^{n} I(0 < x_i) I(x_i < \theta) = I(x_{(n)} < \theta) \prod_{i=1}^{n} I(0 < x_i)$$

where $x_{(n)}$ is the largest of the x values. It follows from Theorem 6.5 that $X_{(n)}$ is sufficient, as had been shown directly in Example 6.3. ||

As a final illustration, consider Example 6.4 from the present point of view.

Example 6.9 Continuation of Example 6.4. If X is distributed as $N(0, \sigma^2)$, the density of X is

$$\frac{1}{\sqrt{2\pi}\sigma} e^{-x^2/2\sigma^2}$$

which depends on x only through x^2, so that (6.3) holds with $T(x) = x^2$. As always, of course, there are many equivalent statistics such as $|X|$, X^4 or e^{X^2}. ||

Quite generally, two statistics, $T = T(X)$ and $T' = T'(X)$, will be said to be equivalent (with respect to a family $\mathcal{P}$ of distributions of X) if each is a function of the other a.e. $\mathcal{P}$, that is, if there exists a $\mathcal{P}$-null set N and functions f and g such that $T(x) = f[T'(x)]$ and $T'(x) = g[T(x)]$ for all $x \in N$. Two such statistics carry the same amount of information.

Example 6.10 Sufficiency of order statistics. Let $X = (X_1, \ldots, X_n)$ be iid according to an unknown continuous distribution F and let $T = (X_{(1)}, \ldots, X_{(n)})$ where $X_{(1)} < \cdots < X_{(n)}$ denotes the ordered observations, the so-called *order statistics*. By the continuity assumptions, the X's are distinct with probability 1. Given T, the only possible values for X are the $n!$ vectors $(X_{(i_1)}, \cdots, X_{(i_n)})$, and by symmetry, each of these has conditional probability $1/n!$ The conditional distribution is thus independent of F, and T is sufficient. In fact, a random vector X' with the same distribution as X can be obtained from T by labeling the n coordinates of T at random. Equivalent to T is the statistic $U = (U_1, \ldots, U_n)$

where $U_1 = \Sigma X_i$, $U_2 = \Sigma X_i X_j$ $(i \neq j), \ldots, U_n = X_1 \cdots X_n$, and also the statistic $V = (V_1, \ldots, V_n)$ where $V_k = X_1^k + \cdots + X_n^k$ (Problem 6.9). ||

Equivalent forms of a sufficient statistic reduce the data to the same extent. There may, however, also exist sufficient statistics which provide different degrees of reduction.

Example 6.11 Different sufficient statistics. Let $X_1, \ldots, X_n$ be iid as $N(0, \sigma^2)$ and consider the statistics

$$T_1(X) = (X_1, \ldots, X_n),$$
$$T_2(X) = (X_1^2, \ldots, X_n^2),$$
$$T_3(X) = (X_1^2 + \cdots + X_m^2, X_{m+1}^2 + \cdots + X_n^2),$$
$$T_4(X) = X_1^2 + \cdots + X_n^2.$$

These are all sufficient (Problem 6.5), with T_i providing increasing reduction of the data as i increases. ||

It follows from the interpretation of sufficiency given at the beginning of this section that if T is sufficient and $T = H(U)$, then U is also sufficient. Knowledge of U implies knowledge of T and hence permits reconstruction of the original data. Furthermore, T provides a greater reduction of the data than U unless H is 1:1, in which case T and U are equivalent. A sufficient statistic T is said to be *minimal* if of all sufficient statistics it provides the greatest possible reduction of the data, that is, if for any sufficient statistic U there exists a function H such that $T = H(U)$ (a.e. $\mathcal{P}$). Minimal sufficient statistics can be shown to exist under weak assumptions (see, for example, Bahadur, 1954), but exceptions are possible (Pitcher 1957, Landers and Rogge 1972). Minimal sufficient statistics exist, in particular if the basic measurable space is Euclidean in the sense of Example 2.2 and the family $\mathcal{P}$ of distributions is dominated (Bahadur 1957).

It is typically fairly easy to construct a minimal sufficient statistic. For the sake of simplicity, we shall restrict attention to the case that the distributions of $\mathcal{P}$ all have the same support (but see Problems 6.11 - 6.17).

Theorem 6.12 *Let $\mathcal{P}$ be a finite family with densities p_i, $i = 0, 1, \ldots, k$, all having the same support. Then, the statistic*

$$T(X) = \left(\frac{p_1(X)}{p_0(X)}, \frac{p_2(X)}{p_0(X)}, \ldots, \frac{p_k(X)}{p_0(X)}\right) \tag{6.6}$$

is minimal sufficient.

The proof is an easy consequence of the following corollary of Theorem 6.5 (Problem 6.6).

Corollary 6.13 *Under the assumptions of Theorem 6.5, a necessary and sufficient condition for a statistic U to be sufficient is that for any fixed θ and θ_0, the ratio $p_\theta(x)/p_{\theta_0}(x)$ is a function only of $U(x)$.*

Proof of Theorem 6.12. The corollary states that U is a sufficient statistic for $\mathcal{P}$ if and only if T is a function of U, and this proves T to be minimal. □

Theorem 6.12 immediately extends to the case that $\mathcal{P}$ is countable. Generalizations to uncountable families are also possible (see Lehmann and Scheffé 1950, Dynkin 1951, and Barndorff-Nielsen, Hoffmann-Jorgensen, and Pedersen 1976), but must contend with measure-theoretic difficulties. In most applications, minimal sufficient statistics can be obtained for uncountable families by combining Theorem 6.12 with the following lemma.

Lemma 6.14 *If $\mathcal{P}$ is a family of distributions with common support and $\mathcal{P}_0 \subset \mathcal{P}$, and if T is minimal sufficient for $\mathcal{P}_0$ and sufficient for $\mathcal{P}$, it is minimal sufficient for $\mathcal{P}$.*

Proof. If U is sufficient for $\mathcal{P}$, it is also sufficient for $\mathcal{P}_0$, and hence T is a function of U. □

Example 6.15 Location families. As an application, let us now determine minimal sufficient statistics for a sample $X_1, \ldots, X_n$ from a location family $\mathcal{P}$, that is, when

$$p_\theta(x) = f(x_1 - \theta) \cdots f(x_n - \theta), \tag{6.7}$$

where f is assumed to be known. By Example 6.10, sufficiency permits the rather trivial reduction to the order statistics for all f. However, this reduction uses only the iid assumption and neither the special structure (6.7) nor the knowledge of f. To illustrate the different possibilities that arise when this knowledge is utilized, we shall take for f the six densities of Table 4.1, each with $b = 1$.

(i) *Normal.* If $\mathcal{P}_0$ consists of the two distributions $N(\theta_0, 1)$ and $N(\theta_1, 1)$, it follows from Theorem 6.12 that the minimal sufficient statistic for $\mathcal{P}_0$ is $T(x) = p_{\theta_1}(X)/p_{\theta_0}(X)$, which is equivalent to $\bar{X}$. Since $\bar{X}$ is sufficient for $\mathcal{P} = \{N(\theta, 1), -\infty < \theta < \infty\}$ by the factorization criterion, it is minimal sufficient.

(ii) *Exponential.* If the X's are distributed as $E(\theta, 1)$, it is easily seen that $X_{(1)}$ is minimal sufficient (Problem 6.17).

(iii) *Uniform.* For a sample from $U(\theta - 1/2, \theta + 1/2)$, the minimal sufficient statistic is $(X_{(1)}, X_{(n)})$ (Problem 6.16).

In these three instances, sufficiency was able to reduce the original n-dimensional data to one or two dimensions. Such extensive reductions are not possible for the remaining three distributions of Table 4.1.

(iv) *Logistic.* The joint density of a sample from $L(\theta, 1)$ is

$$p_\theta(x) = \exp[-\Sigma(x_i - \theta)]/\prod\{1 + \exp[-(x_i - \theta)]\}^2. \tag{6.8}$$

Consider a subfamily $\mathcal{P}_0$ consisting of the distribution (6.8) with $\theta_0 = 0$ and $\theta_1, \ldots, \theta_k$. Then by Theorem 6.12, the minimal sufficient statistic for $\mathcal{P}_0$ is $T(X) = [T_1(X), \ldots, T_k(X)]$, where

$$T_j(\mathbf{x}) = e^{n\theta_j} \prod_{i=1}^{n} \left(\frac{1 + e^{-x_i}}{1 + e^{-x_i + \theta_j}} \right)^2. \tag{6.9}$$

We shall now show that for $k = n + 1$, $T(X)$ is equivalent to the order statistics, that is, that $T(x) = T(y)$ if and only if $x = (x_1, \ldots, x_n)$ and $y = (y_1, \ldots, y_n)$ have the same order statistics, which means that one is a permutation of the other. The equation $T_j(x) = T_j(y)$ is equivalent to

$$\Pi\left(\frac{1+\exp(-x_i)}{1+\exp(-x_i+\theta_j)}\right)^2 = \Pi\left(\frac{1+\exp(-y_i)}{1+\exp(-y_i+\theta_j)}\right)^2$$

and hence $T(x) = T(y)$ to

$$\prod_{i=1}^{n}\frac{1+\xi u_i}{1+u_i} = \prod_{i=1}^{n}\frac{1+\xi v_i}{1+v_i} \quad \text{for } \xi = \xi_1, \ldots, \xi_{n+1}, \tag{6.10}$$

where $\xi_j = e^{\theta_j}$, $u_i = e^{-x_i}$, and $v_i = e^{-y_i}$. Now the left- and right-hand sides of (6.10) are polynomials in ξ of degree n which agree for $n+1$ values of ξ if and only if the coefficients of ξ^r agree for all $r = 0, 1, \ldots, n$. For $r = 0$, this implies $\Pi(1+u_i) = \Pi(1+v_i)$, so that (6.10) reduces to $\Pi(1+\xi u_i) = \Pi(1+\xi v_i)$ for $\xi = \xi_1, \ldots, \xi_{n+1}$, and hence for all ξ. It follows that $\Pi(\eta+u_i) = \Pi(\eta+v_i)$ for all η, so that these two polynomials in η have the same roots. Since this is equivalent to the x's and y's having the same order statistics, the proof is complete.

Similar arguments show that in the Cauchy and double exponential cases, too, the order statistics are minimal sufficient (Problem 6.10). This is, in fact, the typical situation for location families, examples (i) through (iii) being happy exceptions. ||

As a second application of Theorem 6.12 and Lemma 6.1, let us determine minimal sufficient statistics for exponential families.

Corollary 6.16 (Exponential Families) *Let X be distributed with density (5.2). Then, $T = (T_1, \ldots, T_s)$ is minimal sufficient provided the family (5.2) satisfies one of the following conditions:*

(i) It is of full rank.

(ii) The parameter space contains $s+1$ points $\eta^{(j)} (j = 0, \ldots, s)$, which span E_s, in the sense that they do not belong to a proper affine subspace of E_s.

Proof. That T is sufficient follows immediately from Theorem 6.5. To prove minimality under assumption (i), let $\mathcal{P}_0$ be a subfamily consisting of $s+1$ distributions $\eta^{(j)} = (\eta_1^{(j)}, \ldots, \eta_s^{(j)})$, $j = 0, 1, \ldots, s$. Then, the minimal sufficient statistic for $\mathcal{P}_0$ is equivalent to

$$\Sigma(\eta_i^{(1)} - \eta_i^{(0)})T_i(X), \ldots, \Sigma(\eta_i^{(s)} - \eta_i^{(0)})T_i(X),$$

which is equivalent to $T = [T_1(X), \ldots, T_s(X)]$, provided the $s \times s$ matrix $||\eta_i^{(j)} - \eta_i^{(0)}||$ is nonsingular. A subfamily $\mathcal{P}_0$ for which this condition is satisfied exists under the assumption of full rank.

The proof of minimality under assumption (ii) is similar. □

It is seen from this result that the sufficient statistics T of Examples 6.6 and 6.7 are minimal. The following example illustrates the applicability of part (ii).

Example 6.17 Minimal sufficiency in curved exponential families. Let $X_1, X_2, \ldots, X_n$ have joint density (6.4), but, as in Example 5.4, assume that $\xi = \sigma$, so the parameter space is the curve of Figure 10.1 (see Note 10.6). The statistic $T = (\sum X_i, \sum X_i^2)$ is sufficient, and it is also minimal by Corollary 6.16. To see this, recall that the natural parameter is $\eta = (1/\xi, -1/2\xi^2)$, and choose

$$\eta^{(0)} = \left(1, -\tfrac{1}{2}\right), \quad \eta^{(1)} = \left(2, -\tfrac{1}{8}\right), \quad \eta^{(2)} = \left(3, -\tfrac{1}{18}\right)$$

and note that the 2×2 matrix

$$\begin{pmatrix} 2-1 & 3-1 \\ -\frac{1}{8}+\frac{1}{2} & -\frac{1}{18}+\frac{1}{2} \end{pmatrix}$$

has rank 2 and is invertible.

In contrast, suppose that the parameters are restricted according to $\xi = \sigma^2$, another curved exponential family. This defines an affine subspace (with zero curvature) and the sufficient statistic T is no longer minimal (Problem 6.20). ||

Let $X_1, \ldots, X_n$ be iid, each with density (5.2), assumed to be of full rank. Then, the joint distribution of the X's is again full-rank exponential, with $T = (T_1^*, \ldots, T_s^*)$ where $T_i^* = \Sigma_{j=1}^n T_i(X_j)$. This shows that in a sample from the exponential family (5.2), the data can be reduced to an s-dimensional sufficient statistic, regardless of the sample size.

The reduction of a sample to a smaller number of sufficient statistics greatly simplifies the statistical analysis, and it is therefore interesting to ask what other families permit such a reduction. The dimensionality of a sufficient statistic is a property which differs from those considered so far, in that it depends not only on the sets of points of the sample space for which the statistic takes on the same value but it also depends on these values; that is, the dimensionality may not be the same for different representations of a sufficient statistic (see, for example, Denny, 1964, 1969). To make the concept of dimensionality meaningful, let us call T a *continuous s-dimensional sufficient statistic* over a Euclidean sample space $\mathcal{X}$ if the assumptions of Theorem 6.5 hold, if $T(x) = [T_1(x), \ldots, T_s(x)]$ where T is continuous, and if the factorization (6.3) holds not only a.e. but for all $x \in \mathcal{X}$.

Theorem 6.18 *Suppose $X_1, \ldots, X_n$ are real-valued iid according to a distribution with density $f_\theta(x_i)$ with respect to Lebesgue measure, which is continuous in x_i and whose support for all θ is an interval I. Suppose that for the joint density of $X = (X_1, \ldots, X_n)$*

$$p_\theta(x) = f_\theta(x_1) \cdots f_\theta(x_n)$$

there exists a continuous k-dimensional sufficient statistic. Then

(i) if $k = 1$, there exist functions η_1, B and h such that (5.1) holds;

(ii) $k > 1$, and if the densities $f_\theta(x_i)$ have continuous partial derivatives with respect to x_i, then there exist functions η_i, B and h such that (5.1) holds with $s \leq k$.

For a proof of this result, see Barndorff-Nielsen and Pedersen (1968). A corresponding problem for the discrete case is considered by Andersen (1970a).

This theorem states essentially that among "smooth" absolutely continuous families of distributions with fixed support, exponential families are the only ones that permit dimensional reduction of the sample through sufficiency. It is crucial for this result that the support of the distributions P_θ is independent of θ. In the contrary case, a simple example of a family possessing a one-dimensional sufficient statistic for any sample size is provided by the uniform distribution (Example 6.3).

The Dynkin-Ferguson theorem mentioned at the end of the last section and Theorem 6.18 state roughly that (*a*) the only location families which are one-dimensional exponential families are the normal and log of gamma distributions and (*b*) only exponential families permit reduction of the data through sufficiency. Together, these results appear to say that the only location families with fixed support in which a dimensional reduction of the data is possible are the normal and log of gamma families. This is not quite correct, however, because a location family — although it is a one-dimensional family — may also be a curved exponential family.

Example 6.19 Location/curved exponential family. Let $X_1, \ldots, X_n$ be iid with joint density (with respect to Lebesgue measure)

$$(6.11) \qquad C \exp\left[-\sum_{i=1}^{n}(x_i - \theta)^4\right]$$
$$= C\exp(-n\theta^4)\exp(4\theta^3\Sigma x_i - 6\theta^2\Sigma x_i^2 + 4\theta\Sigma x_i^3 - \Sigma x_i^4).$$

According to (5.1), this is a three-dimensional exponential family, and it provides an example of a location family with a three-dimensional sufficient statistic satisfying all the assumptions of Theorem 6.18. This is a curved exponential family with parameter space $\Theta = \{(\theta_1, \theta_2, \theta_3) : \theta_1 = \theta_3^3, \theta_2 = \theta_3^2\}$, a curved subset of three-dimensional space. ||

The tentative conclusion, which had been reached just before Example 6.19 and which was contradicted by this example, is nevertheless basically correct. Typically, a location family with fixed support $(-\infty, \infty)$ will not constitute even a curved exponential family and will, therefore, not permit a dimensional reduction of the data without loss of information.

Example 6.15 shows that the degree of reduction that can be achieved through sufficiency is extremely variable, and an interesting question is, what characterizes the situations in which sufficiency leads to a substantial reduction of the data? The ability of a sufficient statistic to achieve such a reduction appears to be related to the amount of ancillary information it contains. A statistic $V(X)$ is said to be *ancillary* if its distribution does not depend on θ, and *first-order ancillary* if its expectation $E_\theta[V(X)]$ is constant, independent of θ. An ancillary statistic by itself contains no information about θ, but minimal sufficient statistics may still contain much ancillary material. In Example 6.15(iv), for instance, the differences $X_{(n)} - X_{(i)} (i = 1, \ldots, n-1)$ are ancillary despite the fact that they are functions of the minimal sufficient statistics $(X_{(1)}, \ldots, X_{(n)})$.

Example 6.20 Location ancillarity. Example 6.15(iv) is a particular case of a location family. Quite generally, when sampling from any location family, the differences $X_i - X_j, i \neq j$, are ancillary statistics. Similarly, when sampling from scale families, ratios are ancillary. See Problem 6.34 for details. ||

A sufficient statistic T appears to be most successful in reducing the data if no nonconstant function of T is ancillary or even first-order ancillary, that is, if $E_\theta[f(T)] = c$ for all $\theta \in \Omega$ implies $f(t) = c$ (a.e. $\mathcal{P}$). By subtracting c, this condition is seen to be equivalent to

$$(6.12) \qquad E_\theta[f(T)] = 0 \quad \text{for all} \quad \theta \in \Omega \text{ implies } f(t) = 0 \, (\text{a.e. } \mathcal{P})$$

where $\mathcal{P} = \{P_\theta, \theta \in \Omega\}$. A statistic T satisfying (6.12) is said to be *complete*. As will be seen later, completeness brings with it substantial simplifications of the statistical situation.

Since complete sufficient statistics are particularly effective in reducing the data, it is not surprising that a complete sufficient statistic is always minimal. Proofs are given in Lehmann and Scheffé (1950), Bahadur (1957), and Schervish (1995); see also Problem 6.29.

What happens to the ancillary statistics when the minimal sufficient statistic is complete is shown by the following result.

Theorem 6.21 (Basu's Theorem) *If T is a complete sufficient statistic for the family $\mathcal{P} = \{P_\theta, \theta \in \Omega\}$, then any ancillary statistic V is independent of T.*

Proof. If V is ancillary, the probability $p_A = P(V \in A)$ is independent of θ for all A. Let $\eta_A(t) = P(V \in A | T = t)$. Then, $E_\theta[\eta_A(T)] = p_A$ and, hence, by completeness,

$$\eta_A(t) = p_A(a.e. \mathcal{P}).$$

This establishes the independence of V and T. □

We conclude this section by examining some complete and incomplete families through examples.

Theorem 6.22 *If X is distributed according to the exponential family (5.2) and the family is of full rank, then $T = [T_1(X), \ldots, T_s(X)]$ is complete.*

For a proof, see TSH2 Section 4.3, Theorem 1; Barndorff-Nielsen (1978), Lemma 8.2.; or Brown (1986a), Theorem 2.12.

Example 6.23 Completeness in some one-parameter families. We give some examples of complete one-parameter families of distributions.

(i) Theorem 6.22 proves completeness of

(a) X for the binomial family $\{b(p, n), 0 < p < 1\}$

(b) X for the Poisson family $\{P(\lambda), 0 < \lambda\}$

(ii) *Uniform.* Let $X_1, \ldots, X_n$ be iid according to the uniform distribution $U(0, \theta)$, $0 < \theta$. It was seen in Example 6.3 that $T = X_{(n)}$ is sufficient for θ. To see that T is complete, note that

$$P(T \leq t) = t^n/\theta^n, \quad 0 < t < \theta,$$

so that T has probability density

$$(6.13) \qquad p_\theta(t) = nt^{n-1}/\theta^n, \quad 0 < t < \theta.$$

Suppose $E_\theta f(T) = 0$ for all θ, and let f^+ and f^- be its positive and negative parts, respectively. Then,

$$\int_0^\theta t^{n-1} f^+(t)\,dt = \int_0^\theta t^{n-1} f^-(t)\,dt$$

for all θ. It follows that

$$\int_A t^{n-1} f^+(t)\,dt = \int_A t^{n-1} f^-(t)\,dt$$

for all Borel sets A, and this implies $f = 0$ a.e.

(iii) *Exponential.* Let $Y_1, \ldots, Y_n$ be iid according to the exponential distribution $E(\eta, 1)$. If $X_i = e^{-Y_i}$ and $\theta = e^{-\eta}$, then $X_1, \ldots, X_n$ iid as $U(0, \theta)$ (Problem 6.28), and it follows from (ii) that $X_{(n)}$ or, equivalently, $Y_{(1)}$ is sufficient and complete. ||

Example 6.24 Completeness in some two-parameter families.

(i) *Normal* $N(\xi, \sigma^2)$. Theorem 6.22 proves completeness of $(\bar{X}, S^2)$ of Example 6.7 in the normal family $\{N(\xi, \sigma^2), -\infty < \xi < \infty, 0 < \sigma\}$.

(ii) *Exponential* $E(a, b)$. Let $X_1, \ldots, X_n$ be iid according to the exponential distribution $E(a, b), -\infty < a < \infty, 0 < b$, and let $T_1 = X_{(1)}, T_2 = \Sigma[X_i - X_{(1)}]$. Then, (T_1, T_2) are independently distributed as $E(a, b/n)$ and $\frac{1}{2}b\chi^2_{2n-2}$, respectively (Problem 6.18), and they are jointly sufficient and complete. Sufficiency follows from the factorization criterion. To prove completeness, suppose that

$$E_{a,b}[f(T_1, T_2)] = 0 \quad \text{for all } a, b.$$

Then if

$$g(t_1, b) = E_b[f(t_1, T_2)], \tag{6.14}$$

we have that for any fixed b,

$$\int_a^\infty g(t_1, b) e^{-nt_1/b}\,dt_1 = 0 \quad \text{for all } a.$$

It follows from Example 6.23(iii) that

$$g(t_1, b) = 0,$$

except on a set N_b of t_1 values which has Lebesgue measure zero and which may depend on b. Then, by Fubini's theorem, for almost all t_1 we have

$$g(t_1, b) = 0 \text{ a.e. in } b.$$

Since the densities of T_2 constitute an exponential family, $g(t_1, b)$ by (6.14) is a continuous function of b for any fixed t_1. It follows that for almost all t_1, $g(t_1, b) = 0$, not only a.e. but for all b. Applying completeness of T_2 to (6.14), we see that for almost all t_1, $f(t_1, t_2) = 0$ a.e. in t_2. Thus, finally, $f(t_1, t_2) = 0$ a.e. with respect to Lebesgue measure in the (t_1, t_2) plane. [For measurability aspects which have been ignored in this proof, see Lehmann and Scheffé (1955, Theorem 7.1).] ||

Example 6.25 Minimal sufficient but not complete.

(i) *Location uniform.* Let $X_1, \ldots, X_n$ be iid according to $U(\theta - 1/2, \theta + 1/2)$, $-\infty < \theta < \infty$. Here, $T = \{X_{(1)}, X_{(n)}\}$ is minimal sufficient (Problem 6.16). On the other hand, T is not complete since $X_{(n)} - X_{(1)}$ is ancillary. For example, $E_\theta[X_{(n)} - X_{(1)} - (n-1)/(n+1)] = 0$ for all θ.

(ii) *Curved normal family.* In the curved exponential family derived from the $N(\xi, \sigma^2)$ family with $\xi = \sigma$, we have seen (Example 6.17) that the statistic $T = (\sum x_i, \sum x_i^2)$ is minimal sufficient. However, it is not complete since there exists a function $f(T)$ satisfying (6.12). This follows from the fact that we can find unbiased estimators for ξ based on either $\sum X_i$ or $\sum X_i^2$ (see Problem 6.21). ||

We close this section with an illustration of sufficiency and completeness in logit dose-response models.

Example 6.26 Completeness in the logit model. For the model of Example 5.5, where X_i are independent $b(p_i, n_i)$, $i = 1, \ldots, m$, that is,

$$P(X_1 = x_1, \ldots, X_m = x_m) = \prod_{i=1}^{m} \binom{n_i}{x_i} p_i^{x_i}(1-p_i)^{n_i - x_i}, \tag{6.15}$$

it can be shown that $\mathbf{X} = (X_1, \cdots, X_m)$ is minimal sufficient. The natural parameters are the logits $\eta_i = \log[(p_i/(1-p_i)]$, $i = 1, \ldots, m$ [see (5.8)], and if the p_i's are unrestricted, the minimal sufficient statistic is also complete (Problem 6.23). ||

Example 6.27 Dose-response model. Suppose n_i subjects are each given dose level d_i of a drug, $i = 1, 2$, and that $d_1 < d_2$. The response of each subject is either 0 or 1, independent of the others, and the probability of a successful response is $p_i = \eta_\theta(d_i)$. The joint distribution of the response vector $\mathbf{X} = (X_1, X_2)$ is

$$p_\theta(\mathbf{x}) = \prod_{i=1}^{2} \binom{n_i}{x_i} [\eta_\theta(d_i)]^{x_i} [1 - \eta_\theta(d_i)]^{n_i - x_i}. \tag{6.16}$$

Note the similarity to the model (6.15).

The statistic $\mathbf{X}$ is minimal sufficient in the model (6.16), and remains so if $\eta_\theta(d_i)$ has the form

$$\eta_\theta(d_i) = 1 - e^{-\theta d_i}, \quad d_1 = 1, d_2 = 2, n_1 = 2, n_2 = 1. \tag{6.17}$$

However, it is not complete since

$$E_\theta[I(X_1 = 0) - I(X_2 = 0)] = 0. \tag{6.18}$$

If instead of (6.17), we assume that $\eta_\theta(d_i)$ is given by

$$\eta_\theta(d_i) = 1 - e^{-\theta_1 d_i - \theta_2 d_i^2}, \quad i = 1, 2, \tag{6.19}$$

where d_1/d_2 is an irrational number, then $\mathbf{X}$ is a complete sufficient statistic.

These models are special cases of those examined by Messig and Strawderman (1993), who establish conditions for minimal sufficiency and completeness in a large class of dose-response models. ||

Table 7.1. *Convex Functions*

Function ϕ	Interval (a, b)
(i) $\|x\|$	$-\infty < x < \infty$
(ii) x^2	$-\infty < x < \infty$
(iii) x^p, $p \geq 1$	$0 < x$
(iv) $1/x^p$, $p > 0$	$0 < x$
(v) e^x	$-\infty < x < \infty$
(vi) $-\log x$	$0 < x < \infty$

7 Convex Loss Functions

The property of convexity and the associated property of concavity play an important role in point estimation. In particular, the point estimation problem outlined in Section 1 simplifies in a number of ways when the loss function $L(\theta, d)$ is a convex function of d.

Definition 7.1 A real-valued function ϕ defined over an open interval $I = (a, b)$ with $-\infty \leq a < b \leq \infty$ is convex if for any $a < x < y < b$ and any $0 < \gamma < 1$

$$\phi[\gamma x + (1-\gamma)y] \leq \gamma\phi(x) + (1-\gamma)\phi(y). \tag{7.1}$$

The function is said to be *strictly convex* if strict inequality holds in (7.1) for all indicated values of x, y, and γ. A function ϕ is concave on (a, b) if $-\phi$ is convex.

Convexity is a very strong condition which implies, for example, that ϕ is continuous in (a, b) and has a left and right derivative at every point of (a, b). Proofs of these properties and of the other properties of convex functions stated in the following without proof can be found, for example, in Hardy, Littlewood, and Polya (1934), Rudin (1966), Roberts and Varberg (1973), or Dudley (1989).

Determination of whether or not a function is convex is often easy with the help of the following two criteria.

Theorem 7.2

(i) If ϕ is defined and differentiable on (a, b), then a necessary and sufficient condition for ϕ to be convex is that

$$\phi'(x) \leq \phi'(y) \quad \textit{for all } a < x < y < b. \tag{7.2}$$

The function is strictly convex if and only if the inequality (7.2) is strict for all $x < y$.

(ii) If, in addition, ϕ is twice differentiable, then the necessary and sufficient condition (7.2) is equivalent to

$$\phi''(x) \geq 0 \quad \textit{for all } a < x < b \tag{7.3}$$

with strict inequality sufficient (but not necessary) for strict convexity.

Example 7.3 Convex functions. From these criteria, it is easy to see that the functions of Table 7.1 are convex over the indicated intervals: In all these cases, ϕ is strictly convex, except in (i) and in (iii) with $p = 1$. ‖

In general, a convex function is strictly convex unless it is linear over some subinterval of (a, b) (Problems 7.1 and 7.6).

A basic property of convex functions is contained in the following theorem.

Theorem 7.4 *Let ϕ be a convex function defined on $I = (a, b)$ and let t be any fixed point in I. Then, there exists a straight line*

$$y = L(x) = c(x - t) + \phi(t) \tag{7.4}$$

through the point $[t, \phi(t)]$ such that

$$L(x) \leq \phi(x) \quad \textit{for all } x \textit{ in } I. \tag{7.5}$$

By definition, a function ϕ is convex if the value of the function at the weighted average of two points does not exceed the weighted average of its values at these two points. By induction, this is easily generalized to the average of any finite number of points (Problem 7.8). In fact, the inequality also holds for the weighted average of any infinite set of points, and in this general form, it is known as Jensen's inequality.

The weighted average of ϕ with respect to the weight function Λ is represented by

$$\int_I \phi \, d\Lambda \tag{7.6}$$

where Λ is a measure with $\Lambda(I) = 1$. In the particular case that Λ assigns measure γ and $1 - \gamma$ to the points x and y, respectively, this reduces to the right side of (7.1). It is convenient to interpret (7.6) as the expected value of $\phi(X)$, where X is a random variable taking on values in I according to the probability distribution Λ.

Theorem 7.5 (Jensen's Inequality) *If ϕ is a convex function defined over an open interval I, and X is a random variable with $P(X \in I) = 1$ and finite expectation, then*

$$\phi[E(X)] \leq E[\phi(X)]. \tag{7.7}$$

If ϕ is strictly convex, the inequality is strict unless X is a constant with probability 1.

Proof. Let $y = L(x)$ be the equation of the line which satisfies (7.5) and for which $L(t) = \phi(t)$ when $t = E(X)$. Then,

$$E[\phi(X)] \geq E[L(X)] = L[E(X)] = \phi[E(X)], \tag{7.8}$$

which proves (7.7). If ϕ is strictly convex, the inequality in (7.5) is strict for all $x \neq t$, and hence the inequality in (7.8) is strict unless $\phi(X) = E[\phi(X)]$ with probability 1. □

Note that the theorem does not exclude the possibility that $E[\phi(X)] = \infty$.

Corollary 7.6 *If X is a nonconstant positive random variable with finite expectation, then*

$$\frac{1}{E(X)} < E\left(\frac{1}{X}\right) \tag{7.9}$$

and

$$E(\log X) < \log[E(X)]. \tag{7.10}$$

Example 7.7 Entropy distance. For density functions f and g, we define the *entropy distance* between f and g, with respect to f (also known as *Kullback-Leibler Information of* g *at* f or *Kullback-Leibler distance between* g *and* f) as

$$E_f[\log(f(X)/g(X))] = \int \log[f(x)/g(x)]f(x)\,dx. \tag{7.11}$$

Corollary 7.6 shows that

$$\begin{aligned} E_f[\log(f(X)/g(X))] &= -E_f[\log(g(X)/f(X))] \\ &\geq -\log[E_f(g(X)/f(X))] \\ &= 0, \end{aligned} \tag{7.12}$$

and hence that the entropy distance is always non-negative, and equals zero if $f = g$. Note that inequality (7.12) also establishes

$$E_f \log[g(X)] \leq E_f \log[f(X)], \tag{7.13}$$

which plays an important role in the theory of the EM algorithm of Section 6.4.

Entropy distance was explored by Kullback (1968); for an exposition of its properties see, for example, Brown (1986a). Entropy distance has, more recently, found many uses in Bayesian analysis, see e.g., Berger (1985) or Robert (1994a), and Section 4.5. ‖

In Theorem 6.1, it was seen that if T is a sufficient statistic, then for any statistical procedure there exists an equivalent procedure (i.e., having the same risk function) based only on T. We shall now show that in estimation with a strictly convex loss function, a much stronger statement is possible: Given any estimator $\delta(X)$ which is not a function of T, there exists a *better* estimator depending only on T.

Theorem 7.8 (Rao-Blackwell Theorem) *Let X be a random observable with distribution* $P_\theta \in \mathcal{P} = \{P_{\theta'}, \theta' \in \Omega\}$, *and let T be sufficient for* $\mathcal{P}$. *Let* δ *be an estimator of an estimand* $g(\theta)$, *and let the loss function* $L(\theta, d)$ *be a strictly convex function of d. Then, if* δ *has finite expectation and risk,*

$$R(\theta, \delta) = EL[\theta, \delta(X)] < \infty,$$

and if

$$\eta(t) = E[\delta(X)|t], \tag{7.14}$$

the risk of the estimator $\eta(T)$ *satisfies*

$$R(\theta, \eta) < R(\theta, \delta) \tag{7.15}$$

unless $\delta(X) = \eta(T)$ *with probability 1.*

Proof. In Theorem 7.5, let $\phi(d) = L(\theta, d)$, let $\delta = \delta(X)$, and let X have the conditional distribution $P^{X|t}$ of X given $T = t$. Then

$$L[\theta, \eta(t)] < E\{L[\theta, \delta(X)]|t\}$$

unless $\delta(X) = \eta(T)$ with probability 1. Taking the expectation on both sides of this inequality yields (7.15), unless $\delta(X) = \eta(T)$ with probability 1. □

Some points concerning this result are worth noting.

1. Sufficiency of T is used in the proof only to ensure that $\eta(T)$ does not depend on θ and hence is an estimator.
2. If the loss function is convex but not strictly convex, the theorem remains true provided the inequality sign in (7.15) is replaced by $\leq$. Even in that case, the theorem still provides information beyond the results of Section 6 because it shows that the particular estimator $\eta(T)$ is at least as good as $\delta(X)$.
3. The theorem is not true if the convexity assumption is dropped. Examples illustrating this fact will be given in Chapters 2 and 5.

In Section 6, randomized estimators were introduced, and such estimators may be useful, for example, in reducing the maximum risk (see Chapter 5, Example 5.1.8), but this can never be the case when the loss function is convex.

Corollary 7.9 *Given any randomized estimator of $g(\theta)$, there exists a nonrandomized estimator which is uniformly better if the loss function is strictly convex and at least as good when it is convex.*

Proof. Note first that a randomized estimator can be obtained as a nonrandomized estimator $\delta^*(X, U)$, where X and U are independent and U is uniformly distributed on (0, 1). This is achieved by observing $X = x$ and then using U to construct the distribution of Y given $X = x$, where $Y = Y(x)$ is the random variable employed in the definition of a randomized estimator (Problem 7.10). To prove the theorem, we therefore need to show that given any estimator $\delta^*(X, U)$ of $g(\theta)$, there exists an estimator $\delta(X)$, depending on X only, which has uniformly smaller risk. However, this is an immediate consequence of the Rao-Blackwell theorem since for the observations (X, U), the statistic X is sufficient. For $\delta(X)$, one can therefore take the conditional expectation of $\delta^*(X, U)$ given X. □

An estimator δ is said to be *inadmissible* if there exists another estimator δ' which *dominates* it (that is, such that $R(\theta, \delta') \leq R(\theta, \delta)$ for all θ, with strict inequality for some θ) and *admissible* if no such estimator δ' exists. If the loss function L is strictly convex, it follows from Corollary 7.9 that every admissible estimator must be nonrandomized. Another property of admissible estimators in the strictly convex loss case is provided by the following uniqueness result.

Theorem 7.10 *If L is strictly convex and δ is an admissible estimator of $g(\theta)$, and if δ' is another estimator with the same risk function, that is, satisfying $R(\theta, \delta) = R(\theta, \delta')$ for all θ, then $\delta' = \delta$ with probability 1.*

Proof. If $\delta^* = \frac{1}{2}(\delta + \delta')$, then

$$R(\theta, \delta^*) < \frac{1}{2}[R(\theta, \delta) + R(\theta, \delta')] = R(\theta, \delta) \tag{7.16}$$

unless $\delta = \delta'$ with probability 1, and (7.16) contradicts the admissibility of δ. □

The preceding considerations can be extended to the situation in which the estimand $g(\theta) = [g_1(\theta), \ldots, g_k(\theta)]$ and the estimator $\delta(X) = [\delta_1(X), \ldots, \delta_k(X)]$ are vector-valued.

Definition 7.11 For any two points $\mathbf{x} = (x_1, \ldots, x_k)$ and $\mathbf{y} = (y_1, \ldots, y_k)$ in E_k, define $\gamma\mathbf{x} + (1-\gamma)\mathbf{y}$ to be the point with coordinates $\gamma x_i + (1-\gamma)y_i$, $i = 1, \ldots, k$.

(i) A *set* S in E_k is *convex* if for any $\mathbf{x}, \mathbf{y} \in S$, the points

$$\gamma\mathbf{x} + (1-\gamma)\mathbf{y}, \quad 0 < \gamma < 1$$

are also in S. (Geometrically, this means that the line segment connecting any two points in S lies in S.)

(ii) A *real-valued function* ϕ defined over an open convex set S in E_k is *convex* if (7.1) holds with x and y replaced by $\mathbf{x}$ and $\mathbf{y}$; it is strictly convex if the inequality is strict for all $\mathbf{x}$ and $\mathbf{y}$.

Example 7.12 Convex combination. If ϕ_j is a convex function of a real variable defined over an interval I_j for each $j = 1, \ldots, k$, then for any positive constants $a_1, \ldots, a_k$

$$\phi(\mathbf{x}) = \Sigma a_j \phi_j(x_j) \tag{7.17}$$

is a convex function defined over the k-dimensional rectangle with sides $I_1, \ldots, I_k$; it is strictly convex, provided $\phi_1, \ldots, \phi_k$ are all strictly convex. This example implies, in particular, that the loss function

$$L(\theta, \mathbf{d}) = \Sigma a_i [d_i - g_i(\theta)]^2 \tag{7.18}$$

is strictly convex. ||

A useful criterion to determine whether a given function ϕ is convex is the following generalization of (7.3).

Theorem 7.13 *Let ϕ be defined over an open convex set S in E_k and twice differentiable in S. Then, a necessary and sufficient condition for ϕ to be convex is that the $k \times k$ matrix with ijth element $\partial^2\phi(x_1, \ldots, x_k)/\partial x_i \partial x_j$, which is known as the* Hessian matrix, *is positive semidefinite; if the matrix is positive definite, then ϕ is strictly convex.*

Example 7.14 Quadratic loss. Consider the loss function

$$L(\theta, \mathbf{d}) = \Sigma\Sigma a_{ij}[d_i - g_i(\theta)][d_j - g_j(\theta)]. \tag{7.19}$$

Since $\partial^2 L/\partial d_i \partial d_j = a_{ij}$, L is strictly convex, provided the matrix $||a_{ij}||$ is positive definite. ||

Let us now consider some consequences of adopting a convex loss function in a location model. In Section 1, it was pointed out that there exists a unique number a minimizing $\Sigma(x_i - a)^2$, namely $\bar{x}$, and that the minimizing value of $\Sigma_{i=1}^n |x_i - a|$ is either unique (when n is odd) or the minimizing values constitute an interval. This interval structure of the minimizing values does not hold, for example, when minimizing $\Sigma\sqrt{|x_i - a|}$. In the case $n = 2$, for instance, there exist two minimizing values, $a = x_1$ and $a = x_2$ (Problem 7.12). This raises the general question of the set of values a minimizing $\Sigma\rho(x_i - a)$, which, in turn, is a special case of the following problem. Let X be a random variable and $L(\theta, d) = \rho(d - \theta)$ a loss function, with ρ even. Then, what can be said about the set of values a minimizing $E[\rho(X-a)]$? This specializes to the earlier case if X takes on the values $x_1, \ldots, x_n$ with probabilities $1/n$ each.

Theorem 7.15 *Let ρ be a convex function defined on $(-\infty, \infty)$ and X a random variable such that $\phi(a) = E[\rho(X - a)]$ is finite for some a. If ρ is not monotone, $\phi(a)$ takes on its minimum value and the set on which this value is taken is a closed interval. If ρ is strictly convex, the minimizing value is unique.*

The proof is based on the following lemma.

Lemma 7.16 *Let ϕ be a convex function on $(-\infty, \infty)$ which is bounded below and suppose that ϕ is not monotone. Then, ϕ takes on its minimum value; the set S on which this value is taken on is a closed interval and is a single point when ϕ is strictly convex.*

Proof. Since ϕ is convex and not monotone, it tends to ∞ as $x \to \pm\infty$. Since ϕ is also continuous, it takes on its minimizing value. That S is an interval follows from convexity and that it is closed follows from continuity. □

Proof of Theorem 7.15. By the lemma, it is enough to prove that ϕ is (strictly) convex and not monotone. That ϕ is not monotone follows from that fact that $\phi(a) \to \infty$ as $a \to \pm\infty$. This latter property of ϕ is a consequence of the facts that $X - a$ tends in probability to $\mp\infty$ as $a \to \pm\infty$ and that $\rho(t) \to \infty$ as $t \to \pm\infty$. (Strict) convexity of ϕ follows from the corresponding property of ρ. □

Example 7.17 Squared error loss. Let $\rho(t) = t^2$ and suppose that $E(X^2) < \infty$. Since ρ is strictly convex, if follows that $\phi(a)$ has a unique minimizing value. If $E(X) = \mu$, which by assumption is finite, we have, in fact,

$$\phi(a) = E(X - a)^2 = E(X - \mu)^2 + (\mu - a)^2, \tag{7.20}$$

which shows that $\phi(a)$ is a minimum if and only if $a = \mu$. ||

Example 7.18 Absolute error loss. Let $\rho(t) = |t|$ and suppose that $E|X| < \infty$. Since ρ is convex but not strictly convex, it follows from Theorem 7.15 that $\phi(a)$ takes on its minimum value and that the set S of minimizing values is a closed interval. The set S is, in fact, the set of *medians* of X (Problems 1.7 and 1.8). ||

The following is a useful consequence of Theorem 7.15 (see also Problem 7.27).

Corollary 7.19 *Under the assumptions of Theorem 7.15, suppose that ρ is even and X is symmetric about μ. Then, $\phi(a)$ attains its minimum at $a = \mu$.*

Proof. By Theorem 7.15 the minimum is taken on. If $\mu + c$ is a minimizing value, so is $\mu - c$ and so, therefore, are all values a between $\mu - c$ and $\mu + c$, which includes $a = \mu$. □

Now consider an example in which ρ is not convex.

Example 7.20 Nonconvex loss. Let $\rho(t) = 1$ if $|t| \geq k$ and $\rho(t) = 0$ otherwise. Minimizing $\phi(a)$ is then equivalent to maximizing $\psi(a) = P(|X - a| < k)$. Consider the following two special cases (Problem 7.22):

(i) The distribution of X has a probability density (with respect to Lebesgue measure) which is continuous, unimodal, and such that $f(x)$ decreases strictly as x moves away from the mode in either direction. Then, there exists a unique value a for which $f(a - k) = f(a + k)$, and this is the unique maximizing value of $\psi(a)$.

(ii) Suppose that f is even and U-shaped with $f(x)$ attaining its maximum at $x = \pm A$ and $f(x) = 0$ for $|x| > A$. Then, $\psi(a)$ attains its maximum at the two points $a = -A + k$ and $a = A - k$. ||

Convex loss functions have been seen to lead to a number of simplifications of estimation problems. One may wonder, however, whether such loss functions are likely to be realistic. If $L(\theta, d)$ represents not just a measure of inaccuracy but a real (for example, financial) loss, one may argue that all such losses are bounded: once you have lost all, you cannot lose any more. On the other hand, if d can take on all values in $(-\infty, \infty)$ or $(0, \infty)$, no nonconstant bounded function can be convex (Problem 7.18). Unfortunately, bounded loss functions with unbounded d can lead to completely unreasonable estimators (see, for example, Theorem 2.1.15). The reason is roughly that arbitrarily large errors can then be committed with essentially no additional penalty and their leverage used to unfair advantage. Perhaps convex loss functions result in more reasonable estimators because the large penalties they exact for large errors compensate for the unrealistic assumption of unbounded d: They make such values so expensive that the estimator will try hard to avoid them.

The most widely used loss function is squared error

$$L(\theta, d) = [d - g(\theta)]^2 \tag{7.21}$$

or slightly more generally weighted squared error

$$L(\theta, d) = w(\theta)[d - g(\theta)]^2. \tag{7.22}$$

Since these are strictly convex in d, the simplifications represented by Theorem 7.8, Corollary 7.9, and Theorem 7.10 are valid in these cases. The most slowly growing even convex loss function is absolute error

$$L(\theta, d) = |d - g(\theta)|. \tag{7.23}$$

The faster the loss function increases, the more attention it pays to extreme values of the estimators and hence to outlying observations, so that the performance of the resulting estimators is strongly influenced by the tail behavior of the assumed distribution of the observable random variables. As a consequence, fast-growing loss functions lead to estimators that tend to be sensitive to the assumptions made about this tail behavior, and these assumptions typically are based on little information and thus are not very reliable.

It turns out that the estimators produced by squared error loss often are uncomfortably sensitive in this respect. On the other hand, absolute error appears to go too far in leading to estimators which discard all but the central observations. For many important problems, the most appealing results are obtained from the use of loss functions which lie between (7.21) and (7.23). One interesting class of such loss functions, due to Huber (1964), puts

(7.24) $$L(\theta, d) = \begin{cases} [d - g(\theta)]^2 & \text{if } |d - g(\theta)| \leq k \\ 2k|d - g(\theta)| - k^2 & \text{if } |d - g(\theta)| \geq k. \end{cases}$$

This agrees with (7.21) for $|d - g(\theta)| \leq k$, but above k and below $-k$, it replaces the parabola with straight lines joined to the parabola so as to make the function continuous and continuously differentiable (Problem 7.21).

The Huber loss functions are convex but not strictly convex. An alternative family, which also interpolates between (7.21) and (7.23) and which is strictly convex, is

(7.25) $$L(\theta, d) = |d - g(\theta)|^p, \quad 1 < p < 2.$$

It is a disadvantage of both (7.24) and (7.25) that the resulting estimators, even in fairly simple problems, cannot be obtained in closed form and hence are more difficult to grasp intuitively and to interpret. This may account at least in part for the fact that squared error is the most commonly used loss function or measure of accuracy and that the classic estimators in most situations are the ones derived on this basis. As indicated at the end of Section 1, we shall develop here the theory under the more general assumption of convex loss functions (which, in practice, does not appear to be a serious limitation), but we shall work most examples for the conventional squared error loss. The issue of the robustness of the resulting estimators, which requires going outside the assumed model, will not be treated in detail here. References for further study of robustness include Huber (1981), Hampel et al. (1986), and Staudte and Sheather (1990).

With some care, the properties of convex and concave functions generalize to multivariate situations. For example, Theorem 7.4 generalizes to the following supporting hyperplane theorem for convex functions.

Theorem 7.21 *Let ϕ be a convex function defined over an open convex set S in E_k and let $\mathbf{t}$ be any point in S. Then, there exists a hyperplane*

(7.26) $$y = L(\mathbf{x}) = \Sigma c_i(x_i - t_i) + \phi(\mathbf{t})$$

through the point $[\mathbf{t}, \phi(\mathbf{t})]$ *such that*

(7.27) $$L(\mathbf{x}) \leq \phi(\mathbf{x}) \quad \textit{for all } \mathbf{x} \in S.$$

Jensen's inequality (Theorem 7.5) generalizes in the obvious way. The only changes that are needed are replacement of the interval I by an open convex set S, of the random variable X by a random vector $\mathbf{X}$ satisfying $P(\mathbf{X} \in S) = 1$, and of the expectation $E(X)$ by the expectation vector $E(\mathbf{X}) = [E(X_1), \ldots, E(X_k)]$. For the resulting modification of the inequality (7.7) to be meaningful, it is necessary to know that $E(\mathbf{X})$ is in S so that $\phi[E(\mathbf{X})]$ is defined.

Lemma 7.22 *If $\mathbf{X}$ is a random vector with $P(\mathbf{X} \in S) = 1$, where S is an open convex set in E_k, and if $E(\mathbf{X})$ exists, then $E(\mathbf{X}) \in S$.*

A formal proof is given by Ferguson (1967, p. 74). Here, we shall give only a sketch. Suppose that $k = 2$, and suppose that $\xi = E(\mathbf{X})$ is not in S. Then, Theorem 7.21 guarantees the existence of a line $a_1x_1 + a_2x_2 = b$ through the point (ξ_1, ξ_2) such that S lies entirely on one side of the line. By a rotation of the plane, it can be assumed without loss of generality that the equation of the line is $x_2 = \xi_2$ and that S lies above this line so that $P(X_2 > \xi_2) = 1$. It follows that $E(X_2) > \xi_2$, which is a contradiction.

The notions of convexity and concavity can also be extended to the multidimensional case in a slightly different way, one that examines the behavior of the function when it is averaged over spheres instead of over pairs of points.

Definition 7.23 A continuous function $f : R^k \to R$ is *superharmonic* at a point $\mathbf{x}_0 \in R^k$ if, for every $r > 0$, the average of f over the surface of the sphere $S_r(\mathbf{x}_0) = \{\mathbf{x} : ||\mathbf{x} - \mathbf{x}_0|| = r\}$ is less than or equal to $f(\mathbf{x}_0)$. The function f is superharmonic in R^p if it is superharmonic at each $\mathbf{x}_0 \in R^p$. (See Problem 7.15 for an extension.)

If we denote the average of f over the surface of the sphere by $A_{x_0}(f)$, we thus define f to be *superharmonic, harmonic*, or *subharmonic*, depending on whether $A_{x_0}(f)$ is less than or equal to, equal to, or greater than or equal to f, respectively. These definitions are analogous to those of convexity and concavity, but here we take the average over the surface of a sphere. (Note that in one dimension, the sphere reduces to two points, so superharmonic and concave are the same property.) The following characterization of superharmonicity, which is akin to that of Theorem 7.13, is typically easier to check than the definition. (For a proof, see Helms 1969).

Theorem 7.24 *If $f : R^k \to R$ is twice differentiable, then f is superharmonic in R^k if and only if for all $x \in R^k$,*

$$\sum_{i=1}^{k} \frac{\partial^2}{\partial x_i^2} f(x) \leq 0. \tag{7.28}$$

If Equation (7.28) is an equality, then f is harmonic, and if the inequality is reversed, then f is subharmonic.

Example 7.25 Subharmonic functions. Some multivariate analogs of the convex functions in Example 7.3 are subharmonic. For example, if $f(x_1, \ldots, x_k) = \sum_{i=1}^{k} x_i^p$ then

$$\sum_{i=1}^{k} \frac{\partial^2}{\partial x_i^2} f(x) = \sum_{i=1}^{k} p(p-1)x_i^{p-2}.$$

This function is subharmonic if $p \geq 1$ and $x_i > 0$, or if $p \geq 2$ is an even integer. Problem 7.14 considers some other multivariate functions. ||

Example 7.26 Subharmonic loss. The loss function of Example 7.14, given in Equation (7.19), has second derivative $\partial^2 L/\partial d_i^2 = a_{ii}$. Thus, it is subharmonic if, and only if, $\sum_i a_{ii} \geq 0$. This is a weaker condition than that needed for multidimensional convexity. ||

The property of superharmonicity is useful in the theory of minimax point estimation, as will be seen in Section 5.6.

8 Convergence in Probability and in Law

Thus far, our preparations have centered on "small-sample" aspects, that is, we have considered the sample size n as being fixed. However, it is often fruitful to consider a sequence of situations in which n tends to infinity. If the given sample size is sufficiently large, the limit behavior may provide an important complement to the small-sample behavior, and often discloses properties of estimators that are masked by complications inherent in small-sample calculations. In preparation for a study of such large-sample asymptotics in Chapter 6, we here present some of the necessary tools.

In particular, we review the probabilistic foundations necessary to derive the limiting behavior of estimators. It turns out that under rather weak assumptions, the limit distribution of many estimators is normal and hence depends only on a mean and a variance. This mitigates the effect of the underlying assumptions because the results become less dependent on the model and the loss function.

We consider a sample $\mathbf{X} = (X_1, \ldots, X_n)$ as a member of a sequence corresponding to $n = 1, 2$ (or, more generally, $n_0, n_0 + 1, \ldots$) and obtain the limiting behavior of estimator sequences as $n \to \infty$. Mathematically, the results are thus limit theorems.

In applications, the limiting results (particularly the asymptotic variances) are used as approximations to the situation obtaining for the actual finite n. A weakness of this approach is that, typically, no good estimates are available for the accuracy of the approximation. However, we can obtain at least some idea of the accuracy by numerical checks for selected values of n.

Suppose for a moment that $X_1, \ldots, X_n$ are iid according to a distribution P_θ, $\theta \in \Omega$, and that the estimand is $g(\theta)$. As n increases, more and more information about θ becomes available, and one would expect that for sufficiently large values of n, it would typically be possible to estimate $g(\theta)$ very closely. If $\delta_n = \delta_n(X_1, \ldots, X_n)$ is a reasonable estimator, of course, it cannot be expected to be close to $g(\theta)$ for every sample point $(x_1, \ldots, x_n)$ since the values of a particular sample may always be atypical (e.g., a fair coin may fall heads in 1000 successive spins). What one can hope for is that δ_n will be close to $g(\theta)$ with high probability.

This idea is captured in the following definitions, which do not assume the random variables to be iid.

Definition 8.1 A sequence of random variables Y_n defined over sample spaces $(\mathcal{Y}_n, \mathcal{B}_n)$ tends *in probability* to a constant c ($Y_n \xrightarrow{P} c$) if for every $a > 0$

$$P[|Y_n - c| \geq a] \to 0 \quad \text{as } n \to \infty. \tag{8.1}$$

A sequence of estimators δ_n of $g(\theta)$ is *consistent* if for every $\theta \in \Omega$

$$\delta_n \xrightarrow{P_\theta} g(\theta). \tag{8.2}$$

The following condition, which assumes the existence of second moments, frequently provides a convenient method for proving consistency.

Theorem 8.2 *Let $\{\delta_n\}$ be a sequence of estimators of $g(\theta)$ with mean squared error $E[\delta_n - g(\theta)]^2$.*

(i) If

$$E[\delta_n - g(\theta)]^2 \to 0 \quad \textit{for all } \theta, \tag{8.3}$$

then δ_n is consistent for estimating $g(\theta)$.

(ii) Equivalent to (8.3), δ_n is consistent if

$$b_n(\theta) \to 0 \quad \textit{and} \quad \text{var}_\theta(\delta_n) \to 0 \quad \textit{for all } \theta, \tag{8.4}$$

where b_n is the bias of δ_n.

(iii) In particular, δ_n is consistent if it is unbiased for each n and if

$$\text{var}_\theta(\delta_n) \to 0 \quad \textit{for all } \theta. \tag{8.5}$$

The proof follows from Chebychev's Inequality (see Problem 8.1).

Example 8.3 Consistency of the mean. Let $X_1, \ldots, X_n$ be iid with expectation $E(X_i) = \xi$ and variance $\sigma^2 < \infty$. Then, $\bar{X}$ is an unbiased estimator of ξ with variance σ^2/n, and hence is consistent by Theorem 8.2(iii). Actually, it was proved by Khinchin, see, for example, Feller 1968, Chapter X, Section 1,2) that consistency of $\bar{X}$ already follows from the existence of the expectation, so that the assumption of finite variance is not needed. ||

Note. The statement that $\bar{X}$ is consistent is shorthand for the fuller assertion that the sequence of estimators $\bar{X}_n = (X_1 + \cdots + X_n)/n$ is consistent. This type of shorthand is used very common and will be used here. However, the full meaning should be kept in mind.

Example 8.4 Consistency of S^2. Let $X_1, \ldots, X_n$ be iid with finite variance σ^2. Then, the unbiased estimator

$$S_n^2 = \Sigma(X_i - \bar{X})^2/(n-1)$$

is a consistent estimator of σ^2. To see this, assume without loss of generality that $E(X_i) = 0$, and note that

$$S_n^2 = \frac{n}{n-1}\left[\frac{1}{n}\Sigma X_i^2 - \bar{X}^2\right].$$

By Example 8.3, $\Sigma X_i^2/n \xrightarrow{P} \sigma^2$ and $\bar{X}^2 \xrightarrow{P} 0$. Since $n/(n-1) \to 1$, it follows from Problem 8.4 that $S_n^2 \xrightarrow{P} \sigma^2$. (See also Problem 8.5.) ||

Example 8.5 Markov chains. As an illustration of a situation involving dependent random variables, consider a two-state Markov chain. The variables $X_1, X_2, \ldots$ each take on the values 0 and 1, with the joint distribution determined by the initial probability $P(X_1 = 1) = p_1$, and the transition probabilities

$$P(X_{i+1} = 1|X_i = 0) = \pi_0, \quad P(X_{i+1} = 1|X_i = 1) = \pi_1,$$

of which we shall assume $0 < \pi_0, \pi_1 < 1$. For such a chain, the probability

$$p_k = P(X_k = 1)$$

typically depends on k and the initial probability p_1 (but see Problem 8.10). However, as $k \to \infty$, p_k tends to a limit p, which is independent of p_1. It is easy to see what the value of p must be. Consider the recurrence relation

$$(8.6) \qquad p_{k+1} = p_k\pi_1 + (1 - p_k)\pi_0 = p_k(\pi_1 - \pi_0) + \pi_0.$$

If

$$(8.7) \qquad p_k \to p,$$

this implies

$$(8.8) \qquad p = \frac{\pi_0}{1 - \pi_1 + \pi_0}.$$

To prove (8.7), it is only necessary to iterate (8.6) starting with $k = 1$ to find (Problem 8.6).

$$(8.9) \qquad p_k = (p_1 - p)(\pi_1 - \pi_0)^{k-1} + p.$$

Since $|\pi_1 - \pi_0| < 1$, the result follows.

For estimating p, after n trials, the natural estimator is $\bar{X}_n$, the frequency of ones in these trials. Since

$$E(\bar{X}_n) = (p_1 + \cdots + p_n)/n,$$

it follows from (8.7) that $E(\bar{X}_n) \to p$ (Problem 8.7), so that the bias of $\bar{X}_n$ tends to zero. Consistency of $\bar{X}_n$ will therefore follow if we can show that $\text{var}(\bar{X}_n) \to 0$. Now,

$$\text{var}(\bar{X}_n) = \sum_{i=1}^{n}\sum_{j=1}^{n} \text{cov}(X_i, X_j)/n^2.$$

As $n \to \infty$, this average of n^2 terms will go to zero if $\text{cov}(X_i, X_j) \to 0$ sufficiently fast as $|j - i| \to \infty$. The covariance of X_i and X_j can be obtained by a calculation similar to that leading to (8.9) and satisfies

$$(8.10) \qquad |\text{cov}(X_i, X_j)| \leq M|\pi_1 - \pi_0|^{j-i}.$$

From (8.10), one finds that $\text{var}(\bar{X}_n)$ is of order $1/n$ and hence that $\bar{X}_n$ is consistent (Problem 8.11).

Instead of p, one may be interested in estimating π_0 and π_1 themselves. Again, it turns out that the natural estimator $N_{01}/(N_{00} + N_{01})$ for π_0, where N_{0j} is the number of pairs (X_i, X_{i+1}) with $X_i = 0$, $X_{i+1} = j$, $j = 0, 1$, is consistent.

Consider, on the other hand, the estimation of p_1. It does not appear that observations beyond on the first provide any information about p_1, and one would therefore not expect to be able to estimate p_1 consistently. To obtain a formal proof, suppose for a moment that the π's are known, so that p_1 is the only unknown parameter. If a consistent estimator δ_n exists for the original problem, then δ_n will continue to be consistent under this additional assumption. However, when the π's are known, X_1 is a sufficient statistic for p_1 and the problem reduces to that of estimating a success probability from a single trial. That a consistent estimator of p_1 cannot exist under these circumstances follows from the definition of consistency. ||

When $X_1, \ldots, X_n$ are iid according to a distribution $P_\theta, \theta \in \Omega$, consistent estimators of real-valued functions of θ will exist in most of the situations we shall encounter (see, for example, Problem 8.8). There is, however, an important exception. Suppose the X's are distributed according to $F(x_i - \theta)$ where F is $N(\xi, \sigma^2)$, with θ, ξ, and σ^2 unknown. Then, no consistent estimator of θ exists. To see this, note that the X's are distributed as $N(\xi + \theta, \sigma^2)$. Thus, $\bar{X}$ is consistent for estimating $\xi + \theta$, but ξ and θ cannot be estimated separately because they are not uniquely defined, they are *unidentifiable* (see Definition 5.2). More precisely, for $X \sim P_{\theta,\xi}$, there exist pairs (θ_1, ξ_1) and (θ_2, ξ_2) with $\theta_1 \neq \theta_2$ for which $P_{\theta_1,\xi_1} = P_{\theta_2,\xi_2}$, showing the parameter θ to be unidentifiable. A parameter that is unidentifiable cannot be estimated consistently since $\delta(X_1, \ldots, X_n)$ cannot simultaneously be close to both θ_1 and θ_2.

Consistency is too weak a property to be of much interest in itself. It tells us that for large n, the error $\delta_n - g(\theta)$ is likely to be small but not whether the order of the error is $1/n$, $1/\sqrt{n}$, $1/\log n$, and so on. To obtain an idea of the rate of convergence of a consistent estimator δ_n, consider the probability

$$P_n(a) = P\left\{|\delta_n - g(\theta)| \leq \frac{a}{k_n}\right\}. \tag{8.11}$$

If k_n is bounded, then $P_n(a) \to 1$. On the other hand, if $k_n \to \infty$ sufficiently fast, $P_n(a) \to 0$. This suggests that for a given $a > 0$, there might exist an intermediate sequence $k_n \to \infty$ for which $P_n(a)$ tends to a limit strictly between 0 and 1. This will be the case for most of the estimators with which we are concerned. Commonly, there will exist a sequence $k_n \to \infty$ and a limit function H which is a continuous cdf such that for all a

$$P\{k_n[\delta_n - g(\theta)] \leq a\} \to H(a) \quad \text{as } n \to \infty. \tag{8.12}$$

We shall then say that the error $|\delta_n - g(\theta)|$ tends to zero at rate $1/k_n$. The rate, of course, is not uniquely determined by this definition. If $1/k_n$ is a possible rate, so is $1/k'_n$ for any sequence k'_n for which k'_n/k_n tends to a finite nonzero limit. On the other hand, if k'_n tends to ∞ more slowly (or faster) than k_n, that is, if $k'_n/k_n \to 0$ (or ∞), then $k'_n[\delta_n - g(\theta)]$ tends in probability to zero (or ∞) (Problem 8.12).

One can think of the normalizing constants k_n in (8.12) in another way. If δ_n is consistent, the errors $\delta_n - g(\theta)$ tend to zero as $n \to \infty$. Multiplication by constants k_n tending to infinity magnifies these minute errors—it acts as a microscope. If (8.12) holds, then k_n is just the right degree of magnification to give a well-focused picture of the behavior of the errors.

We formalize (8.12) in the following definition.

Definition 8.6 Suppose that $\{Y_n\}$ is a sequence of random variables with cdf

$$H_n(a) = P(Y_n \leq a)$$

and that there exists a cdf H such that

$$H_n(a) \to H(a) \tag{8.13}$$

at all points a at which H is continuous. Then, we shall say that the distribution functions H_n *converge weakly* to H, and that the Y_n have the *limit distribution* H, or *converge in law* to any random variable Y with distribution H. This will be denoted by $Y_n \xrightarrow{\mathcal{L}} Y$ or by $\mathcal{L}(Y_n) \to H$. We may also say that Y_n tends in law to H and write $Y_n \to H$.

The crucial assumption in (8.13) is that $H(-\infty) = 0$ and $H(+\infty) = 1$, that is, that no probability mass escapes to $\pm\infty$ (see Problem 1.37).

The following example illustrates the reason for requiring (8.13) to hold only for the continuity points of H.

Example 8.7 Degenerate limit distribution.

(i) Let Y_n be normally distributed with mean zero and variance σ_n^2 where $\sigma_n \to 0$ as $n \to \infty$.

(ii) Let Y_n be a random variable taking on the value $1/n$ with probability 1.

In both cases, it seems natural to say that Y_n tends in law to a random variable Y which takes on the value 0 with probability 1. The cdf $H(a)$ of Y is zero for $a < 0$ and 1 for $a \geq 0$. The cdf $H_n(a)$ of Y_n in both (i) and (ii) tends to $H(a)$ for all $a \neq 0$, but not for $a = 0$ (Problem 8.14). ||

An important property of weak convergence is given by the following theorem. Its proof, and those of Theorems 8.9-8.12, can be found in most texts on probability theory. See, for example, Billingsley (1995, Section 25).

Theorem 8.8 *The sequence Y_n converges in law to Y if and only if $E[f(Y_n)] \to E[f(Y)]$ for every bounded continuous real-valued function f.*

A basic tool for obtaining the limit distribution of many estimators of interest is the central limit theorem (CLT), of which the following is the simplest case.

Theorem 8.9 (Central Limit Theorem) *Let X_i $(i = 1, \ldots, n)$ be iid with $E(X_i) = \xi$ and* $\text{var}(X_i) = \sigma^2 < \infty$. *Then, $\sqrt{n}(\bar{X} - \xi)$ tends in law to $N(0, \sigma^2)$ and hence $\sqrt{n}(\bar{X} - \xi)/\sigma$ to the standard normal distribution $N(0, 1)$.*

The usefulness of this result is greatly extended by Theorems 8.10 and 8.12 below.

Theorem 8.10 *If $Y_n \xrightarrow{\mathcal{L}} Y$, and A_n and B_n tend in probability to a and b, respectively, then $A_n + B_nY_n \xrightarrow{\mathcal{L}} a + bY$.*

When Y_n converges to a distribution H, it is often required to evaluate probabilities of the form $P(Y_n \leq y_n)$ where $y_n \to y$, and one may hope that these probabilities will tend to $H(y)$.

Corollary 8.11 *If $Y_n \xrightarrow{\mathcal{L}} H$, and y_n converges to a continuity point y of H, then $P(Y_n \leq y_n) \to H(y)$.*

Proof. $P(Y_n \leq y_n) = P[Y_n + (y - y_n) \leq y]$ and the result follows from Theorem 8.10 with $B_n = 1$ and $A_n = y - y_n$. □

The following widely used result is often referred to as the *delta method.*

Theorem 8.12 (Delta Method) *If*

$$\sqrt{n}[T_n - \theta] \xrightarrow{\mathcal{L}} N(0, \tau^2), \tag{8.14}$$

then

$$\sqrt{n}[h(T_n) - h(\theta)] \xrightarrow{\mathcal{L}} N(0, \tau^2[h'(\theta)]^2), \tag{8.15}$$

provided $h'(\theta)$ exists and is not zero.

Proof. Consider the Taylor expansion of $h(T_n)$ around $h(\theta)$:

$$h(T_n) = h(\theta) + (T_n - \theta)[h'(\theta) + R_n], \tag{8.16}$$

where $R_n \to 0$ as $T_n \to \theta$. It follows from (8.14) that $T_n \to \theta$ in probability and hence that $R_n \to 0$ in probability. The result now follows by applying Theorem 8.10 to $\sqrt{n}[h(T_n) - h(\theta)]$. □

Example 8.13 Limit of binomial. Let X_i, $i = 1, 2, \ldots$, be independent Bernoulli (p) random variables and let $T_n = \frac{1}{n}\sum_{i=1}^{n} X_i$. Then by the CLT (Theorem 8.9)

$$\sqrt{n}\,(T_n - p) \to N\,[0, p(1-p)] \tag{8.17}$$

since $E(T_n) = p$ and $\text{var}(T_n) = p(1-p)$.

Suppose now that we are interested in the large sample behavior of the estimate $T_n(1 - T_n)$ of the variance $h(p) = p(1-p)$. Since $h'(p) = 1 - 2p$, it follows from Theorem 8.12 that

$$\sqrt{n}\,[T_n(1-T_n) - p(1-p)] \to N\left[0, (1-2p)^2 p(1-p)\right] \tag{8.18}$$

for $p \neq 1/2$. ||

When the dominant term in the Taylor expansion (8.16) vanishes [as it does at $p = 1/2$ in (8.18)], it is natural to carry the expansion one step further to obtain

$$h(T_n) = h(\theta) + (T_n - \theta)h'(\theta) + \frac{1}{2}(T_n - \theta)^2[h''(\theta) + R_n],$$

where $R_n \to 0$ in probability as $T_n \to \theta$, or, since $h'(\theta) = 0$,

$$h(T_n) - h(\theta) = \frac{1}{2}(T_n - \theta)^2[h''(\theta) + R_n]. \tag{8.19}$$

In view of (8.14), the distribution of $[\sqrt{n}(T_n - \theta)]^2$ tends to a nondegenerate limit distribution, namely (after division by τ^2) to a χ^2-distribution with 1 degree of freedom, and hence

$$n(T_n - \theta)^2 \to \tau^2 \cdot \chi_1^2. \tag{8.20}$$

The same argument as that leading to (8.15), but with $h'(\theta) = 0$ and $h''(\theta) \neq 0$, establishes the following theorem.

Theorem 8.14 *If* $\sqrt{n}[T_n - \theta] \xrightarrow{\mathcal{L}} N(0, \tau^2)$ *and if* $h'(\theta) = 0$, *then*

$$n[h(T_n) - h(\theta)] \to \frac{1}{2}\tau^2 h''(\theta)\chi_1^2 \tag{8.21}$$

provided $h''(\theta)$ *exists and is not zero.*

Example 8.15 Continuation of Example 8.13. For $h(p) = p(1-p)$, we have, at $p = 1/2$, $h'(1/2) = 0$ and $h''(1/2) = -2$. Hence, from Theorem 8.14, at $p = 1/2$,

$$n\left[T_n(1-T_n) - \frac{1}{4}\right] \to -\frac{1}{2}\chi_1^2. \tag{8.22}$$

Although (8.22) might at first appear strange, note that $T_n(1 - T_n) \leq 1/4$, so the left side is always negative. An equivalent form for (8.22) is

$$2n\left[\frac{1}{4} - T_n(1-T_n)\right] \to \chi_1^2. \qquad ||$$

The typical behavior of estimator sequences as sample sizes tend to infinity is that suggested by Theorem 8.12, that is, if δ_n is the estimator of $g(\theta)$ based on n observations, one may expect that $\sqrt{n}[\delta_n - g(\theta)]$ will tend to a normal distribution with mean zero and variance, say $\tau^2(\theta)$. It is in this sense that the large-sample behavior of such estimators can be studied without reference to a specific loss function. The asymptotic behavior of δ_n is governed solely by $\tau^2(\theta)$ since knowledge of $\tau^2(\theta)$ determines the probability of the error $\sqrt{n}[\delta_n - g(\theta)]$ lying in any given interval. In particular, $\tau^2(\theta)$ provides a basis for the large-sample comparison of different estimators.

Contrast this to the finite-sample situation where, for example, if estimators are compared in terms of their risk, one estimator might be best in terms of absolute error, another for squared error, and still another in terms of a higher power of the error or the probability of falling within a stated distance of the true value. This cannot happen here, as $\tau^2(\theta)$ provides the basis for all large-sample evaluations.

It is straightforward to generalize the preceding theorems to functions of several means. The expansion (8.16) is replaced by the corresponding Taylor's theorem in several variables. Although the following theorem starts in a multivariate setting, the conclusion is univariate.

Theorem 8.16 *Let $(X_{1\nu}, \ldots, X_{s\nu})$, $\nu = 1, \ldots, n$, be n independent s-tuples of random variables with $E(X_{i\nu}) = \xi_i$ and $\text{cov}(X_{i\nu}, X_{j\nu}) = \sigma_{ij}$. Let $\bar{X}_i = \Sigma X_{i\nu}/n$, and suppose that h is a real-valued function of s arguments with continuous first partial derivatives. Then,*

$$\sqrt{n}[h(\bar{X}_1, \ldots, \bar{X}_s) - h(\xi_1, \ldots, \xi_s)] \xrightarrow{\mathcal{L}} N(0, v^2), \quad v^2 = \Sigma\Sigma\sigma_{ij}\frac{\partial h}{\partial \xi_i} \cdot \frac{\partial h}{\partial \xi_j},$$

provided $v^2 > 0$.

Proof. See Problem 8.20. □

Example 8.17 Asymptotic distribution of S^2. As an illustration of Theorem 8.16, consider the asymptotic distribution of $S^2 = \Sigma(Z_\nu - \bar{Z})^2/n$ where the Z's are iid. Without loss of generality, suppose that $E(Z_\nu) = 0$, $E(Z_\nu^2) = \sigma^2$. Since $S^2 = (1/n)\Sigma Z_\nu^2 - \bar{Z}^2$, Theorem 8.16 applies with $X_{1\nu} = Z_\nu^2$, $X_{2\nu} = Z_\nu$, $h(x_1, x_2) = x_1 - x_2^2$, $\xi_2 = 0$, and $\xi_1 = \text{var}(Z_\nu) = \sigma^2$. Thus, $\sqrt{n}(S^2 - \sigma^2) \to N(0, v^2)$ where $v^2 = \text{var}(Z_\nu^2)$. ||

We conclude this section by considering the multivariate case and extending some of the basic probability results for random variables to vectors of random variables. The definitions of convergence in probability and in law generalize very naturally as follows.

Definition 8.18 A sequence of random vectors $\mathbf{Y}_n = (Y_{1n}, \ldots, Y_{rn})$, $n = 1, 2, \ldots$, tends *in probability* toward a constant vector $\mathbf{c} = (c_1, \ldots, c_r)$ if $Y_{in} \xrightarrow{P} c_i$ for each $i = 1, \ldots, r$, and it converges *in law* (or *weakly*) to a random vector $\mathbf{Y}$ with cdf H if

$$H_n(\mathbf{a}) \to H(\mathbf{a}) \tag{8.23}$$

at all continuity points $\mathbf{a}$ of H, where

$$H_n(\mathbf{a}) = P[Y_{1n} \le a_1, \ldots, Y_{rn} \le a_r] \tag{8.24}$$

is the cdf of $\mathbf{Y}_n$.

Theorem 8.8 extends to the present case.

Theorem 8.19 *The sequence $\{\mathbf{Y}_n\}$ converges in law to $\mathbf{Y}$ if and only if $E[f(\mathbf{Y}_n)] \to E[f(\mathbf{Y})]$ for every bounded continuous real-valued f.*

[For a proof of this and Theorem 8.20, see Billingsley (1995, Section 29).]

Weak convergence of $\mathbf{Y}_n$ to $\mathbf{Y}$ does not imply

$$P(\mathbf{Y}_n \in A) \to P(\mathbf{Y} \in A) \tag{8.25}$$

for all sets A for which these probabilities are defined since this is not even true for the set A defined by

$$T_1 \le a_1, \ldots, T_r \le a_r$$

unless H is continuous at $\mathbf{a}$.

Theorem 8.20 *The sequence $\{\mathbf{Y}_n\}$ converges in law to $\mathbf{Y}$ if and only if (8.25) holds for all sets A for which the probabilities in question are defined and for which the boundary of A has probability zero under the distribution of $\mathbf{Y}$.*

As in the one-dimensional case, the central limit theorem provides a basic tool for multivariate asymptotic theory.

Theorem 8.21 (Multivariate CLT) *Let $\mathbf{X}_\nu = (X_{1\nu}, \ldots, X_{r\nu})$ be iid with mean vector $\boldsymbol{\xi} = (\xi_1, \ldots, \xi_r)$ and covariance matrix $\Sigma = ||\sigma_{ij}||$, and let $\bar{X}_{in} = (X_{i1} + \cdots + X_{in})/n$. Then,*

$$[\sqrt{n}(\bar{X}_{1n} - \xi_1), \ldots, \sqrt{n}(\bar{X}_{rn} - \xi_r)]$$

tends in law to the multivariate normal distribution with mean vector $\mathbf{0}$ and covariance matrix Σ.

As a last result, we mention a generalization of Theorem 8.16.

Theorem 8.22 *Suppose that*

$$[\sqrt{n}(Y_{1n} - \theta_1), \ldots, \sqrt{n}(Y_{rn} - \theta_r)]$$

tends in law to the multivariate normal distribution with mean vector $\mathbf{0}$ and covariance matrix Σ, and suppose that $h_1, \ldots, h_r$ are r real-valued functions of $\boldsymbol{\theta} = (\theta_1, \ldots, \theta_r)$, defined and continuously differentiable in a neighborhood ω of the parameter point θ and such that the matrix $B = ||\partial h_i/\partial\theta_j||$ of partial derivatives is nonsingular in ω. Then,

$$[\sqrt{n}[h_1(\mathbf{Y}_n) - h_1(\boldsymbol{\theta})], \ldots, \sqrt{n}[h_r(\mathbf{Y}_n) - h_r(\boldsymbol{\theta})]]$$

tends in law to the multivariate normal distribution with mean vector $\mathbf{0}$ and with covariance matrix $B\Sigma B'$.

Proof. See Problem 8.27 □

9 Problems

Section 1

1.1 If $(x_1, y_1), \ldots, (x_n, y_n)$ are n points in the plane, determine the best fitting line $y = \alpha + \beta x$ in the least squares sense, that is, determine the values α and β that minimize $\Sigma[y_i - (\alpha + \beta x_i)]^2$.

1.2 Let $X_1, \ldots, X_n$ be uncorrelated random variables with common expectation θ and variance σ^2. Then, among all linear estimators $\Sigma\alpha_i X_i$ of θ satisfying $\Sigma\alpha_i = 1$, the mean $\bar{X}$ has the smallest variance.

1.3 In the preceding problem, minimize the variance of $\Sigma\alpha_i X_i (\Sigma\alpha_i = 1)$

(a) When the variance of X_i is σ^2/α_i (α_i known).

(b) When the X_i have common variance σ^2 but are correlated with common correlation coefficient ρ.

(For generalizations of these results see, for example, Watson 1967 and Kruskal 1968.)

1.4 Let X and Y have common expectation θ, variances σ^2 and τ^2, and correlation coefficient ρ. Determine the conditions on σ, τ, and ρ under which

(a) $\text{var}(X) < \text{var}[(X + Y)/2]$.

(b) The value of α that minimizes $\text{var}[\alpha X + (1 - \alpha)Y]$ is negative.

Give an intuitive explanation of your results.

1.5 Let X_i $(i = 1, 2)$ be independently distributed according to the Cauchy densities $C(a_i, b_i)$. Then, $X_1 + X_2$ is distributed as $C(a_1 + a_2, b_1 + b_2)$. [*Hint*: Transform to new variables $Y_1 = X_1 + X_2$, $Y_2 = X_2$.]

1.6 If $X_1, \ldots, X_n$ are iid as $C(a, b)$, the distribution of $\bar{X}$ is again $C(a, b)$. [*Hint*: Prove by induction, using Problem 5.]

1.7 A *median* of X is any value m such that $P(X \le m) \ge 1/2$ and $P(X \ge m) \ge 1/2$.

(a) Show that this is equivalent to $P(X < m) \le 1/2$ and $P(X > m) \le 1/2$.

(b) Show that the set of medians is always a closed interval $m_0 \le m \le m_1$.

1.8 If $\phi(a) = E|X - a| < \infty$ for some a, show that $\phi(a)$ is minimized by any median of X. [*Hint*: If $m_0 \le m \le m_1$ (in the notation of Problem 1.7) and $m_1 < c$, then

$$E|X - c| - E|X - m| = (c - m)[P(X \le m) - P(X > m)] + 2\int_{m<x<c} (c - x) dP(x)].$$

1.9 (a) The median of any set of distinct real numbers $x_1, \ldots, x_n$ is defined to be the middle one of the ordered x's when n is odd, and any value between the two middle ordered x's when n is even. Show that this is also the median of the random variable X which takes on each of the values $x_1, \ldots, x_n$ with probability $1/n$.

(b) For any set of distinct real numbers $x_1, \ldots, x_n$, the sum of absolute deviations $\Sigma|x_i - a|$ is minimized by any median of the x's.

(c) For n given points (x_i, y_i), $i = 1, \ldots, n$, find the value b that minimizes $\Sigma|y_i - bx_i|$. [*Hint*: Reduce the problem to a special case of Problem 8.]

1.10 For any set of numbers $x_1, \cdots, x_n$ and a monotone function $h(\cdot)$, show that the value of a that minimizes $\sum_{i=1}^n [h(x_i) - h(a)]^2$ is given by $a = h^{-1}\left(\sum_{i=1}^n h(x_i)/n\right)$. Find functions h that will yield the arithmetic, geometric, and harmonic means as minimizers.

[*Hint*: Recall that the geometric mean of non-negative numbers is $\left(\prod x_i\right)^{1/n}$ and the harmonic mean is $\left[(1/n)\sum(1/x_i)\right]^{-1}$. This problem, and some of its implications, is considered by Casella and Berger (1992).]

1.11 (a) If two estimators δ_1, δ_2 have continuous symmetric densities $f_i(x-\theta)$, $i = 1, 2$, and $f_1(0) > f_2(0)$, then

$$P[|\delta_1 - \theta| < c] > P[|\delta_2 - \theta| < c] \quad \text{for some } c > 0$$

and hence δ_1 will be closer to θ than δ_2 with respect to the measure (1.5).

(b) Let X, Y be independently distributed with common continuous symmetric density f, and let $\delta_1 = X$, $\delta_2 = (X+Y)/2$. The inequality in part (a) will hold provided $2\int f^2(x)\,dx < f(0)$ (Edgeworth 1883, Stigler 1980).

1.12 (a) Let $f(x) = (1/2)(k-1)/(1+|x|)^k$, $k \geq 2$. Show that f is a probability density and that all its moments of order $< k-1$ are finite.

(b) The density of part (a) satisfies the inequality of Problem 1.11(b).

1.13 (a) If X is binomial $b(p, n)$, show that

$$E\left|\frac{x}{n} - p\right| = 2\binom{n-1}{k-1} p^k (1-p)^{n-k+1} \quad \text{for } \frac{k-1}{n} \leq p \leq \frac{k}{n}.$$

(b) Graph the risk function of part (i) for $n = 4$ and $n = 5$.

[*Hint*: For (a), use the identity

$$\binom{n}{x}(x - np) = n\left[\binom{n-1}{x-1}(1-p) - \binom{n-1}{x}p\right], \quad 1 \leq x \leq n.$$

(Johnson 1957–1958, and Blyth 1980).]

Section 2

2.1 If $A_1, A_2, \ldots$ are members of a σ-field $\mathcal{A}$ (the A's need not be disjoint), so are their union and intersection.

2.2 For any $a < b$, the following sets are Borel sets (a) $\{x : a < x\}$ and (b) $\{x : a \leq x \leq b\}$.

2.3 Under the assumptions of Problem 2.1, let

$$\underline{A} = \liminf A_n = \{x : x \in A_n \text{ for all except a finite number of } n\text{'s}\},$$
$$\overline{A} = \limsup A_n = \{x : x \in A_n \text{ for infinitely many } n\}.$$

Then, $\underline{A}$ and $\overline{A}$ are in $\mathcal{A}$.

2.4 Show that

(a) If $A_1 \subset A_2 \subset \cdots$, then $\underline{A} = \overline{A} = \cup A_n$.

(b) If $A_1 \supset A_2 \supset \cdots$, then $\underline{A} = \overline{A} = \cap A_n$.

2.5 For any sequence of real numbers $a_1, a_2, \ldots$, show that the set of all limit points of subsequences is closed. The smallest and largest such limit point (which may be infinite) are denoted by $\liminf a_k$ and $\limsup a_k$, respectively.

2.6 Under the assumptions of Problems 2.1 and 2.3, show that

$$I_{\underline{A}}(x) = \liminf I_{A_k}(x) \quad \text{and} \quad I_{\overline{A}}(x) = \limsup I_{A_k}(x)$$

where $I_A(x)$ denotes the indicator of the set A.

2.7 Let $(\mathcal{X}, \mathcal{A}, \mu)$ be a measure space and let $\mathcal{B}$ be the class of all sets $A \cup C$ with $A \in \mathcal{A}$ and C a subset of a set $A' \in \mathcal{A}$ with $\mu(A') = 0$. Show that $\mathcal{B}$ is a σ-field.

2.8 If f and g are measurable functions, so are (i) $f+g$, and (ii) $\max(f, g)$.

2.9 If f is integrable with respect to μ, so is $|f|$, and $\left|\int f\,d\mu\right| \leq \int |f|\,d\mu$. [*Hint*: Express $|f|$ in terms of f^+ and f^-.]

2.10 Let $\mathcal{X} = \{x_1, x_2, \dots\}$, μ = counting measure on $\mathcal{X}$, and f integrable. Then $\int f d\mu = \Sigma f(x_i)$. [*Hint*: Suppose, first, that $f \geq 0$ and let $s_n(x)$ be the simple function, which is $f(x)$ for $x = x_1, \dots, x_n$, and 0 otherwise.]

2.11 Let $f(x) = 1$ or 0 as x is rational or irrational. Show that the Riemann integral of f does not exist.

Section 3

3.1 Let X have a standard normal distribution and let $Y = 2X$. Determine whether

(a) the cdf $F(x, y)$ of (X, Y) is continuous.

(b) the distribution of (X, Y) is absolutely continuous with respect to Lebesgue measure in the (x, y) plane.

3.2 Show that any function f which satisfies (3.7) is continuous.

3.3 Let X be a *measurable transformation* from $(\mathcal{E}, \mathcal{B})$ to $(\mathcal{X}, \mathcal{A})$ (i.e., such that for any $A \in \mathcal{A}$, the set $\{e : X(e) \in A\}$ is in $\mathcal{B}$), and let Y be a measurable transformation from $(\mathcal{X}, \mathcal{A})$ to $(\mathcal{Y}, \mathcal{C})$. Then, $Y[X(e)]$ is a measurable transformation from $(\mathcal{E}, \mathcal{B})$ to $(\mathcal{Y}, \mathcal{C})$.

3.4 In Example 3.1, show that the support of P is $[a, b]$ if and only if F is strictly increasing on $[a, b]$.

3.5 Let S be the support of a distribution on a Euclidean space $(\mathcal{X}, \mathcal{A})$. Then, (*i*) S is closed; (*ii*) $P(S) = 1$; (*iii*) S is the intersection of all closed sets C with $P(C) = 1$.

3.6 If P and Q are two probability measures over the same Euclidean space which are equivalent, then they have the same support.

3.7 Let P and Q assign probabilities

$$P : P\left(X = \frac{1}{n}\right) = p_n > 0, \quad n = 1, 2, \dots \quad (\Sigma p_n = 1),$$

$$Q : P(X = 0) = \frac{1}{2}; \quad P\left(X = \frac{1}{n}\right) = q_n > 0; \quad n = 1, 2, \dots \quad \left(\Sigma q_n = \frac{1}{2}\right).$$

Then, show that P and Q have the same support but are not equivalent.

3.8 Suppose X and Y are independent random variables with $X \sim E(\lambda, 1)$ and $Y \sim E(\mu, 1)$. It is impossible to obtain direct observations of X and Y. Instead, we observe the random variables Z and W, where

$$Z = \min\{X, Y\} \quad \text{and} \quad W = \begin{cases} 1 & \text{if } Z = X \\ 0 & \text{if } Z = Y. \end{cases}$$

Find the joint distribution of Z and W and show that they are independent. (The X and Y variables are *censored.*, a situation that often arises in medical experiments. Suppose that X measures survival time from some treatment, and the patient leaves the survey for some unrelated reason. We do not get a measurement on X, but only a lower bound.)

Section 4

4.1 If the distributions of a positive random variable X form a scale family, show that the distributions of $\log X$ form a location family.

4.2 If X is distributed according to the uniform distribution $U(0, \theta)$, show that the distribution of $-\log X$ is exponential.

4.3 Let U be uniformly distributed on (0, 1) and consider the variables $X = U^\alpha, 0 < \alpha$. Show that this defines a group family, and determine the density of X.

4.4 Show that a transformation group is a group.

4.5 If g_0 is any element of a group G, show that as g ranges over G so does gg_0.

4.6 Show that for $p = 2$, the density (4.15) specializes to (4.16).

4.7 Show that the family of transformations (4.12) with B nonsingular and lower triangular form a group G.

4.8 Show that the totality of nonsingular multivariate normal distributions can be obtained by the subgroup G of (4.12) described in Problem 4.7.

4.9 In the preceding problem, show that G can be replaced by the subgroup G_0 of lower triangular matrices $B = (b_{ij})$, in which the diagonal elements $b_{11}, \ldots, b_{pp}$ are all positive, but that no proper subgroup of G_0 will suffice.

4.10 Show that the family of all continuous distributions whose support is an interval with positive lower end point is a group family. [*Hint*: Let U be uniformly distributed on the interval (2, 3) and let $X = b[g(U)]^a$ where $\alpha, b > 0$ and where g is continuous and 1:1 from (2, 3) to (2, 3).]

4.11 Find a modification of the transformation group (4.22) which generates a random sample from a population $\{y_1, \ldots, y_N\}$ where the y's, instead of being arbitrary, are restricted to (a) be positive and (b) satisfy $0 < y_i < 1$.

4.12 Generalize the transformation group of Example 4.10 to the case of s populations $\{y_{ij}, j = 1, \ldots, N_i\}, i = 1, \ldots, s$, with a random sample of size n_i being drawn from the ith population.

4.13 Let U be a positive random variable, and let

$$X = bU^{1/c}, \quad b > 0, \quad c > 0.$$

(a) Show that this defines a group family.

(b) If U is distributed as $E(0, 1)$, then X is distributed according to the *Weibull* distribution with density

$$\frac{c}{b}\left(\frac{x}{b}\right)^{c-1} e^{-(x/b)^c}, \quad x > 0.$$

4.14 If F and F_0 are two continuous, strictly increasing cdf's on the real line, and if the cdf of U is F_0 and g is strictly increasing, show that the cdf of $g(U)$ is F if and only if $g = F^{-1}(F_0)$.

4.15 The following two families of distributions are not group families:

(a) The class of binomial distributions $b(p, n)$, with n fixed and $0 < p < 1$.

(b) The class of Poisson distributions $P(\lambda)$, $0 < \lambda$.

[*Hint*: (a) How many 1:1 transformations are there taking the set of integers $\{0, 1, \ldots, n\}$ into itself?]

4.16 Let $X_1, \ldots, X_r$ have a multivariate normal distribution with $E(X_i) = \xi_i$ and with covariance matrix Σ. If X is the column matrix with elements X_i and B is an $r \times r$ matrix of constants, then BX has a multivariate normal distribution with mean $B\xi$ and covariance matrix $B\Sigma B'$.

Section 5

5.1 Determine the natural parameter space of (5.2) when $s = 1$, $T_1(x) = x$, μ is Lebesgue measure, and $h(x)$ is (i) $e^{-|x|}$ and (ii) $e^{-|x|}/(1 + x^2)$.

5.2 Suppose in (5.2), $s = 2$ and $T_2(x) = T_1(x)$. Explain why it is impossible to estimate η_1. [*Hint*: Compare the model with that obtained by putting $\eta_1' = \eta_1 + c$, $\eta_2' = \eta_2 - c$.]

5.3 Show that the distribution of a sample from the p-variate normal density (4.15) constitutes an s-dimensional exponential family. Determine s and identify the functions η_i, T_i, and B of (5.1).

5.4 Efron (1975) gives very general definitions of curvature, which generalize (10.1) and (10.2). For the s-dimensional family (5.1) with covariance matrix Σ_θ, if θ is a scalar, define the *statistical curvature* to be $\gamma_\theta = \left(|M_\theta|/m_{11}^3\right)^{1/2}$ where

$$M_\theta = \begin{pmatrix} m_{11} & m_{12} \\ m_{21} & m_{22} \end{pmatrix} = \begin{pmatrix} \dot{\eta}_\theta' \Sigma_\theta \dot{\eta}_\theta & \dot{\eta}_\theta' \Sigma_\theta \ddot{\eta}_\theta \\ \ddot{\eta}_\theta' \Sigma_\theta \dot{\eta}_\theta & \ddot{\eta}_\theta' \Sigma_\theta \ddot{\eta}_\theta \end{pmatrix},$$

with $\eta(\theta) = \{\eta_i(\theta)\}$, $\dot{\eta}(\theta) = \{\eta_i'(\theta)\}$ and $\ddot{\eta}(\theta) = \{\eta_i''(\theta)\}$. Calculate the curvature of the family (see Example 6.19) $C \exp\left[-\sum_{i=1}^n (x_i - \theta)^m\right]$ for $m = 2, 3, 4$. Are the values of γ_θ ordered in the way you expected them to be?

5.5 Let (X_1, X_2) have a bivariate normal distribution with mean vector $\xi = (\xi_1, \xi_2)$ and identity the covariance matrix. In each of the following situations, verify the curvature, γ_θ of the family.

(a) $\xi = (\theta, \theta)$, $\gamma_\theta = 0$.

(b) $\xi = (\theta_1, \theta_2)$, $\theta_1^2 + \theta_2^2 = r^2$, $\gamma_\theta = 1/r$.

5.6 In the density (5.1)

(a) For $s = 1$ show that $E_\theta[T(X)] = B'(\theta)/\eta'(\theta)$ and $\text{var}_\theta[T(X)] = \frac{B''(\theta)}{[\eta'(\theta)]^2} - \frac{\eta''(\theta)B'(\theta)}{[\eta'(\theta)]^3}$.

(b) For $s > 1$, show that $E_\theta[T(X)] = J^{-1}\nabla B$ where J is the Jacobian matrix defined by $J = \{\frac{\partial \eta_j}{\partial \theta_i}\}$ and ∇B is the gradient vector $\nabla B = \{\frac{\partial}{\partial \theta_i} B(\theta)\}$.

(See Johnson, Ladalla, and Liu (1979) for a general treatment of these identities.)

5.7 Verify the relations (a) (5.22) and (b) (5.26).

5.8 For the binomial distribution (5.28), verify (a) the moment generating function (5.30) and (b) the moments (5.31).

5.9 For the Poisson distribution (5.32), verify the moments (5.35).

5.10 In a Bernoulli sequence of trials with success probability p, let $X + m$ be the number of trials required to achieve m successes.

(a) Show that the distribution of X, the *negative binomial distribution*, is as given in Table 5.1.

(b) Verify that the negative binomial probabilities add up to 1 by expanding $\left(\frac{1}{p} - \frac{q}{p}\right)^{-m} = p^m(1-q)^{-m}$.

(c) Show that the distributions of (a) constitute a one-parameter exponential family.

(d) Show that the moment generating function of X is $M_X(u) = p^m/(1 - qe^u)^m$.

(e) Show that $E(X) = mq/p$ and $\text{var}(X) = mq/p^2$.

(f) By expanding $K_X(u)$, show that the first four cumulants of X are $k_1 = mq/p$, $k_2 = mq/p^2$, $k_3 = mq(1+q)/p^3$, and $k_4 = mq(1 + 4q + q^2)/p^4$.

5.11 In the preceding problem, let $X_i + 1$ be the number of trials required after the $(i-1)$st success has been obtained until the next success occurs. Use the fact that $X = \sum_{i=1}^m X_i$ to find an alternative derivation of the mean and variance in part (e).

5.12 A discrete random variable with probabilities

$$P(X = x) = a(x)\theta^x/C(\theta), \quad x = 0, 1, \ldots; \; a(x) \geq 0; \; \theta > 0,$$

is a *power series distribution*. This is an exponential family (5.1) with $s = 1$, $\eta = \log\theta$, and $T = X$. The moment generating function is $M_X(u) = C(\theta e^u)/C(\theta)$.

5.13 Show that the binomial, negative binomial, and Poisson distributions are special cases of the power series distribution of Problem 5.12, and determine θ and $C(\theta)$.

5.14 The distribution of Problem 5.12 with $a(x) = 1/x$ and $C(\theta) = -\log(1-\theta)$, $x = 1, 2, \ldots; 0 < \theta < 1$, is the *logarithmic series* distribution. Show that the moment generating function is $\log(1-\theta e^u)/\log(1-\theta)$ and determine $E(X)$ and $\text{var}(X)$.

5.15 For the multinomial distribution (5.4), verify the moment formulas (5.16).

5.16 As an alternative to using (5.14) and (5.15), obtain the moments (5.16) by representing each X_i as a sum of n indicators, as was done in (5.5):

5.17 For the gamma distribution (5.41).

(a) verify the formulas (5.42), (5.43), and (5.44);

(b) show that (5.43), with the middle term deleted, holds not only for all positive integers r but for all real $r > -\alpha$.

5.18 (a) Prove Lemma 5.15. (Use integration by parts.)

(b) By choosing $g(x)$ to be x^2 and x^3, use the Stein Identity to calculate the third and fourth moments of the $N(\mu, \sigma^2)$ distribution.

5.19 Using Lemma 5.15:

(a) Derive the form of the identity for $X \sim \text{Gamma}(\alpha, b)$ and use it to verify the moments given in (5.44).

(b) Derive the form of the identity for $X \sim \text{Beta}(a, b)$, and use it to verify that $E(X) = a/(a+b)$ and $\text{var}(X) = ab/(a+b)^2(a+b+1)$.

5.20 As an alternative to the approach of Problem 5.19(b) for calculating the moments of $X \sim B(a, b)$, a general formula for EX^k (similar to equation (5.43)) can be derived. Do so, and use it to verify the mean and variance of X given in Problem 5.19. [*Hint*: Write EX^k as the integral of $x^{c-1}(1-x)^{d-1}$ and use the constant $B(c, d)$ of Table 5.1. Note that a similar approach will work for many other distributions, including the χ^2, Student's t, and F distributions.]

5.21 The Stein Identity can also be applied to discrete exponential families, as shown by Hudson (1978) and generalized by Hwang (1982a). If X takes values in $\mathcal{N} = \{0, 1, \ldots, \}$ with probability function

$$p_\theta(x) = \exp[\theta x - B(\theta)]h(x),$$

then for any $g : \mathcal{N} \to \Re$ with $E_\theta|g(X)| < \infty$, we have the identity

$$Eg(X) = e^{-\theta}E\{t(X)g(X-1)\}$$

where $t(0) = 0$ and $t(x) = h(x-1)/h(x)$ for $x > 0$.

(a) Prove the identity.

(b) Use the identity to calculate the first four moments of the binomial distribution (5.31).

(c) Use the identity to calculate the first four moments of the Poisson distribution (5.35).

5.22 The inverse Gaussian distribution, $IG(\lambda, \mu)$, has density function

$$\sqrt{\frac{\lambda}{2\pi}} e^{(\lambda\mu)^{1/2}} x^{-3/2} e^{-\frac{1}{2}(\frac{\lambda}{x}+\mu x)}, \quad x > 0,\ \lambda, \mu > 0.$$

(a) Show that this density constitutes an exponential family.

(b) Show that this density is a scale family (as defined in Example 4.1).

(c) Show that the statistics $\bar{X} = (1/n)\Sigma x_i$ and $S^* = \Sigma(1/x_i - 1/\bar{x})$ are complete sufficient statistics.

(d) Show that $\bar{X} \sim IG(n\lambda, n\mu)$ and $S^* \sim (1/\lambda)\chi^2_{n-1}$.

Note: Together with the normal and gamma distributions, the inverse Gaussian completes the trio of families that are both an exponential and a group family of distributions. This fact plays an important role in distribution theory based on saddlepoint approximations (Daniels 1983) or likelihood theory (Barndorff-Nielsen 1983).

5.23 In Example 5.14, show that

(a) χ^2_1 is the distribution of Y^2 where Y is distributed as $N(0, 1)$;

(b) χ^2_n is the distribution of $Y_1^2 + \cdots + Y_n^2$ where the Y_i are independent $N(0, 1)$.

5.24 Determine the values α for which the density (5.41) is (a) a decreasing function of x on $(0, \infty)$ and (b) increasing for $x < x_0$ and decreasing for $x > x_0 (0 < x_0)$. In case (b), determine the mode of the density.

5.25 A random variable X has the *Pareto distribution* $P(c, k)$ if its cdf is $1 - (k/x)^c$, $x > k > 0, c > 0$.

(a) The distributions $P(c, 1)$ constitute a one-parameter exponential family (5.2) with $\eta = -c$ and $T = \log X$.

(b) The statistic T is distributed as $E(\log k, 1/c)$.

(c) The family $P(c, k)$ $(0 < k, 0 < c)$ is a group family.

5.26 If (X, Y) is distributed according to the bivariate normal distribution (4.16) with $\xi = \eta = 0$:

(a) Show that the moment generating function of (X, Y) is

$$M_{X,Y}(u_1, u_2) = e^{-[u_1^2\sigma^2 + 2\rho\sigma\tau u_1 u_2 + u_2^2\tau^2]/2}.$$

(b) Use (a) to show that

$$\mu_{12} = \mu_{21} = 0, \quad \mu_{11} = \rho\sigma\tau,$$
$$\mu_{13} = 3\rho\sigma\tau^3, \quad \mu_{31} = 3\rho\sigma^3\tau, \quad \mu_{22} = (1 + 2\rho^2)\sigma^2\tau^2.$$

5.27 (a) If $\mathbf{X}$ is a random column vector with expectation ξ, then the covariance matrix of $\mathbf{X}$ is $\text{cov}(\mathbf{X}) = E[(\mathbf{X}' - \xi)(\mathbf{X}' - \xi')]$.

(b) If the density of $\mathbf{X}$ is (4.15), then $\xi = \mathbf{a}$ and $\text{cov}(\mathbf{X}) = \Sigma$.

5.28 (a) Let X be distributed with density $p_\theta(x)$ given by (5.1), and let A be any fixed subset of the sample space. Then, the distributions of X *truncated* on A, that is, the distributions with density $p_\theta(x)I_A(x)/P_\theta(A)$ again constitute an exponential family.

(b) Give an example in which the natural parameter space of the original exponential family is a proper subset of the natural parameter space of the truncated family.

5.29 If X_i are independently distributed according to $\Gamma(\alpha_i, b)$, show that ΣX_i is distributed as $\Gamma(\Sigma\alpha_i, b)$. [*Hint: Method 1.* Prove it first for the sum of two gamma variables by a transformation to new variables $Y_1 = X_1 + X_2$, $Y_2 = X_1/X_2$ and then use induction. *Method 2.* Obtain the moment generating function of ΣX_i and use the fact that a distribution is uniquely determined by its moment generating function, when the latter exists for at least some $u \neq 0$.]

5.30 When the X_i are independently distributed according to Poisson distributions $P(\lambda_i)$, find the distribution of ΣX_i.

5.31 Let $X_1, \ldots, X_n$ be independently distributed as $\Gamma(\alpha, b)$. Show that the joint distribution is a two-parameter exponential family and identify the functions η_i, T_i, and B of (5.1).

5.32 If Y is distributed as $\Gamma(\alpha, b)$, determine the distribution of $c \log Y$ and show that for fixed α and varying b it defines an exponential family.

5.33 Morris (1982, 1983b) investigated the properties of natural exponential families with quadratic variance functions. There are only six such families: normal, binomial, gamma, Poisson, negative binomial, and the lesser-known generalized hyperbolic secant distribution, which is the density of $X = \frac{1}{\pi}\log(\frac{Y}{1-Y})$ when $Y \sim \text{Beta}(\frac{1}{2} + \frac{\theta}{\pi}, \frac{1}{2} - \frac{\theta}{\pi})$, $|\theta| < \frac{\pi}{2}$.

(a) Find the density of X, and show that it constitutes an exponential family.

(b) Find the mean and variance of X, and show that the variance equals $1 + \mu^2$, where μ is the mean.

Subsequent work on quadratic and other power variance families has been done by Bar-Lev and Enis (1986, 1988), Bar-Lev and Bshouty (1989), and Letac and Mora (1990).

Section 6

6.1 Extend Example 6.2 to the case that $X_1, \ldots, X_r$ are independently distributed with Poisson distributions $P(\lambda_i)$ where $\lambda_i = a_i\lambda$ ($a_i > 0$, known).

6.2 Let $X_1, \ldots, X_n$ be iid according to a distribution F and probability density f. Show that the conditional distribution given $X_{(i)} = a$ of the $i-1$ values to the left of a and the $n-i$ values to the right of a is that of $i-1$ variables distributed independently according to the probability density $f(x)/F(a)$ and $n-i$ variables distributed independently with density $f(x)/[1 - F(a)]$, respectively, with the two sets being (conditionally) independent of each other.

6.3 Let f be a positive integrable function over $(0, \infty)$, and let $p_\theta(x)$ be the density over $(0, \theta)$ defined by $p_\theta(x) = c(\theta)f(x)$ if $0 < x < \theta$, and 0 otherwise. If $X_1, \ldots, X_n$ are iid with density p_θ, show that $X_{(n)}$ is sufficient for θ.

6.4 Let f be a positive integrable function defined over $(-\infty, \infty)$ and let $p_{\xi,\eta}(x)$ be the probability density defined by $p_{\xi,\eta}(x) = c(\xi, \eta)f(x)$ if $\xi < x < \eta$, and 0 otherwise. If $X_1, \ldots, X_n$ are iid with density $p_{\xi,\eta}$, show that $(X_{(1)}, X_{(n)})$ is sufficient for (ξ, n).

6.5 Show that each of the statistics $T_1 - T_4$ of Example 6.11 is sufficient.

6.6 Prove Corollary 6.13.

6.7 Let $X_1, \ldots, X_m$ and $Y_1, \ldots, Y_n$ be independently distributed according to $N(\xi, \sigma^2)$ and $N(\eta, \tau^2)$, respectively. Find the minimal sufficient statistics for these cases:

(a) ξ, η, σ, τ are arbitrary: $-\infty < \xi, \eta < \infty, 0 < \sigma, \tau$.

(b) $\sigma = \tau$ and ξ, η, σ are arbitrary.

(c) $\xi = \eta$ and ξ, σ, τ are arbitrary.

6.8 Let $X_1, \ldots, X_n$ be iid according to $N(\sigma, \sigma^2)$, $0 < \sigma$. Find a minimal set of sufficient statistics.

6.9 (a) If $(x_1, \ldots, x_n)$ and $(y_1, \ldots, y_n)$, have the same elementary symmetric functions $\Sigma x_i = \Sigma y_i$, $\Sigma_{i \neq j} x_i y_j = \Sigma_{i \neq j} y_i y_j, \ldots, x_1 \cdots x_n = y_1 \cdots y_n$, then the y's are a permutation of the x's.

(b) In the notation of Example 6.10, show that U is equivalent to V. [*Hint*: Compare the coefficients and the roots of the polynomials $P(x) = \Pi(x - u_i)$ and $Q(x) = \Pi(x - v_i)$.]

6.10 Show that the order statistics are minimal sufficient for the location family (6.7) when f is the density of

(a) the double exponential distribution $D(0, 1)$.

(b) the Cauchy distribution $C(0, 1)$.

6.11 Prove the following generalization of Theorem 6.12 to families without common support.

Theorem 9.1 *Let $\mathcal{P}$ be a finite family with densities p_i, $i = 0, \ldots, k$, and for any x, let $S(x)$ be the set of pairs of subscripts (i, j) for which $p_i(x) + p_j(x) > 0$. Then, the statistic*

$$T(X) = \left\{ \frac{p_j(X)}{p_i(X)}, \quad i < j \text{ and } (i, j) \in S(X) \right\}$$

is minimal sufficient. Here, $p_j(x)/p_i(x) = \infty$ if $p_i(x) = 0$ and $p_j(x) > 0$.

6.12 In Problem 6.11 it is not enough to replace $p_i(X)$ by $p_0(X)$. To see this let $k = 2$ and $p_0 = U(-1, 0)$, $p_1 = U(0, 1)$, and $p_2(x) = 2x$, $0 < x < 1$.

6.13 Let $k = 1$ and $P_i = U(i, i + 1)$, $i = 0, 1$.

(a) Show that a minimal sufficient statistic for $\mathcal{P} = \{P_0, P_1\}$ is $T(X) = i$ if $i < X < i+1$, $i = 0, 1$.

(b) Let X_1 and X_2 be iid according to a distribution from $\mathcal{P}$. Show that each of the two statistics $T_1 = T(X_1)$ and $T_2 = T(X_2)$ is sufficient for (X_1, X_2).

(c) Show that $T(X_1)$ and $T(X_2)$ are equivalent.

6.14 In Lemma 6.14, show that the assumption of common support can be replaced by the weaker assumption that every $\mathcal{P}_0$-null set is also a $\mathcal{P}$-null set so that (a.e. $\mathcal{P}_0$) is equivalent to (a.e. $\mathcal{P}$).

6.15 Let $X_1, \ldots, X_n$ be iid according to a distribution from $\mathcal{P} = \{U(0, \theta), \theta > 0\}$, and let $\mathcal{P}_0$ be the subfamily of $\mathcal{P}$ for which θ is rational. Show that every $\mathcal{P}_0$-null set in the sample space is also a $\mathcal{P}$-null set.

6.16 Let $X_1, \ldots, X_n$ be iid according to a distribution from a family $\mathcal{P}$. Show that T is minimal sufficient in the following cases:

(a) $\mathcal{P} = \{U(0, \theta), \theta > 0\}$; $T = X_{(n)}$.

(b) $\mathcal{P} = \{U(\theta_1, \theta_2), -\infty < \theta_1 < \theta_2 < \infty\}$; $T = (X_{(1)}, X_{(n)})$.

(c) $\mathcal{P} = \{U(\theta - 1/2, \theta + 1/2), -\infty < \theta < \infty\}$; $T = (X_{(1)}, X_{(n)})$.

6.17 Solve the preceding problem for the following cases:

(a) $\mathcal{P} = \{E(\theta, 1), -\infty < \theta < \infty\}$; $T = X_{(1)}$.

(b) $\mathcal{P} = \{E(0, b), 0 < b\}$; $T = \Sigma X_i$.

(c) $\mathcal{P} = \{E(a, b), -\infty < a < \infty, 0 < b\}$; $T = (X_{(1)}, \Sigma[X_i - X_{(1)}])$.

6.18 Show that the statistics $X_{(1)}$ and $\Sigma[X_i - X_{(1)}]$ of Problem 6.17(c) are independently distributed as $E(a, b/n)$ and bGamma$(n-2, 1)$ respectively.

[*Hint*: If $a = 0$ and $b = 1$, the variables $Y_i = (n-i+1)[X_{(i)} - X_{(i-1)}]$, $i = 2, \ldots, n$, are iid as $E(0, 1)$.]

6.19 Show that the sufficient statistics of (i) Problem 6.3 and (ii) Problem 6.4 are minimal sufficient.

6.20 (a) Show that in the $N(\theta, \theta)$ curved exponential family, the sufficient statistic $T = (\sum x_i, \sum x_i^2)$ is not minimal.

(b) For the density of Example 6.19, show that $T = (\sum x_i, \sum x_i^2, \sum x_i^3)$ is a minimal sufficient statistic.

6.21 For the situation of Example 6.25(ii), find an unbiased estimator of ξ based on $\sum X_i$, and another based on $\sum X_i^2$); hence, deduce that $T = (\sum X_i, \sum X_i^2)$ is not complete.

6.22 For the situation of Example 6.26, show that $\mathbf{X}$ is minimal sufficient and complete.

6.23 For the situation of Example 6.27:

(a) Show that $\mathbf{X} = (X_1, X_2)$ is minimal sufficient for the family (6.16) with restriction (6.17).

(b) Establish (6.18), and hence that the minimal sufficient statistic of part (a) is not complete.

6.24 (Messig and Strawderman 1993) Show that for the general dose-response model

$$p_\theta(\mathbf{x}) = \prod_{i=1}^{m} \binom{n_i}{x_i} [\eta_\theta(d_i)]^{x_i} [1 - \eta_\theta(d_i)]^{n_i - x_i},$$

the statistic $\mathbf{X} = (X_1, X_2, \ldots, X_m)$ is minimal sufficient if there exist vectors $\theta_1, \theta_2, \cdots, \theta_m)$ such that the $m \times m$ matrix

$$P = \left\{ \log \left(\frac{\eta_{\theta_j}(d_i)\left[1 - \eta_{\theta_0}(d_i)\right]}{\eta_{\theta_0}(d_i)\left[1 - \eta_{\theta_j}(d_i)\right]} \right) \right\}$$

is invertible. (Hint: Theorem 6.12.)

6.25 Let (X_i, Y_i), $i = 1, \ldots, n$, be iid according to the uniform distribution over a set R in the (x, y) plane and let $\mathcal{P}$ be the family of distributions obtained by letting R range over a class $\mathcal{R}$ of sets R. Determine a minimal sufficient statistic for the following cases:

(a) $\mathcal{R}$ is the set of all rectangles $a_1 < x < a_2$, $b_1 < y < b_2$, $-\infty < a_1 < a_2 < \infty$, $-\infty < b_1 < b_2 < \infty$.

(b) $\mathcal{R}'$ is the subset of $\mathcal{R}$, for which $a_2 - a_1 = b_2 - b_1$.

(c) $\mathcal{R}''$ is the subset of $\mathcal{R}'$ for which $a_2 - a_1 = b_2 - b_1 = 1$.

6.26 Solve the preceding problem if

(a) $\mathcal{R}$ is the set of all triangles with sides parallel to the x axis, the y axis, and the line $y = x$, respectively.

(b) $\mathcal{R}'$ is the subset of $\mathcal{R}$ in which the sides parallel to the x and y axes are equal.

6.27 Formulate a general result of which Problems 6.25(a) and 6.26(a) are special cases.

6.28 If Y is distributed as $E(\eta, 1)$, the distribution of $X = e^{-Y}$ is $U(0, e^{-\eta})$. (This result is useful in the computer generation of random variables; see Problem 4.4.14.)

6.29 If a minimal sufficient statistic exists, a necessary condition for a sufficient statistic to be complete is for it to be minimal. [*Hint*: Suppose that $T = h(U)$ is minimal sufficient and U is complete. To show that U is equivalent to T, note that otherwise there exists ψ such that $\psi(U) \neq \eta[h(U)]$ with positive probability where $\eta(t) = E[\psi(U)|t]$.]

6.30 Show that the minimal sufficient statistics $T = (X_{(1)}, X_{(n)})$ of Problem 6.16(b) are complete. [*Hint*: Use the approach of Example 6.24.]

6.31 For each of the following problems, determine whether the minimal sufficient statistic is complete: (a) Problem 6.7(a)-(c); (b) Problem 6.25(a)-(c); (c) Problem 6.26(a) and (b).

6.32 (a) Show that if $\mathcal{P}_0, \mathcal{P}_1$ are two families of distributions such that $\mathcal{P}_0 \in \mathcal{P}_1$ and every null set of $\mathcal{P}_0$ is also a null set of $\mathcal{P}_1$, then a sufficient statistic T that is complete for $\mathcal{P}_0$ is also complete for $\mathcal{P}_1$.

(b) Let $\mathcal{P}_0$ be the class of binomial distributions $b(p, n)$, $0 < p < 1$, n = fixed, and let $\mathcal{P}_1 = \mathcal{P}_0 \cup \{Q\}$ where Q is the Poisson distribution with expectation 1. Then $\mathcal{P}_0$ is complete but $\mathcal{P}_1$ is not.

6.33 Let $X_1, \ldots, X_n$ be iid each with density $f(x)$ (with respect to Lebesgue measure), which is unknown. Show that the order statistics are complete.

[*Hint*: Use Problem 6.32(a) with $\mathcal{P}_0$ the class of distributions of Example 6.15(iv). Alternatively, let $\mathcal{P}_0$ be the exponential family with density

$$C(\theta_1, \ldots, \theta_n)e^{-\theta_1 \Sigma x_i - \theta_2 \Sigma x_i^2 - \cdots - \theta_n \Sigma x_i^n - \Sigma x_i^{2n}}.]$$

6.34 Suppose that $X_1, \ldots, X_n$ are an iid sample from a location-scale family with distribution function $F((x-a)/b)$.

(a) If b is known, show that the differences $(X_1 - X_i)/b$, $i = 2, \ldots, n$, are ancillary.

(b) If a is known, show that the ratios $(X_1 - a)/(X_i - a)$, $i = 2, \ldots, n$, are ancillary.

(c) If neither a or b are known, show that the quantities $(X_1 - X_i)/(X_2 - X_i)$, $i = 3, \ldots, n$, are ancillary.

6.35 Use Basu's theorem to prove independence of the following pairs of statistics:

(a) $\overline{X}$ and $\Sigma(X_i - \overline{X})^2$ where the X's are iid as $N(\xi, \sigma^2)$.

(b) $X_{(1)}$ and $\Sigma[X_i - X_{(1)}]$ in Problem 6.18.

6.36 (a) Under the assumptions of Problem 6.18, the ratios $Z_i = [X_{(n)} - X_{(i)}]/X_{(n)} - X_{(n-1)}]$, $i = 1, \ldots, n-2$, are independent of $\{X_{(1)}, \Sigma[X_i - X_{(1)}]\}$.

(b) Under the assumptions of Problems 6.16(b) and 6.30 the ratios $Z_i = [X_{(i)} - X_{(1)}]/X_{(n)} - X_{(1)}]$, $i = 2, \ldots, n-1$, are independent of $\{X_{(1)}, X_{(n)}\}$.

6.37 Under the assumptions of Theorem 6.5, let A be any fixed set in the sample space, P_θ^* the distribution P_θ truncated on A, and $\mathcal{P}^* = \{P_\theta^*, \theta \in \Omega\}$. Then prove

(a) if T is sufficient for $\mathcal{P}$, it is sufficient for $\mathcal{P}^*$.

(b) if, in addition, T is complete for $\mathcal{P}$, it is also complete for $\mathcal{P}^*$.

Generalizations of this result were derived by Tukey in the 1940s and also by Smith (1957). The analogous problem for observations that are *censored* rather than truncated is discussed by Bhattacharyya, Johnson, and Mehrotra (1977).

6.38 If $X_1, \ldots, X_n$ are iid as $B(a, b)$,

(a) Show that $[\Pi X_i, \Pi(1 - X_i)]$ is minimal sufficient for (a, b).

(b) Determine the minimal sufficient statistic when $a = b$.

Section 7

7.1 Verify the convexity of the functions (i)-(vi) of Example 7.3.

7.2 Show that x^p is concave over $(0, \infty)$ if $0 < p < 1$.

7.3 Give an example showing that a convex function need not be continuous on a closed interval.

7.4 If ϕ is convex on (a, b) and ψ is convex and nondecreasing on the range of ϕ, show that the function $\psi[\phi(x)]$ is convex on (a, b).

7.5 Prove or disprove by counterexample each of the following statements. If ϕ is convex on (a, b), then so is (i) $e^{\phi(x)}$ and (ii) $\log \phi(x)$ if $\phi > 0$.

7.6 Show that if equality holds in (7.1) for some $0 < \gamma < 1$, then ϕ is linear on $[x, y]$.

7.7 Establish the following lemma, which is useful in examining the risk functions of certain estimators. (For further discussion, see Casella 1990).

Lemma 9.2 *Let $r : [0, \infty) \to [0, \infty)$ be concave. Then, (i) $r(t)$ is nondecreasing and (ii) $r(t)/t$ is nonincreasing.*

7.8 Prove Jensen's inequality for the case that X takes on the values $x_1, \ldots, x_n$ with probabilities $\gamma_1, \ldots, \gamma_n (\Sigma \gamma_i = 1)$ directly from (7.1) by induction over n.

7.9 A slightly different form of the Rao-Blackwell theorem, which applies only to the variance of an estimator rather than any convex loss, can be established without Jensen's inequality.

(a) For any estimator $\delta(x)$ with $\text{var}[\delta(X)] < \infty$, and any statistic T, show that

$$\text{var}[\delta(X)] = \text{var}[E(\delta(X)|T)] + E[\text{var}(\delta(X)|T)].$$

(b) Based on the identity in part (a), formulate and prove a Rao-Blackwell type theorem for variances.

(c) The identity in part (a) plays an important role in both theoretical and applied statistics. For example, explain how Equation (1.2) can be interpreted as a special case of this identity.

7.10 Let U be uniformly distributed on (0, 1), and let F be a distribution function on the real line.

(a) If F is continuous and strictly increasing, show that $F^{-1}(U)$ has distribution function F.

(b) For arbitrary F, show that $F^{-1}(U)$ continues to have distribution function F.

[*Hint*: Take F^{-1} to be any nondecreasing function such that $F^{-1}[F(x)] = x$ for all x for which there exists no $x' \neq x$ with $F(x') = F(x)$.]

7.11 Show that the k-dimensional sphere $\Sigma_{i=1}^k x_i^2 \leq c$ is convex.

7.12 Show that $f(a) = \sqrt{|x - a|} + \sqrt{|y - a|}$ is minimized by $a = x$ and $a = y$.

7.13 (a) Show that $\phi(\mathbf{x}) = e^{\Sigma x_i}$ is convex by showing that its Hessian matrix is positive semidefinite.

(b) Show that the result of Problem 7.4 remains valid if ϕ is a convex function defined over an open convex set in E_k.

(c) Use (b) to obtain an alternative proof of the result of part (a).

7.14 Determine whether the following functions are super- or subharmonic:

(a) $\sum_{i=1}^k x_i^p, \quad p < 1, x_i > 0$.

(b) $e^{-\sum_{i=1}^k x_i^2}$.

(c) $\log\left(\prod_{i=1}^k x_i\right)$.

7.15 A function is *lower semicontinuous* at the point y if $f(y) \le \liminf_{x\to y} f(x)$. The definition of superharmonic can be extended from continuous to lower semicontinuous functions.

(a) Show that a continuous function is lower semicontinuous.

(b) The function $f(x) = I(a < x < b)$ is superharmonic on $(-\infty, \infty)$.

(c) For an estimator d of θ, show that the loss function

$$L(\theta, d) = \begin{cases} 0 & \text{if } |d-\theta| \le k \\ 2 & \text{if } |d-\theta| > k \end{cases}$$

is subharmonic.

7.16 (a) If $f : \mathfrak{R}^p \to \mathfrak{R}$ is superharmonic, then $\varphi(f(\cdot))$ is also superharmonic, where $\varphi : \mathfrak{R} \to \mathfrak{R}$ is a twice-differentiable increasing concave function.

(b) If h is superharmonic, then $h^*(x) = \int g(x-y)h(y)dy$ is also superharmonic, where $g(\cdot)$ is a density.

(c) If h_γ is superharmonic, then so is $h^*(x) = \int h_\gamma(x)dG(\gamma)$ where $G(\gamma)$ is a distribution function.

(Assume that all necessary integrals exist, and that derivatives may be taken inside the integrals.)

7.17 Use the convexity of the function ϕ of Problem 7.13 to show that the natural parameter space of the exponential family (5.2) is convex.

7.18 Show that if f is defined and bounded over $(-\infty, \infty)$ or $(0, \infty)$, then f cannot be convex (unless it is constant).

7.19 Show that $\phi(x, y) = -\sqrt{xy}$ is convex over $x > 0$, $y > 0$.

7.20 If f and g are real-valued functions such that f^2, g^2 are measurable with respect to the σ-finite measure μ, prove the *Schwarz inequality*

$$\left(\int fg\,d\mu\right)^2 \le \int f^2 d\mu \int g^2 d\mu.$$

[*Hint*: Write $\int fg\,d\mu = E_Q(f/g)$, where Q is the probability measure with $dQ = g^2 d\mu / \int g^2 d\mu$, and apply Jensen's inequality with $\varphi(x) = x^2$.]

7.21 Show that the loss functions (7.24) are continuously differentiable.

7.22 Prove that statements made in Example 7.20(i) and (ii).

7.23 Let f be a unimodal density symmetric about 0, and let $L(\theta, d) = \rho(d-\theta)$ be a loss function with ρ nondecreasing on $(0, \infty)$ and symmetric about 0.

(a) The function $\phi(a) = E[\rho(X - a)]$ defined in Theorem 7.15 takes on its minimum at 0.

(b) If

$$S_a = \{x : [\rho(x+a) - \rho(x-a)][f(x+a) - f(x-a)] \ne 0\},$$

then $\phi(a)$ takes on its unique minimum value at $a = 0$ if and only if there exists a_0 such that $\phi(a_0) < \infty$, and $\mu(S_a) > 0$ for all a. [*Hint*: Note that $\phi(0) \le 1/2[\phi(2a) + \phi(-2a)]$, with strict inequality holding if and only if $\mu(S_a) > 0$ for all a.]

7.24 (a) Suppose that f and ρ satisfy the assumptions of Problem 7.23 and that f is strictly decreasing on $[0, \infty)$. Then, if $\phi(a_0) < \infty$ for some a_0, $\phi(a)$ has a unique minimum at zero unless there exists $c \le d$ such that

$$\rho(0) = c \quad \text{and} \quad \rho(x) = d \quad \text{for all} \quad x \neq 0.$$

(b) If ρ is symmetric about 0, strictly increasing on $[0, \infty)$, and $\phi(a_0) < \infty$ for some a_0, then $\phi(a)$ has a unique minimum at (0) for all symmetric unimodal f.

[Problems 7.23 and 7.24 were communicated by Dr. W.Y. Loh.]

7.25 Let ρ be a real-valued function satisfying

$$0 \le \rho(t) \le M < \infty \text{ and } \rho(t) \to M \text{ as } t \to \pm\infty,$$

and let X be a random variable with a continuous probability density f. Then $\phi(a) = E[\rho(X - 1)]$ attains its minimum. [*Hint*: Show that (a) $\phi(a) \to M$ as $a \to \pm\infty$ and (b) ϕ is continuous. Here, (b) follows from the fact (see, for example, TSH2, Appendix, Section 2) that if $f_n, n = 1, 2, \ldots$, and f are probability densities such that $f_n(x) \to f(x)$ a.e., then $\int \psi f_n \to \int \psi f$ for any bounded ψ.]

7.26 Let ϕ be a strictly convex function defined over an interval I (finite or infinite). If there exists a value a_0 in I minimizing $\phi(a)$, then a_0 is unique.

7.27 Generalize Corollary 7.19 to the case where $\mathbf{X}$ and μ are vectors.

Section 8

8.1 (a) Prove Chebychev's Inequality: For any random variable X and non-negative function $g(\cdot)$,

$$P(g(X) \ge \varepsilon) \le \frac{1}{\varepsilon} Eg(X)$$

for every $\varepsilon > 0$. (In many statistical applications, it is useful to take $g(x) = (x - a)^2/b^2$ for some constants a and b.)

(b) Prove Lemma 9.3. [*Hint*: Apply Chebychev's Inequality.]

Lemma 9.3 *A sufficient condition for Y_n to converge in probability to c is that $E(Y_n - c)^2 \to 0$.*

8.2 To see that the converse of Theorem 8.2 does not hold, let $X_1, \ldots, X_n$ be iid with $E(X_i) = \theta$, $\text{var}(X_i) = \sigma^2 < \infty$, and let $\delta_n = \bar{X}$ with probability $1 - \varepsilon_n$ and $\delta_n = A_n$ with probability ε_n. If ε_n and A_n are constants satisfying

$$\varepsilon_n \to 0 \quad \text{and} \quad \varepsilon_n A_n \to \infty,$$

then δ_n is consistent for estimating θ, but $E(\delta_n - \theta)^2$ does not tend to zero.

8.3 Suppose $\rho(x)$ is an even function, nondecreasing and non-negative for $x \ge 0$ and positive for $x > 0$. Then, $E\{\rho[\delta_n - g(\theta)]\} \to 0$ for all θ implies that δ_n is consistent for estimating $g(\theta)$.

8.4 (a) If A_n, B_n, and Y_n tend in probability to a, b, and y, respectively, then $A_n + B_n Y_n$ tends in probability to $a + by$.

(b) If A_n takes on the constant value a_n with probability 1 and $a_n \to a$, then $A_n \to a$ in probability.

8.5 Referring to Example 8.4, show that $c_n S_n^2 \xrightarrow{P} \sigma^2$ for any sequence of constants $c_n \to 1$. In particular, the MLE $\hat{\sigma}^2 = \frac{n-1}{n} S_n^2$ is a consistent estimator of σ^2.

8.6 Verify Equation (8.9).

8.7 If $\{a_n\}$ is a sequence of real numbers tending to a, and if $b_n = (a_1 + \cdots + a_n)/n$, then $b_n \to a$.

8.8 (a) If δ_n is consistent for θ, and g is continuous, then $g(\delta_n)$ is consistent for $g(\theta)$.

(b) Let $X_1, \ldots, X_n$ be iid as $N(\theta, 1)$, and let $g(\theta) = 0$ if $\theta \neq 0$ and $g(0) = 1$. Find a consistent estimator of $g(\theta)$.

8.9 (a) In Example 8.5, find $\text{cov}(X_i, X_j)$ for any $i \neq j$.

(b) Verify (8.10).

8.10 (a) In Example 8.5, find the value of p_1 for which p_k becomes independent of k.

(b) If p_1 has the value given in (a), then for any integers $i_1 < \cdots < i_r$ and k, the joint distribution of $X_{i_1}, \ldots, X_{i_r}$ is the same as that of $X_{i_1+k}, \ldots, X_{i_r+k}$.

[*Hint*: Do not calculate, but use the definition of the chain.]

8.11 Suppose $X_1, \ldots, X_n$ have a common mean ξ and variance σ^2, and that $\text{cov}(X_i, X_j) = \rho_{j-i}$. For estimating ξ, show that:

(a) $\bar{X}$ is not consistent if $\rho_{j-i} = \rho \neq 0$ for all $i \neq j$;

(b) $\bar{X}$ is consistent if $|\rho_{j-i}| \leq M\gamma^{j-i}$ with $|\gamma| < 1$.

[*Hint*: (a) Note that $\text{var}(\bar{X}) > 0$ for all sufficiently large n requires $\rho \geq 0$, and determine the distribution of $\bar{X}$ in the multivariate normal case.]

8.12 Suppose that $k_n[\delta_n - g(\theta)]$ tends in law to a continuous limit distribution H. Prove that:

(a) If $k_n'/k_n \to d \neq 0$ or ∞, then $k_n'[\delta_n - g(\theta)]$ also tends to a continuous limit distribution.

(b) If $k_n'/k_n \to 0$ or ∞, then $k_n'[\delta_n - g(\theta)]$ tends in probability to zero or infinity, respectively.

(c) If $k_n \to \infty$, then $\delta_n \to g(\theta)$ in probability.

8.13 Show that if $Y_n \to c$ in probability, then it tends in law to a random variable Y which is equal to c with probability 1.

8.14 (a) In Example 8.7(i) and (ii), $Y_n \to 0$ in probability. Show that:

(b) If H_n denotes the distribution function of Y_n in Example 8.7(i) and (ii), then $H_n(a) \to 0$ for all $a < 0$ and $H_n(a) \to 1$ for all $a > 0$.

(c) Determine $\lim H_n(0)$ for Example 8.7(i) and (ii).

8.15 If $T_n > 0$ satisfies $\sqrt{n}[T_n - \theta] \xrightarrow{\mathcal{L}} N(0, \tau^2)$, find the limiting distribution of (a) $\sqrt{T_n}$ and (b) $\log T_n$ (suitably normalized).

8.16 If T_n satisfies $\sqrt{n}[T_n - \theta] \xrightarrow{\mathcal{L}} N(0, \tau^2)$, find the limiting distribution of (a) T_n^2, (b) $\log |T_n|$, (c) $1/T_n$, and (d) e^{T_n} (suitably normalized).

8.17 *Variance stabilizing transformations* are transformations for which the resulting statistic has an asymptotic variance that is independent of the parameters of interest. For each of the following cases, find the asymptotic distribution of the transformed statistic and show that it is variance stabilizing.

(a) $T_n = \frac{1}{n}\sum_{i=1}^n X_i$, $X_i \sim$Poisson(λ), $h(T_n) = \sqrt{T_n}$.

(b) $T_n = \frac{1}{n}\sum_{i=1}^n X_i$, $X_i \sim$Bernoulli(p), $h(T_n) = \arcsin\sqrt{T_n}$.

8.18 (a) The function $v(\cdot)$ is a variance stabilizing transformation if the estimator $v(T_n)$ has asymptotic variance $\tau^2(\theta)[v'(\theta)]^2 = c$, where c is a constant independent of θ.

(b) For any positive integer n, find the variance stabilizing transformation if $\tau^2(\theta) = \theta^n$. In particular, be careful of the important case $n = 2$.

[A variance stabilizing transformation (if it exists) is the solution of a differential equation resulting from the Delta Method approximation of the variance of an estimator (Theorem 8.12) and is not a function of the distribution of the statistic (other than the fact that the distribution will determine the form of the variance). The transformations of part (b) are known as the Box-Cox family of power transformations and play an important role in applied statistics. For more details and interesting discussions, see Bickel and Doksum 1981, Box and Cox 1982, and Hinkley and Runger 1984.]

8.19 Serfling (1980, Section 3.1) remarks that the following variations of Theorem 8.12 can be established. Show that:

(a) If h is differentiable in a neighborhood of θ, and h' is continuous at θ, then $h'(\theta)$ may be replace by $h'(T_n)$ to obtain

$$\sqrt{n}\frac{[h(T_n) - h(\theta)]}{\tau h'(T_n)} \xrightarrow{\mathcal{L}} N(0, 1).$$

(b) Furthermore, if τ^2 is a continuous function of θ, say $\tau^2(\theta)$, it can be replaced by $\tau^2(T_n)$ to obtain

$$\sqrt{n}\frac{[h(T_n) - h(\theta)]}{\tau(T_n)h'(T_n)} \xrightarrow{\mathcal{L}} N(0, 1).$$

8.20 Prove Theorem 8.16.

[*Hint*: Under the assumptions of the theorem we have the Taylor expansion

$$h(x_1, \ldots, x_s) = h(\xi_1, \ldots, \xi_s) + \Sigma(x_i - \xi_i)\left[\frac{\partial h}{\partial \xi_i} + R_i\right]$$

where $R_i \to 0$ as $x_i \to \xi_i$.]

8.21 A sequence of numbers R_n is said to be $o(1/k_n)$ as $n \to \infty$ if $k_n R_n \to 0$ and to be $O(1/k_n)$ if there exist M and n_0 such that $|k_n R_n| < M$ for all $n > n_0$ or, equivalently, if $k_n R_n$ is bounded.

(a) If $R_n = o(1/k_n)$, then $R_n = 0(1/k_n)$.

(b) $R_n = 0(1)$ if and only if R_n is bounded.

(c) $R_n = o(1)$ if and only if $R_n \to 0$.

(d) If R_n is $O(1/k_n)$ and k_n'/k_n tends to a finite limit, then R_n is $O(1/k_n')$.

8.22 (a) If R_n and R_n' are both $O(1/k_n)$, so is $R_n + R_n'$.

(b) If R_n and R_n' are both $o(1/k_n)$, so is $R_n + R_n'$.

8.23 Suppose $k_n'/k_n \to \infty$.

(a) If $R_n = 0(1/k_n)$ and $R_n' = 0(1/k_n')$, then $R_n + R_n' = 0(1/k_n)$.

(b) If $R_n = o(1/k_n)$ and $R_n' = o(1/k_n')$, then $R_n + R_n' = o(1/k_n)$.

8.24 A sequence of random variables Y_n is ***bounded in probability*** if given any $\varepsilon > 0$, there exist M and n_0 such that $P(|Y_n| > M) < \varepsilon$ for all $n > n_0$. Show that if Y_n converges in law, then Y_n is bounded in probability.

8.25 In generalization of the notation o and O, let us say that $Y_n = o_p(1/k_n)$ if $k_n Y_n \to 0$ in probability and that $Y_n = O_p(1/k_n)$ if $k_n Y_n$ is bounded in probability. Show that the results of Problems 8.21 - 8.23 continue to hold if o and O are replaced by o_p and O_p.

8.26 Let (X_n, Y_n) have a bivariate normal distribution with means $E(X_n) = E(Y_n) = 0$, variances $E(X_n^2) = E(Y_n^2) = 1$, and with correlation coefficient ρ_n tending to 1 as $n \to \infty$.

(a) Show that $(X_n, Y_n) \xrightarrow{\mathcal{L}} (X, Y)$ where X is $N(0, 1)$ and $P(X = Y) = 1$.

(b) If $S = \{(x, y) : x = y\}$, show that (8.25) does not hold.

8.27 Prove Theorem 8.22. [*Hint*: Make a Taylor expansion as in the proof of Theorem 8.12 and use Problem 4.16.]

10 Notes

10.1 Fubini's Theorem

Theorem 2.8, called variously Fubini's or Tonelli's theorem, is often useful in mathematical statistics. A variant of Theorem 2.8 allows f to be nonpositive, but requires an integrability condition (Billingsley 1995, Section 18). Dudley (1989) refers to Theorem 2.8 as the Tonelli-Fubini theorem and recounts an interesting history in which Lebesgue played a role. Apparently, Fubini's first published proof of this theorem was incorrect and was later corrected by Tonelli, using results of Lebesgue.

10.2 Sufficiency

The concept of sufficiency is due to Fisher (1920). (For some related history, see Stigler 1973.). In his fundamental paper of 1922, Fisher introduced the term sufficiency and stated the factorization criterion. The criterion was rediscovered by Neyman (1935) and was proved for general dominated families by Halmos and Savage (1949). The theory of minimal sufficiency was initiated by Lehmann and Scheffé (1950) and Dynkin (1951). Further generalizations are given by Bahadur (1954) and Landers and Rogge (1972). Yamada and Morimoto (1992) review the topic. Theorem 7.8 with squared error loss is due to Rao (1945) and Blackwell (1947). It was extended to the pth power of the error ($p \geq 1$) by Barankin (1950) and to arbitrary convex loss functions by Hodges and Lehmann (1950).

10.3 Exponential Families

One-parameter exponential families, as the only (regular) families of distributions for which there exists a one-dimensional sufficient statistic, were also introduced by Fisher (1934). His result was generalized to more than one dimension by Darmois (1935), Koopman (1936), and Pitman (1936). (Their contributions are compared by Barankin and Maitra (1963).) Another discussion of this theorem with reference to the literature is given, for example, by Hipp (1974). Comprehensive treatments of exponential families are provided by Barndorff-Nielsen (1978) and Brown (1986a); a more mathematical treatment is given in Hoffman-Jorgenson (1994). Statistical aspects are emphasized in Johansen (1979).

10.4 Ancillarity

To illustrate his use of ancillary statistics, group families were introduced by Fisher (1934). (For more information on ancillarity, see Buehler 1982, or the review article by Lehmann and Scholtz 1992).)

Ancillary statistics, and more general notions of ancillarity, have played an important role in developing inference in both group families and curved exponential families, the latter having connections to the field of "small-sample asymptotics," where it is shown how to obtain highly accurate asymptotic approximations, based on ancillaries and saddlepoints.

For example, as curved exponential families are not of full rank, it is typical that a minimal sufficient statistic is not complete. One might hope that an s-dimensional sufficient statistic could be split into a d-dimensional sufficient piece and an $s-d$-dimensional ancillary piece. Although this cannot always be done, useful decompositions can be found. Such endeavors lie at the heart of conditional inference techniques.

Good introductions to these topics can be found in Reid (1988), Field and Ronchetti (1990), Hinkley, Reid, and Snell (1991), Barndorff-Nielsen and Cox (1994), and Reid (1995).

10.5 Completeness

Completeness was introduced by Lehmann and Scheffé (1950). Theorem 6.21 is due to Basu (1955b, 1958). Although there is no converse to Basu's theorem as stated here, some alternative definitions and converse results are discussed by Lehmann (1981).

There are alternate versions of Theorem 6.22, which relate completeness in exponential families to having full rank. This is partially due to the fact that a *full* or *full-rank* exponential family can be defined in alternate ways. For example, referring to (5.1), if we define Θ as the index set of the densities $p_\theta(x)$, that is, we consider the family of densities $\{p_\theta(x), \theta \in \Theta\}$, then Brown (1986a, Section 1.1) defines the exponential family to be *full* if $\Theta = \Xi$, where Ξ is the natural parameter space [see (5.3)]. But this property is not needed for completeness. As Brown (1986a, Theorem 2.12) states, as long as the interior of Θ is nonempty (that is, Θ contains an open set), the family $\{p_\theta(x), \theta \in \Theta\}$ is complete. Another definition of a *full exponential model* is given by Barndorff-Nielsen and Cox (1994, Section 1.3), which requires that the statistics $T_1, \ldots, T_s$ not be linearly dependent.

In nonparametric families, the property of completeness, and determination of complete sufficient statistics, continues to be investigated. See, for example, Mandelbaum and Rüschendorf (1987) and Mattner (1992, 1993, 1994). For example, building on the work of Fraser (1954) and Mandelbaum and Rüschendorf (1987), Mattner (1994) showed that the order statistics are complete for the family of densities $\mathcal{P}$, in cases such as

(a) $\mathcal{P}$={all probability measures on the real line with unimodal densities with respect to Lebesgue measure}.

(b) $\mathcal{P} = \{(1-t)P + tQ : P \in \mathcal{P}, Q \in \mathcal{Q}(P), t \in [0, \epsilon]\}$, where ϵ is fixed and, for each $P \in \mathcal{P}$, P is absolutely continuous with respect to the complete and convex family $\mathcal{Q}(P)$.

10.6 Curved Exponential Families

The theory of curved exponential families was initiated by Efron (1975, 1978), who applied the ideas of *plane curvature* and *arc length* to better understand the structure of exponential families. Curved exponential families have been extensively studied since then. (See, for example, Brown 1986a, Chapter 3; Barndorff-Nielsen 1988; McCullagh and Nelder 1989; Barndorff-Nielsen and Cox 1994, Section 2.10.) Here, we give some details in a two-dimensional case; extensions to higher dimensions are reasonably straightforward (Problem 5.4).

For the exponential family (5.1), with $s = 2$, the parameter is $(\eta_1(\theta), \eta_2(\theta))$, where θ is an underlying parameter which is indexing the parameter space. If θ itself is a one-dimensional parameter, then the parameter space is a curve in two dimensions, a subset of the full two-dimensional space. Assuming that the η_i's have at least two derivatives as functions of θ, the parameter space is a one-dimensional *differentiable manifold*, a differentiable curve. (See Amari et al 1987 or Murray and Rice 1993 for an introduction to differential geometry and statistics.)

Figure 10.1. *The curve $\eta(\tau) = (\eta_1(\tau), \eta_2(\tau)) = (\tau, -\frac{1}{2}\tau^2)$. The radius of curvature γ_τ is the instantaneous rate of change of the angle Δa, between the derivatives $\nabla\eta(\tau)$, with respect to the arc length Δs. The vector $\nabla\eta(\tau)$, the tangent vector, and the unit normal vector $N(\tau) = [-\eta_2'(\tau), \eta_1'(\tau)]/[ds_\tau/d\tau]$ provide a moving frame of reference.*

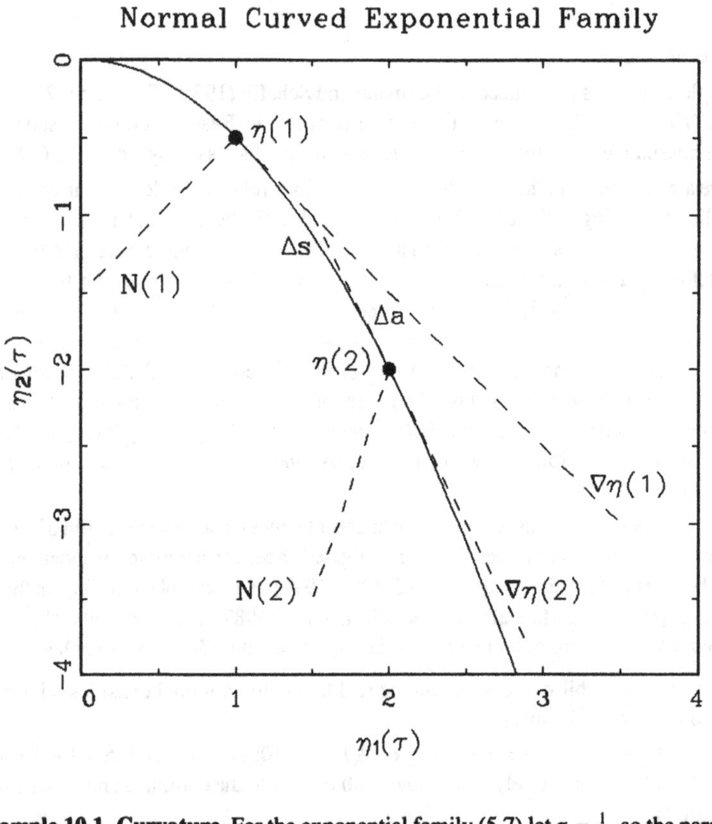

Example 10.1 Curvature. For the exponential family (5.7) let $\tau = \frac{1}{\xi}$, so the parameter space is the curve

$$\eta(\tau) = (\tau, -\frac{1}{2}\tau^2),$$

as shown in Figure 10.1. The direction of the curve $\eta(\tau)$, at any point τ, is measured by the derivative vector (the *gradient*) $\nabla\eta(\tau) = (\eta_1'(\tau), \eta_2'(\tau)) = (1, -\tau)$. At each τ we can assign an angular value

$$\begin{aligned} a(\tau) &= \text{polar angle of normalized gradient vector } \nabla\eta(\tau) \\ &= \text{polar angle of } \frac{(\eta_1'(\tau), \eta_2'(\tau))}{[(\eta_1')^2 + (\eta_2')^2]^{1/2}}, \end{aligned}$$

which measures how the curve "bends." The *curvature*, γ_τ, is a measure of the rate of change of this angle as a function of the *arc length* $s(\tau)$, where $s(\tau) = \int_0^\tau |\nabla\eta(t)|dt$.

Thus,

$$\gamma_\tau = \lim_{\delta \to 0} \frac{a(\tau + \delta\tau) - a(\tau)}{s(\tau + \delta\tau) - s(\tau)} = \frac{da(\tau)}{ds(\tau)}; \tag{10.1}$$

see Figure 10.1. An application of calculus will show that

$$\gamma_\tau = \frac{\eta_1' \eta_2'' - \eta_2' \eta_1''}{[(\eta_1')^2 + (\eta_2')^2]^{3/2}}, \tag{10.2}$$

so for the exponential family (5.7), we have $\gamma_\tau = -(1 + \tau^2)^{3/2}$. ||

For the most part, we are only concerned with $|\gamma_\tau|$, as the sign merely gives the direction of parameterization, and the magnitude gives the degree of curvature. As might be expected, lines have zero curvature and circles have constant curvature. The curvature of a circle is equal to the reciprocal of the radius, which leads to calling $1/|\gamma_\tau|$ the *radius of curvature*. Definitions of arc length, and so forth, naturally extend beyond two dimensions. (See Problems 5.5 and 5.4.)

10.7 Large Deviation Theory

Limit theorems such as Theorem 1.8.12 refer to sequences of situations as $n \to \infty$. However, in a given problem, one is dealing with a specific large value of n. Any particular situation can be embedded in many different sequences, which lead to different approximations.

Suppose, for example, that it is desired to find an approximate value for

$$P(|T_n - g(\theta)| \geq a) \tag{10.3}$$

when $n = 100$ and $a = 0.2$. If $\sqrt{n}[T_n - g(\theta)]$ is asymptotically normally distributed as $N(0, 1)$, one might want to put $a = c/\sqrt{n}$ (so that $c = 2$) and consider (10.3) as a member of the sequence

$$P\left(|T_n - g(\theta)| \geq \frac{2}{\sqrt{n}}\right) \approx 2[1 - \Phi(2)]. \tag{10.4}$$

Alternatively, one could keep $a = 0.2$ fixed and consider (10.3) as a member of the sequence

$$P(|T_n - g(\theta)| \geq 0.2). \tag{10.5}$$

Since $T_n - g(\theta) \to 0$, this sequence of probabilities tends to zero, and in fact does so at a very fast rate. In this approach, the normal approximation is no longer useful (it only tells us that (10.5) $\to 0$ as $n \to \infty$). The study of the limiting behavior of sequences such as (10.5) is called *large deviation theory*. An exposition of large deviation theory is given by Bahadur (1971). Books on large deviation theory include those by Kester (1985) and Bucklew (1990). Much research has been done on this topic, and applications to various aspects of point estimation can be found in Fu (1982), Kester and Kallenberg (1986), Sieders and Dzhaparidze (1987), and Pfanzagl (1990).

We would, of course, like to choose the approximation that comes closer to the true value. It seems plausible that for values of (10.3) not extremely close to 0 and for moderate sample sizes, (10.4) would tend to do better than that obtained from the sequence (10.5). Some numerical comparisons in the context of hypothesis testing can be found in Groeneboom and Oosterhoff (1981); other applications in testing are considered in Barron (1989).

CHAPTER 2

Unbiasedness

1 UMVU Estimators

It was pointed out in Section 1.1 that estimators with uniformly minimum risk typically do not exist, and restricting attention to estimators showing some degree of impartiality was suggested as one way out of this difficulty. As a first such restriction, we shall study the condition of *unbiasedness* in the present chapter.

Definition 1.1 An estimator $\delta(x)$ of $g(\theta)$ is *unbiased* if

(1.1) $$E_\theta[\delta(X)] = g(\theta) \quad \text{for all } \theta \in \Omega.$$

When used repeatedly, an unbiased estimator in the long run will estimate the right value "on the average." This is an attractive feature, but insistence on unbiasedness can lead to problems. To begin with, unbiased estimators of g may not exist.

Example 1.2 Nonexistence of unbiased estimator. Let X be distributed according to the binomial distribution $b(p, n)$ and suppose that $g(p) = 1/p$. Then, unbiasedness of an estimator δ requires

(1.2) $$\sum_{k=0}^{n} \delta(k) \binom{n}{k} p^k q^{n-k} = g(p) \quad \text{for all } 0 < p < 1.$$

That no such δ exists can be seen, for example, for the fact that as $p \to 0$, the left side tends to $\delta(0)$ and the right side to ∞. Yet, estimators of $1/p$ exist which (for n not too small) are close to $1/p$ with high probability. For example, since X/n tends to be close to p, n/X (with some adjustment when $X = 0$) will tend to be close to $1/p$. ||

If there exists an unbiased estimator of g, the estimand g will be called *U-estimable*. (Some authors call such an estimand "estimable," but this conveys the false impression that any g not possessing this property cannot be accurately estimated.) Even when g is U-estimable there is no guarantee that any of its unbiased estimators are desirable in other ways, and one may instead still prefer to use an estimator that does have some bias. On the other hand, a large bias is usually considered a drawback and special methods of bias reduction have been developed for such cases.

Example 1.3 The jackknife. A general method for bias reduction was initiated by Quenouille (1949, 1956) and later named the jackknife by Tukey (1958). Let

$T(\mathbf{x})$ be an estimator of a parameter $\tau(\theta)$ based on a sample $\mathbf{x} = (x_1, \ldots, x_n)$ and satisfying $E[T(\mathbf{x})] = \tau(\theta) + O(\frac{1}{n})$. Define $\mathbf{x}_{(-i)}$ to be the vector of sample values excluding x_i. Then, the jackknifed version of $T(\mathbf{x})$ is

$$T_J(\mathbf{x}) = nT(\mathbf{x}) - \frac{n-1}{n}\sum_{i=1}^{n} T(\mathbf{x}_{(-i)}). \tag{1.3}$$

It can be shown that $E[T_J(\mathbf{x})] = \tau(\theta) + O(\frac{1}{n^2})$, so the bias has been reduced (Stuart and Ord 1991, Section 17.10; see also Problem 1.4). ||

Although unbiasedness is an attractive condition, after a best unbiased estimator has been found, its performance should be investigated and the possibility not ruled out that a slightly biased estimator with much smaller risk might exist (see, for example, Sections 5.5 and 5.6).

The motive for introducing unbiasedness was the hope that within the class of unbiased estimators, there would exist an estimator with uniformly minimum risk. In the search for such an estimator, a natural approach is to minimize the risk for some particular value θ_0 and then see whether the result is independent of θ_0. To this end, the following obvious characterization of the totality of unbiased estimators is useful.

Lemma 1.4 *If δ_0 is any unbiased estimator of $g(\theta)$, the totality of unbiased estimators is given by $\delta = \delta_0 - U$ where U is any unbiased estimator of zero, that is, it satisfies*

$$E_\theta(U) = 0 \quad \textit{for all } \theta \in \Omega.$$

To illustrate this approach, suppose the loss function is squared error. The risk of an unbiased estimator δ is then just the variance of δ. Restricting attention to estimators δ_0, δ, and U with finite variance, we have, if δ_0 is unbiased,

$$\text{var}(\delta) = \text{var}(\delta_0 - U) = E(\delta_0 - U)^2 - [g(\theta)]^2$$

so that the variance of δ is minimized by minimizing $E(\delta_0 - U)^2$.

Example 1.5 Locally best unbiased estimation. Let X take on the values $-1, 0, 1, \ldots$ with probabilities (Problem 1.1)

$$P(X = -1) = p, \quad P(X = k) = q^2 p^k, \quad k = 0, 1, \ldots, \tag{1.4}$$

where $0 < p < 1$ and $q = 1 - p$, and consider the problems of estimating (a) p and (b) q^2. Simple unbiased estimators of p and q^2 are, respectively,

$$\delta_0 = \begin{cases} 1 & \text{if } X = -1 \\ 0 & \text{otherwise} \end{cases} \quad \text{and} \quad \delta_1 = \begin{cases} 1 & \text{if } X = 0 \\ 0 & \text{otherwise.} \end{cases}$$

It is easily checked that U is an unbiased estimator of zero if and only if [Problem 1.1(b)]

$$U(k) = -kU(-1) \quad \text{for } k = 0, 1, \ldots \tag{1.5}$$

or equivalently if $U(k) = ak$ for all $k = -1, 0, 1, \ldots$ and some a. The problem of determining the unbiased estimator which minimizes the variance at p_0 thus reduces to that of determining the value of a which minimizes

$$\Sigma P(X = k)[\delta_i(k) - ak]^2. \tag{1.6}$$

The minimizing values of a are (Problem 1.2)

$$a_0^* = -p_0 \Big/ \left[p_0 + q_0^2 \sum_{k=1}^{\infty} k^2 p_0^k \right] \quad \text{and} \quad a_1^* = 0$$

in cases (a) and (b), respectively. Since a_1^* does not depend on p_0, the estimator $\delta_1^* = \delta_1 - a_1^* X = \delta_1$ minimizes the variance among all unbiased estimators not only when $p = p_0$ but for all values of p. On the other hand, $\delta_0^* = \delta_0 - a_0^* X$ does depend on p_0, and it therefore only minimizes the variance at $p = p_0$. ||

The properties possessed by δ_0^* and δ_1^* are characterized more generally by the following definition.

Definition 1.6 An unbiased estimator $\delta(x)$ of $g(\theta)$ is the *uniform minimum variance unbiased* (UMVU) estimator of $g(\theta)$ if $\text{var}_\theta \delta(x) \le \text{var}_\theta \delta'(x)$ for all $\theta \in \Omega$, where $\delta'(x)$ is any other unbiased estimator of $g(\theta)$. The estimator $\delta(x)$ is *locally minimum variance unbiased* (LMVU) at $\theta = \theta_0$ if $\text{var}_{\theta_0} \delta(x) \le \text{var}_{\theta_0} \delta'(x)$ for any other unbiased estimator $\delta'(x)$.

In terms of Definition 1.6, we have shown in Example 1.5 that δ_1^* is UMVU and that δ_0^* is LMVU. Since δ_0^* depends on p_0, no UMVU estimator exists in this case.

Notice that the definition refers to "the" UMVU estimator, since UMVU estimators are unique (see Problem 1.12). The existence, uniqueness, and characterization of LMVU estimators have been investigated by Barankin (1949) and Stein (1950). Interpreting $E(\delta_0 - U)^2$ as the distance between δ_0 and U, the minimizing U^* can be interpreted as the projection of δ_0 onto the linear space $\mathcal{U}$ formed by the unbiased estimators U of zero. The desired results then follow from the projection theorem of linear space theory (see, for example, Bahadur 1957, and Luenberger 1969).

The relationship of unbiased estimators of $g(\theta)$ with unbiased estimators of zero can be helpful in characterizing and determining UMVU estimators when they exist. Note that if $\delta(X)$ is an unbiased estimator of $g(\theta)$, then so is $\delta(X) + aU(X)$, for any constant a and any unbiased estimator U of zero and that

$$\text{var}_{\theta_0}[\delta(X) + aU(X)] = \text{var}_{\theta_0}\delta(X) + a^2 \text{var}_{\theta_0} U(X) + 2a\text{cov}_{\theta_0}(U(X), \delta(X)).$$

If $\text{cov}_\theta(U(X), \delta(X)) \neq 0$ for some $\theta = \theta_0$, we shall show below that there exists a value of a for which $\text{var}_{\theta_0}[\delta(X) + aU(X)] < \text{var}_{\theta_0}\delta(X)$. As a result, the covariance with unbiased estimators of zero is the key in characterizing the situations in which a UMVU estimator exists. In the statement of the following theorem, attention will be restricted to estimators with finite variance, since otherwise the problem of minimizing the variance does not arise. The class of estimators δ with $E_\theta \delta^2 < \infty$ for all θ will be denoted by Δ.

Theorem 1.7 *Let X have distribution $P_\theta, \theta \in \Omega$, let δ be an estimator in Δ, and let $\mathcal{U}$ denote the set of all unbiased estimators of zero which are in Δ. Then, a* *necessary and sufficient condition for δ to be a UMVU estimator of its expectation $g(\theta)$ is that*

$$E_\theta(\delta U) = 0 \quad \textit{for all } U \in \mathcal{U} \quad \textit{and all } \theta \in \Omega. \tag{1.7}$$

(*Note*: Since $E_\theta(U) = 0$ for all $U \in \mathcal{U}$, it follows that $E_\theta(\delta U) = \text{cov}_\theta(\delta, U)$, so that (1.7) is equivalent to the condition that δ is uncorrelated with every $U \in \mathcal{U}$.)

Proof.

(a) *Necessity.* Suppose δ is UMVU for estimating its expectation $g(\theta)$. Fix $U \in \mathcal{U}, \theta \in \Omega$, and for arbitrary real λ, let $\delta' = \delta + \lambda U$. Then, δ' is also an unbiased estimator of $g(\theta)$, so that

$$\text{var}_\theta(\delta + \lambda U) \geq \text{var}_\theta(\delta) \quad \text{for all } \lambda.$$

Expanding the left side, we see that

$$\lambda^2 \text{var}_\theta U + 2\lambda \text{cov}_\theta(\delta, U) \geq 0 \quad \text{for all } \lambda,$$

a quadratic in λ with real roots $\lambda = 0$ and $\lambda = -2\,\text{cov}_\theta(\delta, U)/\text{var}_\theta(U)$. It will therefore take on negative values unless $\text{cov}_\theta(\delta, U) = 0$.

(b) *Sufficiency.* Suppose $E_\theta(\delta U) = 0$ for all $U \in \mathcal{U}$. To show that δ is UMVU, let δ' be any unbiased estimator of $E_\theta(\delta)$. If $\text{var}_\theta \delta' = \infty$, there is nothing to prove, so assume $\text{var}_\theta \delta' < \infty$. Then, $\delta - \delta' \in \mathcal{U}$ (Problem 1.8) so that

$$E_\theta[\delta(\delta - \delta')] = 0$$

and hence $E_\theta(\delta^2) = E_\theta(\delta\delta')$. Since δ and δ' have the same expectation,

$$\text{var}_\theta \delta = \text{cov}_\theta(\delta, \delta'),$$

and from the covariance inequality (Problem 1.5), we conclude that $\text{var}_\theta(\delta) \leq \text{var}_\theta(\delta')$.

□

The proof of Theorem 1.7 shows that condition (1.7), if required only for $\theta = \theta_0$, is necessary and sufficient for an estimator δ with $E_{\theta_0}(\delta^2) < \infty$ to be LMVU at θ_0. This result also follows from the characterization of the LMVU estimator as $\delta = \delta_0 - U^*$ where δ_0 is any unbiased estimator of g and U^* is the projection of δ_0 onto $\mathcal{U}$. Interpreting the equation $E_{\theta_0}(\delta U) = 0$ as orthogonality of δ_0 and U, the projection of U^* has the property that $\delta = \delta_0 - U^*$ is orthogonal to $\mathcal{U}$, that is, $E_{\theta_0}(\delta U) = 0$ for all $U \in \mathcal{U}$. If the estimator is to be UMVU, this relation must hold for all θ.

Example 1.8 Continuation of Example 1.5. As an application of Theorem 1.7, let us determine the totality of UMVU estimators in Example 1.5. In view of (1.5) and (1.7), a necessary and sufficient condition for δ to be UMVU for its expectation is

$$E_p(\delta X) = 0 \quad \text{for all } p, \tag{1.8}$$

that is, for δX to be in $\mathcal{U}$ and hence to satisfy (1.5). This condition reduces to

$$k\delta(k) = k\delta(-1) \quad \text{for } k = 0, 1, 2, \ldots,$$

which is satisfied provided

$$(1.9) \qquad \delta(k) = \delta(-1) \quad \text{for } k = 1, 2, \ldots$$

with $\delta(0)$ being arbitrary. If we put $\delta(-1) = a$, $\delta(0) = b$, the expectation of such a δ is $g(p) = bq^2 + a(1 - q^2)$ and $g(p)$ is therefore seen to possess a UMVU estimator with finite variance if and only if it is of the form $a + cq^2$. ||

It is interesting to note, although we shall not prove it here, that Theorem 1.7 typically, but not always, holds not only for squared error but for general convex loss functions. This result follows from a theorem of Bahadur (1957). For details, see Padmanabhan (1970) and Linnik and Rukhin (1971).

Constants are always UMVU estimators of their expectations since the variance of a constant is zero. (If δ is a constant, (1.7) is of course trivially satisfied.) Deleting the constants from consideration, three possibilities remain concerning the set of UMVU estimators.

Case 1. No nonconstant U-estimable function has a UMVU estimator.

Example 1.9 Nonexistence of UMVU estimator. Let $X_1, \ldots, X_n$ be a sample from a discrete distribution which assigns probability 1/3 to each of the points $\theta - 1, \theta, \theta + 1$, and let θ range over the integers. Then, no nonconstant function of θ has a UMVU estimator (Problem 1.9). A continuous version of this example is provided by a sample from the uniform distribution $U(\theta - 1/2, \theta + 1/2)$; see Lehmann and Scheffé (1950, 1955, 1956). (For additional examples, see Section 2.3.) ||

Case 2. Some, but not all, nonconstant U-estimable functions have UMVU estimators. Example 1.5 provides an instance of this possibility.

Case 3. Every U-estimable function has a UMVU estimator.

A condition for this to be the case is suggested by (Rao-Blackwell) Theorem 1.7.8. If T is a sufficient statistic for the family $\mathcal{P} = \{P_\theta, \theta \in \Omega\}$ and $g(\theta)$ is U-estimable, then any unbiased estimator δ of $g(\theta)$ which is not a function of T is improved by its conditional expectation given T, say $\eta(T)$. Furthermore, $\eta(T)$ is again an unbiased estimator of $g(\theta)$ since by (6.1), $E_\theta[\eta(T)] = E_\theta[\delta(X)]$.

Lemma 1.10 *Let X be distributed according to a distribution from $\mathcal{P} = \{P_\theta, \theta \in \Omega\}$, and let T be a complete sufficient statistic for $\mathcal{P}$. Then, every U-estimable function $g(\theta)$ has one and only one unbiased estimator that is a function of T. (Here, uniqueness, of course, means that any two such functions agree a.e. $\mathcal{P}$.)*

Proof. That such an unbiased estimator exists was established just preceding the statement of Lemma 1.10. If δ_1 and δ_2 are two unbiased estimators of $g(\theta)$, their difference $f(T) = \delta_1(T) - \delta_2(T)$ satisfies

$$E_\theta f(T) = 0 \quad \text{for all } \theta \in \Omega,$$

and hence by the completeness of T, $\delta_1(T) = \delta_2(T)$ a.e. $\mathcal{P}$, as was to be proved. □

So far, attention has been restricted to squared error loss. However, the Rao-Blackwell theorem applies to any convex loss function, and the preceding argument therefore establishes the following result.

Theorem 1.11 *Let X be distributed according to a distribution in $\mathcal{P} = \{P_\theta, \theta \in \Omega\}$, and suppose that T is a complete sufficient statistic for $\mathcal{P}$.*

(a) For every U-estimable function $g(\theta)$, there exists an unbiased estimator that uniformly minimizes the risk for any loss function $L(\theta, d)$ which is convex in its second argument; therefore, this estimator in particular is UMVU.

(b) The UMVU estimator of (i) is the unique unbiased estimator which is a function of T; it is the unique unbiased estimator with minimum risk, provided its risk is finite and L is strictly convex in d.

It is interesting to note that under mild conditions, the existence of a complete sufficient statistic is not only sufficient but also necessary for Case 3. This result, which is due to Bahadur (1957), will not be proved here.

Corollary 1.12 *If $\mathcal{P}$ is an exponential family of full rank given by (5.1), then the conclusions of Theorem 1.11 hold with $\theta = (\theta_1, \ldots, \theta_s)$ and $T = (T_1, \ldots, T_s)$.*

Proof. This follows immediately from Theorem 1.6.22. □

Theorem 1.11 and its corollary provide best unbiased estimators for large classes of problems, some of which will be discussed in the next three sections. For the sake of simplicity, these estimators will be referred to as being UMVU, but it should be kept in mind that their optimality is not tied to squared error as loss, but, in fact, they minimize the risk for any convex loss function.

Sometimes we happen to know an unbiased estimator δ of $g(\theta)$ which is a function of a complete sufficient statistic. The theorem then states that δ is UMVU. Suppose, for example, that $X_1, \ldots, X_n$ are iid according to $N(\xi, \sigma^2)$ and that the estimand is σ^2. The standard unbiased estimator of σ^2 is then $\delta = \Sigma(X_i - \bar{X})^2/(n-1)$. Since this is a function of the complete sufficient statistic $T = (\Sigma X_i, \Sigma(X_i - \bar{X})^2)$, δ is UMVU. Barring such fortunate accidents, two systematic methods are available for deriving UMVU estimators through Theorem 1.11.

Method One: Solving for δ

If T is a complete sufficient statistic, the UMVU estimator of any U-estimable function $g(\theta)$ is uniquely determined by the set of equations

$$E_\theta \delta(T) = g(\theta) \quad \text{for all } \theta \in \Omega. \tag{1.10}$$

Example 1.13 Binomial UMVU estimator. Suppose that T has the binomial distribution $b(p, n)$ and that $g(p) = pq$. Then, (1.10) becomes

$$\sum_{t=0}^{n} \binom{n}{t} \delta(t) p^t q^{n-t} = pq \quad \text{for all } 0 < p < 1. \tag{1.11}$$

If $\rho = p/q$ so that $p = \rho/(1+\rho)$ and $q = 1/(1+\rho)$, (1.11) can be rewritten as

$$\sum_{t=0}^{n} \binom{n}{t} \delta(t)\rho^t = \rho(1+\rho)^{n-2} = \sum_{t=1}^{n-1} \binom{n-2}{t-1} \rho^t \quad (0 < \rho < \infty).$$

A comparison of the coefficients on the left and right sides leads to

$$\delta(t) = \frac{t(n-t)}{n(n-1)}. \qquad \|$$

Method Two: Conditioning

If $\delta(X)$ is any unbiased estimator of $g(\theta)$, it follows from Theorem 1.11 that the UMVU estimator can be obtained as the conditional expectation of $\delta(X)$ given T. For this derivation, it does not matter which unbiased estimator δ is being conditioned; one can thus choose δ so as to make the calculation of $\delta'(T) = E[\delta(X)|T]$ as easy as possible.

Example 1.14 UMVU estimator for a uniform distribution. Suppose that $X_1, \ldots, X_n$ are iid according to the uniform distribution $U(0, \theta)$ and that $g(\theta) = \theta/2$. Then, $T = X_{(n)}$, the largest of the X's, is a complete sufficient statistic. Since $E(X_1) = \theta/2$, the UMVU estimator of $\theta/2$ is $E[X_1|X_{(n)} = t]$. If $X_{(n)} = t$, then $X_1 = t$ with probability $1/n$, and X_1 is uniformly distributed on $(0, t)$ with the remaining probability $(n-1)/n$ (see Problem 1.6.2). Hence,

$$E[X_1|t] = \frac{1}{n} \cdot t + \frac{n-1}{n} \cdot \frac{t}{2} = \frac{n+1}{n} \cdot \frac{t}{2}.$$

Thus, $[(n+1)/n] \cdot T/2$ and $[(n+1)/n] \cdot T$ are the UMVU estimators of $\theta/2$ and θ, respectively. $\|$

The existence of UMVU estimators under the assumptions of Theorem 1.11 was proved there for convex loss functions. That the situation tends to be very different without convexity of the loss is seen from the following results of Basu (1955a).

Theorem 1.15 *Let the loss function $L(\theta, d)$ for estimating $g(\theta)$ be bounded, say $L(\theta, d) \leq M$, and assume that $L[\theta, g(\theta)] = 0$ for all θ, that is, the loss is zero when the estimated value coincides with the true value. Suppose that g is U-estimable and let θ_0 be an arbitrary value of θ. Then, there exists a sequence of unbiased estimators δ_n for which $R(\theta_0, \delta_n) \to 0$.*

Proof. Since $g(\theta)$ is U-estimable, there exists an unbiased estimator $\delta(X)$. For any $0 < \pi < 1$, let

$$\delta'_\pi(x) = \begin{cases} g(\theta_0) & \text{with probability } 1-\pi \\ \dfrac{1}{\pi}[\delta(x) - g(\theta_0)] + g(\theta_0) & \text{with probability } \pi. \end{cases}$$

Then, δ'_π is unbiased for all π and all θ, since

$$E_\theta(\delta'_\pi) = (1-\pi)g(\theta_0) + \frac{\pi}{\pi}[g(\theta) - g(\theta_0)] + \pi g(\theta_0) = g(\theta).$$

The risk $R(\theta_0, \delta'_\pi)$ at θ_0 is $(1-\pi)\cdot 0$ plus π times the expected loss of $(1/\pi)[\delta(X) - g(\theta_0)] + g(\theta_0)$, so that

$$R(\theta_0, \delta'_\pi) \leq \pi M.$$

As $\pi \to 0$, it is seen that $R(\theta_0, \delta'_\pi) \to 0$. □

This result implies that for bounded loss functions, no uniformly minimum-risk-unbiased or even locally minimum-risk-unbiased estimator exists except in trivial cases, since at each θ_0, the risk can be made arbitrarily small even by unbiased estimators. [Basu (1955a) proved this fact for a more general class of nonconvex loss functions.] The proof lends support to the speculation of Section 1.7 that the difficulty with nonconvex loss functions stems from the possibility of arbitrarily large errors since as $\pi \to 0$, the error $|\delta_\pi(x) - g(\theta_0)| \to \infty$. It is the leverage of these large but relatively inexpensive errors which nullifies the restraining effect of unbiasedness.

This argument applies not only to the limiting case of unbounded errors but also, although to a correspondingly lesser degree, to the case of finite large errors. In the latter situation, convex loss functions receive support from a large-sample consideration. To fix ideas, suppose the observations consist for n iid variables $X_1, \ldots, X_n$. As n increases, the error in estimating a given value $g(\theta)$ will decrease and tend to zero as $n \to \infty$. (See Section 1.8 for a precise statement.) Thus, essentially only the local behavior of the loss function near the true value $g(\theta)$ is relevant. If the loss function is smooth, its Taylor expansion about $d = g(\theta)$ gives

$$L(\theta, d) = a(\theta) + b(\theta)[d - g(\theta)] + c(\theta)[d - g(\theta)]^2 + R,$$

where the remainder R becomes negligible as the error $|d - g(\theta)|$ becomes sufficiently small. If the loss is zero when $d = g(\theta)$, then a must be zero, so that $b(\theta)[d - g(\theta)]$ becomes the dominating term for small errors. The condition $L(\theta, d) \geq 0$ for all θ then implies $b(\theta) = 0$ and hence

$$L(\theta, d) = c(\theta)[d - g(\theta)]^2 + R.$$

Minimizing the risk for large n thus becomes essentially equivalent to minimizing $E[\delta(X) - g(\theta)]^2$, which justifies not only a convex loss function but even squared error. Not only the loss function but also other important aspects of the behavior of estimators and the comparison of different estimators greatly simplify for large samples, as will be discussed in Chapter 6.

The difficulty which bounded loss functions present for the theory of unbiased estimation is not encountered by a different unbiasedness concept, that of median unbiasedness mentioned in Section 1.1. For estimating $g(\theta)$ in a multiparameter exponential family, it turns out that uniformly minimum risk median unbiased estimators exist for any loss function L for which $L(\theta, d)$ is a nondecreasing function of d as d moves in either direction away from $g(\theta)$. A detailed version of this result can be found in Pfanzagl (1979). We shall not discuss the theory of median unbiased estimation here since the methods required belong to the theory of confidence intervals rather than that of point estimation (see TSH2, Section 3.5).

2 Continuous One- and Two-Sample Problems

The problem of estimating an unknown quantity θ from n measurements of θ was considered in Example 1.1.1 as the prototype of an estimation problem. It was formalized by assuming that the n measurements are iid random variables $X_1, \ldots, X_n$ with common distribution belonging to the location family

$$P_\theta(X_i \le x) = F(x - \theta). \tag{2.1}$$

The problem takes different forms according to the assumptions made about F. Some possibilities are the following:

(a) F is completely specified.

(b) F is specified except for an unknown scale parameter. In this case, (2.1) will be replaced by a location-scale family. It will then be convenient to denote the location parameter by ξ rather than θ (to reserve θ for the totality of unknown parameters) and hence to write the family as

$$P_\theta(X_i \le x) = F\left(\frac{x - \xi}{\sigma}\right). \tag{2.2}$$

Here, it will be of interest to estimate both ξ and σ.

(c) The distribution of the X's is only approximately given by Equation (2.1) or (2.2) with a specified F. What is meant by "approximately" leads to the topic of robust estimation

(d) F is known to be symmetric about 0 (so that the X's are symmetrically distributed about θ or ξ) but is otherwise unknown.

(e) F is unknown except that it has finite variance; the estimand is $\xi = E(X_i)$.

In all these models, F is assumed to be continuous.

A treatment of Problems (a) and (b) for an arbitrary known F is given in Chapter 3 from the point of view of equivariance. In the present section, we shall be concerned with unbiased estimation of θ or (ξ, σ) in Problems (a) and (b) and some of their generalizations for some special distributions, particularly for the case that F is normal or exponential. Problems (c), (d), and (e) all fall under the general heading of robust and nonparametric statistics (Huber 1981, Hampel et al. 1986, Staudte and Sheather 1990). We will not attempt a systematic treatment of these topics here, but will touch upon some points through examples. For example, Problem (e) will be considered in Section 2.4.

The following three examples will be concerned with the normal one-sample problems, that is, with estimation problems arising when $X_1, \ldots, X_n$ are distributed with joint density (2.3).

Example 2.1 Estimating polynomials of a normal variance. Let $X_1, \ldots, X_n$ be distributed with joint density

$$\frac{1}{(\sqrt{2\pi}\sigma)^n} \exp\left[-\frac{1}{2\sigma^2}\Sigma(x_i - \xi)^2\right], \tag{2.3}$$

and assume, to begin with, that only one of the parameters is unknown. If σ is known, it follows from Theorem 1.6.22 that the sample mean $\bar{X}$ is a complete sufficient statistic, and since $E(\bar{X}) = \xi$, $\bar{X}$ is the UMVU estimator of ξ. More generally, if $g(\xi)$ is any U-estimable function of ξ, there exists a unique unbiased estimator $\delta(\bar{X})$ based on $\bar{X}$ and it is UMVU. If, in particular, $g(\xi)$ is a polynomial of degree r, $\delta(\bar{X})$ will also be a polynomial of that degree, which can be determined inductively for $r = 2, 3, \ldots$ (Problem 2.1).

If ξ is known, (2.3) is a one-parameter exponential family with $S^2 = \Sigma(X_i - \xi)^2$ being a complete sufficient statistic. Since $Y = S^2/\sigma^2$ is distributed as χ_n^2 independently of σ^2, it follows that

$$E\left(\frac{S^r}{\sigma^r}\right) = \frac{1}{K_{n,r}},$$

where $K_{n,r}$ is a constant, and hence that

$$K_{n,r}S^r \tag{2.4}$$

is UMVU for σ^r. Recall from Example 1.5.14 with $a = n/2$, $b = 2$ and with $r/2$ in place of r that

$$E\left(\frac{S^r}{\sigma^r}\right) = E\left[(\chi_n^2)^{r/2}\right] = \frac{\Gamma[(n+r)/2]}{\Gamma(n/2)}2^{r/2}$$

so that

$$K_{n,r} = \frac{\Gamma(n/2)}{2^{r/2}\Gamma[(n+r)/2]}. \tag{2.5}$$

As a check, note that for $r = 2$, $K_{n,r} = 1/n$, and hence $E(S^2) = n\sigma^2$.

Formula (2.5) is established in Example 1.5.14 only for $r > 0$. It is, however, easy to see (Problem 1.5.19) that it holds whenever

$$n > -r, \tag{2.6}$$

but that the $(r/2)$th moment of χ_n^2 does not exist when $n \leq -r$.

We are now in a position to consider the more realistic case in which both parameters are unknown. Then, by Example 1.6.24, $\bar{X}$ and $S^2 = \sum(X_i - \bar{X})^2$ jointly are complete sufficient statistics for (ξ, σ^2). This shows that $\bar{X}$ continues to be UMVU for ξ. Since $\text{var}(\bar{X}) = \sigma^2/n$, estimation of σ^2 is, of course, also of great importance. Now, S^2/σ^2 is distributed as χ_{n-1}^2 and it follows from (2.4) with n replaced by $n-1$ and the new definition of S^2 that

$$K_{n-1,r}S^r \tag{2.7}$$

is UMVU for σ^r provided $n > -r+1$, and thus in particular $S^2/(n-1)$ is UMVU for σ^2.

Sometimes, it is of interest to measure ξ in σ-units and hence to estimate $g(\xi, \sigma) = \xi/\sigma$. Now $\bar{X}$ is UMVU for ξ and $K_{n-1,-1}/S$ for $1/\sigma$. Since $\bar{X}$ and S are independent, it follows that $K_{n-1,-1}\bar{X}/S$ is unbiased for ξ/σ and hence UMVU, provided $n - 1 > 1$, that is, $n > 2$.

If we next consider calculating the variance of $K_{n-1,-1}\bar{X}/S$ or, more generally, calculating the variance of UMVU estimators of polynomial functions of ξ and σ, we are led to calculating the moments $E(\bar{X}^k)$ and $E(S^k)$ for all $k = 1, 2, \ldots$. This is investigated in Problems 2.4-2.6 ‖

Another class of problems within the framework of the normal one-sample problem relates to the probability

$$p = P(X_1 \le u). \tag{2.8}$$

Example 2.2 Estimating a probability or a critical value. Suppose that the observations X_i denote the performances of past candidates on an entrance examination and that we wish to estimate the cutoff value u for which the probability of a passing performance, $X \ge u$, has a preassigned probability $1 - p$. This is the problem of estimating u in (2.8) for a given value of p. Solving the equation

$$p = P(X_1 \le u) = \Phi\left(\frac{u - \xi}{\sigma}\right) \tag{2.9}$$

(where Φ denotes the cdf of the standard normal distribution) for u shows that

$$u = g(\xi, \sigma) = \xi + \sigma\Phi^{-1}(p).$$

It follows that the UMVU estimator of u is

$$\bar{X} + K_{n-1,1}S\Phi^{-1}(p). \tag{2.10}$$

Consider next the problem of estimating p for a given value of u. Suppose, for example, that a manufactured item is acceptable if some quality characteristic is $\le u$ and that we wish to estimate the probability of an item being acceptable, its *reliability*, given by (2.9).

To illustrate a method which is applicable to many problems of this type, consider, first, the simpler case that $\sigma = 1$. An unbiased estimator δ of p is the indicator of the event $X_1 \le u$. Since $\bar{X}$ is a complete sufficient statistic, the UMVU estimator of $p = P(X_1 \le u) = \Phi(u - \xi)$ is therefore

$$E[\delta|\bar{X}] = P[X_1 \le u|\bar{X}].$$

To evaluate this probability, use the fact that $X_1 - \bar{X}$ is independent of $\bar{X}$. This follows from Basu's theorem (Theorem 1.6.21) since $X_1 - \bar{X}$ is ancillary.[1] Hence,

$$P[X_1 \le u|\bar{x}] = P[X_1 - \bar{X} \le u - \bar{x}|\bar{x}] = P[X_1 - \bar{X} \le u - \bar{x}],$$

and the computation of a conditional probability has been replaced by that of an unconditional one. Now, $X_1 - \bar{X}$ is distributed as $N(0, (n-1)/n)$, so that

$$P[X_1 - \bar{X} \le u - \bar{x}] = \Phi\left[\sqrt{\frac{n}{n-1}}(u - \bar{x})\right], \tag{2.11}$$

which is the UMVU estimator of p.

Closely related to the problem of estimating p, which is the cdf

$$F(u) = P[X_1 \le u] = \Phi(u - \xi)$$

of X_1 evaluated at u, is that of estimating the probability density at $u : g(\xi) = \phi(u - \xi)$. We shall now show that the UMVU estimator of the probability density $g(\xi) = p_\xi^{X_1}(u)$ of X_1 evaluated at u is the conditional density of X_1 given $\bar{X}$

[1] Such applications of Basu's theorem can be simplified when invariance is present. The theory and some interesting illustrations are discussed by Eaton and Morris (1970).

evaluated at u, $\delta(\bar{X}) = p^{X_1|\bar{X}}(u)$. Since this is a function of $\bar{X}$, it is only necessary to check that δ is unbiased. This can be shown by differentiating the UMVU estimator of the cdf after justifying the required interchange of differentiation and integration, or as follows. Note that the joint density of X_1 and $\bar{X}$ is $p^{X_1|\bar{X}}(u)p_\xi^{\bar{X}}(x)$ and that the marginal density is therefore

$$p_\xi^{X_1}(u) = \int_{-\infty}^{\infty} p^{X_1|\bar{X}}(u)p_\xi^{\bar{X}}(x)\,dx.$$

This equation states just that $\delta(\bar{X})$ is an unbiased estimator of $g(\xi)$. Differentiating the earlier equation

$$P[X_1 \le u|\bar{x}] = \Phi\left[\sqrt{\frac{n}{n-1}}(u-\bar{x})\right]$$

with respect to u, we see that the derivative

$$\begin{aligned}\frac{d}{du}[P[X_1 \le u|\bar{X}] &= p^{X_1|\bar{X}}(u)\\ &= \sqrt{\frac{n}{n-1}}\phi\left[\sqrt{\frac{n}{n-1}}(u-\bar{X})\right],\end{aligned}$$

(where ϕ is the standard normal density) is the UMVU estimator of $p_\xi^{X_1}(u)$.

Suppose now that both ξ and σ are unknown. Then, exactly as in the case $\sigma = 1$, the UMVU estimator of $P[X_1 \le u] = \Phi((u-\xi)/\sigma)$ and of the density $p^{X_1}(u) = (1/\sigma)\phi((u-\xi)/\sigma)$ is given, respectively, by $P[X_1 \le u|\bar{X}, S]$ and the conditional density of X_1 given $\bar{X}$ and S evaluated at u, where $S^2 = \Sigma(X_i - \bar{X})^2$. To replace the conditional distribution with an unconditional one, note that $(X_1 - \bar{X})/S$ is ancillary and therefore, by Basu's theorem, independent of $(\bar{X}, S)$. It follows, as in the earlier case, that

$$P[X_1 \le u|\bar{x}, s] = P\left[\frac{X_1 - \bar{X}}{S} \le \frac{u-\bar{x}}{s}\right] \tag{2.12}$$

and that

$$p^{X_1|\bar{x},s}(u) = \frac{1}{s}f\left(\frac{u-\bar{x}}{s}\right) \tag{2.13}$$

where f is the density of $(X_1 - \bar{X})/S$. A straightforward calculation (Problem 2.10) gives

$$f(z) = \frac{\Gamma\left(\frac{n-1}{2}\right)}{\Gamma\left(\frac{1}{2}\right)\Gamma\left(\frac{n-2}{2}\right)}\sqrt{\frac{n}{n-1}}\left(1 - \frac{nz^2}{n-1}\right)^{(n/2)-2} \quad \text{if } 0 < |z| < \sqrt{\frac{n-1}{n}} \tag{2.14}$$

and zero elsewhere. The estimator (2.13) is obtained by substitution of (2.14), and the estimator (2.12) is obtained by integrating the density f. ‖

We shall next consider two extensions of the normal one-sample model. The first extension is concerned with the two-sample problem, in which there are two independent groups of observations, each with a model of this type, but corresponding to different conditions or representing measurements of two different quantities so that the parameters of the two models are not the same. The second extension deals with the multivariate situation of n p-tuples of observations $(X_{1\nu}, \ldots, X_{p\nu})$, $\nu = 1, \ldots, n$, with $(X_{1\nu}, \ldots, X_{p\nu})$ representing measurements of p different characteristics of the νth subject.

Example 2.3 The normal two-sample problem. Let $X_1, \ldots, X_m$ and $Y_1, \ldots, Y_n$ be independently distributed according to normal distributions $N(\xi, \sigma^2)$ and $N(\eta, \tau^2)$, respectively.

(a) Suppose that ξ, η, σ, τ are completely unknown. Then, the joint density

$$\frac{1}{(\sqrt{2\pi})^{m+n}\sigma^m\tau^n} \exp\left[-\frac{1}{2\sigma^2}\Sigma(x_i-\xi)^2 - \frac{1}{2\tau^2}\Sigma(y_j-\eta)^2\right] \tag{2.15}$$

constitutes an exponential family for which the four statistics

$$\bar{X}, \quad \bar{Y}, \quad S_X^2 = \Sigma(X_i - \bar{X})^2, \quad S_Y^2 = \Sigma(Y_j - \bar{Y})^2$$

are sufficient and complete. The UMVU estimators of ξ and σ^r are therefore $\bar{X}$ and $K_{n-1,r}S_X^r$, as in Example 2.1, and those of η and τ^r are given by the corresponding formulas. In the present model, interest tends to focus on comparing parameters from the two distributions. The UMVU estimator of $\eta - \xi$ is $\bar{Y} - \bar{X}$ and that of τ^r/σ^r is the product of the UMVU estimators of τ^r and $1/\sigma^r$.

(b) Sometimes, it is possible to assume that $\sigma = \tau$. Then $\bar{X}, \bar{Y}$, and $S^2 = \Sigma(X_i - \bar{X})^2 + \Sigma(Y_j - \bar{Y})^2$ are complete sufficient statistics [Problem 1.6.35(a)] and the natural unbiased estimators of $\xi, \eta, \sigma^r, \eta-\xi$, and $(\eta-\xi)/\sigma$ are all UMVU (Problem 2.11).

(c) As a third possibility, suppose that $\eta = \xi$ but that σ and τ are not known to be equal, and that it is desired to estimate the common mean ξ. This might arise, for example, when two independent sets of measurements of the same quantity are available. The statistics $T = (\bar{X}, \bar{Y}, S_X^2, S_Y^2)$ are then minimal sufficient (Problem 1.6.7), but they are no longer complete since $E(\bar{Y} - \bar{X}) = 0$.

If $\sigma^2/\tau^2 = \gamma$ is known, the best unbiased linear combination of $\bar{X}$ and $\bar{Y}$ is

$$\delta_\gamma = \alpha\bar{X} + (1-\alpha)\bar{Y}, \quad \text{where } \alpha = \frac{\tau^2}{n}\bigg/\left(\frac{\sigma^2}{m} + \frac{\tau^2}{n}\right)$$

(Problem 2.12). Since, in this case, $T' = (\Sigma X_i^2 + \gamma\Sigma Y_j^2, \Sigma X_i + \gamma\Sigma Y_j)$ is a complete sufficient statistic (Problem 2.12) and δ_γ is a function of T', δ_γ is UMVU. When σ^2/τ^2 is unknown, a UMVU estimator of ξ does not exist (Problem 2.13), but one can first estimate α, and then estimate ξ by $\xi = \hat{\alpha}\bar{X} + (1-\hat{\alpha})\bar{Y}$. It is easy to see that $\hat{\xi}$ is unbiased provided $\hat{\alpha}$ is a function of only S_X^2 and S_Y^2 (Problem 2.13), for example, if σ^2 and τ^2 in α are replaced by $S_X^2/(m-1)$ and $S_Y^2/(n-1)$. The problem of finding a good estimator of ξ has been considered by various authors, among them Graybill and Deal (1959), Hogg (1960), Seshadri (1963), Zacks (1966), Brown and Cohen (1974), Cohen and Sackrowitz (1974), Rubin and Weisberg (1975), Rao (1980), Berry (1987), Kubokawa (1987), Loh (1991), and George (1991). ‖

It is interesting to note that the nonexistence of a UMVU estimator holds not only for ξ but for any U-estimable function of ξ. This fact, for which no easy proof is available, was established by Unni (1978, 1981) using the results of Kagan and Palamadov (1968).

In cases (a) and (b), the difference $\eta - \xi$ provides one comparison between the distributions of the X's and Y's. An alternative measure of the superiority (if large values of the variables are desirable) of the Y's over the X's is the probability $p = P(X < Y)$. The UMVU estimator of p can be obtained as in Example 2.2 as $P(X_1 < Y_1 | \bar{X}, \bar{Y}, S_X^2, S_Y^2)$ and $P(X_1 < Y_1 | \bar{X}, \bar{Y}, S^2)$ in cases (a) and (b), respectively (Problem 2.14). In case (c), the problem disappears since then $p = 1/2$.

Example 2.4 The multivariate normal one-sample problem. Suppose that $(X_i, Y_i, \ldots)$, $i = 1, \ldots, n$, are observations of p characteristics on a random sample of n subjects from a large population, so that the n p-vectors can be assumed to be iid. We shall consider the case that their common distribution is a p-variate normal distribution (Example 1.4.5) and begin with the case $p = 2$.

The joint probability density of the (X_i, Y_i) is then

$$(2.16) \qquad \left(\frac{1}{2\pi\sigma\tau\sqrt{1-\rho^2}}\right)^n \exp\left\{-\frac{1}{2(1-\rho^2)}\left[\frac{1}{\sigma^2}\Sigma(x_i-\xi)^2 - \frac{2\rho}{\sigma\tau}\Sigma(x_i-\xi)(y_i-\eta) + \frac{1}{\tau^2}\Sigma(y_i-\eta)^2\right]\right\}$$

where $E(X_i) = \xi$, $E(Y_i) = \eta$, $\text{var}(X_i) = \sigma^2$, $\text{var}(Y_i) = \tau^2$, and $\text{cov}(X_i, Y_i) = \rho\sigma\tau$, so that ρ is the correlation coefficient between X_i and Y_i. The bivariate family (2.16) constitutes a five-parameter exponential family of full rank, and the set of sufficient statistics $T = (\bar{X}, \bar{Y}, S_X^2, S_Y^2, S_{XY})$ where

$$(2.17) \qquad S_{XY} = \Sigma(X_i - \bar{X})(Y_i - \bar{Y})$$

is therefore complete. Since the marginal distributions of the X_i and Y_i are $N(\xi, \sigma^2)$ and $N(\eta, \tau^2)$, the UMVU estimators of ξ and σ^2 are $\bar{X}$ and $S_X^2/(n-1)$, and those of η and τ^2 are given by the corresponding formulas. The statistic $S_{XY}/(n-1)$ is an unbiased estimator of $\rho\sigma\tau$ (Problem 2.15) and is therefore the UMVU estimator of $\text{cov}(X_i, Y_i)$.

For the correlation coefficient ρ, the natural estimator is the sample correlation coefficient

$$(2.18) \qquad R = S_{XY}\big/\sqrt{S_X^2 S_Y^2}.$$

However, R is not unbiased, since it can be shown [see, for example, Stuart and Ord (1987, Section 16.32)] that

$$(2.19) \qquad E(R) = \rho\left[1 - \frac{(1-\rho^2)}{2n} + O\left(\frac{1}{n^2}\right)\right].$$

By implementing Method One of Section 2.1, together with some results from the theory of Laplace transforms, Olkin and Pratt (1958) derived a function $G(R)$ of

R which is unbiased and hence UMVU. It is given by

$$G(r) = rF\left(\frac{1}{2}, \frac{1}{2}; \frac{n-1}{2}; 1-r^2\right),$$

where $F(a, b; c; x)$ is the *hypergeometric function*

$$F(a, b; c; x) = \sum_{k=0}^{\infty} \frac{\Gamma(a+k)\Gamma(b+k)\Gamma(c)x^k}{\Gamma(a)\Gamma(b)\Gamma(c+k)k!}$$
$$= \frac{\Gamma(c)}{\Gamma(b)\Gamma(c-b)} \int_0^1 \frac{t^{b-1}(1-t)^{c-b-1}}{(1-tx)^a} dt.$$

Calculation of $G(r)$ is facilitated by using a computer algebra program. Alternatively, by substituting in the above series expansion, one can derive the approximation

$$G(r) = r\left[1 + \frac{1-r^2}{2(n-1)} + O\left(\frac{1}{n^2}\right)\right]$$

which is quite accurate.

These results extend easily to the general multivariate case. Let us change notation and denote by $(X_{1\nu}, \ldots, X_{p\nu})$, $\nu = 1, \ldots, n$, a sample from a nonsingular p-variate normal distribution with means $E(X_{i\nu}) = \xi_i$ and covariances $\text{cov}(X_{i\nu}, X_{j\nu}) = \sigma_{ij}$. Then, the density of the X's is

$$\text{(2.20)} \qquad \frac{|\Theta|^{n/2}}{(2\pi)^{pn/2}} \exp\left(-\frac{1}{2}\Sigma\Sigma\theta_{jk}S'_{jk}\right)$$

where

$$\text{(2.21)} \qquad S'_{jk} = \sum_{\nu=1}^{n}(X_{j\nu} - \xi_j)(X_{k\nu} - \xi_k)$$

and where $\Theta = (\theta_{jk})$ is the inverse of the covariance matrix (σ_{jk}). This is a full-rank exponential family, for which the $p + \binom{p}{2} = \frac{1}{2}p(p+1)$ statistics $X_{i\cdot} = \Sigma X_{i\nu}/n$ $(i = 1, \ldots, p)$ and $S_{jk} = \Sigma(X_{j\nu} - X_{j\cdot})(X_{k\nu} - X_{k\cdot})$ are complete.

Since the marginal distributions of the $X_{j\nu}$ and the pair $(X_{j\nu}, X_{k\nu})$ are univariate and bivariate normal, respectively, it follows from Example 2.1 and the earlier part of the present example, that $X_{i\cdot}$ is UMVU for ξ_i and $S_{jk}/(n-1)$ for σ_{jk}. Also, the UMVU estimators of the correlation coefficients $\rho_{jk} = \sigma_{jk}/\sqrt{\sigma_{jj}\sigma_{kk}}$ are just those obtained from the bivariate distribution of the $(X_{j\nu}, X_{k\nu})$. The UMVU estimator of the square of the multiple correlation coefficient of one of the p coordinates with the other $p-1$ was obtained by Olkin and Pratt (1958). The problem of estimating a multivariate normal probability density has been treated by Ghurye and Olkin (1969); see also Gatsonis 1984. ||

Results quite analogous to those found in Examples 2.1–2.3 obtain when the normal density (2.3) is replaced by the exponential density

$$\text{(2.22)} \qquad \frac{1}{b^n} \exp\left[-\frac{1}{b}\Sigma(x_i - a)\right], \quad x_i > a.$$

Despite its name, this two-parameter family does not constitute an exponential family since its support changes with a. However, for fixed a, it constitutes a one-parameter exponential family with parameter $1/b$.

Example 2.5 The exponential one-sample problem. Suppose, first, that b is known. Then, $X_{(1)}$ is sufficient for a and complete (Example 1.6.24). The distribution of $n[X_{(1)} - a]/b$ is the standard exponential distribution $E(0, 1)$ and the UMVU estimator of a is $X_{(1)} - (b/n)$ (Problem 2.17). On the other hand, when a is known, the distribution (2.22) constitutes a one-parameter exponential family with complete sufficient statistic $\Sigma(X_i - a)$. Since $2\Sigma(X_i - a)/b$ is distributed as χ^2_{2n}, it is seen that $\Sigma(X_i - a)/n$ is the UMVU estimator for b (Problem 2.17).

When both parameters are unknown, $X_{(1)}$ and $\Sigma[X_i - X_{(1)}]$ are jointly sufficient and complete (Example 1.6.24). Since they are independently distributed, $n[X_{(1)} - a]/b$ as $E(0, 1)$ and $2\Sigma[X_i - X_{(1)}]/b$ as $\chi^2_{2(n-1)}$ (Problem 1.6.18), it follows that (Problem 2.18)

$$\text{(2.23)} \qquad \frac{1}{n-1}\Sigma[X_i - X_{(1)}] \quad \text{and} \quad X_{(1)} - \frac{1}{n(n-1)}\Sigma[X_i - X_{(1)}]$$

are UMVU for b and a, respectively.

It is also easy to obtain the UMVU estimators of a/b and of the critical value u for which $P(X_1 \le u)$ has a given value p. If, instead, u is given, the UMVU estimator of $P(X_1 \le u)$ can be found in analogy with the normal case (Problems 2.19 and 2.20). Finally, the two-sample problems corresponding to Example 2.3(a) and (b) can be handled very similarly to the normal case (Problems 2.21-2.23). ‖

An important aspect of estimation theory is the comparison of different estimators. As competitors of UMVU estimators, we shall now consider the maximum likelihood estimator (ML estimator, see Section 6.2). This comparison is of interest both because of the widespread use of the ML estimator and because of its asymptotic optimality (which will be discussed in Chapter 6). If a distribution is specified by a parameter θ (which need not be real-valued), the ML estimator of θ is that value $\hat{\theta}$ of θ which maximizes the probability or probability density. The ML estimator of $g(\theta)$ is defined to be $g(\hat{\theta})$.

Example 2.6 Comparing UMVU and ML estimators. Let $X_1, \ldots, X_n$ be iid according to the normal distribution $N(\xi, \sigma^2)$. Then, the joint density of the X's is given by (2.3) and it is easily seen that the ML estimators of ξ and σ^2 are (Problem 2.26)

$$\text{(2.24)} \qquad \hat{\xi} = \bar{X} \quad \text{and} \quad \hat{\sigma}^2 = \frac{1}{n}\sum(X_i - \bar{X})^2.$$

Within the framework of this example, one can illustrate the different possible relationships between UMVU and ML estimators.

(a) When the estimand $g(\xi, \sigma)$ is ξ, then $\bar{X}$ is both the ML estimator and the UMVU estimator, so in this case, the two estimators coincide.

(b) Let σ be known, say $\sigma = 1$, and let $g(\xi, \sigma)$ be the probability $p = \Phi(u - \xi)$ considered in Example 2.2 (see also Example 3.1.13). The UMVU estimator is $\Phi[\sqrt{n/(n-1)}(u - \bar{X})]$, whereas the ML estimator is $\Phi(u - \bar{X})$. Since the

ML estimator is biased (by completeness, there can be only one unbiased function of $\bar{X}$), the comparison should be based on the mean squared error (rather than the variance)

$$R_\delta(\xi, \sigma) = E[\delta - g(\xi, \sigma)]^2 \tag{2.25}$$

as risk. Such a comparison was carried out by Zacks and Even (1966), who found that neither estimator is uniformly better than the other. For $n = 4$, for example, the UMVU estimator is better when $|u - \xi| > 1.3$ or, equivalently, when $p < .1$ or $p > .9$, whereas for the remaining values the ML estimator has smaller mean squared error.

This example raises the question whether there are situations in which the ML estimator is either uniformly better or worse than its UMVU competitor. The following two simple examples illustrate these possibilities.

(c) If ξ and σ^2 are both unknown, the UMVU estimator and the ML estimator of σ^2 are, respectively, $S^2/(n-1)$ and S^2/n, where $S^2 = \sum(X_i - \bar{X})^2$. Consider the general class of estimators cS^2. An easy calculation (Problem 2.28) shows that

$$E(cS^2 - \sigma^2) = \sigma^4\left[(n^2 - 1)c^2 - 2(n-1)c + 1\right]. \tag{2.26}$$

For any given c, this risk function is proportional to σ^4. The risk functions corresponding to different values of c, therefore, do not intersect, but one lies entirely above the other. The right side of (2.26) is minimized by $c = 1/(n+1)$. Since the values $c = 1/(n-1)$ and $c = 1/n$, corresponding to the UMVU and ML estimator, respectively, lie on the same side of $1/(n+1)$ with $1/n$ being closer and the risk function is quadratic, it follows that the ML estimator has uniformly smaller risk than the UMVU estimator, but that the ML estimator, in turn, is dominated by $S^2/(n+1)$. (For further discussion of this problem, see Section 3.3.)

(d) Suppose that σ^2 is known and let the estimand be ξ^2. Then, the ML estimator is $\bar{X}^2$ and the UMVU estimator is $\bar{X}^2 - \sigma^2/n$ (Problem 2.1). That the risk of the ML estimator is uniformly larger follows from the following lemma.

‖

Lemma 2.7 *Let the risk be expected squared error. If δ is an unbiased estimator of $g(\theta)$ and if $\delta^* = \delta + b$, where the bias b is independent of θ, then δ^* has uniformly larger risk than δ, in fact,*

$$R_{\delta^*}(\theta) = R_\delta(\theta) + b^2.$$

For small sample sizes, both the UMVU and ML estimators can be unsatisfactory. One unpleasant possible feature of UMVU estimators is illustrated by the estimation of ξ^2 in the normal case [Problem 2.5; Example 2.6(d)]. The UMVU estimator is $\bar{X}^2 - \sigma^2/n$ when σ is known, and $\bar{X}^2 - S^2/n(n-1)$ when it is unknown. In either case, the estimator can take on negative values although the estimand is known to be non-negative. Except when $\xi = 0$ or n is small, the probability of such values is not large, but when they do occur, they cause some embarrassment. The difficulty can be avoided, and at the same time the risk of the estimator improved, by replacing the estimator by zero whenever it is negative. This idea is developed further in Sections 4.7 and 5.6. It is also the case that most of these problems disappear in large samples, as we will see in Chapter 6.

The examples of this section are fairly typical and suggest that the difference between the two estimators tends to be small. For samples from the exponential families, which constitute the main area of application of UMVU estimation, it has, in fact, been shown under suitable regularity assumptions that the UMVU and ML estimators are asymptotically equivalent as the sample size tends to infinity, so that the UMVU estimator shares the asymptotic optimality of the ML estimator. (For an exact statement and counterexamples, see Portnoy 1977b.)

3 Discrete Distributions

The distributions considered in the preceding section were all continuous. We shall now treat the corresponding problems for some of the basic discrete distributions.

Example 3.1 Binomial UMVU estimators. In the simplest instance of a one-sample problem with qualitative rather than quantitative "measurements," the observations are dichotomous; cure or no cure, satisfactory or defective, yes or no. The two outcomes will be referred to generically as success or failure.

The results of n independent such observations with common success probability p are conveniently represented by random variables X_i which are 1 or 0 as the ith case or "trial" is a success or failure. Then, $P(X_i = 1) = p$, and the joint distribution of the X's is given by

$$(3.1) \qquad P(X_1 = x_1, \ldots, X_n = x_n) = p^{\Sigma x_i} q^{n - \Sigma x_i} \quad (q = 1 - p).$$

This is a one-parameter exponential family, and $T = \Sigma X_i$ —the total number of successes—is a complete sufficient statistic. Since $E(X_i) = E(\bar{X}) = p$ and $\bar{X} = T/n$, it follows that T/n is the UMVU estimator of p. Similarly, $\Sigma(X_i - \bar{X})^2/(n-1) = T(n-T)/n(n-1)$ is the UMVU estimator of $\text{var}(X_i) = pq$ (Problem 3.1; see also Example 1.13).

The distribution of T is the binomial distribution $b(p, n)$, and it was pointed out in Example 1.2 that $1/p$ is not U-estimable on the basis of T, and hence not in the present situation. In fact, it follows from Equation (1.2) that a function $g(p)$ can be U-estimable only if it is a polynomial of degree $\leq n$.

To see that every such polynomial is actually U-estimable, it is enough to show that p^m is U-estimable for every $m \leq n$. This can be established, and the UMVU estimator determined, by Method 1 of Section 1 (Problem 3.2). An alternative approach utilizes Method 2. The quantity p^m is the probability

$$p^m = P(X_1 = \cdots = X_m = 1)$$

and its UMVU estimator is therefore given by

$$\delta(t) = P[X_1 = \cdots = X_m = 1 | T = t].$$

This probability is 0 if $t < m$. For $t \geq m$, $\delta(t)$ is the probability of obtaining m successes in the first m trials and $t - m$ successes in the remaining $n - m$ trials, divided by $P(T = t)$, and hence it is

$$p^m \binom{n-m}{t-m} p^{t-m} q^{n-1} \Big/ \binom{n}{t} p^t q^{n-1},$$

or

$$\delta(T) = \frac{T(T-1)\cdots(T-m+1)}{n(n-1)\cdots(n-m+1)}. \tag{3.2}$$

Since this expression is zero when $T = 0, \ldots, m-1$, it is seen that $\delta(T)$, given by (3.2) for all $T = 0, 1, \ldots, n$, is the UMVU estimator of p^m. This proves that $g(p)$ is U-estimable on the basis of n binomial trials if and only if it is a polynomial of degree $\leq n$. ‖

Consider now the estimation of $1/p$, for which no unbiased estimator exists. This problem arises, for example, when estimating the size of certain animal populations. Suppose that a lake contains an unknown number N of some species of fish. A random sample of size k is caught, tagged, and released again. Somewhat later, a random sample of size n is obtained and the number X of tagged fish in the sample is noted. (This is the *capture-recapture* method. See, for example, George and Robert, 1992.) If, for the sake of simplicity, we assume that each caught fish is immediately returned to the lake (or alternatively that N is very large compared to n), the n fish in this sample constitute n binomial trials with probability $p = k/N$ of success (i.e., obtaining a tagged fish). The population size N is therefore equal to k/p. We shall now discuss a sampling scheme under which $1/p$, and hence k/p, is U-estimable.

Example 3.2 Inverse binomial sampling. Reliable estimation of $1/p$ is clearly difficult when p is close to zero, where a small change of p will cause a large change in $1/p$. To obtain control of $1/p$ for all p, it would therefore seem necessary to take more observations the smaller p is. A sampling scheme achieving this is inverse sampling, which continues until a specified number of successes, say m, have been obtained. Let $Y + m$ denote the required number of trials. Then, Y has the *negative binomial* distribution given by (Problem 1.5.12)

$$P(Y = y) = \binom{m+y-1}{m-1} p^m (1-p)^y, \quad y = 0, 1, \ldots, \tag{3.3}$$

with

$$E(Y) = m(1-p)/p; \quad \text{var}(Y) = m(1-p)/p^2. \tag{3.4}$$

It is seen from (3.4) that

$$\delta(Y) = (Y + m)/m,$$

the reciprocal of the proportion of successes, is an unbiased estimator of $1/p$.

The full data in the present situation are not Y but also include the positions in which the m successes occur. However, Y is a sufficient statistic (Problem 3.6), and it is complete since (3.3) is an exponential family. As a function of Y, $\delta(Y)$ is thus the unique unbiased estimator of $1/p$; based on the full data, it is UMVU.

It is interesting to note that $1/(1-p)$ is not U-estimable with the present sampling scheme, for suppose $\delta(Y)$ is an unbiased estimator so that

$$p^m \sum_{y=0}^{\infty} \delta(y) \binom{m+y-1}{m-1} (1-p)^y = 1/(1-p) \quad \text{for all } 0 < p < 1.$$

The left side is a power series which converges for all $0 < p < 1$, and hence converges and is continuous for all $|p| < 1$. As $p \to 1$, the left side therefore tends to $\delta(0)$ while the right side tends to infinity. Thus, the assumed δ does not exist. (For the estimation of p^r, see Problem 3.4.) ‖

The situations described in Examples 3.1 and 3.2 are special cases of *sequential binomial sampling* in which the number of trials is allowed to depend on the observations. The outcome of such sampling can be represented as a random walk in the plane. The walk starts at (0, 0) and moves a unit to the right or up as the first trial is a success or failure. From the resulting point (1, 0) or (0, 1), it again moves a unit to the right or up, and continues in this way until the sampling plan tells it to stop. A stopping rule is thus defined by a set B of points, a *boundary*, at which sampling stops. We require B to satisfy

$$\sum_{(x,y)\in B} P(x, y) = 1 \tag{3.5}$$

since otherwise there is positive probability that sampling will go on indefinitely. A stopping rule that satisfies (3.5) is called *closed.*

Any particular sample path ending in (x, y) has probability $p^x q^y$, and the probability of a path ending in any particular point (x, y) is therefore

$$P(x, y) = N(x, y) p^x q^y, \tag{3.6}$$

where $N(x, y)$ denotes the number of paths along which the random walk can reach the point (x, y). As illustrations, consider the plans of Examples 3.1 and 3.2.

(a) In Example 3.1, B is the set of points (x, y) satisfying $x+y = n, x = 0, \ldots, n$, and for any $(x, y) \in B$, we have $N(x, y) = \binom{n}{x}$.

(b) In Example 3.2, B is the set of points (x, y) with $x = m; y = 0, 1, \ldots,$ and for any such point

$$N(x, y) = \binom{m+y-1}{y}.$$

The observations in sequential binomial sampling are represented by the sample path, and it follows from (3.6) and the factorization criterion that the coordinates (X, Y) of the stopping point in which the path terminates constitute a sufficient statistic. This can also be seen from the definition of sufficiency, since the conditional probability of any given sample path given that it ends in (x, y) is

$$\frac{p^x q^x}{N(x, y) p^x q^y} = \frac{1}{N(x, y)},$$

which is independent of p.

Example 3.3 Sequential estimation of binomial p. For any closed sequential binomial sampling scheme, an unbiased estimator of p depending only on the sufficient statistic (X, Y) can be found in the following way. A simple unbiased estimator is $\delta = 1$ if the first trial is a success and $\delta = 0$ otherwise. Application of the Rao-Blackwell theorem then leads to

$$\delta'(X, Y) = E[\delta|(X, Y)] = P[1^{\text{st}} \text{ trial} = \text{success}|(X, Y)]$$

as an unbiased estimator depending only on (X, Y). If the point $(1, 0)$ is a stopping point, then $\delta' = \delta$ and nothing is gained. In all other cases, δ' will have a smaller variance than δ. An easy calculation [Problem 3.8(a)] shows that

$$\delta'(x, y) = N'(x, y)/N(x, y) \tag{3.7}$$

where $N'(x, y)$ is the number of paths possible under the sampling schemes which pass through $(1, 0)$ and terminate in (x, y). ||

More generally, if (a, b) is any *accessible point*, that is, if it is possible under the given sampling plan to reach (a, b), the quantity $p^a q^b$ is U-estimable, and an unbiased estimator depending only on (X, Y) is given by (3.7), where $N'(x, y)$ now stands for the number of paths passing through (a, b) and terminating in (x, y) [Problem 3.8(b)].

The estimator (3.7) will be UMVU for any sampling plan for which the sufficient statistic (X, Y) is complete. To describe conditions under which this is the case, let us call an accessible point that is not in B a *continuation point*. A sampling plan is called *simple* if the set of continuation points C_t on each line segment $x + y = t$ is an interval or the empty set. A plan is called *finite* if the number of accessible points is finite.

Example 3.4 Two sampling plans.

(a) Let a, b, and m be three positive integers with $a < b < m$. Continue observation until either a successes or failures have been obtained. If this does not happen during the first m trials, continue until either b successes or failures have been obtained. This sampling plan is simple and finite.

(b) Continue until *both* at least a successes and a failures have been obtained. This plan is neither simple nor finite, but it is closed (Problem 3.10). ||

Theorem 3.5 *A necessary and sufficient condition for a finite sampling plan to be complete is that it is simple.*

We shall here only prove sufficiency. [For a proof of necessity, see Girschick, Mosteller, and Savage 1946.] If the restriction to finite plans is dropped, simplicity is no longer sufficient (Problem 3.9). Another necessary condition in that case is stated in Problem 3.13. This condition, together with simplicity, is also sufficient. (For a proof, see Lehmann and Stein 1950.)

For the following proof it may be helpful to consider a diagram of plan (a) of Example 3.4.

Proof. Proof of sufficiency. Suppose there exists a nonzero function $\delta(X, Y)$ whose expectation is zero for all p $(0 < p < 1)$. Let t_0 be the smallest value of t for which there exists a boundary point (x_0, y_0) on $x + y = t_0$ such that $\delta(x_0, y_0) \neq 0$. Since the continuation points on $x + y = t_0$ (if any) form an interval, they all lie on the same side of (x_0, y_0). Suppose, without loss of generality, that (x_0, y_0) lies to the left and above C_{t_0}, and let (x_1, y_1) be that boundary point on $x + y = t_0$ above C_{t_0} and with $\delta(x, y) \neq 0$, which has the smallest x-coordinate. Then, all boundary points with $\delta(x, y) \neq 0$ satisfy $t \geq t_0$ and $x \geq x_1$. It follows that for all $0 < p < 1$

$$E[\delta(X, Y)] = N(x_1, y_1)\delta(x_1, y_1)p^{x_1}q^{t_0-x_1} + p^{x_1+1}R(p) = 0$$

where $R(p)$ is a polynomial in p. Dividing by p^{x_1} and letting $p \to 0$, we see that $\delta(x_1, y_1) = 0$, which is a contradiction. □

Fixed binomial sampling satisfies the conditions of the theorem, but, there (and for inverse binomial sampling), completeness follows already from the fact that it leads to a full-rank exponential family (5.1) with $s = 1$. An example in which this is not the case is curtailed binomial sampling, in which sampling is continued as long as $X < a$, $Y < b$, and $X + Y < n(a, b < n)$ and is stopped as soon as one of the three boundaries is reached (Problem 3.11). Double sampling and curtailed double sampling provide further applications of the theory. (See Girshick, Mosteller, and Savage 1946; see also Kremers 1986.)

The discrete distributions considered so far were all generated by binomial trials. A large class of examples is obtained by considering one-parameter exponential families (5.2) in which $T(x)$ is integer-valued. Without loss of generality, we shall take $T(x)$ to be x and the distribution of X to be given by

$$P(X = x) = e^{\eta x - B(\eta)}a(x). \tag{3.8}$$

Putting $\theta = e^\eta$, we can write (3.8) as

$$P(X = x) = a(x)\theta^x / C(\theta), \quad x = 0, 1, \ldots, \quad \theta > 0. \tag{3.9}$$

For any function $a(x)$ for which $\Sigma a(x)\theta^x < \infty$ for some $\theta > 0$, this is a family of *power series distributions* (Problems 1.5.14–1.5.16). The binomial distribution $b(p, n)$ is obtained from (3.9) by putting $a(x) = \binom{n}{x}$ for $x = 0, 1, \ldots, n$, and $a(x) = 0$ otherwise; $\theta = p/q$ and $C(\theta) = (\theta + 1)^n$. The negative binomial distribution with $a(x) = \binom{m+x-1}{m-1}$, $\theta = q$, and $C(\theta) = (1 - \theta)^{-m}$ is another example. The family (3.9) is clearly complete. If $a(x) > 0$ for all $x = 0, 1, \ldots$, then θ^r is U-estimable for any positive integer r, and its unique unbiased estimator is obtained by solving the equations

$$\sum_{x=0}^{\infty} \delta(x)a(x)\theta^x = \theta^r \cdot C(\theta) \quad \text{for all } \theta \in \Omega.$$

Since $\Sigma a(x)\theta^x = C(\theta)$, comparison of the coefficients of θ^x yields

$$\delta(x) = \begin{cases} 0 & \text{if } x = 0, \ldots, r-1 \\ a(x-r)/a(x) & \text{if } x \geq r. \end{cases} \tag{3.10}$$

Suppose, next, that $X_1, \ldots, X_n$ are iid according to a power series family (3.9). Then, $X_1 + \cdots + X_n$ is sufficient for θ, and its distribution is given by the following lemma.

Lemma 3.6 *The distribution of* $T = X_1 + \cdots + X_n$ *is the power series family*

$$P(T = t) = \frac{A(t, n)\theta^t}{[C(\theta)]^n}, \tag{3.11}$$

where $A(t, n)$ *is the coefficient of* θ^t *in the power series expansion of* $[C(\theta)]^n$.

Proof. By definition,

$$P(T = t) = \theta^t \sum_t \frac{a(x_1) \cdots a(x_n)}{[C(\theta)]^n}$$

where Σ_t indicates that the summation extends over all n-tuples of integers $(x_1, \ldots, x_n)$ with $x_1 + \cdots + x_n = t$. If

$$B(t, n) = \sum_t a(x_1) \cdots a(x_n), \tag{3.12}$$

the distribution of T is given by (3.11) with $B(t, n)$ in place of $A(t, n)$. On the other hand,

$$[C(\theta)]^n = \left[\sum_{x=0}^{\infty} a(x)\theta^x\right]^n,$$

and for any $t = 0, 1, \ldots$, the coefficient of θ^t in the expansion of the right side as a power series in θ is just $B(t, n)$. Thus, $B(t, n) = A(t, n)$, and this completes the proof. □

It follows from the lemma that T is complete and from (3.10) that the UMVU estimator of θ^r on the basis of a sample of n is

$$\delta(t) = \begin{cases} 0 & \text{if } t = 0, \ldots, r-1 \\ \dfrac{A(t-r, n)}{A(t, n)} & \text{if } t \geq r. \end{cases} \tag{3.13}$$

Consider, next, the problem of estimating the probability distribution of X from a sample $X_1, \ldots, X_n$. The estimand can be written as

$$g(\theta) = P_\theta(X_1 = x)$$

and the UMVU estimator is therefore given by

$$\begin{aligned} \delta(t) &= P[X_1 = x | X_1 + \cdots + X_n = t] \\ &= \frac{P(X_1 = x)P(X_2 + \cdots + X_n = t - x)}{P(T = t)}. \end{aligned}$$

In the present case, this reduces to

$$\delta(t) = \frac{a(x)A(t-x, n-1)}{A(t, n)}, \quad n > 1, \quad 0 \leq x \leq t. \tag{3.14}$$

Example 3.7 Poisson UMVU estimation. The Poisson distribution, shown in Table 1.5.1, arises as a limiting case of the binomial distribution for large n and small p, and more generally as the number of events occurring in a fixed time period when the events are generated by a Poisson process. The distribution $P(\theta)$ of a Poisson variable with expectation θ is given by (3.9) with

$$a(x) = \frac{1}{x!}, \quad C(\theta) = e^{\theta}. \tag{3.15}$$

Thus, $[C(\theta)]^n = e^{n\theta}$ and

$$A(t, n) = \frac{n^t}{t!}. \tag{3.16}$$

The UMVU estimator of θ^r is therefore, by (3.13), equal to

$$\delta(t) = \frac{t(t-1)\cdots(t-r+1)}{n^r} \tag{3.17}$$

for all $t \geq r$. Since the right side is zero for $t = 0, \ldots, r-1$, formula (3.17) holds for all r.

The UMVU estimator of $P_\theta(X = x)$ is given by (3.14), which, by (3.16), becomes

$$\delta(t) = \binom{t}{x}\left(\frac{1}{n}\right)^x \left(\frac{n-1}{n}\right)^{t-x}, \quad x = 0, 1, \ldots, t.$$

For varying x, this is the binomial distribution $b(1/n, t)$.

In some situations, Poisson variables are observed only when they are positive. For example, suppose that we have a sample from a truncated Poisson distribution (truncated on the left at 0) with probability function

$$P(X = x) = \frac{1}{e^{\theta} - 1}\frac{\theta^x}{x!}, \quad x = 1, 2, \ldots. \tag{3.18}$$

This is a power series distribution with

$$a(x) = \frac{1}{x!} \quad \text{if } x \geq 1, \quad a(0) = 0,$$

and

$$C(\theta) = e^{\theta} - 1.$$

For any values of t and n, the UMVU estimator $\delta(t)$ of θ, for example, can now be obtained from (3.13). (See Problems 3.18–3.22; for further discussion, see Tate and Goen 1958.) ||

We next consider some multiparameter situations.

Example 3.8 Multinomial UMVU estimation. Let $(X_0, X_1, \ldots, X_n)$ have the multinomial distribution (5.4). As was seen in Example 1.5.3, this is an s-parameter exponential family, with $(X_1, \ldots, X_s)$ or $(X_0, X_1, \ldots, X_s)$ constituting a complete sufficient statistic. [Recall that $X_0 = n - (X_1 + \cdots + X_s)$.] Since $E(X_i) = np_i$, it follows that X_i/n is the UMVU estimator of p_i. To obtain the UMVU estimator of $p_i p_j$, note that one unbiased estimator is $\delta = 1$ if the first trial results in outcome i and the second trial in outcome j, and $\delta = 0$ otherwise. The UMVU estimator of $p_i p_j$ is therefore

$$E(\delta | X_0, \ldots, X_s) = \frac{(n-2)! X_i X_j}{X_0! \cdots X_s!} \Big/ \frac{n!}{X_0! \cdots X_s!} = \frac{X_i X_j}{n(n-1)}. \quad ||$$

Table 3.1. $I \times J$ *Contingency Table*

	$B_1 \cdots B_J$	Total
A_1	$n_{11} \cdots n_{1J}$	n_{1+}
$\vdots$		$\vdots$
A_I	$n_{I1} \cdots n_{IJ}$	n_{I+}
Total	$n_{+1} \cdots n_{+J}$	n

In the application of multinomial models, the probabilities $p_0, \ldots, p_s$ are frequently subject to additional restrictions, so that the number of independent parameters is less than s. In general, such a restricted family will not constitute a full-rank exponential family, but may be a curved exponential family. There are, however, important exceptions. Simple examples are provided by certain contingency tables.

Example 3.9 Two-way contingency tables. A number n of subjects is drawn at random from a population sufficiently large that the drawings can be considered to be independent. Each subject is classified according to two characteristics: A, with possible outcomes $A_1, \ldots, A_I$, and B, with possible outcomes $B_1, \ldots, B_J$. [For example, students might be classified as being male or female ($I = 2$) and according to their average performance (A, B, C, D, or F; $J = 5$).] The probability that a subject has properties (A_i, B_j) will be denoted by p_{ij} and the number of such subjects in the sample by n_{ij}. The joint distribution of the IJ variables n_{ij} is an unrestricted multinomial distribution with $s = IJ - 1$, and the results of the sample can be represented in an $I \times J$ table, such as Table 3.1. From Example 3.8, it follows that the UMVU estimator of p_{ij} is n_{ij}/n.

A special case of Table 3.1 arises when A and B are independent, that is, when $p_{ij} = p_{i+}p_{+j}$ where $p_{i+} = p_{i1} + \cdots + p_{iJ}$ and $p_{+j} = p_{1j} + \cdots + p_{Ij}$. The joint probability of the IJ cell counts then reduces to

$$\frac{n!}{\prod_{i,j} n_{ij}!} \prod_i p_{i+}^{n_{i+}} \prod_j p_{+j}^{n_{+j}}.$$

This is an $(I + J - 2)$-parameter exponential family with the complete sufficient statistics (n_{i+}, n_{+j}), $i = 1, \ldots, I$, $j = 1, \ldots, J$, or, equivalently, $i = 1, \ldots, I - 1$, $j = 1, \ldots, J - 1$. In fact, $(n_{1+}, \ldots, n_{I+})$ and $(n_{+1}, \ldots, n_{+J})$ are independent, with multinomial distributions $M(p_{1+}, \ldots, p_{I+}; n)$ and $M(p_{+1}, \ldots, p_{+J}; n)$, respectively (Problem 3.27), and the UMVU estimators of p_{i+}, p_{+j} and $p_{ij} = p_{i+}p_{+j}$ are, therefore, n_{i+}/n, n_{+j}/n and $n_{i+}n_{+j}/n^2$, respectively. ||

When studying the relationship between two characteristics A and B, one may find A and B to be dependent although no mechanism appears to exist through which either factor could influence the other. An explanation is sometimes found in the dependence of both factors on a common third factor, C, a phenomenon known as *spurious correlation*. The following example describes a model for this situation.

Example 3.10 Conditional independence in a three-way table. In the situation of Example 3.9, suppose that each subject is also classified according to a third factor C as $C_1, \ldots,$ or C_K. [The third factor for the students of Example 3.9 might be their major (History, Physics, etc.).] Consider this situation under the assumption that conditionally given C_k $(k = 1, \ldots, K)$, the characteristics A and B are independent, so that

$$p_{ijk} = p_{++k}\, p_{i+|k}\, p_{+j|k} \tag{3.19}$$

where $p_{i+|k}$, $p_{+j|k}$, and $p_{ij|k}$ denote the probability of the subject having properties A_i, B_j, or (A_i, B_j), respectively, given that it has property C_k.

After some simplification, the joint probability of the IJK cell counts n_{ijk} is seen to be proportional to (Problem 3.28)

$$\prod_{i,j,k}(p_{++k}\,p_{i+|k}\,p_{+j|k})^{n_{ijk}} = \prod_k\left[p_{++k}^{n_{++k}}\prod_i p_{i+|k}^{n_{i+k}}\prod_j p_{+j|k}^{n_{+jk}}\right]. \tag{3.20}$$

This is an exponential family of dimension

$$(K-1) + K(I+J-2) = K(I+J-1) - 1$$

with complete sufficient statistics $T = \{(n_{++k}, n_{i+k}, n_{+jk}),\ i = 1, \ldots, I,\ j = 1, \ldots, J,\ k = 1, \ldots, K\}$. Since the expectation of any cell count is n times the probability of that cell, the UMVU estimators of p_{++k}, p_{i+k}, and p_{+jk} are n_{++k}/n, n_{i+k}/n, and n_{+jk}/n, respectively. ||

Consider, now, the estimation of the probability p_{ijk}. The unbiased estimator $\delta_0 = n_{ijk}/n$, which is UMVU in the unrestricted model, is not a function of T and hence no longer UMVU. The relationship (3.19) suggests the estimator $\delta_1 = (n_{++k}/n)\cdot(n_{i+k}/n_{++k})\cdot(n_{+jk}/n_{++k})$, which is a function of T. It is easy to see (Problem 3.30) that δ_1 is unbiased and hence is UMVU. (For additional results concerning the estimation of the parameters of this model, see Cohen 1981, or Davis 1989.)

To conclude this section, an example is provided in which the UMVU estimator fails completely.

Example 3.11 Misbehaved UMVU estimator. Let X have the Poisson distribution $P(\theta)$ and let $g(\theta) = e^{-a\theta}$, where a is a known constant. The condition of unbiasedness of an estimator δ leads to

$$\sum \frac{\delta(x)\theta^x}{x!} = e^{(1-a)\theta} = \sum \frac{(1-a)^x\theta^x}{x!}$$

and hence to

$$\delta(X) = (1-a)^X. \tag{3.21}$$

Suppose $a = 3$. Then, $g(\theta) = e^{-3\theta}$, and one would expect an estimator which decreases from 1 to 0 as X goes from 0 to infinity. The ML estimator e^{-3X} meets this expectation. On the other hand, the unique unbiased estimator $\delta(x) = (-2)^x$ oscillates wildly between positive and negative values and appears to bear no relation to the problem at hand. (A possible explanation for this erratic behavior is suggested in Lehmann (1983).) It is interesting to see that the difficulty disappears if the sample size is increased. If $X_1, \cdots, X_n$ are iid according to $P(\theta)$, then $T = \sum X_i$ is a sufficient statistic and has the Poisson $P(n\theta)$ distribution. The condition of unbiasedness now becomes

$$\sum \frac{\delta(t)(n\theta)^t}{t!} = e^{(n-a)\theta} = \sum \frac{(n-a)^t\theta^t}{t!}$$

and the UMVU estimator is

$$\delta(T) = \left(1 - \frac{a}{n}\right)^T. \tag{3.22}$$

This is quite reasonable as soon as $n > a$. ‖

4 Nonparametric Families

Section 2.2 was concerned with continuous parametric families of distributions such as the normal, uniform, or exponential distributions, and Section 2.3 with discrete parametric families such as the binomial and Poisson distributions. We now turn to nonparametric families in which no specific form is assumed for the distribution.

We begin with the one-sample problem in which $X_1, \ldots, X_n$ are iid with distribution $F \in \mathcal{F}$. About the family $\mathcal{F}$, we shall make only rather general assumptions, for example, that it is the family of distributions F which have a density, or are continuous, or have first moments, and so on. The estimand $g(F)$ might, for example, be $E(X_i) = \int x\, dF(x)$, or var X_i, or $P(X_i \leq a) = F(a)$.

It was seen in Problem 1.6.33 that for the family $\mathcal{F}_0$ of all probability densities, the order statistics $X_{(1)} < \cdots < X_{(n)}$ constitute a complete sufficient statistic, and the hint given there shows that this result remains valid if $\mathcal{F}_0$ is further restricted by requiring the existence of some moments.[2] (For an alternative proofs, see TSH2, Section 4.3. Also, Bell, Blackwell, and Breiman (1960) show the result is valid for the family of all continuous distributions.)

An estimator $\delta(X_1, \ldots, X_n)$ is a function of the order statistics if and only if it is symmetric in its n arguments. For families $\mathcal{F}$ for which the order statistics are complete, there can therefore exist at most one symmetric unbiased estimator of any estimand, and this is UMVU. Thus, to find the UMVU estimator of any U-estimable $g(F)$, it suffices to find a symmetric unbiased estimator.

Example 4.1 Estimating the distribution function. Let $g(F) = P(X \leq a) = F(a)$, a known. The natural estimator is the number of X's which are $\leq a$, divided by N. The number of such X's is the outcome of n binomial trials with success probability $F(a)$, so that this estimator is unbiased for $F(a)$. Since it is also symmetric, it is the UMVU estimator. This can be paraphrased by saying that the empirical cumulative distribution function is the UMVU estimator of the unknown true cumulative distribution function.

Note. In the normal case of Section 2.2, it was possible to find unbiased estimators not only of $P(X \leq u)$ but also of the probability density $p_X(u)$ of X. No unbiased

[2] The corresponding problem in which the values of some moments (or expectations of other functions) are given is treated by Hoeffding (1977) and N. Fisher (1982).

estimator of the density exists for the family $\mathcal{F}_0$. For proofs, see Rosenblatt 1956, and Bickel and Lehmann 1969, and for further discussion of the problem of estimating a nonparametric density see Rosenblatt 1971, the books by Devroye and Gyoerfi (1985), Silverman (1986), or Wand and Jones (1995), and the review article by Izenman (1991). Nonparametric density estimation is an example of what Liu and Brown (1993) call *singular problems*, which pose problems for unbiased estimation. See Note 8.3. ||

Example 4.2 Nonparametric UMVU estimation of a mean. Let us now further restrict $\mathcal{F}_0$, the class of all distributions F having a density, by adding the condition $E|X| < \infty$, and let $g(F) = \int xf(x)\,dx$. Since $\bar{X}$ is symmetric and unbiased for $g(F)$, $\bar{X}$ is UMVU. An alternative proof of this result is obtained by noting that X_1 is unbiased for $g(F)$. The UMVU estimator is found by conditioning on the order statistics; $E[X_1|X_{(1)}, \ldots, X_{(n)}]$. But, given the order statistics, X_1 assumes each value with probability $1/n$. Hence, the above conditional expectation is equal to $(1/n)\Sigma X_{(i)} = \bar{X}$.

In Section 2.2, it was shown that $\bar{X}$ is UMVU for estimating $E(X_i) = \xi$ in the family of normal distributions $N(\xi, \sigma^2)$; now it is seen to be UMVU in the family of all distributions that have a probability density and finite expectation. Which of these results is stronger? The uniformity makes the nonparametric result appear much stronger. This is counteracted, however, by the fact that the condition of unbiasedness is much more restrictive in that case. Thus, the number of competitors which the UMVU estimator "beats" for such a wide class of distributions is quite small (see Problem 4.1). It is interesting in this connection to note that, for a family intermediate between the two considered here, the family of all symmetric distributions having a probability density, $\bar{X}$ is *not* UMVU (Problem 4.4; see also Bickel and Lehmann 1975-1979). ||

Example 4.3 Nonparametric UMVU estimation of a variance. Let $g(F) =$ var X. Then $[\Sigma(X_i - \bar{X})^2]/(n-1)$ is symmetric and unbiased, and hence is UMVU. ||

Example 4.4 Nonparametric UMVU estimation of a second moment. Let $g(F) = \xi^2$, where $\xi = EX$. Now, $\sigma^2 = E(X^2) - \xi^2$ and a symmetric unbiased estimator of $E(X^2)$ is $\Sigma X_i^2/n$. Hence, the UMVU estimator of ξ^2 is $\Sigma X_i^2/n - \Sigma(X_i - \bar{X})^2/(n-1)$.

An alternative derivation of this result is obtained by noting that X_1X_2 is unbiased for ξ^2. The UMVU estimator of ξ^2 can thus be found by conditioning: $E[X_1, X_2|X_{(1)}, \ldots, X_{(n)}]$. But, given the order statistics, the pair $\{X_1, X_2\}$ assumes the value of each pair $\{X_{(i)}, X_{(j)}\}$, $i \neq j$, with probability $1/n(n-1)$. Hence, the above conditional expected value is

$$\frac{1}{n(n-1)} \sum_{i \neq j} X_i X_j,$$

which is equivalent to the earlier result. ||

Consider, now, quite generally a function $g(F)$ which is U-estimable in $\mathcal{F}_0$. Then, there exists an integer $m \leq n$ and a function $\delta(X_1, \ldots, X_m)$, which is unbiased for $g(F)$. We can assume without loss of generality that δ is symmetric in its m arguments; otherwise, it can be symmetrized. Then, the estimator

$$(4.1) \qquad \frac{1}{\binom{n}{m}} \sum_{(i_1,\ldots,i_m)} \delta(X_{i_1}, \ldots, X_{i_m})$$

is UMVU for $g(F)$; here, the sum is over all m-tuples $(i_1, \ldots, i_m)$ from the integers $1, 2, \ldots, n$ with $i_1 < \cdots < i_m$. That this estimator is UMVU follows from the facts that it is symmetric and that each of the $\binom{n}{m}$ summands has expectation $g(F)$.

The class of statistics (4.1) called *U-statistics* was studied by Hoeffding (1948) who, in particular, gave conditions for their asymptotic normality; for further work on U-statistics, see Serfling 1980, Staudte and Sheather 1990, Lee 1990, or Koroljuk and Borovskich 1994.

Two problems suggest themselves:

(a) What kind of functions $g(F)$ have unbiased estimators, that is, are *U-estimable*?

(b) If a functional $g(F)$ has an unbiased estimator, what is the smallest number of observations for which the unbiased estimator exists? We shall call this smallest number the *degree* of $g(F)$.

(For the case that F assigns positive probability only to the two values 0 and 1, these equations are answered in the preceding section.)

Example 4.5 Degree of the variance. Let $g(F)$ be the variance σ^2 of F. Then $g(F)$ has an unbiased estimator in the subset $\mathcal{F}_0'$ of $\mathcal{F}_{0_2}$ with $E_F X^2 < \infty$ and $n = 2$ observations, since $\Sigma(X_i - \bar{X})^2/(n-1) = \frac{1}{2}(X_2 - X_1)^2$ is unbiased for σ^2. Hence, the degree of σ^2 is ≤ 2. Furthermore, since in the normal case with unknown mean there is no unbiased estimator of σ^2 based on only one observation (Problem 2.7), there is no such estimator within the class $\mathcal{F}_0'$. It follows that the degree of σ^2 is 2. ||

We shall now give another proof that the degree of σ^2 in this example is greater than 1 to illustrate a method that is of more general applicability for problems of this type.

Let g be any estimand that is of degree 1 in $\mathcal{F}_0'$. Then, there exists δ such that

$$\int \delta(x) dF(x) = g(F), \quad \text{for all } F \in \mathcal{F}_0'.$$

Fix two arbitrary distributions F_1 and F_2 in $cal F_0'$ with $F_1 \neq F_2$, and let $F = \alpha F_1 + (1-\alpha)F_2, 0 \leq \alpha \leq 1$. Then,

$$(4.2) \quad g[\alpha F_1 + (1-\alpha)F_2] = \alpha \int \delta(x) dF_1(x) + (1-\alpha) \int \delta(x) dF_2(x).$$

Then, $\alpha F_1 + (1-\alpha)F_2$ is also in $cal F_0'$, and as a function of α, the right-hand side is linear in α. Thus, the only g's that can be of degree 1 are those for which the left-hand side is linear in α.

Now, consider

$$g(F) = \alpha_F^2 = E(X^2) - [EX]^2.$$

In this case,

$$(4.3) \quad \sigma^2_{\alpha F_1 + (1-\alpha)F_2} = \alpha E(X_1^2) + (1-\alpha)E(X_2^2) - [\alpha EX_1 + (1-\alpha)EX_2]^2$$

where X_i is distributed according to F_i. The coefficient of α^2 on the right-hand side is seen to be $-[E(X_2) - E(X_1)]^2$. Since this is not zero for all $F_1, F_2 \in \mathcal{F}_0'$, the right-hand side is not linear in α, and it follows that σ^2 is not of degree 1. ||

Generalizing (4.2), we see that if $g(F)$ is of degree m, then

$$(4.4) \quad \begin{cases} g[\alpha F_1 + (1-\alpha)F_2] \\ \quad = \int \cdots \int \delta(x_1, \ldots, x_m) d[\alpha F_1(x_1) + (1-\alpha)F_2(x_1)] \cdots \\ \text{is a polynomial of degree at most } m, \end{cases}$$

which is thus a necessary condition for g to be estimable with m observations. Conditions for (4.4) to be also sufficient are given by Bickel and Lehmann (1969).

Condition (4.4) may also be useful for proving that there exists no value of n for which a functional $g(F)$ has an unbiased estimate.

Example 4.6 Nonexistence of unbiased estimator. Let $g(F) = \sigma$. Then $g[\alpha F_1 + (1-\alpha)F_2]$ is the square root of the right-hand side of (4.3). Since this quadratic in α is not a perfect square for all $F_1, F_2 \in \mathcal{F}_0'$, it follows that its square root is not a polynomial. Hence σ does not have an unbiased estimator for any fixed number n of observations. ||

Let us now turn from the one-sample to the two-sample problem. Let $X_1, \ldots, X_m$ and $Y_1, \ldots, Y_n$ be independently distributed according to distributions F and $G \in \mathcal{F}_0$. Then the order statistics $X_{(1)} < \cdots < X_{(m)}$ and $Y_{(1)} < \cdots < Y_{(n)}$ are sufficient and complete (Problem 4.5). A statistic δ is a function of these order statistics if and only if δ is symmetric in the X_i's and separately symmetric in the Y_j's.

Example 4.7 Two-sample UMVU estimator. Let $h(F, G) = E(Y) - E(X)$. Then $\bar{Y} - \bar{X}$ is unbiased for $h(F, G)$. Since it is a function of the complete sufficient statistic, it is UMVU. ||

The concept of degree runs into difficulty in the present case. Smallest values m_0 and n_0 are sought for which a given functional $h(F, G)$ has an unbiased estimator. One possibility is to find the smallest m for which there exists an n such that $h(F, G)$ has an unbiased estimator, and to let m_0 and n_0 be the smallest values so determined. This procedure is not symmetric in m and n. However, it can be shown that if the reverse procedure is used, the same minimum values are obtained. [See Bickel and Lehmann, (1969)].

As a last illustration, let us consider the bivariate nonparametric problem. Let $(X_1, Y_1), \ldots, (X_n, Y_n)$ be iid according to a distribution $F \in \mathcal{F}$, the family of all bivariate distributions having a probability density. In analogy with the order statistics in the univariate case, the set of pairs

$$T = \{[X_{(1)}, Y_{j_1}], \ldots, [X_{(n)}, Y_{j_n}]\}$$

that is, the n pairs (X_i, Y_i), ordered according to the value of their first coordinate, constitute a sufficient statistic. An equivalent statistic is

$$T' = \{[X_{i_1}, Y_{(1)}], \ldots, [X_{i_n}, Y_{(n)}]\},$$

that is, the set of pairs (X_i, Y_i) ordered according to the value of the second coordinate. Here, as elsewhere, the only aspect of T that matters is the set of points to which T assigns a constant value. In the present case, these are the $n!$ points that can be obtained from the given point $[(X_1, Y_1), \ldots, (X_n, Y_n)]$ by permuting the n pairs. As in the univariate case, the conditional probability of each of these permutations given T or T' is $1/n!$. Also, as in the univariate case, T is complete (Problem 4.10).

An estimator δ is a function of the complete sufficient statistic if and only if δ is invariant under permutation of the n pairs. Hence, any such function is the unique UMVU estimator of its expectation.

Example 4.8 ***U*-estimation of covariance.** The estimator $\Sigma(X_i - \bar{X})(Y_i - \bar{Y})/(n-1)$ is UMVU for $\text{cov}(X, Y)$ (Problem 4.8). ||

5 The Information Inequality

The principal applications of UMVU estimators are to exponential families, as illustrated in Sections 2.2–2.3. When a UMVU estimator does not exist, the variance $V_L(\theta_0)$ of the LMVU estimator at θ_0 is the smallest variance that an unbiased estimator can achieve at θ_0. This establishes a useful benchmark against which to measure the performance of a given unbiased estimator δ. If the variance of δ is close to $V_L(\theta)$ for all θ, not much further improvement is possible. Unfortunately, the function $V_L(\theta)$ is usually difficult to determine. Instead, in this section, we shall derive some lower bounds which are typically not sharp [i.e., lie below $V_L(\theta)$] but are much simpler to calculate. One of the resulting inequalities for the variance, the *information inequality*, will be used in Chapter 5 as a tool for minimax estimation. However, its most important role is in Chapter 6, where it provides insight and motivation for the theory of asymptotically efficient estimators.

For any estimator δ of $g(\theta)$ and any function $\psi(x, \theta)$ with a finite second moment, the *covariance inequality* (Problem 1.5) states that

$$\text{var}(\delta) \geq \frac{[\text{cov}(\delta, \psi)]^2}{\text{var}(\psi)}. \tag{5.1}$$

In general, this inequality is not helpful since the right side also involves δ. However, when $\text{cov}(\delta, \psi)$ depends on δ only through $E_\theta(\delta) = g(\theta)$, (5.1) does provide a lower bound for the variance of all unbiased estimators of $g(\theta)$. The following result is due to Blyth (1974).

Theorem 5.1 *A necessary and sufficient condition for* $\text{cov}(\delta, \psi)$ *to depend on* δ *only through* $g(\theta)$ *is that for all* θ

$$cov(U, \psi) = 0 \quad \textit{for all } U \in \mathcal{U}, \tag{5.2}$$

where $\mathcal{U}$ is the class of statistics defined in Theorem 1.1, that is,

$$\mathcal{U} = \{U : E_\theta U = 0, E_\theta U^2 < \infty, \text{ for all } \theta \in \Omega\}.$$

Proof. To say that cov(δ, ψ) depends on δ only through $g(\theta)$ is equivalent to saying that for any two estimators δ_1 and δ_2 with $E_\theta\delta_1 = E_\theta\delta_2$ for all θ, we have $\text{cov}(\delta_1, \psi) = \text{cov}(\delta_2, \psi)$. The proof of the theorem is then easily established by writing

$$\text{(5.3)} \qquad \begin{aligned} \text{cov}(\delta_1, \psi) - \text{cov}(\delta_2, \psi) &= \text{cov}(\delta_1 - \delta_2, \psi) \\ &= \text{cov}(U, \psi) \end{aligned}$$

and noting that therefore, $\text{cov}(\delta_1, \psi) = \text{cov}(\delta_2, \psi)$ for all δ_1 and δ_2 if and only if $\text{cov}(U, \psi) = 0$ for all U. □

Example 5.2 Hammersley-Chapman-Robbins inequality. Suppose X is distributed with density $p_\theta = p(x, \theta)$, and for the moment, suppose that $p(x, \theta) > 0$ for all x. If θ and $\theta + \Delta$ are two values for which $g(\theta) \neq g(\theta + \Delta)$, then the function

$$\text{(5.4)} \qquad \psi(x, \theta) = \frac{p(x, \theta + \Delta)}{p(x, \theta)} - 1$$

satisfies the conditions of Theorem 5.1 since

$$\text{(5.5)} \qquad E_\theta(\psi) = 0$$

and hence

$$\text{cov}(U, \psi) = E(\psi U) = E_{\theta+\Delta}(U) - E_\theta(U) = 0.$$

In fact,

$$\text{cov}(\delta, \psi) = E_\theta(\delta\psi) = g(\theta + \Delta) - g(\theta),$$

so that (5.1) becomes

$$\text{(5.6)} \qquad \text{var}(\delta) \geq [g(\theta + \Delta) - g(\theta)]^2 / E_\theta \left[\frac{p(X, \theta + \Delta)}{p(X, \theta)} - 1\right]^2.$$

Since this inequality holds for all Δ, it also holds when the right side is replaced by its supremum over Δ. The resulting lower bound is due to Hammersley (1950) and Chapman and Robbins (1951). ‖

In this inequality, the assumption of a common support for the distributions p_θ can be somewhat relaxed. If $S(\theta)$ denotes the support of p_θ, (5.6) will be valid provided $S(\theta + \Delta)$ is contained in $S(\theta)$. In taking the supremum over Δ, attention must then be restricted to the values of Δ for which this condition holds.

When certain regularity conditions are satisfied, a classic inequality is obtained by letting $\Delta \to 0$ in (5.4). The inequality (5.6) is unchanged if (5.4) is replaced by

$$\frac{p_{\theta+\Delta} - p_\theta}{\Delta} \frac{1}{p_\theta},$$

which tends to $((\partial/\partial\theta)p_\theta)/p_\theta$ as $\Delta \to 0$, provided p_θ is differentiable with respect to θ. This suggests as an alternative to (5.4)

$$\text{(5.7)} \qquad \psi(x, \theta) = \frac{\partial}{\partial\theta} p(x, \theta) / p(x, \theta).$$

Since for any $U \in \mathcal{U}$, clearly $(d/d\theta)E_\theta(U) = 0$, ψ will satisfy (5.2), provided

$$E_\theta(U) = \int U p_\theta \, d\mu$$

can be differentiated with respect to θ under the integral sign for all $U \in \mathcal{U}$. To obtain the resulting lower bound, let $p'_\theta = (\partial p_\theta / \partial \theta)$ so that

$$\text{cov}(\delta, \psi) = \int \delta p'_\theta \, d\mu.$$

If differentiation under the integral sign is permitted in

$$\int \delta p_\theta \, d\mu = g(\theta),$$

it then follows that

$$\text{cov}(\delta, \psi) = g'(\theta) \tag{5.8}$$

and hence

$$\text{var}(\delta) \geq \frac{[g'(\theta)]^2}{\text{var}\left[\dfrac{\partial}{\partial \theta} \log p(X, \theta)\right]}. \tag{5.9}$$

The assumptions required for this inequality will be stated more formally in Theorem 5.15, where we will pay particular attention to requirements on the estimator. Pitman (1979, Chapter 5) provides an interesting interpretation of the inequality and discussion of the regularity assumptions.

The function ψ defined by (5.7) is the relative rate at which the density p_θ changes at x. The average of the square of this rate is denoted by

$$I(\theta) = E_\theta \left[\frac{\partial}{\partial \theta} \log p(X, \theta)\right]^2 = \int \left(\frac{p'_\theta}{p_\theta}\right)^2 p_\theta \, d\mu. \tag{5.10}$$

It is plausible that the greater this expectation is at a given value θ_0, the easier it is to distinguish θ_0 from neighboring values θ, and, therefore, the more accurately θ can be estimated at $\theta = \theta_0$. (Under suitable assumptions this surmise turns out to be correct for large samples; see Chapter 6.) The quantity $I(\theta)$ is called the *information* (or the Fisher information) that X contains about the parameter θ.

It is important to realize that $I(\theta)$ depends on the particular parametrization chosen. In fact, if $\theta = h(\xi)$ and h is differentiable, the information that X contains about ξ is

$$I^*(\xi) = I[h(\xi)] \cdot [h'(\xi)]^2. \tag{5.11}$$

When different parameterizations are considered in a single problem, the notation $I(\theta)$ is inadequate; however, it suffices for most applications.

To obtain alternative expressions for $I(\theta)$ that sometimes are more convenient, let us make the following assumptions:

(a) Ω is an open interval (finite, infinite, or semi-infinite).

(b) The distributions P_θ have common support, so that without loss of generality the set $A = \{x : p_\theta(x) > 0\}$

(5.12) is independent of θ.

(c) For any x in A and θ in Ω, the derivative $p'_\theta(x) = \partial p_\theta(x)/\partial\theta$ exists and is finite.

Lemma 5.3

(a) If (5.12) holds, and the derivative with respect to θ of the left side of

$$\int p_\theta(x)\, d\mu(x) = 1 \tag{5.13}$$

can be obtained by differentiating under the integral sign, then

$$E_\theta\left[\frac{\partial}{\partial\theta}\log p_\theta(X)\right] = 0 \tag{5.14}$$

and

$$I(\theta) = \text{var}_\theta\left[\frac{\partial}{\partial\theta}\log p_\theta(X)\right]. \tag{5.15}$$

(b) If, in addition, the second derivative with respect to θ of $\log p_\theta(x)$ *exists for all x and θ and the second derivative with respect to θ of the left side of (5.13) can be obtained by differentiating twice under the integral sign, then*

$$I(\theta) = -E_\theta\left[\frac{\partial^2}{\partial\theta^2}\log p_\theta(X)\right]. \tag{5.16}$$

Proof.

(a) Equation (5.14) is derived by differentiating (5.13), and (5.15) follows from (5.10) and (5.14).

(b) We have

$$\frac{\partial^2}{\partial\theta^2}\log p_\theta(x) = \frac{\frac{\partial^2}{\partial\theta^2}p_\theta(x)}{p_\theta(x)} - \left[\frac{\frac{\partial}{\partial\theta}p_\theta(x)}{p_\theta(x)}\right]^2,$$

and the result follows by taking the expectation of both sides. □

Let us now calculate $I(\theta)$ for some of the families discussed in Sections 1.4 and 1.5.

We first look at exponential families with $s = 1$, given in Equation (1.5.1), and derive a relationship between some unbiased estimators and information.

Theorem 5.4 *Let X be distributed according to the exponential family (5.1) with $s = 1$, and let*

$$\tau(\theta) = E_\theta(T), \tag{5.17}$$

the so-called mean-value parameter. Then, T

$$I[\tau(\theta)] = \frac{1}{\text{var}_\theta(T)}. \tag{5.18}$$

Proof. From Equation (5.15), the amount of information that X contains about θ, $I(\theta)$, is

$$\begin{aligned} I(\theta) &= \text{var}_\theta\left[\frac{\partial}{\partial\theta}\log p_\theta(X)\right] \\ &= \text{var}_\theta[\eta'(\theta)T(X) - B'(\theta)] \qquad \text{(from (1.5.1))} \\ &= [\eta'(\theta)]^2\text{var}(T). \end{aligned} \tag{5.19}$$

Now, from (5.11), the information $I[\tau(\theta)]$ that X contains about $\tau(\theta)$, is given by

$$\begin{aligned} I[\tau(\theta)] &= \frac{I(\theta)}{[\tau'(\theta)]^2} \\ &= \left[\frac{\eta'(\theta)}{\tau'(\theta)}\right]^2 \text{var}(T). \end{aligned} \tag{5.20}$$

Finally, using the fact that $\tau(\theta) = B'(\theta)/\eta'(\theta)$ [(Problem 1.5.6)], we have

$$\text{var}(T) = \frac{B''(\theta) - \eta''(\theta)\tau(\theta)}{[\eta'(\theta)]^2} = \left[\frac{\tau'^2(\theta)}{\eta'^2(\theta)}\right]^{1/2} \tag{5.21}$$

and substituting (5.21) into (5.20) yields (5.18). □

If we combine Equations (5.11) and (5.19), then for any differentiable function $h(\theta)$, we have

$$I[h(\theta)] = \left[\frac{\eta'(\theta)}{h'(\theta)}\right]^2 \text{var}(T). \tag{5.22}$$

Example 5.5 Information in a gamma variable. Let $X \sim \text{Gamma}(\alpha, \beta)$, where we assume that α is known. The density is given by

$$\begin{aligned} p_\beta(x) &= \frac{1}{\Gamma(\alpha)\beta^\alpha}x^{\alpha-1}e^{-x/\beta} \\ &= e^{(-1/\beta)x - \alpha\log(\beta)}h(x) \end{aligned} \tag{5.23}$$

with $h(x) = x^{\alpha-1}/\Gamma(\alpha)$. In this parametrization, $\eta(\beta) = -1/\beta$, $T(x) = x$ and $B(\beta) = \alpha\log(\beta)$. Thus, $E(T) = \alpha\beta$, $\text{var}(T) = \alpha\beta^2$, and the information in X about $\alpha\beta$ is $I(\alpha\beta) = 1/\alpha\beta^2$.

If we are instead interested in the information in X about β, then we can reparameterize (5.23) using $\eta(\beta) = -\alpha/\beta$ and $T(x) = x/\alpha$. From (5.22), we have, quite generally, that $I[ch(\theta)] = \frac{1}{c^2}I[h(\theta)]$, so the information in X about β is $I(\beta) = \alpha/\beta^2$. ||

Table 5.1 gives $I[\tau(\theta)]$ for a number of special cases.

Qualitatively, $I[\tau(\theta)]$ given by (5.18) behaves as one would expect. Since T is the UMVU estimator of its expectation $\tau(\theta)$, the variance of T is a measure of the difficulty of estimating $\tau(\theta)$. Thus, the reciprocal of the variance measures the ease with which $\tau(\theta)$ can be estimated and, in this sense, the information X contains about $\tau(\theta)$.

Example 5.6 Information in a normal variable. Consider the case of the $N(\xi, \sigma^2)$ distribution with σ known, when the interest is in estimation of ξ^2. The density is

Table 5.1. $I[\tau(\theta)]$ *for Some Exponential Families*

Distribution	Parameter $\tau(\theta)$	$I(\tau(\theta))$
$N(\xi, \sigma^2)$	ξ	$1/\sigma^2$
$N(\xi, \sigma^2)$	σ^2	$1/2\sigma^4$
$b(p, n)$	p	n/pq
$P(\lambda)$	λ	$1/\lambda$
$\Gamma(\alpha, \beta)$	β	α/β^2

given by

$$p_\xi(x) = \frac{1}{\sqrt{2\pi}\,\sigma} e^{-\frac{1}{2\sigma^2}(x-\xi)^2}$$
$$= e^{\eta(\xi)T(x) - B(\xi)} h(x)$$

with $\eta(\xi) = \xi$, $T(x) = x/\sigma^2$, $B(\xi) = \frac{1}{2}\xi^2/\sigma^2$, and $h(x) = e^{-x^2/2\sigma^2}/\sqrt{2\pi}$. The information in X about $h(\xi) = \xi^2$ is given by

$$I(\xi^2) = \left[\frac{\eta'(\xi)}{h'(\xi)}\right]^2 \text{var}(T) = \frac{1}{4\xi^2\sigma^2}.$$

Note that we could have equivalently defined $\eta(\xi) = \xi/\sigma^2$, $T(x) = x$ and arrived at the same answer. ‖

Example 5.7 Information about a function of a Poisson parameter. Suppose that X has the Poisson (λ) distribution, so that $I[\lambda]$, the information X contains about $\lambda = E(X)$, is $1/\lambda$. For $\eta(\lambda) = \log \lambda$, which is an increasing function of λ, $I[\log \lambda] = \lambda$. Thus, the information in X about λ is inversely proportional to that about $\log \lambda$. In particular, for large values of λ, it seems that the parameter $\log \lambda$ can be estimated quite accurately, although the converse is true for λ. This conclusion is correct and is explained by the fact the $\log \lambda$ changes very slowly when λ is large. Hence, for large λ, even a large error in the estimate of λ will lead to only a small error in $\log \lambda$, whereas the situation is reversed for λ near zero where $\log \lambda$ changes very rapidly. It is interesting to note that there exists a function of λ [namely $h(\lambda) = \sqrt{\lambda}$] whose behavior is intermediate between that of $h(\lambda) = \lambda$ and $h(\lambda) = \log \lambda$, in that the amount of information X contains about it is constant, independent of λ (Problem 5.6). ‖

As a second class of distributions for which to evaluate $I(\theta)$, consider location families with density

(5.24) $$f(x - \theta) \quad (x, \theta \text{ real-valued})$$

where $f(x) > 0$ for all x. Conditions (5.12) are satisfied provided the derivative $f'(x)$ of $f(x)$ exists for all values of x. It is seen that $I(\theta)$ is independent of θ and given by (Problem 5.14)

(5.25) $$I_f = \int_{-\infty}^{\infty} \frac{[f'(x)]^2}{f(x)}\, dx.$$

Table 5.2. I_f *for Some Standard Distributions*

Distribution	$N(0, 1)$	$L(0, 1)$	$C(0, 1)$	$DE(0, 1)$
I_f	1	1/3	1/2	1

Table 5.2 shows I_f for a number of distributions (defined in Table 1.4.1).

Actually, the double exponential density does not satisfy the stated assumptions since $f'(x)$ does not exist at $x = 0$. However, (5.25) is valid under the slightly weaker assumption that f is absolutely continuous [see (1.3.7)] which does hold in the double exponential case. For this and the extensions below, see, for example, Huber 1981, Section 4.4. On the other hand, it does not hold when f is the uniform density on (0, 1) since f is then not continuous and hence, a fortiori, not absolutely continuous. It turns out that whenever f is not absolutely continuous, it is natural to put I_f equal to ∞. For the uniform distribution, for example, it is easier by an order of magnitude to estimate θ (see Problem 5.33) than for any of the distributions listed in Table 5.2, and it is thus reasonable to assign to I_f the value ∞. This should be contrasted with the fact that $f'(x) = 0$ for all $x \neq 0, 1$, so that formal application of (5.25) leads to the incorrect value 0.

When (5.24) is replaced by

$$\frac{1}{b} f\left(\frac{x-\theta}{b}\right), \tag{5.26}$$

the amount of information about θ becomes (Problem 5.14)

$$\frac{I_f}{b^2} \tag{5.27}$$

with I_f given by (5.25).

The information about θ contained in independent observations is, as one would expect, additive. This is stated formally in the following result.

Theorem 5.8 *Let X and Y be independently distributed with densities p_θ and q_θ, respectively, with respect to measures μ and ν satisfying (5.12) and (5.14).*

If $I_1(\theta)$, $I_2(\theta)$, and $I(\theta)$ are the information about θ contained in X, Y, and (X, Y), respectively, then

$$I(\theta) = I_1(\theta) + I_2(\theta). \tag{5.28}$$

Proof. By definition,

$$I(\theta) = E\left[\frac{\partial}{\partial\theta}\log p_\theta(X) + \frac{\partial}{\partial\theta}\log q_\theta(Y)\right]^2,$$

and the result follows from the fact that the cross-product is zero by (5.14). □

Corollary 5.9 *If $X_1, \ldots, X_n$ are iid, satisfy (5.12) and (5.14), and each has information $I(\theta)$, then the information in $X = (X_1, \ldots, X_n)$ is $nI(\theta)$.*

Let us now return to the inequality (5.9), and proceed to a formal statement of when it holds. If (5.12), and hence (5.15), holds, then the denominator of the right side of (5.9) can be replaced by $I(\theta)$. The result is the following version of the *Information Inequality.*

Theorem 5.10 (The Information Inequality) *Suppose p_θ is a family of densities with dominating measure μ for which (5.12) and (5.14) hold, and that $I(\theta) > 0$. Let δ be any statistic with*

$$E_\theta(\delta^2) < \infty \tag{5.29}$$

for which the derivative with respect to θ of $E_\theta(\delta)$ exists and can be differentiated under the integral sign, that is,

$$\frac{d}{d\theta} E_\theta(\delta) = \int \frac{\partial}{\partial\theta} \delta \, p_\theta \, d\mu. \tag{5.30}$$

Then

$$\operatorname{var}_\theta(\delta) \geq \frac{\left[\frac{\partial}{\partial\theta} E_\theta(\delta)\right]^2}{I(\theta)}. \tag{5.31}$$

Proof. The result follows from (5.9)and Lemma 5.3 and is seen directly by differentiating (5.30) and then applying (5.1). □

If δ is an estimator of $g(\theta)$, with

$$E_\theta(\delta) = g(\theta) + b(\theta)$$

where $b(\theta)$ is the bias of δ, then (5.31) becomes

$$\operatorname{var}_\theta(\delta) \geq \frac{[b'(\theta) + g'(\theta)]^2}{I(\theta)}, \tag{5.32}$$

which provides a lower bound for the variance of any estimator in terms of its bias and $I(\theta)$.

If $\delta = \delta(X)$ where $X = (X_1, \ldots, X_n)$ and if the X's are iid, then by Corollary 5.9

$$\operatorname{var}_\theta(\delta) \geq \frac{[b'(\theta) + g'(\theta)]^2}{nI_1(\theta)} \tag{5.33}$$

where $I_1(\theta)$ is the information about θ contained in X_1. Inequalities (5.32) and (5.33) will be useful in Chapter 5.

Unlike $I(\theta)$, which changes under reparametrization, the lower bound (5.31), and hence the bounds (5.32) and (5.33), does not. Let $\theta = h(\xi)$ with h differentiable. Then,

$$\frac{\partial}{\partial\xi} E_{h(\xi)}(\delta) = \frac{\partial}{\partial\theta} E_\theta(\delta) \cdot h'(\xi),$$

and the result follows from (5.11). (See Problem 5.20.)

The lower bound (5.31) for $\operatorname{var}_\theta(\delta)$ typically is not sharp. In fact, under suitable regularity conditions, it is attained if and only if $p_\theta(x)$ is an exponential family (1.5.1) with $s = 1$ and $T(x) = \delta(x)$ (see Problem 5.17). However, (5.1) is based on the Cauchy-Schwarz inequality, which has a well-known condition for equality (see Problems 5.2 and 5.19). The bound (5.31) will be attained by an estimator if and only if

$$\delta = a\left[\frac{\partial}{\partial\theta}\log p_\theta(x)\right] + b \tag{5.34}$$

for some constants a and b (which may depend on θ).

Example 5.11 Binomial attainment of information bound. For the binomial distribution $X \sim b(p, n)$, we have

$$\begin{aligned}\delta = a\left[\frac{\partial}{\partial\theta}\log p_\theta(x)\right] + b &= a\left[\frac{\partial}{\partial p}[x\log p + (n-x)\log(1-p)]\right] + b\\ &= a\left[\frac{x-np}{p(1-p)}\right] + b\end{aligned}$$

with $E\delta = b$ and $\text{var}\,\delta = na^2/p(1-p)$. This form for δ is the only form of function for which the information inequality bound (5.31) can be attained. The function δ is an estimator only if $a = p(1-p)$ and $b = np$. This yields $\delta = X$, $E\delta = np$, and $\text{var}(\delta) = np(1-p)$. Thus, X is the only unbiased estimator that achieves the information inequality bound (5.31). ||

Many authors have presented general necessary and sufficient conditions for attainment of the bound (5.31) (Wijsman 1973, Joshi 1976, Müller-Funk et al., 1989). The following theorem is adapted from Müller-Funk et al.

Theorem 5.12 Attainment. *Suppose (5.12) holds, and δ is a statistic with* $\text{var}_\theta\delta < \infty$ *for all $\theta \in \Omega$. Then δ attains the lower bound*

$$\text{var}_\theta\delta = \left(\frac{\partial}{\partial\theta}E_\theta\delta\right)^2 \Big/ I(\theta)$$

for all $\theta \in \Omega$ if and only if there exists a continuously differentiable function $\varphi(\theta)$ such that

$$p_\theta(x) = C(\theta)e^{\varphi(\theta)\delta(x)}h(x)$$

is a density with respect to a dominating measure $\mu(x)$ for suitably chosen $C(\theta)$ and $h(x)$, i.e., $p_\theta(x)$ constitutes an exponential family.

Moreover, if $E_\theta\delta = g(\theta)$, then δ and g satisfy

$$\begin{aligned}\delta(x) &= \left[\frac{g'(\theta)}{I(\theta)}\right]\frac{\partial}{\partial\theta}\log p_\theta(x) + g(\theta), \\ g(\theta) &= -C'(\theta)/C(\theta)\varphi'(\theta),\end{aligned} \tag{5.35}$$

and $I(\theta) = \varphi'(\theta)g'(\theta)$.

Note that the function δ specified in (5.35) may depend on θ. In such a case, δ is not an estimator, and there is no estimator that attains the information bound.

Example 5.13 Poisson attainment of information bound. Suppose X is a discrete random variable with probability function that is absolutely continuous with respect to μ = counting measure, and satisfies

$$E_\lambda X = \text{var}_\lambda(X) = \lambda.$$

If X attains the Information Inequality bound then $\lambda = [\partial/(\partial\lambda)E_\lambda(X)]^2/I(\lambda)$ so from Theorem 5.12 $\varphi'(\lambda) = 1/\lambda$ and the distribution of X must be

$$p_\lambda(x) = C(\lambda)e^{[\log\lambda]x}h(x).$$

Since $g(\theta) = \lambda = -\lambda C'(\lambda)/C(\lambda)$, it follows that $C(\lambda) = e^{-\lambda}$, which implies $h(x) = x!$, and $p_\lambda(x)$ is the Poisson distribution. ‖

Some improvements over (5.31) are available when the inequality is not attained. These will be briefly mentioned at the end of the next section.

Theorem 5.10 restricts the information inequality to estimators δ satisfying (5.29) and (5.30). The first of these conditions imposes no serious restrictions since any estimator satisfies (5.31) automatically. However, it is desirable to replace (5.30) by a condition (on the densities p_θ) not involving δ, so that (5.31) will then hold for all δ. Such conditions will be given in Theorem 5.15 below, with a more detailed discussion of alternatives given in Note 8.6.

In reviewing the argument leading to (5.9), the conditions that were needed on the estimator $\delta(x)$ were

$$\text{(a)}\quad E_\theta[\delta^2(X)] < \infty \quad \text{for all } \theta$$

$$\text{(5.36)}\qquad \text{(b)}\quad \frac{\partial}{\partial\theta}E_\theta[\delta(X)] = \int \frac{\partial}{\partial\theta}\delta(x)p_\theta(x)\,d\mu(x) = g'(\theta).$$

The key point is to find a way to ensure that $\text{cov}(\delta, \phi) = (\partial/\partial\theta)E_\theta\delta$, and hence (5.30) holds. Consider the following argument, in which one of the steps is not immediately justified. For $q_\theta(x) = \partial \log p_\theta(x)/\partial\theta$, write

$$\begin{aligned}
\text{cov}(\delta, q) &= \int \delta(x)\left[\frac{\partial}{\partial\theta}\log p_\theta(x)\right]p_\theta(x)dx \\
&= \int \delta(x)\left[\lim_{\Delta\to 0}\frac{p_{\theta+\Delta}(x) - p_\theta(x)}{\Delta p_\theta(x)}\right]p_\theta(x)dx \\
\text{(5.37)}\qquad &\overset{?}{=} \lim_{\Delta\to 0}\int \delta(x)\left[\frac{p_{\theta+\Delta}(x) - p_\theta(x)}{\Delta p_\theta(x)}\right]p_\theta(x)dx \\
&= \lim_{\Delta\to 0}\frac{E_{\theta+\Delta}\delta(X) - E_\theta\delta(X)}{\Delta} \\
&= \frac{\partial}{\partial\theta}E_\theta\delta(X)
\end{aligned}$$

Thus (5.30) will hold provided the interchange of limit and integral is valid. A simple condition for this is given in the following lemma.

Lemma 5.14 *Assume that (5.12(a)) and (5.12(b)) hold, and let δ be any estimator for which $E_\theta\delta^2 < \infty$. Let $q_\theta(x) = \partial \log p_\theta(x)/\partial\theta$ and, for some $\varepsilon > 0$, let b_θ be a function that satisfies*

$$\text{(5.38)}\qquad E_\theta b_\theta^2(X) < \infty \text{ and } \left|\frac{p_{\theta+\Delta}(x) - p_\theta(x)}{\Delta p_\theta(x)}\right| \le b_\theta(x) \text{ for all } |\Delta| < \varepsilon.$$

Then $E_\theta q_\theta(X) = 0$ and

$$\text{(5.39)}\qquad \frac{\partial}{\partial\theta}E_\theta\delta(X) = E\delta(X)q_\theta(X) = \text{cov}_\theta(\delta, q_\theta),$$

and thus (5.30) holds.

Proof. Since

$$\left|\delta(x)\frac{p_{\theta+\Delta}(x)-p_\theta(x)}{\Delta p_\theta(x)}\right| \le |\delta(x)|b(x)|,$$

and

$$E_\theta[|\delta(x)|b(x)] \le \{E_\theta[\delta(x)^2]\}^{1/2}\{E_\theta[b(x)^2]\}^{1/2} << \infty,$$

it follows from the Dominated Convergence Theorem (Theorem 1.2.5) that the interchange of limit and integral in (5.37) is valid. □

An immediate consequence of Lemma 5.14 is the following theorem.

Theorem 5.15 *Suppose $p_\theta(x)$ is a family of densities with dominating measure $\mu(x)$ satisfying (5.12), $I(\theta) > 0$, and there exists a function b_θ and $\varepsilon > 0$ for which (5.38) holds, If δ is any statistic for which $E_\theta(\delta^2) << \infty$, then the information inequality (5.31) will hold.*

We note that condition (5.38) is similar to what is known as a *Lipschitz condition*, which imposes a smoothness constraint on a function by bounding the left side of (5.38) by a constant. It is satisfied for many families of densities (see Problem 5.27), including of course the exponential family. We give one illustration here.

Example 5.16 Integrability. Suppose that $X \sim f(x-\theta)$, where $f(x-\theta)$ is Students t distribution with m degrees of freedom. It is not immediately obvious that this family of densities satisfies (5.14), so we cannot directly apply Theorem 5.10. We leave the general case to Problem 5.27(b), and show here that the Cauchy family ($m = 1$), with density $p_\theta(x) = \frac{1}{\pi}\frac{1}{1+(x-\theta)^2}$, satisfies (5.38). The left side of (5.38) is

$$\begin{aligned}
&\left|\frac{1}{\Delta}\left(\frac{1+(x-\theta)^2}{1+(x-\Delta-\theta)^2}-1\right)\right| \\
&= \left|\frac{1}{\Delta}\frac{1+(x-\theta)^2-1-(x-\Delta-\theta)^2}{1+(x-\Delta-\theta)^2}\right| \\
&= \left|\frac{1}{\Delta}\frac{2\Delta(x-\theta)-\Delta^2}{1+(x-\Delta-\theta)^2}\right| \\
&\le 2\,\frac{|x-\Delta-\theta|}{1+(x-\Delta-\theta)^2}+\frac{|\Delta|}{1+(x-\Delta-\theta)^2} \\
&\le 2+\varepsilon.
\end{aligned}$$

Here the last inequality follows from the facts that $|\Delta| << \varepsilon$ and $|x|/(1+x^2) \le 1$ for all x . Condition (5.38) therefore holds with $b_\theta(x) = 2+\varepsilon$, which verifies the information inequality (5.31) for the Cauchy case. ||

As a consequence of Theorem 5.15, note

Corollary 5.17 *If (5.38) holds, then (5.14) is valid.*

Proof. Putting $\delta(x) = 1$ in (5.29), we have that

$$0 = \frac{d}{d\theta}(1) = \int \frac{\partial}{\partial\theta}p_\theta d\mu = E_\theta\left[\frac{\partial}{\partial\theta}\log p_\theta(X)\right]$$

□

6 The Multiparameter Case and Other Extensions

In discussing the information inequality, we have so far assumed that θ is real-valued. To extend the inequalities of the preceding section to the multiparameter case, we begin by generalizing the inequality (5.1) to one involving several functions ψ_i $(i = 1, \ldots, r)$. This extension also provides a tool for sharpening the inequality (5.31).

Theorem 6.1 *For any unbiased estimator δ of $g(\theta)$ and any functions $\psi_i(x, \theta)$ with finite second moments, we have*

$$\text{var}(\delta) \geq \gamma' C^{-1} \gamma, \tag{6.1}$$

where $\gamma' = (\gamma_1 \cdots \gamma_r)$ and $C = ||C_{ij}||$ are defined by

$$\gamma_i = cov(\delta, \psi_i), \quad C_{ij} = cov(\psi_i, \psi_j). \tag{6.2}$$

The right side of (6.1) will depend on δ only through $g(\theta) = E_\theta(\delta)$, provided each of the functions ψ_i satisfies (5.2).

Proof. For any constants $a_1, \ldots, a_r$, it follows from (5.1) that

$$\text{var}(\delta) \geq \frac{[\text{cov}(\delta, \Sigma a_i \psi_i)]^2}{\text{var}(\Sigma a_i \psi_i)}, \tag{6.3}$$

and direct calculation shows

$$\text{cov}(\delta, \Sigma a_i \psi_i) = \Sigma a_i \gamma_i = a'\gamma, \qquad \text{var}(\Sigma a_i \psi_i) = a'Ca. \tag{6.4}$$

Since (6.3) is true for any vector a, from (6.4) and (5.1) we have

$$\text{var}(\delta) \geq \max_a \frac{[a'\gamma]^2}{a'Ca} = \gamma' C^{-1} \gamma,$$

where we use the fact (see Problem 6.2) that if P is an $r \times r$ matrix and p an $r \times 1$ column vector such that $P = pp'$, then

$$\max_a \frac{a'Pa}{a'Qa} = \text{largest eigenvalue of } Q^{-1}P \tag{6.5}$$
$$= p'Q^{-1}p.$$

□

As the first and principal application of (6.1), we shall extend the information inequality (5.31) to the multiparameter case. Let X be distributed with density $p_\theta, \theta \in \Omega$, with respect to μ where θ is vector-valued, say $\theta = (\theta_1, \ldots, \theta_s)$. Suppose that

(6.6) (5.12)(a) and (b) hold, and in addition
(c) For any x in A, θ in Ω, and $i = 1, \ldots, s$,
the derivative $\partial p_\theta(x)/\partial \theta_i$ exists and is finite.

In a generalization of (5.10), define the *information matrix* as the $s \times s$ matrix

$$I(\theta) = ||I_{ij}(\theta)|| \tag{6.7}$$

where

$$I_{ij}(\theta) = E_\theta\left[\frac{\partial}{\partial\theta_i}\log p_\theta(X)\cdot\frac{\partial}{\partial\theta_j}\log p_\theta(X)\right]. \tag{6.8}$$

If (6.6) holds and the derivative with respect to each θ_i of the left side of (5.13) can be obtained by differentiating under the integral sign, one obtains, as in Lemma 5.3,

$$E\left[\frac{\partial}{\partial\theta_i}\log p_\theta(X)\right] = 0 \tag{6.9}$$

and

$$I_{ij}(\theta) = \text{cov}\left[\frac{\partial}{\partial\theta_i}\log p_\theta(X), \frac{\partial}{\partial\theta_j}\log p_\theta(X)\right]. \tag{6.10}$$

Being a covariance matrix, $I(\theta)$ is positive semidefinite and positive definite unless the $(\partial/\partial\theta_i)\log p_\theta(X)$, $i = 1, \ldots, s$, are affinely dependent (and hence, by (6.9), linearly dependent).

If, in addition to satisfying (6.6) and (6.9), the density p_θ also has second derivatives $\partial^2 p_\theta(x)/\partial\theta_i\partial\theta_j$ for all i and j, there is in generalization of (5.16), an alternative expression for $I_{ij}(\theta)$ which is often more convenient (Problem 6.4),

$$I_{ij}(\theta) = -E\left[\frac{\partial^2}{\partial\theta_i\partial\theta_j}\log p_\theta(X)\right]. \tag{6.11}$$

In the multiparameter situation with $\theta = (\theta_i, \ldots, \theta_s)$, Theorem 5.8 and Corollary 5.9 continue to hold with only the obvious changes, that is, information matrices for independent observations are additive.

To see how an information matrix changes under reparametrization, suppose that

$$\theta_i = h_i(\xi_1, \ldots, \xi_s), \quad i = 1, \ldots, s, \tag{6.12}$$

and let J be the matrix

$$J = \left\|\frac{\partial\theta_j}{\partial\xi_i}\right\|. \tag{6.13}$$

Let the information matrix for $(\xi_1, \ldots, \xi_s)$ be $I^*(\xi) = ||I^*_{ij}(\xi)||$ where

$$I^*_{ij}(\xi) = E\left[\frac{\partial}{\partial\xi_i}\log p_{\theta(\xi)}(X)\cdot\frac{\partial}{\partial\xi_j}\log p_{\theta(\xi)}(X)\right]. \tag{6.14}$$

Then, it is seen from the chain rule for differentiating a function of several variables that (Problem 6.7)

$$I^*_{ij}(\xi) = \sum_k\sum_l I_{kl}(\theta)\frac{\partial\theta_k}{\partial\xi_i}\frac{\partial\theta_l}{\partial\xi_j} \tag{6.15}$$

and hence that

$$I^*(\xi) = JIJ'. \tag{6.16}$$

In generalization of Theorem 5.4, let us now calculate $I(\theta)$ for multiparameter exponential families.

Theorem 6.2 *Let X be distributed according to the exponential family (1.5.1) and let*

(6.17) $$\tau_i = ET_i(X), \quad i = 1, \ldots, s,$$

the mean-value parametrization. Then,

(6.18) $$I(\tau) = C^{-1}$$

where C is the covariance matrix of $(T_1, \ldots, T_s)$.

Proof. It is easiest to work with the natural parametrization (1.5.2), which is equivalent. By (6.10) and (1.5.15), the information in X about the natural parameter η is

$$I^*(\eta) = \left\| \frac{\partial^2}{\partial\eta_j \partial\eta_k} A(\eta) \right\| = \text{cov}(T_j, T_k) = C.$$

Furthermore, (1.5.14) shows that $\tau_j = \partial/\partial\eta_j A(\eta)$ and, hence, (6.13) shows that

$$J = \left\| \frac{\partial\tau_j}{\partial\eta_i} \right\| = C.$$

Thus, from (6.16)

$$C = I^*(\eta) = JI(\tau)J' = CI(\tau)C,$$

which implies (6.18). □

Example 6.3 Multivariate normal information matrix. Let $(X_1, \ldots, X_p)$ have a multivariate normal distribution with mean 0 and covariance matrix $\Sigma = ||\sigma_{ij}||$, so that by (1.4.15), the density is proportional to

$$e^{-\Sigma\Sigma\eta_{ij}x_ix_j/2}$$

where $||\eta_{ij}|| = \Sigma^{-1}$. Since $E(X_iX_j) = \sigma_{ij}$, we find that the information matrix of the σ_{ij} is

(6.19) $$I(\Sigma) = \Sigma^{-1}.$$

||

Example 6.4 Exponential family information matrices. Table 6.1 gives $I(\theta)$ for three two-parameter exponential families, where $\psi(\alpha) = \Gamma'(\alpha)/\Gamma(\alpha)$ and $\psi'(\alpha) = d\psi(\alpha)/d\alpha$ are, respectively, the digamma and trigamma function (Problem 6.5). ||

Example 6.5 Information in location-scale families. For the location-scale families with density $(1/\theta_2)f((x-\theta_1)/\theta_2)$, $\theta_2 > 0$, $f(x) > 0$ for all x, the elements of the information matrix are (Problem 6.5)

(6.20) $$I_{11} = \frac{1}{\theta_2^2} \int \left[\frac{f'(y)}{f(y)}\right]^2 f(y)\,dy,$$

$$I_{22} = \frac{1}{\theta_2^2} \int \left[\frac{yf'(y)}{f(y)} + 1\right]^2 f(y)\,dy$$

and

(6.21) $$I_{12} = \frac{1}{\theta_2^2} \int y \left[\frac{f'(y)}{f(y)}\right]^2 f(y)\,dy.$$

Table 6.1. *Three Information Matrices*

$N(\xi, \sigma^2)$	$\Gamma(\alpha, \beta)$
$I(\xi, \sigma) = \begin{pmatrix} 1/\sigma^2 & 0 \\ 0 & 2/\sigma^2 \end{pmatrix}$	$I(\alpha, \beta) = \begin{pmatrix} \psi'(\alpha) & 1/\beta \\ 1/\beta & \alpha/\beta^2 \end{pmatrix}$
$B(\alpha, \beta)$	
$I(\alpha, \beta) = \begin{pmatrix} \psi'(\alpha) - \psi'(\alpha+\beta) & -\psi'(\alpha+\beta) \\ -\psi'(\alpha+\beta) & \psi'(\beta) - \psi'(\alpha+\beta) \end{pmatrix}$	

The covariance term I_{12} is zero whenever f is symmetric about the origin. ||

Let us now generalize Theorems 5.10 and 5.15 to the multiparameter case in which $\boldsymbol{\theta} = (\theta_1, \ldots, \theta_s)$. For convenience, we state the generalizations in one theorem.

Theorem 6.6 (Multiparameter Information Inequality) *Suppose that (6.6) holds, and $I(\boldsymbol{\theta})$ is positive definite. Let δ be any statistic for which $E_{\boldsymbol{\theta}}(|\delta|^2) < \infty$ and either*

(i) For $i = 1, \ldots, s$, $(\partial/\partial\theta_i)E_{\boldsymbol{\theta}}\delta$ exists and can be obtained by differentiating under the integral sign,

or

(ii) There exist functions $b_{\boldsymbol{\theta}}^{(i)}$, $i = 1, \ldots, s$, with $E_{\boldsymbol{\theta}} b_{\boldsymbol{\theta}}^{(i)}(X)^2 < \infty$ that satisfy

$$\left| \frac{p_{\boldsymbol{\theta}+\Delta\epsilon_i}(x) - p_{\boldsymbol{\theta}}(x)}{\Delta p_{\boldsymbol{\theta}}(x)} \right| \le b_{\boldsymbol{\theta}}^{(i)}(x) \text{ for all } \Delta,$$

where $\epsilon_i \in R^s$ is the unit vector with 1 in the ith position and zero elsewhere.

Then, $E_{\boldsymbol{\theta}}(\partial/\partial\theta_i)\log p_{\boldsymbol{\theta}}(\mathbf{X}) = 0$ and

$$\text{var}_\theta(\delta) \ge \alpha' I^{-1}(\theta)\alpha \tag{6.22}$$

where α' is the row matrix with ith element

$$\alpha_i = \frac{\partial}{\partial\theta_i} E_\theta[\delta(X)]. \tag{6.23}$$

Proof. If the functions ψ_i of Theorem 6.1 are taken to be $\psi_i = (\partial/\partial\theta_i)\log p_\theta(X)$, (6.22) follows from (6.1) and (6.10). □

If δ is an estimator of $g(\theta)$ and $b(\theta)$ is its bias, then (6.23) reduces to

$$\alpha_i = \frac{\partial}{\partial\theta_i}[b(\theta) + g(\theta)]. \tag{6.24}$$

It is interesting to compare the lower bound (6.22) with the corresponding bound when the θ's other than θ_i are known. By Theorem 5.15, the latter is equal to $[(\partial/\partial\theta_i)E_\theta(\delta)]^2/I_{ii}(\theta)$. This is the bound obtained by setting $a = \epsilon_i$ in (6.4), where ϵ_i is the ith unit vector. For example, if the θ's other than θ_i are zero, then the only nonzero element of the vector α of (6.22) is α_i. Since (6.22) was obtained by maximizing (6.4), comparing the two bounds shows

$$I_{ii}^{-1}(\theta) \leq ||I^{-1}(\theta)||_{ii}. \tag{6.25}$$

(See Problem 6.10 for a different derivation.) The two sides of (6.25) are equal if

$$I_{ij}(\theta) = 0 \quad \text{for all } j \neq i, \tag{6.26}$$

as is seen from the definition of the inverse of a matrix, and, in fact, (6.26) is also necessary for equality in (6.25) (Problem 6.10). In this situation, when (6.26) holds, the parameters are said to be orthogonal. This is illustrated by the first matrix in Table 6.1. There, the information bound for one of the parameters is independent of whether the other parameter is known. This is not the case, however, in the second and third situations in Table 6.1, where the value of one parameter affects the information for another. Some implications of these results for estimation will be taken up in Section 6.6. (Cox and Reid (1987) discuss methods for obtaining parameter orthogonality, and some of its consequences; see also Barndorff-Nielsen and Cox 1994.)

In a manner analogous to the one-parameter case, it can be shown that the information inequality bound is attained only if $\delta(x)$ has the form

$$\delta(x) = g(\theta) + [\nabla g(\theta)]' I(\theta)^{-1} [\nabla \log p_\theta(x)], \tag{6.27}$$

where $E\delta = g(\theta)$, $\nabla g(\theta) = \{(\partial/\partial\theta_i) g(\theta), i = 1, 2, \ldots, s\}$, $\nabla \log p_\theta(x) = \{(\partial/\partial\theta_i) \log p_\theta(x), i = 1, 2, \ldots, s\}$. It is also the case, analogous to Theorem 5.12, that if the bound is attainable then the underlying family of distributions constitutes an exponential family (Joshi 1976, Fabian and Hannan, 1977; Müller-Funk et al. 1989).

The information inequalities (5.31) and (6.22) have been extended in a number of directions, some of which are briefly sketched in the following.

(a) When the lower bound is not sharp, it can usually be improved by considering not only the derivatives ψ_i but also higher derivatives:

$$\psi_{i_1,\ldots,i_s} = \frac{1}{p_\theta(x)} \frac{\partial^{i_1+\cdots+i_s} p_\theta(x)}{\partial\theta_1^{i_1} \cdots \partial\theta_s^{i_s}}. \tag{6.28}$$

It is then easy to generalize (5.31) and (5.24) to obtain a lower bound based on any given set S of the ψ's. Assume (6.6) with (c) replaced by the corresponding assumption for all the derivatives needed for the set S, and suppose that the covariance matrix $K(\theta)$ of the given set of ψ's is positive definite. Then, (6.1) yields the *Bhattacharyya inequality*

$$\text{var}_\theta(\delta) \geq \alpha' K^{-1}(\theta) \alpha \tag{6.29}$$

where α' is the row matrix with elements

$$\frac{\partial^{i_1+\cdots+i_s}}{\partial\theta_1^{i_1} \cdots \partial\theta_s^{i_s}} E_\theta \delta(X) = \text{cov}(\delta, \psi_{i_1,\ldots,i_s}). \tag{6.30}$$

It is also seen that equality holds in (6.29) if and only if δ is a linear function of the ψ's in S (Problem 6.12). The problem of whether the Bhattacharyya bounds become sharp as $s \to \infty$ has been investigated for some one-parameter cases by Blight and Rao (1974).

(b) A different kind of extension avoids the need for regularity conditions by considering differences instead of derivatives. (See Hammersley 1950, Chapman and Robbins 1951, Kiefer 1952, Fraser and Guttman 1952, Fend 1959, Sen and Ghosh 1976, Chatterji 1982, and Klaassen 1984, 1985.)

(c) Applications of the inequality to the sequential case in which the number of observations is not a fixed integer but a random variable, say N, determined from the observations is provided by Wolfowitz (1947), Blackwell and Girshick (1947), and Seth (1949). Under suitable regularity conditions, (6.23) then continues to hold with n replaced by $E_\theta(N)$; see also Simons 1980, Govindarajulu and Vincze 1989, and Stefanov 1990.

(d) Other extensions include arbitrary convex loss functions (Kozek 1976); weighted loss functions (Mikulski and Monsour 1988); to the case that g and δ are vector-valued (Rao 1945, Cramér 1946b, Seth 1949, Shemyakin 1987, and Rao 1992); to nonparametric problems (Vincze 1992); location problems (Klaassen 1984); and density estimation (Brown and Farrell 1990).

7 Problems

Section 1

1.1 Verify (a) that (1.4) defines a probability distribution and (b) condition (1.5).

1.2 In Example 1.5, show that a_i^* minimizes (1.6) for $i = 0, 1$, and simplify the expression for a_0^*. [*Hint*: $\Sigma \kappa p^{\kappa-1}$ and $\Sigma \kappa(\kappa-1)p^{\kappa-2}$ are the first and second derivatives of $\Sigma p^\kappa = 1/q$.]

1.3 Let X take on the values $-1, 0, 1, 2, 3$ with probabilities $P(X = -1) = 2pq$ and $P(X = k) = p^k q^{3-k}$ for $k = 0, 1,2,3$.

(a) Check that this is a probability distribution.

(b) Determine the LMVU estimator at p_0 of (i) p, and (ii) pq, and decide for each whether it is UMVU.

1.4 For a sample of size n, suppose that the estimator $T(\mathbf{x})$ of $\tau(\theta)$ has expectation

$$E[T(\mathbf{X})] = \tau(\theta) + \sum_{k=1}^{\infty} \frac{a_k}{n^k},$$

where a_k may depend on θ but not on n.

(a) Show that the expectation of the jackknife estimator T_J of (1.3) is

$$E[T_J(\mathbf{X})] = \tau(\theta) - \frac{a_2}{n^2} + O(1/n^3).$$

(b) Show that if var $T \sim c/n$ for some constant c, then var $T_J \sim c'/n$ for some constant c'. Thus, the jackknife will reduce bias and not increase variance.

A *second-order jackknife* can be defined by jackknifing T_J, and this will result in further bias reduction, but may not maintain a variance of the same order (Robson and Whitlock 1964; see also Thorburn 1976 and Note 8.3).

1.5 (a) Any two random variables X and Y with finite second moments satisfy the covariance inequality $[\text{cov}(X, Y)]^2 \le \text{var}(X) \cdot \text{var}(Y)$.

(b) The inequality in part (a) is an equality if and only if there exist constants a and b for which $P(X = aY + b) = 1$.

[*Hint*: Part (a) follows from the Schwarz inequality (Problem 1.7.20) with $f = X - E(X)$ and $g = Y - E(Y)$.]

1.6 An alternative proof of the Schwarz inequality is obtained by noting that

$$\int (f + \lambda g)^2 dP = \int f^2 dP + 2\lambda \int fg\, dP + \lambda^2 \int g^2 dP \ge 0 \quad \text{for all } \lambda,$$

so that this quadratic in λ has at most one root.

1.7 Suppose X is distributed on (0, 1) with probability density $p_\theta(x) = (1 - \theta) + \theta/2\sqrt{x}$ for all $0 < x < 1, 0 \le \theta \le 1$. Show that there does not exist an LMVU estimator of θ. [*Hint*: Let $\delta(x) = a[x^{-1/2} + b]$ for $c < x < 1$ and $\delta(x) = 0$ for $0 < x < c$. There exist values a and b, and c such that $E_0(\delta) = 0$ and $E_1(\delta) = 1$ (and δ is unbiased) and that $E_0(\delta^2)$ is arbitrarily close to zero (Stein 1950).]

1.8 If δ and δ' have finite variance, so does $\delta' - \delta$. [*Hint*: Problem 1.5.]

1.9 In Example 1.9, (a) determine all unbiased estimators of zero; (b) show that no nonconstant estimator is UMVU.

1.10 If estimators are restricted to the class of linear estimators, characterization of best unbiased estimators is somewhat easier. Although the following is a consequence of Theorem 1.7, it should be established without using that theorem.

Let $\mathbf{X}_{p\times 1}$ satisfy $E(\mathbf{X}) = B\psi$ and $\text{var}(\mathbf{X}) = I$, where $B_{p\times r}$ is known, and $\psi_{r\times 1}$ is unknown. A *linear estimator* is an estimator of the form $a'\mathbf{X}$, where $a_{p\times 1}$ is a known vector. We are concerned with the class of estimators

$$\mathcal{D} = \{\delta(\mathbf{x}) : \delta(\mathbf{x}) = a'\mathbf{x}, \text{ for some known vector } a\}.$$

(a) For a known vector c, show that the estimators in $\mathcal{D}$ that are unbiased estimators of $c'\psi$ satisfy $a'B = c'$.

(b) Let $\mathcal{D}_c = \{\delta(\mathbf{x}) : \delta(\mathbf{x}) = a'\mathbf{x}, a'B = c'\}$ be the class of linear unbiased estimators of $c'\psi$. Show that the *best linear unbiased estimator* (BLUE) of $c'\psi$, the linear unbiased estimator with minimum variance, is $\delta^*(\mathbf{x}) = a^{*\prime}\mathbf{x}$, where $a^{*\prime} = a'B(B'B)^{-1}B'$ and $a^{*\prime}B = c$ with variance $\text{var}(\delta^*) = c'c$.

(c) Let $\mathcal{D}_0 = \{\delta(\mathbf{x}) : \delta(\mathbf{x}) = a'\mathbf{x}, a'B = 0.\}$ be the class of linear unbiased estimators of zero. Show that if $\delta \in \mathcal{D}_0$, then $\text{cov}(\delta, \delta^*) = 0$.

(d) Hence, establish the analog of Theorem 1.7 for linear estimators:

Theorem. *An estimator* $\delta^* \in \mathcal{D}_c$ *satisfies* $\text{var}(\delta^*) = \min_{\delta \in \mathcal{D}_c} \text{var}(\delta)$ *if and only if* $\text{cov}(\delta^*, U) = 0$, *where* U *is any estimator in* $\mathcal{D}_0$.

(e) Show that the results here can be directly extended to the case of $\text{var}(\mathbf{X}) = \Sigma$, where $\Sigma_{p\times p}$ is a known matrix, by considering the transformed problem with $\mathbf{X}^* = \Sigma^{1/2}\mathbf{X}$ and $B^* = \Sigma^{1/2}B$.

1.11 Use Theorem 1.7 to find UMVU estimators of some of the $\eta_\theta(d_i)$ in the dose-response model (1.6.16), with the restriction (1.6.17) (Messig and Strawderman 1993). Let the classes Δ and $\mathcal{U}$ be defined as in Theorem 1.7.

(a) Show that an estimator $U \in \mathcal{U}$ if and only if $U(x_1, x_2) = a[I(x_1 = 0) - I(x_2 = 0)]$ for an arbitrary constant $a < \infty$.

(b) Using part (a) and (1.7), show that an estimator δ is UMVU for its expectation only if it is of the form $\delta(x_1, x_2) = aI_{(0,0)}(x_1, x_2) + bI_{(0,1),(1,0),(2,0)}(x_1, x_2) + cI_{(1,1)}(x_1, x_2) + dI_{(2,1)}(x_1, x_2)$ where a, b, c, and d are arbitrary constants.

(c) Show that there does not exist a UMVU estimator of $\eta_\theta(d_1) = 1 - e^{-\theta}$, but the UMVU estimator of $\eta_\theta(d_2) = 1 - e^{-2\theta}$ is $\delta(x_1, x_2) = 1 - \frac{1}{2}[I(x_1 = 0) + I(x_2 = 0)]$.

(d) Show that the LMVU estimator of $1 - e^{-\theta}$ is $\delta(x_1, x_2) = \frac{x_1}{2} + \frac{1}{2(1+e^{-\theta})}[I(x_1 = 0) - I(x_2 = 0)]$.

1.12 Show that if $\delta(X)$ is a UMVU estimator of $g(\theta)$, it is the unique UMVU estimator of $g(\theta)$. (Do not assume completeness, but rather use the covariance inequality and the conditions under which it is an equality.)

1.13 If δ_1 and δ_2 are in Δ and are UMVU estimators of $g_1(\theta)$ and $g_2(\theta)$, respectively, then $a_1\delta_1 + a_2\delta_2$ is also in Δ and is UMVU for estimating $a_1g_1(\theta) + a_2g_2(\theta)$, for any real a_1 and a_2.

1.14 Completeness of T is not only sufficient but also necessary so that every $g(\theta)$ that can be estimated unbiasedly has only one unbiased estimator that is a function of T.

1.15 Suppose $X_1, \ldots, X_n$ are iid Poisson (λ).

(a) Show that $\bar{X}$ is the UMVU estimator for λ.

(b) For $S^2 = \sum_{i=1}^{n}(X_i - \bar{X})^2/(n-1)$, we have that $ES^2 = E\bar{X} = \lambda$. To directly establish that var $S^2 >$ var $\bar{X}$, prove that $E(S^2|\bar{X}) = \bar{X}$.

Note: The identity $E(S^2|\bar{X}) = \bar{X}$ shows how completeness can be used in calculating conditional expectations.

1.16 (a) If $X_1, \ldots, X_n$ are iid (not necessarily normal) with var$(X_i) = \sigma^2 < \infty$, show that $\delta = \Sigma(X_i - \bar{X})^2/(n-1)$ is an unbiased estimator of σ^2.

(b) If the X_i take on the values 1 and 0 with probabilities p and $q = 1 - p$, the estimator δ of (a) depends only on $T = \Sigma X_i$ and hence is UMVU for estimating $\sigma^2 = pq$. Compare this result with that of Example 1.13.

1.17 If T has the binomial distribution $b(p, n)$ with $n > 3$, use Method 1 to find the UMVU estimator of p^3.

1.18 Let $X_1, \ldots, X_n$ be iid according to the Poisson distribution $P(\lambda)$. Use Method 1 to find the UMVU estimator of (a) λ^k for any positive integer k and (b) $e^{-\lambda}$.

1.19 Let $X_1, \ldots, X_n$ be distributed as in Example 1.14. Use Method 1 to find the UMVU estimator of θ^k for any integer $k > -n$.

1.20 Solve Problem 1.18(b) by Method 2, using the fact that an unbiased estimator of $e^{-\lambda}$ is $\delta = 1$ if $X_1 = 0$, and $\delta = 0$ otherwise.

1.21 In n Bernoulli trials, let $X_i = 1$ or 0 as the ith trial is a success or failure, and let $T = \Sigma X_i$. Solve Problem 1.17 by Method 2, using the fact that an unbiased estimator of p^3 is $\delta = 1$ if $X_1 = X_2 = X_3 = 1$, and $\delta = 0$ otherwise.

1.22 Let X take on the values 1 and 0 with probability p and q, respectively, and assume that $1/4 < p < 3/4$. Consider the problem of estimating p with loss function $L(p, d) = 1$ if $|d - p| \geq 1/4$, and 0 otherwise. Let δ^* be the randomized estimator which is Y_0 or Y_1 when $X = 0$ or 1 where Y_0 and Y_1 are distributed as $U(-1/2, 1/2)$ and $U(1/2, 3/2)$, respectively.

(a) Show that δ^* is unbiased.

(b) Compare the risk function of δ^* with that of X.

Section 2

2.1 If $X_1, \ldots, X_n$ are iid as $N(\xi, \sigma^2)$ with σ^2 known, find the UMVU estimator of (a) ξ^2, (b) ξ^3, and (c) ξ^4. [*Hint*: To evaluate the expectation of $\bar{X}^k$, write $\bar{X} = Y + \xi$, where Y is $N(0, \sigma^2/n)$ and expand $E(Y + \xi)^k$.]

2.2 Solve the preceding problem when σ is unknown.

2.3 In Example 2.1 with σ known, let $\delta = \Sigma c_i X_i$ be any linear estimator of ξ. If δ is biased, show that its risk $E(\delta - \xi)^2$ is unbounded. [*Hint*: If $\Sigma c_i = 1 + k$, the risk is $\geq k^2\xi^2$.]

2.4 Suppose, as in Example 2.1, that $X_1, \ldots, X_n$ are iid as $N(\xi, \sigma^2)$, with one of the parameters known, and that the estimand is a polynomial in ξ or σ. Then, the UMVU estimator is a polynomial in $\bar{X}$ or $S^2 = \sum(X_i - \xi)^2$. The variance of any such polynomial can be estimated if one knows the moments $E(\bar{X}^k)$ and $E(S^k)$ for all $k = 1, 2, \ldots$. To determine $E(\bar{X}^k)$, write $\bar{X} = Y + \xi$, where Y is distributed as $N(0, \sigma^2/n)$. Show that

(a)

$$E(\bar{X}^k) = \sum_{r=0}^{k} \binom{k}{r} \xi^{k-r} E(Y^r)$$

with

$$E(Y^r) = \begin{cases} (r-1)(r-3)\cdots 3 \cdot 1(\sigma^2/n)^{r/2} & \text{when } r \geq 2 \text{ is even} \\ 0 & \text{when } r \text{ is odd}. \end{cases}$$

(b) As an example, consider the UMVU estimator S^2/n of σ^2. Show that $E(S^4) = n(n+2)\sigma^2$ and $\text{var}\left(\frac{S^2}{n}\right) = \frac{2\sigma^4}{n}$ and that the UMVU estimator of this variance is $2S^4/n^2(n+2)$.

2.5 In Example 2.1, when both parameters are unknown, show that the UMVU estimator of ξ^2 is given by $\delta = \bar{X}^2 - \frac{S^2}{n(n-1)}$ where now $S^2 = \sum(X_i - \bar{X})^2$.

2.6 (a) Determine the variance of the estimator Problem 2.5.

(b) Find the UMVU estimator of the variance in part (a).

2.7 If X is a single observation from $N(\xi, \sigma^2)$, show that no unbiased estimator δ of σ^2 exists when ξ is unknown. [*Hint*: For fixed $\sigma = a$, X is a complete sufficient statistic for ξ, and $E[\delta(X)] = a^2$ for all ξ implies $\delta(x) = a^2$ a.e.]

2.8 Let X_i, $i = 1, \ldots, n$, be independently distributed as $N(\alpha + \beta t_i, \sigma^2)$ where α, β, and σ^2 are unknown, and the t's are known constants that are not all equal. Find the UMVU estimators of α and β.

2.9 In Example 2.2 with $n = 1$, the UMVU estimator of p is the indicator of the event $X_1 \leq u$ whether σ is known or unknown.

2.10 Verify Equation (2.14), the density of $(X_1 - \bar{X})/S$ in normal sampling. [The UMVU estimator in (2.13) is used by Kiefer (1977) as an example of his estimated confidence approach.]

2.11 Assuming (2.15) with $\sigma = \tau$, determine the UMVU estimators of σ^2 and $(\eta - \xi)/\sigma$.

2.12 Assuming (2.15) with $\eta = \xi$ and $\sigma^2/\tau^2 = \gamma$, show that when γ is known:

(a) T' defined in Example 2.3(iii) is a complete sufficient statistic;

(b) δ_γ is UMVU for ξ.

2.13 Show that in the preceding problem with γ unknown,

(a) a UMVU estimator of ξ does not exist;

(b) the estimator $\hat{\xi}$ is unbiased under the conditions stated in Example 2.3. [*Hint*: (i) Problem 2.12(b) and the fact that δ_γ is unbiased for ξ even when $\sigma^2/\tau^2 \neq \gamma$. (ii) Condition on (S_X, S_Y).]

2.14 For the model (2.15) find the UMVU estimator of $P(X_1 < Y_1)$ when (a) $\sigma = \tau$ and (b) when σ and τ are arbitrary. [*Hint*: Use the conditional density (2.13) of X_1 given $\bar{X}$, S_X^2 and that of Y_1 given $\bar{Y}$, S_Y^2 to determine the conditional density of $Y_1 - X_1$ given $\bar{X}, \bar{Y}, S_X^2$, and S_Y^2.]

2.15 If $(X_1, Y_1), \ldots, (X_n, Y_n)$ are iid according to any bivariate distribution with finite second moments, show that $S_{XY}/(n-1)$ given by (2.17) is an unbiased estimator of $\text{cov}(X_i, Y_i)$.

2.16 In a sample size $N = n + k + 1$, some of the observations are missing. Assume that (X_i, Y_i), $i = 1, \ldots, n$, are iid according to the bivariate normal distribution (2.16), and that $U_1, \ldots, U_k$ and $V_1, \ldots, V_l$ are independent $N(\xi, \sigma^2)$ and $N(\eta, \tau^2)$, respectively.

(a) Show that the minimal sufficient statistics are complete when ξ and η are known but not when they are unknown.

(b) When ξ and η are known, find the UMVU estimators for σ^2, τ^2, and $\rho\sigma\tau$, and suggest reasonable unbiased estimators for these parameters when ξ and η are unknown.

2.17 For the family (2.22), show that the UMVU estimator of a when b is known and the UMVU estimator of b is known are as stated in Example 2.5. [*Hint*: Problem 6.18.]

2.18 Show that the estimators (2.23) are UMVU. [*Hint*: Problem 1.6.18.].

2.19 For the family (2.22) with $b = 1$, find the UMVU estimator of $P(X_1 \geq u)$ and of the density $e^{-(u-a)}$ of X_1 at u. [*Hint*: Obtain the estimator $\delta(X_{(1)})$ of the density by applying Method 2 of Section 2.1 and then the estimator of the probability by integration. Alternatively, one can first obtain the estimator of the probability as $P(X_1 \geq u | X_{(1)})$ using the fact that $X_1 - X_{(1)}$ is ancillary and that given $X_{(1)}$, X_1 is either equal to $X_{(1)}$ or distributed as $E(X_{(1)}, 1)$.]

2.20 Find the UMVU estimator of $P(X_1 \geq u)$ for the family (2.22) when both a and b are unknown.

2.21 Let $X_1, \ldots, X_m$ and $Y_1, \ldots, Y_n$ be independently distributed as $E(a, b)$ and $E(a', b')$, respectively.

(a) If a, b, a', and b' are completely unknown, $X_{(1)}$, $Y_{(1)}$, $\Sigma[X_i - X_{(1)}]$, and $\Sigma[Y_j - Y_{(1)}]$ jointly are sufficient and complete.

(b) Find the UMVU estimators of $a' - a$ and b'/b.

2.22 In the preceding problem, suppose that $b' = b$.

(a) Show that $X_{(1)}$, $Y_{(1)}$, and $\Sigma[X_i - X_{(1)}] + \Sigma[Y_j - Y_{(1)}]$ are sufficient and complete.

(b) Find the UMVU estimators of b and $(a' - a)/b$.

2.23 In Problem 2.21, suppose that $a' = a$.

(a) Show that the complete sufficient statistic of Problem 2.21(a) is still minimal sufficient but no longer complete.

(b) Show that a UMVU estimator for $a' = a$ does not exist.

(c) Suggest a reasonable unbiased estimator for $a' = a$.

2.24 Let $X_1, \ldots, X_n$ be iid according to the uniform distribution $U(\xi - b, \xi + b)$. If ξ, b are both unknown, find the UMVU estimators of ξ, b, and ξ/b. [*Hint*: Problem 1.6.30.]

2.25 Let $X_1, \ldots, X_m$ and $Y_1, \ldots, Y_n$ be iid as $U(0, \theta)$ and $U(0, \theta')$, respectively. If $n > 1$, determine the UMVU estimator of θ/θ'.

2.26 Verify the ML estimators given in (2.24).

2.27 In Example 2.6(b), show that

(a) The bias of the ML estimator is 0 when $\xi = u$.

(b) At $\xi = u$, the ML estimator has smaller expected squared error than the UMVU estimator.

[Hint: In (b), note that $u - \bar{X}$ is always closer to 0 than $\sqrt{\frac{n}{n-1}}(u - \bar{X})$.]

2.28 Verify (2.26).

2.29 Under the assumptions of Lemma 2.7, show that:

(a) If b is replaced by any random variable B which is independent of X and not 0 with probability 1, then $R_\delta(\theta) < R_{\delta^*}(\theta)$.

(b) If squared error is replaced by any loss function of the form $L(\theta, \delta) = \rho(d - \theta)$ and δ is risk unbiased with respect to L, then $R_\delta(\theta) < R_{\delta^*}(\theta)$.

Section 3

3.1 (a) In Example 3.1, show that $\Sigma(X_i - \bar{X})^2 = T(n - T)/n$.

(b) The variance of $T(n-T)/n(n-1)$ in Example 3.1 is $(pq/n)[(q-p)^2 + 2pq/(n-1)]$.

3.2 If T is distributed as $b(p, n)$, find an unbiased estimator $\delta(T)$ of p^m $(m \leq n)$ by Method 1, that is, using (1.10). [*Hint*: Example 1.13.]

3.3 (a) Use the method leading to (3.2) to find the UMVU estimator $\pi_k(T)$ of $P[X_1 + \cdots + X_m = k] = \binom{m}{k} p^k q^{m-k}$ $(m \leq n)$.

(b) For fixed t and varying k, show that the $\pi_k(t)$ are the probabilities of a hypergeometric distribution.

3.4 If Y is distributed according to (3.3), use Method 1 of Section 2.1

(a) to show that the UMVU estimator of p^r $(r < m)$ is

$$\delta(y) = \frac{(m - r + y - 1)(m - r + y - 2)\ldots(m - r)}{(m + y - 1)(m + y - 2)\cdots m},$$

and hence in particular that the UMVU estimator of $1/p$, $1/p^2$ and p are, respectively, $(m + y)/m$, $(m + y)(m + y + 1)/m(m + 1)$, and $(m - 1)/(m + y - 1)$;

(b) to determine the UMVU estimator of $\text{var}(Y)$;

(c) to show how to calculate the UMVU estimator δ of $\log p$.

3.5 Consider the scheme in which binomial sampling is continued until at least a successes *and* b failures have been obtained. Show how to calculate a reasonable estimator of $\log(p/q)$. [*Hint*: To obtain an unbiased estimator of $\log p$, modify the UMVU estimator δ of Problem 3.4(c).]

3.6 If binomial sampling is continued until m successes have been obtained, let X_i $(i = 1, \ldots, m)$ be the number of failures between the $(i - 1)$st and ith success.

(a) The X_i are iid according to the *geometric distribution* $P(X_i = x) = pq^x$, $x = 0, 1, \ldots$.

(b) The statistic $Y = \Sigma X_i$ is sufficient for $(X_1, \ldots, X_m)$ and has the distribution (3.3).

3.7 Suppose that binomial sampling is continued until the number of successes equals the number of failures.

(a) This rule is closed if $p = 1/2$ but not otherwise.

(b) If $p = 1/2$ and N denotes the number of trials required, $E(N) = \infty$.

3.8 Verify Equation (3.7) with the appropriate definition of $N'(x, y)$ (a) for the estimation of p and (b) for the estimation of $p^a q^b$.

3.9 Consider sequential binomial sampling with the stopping points $(0, 1)$ and $(2, y)$, $y = 0, 1, \ldots$. (a) Show that this plan is closed and simple. (b) Show that (X, Y) is not complete by finding a nontrivial unbiased estimator of zero.

3.10 In Example 3.4(ii), (a) show that the plan is closed but not simple, (b) show that (X, Y) is not complete, and (c) evaluate the unbiased estimator (3.7) of p.

3.11 *Curtailed single sampling.* Let $a, b < n$ be three non-negative integers. Continue observation until either a successes, b failures, or n observations have been obtained. Determine the UMVU estimator of p.

3.12 For any sequential binomial sampling plan, the coordinates (X, Y) of the end point of the sample path are minimal sufficient.

3.13 Consider any closed sequential binomial sampling plan with a set B of stopping points, and let B' be the set $B \cup \{(x_0, y_0)\}$ where (x_0, y_0) is a point not in B that has positive probability of being reached under plan B. Show that the sufficient statistic $T = (X, Y)$ is not complete for the sampling plan which has B' as its set of stopping points. [*Hint*: For any point $(x, y) \in B$, let $N(x, y)$ and $N'(x, y)$ denote the number of paths to (x, y) when the set of stopping points is B and B', respectively, and let $N(x_0, y_0) = 0$, $N'(x_0, y_0) = 1$. Then, the statistic $1 - [N(X, Y)/N'(X, Y)]$ has expectation 0 under B' for all values of p.]

3.14 For any sequential binomial sampling plan under which the point $(1, 1)$ is reached with positive probability but is not a stopping point, find an unbiased estimator of pq depending only on (X, Y). Evaluate this estimator for

(a) taking a sample of fixed size $n > 2$;

(b) inverse binomial sampling.

3.15 Use (3.3) to determine $A(t, n)$ in (3.11) for the negative binomial distribution with $m = n$, and evaluate the estimators (3.13) of q^r, and (3.14).

3.16 Consider n binomial trials with success probability p, and let r and s be two positive integers with $r + s < n$. To the boundary $x + y = n$, add the boundary point (r, s), that is, if the number of successes in the first $r + s$ trials is exactly r, the process is stopped and the remaining $n - (r + s)$ trials are not performed.

(a) Show that U is an unbiased estimator of zero if and only if $U(k, n - k) = 0$ for $k = 0, 1, \ldots, r - 1$ and $k = n - s + 1, n - s + 2, \ldots, n$, and $U(k, n - k) = c_k U(r, s)$ for $k = r, \ldots, n - s$, where the c's are given constants $\neq 0$.

(b) Show that δ is the UMVU estimator of its expectation if and only if

$$\delta(k, n - k) = \delta(r, s) \quad \text{for } k = r, \ldots, n - s.$$

3.17 Generalize the preceding problem to the case that two points (r_1, s_1) and (r_2, s_2) with $r_i + s_i < n$ are added to the boundary. Assume that these two points are such that all $n + 1$ points $x + y = n$ remain boundary points. [*Hint*: Distinguish the three cases that the intervals (r_1, s_1) and (r_2, s_2) are (i) mutually exclusive, (ii) one contained in the other, and (iii) overlapping but neither contained in the other.]

3.18 If X has the Poisson distribution $P(\theta)$, show that $1/\theta$ does not have an unbiased estimator.

3.19 If $X_1, \ldots, X_n$ are iid according to (3.18), the Poisson distribution truncated on the left at 0, find the UMVU estimator of θ when (a) $n = 1$ and (b) $n = 2$.

3.20 Let $X_1, \ldots, X_n$ be a sample from the Poisson distribution truncated on the left at 0, i.e., with sample space $\mathcal{X} = \{1, 2, 3, \ldots\}$.

(a) For $t = \Sigma x_i$, the UMVU estimator of λ is (Tate and Goen 1958) $\hat{\lambda} = \frac{C^n_{t-1}}{C^n_t} t$ where $C^n_t = \frac{(-1)^n}{n!} \sum_{k=0}^{\infty} \binom{n}{k} (-1)^k k^t$ is a *Stirling number of the second kind.*

(b) An alternate form of the UMVU estimator is $\hat{\lambda} = \frac{t}{n}\left(1 - \frac{C^{n-1}_{t-1}}{C^n_t}\right)$. [Hint: Establish the identity $C^n_t = C^{n-1}_{t-1} + nC^n_{t-1}$.]

(c) The Cramér-Rao lower bound for the variance of unbiased estimators of λ is $\lambda(1 - e^{-\lambda})^2/[n(1 - e^{-\lambda} - \lambda e^{-\lambda})]$, and it is not attained by the UMVU estimator. (It is, however, the asymptotic variance of the ML estimator.)

3.21 Suppose that X has the Poisson distribution truncated on the right at a, so that it has the conditional distribution of Y given $Y \le a$, where Y is distributed as $P(\lambda)$. Show that λ does not have an unbiased estimator.

3.22 For the negative binomial distribution truncated at zero, evaluate the estimators (3.13) and (3.14) for $m = 1, 2$, and 3.

3.23 If $X_1, \ldots, X_n$ are iid $P(\lambda)$, consider estimation of $e^{-b\lambda}$, where b is known.

(a) Show that $\delta^* = (1 - b/n)^t$ is the UMVU estimator of $e^{-b\lambda}$.

(b) For $b > n$, describe the behavior of δ^*, and suggest why it might not be a reasonable estimator.

(The probability $e^{-b\lambda}$, for $b > n$, is that of an "unobservable" event, in that it can be interpreted as the probability of no occurrence in a time interval of length b. A number of such situations are described and analyzed in Lehmann (1983), where it is suggested that, in these problems, no reasonable estimator may exist.)

3.24 If $X_1, \ldots, X_n$ are iid according to the logarithmic series distribution of Problem 1.5.14, evaluate the estimators (3.13) and (3.14) for $n = 1, 2$, and 3.

3.25 For the multinomial distribution of Example 3.8,

(a) show that $p_0^{r_0} \cdots p_s^{r_s}$ has an unbiased estimator provided $r_0, \ldots, r_s$ are nonnegative integers with $\Sigma r_i \le n$;

(b) find the totality of functions that can be estimated unbiasedly;

(c) determine the UMVU estimator of the estimand of (a).

3.26 In Example 3.9 when $p_{ij} = p_{i+}p_{+j}$, determine the variances of the two unbiased estimators $\delta_0 = n_{ij}/n$ and $\delta_1 = n_{i+}n_{+j}/n^2$ of p_{ij}, and show directly that $\text{var}(\delta_0) > \text{var}(\delta_1)$ for all $n > 1$.

3.27 In Example 3.9, show that independence of A and B implies that $(n_{1+}, \ldots, n_{I+})$ and $(n_{+1}, \ldots, n_{+J})$ are independent with multinomial distributions as stated.

3.28 Verify (3.20).

3.29 Let X, Y, and g be such that $E[g(X, Y)|y]$ is independent of y. Then, $E[f(Y)g(X, Y)] = E[f(Y)]E[g(X, Y)]$, and hence $f(Y)$ and $g(X, Y)$ are uncorrelated, for all f.

3.30 In Example 3.10, show that the estimator δ_1 of p_{ijk} is unbiased for the model (3.20). [*Hint*: Problem 3.29.]

Section 4

4.1 Let $X_1, \ldots, X_n$ be iid with distribution F.

(a) Characterize the totality of functions $f(X_1, \ldots, X_n)$ which are unbiased estimators of zero for the class $\mathcal{F}_0$ of all distributions F having a density.

(b) Give one example of a nontrivial unbiased estimator of zero when (i) $n = 2$ and (ii) $n = 3$.

4.2 Let $\mathcal{F}$ be the class of all univariate distribution functions F that have a probability density function f and finite mth moment.

(a) Let $X_1, \ldots, X_n$ be independently distributed with common distribution $F \in \mathcal{F}$. For $n \geq m$, find the UMVU estimator of ξ^m where $\xi = \xi(F) = EX_i$.

(b) Show that for the case that $P(X_i = 1) = p$, $P(X_i = 0) = q$, $p + q = 1$, the estimator of (a) reduces to (3.2).

4.3 In the preceding problem, show that $1/\text{var}_F X_i$ does not have an unbiased estimator for any n.

4.4 Let $X_1, \ldots, X_n$ be iid with distribution $F \in \mathcal{F}$ where $\mathcal{F}$ is the class of all symmetric distributions with a probability density. There exists no UMVU estimator of the center of symmetry θ of F (if unbiasedness is required only for the distributions F for which the expectation of the estimator exists). [*Hint*: The UMVU estimator of θ when F is $U(\theta - 1/2, \theta + 1/2)$, which was obtained in Problem 2.24, is unbiased for all $F \in \mathcal{F}$; so is $\bar{X}$.]

4.5 If $X_1, \ldots, X_m$ and $Y_1, \ldots, Y_n$ are independently distributed according to F and $G \in \mathcal{F}_0$, defined in Problem 4.1, the order statistics $X_{(1)} < \cdots < X_{(m)}$ and $Y_{(1)} < \cdots < Y_{(n)}$ are sufficient and complete. [*Hint*: For completeness, generalize the second proof suggested in Problem 6.33.]

4.6 Under the assumptions of the preceding problem, find the UMVU estimator of $P(X_i < Y_j)$.

4.7 Under the assumptions of Problem 4.5, let $\xi = EX_i$ and $\eta = EY_j$. Show that $\xi^2\eta^2$ possesses an unbiased estimator if and only if $m \geq 2$ and $n \geq 2$.

4.8 Let $(X_1, Y_1), \ldots, (X_n, Y_n)$ be iid $F \in \mathcal{F}$, where $\mathcal{F}$ is the family of all distributions with probability density and finite second moments. Show that $\delta(X, Y) = \sum(X_i - \bar{X})(Y_i - \bar{Y})/(n-1)$ is UMVU for $\text{cov}(X, Y)$.

4.9 Under the assumptions of the preceding problem, find the UMVU estimator of

(a) $P(X_i \leq Y_i)$;

(b) $P(X_i \leq X_j \text{ and } Y_i \leq Y_j), i \neq j$.

4.10 Let $(X_1, Y_1), \ldots, (X_n, Y_n)$ be iid with $F \in \mathcal{F}$, where $\mathcal{F}$ is the family of all bivariate densities. Show that the sufficient statistic T, which generalizes the order statistics to the bivariate case, is complete. [*Hint*: Generalize the second proof suggested in Problem 6.33. As an exponential family for (X, Y), take the densities proportional to $e^{Q(x,y)}$ where

$$Q(x, y) = (\theta_{01}x + \theta_{10}y) + (\theta_{02}x^2 + \theta_{11}xy + \theta_{20}y^2) + \cdots$$
$$+(\theta_{0n}x^n + \cdots + \theta_{n0}y^n) - x^{2n} - y^{2n}.]$$

Section 5

5.1 Under the assumptions of Problem 1.3, determine for each p_1, the value $L_V(p_1)$ of the LMVU estimator of p at p_1 and compare the function $L_V(p)$, $0 < p < 1$ with the variance $V_{p_0}(p)$ of the estimator which is LMVU at (a) $p_0 = 1/3$ and (b) $p_0 = 1/2$.

5.2 Determine the conditions under which equality holds in (5.1).

5.3 Verify $I(\theta)$ for the distributions of Table 5.1.

5.4 If X is normal with mean zero and standard deviation σ, determine $I(\sigma)$.

5.5 Find $I(p)$ for the negative binomial distribution.

5.6 If X is distributed as $P(\lambda)$, show that the information it contains about $\sqrt{\lambda}$ is independent of λ.

5.7 Verify the following statements, asserted by Basu (1988, Chapter 1), which illustrate the relationship between information, sufficiency, and ancillarity. Suppose that we let $I(\theta) = E_\theta\left[-\partial^2/\partial\theta^2 \log f(x|(\theta)\right]$ be the information in X about θ and let $J(\theta) = E_\theta\left[-\partial^2/\partial\theta^2 \log g(T|\theta)\right]$ be the information about θ contained in a statistic T, where $g(\cdot|\theta)$ is the density function of T. Define $\lambda(\theta) = I(\theta) - J(\theta)$, a measure of information lost by using T instead of X. Under suitable regularity conditions, show that

(a) $\lambda(\theta) \geq 0$ for all θ

(b) $\lambda(\theta) = 0$ if and only if T is sufficient for θ.

(c) If Y is ancillary but (T, Y) is sufficient, then $I(\theta) = E_\theta[J(\theta|Y)]$, where

$$J(\theta|y) = E_\theta\left[-\frac{\partial^2}{\partial\theta^2}\log h(T|y,\theta)|Y = y\right]$$

and $h(t|y, \theta)$ is the conditional density of T given $Y = y$.

(Basu's "regularity conditions" are mainly concerned with interchange of integration and differentiation. Assume any such interchanges are valid.)

5.8 Find a function of θ for which the amount of information is independent of θ:

(a) for the gamma distribution $\Gamma(\alpha, \beta)$ with α known and with $\theta = \beta$;

(b) for the binomial distribution $b(p, n)$ with $\theta = p$.

5.9 For inverse binomial sampling (see Example 3.2):

(a) Show that the best unbiased estimator of p is given by $\delta^*(Y) = (m-1)/(Y+m-1)$.

(b) Show that the information contained in Y about P is $I(p) = m/p^2(1-p)$.

(c) Show that $\text{var}\delta^* > 1/I(p)$.

(The estimator δ^* can be interpreted as the success rate if we ignore the last trial, which we know must be a success.)

5.10 Show that (5.13) can be differentiated by differentiating under the integral sign when $p_\theta(x)$ is given by (5.24), for each of the distributions of Table 5.2. [*Hint*: Form the difference quotient and apply the dominated convergence theorem.]

5.11 Verify the entries of Table 5.2.

5.12 Evaluate (5.25) when f is the density of Student's t-distribution with ν degrees of freedom. [*Hint*: Use the fact that

$$\int_{-\infty}^{\infty} \frac{dx}{(1+x^2)^k} = \frac{\Gamma(1/2)\Gamma(k-1/2)}{\Gamma(k)}.\Big]$$

5.13 For the distribution with density (5.24), show that $I(\theta)$ is independent of θ.

5.14 Verify (a) formula (5.25) and (b) formula (5.27).

5.15 For the location t density, calculate the information inequality bound for unbiased estimators of θ.

5.16 (a) For the scale family with density $(1/\theta)f(x/\theta), \theta > 0$, the amount of information a single observation X has about θ is

$$\frac{1}{\theta^2}\int\left[\frac{yf'(y)}{f(y)}+1\right]^2 f(y)\,dy.$$

(b) Show that the information X contains about $\xi = \log\theta$ is independent of θ.

(c) For the Cauchy distribution $C(0, \theta)$, $I(\theta) = 1/(2\theta^2)$.

5.17 If $p_\theta(x)$ is given by 1.5.1 with $s = 1$ and $T(x) = \delta(x)$, show that var$[\delta(X)]$ attains the lower bound (5.31) and is the only estimator to do so. [*Hint*: Use (5.18) and (1.5.15).]

5.18 Show that if a given function $g(\theta)$ has an unbiased estimator, there exists an unbiased estimator δ which for all θ values attains the lower bound (5.1) for some $\psi(x, \theta)$ satisfying (5.2) if and only if $g(\theta)$ has a UMVU estimator δ_0. [*Hint*: By Theorem 5.1, $\psi(x, \theta) = \delta_0(x)$ satisfies (5.2). For any other unbiased δ, $\text{cov}(\delta - \delta_0, \delta_0) = 0$ and hence $\text{var}(\delta_0) = [\text{cov}(\delta, \delta_0)]^2/\text{var}(\delta_0)$, so that $\psi = \delta_0$ provides an attainable bound.] (Blyth 1974).

5.19 Show that if $E_\theta\delta = g(\theta)$, and var$(\delta)$ attains the information inequality bound (5.31), then

$$\delta(x) = g(\theta) + \frac{g'(\theta)}{I(\theta)}\frac{\partial}{\partial\theta}p_\theta(x).$$

5.20 If $E_\theta\delta = g(\theta)$, the information inequality lower bound is $IB(\theta) = [g'(\theta)]^2/I(\theta)$. If $\theta = h(\xi)$ where h is differentiable, show that $IB(\xi) = IB(\theta)$.

5.21 (Liu and Brown 1993) Let X be an observation from the normal mixture density

$$p_\theta(x) = \frac{1}{2\sqrt{2\pi}}\left\{e^{-(1/2)(x-\theta)^2} + e^{-(1/2)(x+\theta)^2}\right\}, \quad \theta \in \Omega,$$

where Ω is any neighborhood of zero. Thus, the random variable X is either $N(\theta, 1)$ or $N(-\theta, 1)$, each with probability 1/2. Show that $\theta = 0$ is a *singular point*, that is, if there exists an unbiased estimator of θ it will have infinite variance at $\theta = 0$.

5.22 Let $X_1, \ldots, X_n$ be a sample from the Poisson (λ) distribution truncated on the left at 0, i.e., with sample space $\mathcal{X} = \{1, 2, 3, \ldots\}$ (see Problem 3.20). Show that the Cramér-Rao lower bound for the variance of unbiased estimators of λ is

$$\frac{\lambda(1-e^{-\lambda})^2}{n(1-e^{-\lambda}-\lambda e^{-\lambda})}$$

and is not attained by the UMVU estimator. (It is, however, the asymptotic variance of the ML estimator.)

5.23 Let $X_1, \ldots, X_n$ be iid according to a density $p(x, \theta)$ which is positive for all x. Then, the variance of any unbiased estimator δ of θ satisfies

$$\text{var}_{\theta_0}(\delta) \geq \frac{(\theta-\theta_0)^2}{\left\{\int_{-\infty}^{\infty}\frac{[p(x,\theta)]^2}{p(x,\theta_0)}\right\}^n - 1}, \quad \theta \neq \theta_0.$$

[*Hint*: Direct consequence of (5.6).]

5.24 If $X_1, \ldots, X_n$ are iid as $N(\theta, \sigma^2)$ where σ is known and θ is known to have one of the values $0, \pm 1, \pm 2, \ldots$, the inequality of the preceding problem shows that any unbiased estimator δ of the restricted parameter θ satisfies

$$\text{var}_{\theta_0}(\delta) \geq \frac{\Delta^2}{e^{n\Delta^2/\sigma^2} - 1}, \quad \Delta \neq 0,$$

where $\Delta = \theta - \theta_0$, and hence $\sup_{\Delta \neq 0} \text{var}_{\theta_0}(\delta) \geq 1/[e^{n/\sigma^2} - 1]$.

5.25 Under the assumptions of the preceding problem, let $\bar{X}^*$ be the integer closest to $\bar{X}$.

(a) The estimator $\bar{X}^*$ is unbiased for the restricted parameter θ.

(b) There exist positive constants a and b such that for all sufficiently large n, $\text{var}_\theta(\bar{X}^*) \leq ae^{-bn}$ for all integers θ.

[*Hint*: (b) One finds $P(\bar{X}^* = k) = \int_{I_k} \phi(t)\,dt$, where I_k is the interval $((k - \theta - 1/2)\sqrt{n}/\sigma, (k - \theta + 1/2)\sqrt{n}/\sigma)$, and hence

$$\text{var}(\bar{X}^*) \leq 4 \sum_{k=1}^{\infty} k \left\{ 1 - \Phi\left[\frac{\sqrt{n}}{\sigma} \left(k - \frac{1}{2} \right) \right] \right\}.$$

The result follows from the fact that for all $y > 0$, $1 - \Phi(y) \leq \phi(y)/y$. See, for example, Feller 1968, Chapter VII, Section 1. Note that $h(y) = \phi(y)/(1 - \Phi(y))$ is the *hazard function* for the standard normal distribution, so we have $h(y) \geq y$ for all $y > 0$. $(1 - \Phi(y))/\phi(y)$ is also known as *Mill's ratio* (see Stuart and Ord, 1987, Section 5.38.) Efron and Johnstone (1990) relate the hazard function to the information inequality].

Note. The surprising results of Problems 5.23–5.25 showing a lower bound and variance which decrease exponentially are due to Hammersley (1950), who shows that, in fact,

$$\text{var}(\bar{X}^*) \sim \sqrt{\frac{8\sigma^2}{\pi n}} e^{-n/8\sigma^2} \quad \text{as} \quad \frac{n}{\sigma^2} \to \infty.$$

Further results concerning the estimation of restricted parameters and properties of $\bar{X}^*$ are given in Khan (1973), Ghosh (1974), Ghosh and Meeden (1978), and Kojima, Morimoto, and Takeuchi (1982).

5.26 *Kiefer inequality*.

(a) Let X have density (with respect to μ) $p(x, \theta)$ which is > 0 for all x, and let Λ_1 and Λ_2 be two distributions on the real line with finite first moments. Then, any unbiased estimator δ of θ satisfies

$$\text{var}(\delta) \geq \frac{[\int \Delta d\Lambda_1(\Delta) - \int \Delta d\Lambda_2(\Delta)]^2}{\int \psi^2(x, \theta) p(x, \theta)\,d\mu(x)}$$

where

$$\psi(x, \theta) = \frac{\int_{\Omega_\theta} p(x, \theta + \Delta)[d\Lambda_1(\Delta) - d\Lambda_2(\Delta)]}{p(x, \theta)}$$

with $\Omega_\theta = \{\Delta : \theta + \Delta_\varepsilon \Omega\}$.

(b) If Λ_1 and Λ_2 assign probability 1 to $\Delta = 0$ and Δ, respectively, the inequality reduces to (5.6) with $g(\theta) = \theta$. [*Hint*: Apply (5.1).] (Kiefer 1952.)

5.27 Verify directly that the following families of densities satisfy (5.38).

(a) The exponential family of (1.5.1),

$$p_\eta(x) = h(x)e^{\eta T(x) - A(\eta)}.$$

(b) The location t family of Example 5.16.

(c) The logistic density of Table 1.4.1.

5.28 Extend condition (5.38) to vector-valued parameters, and show that it is satisfied by the exponential family (1.5.1) for $s > 1$.

5.29 Show that the assumption (5.36(b)) implies (5.38), so Theorem 5.15 is, in fact, a corollary of Theorem 5.10.

5.30 Show that (5.38) is satisfied if either of the following is true:

(a) $|\partial \log p_\theta / \partial\theta|$ is bounded.

(b) $[p_{\theta+\Delta}(x) - p_\theta(x)]/\Delta \to \partial \log p_\theta/\partial\theta$ uniformly.

5.31 (a) Show that if (5.38) holds, then the family of densities is *strongly differentiable* (see Note 8.6).

(b) Show that *weak differentiability* is implied by strong differentiability.

5.32 Brown and Gajek (1990) give two different sufficient conditions for (8.2) to hold, which are given below. Show that each implies (8.2). (Note that, in the progression from (a) to (b) the conditions become weaker, thus more widely applicable and harder to check.)

(a) For some $B < \infty$,

$$E_{\theta_0}\left[\frac{\partial^2}{\partial\theta^2} p_\theta(X)/p_{\theta_0}(X)\right]^2 < B$$

for all θ in a neighborhood of θ.

(b) If $p_t^*(x) = \partial/\partial\theta p_\theta(x)|_{\theta=t}$, then

$$\lim_{\Delta\to 0} E_{\theta_0}\left[\frac{p^*_{\theta_0+\Delta}(X) - p^*_{\theta_0}(X)}{p_{\theta_0}(X)}\right]^2 = 0.$$

5.33 Let $\mathcal{F}$ be the class of all unimodal symmetric densities or, more generally, densities symmetric around zero and satisfying $f(x) \le f(0)$ for all x. Show that

$$\min_{f\in\mathcal{F}} \int x^2 f(x)dx = \frac{1}{12},$$

and that the minimum is attained by the uniform$(-\frac{1}{2}, \frac{1}{2})$ distribution. Thus, the uniform distribution has minimum variance among symmetric unimodal distributions. (See Example 4.8.6 for large-sample properties of the scale uniform.) [Hint: The side condition $\int f(x)dx = 1$, together with the method of undetermined multipliers, yields an equivalent problem, minimization of $\int(x^2 - a^2)f(x)dx$, where a is chosen to satisfy the constraint. A Neyman-Pearson type argument will now work.]

Section 6

6.1 For any random variables $(\psi_1, \ldots, \psi_s)$, show that the matrices $||E\psi_i\psi_j||$ and $C = ||\text{cov}(\psi_i, \psi_j)||$ are positive semidefinite.

6.2 In this problem, we establish some facts about eigenvalues and eigenvectors of square matrices. (For a more general treatment, see, for example, Marshall and Olkin 1979, Chapter 20.)

We use the facts that a scalar $\lambda > 0$ is an *eigenvalue* of the $n \times n$ symmetric matrix A if there exists an $n \times 1$ vector p, the corresponding *eigenvector*, satisfying $Ap = \lambda p$. If A is nonsingular, there are n eigenvalues with corresponding linearly independent eigenvectors.

(a) Show that $A = P'D_\lambda P$, where D_λ is a diagonal matrix of eigenvalues of A and P is and $n \times n$ matrix whose rows are the corresponding eigenvalues that satisfies $P'P = PP' = I$, the identity matrix.

(b) Show that $\max_x \frac{x'Ax}{x'x}$ = largest eigenvalue of A.

(c) If B is a nonsingular symmetric matrix with eigenvector-eigenvalue representation $B = Q'D_\beta Q$, then $\max_x \frac{x'Ax}{x'Bx}$ = largest eigenvalue of A^*, where $A^* = D_\beta^{-1/2} QAQ' D_\beta^{-1/2}$ and $D_\beta^{-1/2}$ is a diagonal matrix whose elements are the reciprocals of the square roots of the eigenvalues of B.

(d) For any square matrices C and D, show that the eigenvalues of the matrix CD are the same as the eigenvalues of the matrix DC, and hence that $\max_x \frac{x'Ax}{x'Bx}$ = largest eigenvalue of AB^{-1}.

(e) If $A = aa'$, where a is a $n \times 1$ vector (A is thus a rank-one matrix), then $\max_x \frac{x'aa'x}{x'Bx} = a'B^{-1}a$.

[Hint: For part (b) show that $\frac{x'Ax}{x'x} = \frac{y'D_\lambda y}{y'y} = \frac{\sum_i \lambda_i y_i^2}{\sum_i y_i^2}$, where $y = Px$, and hence the maximum is achieved at the vector y that is 1 at the coordinate of the largest eigenvalue and zero everywhere else.]

6.3 An alternate proof of Theorem 6.1 uses the method of Lagrange (or undetermined) multipliers. Show that, for fixed γ, the maximum value of $a'\gamma$, subject to the constraint that $a'Ca = 1$, is obtained by the solutions to

$$\frac{\partial}{\partial a_i}\left\{a'\gamma - \frac{1}{2}\lambda[a'Ca - 1]\right\} = 0,$$

where λ is the undetermined multiplier. (The solution is $a = \pm C^{-1}\gamma / \sqrt{\gamma' C^{-1}\gamma}$.)

6.4 Prove (6.11) under the assumptions of the text.

6.5 Verify (a) the information matrices of Table 6.1 and (b) Equations (6.15) and (6.16).

6.6 If $p(x) = (1-\varepsilon)\phi(x-\xi) + (\varepsilon/\tau)\phi[(x-\xi)/\tau]$ where ϕ is the standard normal density, find $I(\varepsilon, \xi, \tau)$.

6.7 Verify the expressions (6.20) and (6.21).

6.8 Let $A = \begin{pmatrix} A_{11} & A_{12} \\ A_{21} & A_{22} \end{pmatrix}$ be a partitioned matrix with A_{22} square and nonsingular, and let

$$B = \begin{pmatrix} I & -A_{12}A_{22}^{-1} \\ 0 & I \end{pmatrix}.$$

Show that $|A| = |A_{11} - A_{12}A_{22}^{-1}A_{21}| \cdot |A_{22}|$.

6.9 (a) Let

$$A = \begin{pmatrix} a & b' \\ b & C \end{pmatrix}$$

where a is a scalar and b a column matrix, and suppose that A is positive definite. Show that $|A| \leq a|C|$ with equality holding if and only if $b = 0$.

(b) More generally, if the matrix A of Problem 6.8 is positive definite, show that $|A| \le |A_{11}| \cdot |A_{22}|$ with equality holding if and only if $A_{12} = 0$.

[*Hint*: Transform A_{11} and the positive semidefinite $A_{12}A_{22}^{-1}A_{21}$ simultaneously to diagonal form.]

6.10 (a) Show that if the matrix A is nonsingular, then for any vector x, $(x'Ax)(x'A^{-1}x) > (x'x)^2$.

(b) Show that, in the notation of Theorem 6.6 and the following discussion,

$$\frac{\left[\frac{\partial}{\partial\theta_i} E_\theta \delta\right]^2}{I_{ii}(\theta)} = \frac{(\varepsilon_i'\alpha)^2}{\varepsilon_i' I(\theta)\varepsilon_i},$$

and if $\alpha = (0, \ldots, 0, \alpha_i, 0, \ldots 0)$, $\alpha' I(\theta)^{-1}\alpha = (\varepsilon_i'\alpha)^2 \varepsilon_i' I(\theta)^{-1}\varepsilon_i$, and hence establish (6.25).

6.11 Prove that (6.26) is necessary for equality in (6.25). [*Hint*: Problem 6.9(a).]

6.12 Prove the Bhattacharyya inequality (6.29) and show that the condition of equality is as stated.

8 Notes

8.1 Unbiasedness and Information

The concept of unbiasedness as "lack of systematic error" in the estimator was introduced by Gauss (1821) in his work on the theory of least squares. It has continued as a basic assumption in the developments of this theory since then.

The amount of information that a data set contains about a parameter was introduced by Edgeworth (1908, 1909) and was developed more systematically by Fisher (1922 and later papers). The first version of the information inequality, and hence connections with unbiased estimation, appears to have been given by Fréchet (1943). Early extensions and rediscoveries are due to Darmois (1945), Rao (1945), and Cramér (1946b). The designation "information inequality," which replaced the earlier "Cramér-Rao inequality," was proposed by Savage (1954).

8.2 UMVU Estimators

The first UMVU estimators were obtained by Aitken and Silverstone (1942) in the situation in which the information inequality yields the same result (Problem 5.17). UMVU estimators as unique unbiased functions of a suitable sufficient statistic were derived in special cases by Halmos (1946) and Kolmogorov (1950) and were pointed out as a general fact by Rao (1947). An early use of Method 1 for determining such unbiased estimators is due to Tweedie (1947). The concept of completeness was defined, its implications for unbiased estimation developed, and Theorem 1.7 obtained, in Lehmann and Scheffé (1950, 1955, 1956).

Theorem 1.11 has been used to determine UMVU estimators in many special cases. Some applications include those of Abbey and David (1970, exponential distribution), Ahuja (1972, truncated Poisson), Bhattacharyya et al. (1977, censored), Bickel and Lehmann (1969, convex), Varde and Sathe (1969, truncated exponential), Brown and Cohen (1974, common mean), Downton (1973, $P(X \le Y)$), Woodward and Kelley (1977, $P(X \le Y)$), Iwase (1983, inverse Gaussian), and Kremers (1986, sum-quota sampling).

Figure 8.1. *Illustration of the information inequality*

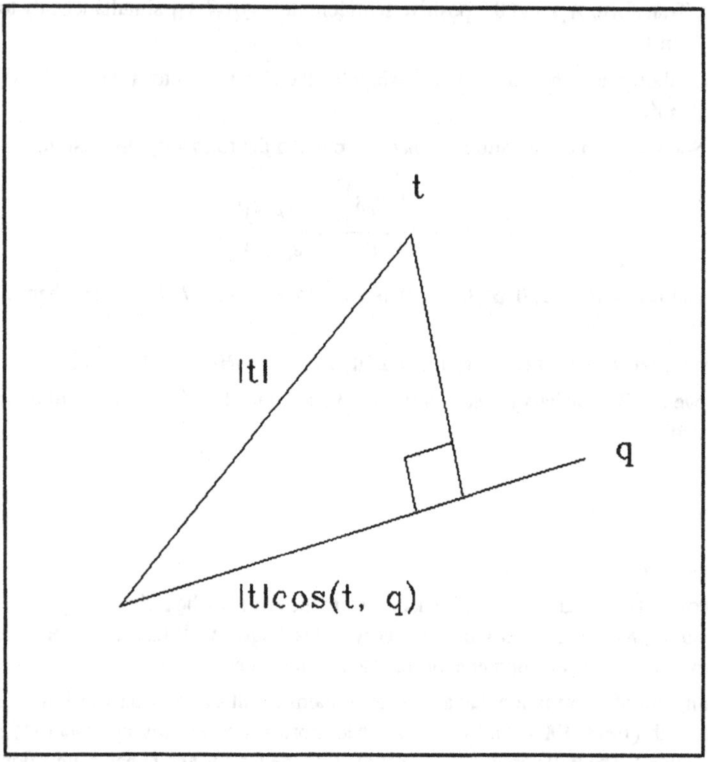

8.3 Existence of Unbiased Esimators

Doss and Sethuraman (1989) show that the process of bias reduction may not always be the wisest course. If an estimand $g(\theta)$ does not have an unbiased estimator, and one tries to reduce the bias in a biased estimator δ, they show that as the bias goes to zero, $\text{var}(\delta) \to \infty$ (see Problem 1.4).

This result has implications for bias-reduction procedures such as the jackknife and the bootstrap. (For an introduction to the jackknife and the bootstrap, see Efron and Tibshirani 1993 or Shao and Tu 1995.) In particular, Efron and Tibshirani (1993, Section 10.6) discuss some practical implications of bias reduction, where they urge caution in its use, as large increases in standard errors can result.

Liu and Brown (1993) call a problem *singular* if there exists no unbiased estimator with finite variance. More precisely, if $\mathcal{F}$ is a family of densities, then if a problem is singular, there will be at least one member of $\mathcal{F}$, called a *singular point*, where any unbiased estimator of a parameter (or functional) will have infinite variance. There are many examples of singular problems, both in parametric and nonparametric estimation, with nonparametric density estimation being, perhaps, the best known. Two particularly simple examples of singular problems are provided by Example 1.2 (estimation of $1/p$ in a binomial problem) and Problem 5.21 (a mixture estimation problem).

8.4 Geometry of the Information Inequality

The information inequality can be interpreted as, and a proof can be based on, the fact that the length of the hypotenuse of a right triangle exceeds the length of each side. For two vectors a and b, define $< t, q > = t'q$, with $< t, t >^2 = |t|^2$. For the triangle in Figure 8.1, using the fact that the cosine of the angle between t and q is $\cos(t, q) = t'q/|t||q|$ and the fact that the hypotenuse is the longest side, we have

$$|t| > |t|\cos(t, q) = |t|\left[\frac{< t, q >}{|t||q|}\right] = \frac{< t, q >}{|q|}.$$

If we define $< X, Y > = E[(X - EX)(Y - EY)]$ for random variables X and Y, applying the above inequality with this definition results in the covariance inequality (5.1), which, in turn, leads to the information inequality. See Fabian and Hannan (1977) for a rigorous development.

8.5 Fisher Information and the Hazard Function

Efron and Johnstone (1990) investigate an identity between the Fisher information number and the *hazard function*, h, defined by

$$h_\theta(x) = \lim_{\Delta\to 0} \Delta^{-1} P(x \le X < x + \Delta | X \ge x) = \frac{f_\theta(x)}{1 - F_\theta(x)}$$

where f_θ and F_θ are the density and distribution function of the random variable X, respectively. The hazard function, $h(x)$, represents the conditional survival rate given survival up to time x and plays and important role in survival analysis. (See, for example, Kalbfleish and Prentice 1980, Cox and Oakes 1984, Fleming and Harrington 1991.)

Efron and Johnstone show that

$$I(\theta) = \int_{-\infty}^{\infty} \frac{\partial}{\partial\theta} \log[f_\theta(x)]^2 f_\theta(x)dx = \int_{-\infty}^{\infty} \frac{\partial}{\partial\theta} \log[h_\theta(x)]^2 f_\theta(x)dx.$$

They then interpret this identity and discuss its implications to, and connections with, survival analysis and statistical curvature of hazard models, among other things. They also note that this identity can be derived as a consequence of the more general result of James (1986), who showed that if $b(\cdot)$ is a continuous function of the random variable X, then

$$\text{var}[b(X)] = E[b(X) - \bar{b}(X)]^2, \quad \text{where} \quad \bar{b}(x) = E[b(X)|b(X) > x],$$

as long as the expectations exist.

8.6 Weak and Strong Differentiability

Research into determining necessary and sufficient conditions for the applicability of the Information Inequality bound has a long history (see, for example, Blyth and Roberts 1972, Fabian and Hannan 1977, Ibragimov and Has'minskii 1981, Section 1.7, Müller-Funk et al. 1989, Brown and Gajek 1990). What has resulted is a condition on the density sufficient to ensure (5.29).

The precise condition needed was presented by Fabian and Hannan (1977), who call it *weak differentiability*. The function $p_{\theta+\Delta}(x)/p_\theta(x)$ is *weakly differentiable* at θ if there is a measurable function q such that

$$\lim_{\Delta\to 0} \int h(x)\left\{\left[\Delta^{-1}\left(\frac{p_{\theta+\Delta}(x)}{p_\theta(x)} - 1\right)\right] - q(x)\right\} p_\theta(x)\,d\mu(x) = 0 \tag{8.1}$$

for all $h(\cdot)$ such that $\int h^2(x)p_\theta(x)\,d\mu(x) < \infty$. Weak differentiability is actually equivalent (necessary and sufficient) to the existence of a function $q_\theta(x)$ such that $(\partial/\partial\theta)E_\theta\delta = E\delta q$. Hence, it can replace condition (5.38) in Theorem 5.15.

Since weak differentiability is often difficult to verify, Brown and Gajek (1990) introduce the more easily verifiable condition of *strong differentiability*, which implies weak differentiability, and thus can also replace condition (5.38) in Theorem 5.15 (Problem 5.31). The function $p_{\theta+\Delta}(x)/p_\theta(x)$ is *strongly differentiable at* $\theta = \theta_0$ *with derivative* $q_{\theta_0}(x)$ if

$$\lim_{\Delta\to 0}\int\left\{\left[\Delta^{-1}\left(\frac{p_{\theta+\Delta}(x)}{p_\theta(x)}-1\right)\right]-q_{\theta_0}(x)\right\}^2 p_{\theta_0}(x)\,d\mu(x)=0. \tag{8.2}$$

These variations of the usual definition of differentiability are well suited for the information inequality problem. In fact, consider the expression in the square brackets in (8.1). If the limit of this expression exists, it is $q_\theta(x) = \partial \log p_\theta(x)/\partial\theta$. Of course, existence of this limit does not, by itself, imply condition (8.2); such an implication requires an integrability condition.

Brown and Gajek (1990) detail a number of easier-to-check conditions that imply (8.2). (See Problem 5.32.) Fabian and Hannan (1977) remark that if (8.1) holds and $\partial \log p_\theta(x)/\partial\theta$ exists, then it must be the case that $q_\theta(x) = \partial \log p_\theta(x)/\partial\theta$. However, the existence of one does not imply the existence of the other.

CHAPTER 3

Equivariance

1 First Examples

In Section 1.1, the principle of unbiasedness was introduced as an impartiality restriction to eliminate estimators such as $\delta(X) \equiv g(\theta_0)$, which would give very low risk for some parameter values at the expense of very high risk for others. As was seen in Sections 2.2–2.4, in many important situations there exists within the class of unbiased estimators a member that is uniformly better for any convex loss function than any other unbiased estimator.

In the present chapter, we shall use symmetry considerations as the basis for another such impartiality restriction with a somewhat different domain of applicability.

Example 1.1 Estimating binomial p. Consider n binomial trials with unknown probability p $(0 < p < 1)$ of success which we wish to estimate with loss function $L(p, d)$, for example, $L(p, d) = (d - p)^2$ or $L(p, d) = (d - p)^2/p(1 - p)$. If X_i, $i = 1, \ldots, n$ is 1 or 0 as the ith trial is a success or failure, the joint distribution of the X's is

$$P(x_1, \ldots, x_n) = p^{\Sigma x_i}(1 - p)^{\Sigma(1-x_i)}.$$

Suppose now that another statistician interchanges the definition of success and failure. For this worker, the probability of success is

$$p' = 1 - p \tag{1.1}$$

and the indicator of success and failure on the ith trial is

$$X_i' = 1 - X_i. \tag{1.2}$$

The joint distribution of the X_i' is

$$P(x_1', \cdots, x_n') = p'^{\Sigma x_i'}(1 - p')^{\Sigma(1-x_i')}$$

and hence satisfies

$$P(x_i', \ldots, x_n') = P(x_1, \ldots, x_n). \tag{1.3}$$

In the new terminology, the estimated value d' of p' is

$$d' = 1 - d, \tag{1.4}$$

and the loss resulting from its use is $L(p', d')$. The loss functions suggested at the beginning of the example (and, in fact, most loss functions that we would want to employ in this situation) satisfy

$$L(p, d) = L(p', d'). \tag{1.5}$$

Under these circumstances, the problem of estimating p with loss function L is said to be *invariant* under the transformations (1.1), (1.2), and (1.4). This invariance is an expression of the complete symmetry of the estimation problem with respect to the outcomes of success and failure.

Suppose now that in the above situation, we had decided to use $\delta(\mathbf{x})$, where $\mathbf{x} = (x_1, \ldots, x_n)$ as an estimator of p. Then, the formal identity of the primed and unprimed problem suggests that we should use

$$\delta(\mathbf{x}') = \delta(1 - x_1, \ldots, 1 - x_n) \tag{1.6}$$

to estimate $p' = 1 - p$. On the other hand, it is natural to estimate $1 - p$ by 1 minus the estimator of p, i.e., by

$$1 - \delta(\mathbf{x}). \tag{1.7}$$

It seems desirable that these two estimators should agree and hence that

$$\delta(\mathbf{x}') = 1 - \delta(\mathbf{x}). \tag{1.8}$$

An estimator satisfying (1.8) will be called equivariant under the transformations (1.1), (1.2), and (1.4). Note that the standard estimate $\Sigma X_i/n$ satisfies (1.8).

The arguments for (1.6) and (1.7) as estimators of $1 - p$ are of a very different nature. The appropriateness of (1.6) depends entirely on the symmetry of the situation. It would continue to be suitable if it were known, for example, that $\frac{1}{3} < p < \frac{2}{3}$ but not if, say, $\frac{1}{4} < p < \frac{1}{2}$. In fact, in the latter case, $\delta(\mathbf{X})$ would typically be chosen to be $< \frac{1}{2}$ for all $\mathbf{X}$, and hence $\delta(\mathbf{X}')$ would be entirely unsuitable as an estimator of $1 - p$, which is known to be $> \frac{1}{2}$. More generally, (1.6) would cease to be appropriate if any prior information about p is available which is not symmetric about $\frac{1}{2}$. In contrast, the argument leading to (1.7) is quite independent of any symmetry assumptions, but simply reflects the fact that if $\delta(\mathbf{X})$ is a reasonable estimator of a parameter θ (that is, is likely to be close to θ), then $1 - \delta(\mathbf{X})$ is reasonable as an estimator of $1 - \theta$. ‖

We shall postpone giving a general definition of equivariance to the next section, and in the remainder of the present section, we formulate this concept and explore its implications for the special case of location problems.

Let $\mathbf{X} = (X_1, \ldots, X_n)$ have joint distribution with probability density

$$f(\mathbf{x} - \xi) = f(x_1 - \xi, \ldots, x_n - \xi), \qquad -\infty < \xi < \infty, \tag{1.9}$$

where f is known and ξ is an unknown *location parameter*. Suppose that for the problem of estimating ξ with loss function $L(\xi, d)$, we have found a satisfactory estimator $\delta(\mathbf{X})$.

In analogy with the transformations (1.2) and (1.1) of the observations X_i and the parameter p in Example 1.1, consider the transformations

$$X_i' = X_i + a \tag{1.10}$$

and

$$\xi' = \xi + a. \tag{1.11}$$

The joint density of $\mathbf{X}' = (X_1', \ldots, X_n')$ can be written as

$$f(\mathbf{x}' - \xi') = f(x_1' - \xi', \cdots, x_n' - \xi')$$

so that in analogy with (1.3) we have by (1.10) and (1.11)

$$f(\mathbf{x}' - \xi') = f(\mathbf{x} - \xi) \text{ for all } \mathbf{x} \text{ and } \xi. \tag{1.12}$$

The estimated value d' of ξ' is

$$d' = d + a \tag{1.13}$$

and the loss resulting from its use is $L(\xi', d')$.

In analogy with (1.5), we require L to satisfy $L(\xi', d') = L(\xi, d)$ and hence

$$L(\xi + a, d + a) = L(\xi, d). \tag{1.14}$$

A loss function L satisfies (1.14) for all values of a if and only if it depends only on the difference $d - \xi$, that is, it is of the form

$$L(\xi, d) = \rho(d - \xi). \tag{1.15}$$

That (1.15) implies (1.14) is obvious. The converse follows by putting $a = -\xi$ in (1.14) and letting $\rho(d - \xi) = L(0, d - \xi)$.

We can formalize these considerations in the following definition.

Definition 1.2 A family of densities $f(x|\xi)$, with parameter ξ, and a loss function $L(\xi, d)$ are *location invariant* if, respectively, $f(x'|\xi') = f(x|\xi)$ and $L(\xi, d) = L(\xi', d')$ whenever $\xi' = \xi + a$ and $d' = d + a$. If both the densities and the loss function are location invariant, the problem of estimating ξ is said to be *location invariant* under the transformations (1.10), (1.11), and (1.13).

As in Example 1.1, this invariance is an expression of symmetry. Quite generally, symmetry in a situation can be characterized by its lack of change under certain transformations. After a transformation, the situation looks exactly as it did before. In the present case, the transformations in question are the shifts (1.10), (1.11), and (1.13), which leave both the density (1.12) and the loss function (1.14) unchanged.

Suppose now that in the original (unprimed) problem, we had decided to use $\delta(\mathbf{X})$ as an estimator of ξ. Then, the formal identity of the primed and unprimed problem suggest that we should use

$$\delta(\mathbf{X}') = \delta(X_1 + a, \ldots, X_n + a) \tag{1.16}$$

to estimate $\xi' = \xi + a$. On the other hand, it is natural to estimate $\xi + a$ by adding a to the estimator of ξ, i.e., by

$$\delta(\mathbf{X}) + a. \tag{1.17}$$

As before, it seems desirable that these two estimators should agree and hence that

$$\delta(X_1 + a, \ldots, X_n + a) = \delta(X_1, \ldots, X_n) + a \quad \text{for all } a. \tag{1.18}$$

Definition 1.3 An estimator satisfying (1.18) will be called equivariant under the transformations (1.10), (1.11), and (1.13), or *location equivariant.*[1]

All the usual estimators of a location parameter are location equivariant. This is the case, for example, for the mean, the median, or any weighted average of the order statistics (with weights adding up to one). The MLE $\hat{\xi}$ is also equivariant since, if $\hat{\xi}$ maximizes $f(\mathbf{x}-\xi)$, $\hat{\xi}+a$ maximizes $f(\mathbf{x}-\xi-a)$.

As was the case in Example 1.1, the arguments for (1.16) and (1.17) as estimators of $\xi+a$ are of a very different nature. The appropriateness of (1.16) results from the invariance of the situation under shift. It would not be suitable for an estimator of $\xi+a$, for example, if it were known that $0<\xi<1$. Then, $\delta(\mathbf{X})$ would typically only take values between 0 and 1, and hence $\delta(\mathbf{X}')$ would be disastrous as an estimate of $\xi+a$ if $a>1$. In contrast, the argument leading to (1.17) is quite independent of any equivariance arguments, but simply reflects the fact that if $\delta(\mathbf{X})$ is a reasonable estimator of a parameter ξ, then $\delta(\mathbf{X})+a$ is reasonable for estimating $\xi+a$.

The following theorem states an important set of properties of location equivariant estimators.

Theorem 1.4 *Let* $\mathbf{X}$ *be distributed with density (1.9), and let* δ *be equivariant for estimating* ξ *with loss function (1.15). Then, the bias, risk, and variance of* δ *are all constant (i.e., do not depend on* ξ*).*

Proof. Note that if $\mathbf{X}$ has density $f(\mathbf{x})$ (i.e., $\xi=0$), then $\mathbf{X}+\xi$ has density (1.9). Thus, the bias can be written as

$$b(\xi)=E_\xi[\delta(\mathbf{X})]-\xi=E_0[\delta(\mathbf{X}+\xi)]-\xi=E_0[\delta(\mathbf{X})],$$

which does not depend on ξ.

The proofs for risk and variance are analogous (Problem 1.1). □

Theorem 1.4 has an important consequence. Since the risk of any equivariant estimator is independent of ξ, the problem of uniformly minimizing the risk within this class of estimators is replaced by the much simpler problem of determining the equivariant estimator for which this constant risk is smallest.

Definition 1.5 In a location invariant estimation problem, if a location equivariant estimator exists which minimizes the constant risk, it is called the *minimum risk equivariant* (MRE) estimator.

Such an estimator will typically exist, and is often unique, although in rare cases there could be a sequence of estimators whose risks decrease to a value not assumed. To derive an explicit expression for the MRE estimator, let us begin by finding a representation of the most general location equivariant estimator.

Lemma 1.6 *If* δ_0 *is any equivariant estimator, then a necessary and sufficient condition for* δ *to be equivariant is that*

$$\delta(\mathbf{x})=\delta_0(\mathbf{x})+u(\mathbf{x}) \tag{1.19}$$

[1] Some authors have called such estimators *invariant*, which could suggest that the estimator remains unchanged, rather than changing in a prescribed way. We will reserve that term for functions that do remain unchanged, such as those satisfying (1.20).

where $u(\mathbf{x})$ *is any function satisfying*

$$u(\mathbf{x}+a) = u(\mathbf{x}), \quad \textit{for all} \quad \mathbf{x},\ a. \tag{1.20}$$

Proof. Assume first that (1.19) and (1.20) hold. Then, $\delta(\mathbf{x}+a) = \delta_0(\mathbf{x}+a)+u(\mathbf{x}+a) = \delta_0(\mathbf{x}) + a + u(\mathbf{x}) = \delta(\mathbf{x}) + a$, so that δ is equivariant.

Conversely, if δ is equivariant, let

$$u(\mathbf{x}) = \delta(\mathbf{x}) - \delta_0(\mathbf{x}).$$

Then

$$\begin{aligned} u(\mathbf{x}+a) &= \delta(\mathbf{x}+a) - \delta_0(\mathbf{x}+a) \\ &= \delta(\mathbf{x}) + a - \delta_0(\mathbf{x}) - a = u(\mathbf{x}) \end{aligned}$$

so that (1.19) and (1.20) hold. □

To complete the representation, we need a characterization of the functions u satisfying (1.20).

Lemma 1.7 *A function u satisfies (1.20) if and only if it is a function of the differences* $y_i = x_i - x_n$ $(i = 1, \ldots, n-1)$, $n \geq 2$; *for* $n = 1$, *if and only if it is a constant.*

Proof. The proof is essentially the same as that of (1.15). □

Note that the function $u(\cdot)$, which is invariant, is only a function of the ancillary statistic $(y_1, \ldots, y_{n-1})$ (see Section 1.6). Hence, by itself, it does not carry any information about the parameter ξ. The connection between invariance and ancillarity is not coincidental. (See Lehmann and Scholz 1992, and Problems 2.11 and 2.12.)

Combining Lemmas 1.6 and 1.7 gives the following characterization of equivariant estimators.

Theorem 1.8 *If* δ_0 *is any equivariant estimator, then a necessary and sufficient condition for* δ *to be equivariant is that there exists a function v of* $n-1$ *arguments for which*

$$\delta(\mathbf{x}) = \delta_0(\mathbf{x}) - v(\mathbf{y}) \quad \textit{for all} \quad \mathbf{x}. \tag{1.21}$$

Example 1.9 Location equivariant estimators based on one observation. Consider the case $n = 1$. Then, it follows from Theorem 1.8 that the only equivariant estimators are $X + c$ for some constant c. ||

We are now in a position to determine the equivariant estimator with minimum risk.

Theorem 1.10 *Let* $\mathbf{X} = (X_1, \ldots, X_n)$ *be distributed according to (1.9), let* $Y_i = X_i - X_n$ $(i = 1, \ldots, n-1)$ *and* $\mathbf{Y} = (Y_1, \ldots, Y_{n-1})$. *Suppose that the loss function is given by (1.15) and that there exists an equivariant estimator* δ_0 *of* ξ *with finite risk. Assume that for each* $\mathbf{y}$ *there exists a number* $v(\mathbf{y}) = v^*(\mathbf{y})$ *which minimizes*

$$E_0\{\rho[\delta_0(\mathbf{X}) - v(\mathbf{y})] | \mathbf{y}\}. \tag{1.22}$$

Then, a location equivariant estimator δ of ξ with minimum risk exists and is given by

$$\delta^*(\mathbf{X}) = \delta_0(\mathbf{X}) - v^*(\mathbf{Y}).$$

Proof. By Theorem 1.8, the MRE estimator is found by determining v so as to minimize

$$R_\xi(\delta) = E_\xi\{\rho[\delta_0(\mathbf{X}) - v(\mathbf{Y}) - \xi]\}.$$

Since the risk is independent of ξ, it suffices to minimize

$$\begin{aligned} R_0(\delta) &= E_0\{\rho[\delta_0(\mathbf{X}) - v(\mathbf{Y})]\} \\ &= \int E_0\{\rho[\delta_0(\mathbf{X}) - v(\mathbf{y})]|\mathbf{y}\}\, dP_0(\mathbf{y}). \end{aligned}$$

The integral is minimized by minimizing the integrand, and hence (1.22), for each $\mathbf{y}$. Since δ_0 has finite risk $E_0\{\rho[\delta_0(\mathbf{X})]|y\} < \infty$ (a.e. P_0), the minimization of (1.22) is meaningful. The result now follows from the assumptions of the theorem. □

Corollary 1.11 *Under the assumptions of Theorem 1.10, suppose that ρ is convex and not monotone. Then, an MRE estimator of ξ exists; it is unique if ρ is strictly convex.*

Proof. Theorems 1.10 and 1.7.15. □

Corollary 1.12 *Under the assumptions of Theorem 1.10:*

(i) if $\rho(d-\xi) = (d-\xi)^2$, then

$$v^*(\mathbf{y}) = E_0[\delta_0(\mathbf{X})|\mathbf{y}]; \tag{1.23}$$

(ii) if $\rho(d-\xi) = |d-\xi|$, then $v^*(\mathbf{y})$ is any median of $\delta_0(\mathbf{X})$ under the conditional distribution of $\mathbf{X}$ given $\mathbf{y}$.

Proof. Examples 1.7.17 and 1.7.18 □

Example 1.13 Continuation of Example 1.9. For the case $n = 1$, if X has finite risk, the arguments of Theorem 1.10 and Corollary 1.11 show that the MRE estimator is $X - v^*$ where v^* is any value minimizing

$$E_0[\rho(X - v)]. \tag{1.24}$$

In particular, the MRE estimator is $X - E_0(X)$ and $X - \text{med}_0(X)$ when the loss is squared error and absolute error, respectively.

Suppose, now, that X is symmetrically distributed about ξ. Then, for any ρ which is convex and even, if follows from Corollary 1.7.19 that (1.24) is minimized by $v = 0$, so that X is MRE. Under the same assumptions, if $n = 2$, the MRE estimator is $(X_1 + X_2)/2$. (Problem 1.3). ||

The existence of MRE estimators is, of course, not restricted to convex loss functions. As an important class of nonconvex loss functions, consider the case that ρ is bounded.

Corollary 1.14 *Under the assumptions of Example 1.13, suppose that $0 \le \rho(t) \le M$ for all values of t, that $\rho(t) \to M$ as $t \to \pm\infty$, and that the density f of X is continuous a.e. Then, an MRE estimator of ξ exists.*

Proof. See Problem 1.8. □

Example 1.15 MRE under 0 − 1 loss. Suppose that

$$\rho(d-\xi)=\begin{cases}1 & \text{if } |d-\xi|>k\\ 0 & \text{otherwise.}\end{cases}$$

Then, v will minimize (1.24), provided it maximizes

$$P_0\{|X-v|\le k\}. \tag{1.25}$$

Suppose that the density f is symmetric about 0. If f is unimodal, then $v=0$ and the MRE estimator of ξ is X. On the other hand, suppose that f is U-shaped, say $f(x)$ is zero for $|x|>c>k$ and is strictly increasing for $0<x<c$. Then, there are two values of v maximizing (1.25), namely $v=c-k$ and $v=-c+k$, hence, $X-c+k$ and $X+c-k$ are both MRE. ||

Example 1.16 Normal. Let $X_1,\ldots,X_n$ be iid according to $N(\xi,\sigma^2)$, where σ is known. If $\delta_0=\bar{X}$ in Theorem 1.10, it follows from Basu's theorem that δ_0 is independent of $\mathbf{Y}$ and hence that $v(\mathbf{y})=v$ is a constant determined by minimizing (1.24) with $\bar{X}$ in place of X. Thus $\bar{X}$ is MRE for all convex and even ρ. It is also MRE for many nonconvex loss functions including that of Example 1.15. ||

This example has an interesting implication concerning a "least favorable" property of the normal distribution.

Theorem 1.17 *Let $\mathcal{F}$ be the class of all univariate distributions F that have a density f (w.r.t. Lebesgue measure) and fixed finite variance, say $\sigma^2=1$. Let $X_1,\ldots,X_n$ be iid with density $f(x_i-\xi)$, $\xi=E(X_i)$, and let $r_n(F)$ be the risk of the MRE estimator of ξ with squared error loss. Then, $r_n(F)$ takes on its maximum value over $\mathcal{F}$ when F is normal.*

Proof. The MRE estimator in the normal case is $\bar{X}$ with risk $E(\bar{X}-\xi)^2=1/n$. Since this is the risk of $\bar{X}$, regardless of F, the MRE estimator for any other F must have risk $\le 1/n$, and this completes the proof. □

For $n\ge 3$, the normal distribution is, in fact, the only one for which $r_n(F)=1/n$. Since the MRE estimator is unique, this will follow if the normal distribution can be shown to be the only one whose MRE estimator is $\bar{X}$. From Corollary 1.12, it is seen that the MRE estimator is $\bar{X}-E_0[\bar{X}|\mathbf{Y}]$ and, hence, is $\bar{X}$ if and only if $E_0[\bar{X}|\mathbf{Y}]=0$. It was proved by Kagan, Linnik, and Rao (1965, 1973) that this last equation holds if and only if F is normal.

Example 1.18 Exponential. Let $X_1,\ldots,X_n$ be iid according to the exponential distribution $E(\xi,b)$ with b known. If $\delta_0=X_{(1)}$ in Theorem 1.10, it again follows from Basu's theorem that δ_0 is independent of $\mathbf{Y}$ and hence that $v(\mathbf{y})=v$ is determined by minimizing

$$E_0[\rho(X_{(1)}-v)]. \tag{1.26}$$

(a) If the loss is squared error, the minimizing value is $v=E_0[X_{(1)}]=b/n$, and hence the MRE estimator is $X_{(1)}-(b/n)$.

(b) If the loss is absolute error, the minimizing value is $v=b(\log 2)/n$ (Problem 1.4).

(c) If the loss function is that of Example 1.15, then v is the center of the interval I of length $2k$ which maximizes $P_{\xi=0}[X_{(1)} \varepsilon I]$. Since for $\xi = 0$, the density of $X_{(1)}$ is decreasing on $(0, \infty)$, $v = k$, and the MRE estimator is $X_{(1)} - k$.

See Problem 1.5 for another comparison. ||

Example 1.19 Uniform. Let $X_1, \ldots, X_n$ be iid according to the uniform distribution $U(\xi - 1/2b, \xi + 1/2b)$, with b known, and suppose the loss function ρ is convex and even. For δ_0, take $[X_{(1)} + X_{(n)}]/2$ where $X_{(1)} < \cdots < X_{(n)}$ denote the ordered X's To find $v(\mathbf{y})$ minimizing (1.22), consider the conditional distribution of δ_0 given $\mathbf{y}$. This distribution depends on $\mathbf{y}$ only through the differences $X_{(i)} - X_{(1)}$, $i = 2, \ldots, n$. By Basu's theorem, the pair $(X_{(1)}, X_{(n)})$ is independent of the ratios $Z_i = [X_{(i)} - X_{(1)}]/X_{(n)} - X_{(1)}]$, $i = 2, \ldots, n-1$ (Problem 1.6.36(b)). Therefore, the conditional distribution of δ_0 given the differences $X_{(i)} - X_{(1)}$, which is equivalent to the conditional distribution of δ_0 given $X_{(n)} - X_{(1)}$ and the Z's, depends only on $X_{(n)} - X_{(1)}$. However, the conditional distribution of δ_0 given $V = X_{(n)} - X_{(1)}$ is symmetric about 0 (when $\xi = 0$; Problem 1.2). It follows, therefore, as in Example 1.13 that the MRE estimator of ξ is $[X_{(1)} + X_{(n)}]/2$, the midrange. ||

When loss is squared error, the MRE estimator

$$\delta^*(\mathbf{X}) = \delta_0(\mathbf{X}) - E[\delta_0(\mathbf{X})|\mathbf{Y}] \tag{1.27}$$

can be evaluated more explicitly.

Theorem 1.20 *Under the assumptions of Theorem 1.15, with $L(\xi, d) = (d - \xi)^2$, the estimator (1.27) is given by*

$$\delta^*(\mathbf{x}) = \frac{\int_{-\infty}^{\infty} u f(x_1 - u, \ldots, x_n - u)\, du}{\int_{-\infty}^{\infty} f(x_1 - u, \ldots, x_n - u)\, du}, \tag{1.28}$$

and in this form, it is known as the Pitman estimator of ξ.

Proof. Let $\delta_0(\mathbf{X}) = X_n$. To compute $E_0(X_n|\mathbf{y})$ (which exists by Problem 1.21), make the change of variables

$$y_i = x_i - x_n \quad (i = 1, \ldots, n-1); \quad y_n = x_n.$$

The Jacobian of the transformation is 1. The joint density of the Y's is therefore

$$p_Y(y_1, \ldots, y_n) = f(y_1 + y_n, \ldots, y_{n-1} + y_n, y_n),$$

and the conditional density of Y_n given $\mathbf{y} = (y_1, \ldots, y_{n-1})$ is

$$\frac{f(y_1 + y_n, \ldots, y_{n-1} + y_n, y_n)}{\int f(y_1 + t, \ldots, y_{n-1} + t, t)\, dt}.$$

It follows that

$$E_0[X_n|\mathbf{y}] = E_0[Y_n|\mathbf{y}] = \frac{\int t f(y_1 + t, \ldots, y_{n-1} + t, t)\, dt}{\int f(y_1 + t, \ldots, y_{n-1} + t, t)\, dt}.$$

This can be reexpressed in terms of the x's as

$$E_0[X_n|\mathbf{y}] = \frac{\int t f(x_1 - x_n + t, \ldots, x_{n-1} - x_n + t, t)\, dt}{\int f(x_1 - x_n + t, \ldots, x_{n-1} - x_n + t, t)\, dt}$$

or, finally, by making the change of variables $u = x_n - t$ as

$$E_0[X_n|\mathbf{y}] = x_n - \frac{\int u f(x_1 - u, \ldots, x_n - u)\,du}{\int f(x_1 - u, \ldots, x_n - u)\,du}.$$

This completes the proof. □

Example 1.21 (Continuation of Example 1.19). As an illustration of (1.28), let us apply it to the situation of Example 1.19. Then

$$f(x_1 - \xi, \ldots, x_n - \xi) = \begin{cases} b^{-n} \text{ if } \xi - \frac{b}{2} \le X_{(1)} \le X_{(n)} \le \xi + \frac{b}{2} \\ 0 \quad \text{otherwise} \end{cases}$$

where b is known. The Pitman estimator is therefore given by

$$\delta^*(\mathbf{x}) = \int_{x_{(n)}-b/2}^{x_{(1)}+b/2} u\,du \left(\int_{x_{(n)}-b/2}^{x_{(1)}+b/2} du \right)^{-1} = \frac{1}{2}[x_{(1)} + x_{(n)}],$$

which agrees with the result of Example 1.19. ||

For most densities, the integrals in (1.28) are difficult to evaluate. The following example illustrates the MRE estimator for one more case.

Example 1.22 Double exponential. Let $X_1, \ldots, X_n$ be iid with double exponential distribution $DE(\xi, 1)$, so that their joint density is $(1/2^n) \times \exp(-\Sigma|x_i - \xi|)$. It is enough to evaluate the integrals in (1.28) over the set where $x_1 < \cdots < x_n$. If $x_k < \xi < x_{k+1}$,

$$\begin{aligned} \Sigma|x_i - \xi| &= \sum_{k+1}^{n}(x_i - \xi) - \sum_{1}^{k}(x_i - \xi) \\ &= \sum_{k+1}^{n} x_i - \sum_{1}^{k} x_i + (2k - n)\xi. \end{aligned}$$

The integration then leads to two sums, both in numerator and denominator of the Pitman estimator. The resulting expression is the desired estimator. ||

So far, the estimator δ has been assumed to be nonrandomized. Let us now consider the role of randomized estimators for equivariant estimation. Recall from the proof of Corollary 1.7.9 that a randomized estimator can be obtained as a nonrandomized estimator $\delta(\mathbf{X}, W)$ depending on $\mathbf{X}$ and an independent random variable W with known distribution. For such an estimator, the equivariance condition (1.18) becomes

$$\delta(\mathbf{X} + a, W) = \delta(\mathbf{X}, W) + a \quad \text{for all } a.$$

There is no change in Theorem 1.4, and Lemma 1.6 remains valid with (1.20) replaced by

$$u(\mathbf{x} + a, w) = u(\mathbf{x}, w) \quad \text{for all } \mathbf{x},\ w, \text{ and } a.$$

The proof of Lemma 1.7 shows that this condition holds if and only if u is a function only of $\mathbf{y}$ and w, so that, finally, in generalization of (1.21), an estimator

$\delta(\mathbf{X}, W)$ is equivariant if and only if it is of the form

$$\delta(\mathbf{X}, W) = \delta_0(\mathbf{X}, W) - v(\mathbf{Y}, W). \tag{1.29}$$

Applying the proof of Theorem 1.10 to (1.29), we see that the risk is minimized by choosing for $v(\mathbf{y}, w)$ the function minimizing

$$E_0\{\rho[\delta_0(\mathbf{X}, w) - v(\mathbf{y}, w)] | \mathbf{y}, w\}.$$

Since the starting δ_0 can be any equivariant estimator, let it be nonrandomized, that is, not dependent on W. Since $\mathbf{X}$ and W are independent, it then follows that the minimizing $v(\mathbf{y}, w)$ will not involve w, so that the MRE estimator (if it exists) will be nonrandomized.

Suppose now that T is a sufficient statistic for ξ. Then, $\mathbf{X}$ can be represented as (T, W), where W has a known distribution (see Section 1.6), and any estimator $\delta(\mathbf{X})$ can be viewed as a randomized estimator based on T. The above argument then suggests that a MRE estimator can always be chosen to depend on T only. However, the argument does not apply since the family $\{P_\xi^T, -\infty < \xi < \infty\}$ no longer needs be a location family. Let us therefore add the assumption that $T = (T_1, \ldots, T_r)$ where $T_i = T_i(\mathbf{X})$ are real-valued and equivariant, that is, satisfy

$$T_i(\mathbf{x} + a) = T_i(\mathbf{x}) + a \quad \text{for all } \mathbf{x} \text{ and } a. \tag{1.30}$$

Under this assumption, the distributions of T do constitute a location family. To see this, let $\mathbf{V} = \mathbf{X} - \xi$ so that $\mathbf{V}$ is distributed with density $f(v_1, \ldots, v_n)$. Then, $T_i(\mathbf{X}) = T_i(\mathbf{V} + \xi) = T_i(\mathbf{V}) + \xi$, and this defines a location family. The earlier argument therefore applies, and under assumption (1.30), an MRE estimator can be found which depends only on T. (For a general discussion of the relationship of invariance and sufficiency, see Hall, Wijsman, and Ghosh 1965, Basu 1969, Berk 1972a, Landers and Rogge 1973, Arnold 1985, Kariya 1989, and Ramamoorthi 1990.)

In Examples 1.16, 1.18 and 1.19, the sufficient statistics $\bar{X}$, $X_{(1)}$, and $(X_{(1)}, X_{(n)})$, respectively, satisfy (1.30), and the previous remark provides an alternative derivation for the MRE estimators in these examples.

It is interesting to compare the results of the present section with those on unbiased estimation in Chapter 2. It was found there that when a UMVU estimator exists, it typically minimizes the risk for all convex loss functions, but that for bounded loss functions not even a locally minimum risk unbiased estimator can be expected to exist. In contrast:

(a) An MRE estimator typically exists not only for convex loss functions but even when the loss function is not so restricted.

(b) On the other hand, even for convex loss functions, the MRE estimator often varies with the loss function.

(c) Randomized estimators need not be considered in equivariant estimation since there are always uniformly better nonrandomized ones.

(d) Unlike UMVU estimators which are frequently inadmissible, the Pitman estimator is admissible under mild assumptions (Stein 1959, and Section 5.4).

(e) The principal area of application of UMVU estimation is that of exponential families, and these have little overlap with location families (see Section 1.5).

(f) For location families, UMVU estimators typically do not exist. (For specific results in this direction, see Bondesson 1975.)

Let us next consider whether MRE estimators are unbiased.

Lemma 1.23 *Let the loss function be squared error.*

(a) When $\delta(\mathbf{X})$ is any equivariant estimator with constant bias b, then $\delta(\mathbf{X}) - b$ is equivariant, unbiased, and has smaller risk than $\delta(\mathbf{X})$.

(b) The unique MRE estimator is unbiased.

(c) If a UMVU estimator exists and is equivariant, it is MRE.

Proof. Part (a) follows from Lemma 2.2.7; (b) and (c) are immediate consequences of (a). □

That an MRE estimator need not be unbiased for general loss functions is seen from Example 1.18 with absolute error as loss. Some light is thrown on the possible failure of MRE estimators to be unbiased by considering the following decision-theoretic definition of unbiasedness, which depends on the loss function L.

Definition 1.24 An estimator δ of $g(\theta)$ is said to be *risk-unbiased* if it satisfies

$$E_\theta L[\theta, \delta(\mathbf{X})] \le E_\theta L[\theta', \delta(\mathbf{X})] \quad \text{for all } \theta' \neq \theta, \tag{1.31}$$

If one interprets $L(\theta, d)$ as measuring how far the estimated value d is from the estimand $g(\theta)$, then (1.31) states that, on the average, δ is at least as close to the true value $g(\theta)$ as it is to any false value $g(\theta')$.

Example 1.25 Mean-unbiasedness. If the loss function is squared error, (1.31) becomes

$$E_\theta[\delta(X) - g(\theta')]^2 \ge E_\theta[\delta(X) - g(\theta)]^2 \quad \text{for all } \theta' \neq \theta. \tag{1.32}$$

Suppose that $E_\theta(\delta^2) < \infty$ and that $E_\theta(\delta) \in \Omega_g$ for all θ, where $\Omega_g = \{g(\theta) : \theta \in \Omega\}$. [The latter condition is, of course, automatically satisfied when $\Omega = (-\infty, \infty)$ and $g(\theta) = \theta$, as is the case when θ is a location parameter.] Then, the left side of (1.32) is minimized by $g(\theta') = E_\theta\delta(X)$ (Example 1.7.17) and the condition of risk-unbiasedness, therefore, reduces to the usual unbiasedness condition

$$E_\theta\delta(X) = g(\theta). \tag{1.33}$$

‖

Example 1.26 Median-unbiasedness. If the loss function is absolute error, (1.31) becomes

$$E_\theta|\delta(X) - g(\theta')| \ge E_\theta|\delta(X) - g(\theta)| \quad \text{for all } \theta' \neq \theta. \tag{1.34}$$

By Example 1.7.18, the left side of (1.34) is minimized by any median of $\delta(X)$. It follows that (1.34) reduces to the condition

$$\text{med}_\theta\delta(X) = g(\theta), \tag{1.35}$$

that is, $g(\theta)$ is a median of $\delta(X)$, provided $E_\theta|\delta| < \infty$ and Ω_g contains a median of $\delta(X)$ for all θ. An estimator δ satisfying (1.35) is called *median-unbiased.* ||

Theorem 1.27 *If δ is MRE for estimating ξ in model (1.9) with loss function (1.15), then it is risk-unbiased.*

Proof. Condition (1.31) now becomes

$$E_\xi \rho[\delta(\mathbf{X}) - \xi'] \geq E_\xi \rho[\delta(\mathbf{X}) - \xi] \quad \text{for all } \xi' \neq \xi,$$

or, if without loss of generality we put $\xi = 0$,

$$E_0 \rho[\delta(\mathbf{X}) - a] \geq E_0 \rho[\delta(\mathbf{X})] \quad \text{for all } a.$$

□

That this holds is an immediate consequence of the fact that $\delta(\mathbf{X}) = \delta_0(\mathbf{X}) - v^*(\mathbf{Y})$ where $v^*(\mathbf{y})$ minimizes (1.22).

2 The Principle of Equivariance

In the present section, we shall extend the invariance considerations of the binomial situation of Example 1.1 and the location families (1.9) to the general situation in which the probability model remains invariant under a suitable group of transformations.

Let X be a random observable taking on values in a sample space $\mathcal{X}$ according to a probability distribution from the family

$$\mathcal{P} = \{P_\theta, \theta \in \Omega\}. \tag{2.1}$$

Denote by $\mathcal{C}$ a class of 1 : 1 transformations g of the sample space onto itself.

Definition 2.1

(i) If g is a 1 : 1 transformation of the sample space onto itself, if for each θ the distribution of $X' = gX$ is again a member of $\mathcal{P}$, say $P_{\theta'}$, and if as θ traverses Ω, so does θ', then the *probability model (2.1) is invariant* under the transformation g.

(ii) If (i) holds for each member of a class of transformations $\mathcal{C}$, then the model (2.1) is *invariant under* $\mathcal{C}$.

A class of transformations that leave a probability model invariant can always be assumed to be a group. To see this, let $G = G(\mathcal{C})$ be the set of all compositions (defined in Section 1.4) of a finite number of transformations $g_1^{\pm 1} \cdots g_m^{\pm 1}$ with $g_1, \ldots, g_m \in \mathcal{C}$, where each of the exponents can be +1 or −1 and where the elements $g_1, \ldots, g_m$ need not be distinct. Then, any element $g \in G$ leaves (2.1) invariant, and G is a group (Problem 2.1), the group *generated* by $\mathcal{C}$.

Example 2.2 Location family.

(a) Consider the location family (1.9) and the group of transformations $\mathbf{X}' = \mathbf{X}+a$, which was already discussed in (1.10) and Example 4.1. It is seen from (1.12) that if $\mathbf{X}$ is distributed according to (1.9) with $\theta = \xi$, then $\mathbf{X}' = \mathbf{X} + a$ has the density (1.9) with $\theta' = \xi' = \xi + a$, so that the model (1.9) is preserved under these transformations.

(b) Suppose now that, in addition, f has the symmetry property

$$f(-\mathbf{x}) = f(\mathbf{x}) \tag{2.2}$$

where $-\mathbf{x} = (-x_1, \ldots, -x_n)$, and consider the transformation $\mathbf{x}' = -\mathbf{x}$. The density of $\mathbf{X}'$ is

$$f(-x_1' - \xi, \ldots, -x_n' - \xi) = f(x_1' - \xi', \ldots, x_n' - \xi')$$

if $\xi' = -\xi$. Thus, model (1.9) is invariant under the transformations $\mathbf{x}' = -\mathbf{x}$, $\xi' = -\xi$, and hence under the group consisting of this transformation and the identity (Problem 2.2). This is not true, however, if f does not satisfy (1.10). If, for example, $X_1, \ldots, X_n$ are iid according to the exponential distribution $E(\xi, 1)$, then the variables $-X_1, \ldots, -X_n$ no longer have an exponential distribution. ||

Let $\{gX, g \in G\}$ be a group of transformations of the sample space which leave the model invariant. If gX has the distribution $P_{\theta'}$, then $\theta' = \bar{g}\theta$ is a function which maps Ω onto Ω, and the transformation $\bar{g}\theta$ is 1 : 1, provided the distributions P_θ, $\theta \in \Omega$ are distinct (Problem 2.3). It is easy to see that the transformations $\bar{g}$ then also form a group which will be denoted by $\bar{G}$ (Problem 2.4). From the definition of $\bar{g}\theta$, it follows that

$$P_\theta(gX \in A) = P_{\bar{g}\theta}(X \in A) \tag{2.3}$$

where the subscript on the left side indicates the distribution of X, not that of gX. More generally, for a function ψ whose expectation is defined,

$$E_\theta[\psi(gX)] = E_{\bar{g}\theta}[\psi(X)]. \tag{2.4}$$

We have now generalized the transformations (1.10) and (1.11), and it remains to consider (1.13). This last generalization is most easily introduced by an example.

Example 2.3 Two-sample location family. Let $\mathbf{X} = (X_1, \ldots, X_m)$ and $\mathbf{Y} = (Y_1, \ldots, Y_n)$ and suppose that $(\mathbf{X}, \mathbf{Y})$ has the joint density

$$f(\mathbf{x} - \xi, \mathbf{y} - \eta) = f(x_1 - \xi, \ldots, x_m - \xi, y_1 - \eta, \ldots, y_n - \eta). \tag{2.5}$$

This model remains invariant under the transformations

$$g(\mathbf{x}, \mathbf{y}) = (\mathbf{x} + a, \mathbf{y} + b), \quad \bar{g}(\xi, \eta) = (\xi + a, \eta + b). \tag{2.6}$$

Consider the problem of estimating

$$\Delta = \eta - \xi. \tag{2.7}$$

If the transformed variables are denoted by

$$\mathbf{x}' = \mathbf{x} + a, \quad \mathbf{y}' = \mathbf{y} + b, \quad \xi' = \xi + a, \quad \eta' = \eta + b,$$

then Δ is transformed into $\Delta' = \Delta + (b - a)$. Hence, an estimated value d, when expressed in the new coordinates, becomes

$$d' = d + (b - a). \tag{2.8}$$

For the problem to remain invariant, we require, analogously to (1.14), that the loss function $L(\xi, \eta; d)$ satisfies

$$L[\xi + a, \eta + b; d + (b - a)] = L(\xi, \eta; d). \tag{2.9}$$

It is easy to see (Problem 2.5) that this is the case if and only if L depends only on the difference $(\eta - \xi) - d$, that is, if

$$L(\xi, \eta; d) = \rho(\Delta - d). \tag{2.10}$$

Suppose, next, that instead of estimating $\eta - \xi$, the problem is that of estimating

$$h(\xi, \eta) = \xi^2 + \eta^2.$$

Under the transformations (2.6), $h(\xi, \eta)$ is transformed into $(\xi + a)^2 + (\eta + b)^2$. This does not lead to an analog of (2.8) since the transformed value does not depend on (ξ, η) only though $h(\xi, \eta)$. Thus, the form of the function to be estimated plays a crucial role in invariance considerations.

‖

Now, consider the general problem of estimating $h(\theta)$ in model (2.1), which is assumed to be invariant under the transformations $X' = gX$, $\theta' = \bar{g}\theta$, $g \in G$. The additional assumption required is that for any given $\bar{g}$, $h(\bar{g}\theta)$ depends on θ only through $h(\theta)$, that is,

$$h(\theta_1) = h(\theta_2) \quad \text{implies} \quad h(\bar{g}\theta_1) = h(\bar{g}\theta_2). \tag{2.11}$$

The common value of $h(\bar{g}\theta)$ for all θ's to which h assigns the same value will then be denoted by

$$h(\bar{g}\theta) = g^* h(\theta). \tag{2.12}$$

If $\mathcal{H}$ is the set of values taken on by $h(\theta)$ as θ ranges over Ω, the transformations g^* are 1 : 1 from $\mathcal{H}$ onto itself. [Problem 2.8(a)]. As $\bar{g}$ ranges over $\bar{G}$, the transformations g^* form a group G^* (Problem 2.6).

The estimated value d of $h(\theta)$ when expressed in the new coordinates becomes

$$d' = g^* d. \tag{2.13}$$

Since the problems of estimating $h(\theta)$ in terms of (X, θ, d) or $h(\theta')$ in terms of (X', θ', d') represent the same physical situation expressed in a new coordinate system, the loss function should satisfy $L(\theta', d') = L(\theta, d)$.

This leads to the following definition.

Definition 2.4 If the probability model (2.1) is invariant under g, the loss function L satisfies

$$L(\bar{g}\theta, g^* d) = L(\theta, d), \tag{2.14}$$

and $h(\theta)$ satisfies (2.11), the problem of estimating $h(\theta)$ with loss function L is *invariant under* g.

In this discussion, it was tacitly assumed that the set $\mathcal{D}$ of possible decisions coincides with $\mathcal{H}$. This need not, however, be the case. In Chapter 2, for example, estimators of a variance were permitted (with some misgiving) to take on negative values. In the more general case that $\mathcal{H}$ is a subset of $\mathcal{D}$, one can take the condition that (2.14) holds for all θ as the definition of g^*d. If $L(\theta, d) = L(\theta, d')$ for all θ implies $d = d'$, as is typically the case, g^*d is uniquely defined by the above condition, and g^* is 1 : 1 from $\mathcal{D}$ onto itself [Problem 2.8(b)].

In an invariant estimation problem, if δ is the estimator that we would like to use to estimate $h(\theta)$, there are two natural ways of estimating $g^*h(\theta)$, the estimand $h(\theta)$ expressed in the transformed system. One of these generalizes the estimators (1.6) and (1.16), and the other the estimators (1.6) and (1.17) of the preceding section.

1. Functional Equivariance. Quite generally, if we have decided to use $\delta(X)$ to estimate $h(\theta)$, it is natural to use

$$\phi[\delta(X)] \text{ as the estimator of } \phi[\bar{h}(\theta)],$$

for any function ϕ. If, for example, $\delta(X)$ is used to estimate the length θ of the edge of a cube, it is natural to estimate the volume θ^3 of the cube by $[\delta(X)]^3$. Hence, if d is the estimated value of $h(\theta)$, then g^*d should be the estimated value of $g^*h(\theta)$.

Applying this to $\phi = g^*$ leads to

$$g^*\delta(X) \text{ as the estimator of } g^*h(\theta) \tag{2.15}$$

when $\delta(X)$ is used to estimate $h(\theta)$.

2. Formal Invariance. Invariance under transformations g, $\bar{g}$, and g^* of the estimation of $h(\theta)$ means that the problem of estimating $h(\theta)$ in terms of X, θ, and d and that of estimating $g^*h(\theta)$ in terms of X', θ', and d' are formally the same, and should therefore be treated the same. In generalization of (1.6) and (1.16), this means that we should use

$$\delta(X') = \delta(gX) \text{ to estimate } g^*[\bar{h}(\theta)] = h(\bar{g}\theta). \tag{2.16}$$

It seems desirable that these two principles should lead to the same estimator and hence that

$$\delta(gX) = g^*\delta(X). \tag{2.17}$$

Definition 2.5 In an invariant estimation problem, an estimator $\delta(X)$ is said to be *equivariant* if it satisfies (2.17) for all $g \in G$.

As was discussed in Section 1, the arguments for (2.15) and (2.16) are of a very different nature. The appropriateness of (2.16) results from the symmetries exhibited by the situation and represented mathematically by the invariance of the problem under the transformations $g \in G$. It gives expression to the idea that if some symmetries are present in an estimation problem, the estimators should possess the corresponding symmetries. It follows that (2.16) is no longer appropriate if the symmetry is invalidated by asymmetric prior information; if, for example, θ is known to be restricted to a subset ω of the parameter space Ω, for which $\bar{g}\omega \neq \omega$, as was the case mentioned at the end of Example 1.1.1 and after Definition 1.3. In contrast, the argument leading to (2.15) is quite independent of any symmetry assumptions and simply reflects the fact that if $\delta(X)$ is a reasonable estimator of, say, θ then $\phi[\delta(X)]$ is a reasonable estimator of $\phi(\theta)$.

Example 2.6 Continuation of Example 2.3. In Example 2.3, $h(\xi, \eta) = \eta - \xi$, and by (2.8), $g^*d = d + (b - a)$. It follows that (2.17) becomes

$$\delta(\mathbf{x} + a, \mathbf{y} + b) = \delta(\mathbf{x}, \mathbf{y}) + b - a. \tag{2.18}$$

If $\delta_0(\mathbf{X})$ and $\delta_0'(\mathbf{Y})$ are location equivariant estimators of ξ and η, respectively, then $\delta(\mathbf{X}, \mathbf{Y}) = \delta_0'(\mathbf{Y}) - \delta_0(\mathbf{X})$ is an equivariant estimator of $\eta - \xi$. ||

The following theorem generalizes Theorem 1.4 to the present situation.

Theorem 2.7 *If δ is an equivariant estimator in a problem which is invariant under a transformation g, then the risk function of δ satisfies*

$$R(\bar{g}\theta, \delta) = R(\theta, \delta) \quad \textit{for all} \quad \theta. \tag{2.19}$$

Proof. By definition

$$R(\bar{g}\theta, \delta) = E_{\bar{g}\theta} L[\bar{g}\theta, \delta(X)].$$

It follows from (2.4) that the right side is equal to

$$E_\theta L[\bar{g}\theta, \delta(gX)] = E_\theta L[\bar{g}\theta, g^*\delta(X)] = R(\theta, \delta).$$

□

Looking back on Section 1, we see that the crucial fact underlying the success of the invariance approach was the constancy of the risk function of any equivariant estimator. Theorem 2.7 suggests the following simple condition for this property to obtain.

A group G of transformations of a space is said to be *transitive* if for any two points there is a transformation in G taking the first point into the second.

Corollary 2.8 *Under the assumptions of Theorem 2.7, if $\bar{G}$ is transitive over the parameter space Ω, then the risk function of any equivariant estimator is constant, that is, independent of θ.*

When the risk function of every equivariant estimator is constant, the best equivariant estimator (MRE) is obtained by minimizing that constant, so that a uniformly minimum risk equivariant estimator will then typically exist. In such problems, alternative characterizations of the best equivariant estimator can be obtained. (See Problems 2.11 and 2.12.) Berk (1967a) and Kariya (1989) provide a rigorous treatment, taking account of the associated measurability problems. A Bayesian approach to the derivation of best equivariant estimators is treated in Section 4.4.

Example 2.9 Conclusion of Example 2.3. In this example, $\theta = (\xi, \eta)$ and $\bar{g}\theta = (\xi + a, \eta + b)$. This group of transformations is transitive over Ω since, given any two points (ξ, η) and (ξ', η'), a and b exist such that $\xi + a = \xi'$, and $\eta + b = \eta'$. The MRE estimator can now be obtained in exact analogy to Section 3.1 (Problems 1.13 and 1.14). ||

The estimation problem treated in Section 1 was greatly simplified by the fact that it was possible to dispense with randomized estimators. The corresponding result holds quite generally when $\bar{G}$ is transitive. If an estimator δ exists which is MRE among all nonrandomized estimators, it is then also MRE when randomization is permitted. To see this, note that a randomized estimator can be represented as $\delta'(X, W)$ where W is independent of X and has a known distribution and that it is equivariant if $\delta'(gX, W) = g^*\delta'(X, W)$. Its risk is again constant, and for any $\theta = \theta_0$, it is equal to $E[h(W)]$ where

$$h(w) = E_{\theta_0}\{L[\delta'(X, w), \theta_0]\}.$$

This risk is minimized by minimizing $h(w)$ for each w. However, by assumption, $\delta'(X, w) = \delta(X)$ minimizes $h(w)$, and hence the MRE estimator can be chosen to be nonrandomized. The corresponding result need not hold when $\bar{G}$ is not transitive. A counterexample is given in Example 5.1.8.

Definition 2.10 For a group $\mathcal{G}$ of transformations of Ω, two points $\theta_1, \theta_2 \in \Omega$ are *equivalent* if there exists a $g \in \mathcal{G}$ such that $g\theta_1 = \theta_2$. The totality of points equivalent to a given point (and hence to each other) is called an *orbit* of $\mathcal{G}$. The group $\mathcal{G}$ is *transitive* over Ω if it has only one orbit.

For the most part, we will consider transitive groups; however, there are some groups of interest that are not transitive.

Example 2.11 Binomial transformation group. Let $X \sim$ binomial(n, p), $0 < p < 1$, and consider the group of transformations.

$$gX = n - X,$$
$$\bar{g}p = 1 - p.$$

The orbits are the pairs $(p, 1 - p)$. The group is not transitive. ||

Example 2.12 Orbits of a scale group. Let $X_1, \ldots, X_n$ be iid $N(\mu, \sigma^2)$, both unknown, and consider estimation of σ^2. The model remains invariant under the scale group

$$gX_i = aX_i,$$
$$\bar{g}(\mu, \sigma^2) = (a\mu, a^2\sigma^2), \quad a > 0.$$

We shall now show that (μ_1, σ_1^2) and (μ_2, σ_2^2) lie on the same orbit if and only if $\mu_1/\sigma_1 = \mu_2/\sigma_2$.

On the one hand, suppose that $\mu_1/\sigma_1 = \mu_2/\sigma_2$. Then, $\mu_2/\mu_1 = \sigma_2/\sigma_1 = a$, say, and $\mu_2 = a\mu_1$; $\sigma_2^2 = a^2\sigma_1^2$. On the other hand, if $\mu_2 = a\mu_1$ and $\sigma_2^2 = a^2\sigma_1^2$, then $\mu_2/\mu_1 = a$ and $\sigma_2^2/\sigma_1^2 = a$. Thus, the values of $\tau = \mu/\sigma$ can be used to label the orbits of $\mathcal{G}$. ||

The following corollary is a straightforward consequence of Theorem 2.7.

Corollary 2.13 *Under the assumptions of Theorem 2.7, the risk function of any equivariant estimator is constant on the orbits of $\mathcal{G}$.*

Proof. See Problem 2.15. □

In Section 1.4, *group families* were introduced as families of distributions generated by subjecting a random variable with a fixed distribution to a group of transformations. Consider now a family of distributions $\mathcal{P} = \{P_\theta, \theta \in \Omega\}$ which remains invariant under a group G for which $\bar{G}$ is transitive over Ω and $g_1 \neq g_2$

implies $\bar{g}_1 \neq \bar{g}_2$. Let θ_0 be any fixed element of Ω. Then $\mathcal{P}$ is exactly the group family of distributions of $\{gX, g \in G\}$ when X has distribution P_{θ_0}.

Conversely, let $\mathcal{P}$ be the group family of the distributions of gX as g varies over G, when X has a fixed distribution P, so that $\mathcal{P} = \{P_g, g \in G\}$. Then, g can serve as the parameter θ and G as the parameter space. In this notation, the starting distribution P becomes P_e, where e is the identity transformation. Thus, a family of distributions remains invariant under a transitive group of transformations of the sample space if and only if it is a group family.

When an estimation problem is invariant under a group of transformations and an MRE estimator exists, this seems the natural estimator to use—of the various principles we shall consider, equivariance, where it applies, is perhaps the most convincing. Yet, even this principle can run into difficulties. The following example illustrates the possibility of a problem remaining invariant under two different groups, G_1 and G_2, which lead to two different MRE estimators δ_1 and δ_2.

Example 2.14 Counterexample. Let the pairs (X_1, X_2) and (Y_1, Y_2) be independent, each with a bivariate normal distribution with mean zero. Let their covariance matrices be $\Sigma = [\sigma_{ij}]$ and $\Delta\Sigma = [\Delta\sigma_{ij}]$, $\Delta > 0$, and consider the problem of estimating Δ.

Let G_1 be the group of transformations

(2.20) $$X_1' = a_1X_1 + a_2X_2 \qquad Y_1' = c(a_1Y_i + a_2Y_2)$$ $$X_2' = bX_2 \qquad Y_2' = cbY_2\,.$$

Then, (X_1', X_2') and (Y_1', Y_2') will again be independent and each will have a bivariate normal distribution with zero mean. If the covariance matrix of (X_1', X_2') is Σ', that of (Y_1', Y_2') is $\Delta'\Sigma'$ where $\Delta' = c^2\Delta$ (Problem 2.16). Thus, G_1 leaves the model invariant.

If $h(\Sigma, \Delta) = \Delta$, (2.11) clearly holds, (2.12) and (2.13) become

(2.21) $$\Delta' = c^2\Delta, \quad d' = c^2 d,$$

respectively, and a loss function $L(\Delta, d)$ satisfies (2.14) provided $L(c^2\Delta, c^2 d) = L(\Delta, d)$. This condition holds if and only if L is of the form

(2.22) $$L(\Delta, d) = \rho(d/\Delta).$$

[For the necessity of (2.22), see Problem 2.10.]

An estimator δ of Δ is equivariant under the above transformation if

(2.23) $$\delta(\mathbf{x}', \mathbf{y}') = c^2\delta(\mathbf{x}, \mathbf{y}).$$

We shall now show that (2.23) holds if and only if

(2.24) $$\delta(\mathbf{x}, \mathbf{y}) = \frac{ky_2^2}{x_2^2} \quad \text{for some value of } k \text{ a.e.}$$

It is enough to prove this for the reduced sample space in which the matrix $\begin{pmatrix} x_1 x_2 \\ y_1 y_2 \end{pmatrix}$ is nonsingular and in which both x_2 and y_2 are $\neq 0$, since the rest of the sample space has probability zero.

Let G_1' be the subgroup of G_1 consisting of the transformations (2.20) with $b = c = 1$. The condition of equivariance under these transformations reduces to

$$\delta(\mathbf{x}', \mathbf{y}') = \delta(\mathbf{x}, \mathbf{y}). \tag{2.25}$$

This is satisfied whenever δ depends only on x_2 and y_2 since $x_2' = x_2$ and $y_2' = y_2$. To see that this condition is also necessary for (2.25), suppose that δ satisfies (2.25) and let $(x_1', x_2; y_1', y_2)$ and $(x_1, x_2; y_1, y_2)$ be any two points in the reduced sample space which have the same second coordinates. Then, there exist a_1 and a_2 such that

$$x_1' = a_1x_1 + a_2x_2;\ y_1' = a_1y_1 + a_2y_2,$$

that is, there exists $g \in G_1'$ for which $g(\mathbf{x}, \mathbf{y}) = (\mathbf{x}'\mathbf{y}')$, and hence δ depends only on x_2, y_2.

Consider now any $\delta'(x_2, y_2)$. To be equivariant under the full group G_1, δ' must satisfy

$$\delta'(bx_2, cby_2) = c^2\delta'(x_2, y_2). \tag{2.26}$$

For $x_2 = y_2 = 1$, this condition becomes

$$\delta'(b, cb) = c^2\delta'(1, 1)$$

and hence reduces to (2.24) with $x_2 = b$, $y_2 = bc$, and $k = \delta'(1, 1)$. This shows that (2.24) is necessary for δ to be equivariant; that it is sufficient is obvious.

The best equivariant estimator under G_1 is thus $k^*Y_2^2/X_2^2$ where k^* is a value which minimizes

$$E_\Delta \rho\left(\frac{kY_2^2}{\Delta X_2^2}\right) = E_1\rho\left(\frac{kY_2^2}{X_2^2}\right).$$

Such a minimizing value will typically exist. Suppose, for example, that the loss is 1 if $|d - \Delta|/\Delta > 1/2$ and zero otherwise. Then, k^* is obtained by maximizing

$$P_1\left(\left|k\frac{Y_2^2}{X_2^2} - 1\right| < \frac{1}{2}\right) = P_1\left(\frac{1}{2k} < \frac{Y_2^2}{X_2^2} < \frac{3}{2k}\right).$$

As $k \to 0$ or ∞, this probability tends to zero, and a maximizing value therefore exists and can be determined from the distribution of Y_2^2/X_2^2 when $\Delta = 1$.

Exactly the same argument applies if G_1 is replaced by the transformations G_2

$$\begin{array}{l|l} X_1' = bX_1 & Y_1' = cbY_1 \\ X_2' = a_1X_1 + a_2X_2 & Y_2' = c(a_1Y_1 + a_2Y_2) \end{array}$$

and leads to the MRE estimator $k^*Y_1^2/X_1^2$. See Problems 2.19 and 2.20. ‖

In the location case, it turned out (Theorem 1.27) that an MRE estimator is always risk-unbiased. The extension of this result to the general case requires some assumptions.

Theorem 2.15 *If $\bar{G}$ is transitive and G^* commutative, then an MRE estimator is risk-unbiased.*

Proof. Let δ be MRE and $\theta, \theta' \in \Omega$. Then, by the transitivity of $\bar{G}$, there exists $\bar{g} \in \bar{G}$ such that $\theta = \bar{g}\theta'$, and hence

$$E_\theta L[\theta', \delta(X)] = E_\theta L[\bar{g}^{-1}\theta, \delta(X)] = E_\theta L[\theta, g^*\delta(X)].$$

Now, if $\delta(X)$ is equivariant, so is $g^*\delta(X)$ (Problem 2.18), and, therefore, since δ is MRE,

$$E_\theta L[\theta, g^*\delta(X)] \geq E_\theta L[\theta, \delta(X)],$$

which completes the proof. □

Transitivity of $\bar{G}$ will usually [but not always, see Example 2.14(a) below] hold when an MRE estimator exists. On the other hand, commutativity of G^* imposes a severe restriction. That the theorem need not be valid if either condition fails is shown by the following example.

Example 2.16 Counterexample. Let X be $N(\xi, \sigma^2)$ with both parameters unknown, let the estimand be ξ and the loss function be

$$L(\xi, \sigma; d) = (d - \xi)^2/\sigma^2. \tag{2.27}$$

(a) The problem remains invariant under the group G_1; $gx = x + c$. It follows from Section 1 that X is MRE under G_1. However, X is not risk-unbiased (Problem 2.19). Here, $\bar{G}_1$ is the group of transformations

$$\bar{g}(\xi, \sigma) = (\xi + c, \sigma),$$

which is clearly not transitive.

If the loss function is replaced by $(d - \xi)^2$, the problem will remain invariant under G_1; X remains equivariant but is now risk-unbiased by Example 1.25. Transitivity of $\bar{G}$ is thus not necessary for the conclusion of Theorem 2.15.

(b) When the loss function is given by (2.27), the problem also remains invariant under the larger group $G_2 : ax+c, 0 < a$. Since X is equivariant under G_2 and MRE under G_1, it is also MRE under G_2. However, as stated in (i), X is not risk-unbiased with respect to (1.35). Here, G_2^* is the group of transformations $g^*d = ad + c$, and this is not commutative (Problem 2.19). ||

The location problem considered in Section 1 provides an important example in which the assumptions of Theorem 2.15 are satisfied, and Theorem 1.27 is the specialization of Theorem 2.15 to that case. The scale problem, which will be considered in Section 3, can also provide another illustration.

We shall not attempt to generalize to the present setting the characterization of equivariant estimators which was obtained for the location case in Theorem 1.8. Some results in this direction, taking account also of the associated measurability problems, can be found in Eaton (1989) or Wijsman (1990). Instead, we shall consider in the next section some other extensions of the problem treated in Section 1.

We close this section by exhibiting a family $\mathcal{P}$ of distributions for which there exists no group leaving $\mathcal{P}$ invariant (except the trivial group consisting of the identity only).

Theorem 2.17 *Let X be distributed according to the power series distribution [see (2.3.9)]*

$$P(X = k) = c_k \theta^k h(\theta); \quad k = 0, 1, \ldots, 0 < \theta < \infty. \tag{2.28}$$

If $c_k > 0$ for all k, then there does not exist a transformation $gx = g(x)$ leaving the family (2.28) invariant except the identity transformation $g(x) = x$ for all x.

Proof. Suppose $Y = g(X)$ is a transformation leaving (2.28) invariant, and let $g(k) = a_k$ and $\bar{g}\theta = \mu$. Then, $P_\theta(X = k) = P_\mu(Y = a_k)$ and hence

$$c_k \theta^k h(\theta) = c_{a_k} \mu^{a_k} h(\mu). \tag{2.29}$$

Replacing k by $k+1$ and dividing the resulting equation by (2.29), we see that

$$\frac{c_{k+1}}{c_k}\,\theta = \frac{c_{a_{k+1}}}{c_{a_k}}\,\mu^{a_{k+1}-a_k}. \tag{2.30}$$

Replacing k by $k+1$ in (2.30) and dividing the resulting equation by (2.30) shows that

$$\mu^{a_{k+2}-a_{k+1}} \text{ is proportional to } \mu^{a_{k+1}-a_k} \text{ for all } 0 < \mu < m$$

and hence that

$$a_{k+2} - a_{k+1} = a_{k+1} - a_k.$$

If we denote this common value by Δ, we get

$$a_k = a_0 + k\Delta \qquad \text{for} \qquad k = 0, 1, 2, \ldots. \tag{2.31}$$

Invariance of the model requires the set (2.31) to be a permutation of the set $\{0, 1, 2, \ldots\}$. This implies that $\Delta > 0$ and hence that $a_0 = 0$ and $\Delta = 1$, i.e., that $a_k = k$ and g is the identity. □

Example 2.11 shows that this result no longer holds if $c_k = 0$ for k exceeding some k_0; see Problem 2.28.

3 Location-Scale Families

The location model discussed in Section 1 provides a good introduction to the ideas of equivariance, but it is rarely realistic. Even when it is reasonable to assume the form of the density f in (1.9) to be known, it is usually desirable to allow the model to contain an unknown scale parameter. The standard normal model according to which $X_1, \ldots, X_n$ are iid as $N(\xi, \sigma^2)$ is the most common example of such a location-scale model. In this section, we apply some of the general principles developed in Section 2 to location-scale models, as well as some other group models. As preparation for the analysis of these models, we begin with the case, which is of interest also in its own right, in which the only unknown parameter is scale parameter.

Let $\mathbf{X} = (X_1, \ldots, X_n)$ have a joint probability density

$$\frac{1}{\tau^n} f\left(\frac{\mathbf{x}}{\tau}\right) = \frac{1}{\tau^n} f\left(\frac{x_1}{\tau}, \ldots, \frac{x_n}{\tau}\right), \quad \tau > 0, \tag{3.1}$$

where f is known and τ is an unknown *scale parameter*. This model remains invariant under the transformations

$$X_i' = bX_i, \quad \tau' = b\tau \quad \text{for } b > 0. \tag{3.2}$$

The estimand of primary interest is $h(\tau) = \tau^r$. Since h is strictly monotone, (2.11) is vacuously satisfied. Transformations (3.2) induce the transformations

$$h(\tau) \to b^r\tau^r = b^r h(\tau) \quad \text{and} \quad d' = b^r d, \tag{3.3}$$

and the loss function L is invariant under these transformations, provided

$$L(b\tau, b^r d) = L(\tau, d). \tag{3.4}$$

This is the case if and only if it is of the form (Problem 3.1)

$$L(\tau, d) = \gamma\left(\frac{d}{\tau^r}\right). \tag{3.5}$$

Examples are

$$L(\tau, d) = \frac{(d - \tau^r)^2}{\tau^{2r}} \quad \text{and} \quad L(\tau, d) = \frac{|d - \tau^r|}{\tau^r} \tag{3.6}$$

but not squared error.

An estimator δ of τ^r is equivariant under (3.2), or *scale equivariant*, provided

$$\delta(b\mathbf{X}) = b^r\delta(\mathbf{X}). \tag{3.7}$$

All the usual estimators of τ are scale equivariant; for example, the standard deviation $\sqrt{\Sigma\left(X_i - \bar{X}\right)^2/(n-1)}$, the mean deviation $\Sigma|X_i - \bar{X}|/n$, the range, and the maximum likelihood estimator [Problem 3.1(b)].

Since the group $\bar{G}$ of transformations $\tau' = b\tau$, $b > 0$, is transitive over Ω, the risk of any equivariant estimator is constant by Corollary 2.8, so that one can expect an MRE estimator to exist. To derive it, we first characterize the totality of equivariant estimators.

Theorem 3.1 *Let* $\mathbf{X}$ *have density (3.1) and let* $\delta_0(\mathbf{X})$ *be any scale equivariant estimator of* τ^r. *Then, if*

$$z_i = \frac{x_i}{x_n} \quad (i = 1, \ldots, n-1) \quad \text{and} \quad z_n = \frac{x_n}{|x_n|} \tag{3.8}$$

and if $\mathbf{z} = (z_1, \ldots, z_n)$, *a necessary and sufficient condition for* δ *to satisfy (3.7) is that there exists a function* $w(\mathbf{z})$ *such that*

$$\delta(\mathbf{x}) = \frac{\delta_0(\mathbf{x})}{w(\mathbf{z})}.$$

Proof. Analogous to Lemma 1.6, a necessary and sufficient condition for δ to satisfy (3.7) is that it is of the form $\delta(\mathbf{x}) = \delta_0(\mathbf{x})/u(\mathbf{x})$ where (Problem 3.4)

$$u(b\mathbf{x}) = u(\mathbf{x}) \quad \text{for all } \mathbf{x} \quad \text{and all } b > 0. \tag{3.9}$$

It remains to show that (3.9) holds if and only if u depends on $\mathbf{x}$ only through $\mathbf{z}$. Note here that $\mathbf{z}$ is defined when $x_n \neq 0$ and, hence, with probability 1. That any function of $\mathbf{z}$ satisfies (3.9) is obvious. Conversely, if (3.9) holds, then

$$u(x_1, \ldots, x_n) = u\left(\frac{x_1}{x_n}, \ldots, \frac{x_{n-1}}{x_n}, \frac{x_n}{|x_n|}\right);$$

hence, u does depend only on $\mathbf{z}$, as was to be proved. □

Example 3.2 Scale equivariant estimator based on one observation. Suppose that $n = 1$. Then, the most general estimator satisfying (3.7) is of the form $X^r/w(Z)$ where $Z = X/|X|$ is ± 1 as X is $>$ or < 0, so that

$$\delta(X) = \begin{cases} AX^r & \text{if } X > 0 \\ -BX^r & \text{if } X < 0, \end{cases}$$

A, B being two arbitrary constants. ||

Let us now determine the MRE estimator for a general scale family.

Theorem 3.3 *Let* $\mathbf{X}$ *be distributed according to (3.1) and let* $\mathbf{Z}$ *be given by (3.8). Suppose that the loss function is given by (3.5) and that there exists an equivariant estimator* δ_0 *of* τ^r *with finite risk. Assume that for each* $\mathbf{z}$, *there exists a number* $w(\mathbf{z}) = w^*(\mathbf{z})$ *which minimizes*

$$E_1\{\gamma[\delta_0(\mathbf{X})/w(\mathbf{z})]|\mathbf{z}\}. \tag{3.10}$$

Then, an MRE estimator δ^* *of* τ^r *exists and is given by*

$$\delta^*(\mathbf{X}) = \frac{\delta_0(\mathbf{X})}{w^*(\mathbf{X})}. \tag{3.11}$$

The proof parallels that of Theorem 1.10.

Corollary 3.4 *Under the assumptions of Theorem 3.3, suppose that* $\rho(v) = \gamma(e^v)$ *is convex and not monotone. Then, an MRE estimator of* τ^r *exists; it is unique if* ρ *is strictly convex.*

Proof. By replacing $\gamma(w)$ by $\rho(\log w)$ [with $\rho(-\infty) = \gamma(0)$], the result essentially reduces to that of Corollary 1.11. This argument requires that $\delta \geq 0$, which can be assumed without loss of generality (Problem 3.2). □

Example 3.5 Standardized power loss. Consider the loss function

$$L(\tau, d) = \frac{|d - \tau^r|^p}{\tau^{pr}} = \left|\frac{d}{\tau^r} - 1\right|^p = \gamma\left(\frac{d}{\tau^r}\right) \tag{3.12}$$

with $\gamma(v) = |v - 1|^p$. Then, ρ is strictly convex for $v > 0$, provided $p \geq 1$ (Problem 3.5). Under the assumptions of Theorem 3.3, if we set

$$\gamma\left(\frac{d}{\tau^r}\right) = \frac{(d - \tau^r)^2}{\tau^{2r}}, \tag{3.13}$$

then (Problem 3.10)

$$\delta^*(\mathbf{X}) = \frac{\delta_0(\mathbf{X}) E_1[\delta_0(\mathbf{X})|\mathbf{Z}]}{E_1[\delta_0^2(\mathbf{X})|\mathbf{Z}]}; \tag{3.14}$$

if

$$\gamma\left(\frac{d}{\tau^r}\right) = \frac{|d - \tau^r|}{\tau^r}, \tag{3.15}$$

then $\delta^*(\mathbf{X})$ is given by (3.11), with $w^*(\mathbf{Z})$ any *scale median* of $\delta_0(\mathbf{X})$ under the conditional distribution of $\mathbf{X}$ given $\mathbf{Z}$ and with $\tau = 1$, that is, $w^*(\mathbf{z})$ satisfies

$$E(\mathbf{X}|\mathbf{Z}) I(X \geq w^*(\mathbf{Z})) = E(\mathbf{X}|\mathbf{Z}) I(X \leq w^*(\mathbf{Z})) \tag{3.16}$$

(Problems 3.7 and 3.10). ||

Example 3.6 Continuation of Example 3.2. Suppose that $n = 1$, and $X > 0$ with probability 1. Then, the arguments of Theorem 3.3 and Example 3.5 show that if X^r has finite risk, the MRE estimator of τ^r is X^r/w^* where w^* is any value minimizing

$$E_1[\gamma(X^r/w)]. \tag{3.17}$$

In particular, the MRE estimator is

$$X^r E_1(X^r)/E_1(X^{2r}) \tag{3.18}$$

when the loss is (3.13), and it is X^r/w^*, where w^* is any scale median of X^r for $\tau = 1$, when the loss is (3.15). ||

Example 3.7 MRE for normal variance, known mean. Let $X_1, \ldots, X_n$ be iid according to $N(0, \sigma^2)$ and consider the estimation of σ^2. For $\delta_0 = \Sigma X_i^2$, it follows from Basu's theorem that δ_0 is independent of $\mathbf{Z}$ and hence that $w^*(\mathbf{z}) = w^*$ is a constant determined by minimizing (3.17) with ΣX_i^2 in place of X^r. For the loss function (3.13) with $r = 2$, the MRE estimator turns out to be $\Sigma X_i^2/(n+2)$ [Equation (2.2.26) or Problem 3.7]. ||

Quite generally, when the loss function is (3.13), the MRE estimator of τ^r is given by

$$\delta^*(\mathbf{x}) = \frac{\int_0^\infty v^{n+r-1} f(vx_1, \ldots, vx_n)\,dv}{\int_0^\infty v^{n+2r-1} f(vx_1, \ldots, vx_n)\,dv}, \tag{3.19}$$

and in this form, it is known as the *Pitman estimator* of τ^r. The proof parallels that of Theorem 1.20 (Problem 3.16).

The loss function (3.13) satisfies

$$\lim_{d\to\infty} L(\tau, d) = \infty \text{ but } \lim_{d\to 0} L(\tau, d) = 1,$$

so that it assigns much heavier penalties to overestimation than to underestimation. An alternative to the loss function (3.13) and (3.15), first introduced by Stein (James and Stein, 1961), and known as *Stein's loss*, is given by

$$L_s(\tau, d) = (d/\tau^r) - \log(d/\tau^r) - 1. \tag{3.20}$$

For this loss, $\lim_{d\to\infty} L_s(\tau, d) = \lim_{d\to 0} L_s(\tau, d) = \infty$; it is thus somewhat more evenhanded. For another justification of (3.20), see Brown 1968, 1990b and also Dey and Srinivasan 1985.

The change in the estimator (3.14) if (3.13) is replaced by (3.20) is shown in the following corollary.

Corollary 3.8 *Under the assumptions of Theorem 3.3, if the loss function is given by (3.20), the MRE estimator δ^* of τ^r is uniquely given by*

$$\delta_s^* = \delta_0(\mathbf{X})/E_1(\delta_0(\mathbf{X})|\mathbf{z}). \tag{3.21}$$

Proof. Problem 3.19. □

In light of the above discussion about skewness of the loss function, it is interesting to compare δ_s^* of (3.21) with δ^* of (3.14). It is clear that $\delta_s^* \geq \delta^*$ if and only if $E_1(\delta_0^2(\mathbf{X})|\mathbf{Z}) \geq [E_1(\delta_0(\mathbf{X})|Z)]^2$, which will always be the case. Thus, L_s results in an estimator which is larger.

Example 3.9 Normal scale estimation under Stein's loss. For the situation of Example 3.7, with $r = 2$, the MLE is $\delta_s^*(\mathbf{x}) = \Sigma X_i^2/n$ which is always larger than $\delta^* = \Sigma X_i^2/(n+2)$, the MRE estimator under $L_2(\tau, d)$. Brown (1968) explores the loss function L_s further, and shows that it is the only scale invariant loss function for which the UMVU estimator is also the MRE estimator. ||

So far, the estimator δ has been assumed to be nonrandomized. Since $\bar{G}$ is transitive over Ω, it follows from the result proved in the preceding section that randomized estimators need not be considered. It is further seen, as for the corresponding result in the location case, that if a sufficient statistic T exists which permits a representation $T = (T_1, \ldots, T_r)$ with

$$T_i(b\mathbf{X}) = bT_i(\mathbf{X}) \quad \text{for all } b > 0,$$

then an MRE estimator can be found which depends only on T. Illustrations are provided by Example 3.7 and Problem 3.12, with $T = (\Sigma X_i^2)^{1/2}$ and $T = X_{(n)}$, respectively. When the loss function is (3.13), it follows from the factorization criterion that the MRE estimator (3.19) depends only on T.

Since the group $\tau' = b\tau, b > 0$, is transitive and the group $d' = \tau^r d$ is commutative, Theorem 3.3 applies and an MRE estimator is always risk-unbiased, although the MRE estimators of Examples 3.7 and 3.9 are not unbiased in the sense of Chapter 2. See also Problem 3.12.

Example 3.10 Risk-unbiasedness. If the loss function is (3.13), the condition of risk-unbiasedness reduces to

$$E_\tau[\delta^2(\mathbf{X})] = \tau^r E_\tau[\delta(\mathbf{X})]. \tag{3.22}$$

Given any scale equivariant estimator $\delta_0(\mathbf{X})$ of τ^r, there exists a value of c for which $c\delta_0(\mathbf{X})$ satisfies (3.22), and for this value, $c\delta_0(\mathbf{X})$ has uniformly smaller risk than $\delta_0(\mathbf{X})$ unless $c = 1$ (Problem 3.21).

If the loss function is (3.15), the condition of risk-unbiasedness requires that $E_\tau|\delta(\mathbf{X}) - a|/a$ be minimized by $a = \tau^r$. From Example 3.5, for this loss function, risk-unbiasedness is equivalent to the condition that the estimand τ^r is equal to the scale median of $\delta(\mathbf{X})$. ||

Let us now turn to location-scale families, where the density of $\mathbf{X} = (X_1, \ldots, X_n)$ is given by

$$\frac{1}{\tau^n} f\left(\frac{x_1 - \xi}{\tau}, \ldots, \frac{x_n - \xi}{\tau}\right) \tag{3.23}$$

with both parameters unknown. Consider first the estimation of τ^r with loss function (3.5). This problem remains invariant under the transformations

$$X_i' = a + bX_i, \quad \xi' = a + b\xi, \quad \tau' = b\tau \quad (b > 0), \tag{3.24}$$

and $d' = b^r d$, and an estimator δ of τ^r is equivariant under this group if

$$\delta(a + b\mathbf{X}) = b^r \delta(\mathbf{X}). \tag{3.25}$$

Consider first only a change in location,

$$X_i' = X_i + a, \tag{3.26}$$

which takes ξ into $\xi' = \xi + a$ but leaves τ unchanged. By (3.25), δ must then satisfy

$$\delta(\mathbf{x}+a) = \delta(\mathbf{x}), \tag{3.27}$$

that is, remain *invariant*. By Lemma 1.7, condition (3.27) holds if and only if δ is a function only of the differences $y_i = x_i - x_n$. The joint density of the Y's is

$$\begin{aligned} &\frac{1}{\tau^n}\int_{-\infty}^{\infty} f\left(\frac{y_1+t}{\tau}, \ldots, \frac{y_{n-1}+t}{\tau}, \frac{t}{\tau}\right) dt \\ &= \frac{1}{\tau^{n-1}}\int_{-\infty}^{\infty} f\left(\frac{y_1}{\tau}+u, \ldots, \frac{y_{n-1}}{\tau}+u, u\right) du. \end{aligned} \tag{3.28}$$

Since this density has the structure (3.1) of a scale family, Theorem 3.3 applies and provides the estimator that uniformly minimizes the risk among all estimators satisfying (3.25).

It follows from Theorem 3.3 that such an MRE estimator of τ^r is given by

$$\delta(\mathbf{X}) = \frac{\delta_0(\mathbf{Y})}{w^*(\mathbf{Z})} \tag{3.29}$$

where $\delta_0(\mathbf{Y})$ is any finite risk scale equivariant estimator of τ^r based on $\mathbf{Y} = (Y_1, \ldots, Y_{n-1})$, where $\mathbf{Z} = (Z_1, \ldots, Z_{n-1})$ with

$$Z_i = \frac{Y_i}{Y_{n-1}} \quad (i = 1, \ldots, n-2) \quad \text{and} \quad Z_{n-1} = \frac{Y_{n-1}}{|Y_{n-1}|}, \tag{3.30}$$

and where $w^*(\mathbf{Z})$ is any number minimizing

$$E_{\tau=1}\{\gamma[\delta_0(\mathbf{Y})/w(\mathbf{Z})|\mathbf{Z}]\}. \tag{3.31}$$

Example 3.11 MRE for normal variance, unknown mean. Let $X_1, \ldots, X_n$ be iid according to $N(\xi, \sigma^2)$ and consider the estimation of σ^2 with loss function (3.13), $r = 2$. By Basu's theorem, $(\bar{X}, \Sigma(X_i - \bar{X})^2)$ is independent of $\mathbf{Z}$. If $\delta_0 = \Sigma(X_i - \bar{X})^2$, then δ_0 is equivariant under (3.24) and independent of $\mathbf{Z}$. Hence, $w^*(\mathbf{z}) = w^*$ in (3.29) is a constant determined by minimizing (3.17) with $\Sigma(X_i - \bar{X})^2$ in place of X^r. Since $\Sigma(X_i - \bar{X})^2$ has the distribution of δ_0 of Example 3.7 with $n-1$ in place of n, the MRE estimator for the loss function (3.13) with $r = 2$ is $\Sigma(X_i - \bar{X})^2/(n+1)$. ||

Example 3.12 Uniform. Let $X_1, \ldots, X_n$ be iid according to $U(\xi - \frac{1}{2}\tau, \xi + \frac{1}{2}\tau)$, and consider the problem of estimating τ with loss function (3.13), $r = 1$. By Basu's theorem, $(X_{(1)}, X_{(n)})$ is independent of $\mathbf{Z}$. If δ_0 is the range $R = X_{(n)} - X_{(1)}$, it is equivariant under (3.24) and independent of $\mathbf{Z}$. It follows from (3.18) with $r = 1$ that (Problem 3.22) $\delta^*(\mathbf{X}) = [(n+2)/n]R$. ||

Since the group $\xi' = a + b\xi, \tau' = b\tau$ is transitive and the group $d' = b^r d$ is commutative, it follows (as in the pure scale case) that an MRE estimator is always risk-unbiased.

The principle of equivariance seems to suggest that we should want to invoke as much invariance as possible and hence use the largest group G of transformations leaving the problem invariant. Such a group may have the disadvantage of restricting the class of eligible estimators too much. (See, for example, Problem

2.7.) To increase the number of available estimators, we may then want to restrict attention to a subgroup G_0 of G. Since estimators that are equivariant under G are automatically also equivariant under G_0, invariance under G_0 alone will leave us with a larger choice, which may enable us to obtain improved risk performance.

For estimating the scale parameter in a location-scale family, a natural subgroup of (3.24) is obtained by setting $a = 0$, which reduces (3.24) to the scale group

$$X_i' = bX_i, \quad \xi_i' = b\xi_i, \quad \tau' = b\tau \quad (b > 0), \tag{3.32}$$

and $d' = b^r d$. An estimator δ of τ^r is equivariant under this group if $\delta(bX) = b^r\delta(X)$, as in (3.7). Application of Theorem 3.1 shows that the equivariant estimators are of the form

$$\delta(\mathbf{x}) = \frac{\delta_0(\mathbf{x})}{\omega(\mathbf{x})} \tag{3.33}$$

where δ_0 is any scale equivariant estimator and $w(\mathbf{z})$ is a function of $z_i = X_i/X_n$, $i = 1, \ldots, n-1$, and $z_n = x_n/|x_n|$. However, we cannot now apply Theorem 3.3 to obtain the MRE estimator, because the group is no longer transitive (Example 2.14), and the risk of equivariant estimators is no longer constant.

We can, however, go further in special cases, such as in the following example.

Example 3.13 More normal variance estimation. If $X_1, \ldots, X_n$ are iid as $N(\xi, \tau^2)$, with both parameters unknown, then it was shown in Example 3.11 that $\delta_0(\mathbf{x}) = \Sigma(x_i - \bar{x})^2/(n+1) = S^2/(n+1)$ is MRE under the location-scale group (3.24) for the loss function (3.13) with $r = 2$.

Now consider the scale group (3.32). Of course, δ_0 is equivariant under this group, but so are the estimators

$$\delta(\mathbf{x}) = \varphi(\bar{x}/s)s^2$$

for some function $\varphi(\cdot)$ (Problem 3.24). Stein (1964) showed that $\varphi(\bar{x}/s) = \min\{(n+1)^{-1}, (n+2)^{-1}(1 + n\bar{x}^2/s^2)\}$ produces a uniformly better estimator than δ_0, and Brewster and Zidek (1974) found the best scale equivariant estimator. See Example 5.2.15 and Problem 5.2.14 for more details. ‖

In the location-scale family (3.23), we have so far considered only the estimation of τ^r; let us now take up the problem of estimating the location parameter ξ. The transformations (3.24) relating to the sample space and parameter space remain the same, but the transformations of the decision space now become $d' = a + bd$. A loss function $L(\xi, \tau; d)$ is invariant under these transformations if and only if it is of the form

$$L(\xi, \tau; d) = \rho\left(\frac{d - \xi}{\tau}\right). \tag{3.34}$$

That any such loss function is invariant is obvious. Conversely, suppose that L is invariant and that $(\xi, \tau; d)$ and $(\xi', \tau'; d')$ are two points with $(d' - \xi')/\tau' = (d - \xi)/\tau$. Putting $b = \tau'/\tau$ and $\xi' - a = b\xi$, one has $d' = a + bd$, $\xi' = a + b\xi$, and $\tau' = b\tau$, hence $L(\xi', \tau'; d') = L(\xi, \tau; d)$, as was to be proved.

Equivariance in the present case becomes

$$\delta(a + b\mathbf{x}) = a + b\delta(\mathbf{x}), \quad b > 0. \tag{3.35}$$

Since $\bar{G}$ is transitive over the parameter space, the risk of any equivariant estimator is constant so that an MRE estimator can be expected to exist. In some special cases, the MRE estimator reduces to that derived in Section 1 with τ known, as follows.

For fixed τ, write

$$g_\tau(x_1, \ldots, x_n) = \frac{1}{\tau^n} f\left(\frac{x_1}{\tau}, \ldots, \frac{x_n}{\tau}\right) \tag{3.36}$$

so that (3.23) becomes

$$g_\tau(x_1 - \xi, \ldots, x_n - \xi). \tag{3.37}$$

Lemma 3.14 *Suppose that for the location family (3.37) and loss function (3.34), there exists an MRE estimator δ^* of ξ with respect to the transformations (1.10) and (1.11) and that*

(a) δ^* is independent of τ, and

(b) δ^* satisfies (3.35).

Then δ^* minimizes the risk among all estimators satisfying (3.35).

Proof. Suppose δ is any other estimator which satisfies (3.35) and hence, a fortiori, is equivariant with respect to the transformations (1.10) and (1.11), and that the value τ of the scale parameter is known. It follows from the assumptions about δ^* that for this τ, the risk of δ^* does not exceed the risk of δ. Since this is true for all values of τ, the result follows. □

Example 3.15 MRE for normal mean. Let $X_1, \ldots, X_n$ be iid as $N(\xi, \tau^2)$, both parameters being unknown. Then, it follows from Example 1.15 that $\delta^* = \bar{X}$ for any loss function $\rho[(d - \xi)/\tau]$ for which ρ satisfies the assumptions of Example 1.15. Since (i) and (ii) of Lemma 3.14 hold for this δ^*, it is the MRE estimator of ξ under the transformations (3.24). ||

Example 3.16 Uniform location parameter. Let $X_1, \ldots, X_n$ be iid as $U(\xi - \frac{1}{2}\tau, \xi + \frac{1}{2}\tau)$. Then, analogous to Example 3.15, it follows from Example 1.19 that $[X_{(1)} + X_{(n)}]/2$ is MRE for the loss functions of Example 3.15. ||

Unfortunately, the MRE estimators of Section 1 typically do not satisfy the assumptions of Lemma 3.14. This is the case, for instance, with the estimators of Examples 1.18 and 1.22. To derive the MRE estimator without these assumptions, let us first characterize the totality of equivariant estimators.

Theorem 3.17 *Let δ_0 be any estimator ξ satisfying (3.35) and δ_1 any estimator of τ taking on positive values only and satisfying*

$$\delta_1(a + b\mathbf{x}) = b\delta_1(\mathbf{x}) \quad \textit{for all } b > 0 \textit{ and all } a. \tag{3.38}$$

Then, δ satisfies (3.35) if and only if it is of the form

$$\delta(\mathbf{x}) = \delta_0(\mathbf{x}) - w(\mathbf{z})\delta_1(\mathbf{x}) \tag{3.39}$$

where $\mathbf{z}$ is given by (3.30).

Proof. Analogous to Lemma 1.6, it is seen that δ satisfies (3.35) if and only if it is of the form

$$\delta(\mathbf{x}) = \delta_0(\mathbf{x}) - u(\mathbf{x})\delta_1(\mathbf{x}), \tag{3.40}$$

where

$$u(a + b\mathbf{x}) = u(\mathbf{x}) \quad \text{for all } b > 0 \text{ and all } a \tag{3.41}$$

(Problem 3.26). That (3.40) holds if and only if u depends on $\mathbf{x}$ only through $\mathbf{z}$ follows from Lemma 1.7 and Theorem 3.1.

An argument paralleling that of Theorem 1.10 now shows that the MRE estimator of ξ is

$$\delta(\mathbf{X}) = \delta_0(\mathbf{X}) - w^*(\mathbf{Z})\delta_1(\mathbf{X})$$

where for each $\mathbf{z}$, $w^*(\mathbf{z})$ is any number minimizing

$$E_{0,1}\{\rho[\delta_0(\mathbf{X}) - w^*(\mathbf{z})\delta_1(\mathbf{X})]|\mathbf{z}\}. \tag{3.42}$$

Here, $E_{0,1}$ indicates that the expectation is evaluated at $\xi = 0$, $\tau = 1$. □

If, in particular,

$$\rho\left(\frac{d-\xi}{\tau}\right) = \frac{(d-\xi)^2}{\tau^2}, \tag{3.43}$$

it is easily seen that $w^*(\mathbf{z})$ is

$$w^*(\mathbf{z}) = E_{0,1}[\delta_0(\mathbf{X})\delta_1(\mathbf{X})|\mathbf{z}]/E_{0,1}[\delta_1^2(\mathbf{X})|\mathbf{z}]. \tag{3.44}$$

Example 3.18 Exponential. Let $X_1, \ldots, X_n$ be iid according to the exponential distribution $E(\xi, \tau)$. If $\delta_0(\mathbf{X}) = X_{(1)}$ and $\delta_1(\mathbf{X}) = \Sigma[X_i - X_{(1)}]$, it follows from Example 1.6.24 that (δ_0, δ_1) are jointly independent of $\mathbf{Z}$ and are also independent of each other. Then (Problem 3.25),

$$w^*(\mathbf{z}) - w^* = E\frac{[\delta_0(\mathbf{X})\delta_1(\mathbf{X})]}{E[\delta_1^2(\mathbf{X})]} = \frac{1}{n^2},$$

and the MRE estimator of ξ is therefore

$$\delta^*(\mathbf{X}) = X_{(1)} - \frac{1}{n^2}\Sigma[X_i - X_{(1)}].$$

When the best location equivariant estimate is not also scale equivariant, its risk is, of course, smaller than that of the MRE under (3.35). Some numerical values of the increase that results from the additional requirement are given for a number of situations by Hoaglin (1975). ∥

For the loss function (3.43), no risk-unbiased estimator δ exists, since this would require that for all ξ, ξ', τ, and τ'

$$\frac{1}{\tau^2}E_{\xi,\tau}[\delta(\mathbf{X}) - \xi]^2 \le \frac{1}{\tau'^2}E_{\xi,\tau}[\delta(\mathbf{X}) - \xi']^2, \tag{3.45}$$

which is clearly impossible. Perhaps (3.45) is too strong and should be required only when $\tau' = \tau$. It then reduces to (1.32) with $\theta = (\xi, \tau)$ and $g(\theta) = \xi$, and this weakened form of (3.45) reduces to the classical unbiasedness condition $E_{\xi,\tau}[\delta(\mathbf{X})] = \xi$. A UMVU estimator of ξ exists in Example 3.18 (Problem

2.2.18), but it is

$$\delta(\mathbf{X}) = X_{(1)} - \frac{1}{n(n-1)}\Sigma[X_i - X_{(1)}]$$

rather than $\delta^*(\mathbf{X})$, and the latter is not unbiased (Problem 3.27).

4 Normal Linear Models

Having developed the theory of unbiased estimation in Chapter 2 and of equivariant estimation in the first three sections of the present chapter, we shall now apply these results to some important classes of statistical models. One of the most widely used bodies of statistical techniques, comprising particularly the analysis of variance, regression, and the analysis of covariance, is formalized in terms of linear models, which will be defined and illustrated in the following. The examples, however, are not enough to give an idea of the full richness of the applications. For a more complete treatment, see, for example, the classic book by Scheffé (1959), or Seber (1977), Arnold (1981), Searle (1987), or Christensen (1987).

Consider the problem of investigating the effect of a number of different factors on a response. Typically, each factor can occur in a number of different forms or at a number of different levels. Factor levels can be qualitative or quantitative. Three possibilities arise, corresponding to three broad categories of linear models:

(a) All factor levels qualitative.

(b) All factor levels quantitative.

(c) Some factors of each kind.

Example 4.1 One-way layout. A simple illustration of category (a) is provided by the *one-way layout* in which a single factor occurs at a number of qualitatively different levels. For example, we may wish to study the effect on performance of a number of different textbooks or the effect on weight loss of a number of diets. If X_{ij} denotes the response of the jth subject receiving treatment i, it is often reasonable to assume that the X_{ij} are independently distributed as

$$X_{ij} : N(\xi_i, \sigma^2), \quad j = 1, \ldots, n_i; \quad i = 1, \ldots, s. \tag{4.1}$$

Estimands that may be of interest are ξ_i and $\xi_i - (1/s)\Sigma_{j=1}^s \xi_j$. ||

Example 4.2 A simple regression model. As an example of type (b), consider the time required to memorize a list of words. If the number of words presented to the ith subject and the time it takes the subject to learn the words are denoted by t_i and X_i, respectively, one might assume that for the range of t's of interest, the X's are independently distributed as

$$X_i : N(\alpha + \beta t_i + \gamma t_i^2, \sigma^2) \tag{4.2}$$

where α, β, and γ are the unknown regression coefficients, which are to be estimated.

This would turn into an example of the third type if there were several groups of subjects. One might, for example, wish to distinguish between women and men or to see how learning ability is influenced by the form of the word list (whether it is handwritten, typed, or printed). The model might then become

$$X_{ij} : N(\alpha_i + \beta_i t_{ij} + \gamma_i t_{ij}^2, \sigma^2) \tag{4.3}$$

where X_{ij} is the response of the jth subject in the ith group. Here, the group is a qualitative factor and the length of the list a quantitative one.

||

The general *linear model*, which covers all three cases, assumes that

$$X_i \text{ is distributed as } N(\xi_i, \sigma^2), \quad i = 1, \ldots, n, \tag{4.4}$$

where the X_i are independent and $(\xi_i, \ldots, \xi_n) \in \prod_\Omega$, an s-dimensional linear subspace of $E_n (s < n)$.

It is convenient to reduce this model to a canonical form by means of an orthogonal transformation

$$\mathbf{Y} = \mathbf{X}C \tag{4.5}$$

where we shall use $\mathbf{Y}$ to denote both the vector with components $(Y_1, \ldots, Y_n)$ and the row matrix $(Y_1, \ldots, Y_n)$. If $\eta_i = E(Y_i)$, the η's and ξ's are related by

$$\eta = \xi C \tag{4.6}$$

where $\eta = (\eta_1, \ldots, \eta_n)$ and $\xi = (\xi_1, \ldots \xi_n)$.

To find the distribution of the Y's, note that the joint density of $X_1, \ldots, X_n$ is

$$\frac{1}{(\sqrt{2\pi}\sigma)^n} \exp\left[-\frac{1}{2\sigma^2}\Sigma(x_i - \xi_i)^2\right],$$

that

$$\Sigma(x_i - \xi_i)^2 = \Sigma(y_i - \eta_i)^2,$$

since C is orthogonal, and that the Jacobian of the transformation is 1. Hence, the joint density of $Y_1, \ldots, Y_n$ is

$$\frac{1}{(\sqrt{2\pi}\sigma)^n} \exp\left[-\frac{1}{2\sigma^2}\Sigma(y_i - \eta_i)^2\right].$$

The Y's are therefore independent normal with $Y_i \sim N(\eta_i, \sigma^2)$, $i = 1, \ldots, n$. If $\mathbf{c}_i'$ denotes the ith column of C, the desired form is obtained by choosing the $\mathbf{c}_i$ so that the first s columns $\mathbf{c}_1', \ldots, \mathbf{c}_s'$ span $\prod_\Omega$. Then,

$$\xi \in \prod_\Omega \iff \xi \text{ is orthogonal to the last } n - s \text{ columns of } C.$$

Since $\eta = \xi C$, it follows that

$$\xi \in \prod_\Omega \iff \eta_{s+1} = \cdots = \eta_n = 0. \tag{4.7}$$

In terms of the Y's, the model (4.4) thus becomes

$$Y_i : N(\eta_i, \sigma^2), \quad i = 1, \ldots, s, \quad \text{and } Y_j : N(0, \sigma^2), \quad j = s + 1, \ldots, n. \tag{4.8}$$

As $(\xi_1, \ldots, \xi_n)$ varies over $\prod_\Omega$, $(\eta_1, \ldots, n_s)$ varies unrestrictedly over E_s while $\eta_{s+1} = \cdots = \eta_n = 0$.

In this canonical model, $Y_1, \ldots, Y_s$ and $S^2 = \Sigma_{j=s+1}^n Y_j^2$ are complete sufficient statistics for $(\eta_1, \ldots, \eta_s, \sigma^2)$.

Theorem 4.3

(a) The UMVU estimators of $\Sigma_{i=1}^s \lambda_i \eta_i$ (where the λ's are known constants) and σ^2 are $\Sigma_{i=1}^s \lambda_i Y_i$ and $S^2/(n-s)$, respectively. (Here, UMVU is used in the strong sense of Section 2.1.)

(b) Under the transformations

$$Y_i' = Y_i + a_i \ (i = 1, \ldots, s); \quad Y_j' = Y_j \ (j = s+1, \ldots, n)$$

$$\eta_i' = \eta_i + a_i \ (i = 1, \ldots, s); \quad \text{and } d' = d + \sum_{i=1}^s a_i \lambda_i$$

and with loss function $L(\eta, d) = \rho(d - \Sigma \lambda_i \eta_i)$ where ρ is convex and even, the UMVU estimator $\Sigma_{i=1}^s \lambda_i Y_i$ is also the MRE estimator of $\Sigma_{i=1}^s \lambda_i \eta_i$.

(c) Under the loss function $(d-\sigma^2)^2/\sigma^4$, the MRE estimator of σ^2 is $S^2/(n-s+2)$.

Proof.

(a) Since $\Sigma_{i=1}^s \lambda_i Y_i$ and $S^2/(n-s)$ are unbiased and are functions of the complete sufficient statistics, they are UMVU.

(b) The condition of equivariance is that

$$\delta(Y_1 + c_1, \ldots, Y_s + c_s, Y_{s+1}, \ldots, Y_n)$$
$$= \delta(Y_1, \ldots, Y_s, Y_{s+1}, \ldots, Y_n) + \sum_{i=1}^s \lambda_i c_i$$

and the result follows from Problem 2.27.

(c) This follows essentially from Example 3.7 (see Problem 4.3) .

□

It would be more convenient to have the estimator expressed in terms of the original variables $X_1, \ldots, X_n$, rather than the transformed variables $Y_1, \ldots, Y_n$. For this purpose, we introduce the following definition.

Let $\xi = (\xi_1, \ldots, \xi_n)$ be any vector in $\prod_\Omega$. Then, the *least squares estimators* (LSE) $(\hat{\xi}_1, \ldots, \hat{\xi}_n)$ of $(\xi_1, \ldots, \xi_n)$ are those estimators which minimize $\Sigma_{i=1}^n (X_i - \xi_i)^2$ subject to the condition $\xi \in \prod_\Omega$.

Theorem 4.4 *Under the model (4.4), the UMVU estimator of $\Sigma_{i=1}^n \gamma_i \xi_i$ is $\Sigma_{i=1}^n \gamma_i \hat{\xi}_1$.*

Proof. By Theorem 4.3 (and the completeness of $Y_1, \ldots, Y_s$ and S^2), it suffices to show that $\Sigma_{i=1}^n \gamma_i \hat{\xi}_i$ is a linear function of $Y_1, \ldots, Y_s$, and that it is unbiased for $\Sigma_{i=1}^n \gamma_i \xi_i$. Now,

$$\sum_{i=1}^n (X_i - \xi_i)^2 = \sum_{i=1}^n [Y_i - E(Y_i)]^2 = \sum_{i=1}^s (Y_i - \eta_i)^2 + \sum_{j=s+1}^n Y_j^2. \tag{4.9}$$

The right side is minimized by $\hat{\eta}_i = Y_i$ $(i = 1, \ldots, s)$, and the left side is minimized by $\hat{\xi}_1, \ldots, \hat{\xi}_n$. Hence,

$$(Y_1 \cdots Y_s 0 \cdots 0) = (\hat{\xi}_1 \cdots \hat{\xi}_n) C = \hat{\xi} C$$

so that

$$\hat{\boldsymbol{\xi}} = (Y_1 \cdots Y_s\, 0 \cdots 0)C^{-1}.$$

It follows that each $\hat{\xi}_i$ and, therefore, $\Sigma_{i=1}^n \gamma_i \hat{\xi}_i$ is a linear function of $Y_1, \ldots, Y_s$. Furthermore,

$$E(\hat{\boldsymbol{\xi}}) = E[(Y_1 \cdots Y_s\, 0 \cdots 0)C^{-1}] = (\eta_1 \cdots \eta_s\, 0 \cdots 0)C^{-1} = \boldsymbol{\xi}.$$

Thus, each $\hat{\xi}_i$ is unbiased for ξ_i; consequently, $\Sigma_{i=1}^n \gamma_i \hat{\xi}_i$ is unbiased for $\Sigma_{i=1}^n \gamma_i \xi_i$. □

It is interesting to note that each of the two quite different equations

$$\mathbf{X} = (Y_1 \cdots Y_n)C^{-1} \quad \text{and} \quad \hat{\boldsymbol{\xi}} = (Y_1 \cdots Y_s\, 0 \cdots 0)C^{-1}$$

leads to $\boldsymbol{\xi} = (\eta_1, \ldots, \eta_s\, 0 \cdots 0)C^{-1}$ by taking expectations.

Let us next reinterpret the equivariance considerations of Theorem 4.3 in terms of the original variables. It is necessary first to specify the group of transformations leaving the problem invariant. The transformations of Y-space defined in Theorem 4.3(b), in terms of the X's become $X_i' = X_i + b_i$, $i = 1, \ldots, n$, but the b_i are not arbitrary since the problem remains invariant only if $\boldsymbol{\xi}' = \xi + \mathbf{b} \in \prod_\Omega$; that is, the b_i must satisfy $\mathbf{b} = (b_1, \ldots, b_n) \in \prod_\Omega$. Theorem 4.3(ii) thus becomes the following corollary.

Corollary 4.5 *Under the transformations*

$$\mathbf{X}' = \mathbf{X} + b \quad with \;\; \mathbf{b} \in \prod_\Omega, \tag{4.10}$$

$\Sigma_{i=1}^n \gamma_i \hat{\xi}_i$ is MRE for estimating $\Sigma_{i=1}^n \gamma_i \xi_i$ with the loss function $\rho(d - \Sigma \gamma_i \xi_i)$ provided ρ is convex and even.

To obtain the UMVU and MRE estimators of σ^2 in terms of the X's, it is only necessary to reexpress S^2. From the minimization of the two sides of (4.9), it is seen that

$$\sum_{i=1}^n (X_i - \hat{\xi}_i)^2 = \sum_{j=s+1}^n Y_j^2 = S^2. \tag{4.11}$$

The UMVU and MRE estimators of σ^2 given in Theorem 4.3, in terms of the X's are therefore $\Sigma(X_i - \hat{\xi}_i)^2/(n-s)$ and $\Sigma(X_i - \hat{\xi}_i)^2/(n-s+2)$, respectively.

Let us now illustrate these results.

Example 4.6 Continuation of Example 4.1. Let X_{ij} be independent $N(\xi_i, \sigma^2)$, $j = 1, \ldots, n_i$, $i = 1, \ldots, s$. To find the UMVU or MRE estimator of a linear function of the ξ_i, it is only necessary to find the least squares estimators $\hat{\xi}_i$. Minimizing

$$\sum_{i=1}^s \sum_{j=1}^{n_i} (X_{ij} - \xi_i)^2 = \sum_{i=1}^s \left[\sum_{j=1}^{n_i} (X_{ij} - X_{i\cdot})^2 + n_i (X_{i\cdot} - \xi_i)^2 \right],$$

we see that

$$\hat{\xi}_i = X_{i\cdot} = \frac{1}{n_i} \sum_{j=1}^{n_i} X_{ij}.$$

From (4.11), the UMVU estimator of σ^2 in the present case is seen to be

$$\hat{\sigma}^2 = \sum_{i=1}^{s}\sum_{j=1}^{n_i}(X_{ij} - X_{i\cdot})^2/(\Sigma n_i - s). \qquad \|$$

Example 4.7 Simple linear regression. Let X_i be independent $N(\xi_i, \sigma^2)$, $i = 1, \ldots, n$, with $\xi_i = \alpha + \beta t_i$, t_i known and not all equal. Here, $\prod_\Omega$ is spanned by the vectors $(1, \ldots, 1)$ and $(t_1, \ldots, t_n)$ so that the dimension of $\prod_\Omega$ is $s = 2$. The least squares estimators of ξ_i are obtained by minimizing $\sum_{i=1}^n (X_i - \alpha - \beta t_i)^2$ with respect to α and β. It is easily seen that for any i and j with $t_i \neq t_j$,

$$\beta = \frac{\xi_j - \xi_i}{t_j - t_i}, \quad \alpha = \frac{t_j\xi_i - t_i\xi_j}{t_j - t_i} \tag{4.12}$$

and that $\hat{\beta}$ and $\hat{\alpha}$ are given by the same functions of $\hat{\xi}_i$ and $\hat{\xi}_j$ (Problem 4.4). Hence, $\hat{\alpha}$ and $\hat{\beta}$ are the best unbiased and equivariant estimators of α and β, respectively.

Note that the representation of α and β in terms of the ξ_i's is not unique. Any two ξ_i and ξ_j values with $t_i \neq t_j$ determine α and β and thus all the ξ's. The reason, of course, is that the vectors $(\xi_1, \ldots, \xi_n)$ lie in a two-dimensional linear subspace of n-space. $\|$

Example 4.7 is a special case of the model specified by the equation

$$\boldsymbol{\xi} = \boldsymbol{\theta} A \tag{4.13}$$

where $\boldsymbol{\theta} = (\theta_1 \cdots \theta_s)$ are s unknown parameters and A is a known $s \times n$ matrix of rank s, the so-called *full-rank model*. In Example 4.7,

$$\boldsymbol{\theta} = (\alpha, \beta) \quad \text{and} \quad A = \begin{pmatrix} 1 \cdots 1 \\ t_1 \cdots t_n \end{pmatrix}.$$

The least squares estimators of the ξ_i in (4.13) are obtained by minimizing

$$\sum_{i=1}^{n}[X_i - \xi_i(\boldsymbol{\theta})]^2$$

with respect to $\boldsymbol{\theta}$. The minimizing values $\hat{\theta}_i$ are the LSEs of θ_i, and the LSEs of the ξ_i are given by

$$\hat{\boldsymbol{\xi}} = \hat{\boldsymbol{\theta}} A. \tag{4.14}$$

Theorems 4.3 and 4.4 establish that the various optimality results apply to the estimators of the ξ_i and their linear combinations. The following theorem shows that they also apply to the estimators of the $\boldsymbol{\theta}$'s and their linear functions.

Theorem 4.8 *Let $X_i \sim N(\xi_i, \sigma^2)$, $i = 1, \ldots, n$, be independent, and let $\boldsymbol{\xi}$ satisfy (4.13) with A of rank s. Then, the least squares estimator $\hat{\boldsymbol{\theta}}$ of $\boldsymbol{\theta}$ is a linear function of the $\hat{\xi}_i$ and hence has the optimality properties established in Theorems 4.3 and 4.4 and Corollary 4.5.*

Proof. It need only be shown that $\boldsymbol{\theta}$ is a linear function of $\boldsymbol{\xi}$; then, by (4.13) and (4.14), $\hat{\boldsymbol{\theta}}$ is the corresponding linear function of $\hat{\boldsymbol{\xi}}$.

Assume without loss of generality that the first s columns of A are linearly independent, and form the corresponding nonsingular $s \times s$ submatrix A^*. Then,

$$(\xi_1 \cdots \xi_s) = (\theta_1 \cdots \theta_s)A^*,$$

so that

$$(\theta_1 \cdots \theta_s) = (\xi_1 \cdots \xi_s)A^{*-1},$$

and this completes the proof. □

Typical examples in which $\boldsymbol{\xi}$ is given in terms of (4.13) are polynomial regressions such as

$$\xi_i = \alpha + \beta t_i + \gamma t_i^2$$

or regression in more than one variable such as

$$\xi_i = \alpha + \beta t_i + \gamma u_i$$

where the t's and u's are given, and α, β, and γ are the unknown parameters. Or there might be several regression lines with a common slope, say

$$\xi_{ij} = \alpha_i + \beta t_{ij} \quad (j = 1, \ldots, n_i;\ i = 1, \ldots, a),$$

and so on.

The full-rank model does not always provide the most convenient parametrization; for reasons of symmetry, it is often preferable to use the model (4.13) with more parameters than are needed. Before discussing such models more fully, let us illustrate the resulting difficulties on a trivial example. Suppose that $\xi_i = \xi$ for all i and that we put $\xi_i = \lambda + \mu$. Such a model does not define λ and μ uniquely but only their sum. One can then either let this ambiguity remain but restrict attention to clearly defined functions such as $\lambda + \mu$, or one can remove the ambiguity by placing an additional restriction on λ and μ, such as $\mu - \lambda = 0$, $\mu = 0$, or $\lambda = 0$.

More generally, let us suppose that the model is given by

$$\boldsymbol{\xi} = \boldsymbol{\theta} A \tag{4.15}$$

where A is a $t \times n$ matrix of rank $s < t$. To define the $\boldsymbol{\theta}$'s uniquely, (4.15) is supplemented by side conditions

$$\boldsymbol{\theta} B = 0 \tag{4.16}$$

chosen so that the set of equations (4.15) and (4.16) has a unique solution $\boldsymbol{\theta}$ for every $\boldsymbol{\xi} \in \prod_\Omega$.

Example 4.9 Unbalanced one-way layout. Consider the one-way layout of Example 4.1, with X_{ij} ($j = 1, \ldots, n_i$; $i = 1, \ldots, s$) independent normal variables with means ξ_i and variance σ^2. When the principal concern is a comparison of the s treatments or populations, one is interested in the differences of the ξ's and may represent these by means of the differences between the ξ_i and some mean value μ, say $\alpha_i = \xi_i - \mu$. The model then becomes

$$\xi_i = \mu + \alpha_i, \quad i = 1, \ldots, s, \tag{4.17}$$

which expresses the s ξ's in terms of $s+1$ parameters. To specify the parameters, an additional restriction is required, for example,

$$\Sigma\alpha_i = 0. \tag{4.18}$$

Adding the s equations (4.17) and using (4.18), one finds

$$\mu = \Sigma\frac{\xi_i}{s} = \bar{\xi} \tag{4.19}$$

and hence

$$\alpha_i = \xi_i - \bar{\xi}. \tag{4.20}$$

The quantity α_i measures the effect of the ith treatment. Since $X_{i\cdot}$ is the least squares estimator of ξ_i, the UMVU estimators of μ and the α's are

$$\hat{\mu} = \Sigma\frac{X_{i\cdot}}{s} = \Sigma\Sigma\frac{X_{ij}}{sn_i} \quad \text{and} \quad \hat{\alpha}_i = X_{i\cdot} - \hat{\mu}. \tag{4.21}$$

When the sample sizes n_i are not all equal, a possible disadvantage of this representation is that the vectors of the coefficients of the X_{ij} in the $\hat{\alpha}_i$ are not orthogonal to the corresponding vector of coefficients of $\hat{\mu}$ [Problem 4.7(a)]. As a result, $\hat{\mu}$ is not independent of the $\hat{\alpha}_i$. Also, when the α_i are known to be zero, the estimator of μ is no longer given by (4.21) (Problem 4.8).

For these reasons, the side condition (4.18) is sometimes replaced by

$$\Sigma n_i\alpha_i = 0, \tag{4.22}$$

which leads to

$$\mu = \Sigma\frac{n_i\xi_i}{N} = \tilde{\xi} \quad (N = \Sigma n_i) \tag{4.23}$$

and hence

$$\alpha_i = \xi_i - \tilde{\xi}. \tag{4.24}$$

Although the α_i of (4.22) seems to be a less natural measure of the effect of the ith treatment, the resulting UMVU estimators $\hat{\hat{\alpha}}_i$ and $\hat{\hat{\mu}}$ have the orthogonality property not possessed by the estimators (4.21) [Problem 4.7(b)]. The side conditions (4.18) and (4.22), of course, agree when the n_i are all equal. ||

The following theorem shows that the conclusion of Theorem 4.8 continues to hold when the θ's are defined by (4.15) and (4.16) instead of (4.13).

Theorem 4.10 *Let X_i be independent $N(\xi_i, \sigma^2)$, $i = 1, \ldots, n$, with $\boldsymbol{\xi} \in \prod_\Omega$, an s-dimensional linear subspace of E_n. Suppose that $(\theta_1, \ldots, \theta_t)$ are uniquely determined by (4.15) and (4.16), where A is of rank $s < t$ and B of rank k. Then, $k = t - s$, and the optimality results of Theorem 4.4 and Corollary 4.5 apply to the parameters $\theta_1, \ldots, \theta_t$ and their least squares estimators $\hat{\theta}_1, \ldots, \hat{\theta}_t$.*

Proof. Let $\hat{\theta}_1, \ldots, \hat{\theta}_t$ be the LSEs of $\theta_1, \ldots, \theta_t$, that is, the values that minimize

$$\sum_{i=1}^{n}[X_i - \xi_i(\theta)]^2$$

subject to (4.15) and (4.16). It must be shown, as in the proof of Theorem 4.8, that the $\hat{\theta}_i$'s are linear functions of $\hat{\xi}_1, \ldots, \hat{\xi}_n$, and that the θ_i's are the same functions of $\xi_1, \ldots, \xi_n$.

Without loss of generality, suppose that the θ's are numbered so that the last k columns of B are linearly independent. Then, one can solve for $\theta_{t-k+1}, \ldots, \theta_t$ in terms of $\theta_1, \ldots, \theta_{t-k}$, obtaining the unique solution

$$\theta_j = L_j(\theta_1, \ldots, \theta_{t-k}) \quad \text{for } j = t-k+1, \ldots, t. \tag{4.25}$$

Substituting into $\boldsymbol{\xi} = \theta A$ gives

$$\boldsymbol{\xi} = (\theta_1 \cdots \theta_{t-k})A^*$$

for some matrix A^*, with $(\theta_1, \ldots, \theta_{t-k})$ varying freely in E_{t-k}. Since each $\boldsymbol{\xi} \in \prod_\Omega$ uniquely determines θ, in particular the value $\boldsymbol{\xi} = \mathbf{0}$ has the unique solution $\theta = \mathbf{0}$, so that $(\theta_1 \cdots \theta_{t-k})A^* = \mathbf{0}$ has a unique solution. This implies that A^* has rank $t-k$. On the other hand, since $\boldsymbol{\xi}$ ranges over a linear space of dimension s, it follows that $t-k = s$ and, hence, that $k = t-s$.

The situation is now reduced to that of Theorem 4.8 with $\boldsymbol{\xi}$ a linear function of $t-k = s$ freely varying θ's, so the earlier result applies to $\theta_1, \ldots, \theta_{t-k}$. Finally, the remaining parameters $\theta_{t-k+1}, \ldots, \theta_t$ and their LSEs are determined by (4.25), and this completes the proof. □

Example 4.11 Two-way layout. A typical illustration of the above approach is provided by a two-way layout. This arises in the investigation of the effect of two factors on a response. In a medical situation, for example, one of the factors might be the kind of treatment (e.g., surgical, nonsurgical, or no treatment at all), the other the severity of the disease. Let X_{ijk} denote the response of the kth subject to which factor 1 is applied at level i and factor 2 at level j. We assume that the X_{ijk} are independently, normally distributed with means ξ_{ij} and common variance σ^2. To avoid the complications of Example 4.9, we shall suppose that each treatment combination (i, j) is applied to the same number of subjects. If the number of levels of the two factors is a and b, respectively, the model is thus

$$X_{ijk} : N(\xi_{ij}, \sigma^2), \quad i = 1, \ldots, I; \quad j = 1, \ldots, J; \quad k = 1, \ldots, m. \tag{4.26}$$

This model is frequently parametrized by

$$\xi_{ij} = \mu + \alpha_i + \beta_j + \gamma_{ij} \tag{4.27}$$

with the side conditions

$$\sum_i \alpha_i = \sum_j \beta_j = \sum_i \gamma_{ij} = \sum_j \gamma_{ij} = 0. \tag{4.28}$$

It is easily seen that (4.27) and (4.28) uniquely determine μ and the α's, β's, and γ's. Using a dot to denote averaging over the indicated subscript, we find by averaging (4.27) over both i and j and separately over i and over j that

$$\xi_{\cdot\cdot} = \mu, \quad \xi_{i\cdot} = \mu + \alpha_i, \quad \xi_{\cdot j} = \mu + \beta_j$$

and hence that

$$\mu = \xi_{\cdot\cdot}, \quad \alpha_i = \xi_{i\cdot} - \xi_{\cdot\cdot}, \quad \beta_j = \xi_{\cdot j} - \xi_{\cdot\cdot}, \tag{4.29}$$

and

$$\gamma_{ij} = \xi_{ij} - \xi_{i\cdot} - \xi_{\cdot j} + \xi_{\cdot\cdot}\,. \tag{4.30}$$

||

Thus, α_i is the average effect (averaged over the levels of the second factor) of the first factor at level i, and β_j is the corresponding effect of the second factor at level j. The quantity γ_{ij} can be written as

$$\gamma_{ij} = (\xi_{ij} - \xi_{\cdot\cdot}) - [(\xi_{i\cdot} - \xi_{\cdot\cdot}) + (\xi_{j\cdot} - \xi_{\cdot\cdot})]. \tag{4.31}$$

It is therefore the difference between the joint effect of the two treatments at levels i and j, respectively, and the sum of the separate effects $\alpha_i + \beta_j$. The quantity γ_{ij} is called the *interaction* of the two factors when they are at levels i and j, respectively.

The UMVU estimators of these various effects follow immediately from Theorem 4.3 and Example 4.6. This example shows that the UMVU estimator of ξ_{ij} is X_{ij} and the associated estimators of the various parameters are thus

$$\hat{\mu} = X_{\cdots}, \quad \hat{\alpha}_i = X_{i\cdot\cdot} - X_{\cdots}, \quad \hat{\beta}_j = X_{\cdot j\cdot} - X_{\cdots}, \tag{4.32}$$

and

$$\hat{\gamma}_{ij} = X_{ij\cdot} - X_{i\cdot\cdot} - X_{\cdot j\cdot} + X_{\cdots}\,. \tag{4.33}$$

The UMVU estimator of σ^2 is

$$\frac{1}{IJ(m-1)} \Sigma\Sigma\Sigma(X_{ijk} - X_{ij\cdot})^2. \tag{4.34}$$

These results for the two-way layout easily generalize to other *factorial experiments*, that is, experiments concerning the joint effect of several factors, provided the numbers of observations at the various combinations of factor levels are equal. Theorems 4.8 and 4.10, of course, apply without this restriction, but then the situation is less simple.

Model (4.4) assumes that the random variables X_i are independently normally distributed with common unknown variance σ^2 and means ξ_i, which are subject to certain linear restrictions. We shall now consider some models that retain the linear structure but drop the assumption of normality.

(i) A very simple treatment is possible if one is willing to restrict attention to unbiased estimators that are linear functions of the X_i and to squared error loss. Suppose we retain from (4.4) only the assumptions about the first and second moments of the X_i, namely

$$E(X_i) = \xi_i, \quad \boldsymbol{\xi} \in \prod_{\Omega}, \tag{4.35}$$

$$\text{var}(X_i) = \sigma^2, \quad \text{cov}(X_i, X_j) = 0 \quad \text{for } i \neq j.$$

Thus, both the normality and independence assumptions are dropped.

Theorem 4.12 *(***Gauss' Theorem on Least Squares***) Under assumptions (4.35), $\Sigma_{i=1}^n \gamma_i \hat{\xi}_i$ of Theorem 4.4 is UMVU among all linear estimators of $\Sigma_{i=1}^n \gamma_i \xi_i$.*

Proof. The estimator is still unbiased, since the expectations of the X_i are the same under (4.35) as under (4.4). Let $\Sigma_{i=1}^n c_i X_i$ be any other linear unbiased estimator of $\Sigma_{i=1}^n \gamma_i \xi_i$. Since $\Sigma_{i=1}^n \gamma_i \hat{\xi}_i$ is UMVU in the normal case, and variances of linear functions of the X_i depend only on first and second moments, it follows that $\text{var}\ \Sigma_{i=1}^n \gamma_i \hat{\xi}_i \leq \text{var}\ \Sigma_{i=1}^n c_i X_i$. Hence, $\Sigma_{i=1}^n \gamma_i \hat{\xi}_i$ is UMVU among linear unbiased estimators. □

Corollary 4.13 *Under the assumptions (4.35) and with squared error loss, $\Sigma_{i=1}^n \gamma_i \hat{\xi}_i$ is MRE with respect to the transformations (4.10) among all linear equivariant estimators of $\Sigma_{i=1}^n \gamma_i \xi_i$.*

Proof. This follows from the argument of Lemma 1.23, since $\Sigma_{i=1}^n \gamma_i \hat{\xi}_i$ is UMVU and equivariant. □

Theorem 4.12, which is also called the Gauss-Markov theorem, has been extensively generalized (see, for example, Rao 1976, Harville 1976, 1981, Kariya, 1985). We shall consider some extensions of this theorem in the next section. On the other hand, the following result, due to Shaffer (1991), shows a direction in which the theorem does not extend. If, in (4.35), we adopt the parametrization $\xi = \theta A$ for some $s \times n$ matrix A, there are some circumstances in which it is reasonable to assume that A also has a distribution (for example, if the data (X, A) are obtained from a sample of units, rather than A being a preset design matrix as is the case in many experiments). The properties of the resulting least squares estimator, however, will vary according to what is assumed about both the distribution of A and the distribution of X. Note that in the following theorem, all expectations are over the joint distribution of X and A.

Theorem 4.14 *Under assumptions (4.35), with $\xi = \theta A$, the following hold.*

(a) If (X, A) are jointly multivariate normal with all parameters unknown, then the least squares estimator $\Sigma \gamma_i \hat{\xi}_i$ is the UMVU estimator of $\Sigma \gamma_i \xi_i$.

(b) If the distribution of A is unknown, then the least squares estimator $\Sigma \gamma_i \hat{\xi}_i$ is UMVU among all linear estimators of $\Sigma \gamma_i \xi_i$.

(c) If $E(AA')$ is known, no best linear unbiased estimator of $\Sigma \gamma_i \xi_i$ exists.

Proof. Part (a) follows from the fact that the least squares estimator is a function of the complete sufficient statistic. Part (b) can be proved by showing that if $\Sigma \gamma_i \hat{\xi}_i$ is unconditionally unbiased then it is conditionally unbiased, and hence Theorem 4.12 applies. For this purpose, one can use a variation of Problem 1.6.33, where it was shown that the order statistics are complete sufficient. Finally, part (c) follows from the fact that the extra information about the variance of A can often be used to improve any unbiased estimator. See Problems 4.16–4.18 for details. □

The formulation of the regression problem in Theorem 4.14, in which the p rows of A are sometimes referred to as "random regressors," has other interesting implications. If A is *ancillary*, the distribution of A and hence $E(A'A)$ are known and so we have a situation where the distribution of an ancillary statistic will affect the properties of an estimator. This paradox was investigated by Brown (1990a), who established some interesting relationships between ancillarity and admissibility (see Problems 5.7.31 and 5.7.32).

For estimating σ^2, it is natural to restrict attention to unbiased quadratic (rather than linear) estimators Q of σ^2. Among these, does the estimator $S^2/(n-s)$ which is UMVU in the normal case continue to minimize the variance? Under mild additional restrictions—for example, invariance under the transformations (4.10) or restrictions to Q's taking on only positive values—it turns out that this is true in some cases (for instance, in Example 4.15 below when the n_i are equal) but not in others. For details, see Searle et al. (1992, Section 11.3).

Example 4.15 Quadratic unbiased estimators. Let X_{ij} ($j = 1, \ldots, n_i$; $i = 1, \ldots, s$) be independently distributed with means $E(X_{ij}) = \xi_i$ and common variance and fourth moment

$$\sigma^2 = E(X_{ij} - \xi_i)^2 \quad \text{and} \quad \beta = E(X_{ij} - \xi_i)^4/\sigma^4,$$

respectively. Consider estimators of σ^2 of the form $Q = \Sigma\lambda_i S_i^2$ where $S_i^2 = \Sigma(X_{ij} - X_{i\cdot})^2$ and $\Sigma\lambda_i(n_i - 1) = 1$ so that Q is an unbiased estimator of σ^2. Then, the variance of Q is minimized (Problem 4.19) when the λ's are proportional to $1/(\alpha_i + 2)$ where $\alpha_i = [(n_i - 1)/n_i](\beta - 3)$. The standard choice of the λ_i (which is to make them equal) is, therefore, best if either the n_i are equal or $\beta = 3$, which is the case when the X_{ij} are normal.

(ii) Let us now return to the model obtained from (4.4) by dropping the assumption of normality but without restricting attention to linear estimators. More specifically, we shall assume that $X_1, \ldots, X_n$ are random variables such that

(4.36) the variables $X_i - \xi_i$ are iid with a common distribution F which has expectation zero and an otherwise unknown probability density f,

and such that (4.13) holds with A an $n \times n$ matrix of rank s. ||

In Section 2.4, we found that for the case $\xi_i = \theta$, the LSE $\bar{X}$ of θ is UMVU in this nonparametric model. To show that the corresponding result does not generally hold when $\boldsymbol{\xi}$ is given by (4.13), consider the two-way layout of Example 4.11 and the estimation of

$$\alpha_i = \xi_{i\cdot} - \xi_{\cdot\cdot} = \frac{1}{IJ}\sum_{j=1}^{I}\sum_{k=1}^{J}(\xi_{ik} - \xi_{jk}). \tag{4.37}$$

To avoid calculations, suppose that F is t_2, the t-distribution with 2 degrees of freedom. Then, the least squares estimators have infinite variance. On the other hand, let $\tilde{X}_{ij}$ be the median of the observations X_{ijv}, $v = 1, \ldots, m$. Then $\tilde{X}_{ik} - \tilde{X}_{jk}$ is an unbiased estimator of $\xi_{ik} - \xi_{jk}$ so that $\delta = (1/ab)\Sigma\Sigma(\tilde{X}_{ik} - \tilde{X}_{jk})$ is an unbiased estimator of α_i. Furthermore, if $m \geq 3$, the $\tilde{X}_{ij}$ have finite variance and so, therefore, does δ. (A sum of random variables with finite variance has finite variance.) This shows that the least squares estimators of the α_i are not UMVU when F is unknown. The same argument applies to the β's and γ's.

The situation is quite different for the estimation of μ. Let $\mathcal{U}$ be the class of unbiased estimators of μ in model (4.27) with F unknown, and let $\mathcal{U}'$ be the corresponding class of unbiased estimators when the α's, β's, and γ's are all zero. Then, clearly, $\mathcal{U} \subset \mathcal{U}'$; furthermore, it follows from Section 2.4 that $X_{\dots}$ uniformly minimizes the variance within $\mathcal{U}'$. Since $X_{\dots}$ is a member of $\mathcal{U}$, it uniformly minimizes the variance within $\mathcal{U}$ and, hence, is UMVU for μ in model (4.27) when F is unknown.

For a more detailed discussion of this problem, see Anderson (1962).

(iii) Instead of assuming the density f in (4.36) to be unknown, we may be interested in the case in which f is known but not normal. The model then remains invariant under the transformations

$$X'_v = X_v + \sum_{j=1}^{s} a_{jv}\gamma_j, \quad -\infty < \gamma_1, \dots, \gamma_s < \infty. \tag{4.38}$$

Since $E(X'_v) = \Sigma a_{jv}(\theta_j + \gamma_j)$, the induced transformations in the parameter space are given by

$$\theta'_j = \theta_j + \gamma_j \quad (j = 1, \dots, s). \tag{4.39}$$

The problem of estimating θ_j remains invariant under the transformations (4.38), (4.39), and

$$d' = d + \gamma_j \tag{4.40}$$

for any loss function of the form $\rho(d - \theta_j)$, and an estimator δ of θ_j is equivariant with respect to these transformations if it satisfies

$$\delta(\mathbf{X}') = \delta(\mathbf{X}) + \gamma_j. \tag{4.41}$$

Since (4.39) is transitive over Ω, the risk of any equivariant estimator is constant, and an MRE estimator of θ_j can be found by generalizing Theorems 1.8 and 1.10 to the present situation (see Verhagen 1961).

(iv) Important extensions to random and mixed effects models, and to general exponential families, will be taken up in the next two sections.

5 Random and Mixed Effects Models

In many applications of linear models, the effects of the various factors $A, B, C, \dots$ which were considered to be unknown constants in Section 3.4 are, instead, random. One then speaks of a *random effects* model (or Model II); in contrast, the corresponding model of Section 3.4 is a *fixed effects* model (or Model I). If both fixed and random effects occur, the model is said to be *mixed*.

Example 5.1 Random effects one-way layout. Suppose that, as a measure of quality control, an auto manufacturer tests a sample of new cars, observing for each car, the mileage achieved on a number of occasions on a gallon of gas. Suppose X_{ij} is the mileage of the ith car on the jth occasion, at time t_{ij}, with all the t_{ij} being selected at random and independently of each other. This would have been modeled in Example 4.1 as

$$X_{ij} = \mu + \alpha_i + U_{ij}$$

where the U_{ij} are independent $N(0, \sigma^2)$. Such a model would be appropriate if these particular cars were the object of study and a replication of the experiment thus consisted of a number of test runs by the same cars. However, the manufacturer is interested in the performance of the thousands of cars to be produced that year and, for this reason, has drawn a random sample of cars for the test. A replication of the experiment would start by drawing a new sample. The effect of the ith car is therefore a random variable, and the model becomes

$$X_{ij} = \mu + A_i + U_{ij} \quad (j = 1, \ldots, n_i;\ i = 1, \ldots, s). \tag{5.1}$$

Here and following, the populations being sampled are assumed to be large enough so that independence and normality of the unobservable random variables A_i and U_{ij} can be assumed as a reasonable approximation. Without loss of generality, one can put $E(A_i) = E(U_{ij}) = 0$ since the means can be absorbed into μ. The variances will be denoted by $\text{var}(A_i) = \sigma_A^2$ and $\text{var}(U_{ij}) = \sigma^2$.

The X_{ij} are dependent, and their joint distribution, and hence the estimation of σ_A^2 and σ^2, is greatly simplified if the model is assumed to be *balanced*, that is, to satisfy $n_i = n$ for all i. In that case, in analogy with the transformation (4.5), let each set $(X_{i1}, \ldots, X_{in})$ be subjected to an orthogonal transformation to $(Y_{i1}, \ldots, Y_{in})$ such that $Y_{i1} = \sqrt{n}\, X_{i\cdot}$. An additional orthogonal transformation is made from $(Y_{11}, \ldots, Y_{s1})$ to $(Z_{11}, \ldots, Z_{s1})$ such that $Z_{11} = \sqrt{s}\, Y_{\cdot 1}$, whereas for $i > 1$, we put $Z_{ij} = Y_{ij}$. Unlike the X_{ij}, the Y_{ij} and Z_{ij} are all independent (Problem 5.1). They are normal with means

$$E(Z_{11}) = \sqrt{sn}\,\mu, \quad E(Z_{ij}) = 0 \quad \text{if } i > 1 \text{ or } j > 1$$

and variances

$$\text{var}(Z_{i1}) = \sigma^2 + n\sigma_A^2, \quad \text{var}(Z_{ij}) = \sigma^2 \quad \text{for } j > 1,$$

so that the joint density of the Z's is proportional to

$$\exp\left\{-\frac{1}{2(\sigma^2 + n\sigma_A^2)}\left[(Z_{11} - \sqrt{sn}\,\mu)^2 + S_A^2\right] - \frac{1}{2\sigma^2}S^2\right\} \tag{5.2}$$

with

$$S_A^2 = \sum_{i=2}^{s} Z_{i1}^2 = n\Sigma(X_{i\cdot} - X_{\cdot\cdot})^2, \quad S^2 = \sum_{i=1}^{s}\sum_{j=2}^{n} Z_{ij}^2 = \sum_{i=1}^{s}\sum_{j=1}^{n}(X_{ij} - X_{i\cdot})^2.$$

This is a three-parameter exponential family with

$$\eta_1 = \frac{\mu}{\sigma^2 + n\sigma_A^2}, \quad \eta_2 = \frac{1}{\sigma^2 + n\sigma_A^2}, \quad \eta_3 = \frac{1}{\sigma^2}. \tag{5.3}$$

The variance of X_{ij} is $\text{var}(X_{ij}) = \sigma^2 + \sigma_A^2$, and we are interested in estimating the *variance components* σ_A^2 and σ^2. Since

$$E\left(\frac{S_A^2}{s-1}\right) = \sigma^2 + n\sigma_A^2 \quad \text{and} \quad E\left(\frac{S^2}{s(n-1)}\right) = \sigma^2,$$

it follows that

$$\hat{\sigma}^2 = \frac{S^2}{s(n-1)} \quad \text{and} \quad \hat{\sigma}_A^2 = \frac{1}{n}\left[\frac{S_A^2}{s-1} - \frac{S^2}{s(n-1)}\right] \tag{5.4}$$

are UMVU estimators of σ^2 and σ_A^2, respectively. The UMVU estimator of the ratio is σ_A^2/σ^2 is

$$\frac{1}{n}\left[\frac{K_{f,-2}}{s(n-1)}\frac{\hat{\sigma}^2+n\hat{\sigma}_A^2}{\hat{\sigma}^2}-1\right],$$

where $K_{f,-2}$ is given by (2.2.5) with $f = s(n-1)$ (Problem 5.3). Typically, the only linear subspace of the η's of interest here is the trivial one defined by $\sigma_A^2 = 0$, which corresponds to $\eta_2 = \eta_3$ and to the case in which the sn X_{ij} are iid as $N(\mu, \sigma^2)$. ||

Example 5.2 Random effects two-way layout. In analogy to Example 4.11, consider the random effects two-way layout.

$$X_{ijk} = \mu + A_i + B_j + C_{ij} + U_{ijk} \tag{5.5}$$

where the unobservable random variables A_i, B_j, C_{ij}, and U_{ijk} are independently normally distributed with zero mean and with variances $\sigma_A^2, \sigma_B^2, \sigma_C^2$, and σ^2, respectively. We shall restrict attention to the balanced case $i = 1, \ldots, I$, $j = 1, \ldots, J$, and $k = 1, \ldots n$. As in the preceding example, a linear transformation leads to independent normal variables Z_{ijk} with means $E(Z_{111}) = \sqrt{IJn}\,\mu$ and 0 for all other Z 's and with variances

$$\begin{aligned}
&\text{var}(Z_{111}) = nJ\sigma_A^2 + nI\sigma_B^2 + n\sigma_C^2 + \sigma^2,\\
&\text{var}(Z_{i11}) = nJ\sigma_A^2 + n\sigma_C^2 + \sigma^2, \quad i > 1,\\
&\text{var}(Z_{1j1}) = nI\sigma_B^2 + n\sigma_C^2 + \sigma^2, \quad j > 1,\\
&\text{var}(Z_{ij1}) = n\sigma_C^2 + \sigma^2, \quad i, j > 1,\\
&\text{var}(Z_{ijk}) = \sigma^2, \quad k > 1.
\end{aligned} \tag{5.6}$$

As an example in which such a model might arise, consider a reliability study of blood counts, in which blood samples from each of J patients are divided into nI subsamples of which n are sent to each of I laboratories. The study is not concerned with these particular patients and laboratories, which, instead, are assumed to be random samples from suitable patient and laboratory populations. From (5.5) it follows that $\text{var}(X_{ijk}) = \sigma_A^2 + \sigma_B^2 + \sigma_C^2 + \sigma^2$. The terms on the right are the variance components due to laboratories, patients, the interaction between the two, and the subsamples from a patient.

The joint distribution of the Z_{ijk} constitutes a five-parameter exponential family with the complete set of sufficient statistics (Problem 5.9)

$$\begin{aligned}
S_A^2 &= \sum_{i=2}^{I} Z_{i11}^2 = nJ\sum_{i=1}^{I}(X_{i\cdot\cdot} - X_{\cdot\cdot\cdot})^2,\\
S_B^2 &= \sum_{j=2}^{J} Z_{1j1}^2 = nI\sum_{j=1}^{J}(X_{\cdot j\cdot} - X_{\cdot\cdot\cdot})^2,\\
S_C^2 &= \sum_{i=2}^{I}\sum_{j=2}^{J} Z_{ij1}^2 = n\sum_{i=1}^{I}\sum_{j=1}^{J}(X_{ij\cdot} - X_{i\cdot\cdot} - X_{\cdot j\cdot} + X_{\cdot\cdot\cdot})^2,\\
S^2 &= \sum_{i=1}^{I}\sum_{j=1}^{J}\sum_{k=2}^{n} Z_{ijk}^2 = \sum_{i=1}^{I}\sum_{j=1}^{J}\sum_{k=1}^{n}(X_{ijk} - X_{ij\cdot})^2,
\end{aligned} \tag{5.7}$$

$$Z_{111} = \sqrt{IJn}X_{...}.$$

From the expectations of these statistics, one finds the UMVU estimators of the variance components σ^2, σ_C^2, σ_A^2, and σ_B^2 to be

$$\hat{\sigma}^2 = \frac{S^2}{IJ(n-1)}, \quad \hat{\sigma}_C^2 = \frac{1}{n}\left[\frac{S_C^2}{(I-1)(J-1)} - \hat{\sigma}^2\right],$$
$$\hat{\sigma}_A^2 = \frac{1}{nJ}\left[\frac{S_A^2}{I-1} - n\hat{\sigma}_C^2 - \hat{\sigma}^2\right], \quad \hat{\sigma}_B^2 = \frac{1}{nI}\left[\frac{S_B^2}{J-1} - n\hat{\sigma}_C^2 - \hat{\sigma}^2\right].$$

A submodel of (5.5), which is sometimes appropriate, is the *additive* model corresponding to the absence of the interaction terms C_{ij} and hence to the assumption $\sigma_C^2 = 0$. If $\eta_1 = \mu/\text{var}(Z_{111})$, $1/\eta_2 = nJ\sigma_A^2 + n\sigma_C^2 + \sigma^2$, $1/\eta_3 = nI\sigma_B^2 + n\sigma_C^2 + \sigma^2$, $1/\eta_4 = n\sigma_C^2 + \sigma^2$, and $1/\eta_5 = \sigma^2$, this assumption is equivalent to $\eta_4 = \eta_5$ and thus restricts the η's to a linear subspace. The submodel constitutes a four-parameter exponential family, with the complete set of sufficient statistics Z_{111}, S_A^2, S_B^2, and $S'^2 = S_C^2 = \Sigma\Sigma\Sigma(X_{ijk} - X_{i..} - X_{.j.} + X_{...})^2$. The UMVU estimators of the variance components σ_A^2, σ_B^2, and σ^2 are now easily obtained as before (Problem 5.10).

Another submodel of (5.5) which is of interest is obtained by setting $\sigma_B^2 = 0$, thus eliminating the B_j terms from (5.5). However, this model, which corresponds to the linear subspace $\eta_3 = \eta_4$, does not arise naturally in the situations leading to (5.5), as illustrated by the laboratory example. These situations are characterized by a *crossed* design in which each of the IA units (laboratories) is observed in combination with each of the JB units (patients). On the other hand, the model without the B terms arises naturally in the very commonly occurring *nested* design illustrated in the following example. ||

Example 5.3 Two nested random factors. For the two factors A and B, suppose that each of the units corresponding to different values of i (i.e., different levels of A) is itself a collection of smaller units from which the values of B are drawn. Thus, the A units might be hospitals, schools, or farms that constitute a random sample from a population of such units from each of which a random sample of patients, students, or trees is drawn. On each of the latter, a number of observations is taken (for example, a number of blood counts, grades, or weights of a sample of apples). The resulting model [with a slight change of notation from (5.5)] may be written as

$$X_{ijk} = \mu + A_i + B_{ij} + U_{ijk}. \tag{5.8}$$

Here, the A's, B's, and U's are again assumed to be independent normal with zero means and variances σ_A^2, σ_B^2, and σ^2, respectively. In the balanced case ($i = 1, \ldots, I$, $j = 1, \ldots, J$, $k = 1, \ldots, n$), a linear transformation produces independent variables with means $E(Z_{111}) = \sqrt{IJn}\,\mu$ and $= 0$ for all other Z's and variances

$$\text{var}(Z_{i11}) = \sigma^2 + n\sigma_B^2 + Jn\sigma_A^2 \quad (i = 1, \ldots, 1),$$
$$\text{var}(Z_{ij1}) = \sigma^2 + n\sigma_B^2 \quad (j > 1),$$
$$\text{var}(Z_{ijk}) = \sigma^2 \quad (k > 1).$$

The joint distribution of the Z's constitutes a four-parameter exponential family with the complete set of sufficient statistics

$$
\begin{aligned}
S_A^2 &= \sum_{i=2}^{I} Z_{i11}^2 = Jn\Sigma(X_{i\cdot\cdot} - X_{\cdots})^2, \\
S_B^2 &= \sum_{j=2}^{J} Z_{1j1}^2 = n\Sigma\Sigma(X_{ij\cdot} - X_{i\cdot\cdot})^2, \qquad (5.9)\\
S^2 &= \sum_{i=1}^{I}\sum_{j=1}^{J}\sum_{k=2}^{n} Z_{ijk}^2 = \sum_{i=1}^{I}\sum_{j=1}^{J}\sum_{k=1}^{n}(X_{ijk} - X_{ij\cdot})^2, \\
Z_{111} &= \sqrt{IJn}X_{\cdots},
\end{aligned}
$$

and the UMVU estimators of the variance components can be obtained as before (Problem 5.12). ||

The models illustrated in Examples 5.2 and 5.3 extend in a natural way to more than two factors, and in the balanced cases, the UMVU estimators of the variance components are easily derived.

The estimation of variance components described above suffers from two serious difficulties.

(i) The UMVU estimators of all the variance components except σ^2 can take on negative values with probabilities as high as .5 and even in excess of that value (Problem 5.5–5.7) (and, correspondingly, their expected squared errors are quite unsatisfactory; see Klotz, Milton, and Zacks 1969).

The interpretation of such negative values either as indications that the associated components are negligible (which is sometimes formalized by estimating them to be zero) or that the model is incorrect is not always convincing because negative values do occur even when the model is correct and the components are positive. An alternative possibility, here and throughout this section, is to fall back on maximum likelihood estimation or restricted MLE's (REML estimates) obtained by maximizing the likelihood after first reducing the data through location invariance (Thompson, 1962; Corbeil and Searle, 1976). Although these methods have no small-sample justification, they are equivalent to a noninformative prior Bayesian solution (Searle et al. 1992; see also Example 2.7). Alternatively, there is an approach due to Hartung (1981), who minimizes bias subject to non-negativity, or Pukelsheim (1981) and Mathew (1984), who find non-negative unbiased estimates of variance.

(ii) Models as simple as those obtained in Examples 5.1–5.3 are not available when the layout is not balanced.

The joint density of the X's can then be obtained by noting that they are linear functions of normal variables and thus have a joint multivariate normal distribution. To obtain it, one only need write down the covariance matrix of the X's and invert it. The result is an exponential family which typically is not complete unless the model is balanced. (This is illustrated for the one-way layout in Problem 5.4.) UMVU estimators cannot be expected in this case (see Pukelsheim 1981). A characterization of U-estimable functions permitting UMVU estimators is given by Unni (1978). Two general methods for the estimation of variance components have been developed in some detail; these are maximum and restricted maximum likelihood, and the minimum norm quadratic unbiased estimation (Minque) introduced by Rao (1970). Surveys of the area are given by Searle (1971b), Harville (1977), and Kleffe (1977). More detailed introductions can be found, for example, in the books by Rao and Kleffe (1988), Searle et al. (1992), and Burdick and Graybill (1992).

So far, the models we have considered have had factors that were either all fixed or all random. We now look at an example of a mixed model, which contains both types of factors.

Example 5.4 Mixed effects model. In Example 5.3, it was assumed that the hospital, schools, or farms were obtained as a random sample from a population of such units. Let us now suppose that it is only these particular hospitals that are of interest (perhaps it is the set of all hospitals in the city), whereas the patients continue to be drawn at random from these hospitals. Instead of (5.8), we shall assume that the observations are given by

$$X_{ijk} = \mu + \alpha_i + B_{ij} + U_{ijk} \quad (\Sigma\alpha_i = 0). \tag{5.10}$$

A transformation very similar to the earlier one (Problem 5.14) now leads to independent normal variables W_{ijk} with joint density proportional to

$$\exp\left\{-\frac{1}{2(\sigma^2 + n\sigma_B^2)}[\Sigma(w_{i11} - \mu - \alpha_i)^2 + S_B^2] - \frac{1}{2\sigma^2}S^2\right\} \tag{5.11}$$

with S_B^2 and S^2 given by (5.9), and with $W_{i11} = \sqrt{Jn}\, X_{i\cdot\cdot}$. This is an exponential family with the complete set of sufficient statistics $X_{i\cdot\cdot}$, S_B^2, and S^2. The UMVU estimators of σ_B^2 and σ^2 are the same as in Example 5.3, whereas the UMVU estimator of α_i is $X_{i\cdot\cdot} - X_{\cdot\cdot\cdot}$, as it would be if the B's were fixed. ||

Thus far in this section, our focus has been the estimation of the variance components in random and mixed effects models. There is, however, another important estimation target in these models, the random effects themselves. This presents a somewhat different problem than is considered in the rest of this book, as the estimand is now a random variable rather than a fixed parameter. However, the theory of UMVU estimation has a fairly straightforward extension to the present case. We illustrate this in the following example.

Example 5.5 Best prediction of random effects. Consider, once more, the random effects model (5.1), where the value α_i of A_i, the effect on gas mileage, could itself be of interest.

Since α_i is the realized value of a random variable rather than a fixed parameter, it is common to speak of *prediction* of α_i rather than estimation of α_i. To avoid identifiability problems, we will, in fact, predict $\mu + \alpha_i$ rather than α_i. If $\delta(\mathbf{X})$ is a predictor, then under squared error loss we have

$$\begin{aligned} E[\delta(\mathbf{X}) - (\mu+\alpha_i)]^2 &= E[\delta(\mathbf{X}) \pm E(\mu+\alpha_i|\mathbf{X}) - (\mu+\alpha_i)]^2 \\ &= E[\delta(\mathbf{X}) - E(\mu+\alpha_i|\mathbf{X})]^2 \end{aligned} \tag{5.12}$$

$$+E[E(\mu+\alpha_i|\mathbf{X})-(\mu+\alpha_i)]^2.$$

As we have no control over the second term on the right side of (5.12), we only need be concerned with minimization of the first term. (In this sense, prediction of a random variable is the same as estimation of its conditional expected value.)

Under the normality assumptions of Example 5.1,

$$E(\mu+\alpha_i|\mathbf{X})=\frac{n\sigma_A^2}{n\sigma_A^2+\sigma^2}\bar{X}_i+\frac{\sigma^2}{n\sigma_A^2+\sigma^2}\mu. \tag{5.13}$$

Assuming the variances known, we set

$$\delta(\mathbf{X})=\frac{n\sigma_A^2}{n\sigma_A^2+\sigma^2}\bar{X}_i+\frac{\sigma^2}{n\sigma_A^2+\sigma^2}\delta'(\mathbf{X})$$

and choose $\delta'(\mathbf{X})$ to minimize $E[\delta'(\mathbf{X})-\mu]^2$. The UMVU estimator of μ is $\bar{X}$, and the UMVU predictor of $\mu+\alpha_i$ is

$$\frac{n\sigma_A^2}{n\sigma_A^2+\sigma^2}\bar{X}_i+\frac{\sigma^2}{n\sigma_A^2+\sigma^2}\bar{X}. \tag{5.14}$$

As we will see in Chapter 4, this predictor is also a Bayes estimator in a hierarchical model (which is another way of thinking of the model (5.1); see Searle et al. 1992, Chapter 9, and Problem 4.7.15).

Although we have assumed normality, optimality of (5.14) continues if the distributional assumptions are relaxed, similar to (4.35). Under such relaxed assumptions, (5.14) continues to be best among linear unbiased predictors (Problem 5.17). Harville (1976) has formulated and proved a Gauss-Markov-type theorem for a general mixed model. ||

6 Exponential Linear Models

The great success of the linear models described in the previous sections suggests the desirability of extending these models beyond the normal case. A natural generalization combines a general exponential family with the structure of a linear model and will often result in exponential linear models in terms of new parameters [see, for example, (5.2) and (5.3)]. However, the models in this section are discrete and do not arise from normal theory.

Equivariance tends to play little role in the resulting models; they are therefore somewhat out of place in this chapter. But certain analogies with normal linear models make it convenient to present them here.

(i) Contingency Tables

Suppose that the underlying exponential family is the set of multinomial distributions (1.5.4), which may be written as

$$\exp\left(\sum_{i=0}^{s}x_i\log p_i\right)h(x), \tag{6.1}$$

and that a linear structure is imposed on the parameters $\eta_i = \log p_i$. Expositions of the resulting theory of *log linear models* can be found in the books by Agresti (1990), Christensen (1990), Santner and Duffy (1990), and Everitt (1992). Diaconis (1988, Chapter 9) shows how a combination of exponential family theory and group representations lead naturally to log linear models.

The models have close formal similarities with the corresponding normal models, and a natural linear subspace of the log p_i often corresponds to a natural restriction on the p's. In particular, since sums of the log p's correspond to products of the p's, a subspace defined by setting suitable interaction terms equal to zero often is equivalent to certain independence properties in the multinomial model.

The exponential family (6.1) is not of full rank since the p's must add up to 1. A full-rank form is

$$\left[\exp \sum_{i=1}^{s} x_i \log(p_i/p_0)\right] h(x). \tag{6.2}$$

If we let

$$\eta_i' = \log \frac{p_i}{p_0} = \eta_i - \eta_0, \tag{6.3}$$

we see that arbitrary linear functions of the η_i' correspond to arbitrary contrasts (i.e., functions of the differences) of the η_i. From Example 2.3.8, it follows that $(X_1, \ldots, X_s)$ or $(X_0, X_1, \ldots, X_s)$ is sufficient and complete for (6.2) and hence also for (6.1). In applications, we shall find (6.1) the more convenient form to use.

If the η's are required to satisfy r independent linear restrictions $\Sigma a_{ij}\eta_j = b_i (i = 1, \ldots, r)$, the resulting distributions will form an exponential family of rank $s - r$, and the associated minimal sufficient statistics T will continue to be complete. Since $E(X_i/n) = p_i$, the probabilities p_i are always U-estimable; their UMVU estimators can be obtained as the conditional expectations of X_i/n given T. If $\hat{p}_i$ is the UMVU estimator of p_i, a natural estimator of η_i is $\hat{\eta}_i = \log \hat{p}_i$, but, of course, this is no longer unbiased. In fact, no unbiased estimator of η_i exists because only polynomials of the p_i can be U-estimable (Problem 2.3.25). When $\hat{p}_i$ is also the MLE of p_i, $\hat{\eta}_i$ is the MLE of η_i. However, the MLE $\hat{\hat{p}}_i$ does not always coincide with the UMVU estimator $\hat{p}_i$. An example of this possibility with $\log p_i = \alpha + \beta t_i$ (t's known; α and β unknown) is given by Haberman (1974, Example 1.16, p. 29; Example 3.3, p. 60). It is a disadvantage of the $\hat{p}_i$ in this case that, unlike $\hat{\hat{p}}_i$, they do not always satisfy the restrictions of the model, that is, for some values of the X's, no α and β exist for which $\log \hat{p}_i = \alpha + \beta t_i$. Typically, if $\hat{p}_i \neq \hat{\hat{p}}_i$, the difference between the two is moderate.

For estimating the η_i, Goodman (1970) has recommended in some cases applying the estimators not to the cell frequencies X_i/n but to $X_i/n + 1/2$, in order to decrease the bias of the MLE. This procedure also avoids difficulties that may arise when some of the cell counts are zero. (See also Bishop, Fienberg, and Holland 1975, Chapter 12.)

Example 6.1 Two-way contingency table. Consider the situation of Example 2.3.9 in which n subjects are classified according to two characteristics A and B with possible outcomes $A_1, \ldots, A_I$ and $B_1, \ldots, B_J$. If n_{ij} is the number of subjects with properties A_i and B_j, the joint distribution of the n_{ij} can be written as

$$\frac{n!}{\Pi_{i,j}(n_{ij})!}\exp\Sigma\Sigma n_{ij}\xi_{ij}, \quad \xi_{ij}=\log p_{ij}.$$

Write $\xi_{ij}=\mu+\alpha_i+\beta_j+\gamma_{ij}$ as in Example 4.11, with the side conditions (4.28). This implies no restrictions since any IJ numbers ξ_{ij} can be represented in this form. The p_{ij} must, of course, satisfy $\Sigma\Sigma p_{ij}=1$ and the ξ_{ij} must therefore satisfy $\Sigma\exp\xi_{ij}=1$. This equation determines μ as a function of the α's, β's, and γ's which are free, subject only to (4.28). The UMVU estimators of the p_{ij} were seen in Example 2.3.9 to be n_{ij}/n. ||

In Example 4.11 (normal two-way layout), it is sometimes reasonable to suppose that all the γ_{ij}s (the interactions) are zero. In the present situation, this corresponds exactly to the assumption that the characteristics A and B are independent, that is, that $p_{ij}=p_{i+}p_{+j}$ (Problem 6.1). The UMVU estimator of p_{ij} is now $n_{i+}n_{+j}/n^2$.

Example 6.2 Conditional independence in a three-way table. In Example 2.3.10, it was assumed that the subjects are classified according to three characteristics A, B, and C and that conditionally, given outcome C, the two characteristics A and B are independent. If $\xi_{ijk}=\log p_{ijk}$ and ξ_{ijk} is written as

$$\xi_{ijk}=\mu+\alpha_i^A+\alpha_j^B+\alpha_k^C+\alpha_{ij}^{AB}+\alpha_{ik}^{AC}+\alpha_{jk}^{BC}+\alpha_{ijk}^{ABC}$$

with the α's subject to the usual restrictions and with μ determined by the fact that the p_{ijk} add up to 1, it turns out that the conditional independence of A and B given C is equivalent to the vanishing of both the three-way interactions α_{ijk}^{ABC} and the A, B interactions α_{ij}^{AB} (Problem 6.2). The UMVU estimators of the p_{ijk} in this model were obtained in Example 2.3.10. ||

(ii) Independent Binomial Experiments

The submodels considered in Example 5.2–5.4 and 6.1–6.2 corresponded to natural assumptions about the variances or probabilities in question. However, in general, the assumption of linearity in the η's made at the beginning of this section is rather arbitrary and is dictated by mathematical convenience rather than by meaningful structural assumptions. We shall now consider a particularly simple class of problems, in which this linearity assumption is inconsistent with more customary assumptions. Agreement with these assumptions can be obtained by not insisting on a linear structure for the parameters η_i themselves but permitting a linear structure for a suitable function of the η's.

The problems are concerned with a number of independent random variables X_i having the binomial distributions $b(p_i, n_i)$. Suppose the X's have been obtained from some unobservable variables Z_i distributed independently as $N(\zeta_i, \sigma^2)$ by setting

(6.4)
$$X_i=\begin{cases}0 & \text{if } Z_i\le u\\ 1 & \text{if } Z_i>u.\end{cases}$$

Then

(6.5)
$$p_i=P(Z_i>u)=\Phi\left(\frac{\zeta_i-u}{\sigma}\right)$$

and hence

$$(6.6) \qquad \zeta_i = u + \sigma\Phi^{-1}(p_i).$$

Now consider a two-way layout for the Z's in which the effects are additive, as in Example 4.11. The subspace of the ζ_{ij} ($i = 1, \ldots, a$, $j = 1, \ldots, b$) defining this model is characterized by the fact that the interactions satisfy

$$(6.7) \qquad \gamma_{ij} = \zeta_{ij} - \zeta_{i\cdot} - \zeta_{\cdot j} + \zeta_{\cdot\cdot} = 0$$

which, by (6.6), implies that

$$(6.8) \qquad \Phi^{-1}(p_{ij}) - \frac{1}{J}\sum_j \Phi^{-1}(p_{ij}) - \frac{1}{I}\sum_i \Phi^{-1}(p_{ij}) + \frac{1}{IJ}\sum_i\sum_j \Phi^{-1}(p_{ij}) = 0.$$

The "natural" linear subspace of the parameter space for the Z's thus translates into a linear subspace in terms of the parameters $\Phi^{-1}(p_{ij})$ for the X's, and the corresponding fact by (6.6) is true quite, generally, for subspaces defined in terms of differences of the ζ's. On the other hand, the joint distribution of the X's is proportional to

$$(6.9) \qquad \exp\left[\Sigma x_i \log \frac{p_i}{q_i}\right] h(x),$$

and the natural parameters of this exponential family are $\eta_i = \log(p_i/q_i)$. The restrictions (6.8) are not linear in the η's, and the minimal sufficient statistics for the exponential family (6.9) with the restrictions (6.8) are not complete.

It is interesting to ask whether there exists a distribution F for the underlying variables Z_i such that a linear structure for the ζ_i will result in a linear structure for $\eta_i = \log(p_i/q_i)$ when the p_i and the ζ_i are linked by the equation

$$(6.10) \qquad q_i = P(Z_i \le u) = F(u - \zeta_i)$$

instead of by (6.5). Then, $\zeta_i = u - F^{-1}(q_i)$ so that linear functions of the ζ_i correspond to linear functions of the $F^{-1}(q_i)$ and hence of $\log(p_i/q_i)$, provided

$$(6.11) \qquad F^{-1}(q_i) = a - b\log\frac{p_i}{q_i}.$$

Suppressing the subscript i and putting $x = a - b\log(p/q)$, we see that (6.11) is equivalent to

$$(6.12) \qquad q = F(x) = \frac{1}{1 + e^{-(x-a)/b}},$$

which is the cdf of the logistic distribution $L(a, b)$ whose density is shown in Table 2.3.1.

Inferences based on the assumption of linearity in $\Phi^{-1}(p_i)$ and $\log(p_i/q_i) = F^{-1}(q_i)$ with F given by (6.12) where, without loss of generality, we can take $a = 0$, $b = 1$, are known as *probit* and *logit analysis*, respectively, and are widely used analysis techniques. For more details and many examples, see Cox 1970, Bishop, Fienberg, and Holland 1975, or Agresti 1990. As is shown by Cox (p. 28), the two analyses may often be expected to give very similar results, provided the

p's are not too close to 0 or 1. The probit model can also be viewed as a special case of a *threshold model*, a model in which it is only observed whether a random variable exceeds a threshold (Finney 1971). For the calculation of the MLEs in this model see Problem 6.4.16.

The outcomes of s independent binomial experiments can be represented by a $2 \times s$ contingency table, as in Table 3.3.1, with $I = 2$ and $J = s$, and the outcomes A_1 and A_2 corresponding to success and failure, respectively. The column totals $n_{+1}, \ldots, n_{+s}$ are simply the s sample sizes and are, therefore, fixed in the present model. In fact, this is the principal difference between the present model and that assumed for a $2 \times J$ table in Example 2.3.9. The case of s independent binomials arises in the situation of that example, if the n subjects, instead of being drawn at random from the population at large, are obtained by drawing n_{+j} subjects from the subpopulation having property B_j for $j = 1, \ldots, s$.

A $2 \times J$ contingency table, with fixed column totals and with the distribution of the cell counts given by independent binomials, occurs not only in its own right through the sampling of $n_{+1}, \ldots, n_{+J}$ subjects from categories $B_1, \ldots, B_J$, respectively, but also in the multinomial situation of Example 6.1 with $I = 2$, as the conditional distribution of the cell counts given the column totals. This relationship leads to an apparent paradox. In the conditional model, the UMVU estimator of the probability $p_j = p_{1j}/(p_{1j} + p_{2j})$ of success, given that the subject is in B_j, is $\delta_j = n_{1j}/n_{+j}$. Since δ_j satisfies

$$E(\delta_j | B_j) = p_j, \tag{6.13}$$

it appears also to satisfy $E(\delta_j) = p_j$ and hence to be an unbiased estimator of $p_{1j}/(p_{1j} + p_{2j})$ in the original multinomial model. On the other hand, an easy extension of the argument of Example 3.3.1 (see Problem 2.3.25) shows that, in this model, only polynomials in the p_{ij} can be U-estimable, and the ratio in question clearly is not a polynomial.

The explanation lies in the tacit assumption made in (6.13) that $n_{+j} > 0$ and in the fact that δ_j is not defined when $n_{+j} = 0$. To ensure at least one observation in B_j, one needs a sampling scheme under which an arbitrarily large number of observations is possible. For such a scheme, the U-estimability of $p_{1j}/(p_{1j} + p_{2j})$ would no longer be surprising.

It is clear from the discussion leading to (6.8) that the generalization of normal linear models to models linear in the natural parameters η_i of an exponential family is too special and that, instead, linear spaces in suitable functions of the η_i should be permitted. Because in exponential families the parameters of primary interest often are the expectations $\theta_i = E(T_i)$ [for example in (6.9), the $p_i = E(X_i)$], generalized linear models are typically defined by restricting the parameters to lie in a space defined by linear conditions on $v(\theta_i)$ [or in some cases $v_i(\theta_i)$] for a suitable *link function* v (linking the θ's with the linear space). A theory of such models was developed Dempster (1971) and Nelder and Wedderburn (1972), who, in particular, discuss maximum likelihood estimation of the parameters. Further aspects are treated in Wedderburn (1976) and in Pregibon (1980). For a comprehensive treatment of these *generalized linear models*, see the book by McCullagh and Nelder (1989), an essential reference on this topic; an introductory treatment is provided by Dobson (1990). A generalized linear interactive modeling (GLIM) package has been developed by Baker and Nelder (1983ab). The GLIM package has proved invaluable in implementing these methods and has been in the center of much of the research and modeling (see, for example, Aitken et al. 1989).

7 Finite Population Models

In the location-scale models of Sections 3.1 and 3.3, and the more general linear models of Section 4.4 and 4.5, observations are measurements that are subject to random errors. The parameters to be estimated are the true values of the quantities being measured, or differences and other linear functions of these values, and the variance of the measurement errors. We shall now consider a class of problems in which the measurements are assumed to be without error, but in which the observations are nevertheless random because the subjects (or objects) being observed are drawn at random from a finite population.

Problems of this kind occur whenever one wishes to estimate the average income, days of work lost to illness, reading level, or the proportion of a population supporting some measure or candidate. The elements being sampled need not be human but may be trees, food items, financial records, schools, and so on. We shall consider here only the simplest sampling schemes. For a fuller account of the principal methods of sampling, see, for example, Cochran (1977); a systematic treatment of the more theoretical aspects is given by Cassel, Särndal, and Wretman (1977) and Särndal, Swensson, and Wretman (1992).

The prototype of the problems to be considered is the estimation of a population average on the basis of a simple random sample from that population. In order to draw a random sample, one needs to be able to identify the members of the population. Telephone subscribers, for example, can conveniently be identified by the page and position on the page, trees by their coordinates, and students in a class by their names or by the row and number of their seat. In general, a list or other identifying description of the members of the population is called a *frame*. To represent the sampling frame, suppose that N population elements are labeled $1, \ldots, N$; in addition, a value a_i (the quantity of interest) is associated with the element i. (This notation is somewhat misleading because, in any realization of the model, the a's will simply be N real numbers without identifying subscripts.) For the purpose of estimating $\bar{a} = \Sigma_{i=1}^{N} a_i/N$, a sample of size n is drawn in order, one element after another, without replacement. It is a *simple random sample* if all $N(N-1)\ldots(N-n+1)$ possible n-tuples are equally likely.

The data resulting from such a sampling process consist of the n labels of the sampled elements and the associated a values, in the order in which they were drawn, say

$$X = \{(I_1, Y_1), \ldots, (I_n, Y_n)\} \tag{7.1}$$

where the I's denote the labels and the Y's the associated a values, $Y_k = a_{I_k}$. The unknown aspect of the situation, which as usual we shall denote by θ, is the set of population a values of the N elements,

$$\theta = \{(1, a_1), \ldots, (N, a_N)\}. \tag{7.2}$$

In the classic approach to sampling, the labels are discarded. Let us for a moment follow this approach, so that what remains of the data is the set of n observed a values: $Y_1, \ldots, Y_n$. Under simple random sampling, the order statistics $Y_{(1)} \le \cdots \le Y_{(n)}$ are then sufficient. To obtain UMVU estimators of $\bar{a}$ and other functions of the a's, one needs to know whether this sufficient statistic is complete. The answer depends on the parameter space Ω, which we have not yet specified.

It frequently seems reasonable to assume that the set V of possible values is the same for each of the a's and does not depend on the values taken on by the other a's. (This would not be the case, for example, if the a's were the grades obtained by the students in a class which is being graded "on the curve.") The parameter space is then the set Ω of all θ's given by (7.2) with $(a_1, \ldots, a_N)$ in the Cartesian product

$$V \times V \times \cdots \times V. \tag{7.3}$$

Here, V may, for example, be the set of all real numbers, all positive real numbers, or all positive integers. Or it may just be the set $V = \{0, 1\}$ representing a situation in which there are only two kinds of elements—those who vote yes or no, which are satisfactory or defective, and so on.

Theorem 7.1 *If the parameter space is given by (7.3), the order statistics $Y_{(1)}, \ldots, Y_{(n)}$ are complete.*

Proof. Denote by s an unordered sample of n elements and by $Y_{(1)}(s, \theta), \ldots, Y_{(n)}(s, \theta)$ its a values in increasing size. Then, the expected value of any estimator δ depending only on the order statistics is

$$E_\theta\{\delta[Y_{(1)}, \ldots, Y_{(n)}]\} = \Sigma P(s)\delta[Y_{(1)}(s, \theta), \ldots, Y_{(n)}(s, \theta)], \tag{7.4}$$

where the summation extends over all $\binom{N}{n}$ possible samples, and where for simple random sampling, $P(s) = 1/\binom{N}{n}$ for all s. We need to show that

$$E_\theta\{\delta[Y_{(1)}, \ldots, Y_{(n)}]\} = 0 \quad \text{for all} \quad \theta \in \Omega \tag{7.5}$$

implies that $\delta[y_{(1)}, \ldots, y_{(n)}] = 0$ for all $y_{(1)} \le \cdots \le y_{(n)}$.

Let us begin by considering (7.5) for all parameter points θ for which $(a_1, \ldots, a_N)$ is of the form $(a, \ldots, a)$, $a \in V$. Then, (7.5) reduces to

$$\sum_s P(s)\delta(a, \ldots, a) = 0 \quad \text{for all} \quad a,$$

which implies $\delta(a, \ldots, a) = 0$. Next, suppose that $N - 1$ elements in θ are equal to a, and one is equal to $b > a$. Now, (7.5) will contain two kinds of terms: those corresponding to samples consisting of n a's and those in which the sample contains b, and (7.5) becomes

$$p\,\delta(a, \ldots, a) + q\delta(a, \ldots, a, b) = 0$$

where p and q are known numbers $\neq 0$. Since the first term has already been shown to be zero, it follows that $\delta(a, \ldots, a, b) = 0$. Continuing inductively, we see that $\delta(a, \ldots, a, b, \ldots, b) = 0$ for any k a's and $n - k$ b's, $k = 0, \ldots, n$.

As the next stage in the induction argument, consider θ's of the form $(a, \ldots, a, b, c)$ with $a < b < c$, then θ's of the form $(a, \ldots, a, b, b, c)$, and so on, showing successively that $\delta(a, \ldots, a, b, c), \delta(a, \ldots, a, b, b, c), \ldots$ are equal to zero. Continuing in this way, we see that $\delta[y_{(1)}, \ldots, y_{(n)}] = 0$ for all possible $(y_{(1)}, \ldots, y_{(n)})$, and this proves completeness. □

It is interesting to note the following:

(a) No use has been made of the assumption of simple random sampling, so that the result is valid also for other sampling methods for which the probabilities $P(s)$ are known and positive for all s.

(b) The result need not be true for other parameter spaces Ω (Problem 7.1).

Corollary 7.2 *On the basis of the sample values $Y_1, \ldots, Y_n$, a UMVU estimator exists for any U-estimable function of the a's, and it is the unique unbiased estimator $\delta(Y_1, \ldots, Y_n)$ that is symmetric in its n arguments.*

Proof. The result follows from Theorem 2.1.11 and the fact that a function of $y_1, \ldots, y_n$ depends only on $y_{(1)}, \ldots, y_{(n)}$ if and only if it is symmetric in its n arguments (see Section 2.4). □

Example 7.3 UMVU estimation in simple random sampling. If the sampling method is simple random sampling and the estimand is $\bar{a}$, the sample mean $\bar{Y}$ is clearly unbiased since $E(Y_i) = \bar{a}$ for all i (Problem 7.2). Since $\bar{Y}$ is symmetric in $Y_1, \ldots, Y_n$, it is UMVU and among unbiased estimators, it minimizes the risk for any convex loss function. The variance of $\bar{Y}$ is (Problem 7.3)

$$\text{var}(\bar{Y}) = \frac{N-n}{N-1} \cdot \frac{1}{n}\tau^2 \tag{7.6}$$

where

$$\tau^2 = \frac{1}{N}\Sigma(a_i - a)^2 \tag{7.7}$$

is the *population variance*. To obtain an unbiased estimator of τ^2, note that (Problem 7.3)

$$E\left[\frac{1}{n-1}\Sigma(Y_i - \bar{Y})^2\right] = \frac{N}{N-1}\tau^2. \tag{7.8}$$

Thus, $[(N-1)/N(n-1)]\Sigma_{i=1}^n(Y_i - \bar{Y})^2$ is unbiased for τ^2, and because it is symmetric in its n arguments, it is UMVU. ||

If the sampling method is sequential, the stopping rule may add an additional complication.

Example 7.4 Sum-quota sampling. Suppose that each Y_i has associated with it a cost C_i, a positive random variable, and sampling is continued until ν observations are taken, where $\sum_1^{\nu-1} C_i < Q < \sum_1^{\nu} C_i$, with Q being a specified quota. (Note the similarity to inverse binomial sampling, as discussed in Example 2.3.2.) Under this sampling scheme, Pathak (1976) showed that $\bar{Y}_{\nu-1} = \frac{1}{\nu-1}\sum_1^{\nu-1} Y_i$ is an unbiased estimator of the population average $\bar{a}$ (Problem 7.4).

Note that Pathak's estimator drops the terminal observation Y_ν, which tends to be upwardly biased. As a consequence, Pathak's estimator can be improved upon. This was done by Kremers (1986), who showed the following:

(a) $T = \{(C_1, Y_1), \dots, (C_\nu, Y_\nu)\}$ is complete sufficient.

(b) Conditional on T, $\{(C_i, Y_1), \dots, (C_{\nu-1}, Y_{\nu-1})\}$ are *exchangeable* (Problem 7.5).

Under these conditions, the estimator

$$\hat{a} = \bar{Y} - (\bar{Y}_{[\nu]} - \bar{Y})/(\nu - 1) \tag{7.9}$$

is UMVU if $\nu > 1$, where $\bar{Y}_{[\nu]}$ is the mean of all of the observations that *could have been* the terminal observation; that is, $\bar{Y}_{[\nu]}$ is the mean of all the Y_i's in the set

$$\{(c_j, y_j) : \sum_{i \neq j} c_i < Q, j = 1, \dots, \nu\}. \tag{7.10}$$

See Problem 7.6. ||

So far, we have ignored the labels. That Theorem 7.1 and Corollary 7.2 no longer hold when the labels are included in the data is seen by the following result.

Theorem 7.5 *Given any sampling scheme of fixed size n which assigns to the sample* s *a known probability $P(s)$ (which may depend on the labels but not on the a values of the sample), given any U-estimable function $g(\theta)$, and given any preassigned parameter point $\theta_0 = \{(1, a_{10}), \dots, (N, a_{N0})\}$, there exists an unbiased estimator δ^* of $g(\theta)$ with variance* $\text{var}_{\theta_0}(\delta^*) = 0$.

Proof. Let δ be any unbiased estimator of $g(\theta)$, which may depend on both labels and y values, say

$$\delta(s) = \delta[(i_1, y_1), \dots, (i_n, y_n)],$$

and let

$$\delta_0(s) = \delta[(i_1, a_{i_1 0}), \dots, (i_n, a_{i_n 0})].$$

Note that δ_0 depends on the labels whether or not δ does and thus would not be available if the labels had been discarded. Let

$$\delta^*(s) = \delta(s) - \delta_0(s) + g(\theta_0).$$

Since

$$E_\theta(\delta) = g(\theta) \quad \text{and} \quad E_\theta(\delta_0) = g(\theta_0),$$

it is seen that δ^* is unbiased for estimating $g(\theta)$. When $\theta = \theta_0$, $\delta^* = g(\theta_0)$ and is thus a constant. Its variance is therefore zero, as was to be proved. □

To see under what circumstances the labels are likely to be helpful and when it is reasonable to discard them, let us consider an example.

Example 7.6 Informative labels. Suppose the population is a class of several hundred students. A random sample is drawn and each of the sampled students is asked to provide a numerical evaluation of the instructor. (Such a procedure may be more accurate than distributing reaction sheets to the whole class, if for the much smaller sample it is possible to obtain a considerably higher rate of response.) Suppose that the frame is an alphabetically arranged class list and that the label is the number of the student on this list. Typically, one would not expect this label to carry any useful information since the place of a name in the alphabet does not usually shed much light on the student's attitude toward the instructor. (Of course, there may be exceptional circumstances that vitiate this argument.) On the other hand, suppose the students are seated alphabetically. In a large class, the students sitting in front may have the advantage of hearing and seeing better, receiving more attention from the instructor, and being less likely to read the campus newspaper or fall asleep. Their attitude could thus be affected by the place of their name in the alphabet, and thus the labels could carry some information. ||

We shall discuss two ways of formalizing the idea that the labels can reasonably be discarded if they appear to be unrelated to the associated a values.

(i) *Invariance*. Consider the transformations of the parameter and sample space obtained by an arbitrary permutation of the labels:

$$\bar{g}\theta = \{(j(1), a_1), \ldots, (j(N), a_N)\},$$
$$gX = \{(j(I_1), Y_1), \ldots, (j(I_n), Y_n)\}. \tag{7.11}$$

The estimand $\bar{a}$ [or, more generally, any function $h(a_1, \ldots, a_N)$ that is symmetric in the a's] is unchanged by these transformations, so that $g^*d = d$ and a loss function $L(\theta, d)$ is invariant if it depends on θ only through the a's (in fact, as a symmetric function of the a's) and not the labels. [For estimating $\bar{a}$, such a loss function would be typically of the form $\rho(d - \bar{a})$.] Since $g^*d = d$, an estimator δ is equivariant if it satisfies the condition

$$\delta(gX) = \delta(X) \quad \text{for all } g \text{ and } X. \tag{7.12}$$

In this case, equivariance thus reduces to invariance. Condition (7.12) holds if and only if the estimator δ depends only on the observed Y values and not on the labels. Combining this result with Corollary 7.2, we see that for any U-estimable function $h(a_1, \ldots, a_N)$, the estimator of Corollary 7.2 uniformly minimizes the risk for any convex loss function that does not depend on the labels among all estimators of h which are both unbiased and invariant.

The appropriateness of the principle of equivariance, which permits restricting consideration to equivariant (in the present case, invariant) estimators, depends on the assumption that the transformations (7.11) leave the problem invariant. This is clearly not the case when there is a relationship between the labels and the associated a values, for example, when low a values tend to be associated with low labels and high a values with high labels, since permutation of the labels will destroy this relationship. Equivariance considerations therefore justify discarding the labels if, in our judgment, the problem is symmetric in the labels, that is, unchanged under any permutation of the labels.

(ii) *Random labels*. Sometimes, it is possible to adopt a slightly different formulation of the model which makes an appeal to equivariance unnecessary. Suppose that the labels have been assigned at random, that is, so that all $N!$ possible assignments are equally likely. Then, the observed a values $Y_1, \ldots, Y_n$ are sufficient. To see this, note that given these values, any n labels $(I_1, \ldots, I_n)$ associated with them are equally likely, so that the conditional distribution of X given $(Y_1, \ldots, Y_n)$ is independent of θ. In this model, the estimators of Corollary 7.2 are, therefore, UMVU without any further restriction.

Of course, the assumption of random labeling is legitimate only if the labels really were assigned at random rather than in some systematic way such as alphabetically or first come, first labeled. In the latter cases, rather than incorporating a very shaky assumption into the model, it seems preferable to invoke equivariance when it comes to the analysis of the data with the implied admission that we believe the labels to be unrelated to the a values but without denying that a hidden relationship may exist.

Simple random sampling tends to be inefficient unless the population being sampled is fairly homogeneous with respect to the a's. To see this, suppose that $a_1 = \cdots = a_{N_1} = a$ and $a_{N_1+1} = \cdots = a_{N_1+N_2} = b (N_1 + N_2 = N)$. Then (Problem 7.3)

$$\text{var}(\bar{Y}) = \frac{N-n}{N-1} \cdot \frac{\gamma(1-\gamma)}{n}(b-a)^2 \tag{7.13}$$

where $\gamma = N_1/N$. On the other hand, suppose that the subpopulations Π_i consisting of the a's and b's, respectively, can be identified and that one observation X_i is taken from each of the Π_i $(i = 1, 2)$. Then $X_1 = a$ and $X_2 = b$ and $(N_1X_1+N_2X_2)/N = \bar{a}$ is an unbiased estimator of $\bar{a}$ with variance zero.

This suggests that rather than taking a simple random sample from a heterogeneous population Π, one should try to divide Π into more homogeneous subpopulations Π_i, called *strata*, and sample each of the strata separately. Human populations are frequently stratified by such factors as age, gender, socioeconomic background, severity of disease, or by administrative units such as schools, hospitals, counties, voting districts, and so on.

Suppose that the population Π has been partitioned into s strata $\Pi_1, \ldots, \Pi_s$ of sizes $N_1, \ldots, N_s$ and that independent simple random samples of size n_i are taken from each Π_i $(i = 1, \ldots, s)$. If a_{ij} $(j = 1, \ldots, N_i)$ denote the a values in the ith stratum, the parameter is now $\theta = (\theta_1, \ldots, \theta_s)$, where

$$\theta_i = \{(1, a_{i1}), \ldots, (N_i, a_{iN_i}); i\},$$

and the observations are $X = (X_1, \ldots, X_s)$, where

$$X_i = \{(K_{i1}, Y_{i1}), \ldots, (K_{in_i}, Y_{in_i}); i\}.$$

Here, K_{ij} is the label of the jth element drawn from Π_i and Y_{ij} is its a value.

It is now easy to generalize the optimality results for simple random sampling to stratified sampling.

Theorem 7.7 *Let the Y_{ij} $(j = 1, \ldots, n_i)$, ordered separately for each i, be denoted by $Y_{i(1)} < \cdots < Y_{i(n_i)}$. On the basis of the Y_{ij} (i.e., without the labels), these ordered sample values are sufficient. They are also complete if the parameter space Ω_i for θ_i is of the form $V_i \times \cdots \times V_i$ (N_i factors) and the overall parameter space is $\Omega = \Omega_1 \times \cdots \times \Omega_s$. (Note that the value sets V_i may be different for different strata.)*

The proof is left to the reader (Problem 7.9).

It follows from Theorem 7.7 that on the basis of the Y's, a UMVU estimator exists for any U-estimator function of the a's and that it is the unique unbiased estimator

$\delta(Y_{i1}, \ldots, Y_{1n_i}; Y_{21}, \ldots, Y_{2n_2}; \ldots)$ which is symmetric in its first n_1 arguments, symmetric in its second set of n_2 arguments, and so forth.

Example 7.8 UMVU estimation in stratified random sampling. Suppose that we let $a_{..} = \Sigma\Sigma a_{ij}/N$ be the average of the a's for the population Π. If $a_{i.}$ is the average of the a's in Π_i, $Y_{i.}$ is unbiased for estimating $a_{i.}$ and hence

$$(7.14) \qquad \delta = \Sigma \frac{N_i Y_{i.}}{N}$$

is an unbiased estimator of $a_{..}$. Since δ is symmetric for each of the s subsamples, it is UMVU for $a_{..}$ on the basis of the Y's. From (7.6) and the independence of the Y_i's, it is seen that

$$(7.15) \qquad \text{var}(\delta) = \Sigma \frac{N_i^2}{N^2} \cdot \frac{N_i - n_i}{N_i - 1} \cdot \frac{1}{n_i} \tau_i^2,$$

where τ_i^2 is the population variance of Π_i, and from (7.8), one can read off the UMVU estimator of (7.15). ||

Discarding the labels within each stratum (but not the strata labels) can again be justified by invariance considerations if these labels appear to be unrelated to the associated a values. Permutation of the labels within each stratum then leaves the problem invariant, and the condition of equivariance reduces to the invariance condition (7.12). In the present situation, an estimator again satisfies (7.12) if and only if it does not depend on the within-strata labels. The estimator (7.14), and other estimators which are UMVU when these labels are discarded, are therefore also UMVU invariant without this restriction.

A central problem in stratified sampling is the choice of the sample sizes n_i. This is a design question and hence outside the scope of this book (but see Hedayat and Sinha 1991). We only mention that a natural choice is *proportional allocation*, in which the sample sizes n_i are proportional to the population sizes N_i. If the τ_i are known, the best possible choice in the sense of minimizing the approximate variance

$$(7.16) \qquad \Sigma(N_i^2 \tau_i^2 / n_i N^2)$$

is the *Tschuprow-Neyman* allocation with n_i proportional to $N_i \tau_i$ (Problem 7.11).

Stratified sampling, in addition to providing greater precision for the same total sample size than simple random sampling, often has the advantage of being administratively more convenient, which may mean that a larger sample size is possible on the same budget. Administrative convenience is the principal advantage of a third sampling method, *cluster sampling*, which we shall consider next. The population is divided into K clusters of sizes $M_1, \ldots, M_K$. A single random sample of k clusters is taken and the a values of all the elements in the sampled clusters are obtained. The clusters might, for example, be families or city blocks. A field worker obtaining information about one member of a family can often obtain the same information for all the members at relatively little additional cost.

An important special case of cluster sampling is *systematic sampling*. Suppose the items on a conveyor belt or the cards in a card catalog are being sampled. The easiest way of drawing a sample in these cases and in many situations in which the sampling is being done in the field is to take every rth element, where r is some positive number. To inject some randomness into the process, the starting point is chosen at random. Here, there are r clusters consisting of the items labeled

$$\{1, r+1, 2r+1, \ldots\}, \{2, r+2, 2r+2, \ldots\}, \ldots, \{r, 2r, 3r, \ldots\},$$

of which one is chosen at random, so that $K = r$ and $k = 1$. In general, let the elements of the ith cluster be $\{a_{i1}, \ldots, a_{iM_i}\}$ and let $u_i = \Sigma_{j=1}^{M_i} a_{ij}$ be the total for the ith cluster. We shall be interested in estimating some function of the u's such as the population average $a_{..} = \Sigma u_i / \Sigma M_i$. Of the a_{ij}, we shall assume that the vector of values $(a_{i1}, \ldots, a_{iM_i})$ belongs to some set W_i (which may, but need not be, of the form $V \times \cdots \times V$) and that $(a_{11}, \ldots, a_{1M_1}; a_{21}, \ldots, a_{2M_2}; \ldots) \in W_1 \times \cdots \times W_K$. The observations consist of the labels of the clusters included in the sample together with the full set of labels and values of the elements of each such cluster:

$$X = \{[i_1; (1, a_{i_1,1}), (2, a_{i_1,2}), \ldots]; [i_2; (1, a_{i_2,1}), (2, a_{i_2,2}), \ldots]; \ldots\}.$$

Let us begin the reduction of the statistical problem with invariance considerations. Clearly, the problem remains invariant under permutations of the labels within each cluster, and this reduces the observation to

$$X' = \{[i_1, (a_{i_1,1}, \ldots, a_{i_1,M_{i_1}})]; [i_2, (a_{i_2,1}, \ldots, a_{i_2,M_{i_2}})]; \ldots\}$$

in the sense that an estimator is invariant under these permutations if and only if it depends on X only through X'.

The next group is different from any we have encountered so far. Consider any transformation taking $(a_{i1}, \ldots, a_{iM_i})$ into $(a'_{i1}, \ldots, a'_{iM_i})$, $i = 1, \ldots, K$, where the a'_{ij} are arbitrary, except that they must satisfy

(a) $$(a'_{i1}, \ldots, a'_{iM_i}) \in W_i$$

and

(b) $$\sum_{j=1}^{M_i} a'_{ij} = u_i.$$

Note that for some vectors $(a_{i1}, \ldots, a_{iM_i})$, there may be no such transformations except the identity; for others, there may be just the identity and one other, and so on, depending on the nature of W_i.

It is clear that these transformations leave the problem invariant, provided both the estimand and the loss function depend on the a's only through the u's. Since the estimand remains unchanged, the same should then be true for δ, which, therefore, should satisfy

(7.17) $$\delta(gX') = \delta(X')$$

for all these transformations. It is easy to see (Problem 7.17) that δ satisfies (7.17) if and only if δ depends on X' only through the observed cluster labels, cluster sizes, and the associated cluster totals, that is, only on

(7.18) $$X'' = \{(i_i, u_{i_1}, M_{i_1}), \ldots, (i_k, u_{i_k}, M_{i_k})\}$$

and the order in which the clusters were drawn.

This differs from the set of observations we would obtain in a simple random sample from the collection

$$\{(1, u_1), \ldots, (K, u_K)\} \tag{7.19}$$

through the additional observations provided by the cluster sizes. For the estimation of the population average or total, this information may be highly relevant and the choice of estimator must depend on the relationship between M_i and u_i. The situation does, however, reduce to that of simple random sampling from (7.19) under the additional assumption that the cluster sizes M_i are equal, say $M_i = M$, where M can be assumed to be known. This is the case, either exactly or as a very close approximation, for systematic sampling, and also in certain applications to industrial, commercial, or agricultural sampling—for example, when the clusters are cartons of eggs of other packages or boxes containing a fixed number of items. From the discussion of simple random sampling, we know that the average $\bar{Y}$ of the observed u values is then the UMVU invariant estimator $\bar{u} = \Sigma u_i/K$ and hence that $\bar{Y}/M$ is UMVU invariant for estimating $a_{..}$. The variance of the estimator is easily obtained from (7.6) with $\tau^2 = \Sigma(u_i - \bar{u})^2/K$.

In stratified sampling, it is desirable to have the strata as homogeneous as possible: The more homogeneous a stratum, the smaller the sample size it requires. The situation is just the reverse in cluster sampling, where the whole cluster will be observed in any case. The more homogeneous a cluster, the less benefit is derived from these observations: "If you have seen one, you have seen them all." Thus, it is desirable to have the clusters as heterogeneous as possible. For example, families, for some purposes, constitute good clusters by being both administratively convenient and heterogeneous with respect to age and variables related to age. The advantages of stratified sampling apply not only to the sampling of single elements but equally to the sampling of clusters. *Stratified cluster sampling* consists of drawing a simple random sample of clusters from each stratum and combining the estimates of the strata averages or totals in the obvious way. The resulting estimator is again UMVU invariant, provided the cluster sizes are constant within each stratum, although they may differ from one stratum to the next. (For a more detailed discussion of stratified cluster sampling, see, for example, Kish 1965.)

To conclude this section, we shall briefly indicate two ways in which the equivariance considerations in the present section differ from those in the rest of the chapter.

(i) In all of the present applications, the transformations leave the estimand unchanged rather than transforming it into a different value, and the condition of equivariance then reduces to the invariance condition: $\delta(gX) = \delta(X)$. Correspondingly, the group $\bar{G}$ is not transitive over the parameter space and a UMRE estimator cannot be expected to exist. To obtain an optimal estimator, one has to invoke unbiasedness in addition to invariance. (For an alternative optimality property, see Section 5.4.)

(ii) Instead of starting with transformations of the sample space which would then induce transformations of the parameter space, we inverted the order and began by transforming θ, thereby inducing transformations of X. This does not involve a new approach but was simply more convenient than the usual order. To see how to present the transformations in the usual order, let us consider the sample space as the totality of possible samples s together with the labels and values of their elements. Suppose, for example, that the transformations are permutations of the labels. Since the same elements appear in many different samples, one must ensure that the transformations g of the samples are consistent, that is, that the transform of an element is independent of the particular sample in which it appears. If a transformation has this property, it will define a permutation of all the labels in the population and hence a transformation $\bar{g}$ of θ'. Starting with g or $\bar{g}$ thus leads to the same result; the latter is more convenient because it provides the required consistency property automatically.

8 Problems

Section 1

1.1 Prove the parts of Theorem 1.4 relating to (a) risk and (b) variance.

1.2 In model (1.9), suppose that $n = 2$ and that f satisfies $f(-x_1, -x_2) = f(x_2, x_1)$. Show that the distribution of $(X_1 + X_2)/2$ given $X_2 - X_1 = y$ is symmetric about 0. Note that if X_1 and X_2 are iid according to a distribution which is symmetric about 0, the above equation holds.

1.3 If X_1 and X_2 are distributed according to (1.9) with $n = 2$ and f satisfying the assumptions of Problem 1.2, and if ρ is convex and even, then the MRE estimator of ξ is $(X_1 + X_2)/2$.

1.4 Under the assumptions of Example 1.18, show that (a) $E[X_{(1)}] = b/n$ and (b) $\text{med}[X_{(1)}] = b \log 2/n$.

1.5 For each of the three loss functions of Example 1.18, compare the risk of the MRE estimator to that of the UMVU estimator.

1.6 If T is a sufficient statistic for the family (1.9), show that the estimator (1.28) is a function of T only. [*Hint*: Use the factorization theorem.]

1.7 Let $X_i (i = 1, 2, 3)$ be independently distributed with density $f(x_i - \xi)$ and let $\delta = X_1$ if $X_3 > 0$ and $= X_2$ if $X_3 \le 0$. Show that the estimator δ of ξ has constant risk for any invariant loss function, but δ is not location equivariant.

1.8 Prove Corollary 1.14. [*Hint*: Show that (a) $\phi(v) = E_0\rho(X - v) \to M$ as $v \to \pm\infty$ and (b) that ϕ is continuous; (b) follows from the fact (see TSH2, Appendix Section 2) that if $f_n, n = 1, 2, \ldots$ and f are probability densities such that $f_n(x) \to f(x)$ a.e., then $\int \psi f_n \to \int \psi f$ for any bounded ψ.]

1.9 Let $X_1, \ldots, X_n$ be distributed as in Example 1.19 and let the loss function be that of Example 1.15. Determine the totality of MRE estimators and show that the midrange is one of them.

1.10 Consider the loss function

$$\rho(t) = \begin{cases} -At & \text{if } t < 0 \\ Bt & \text{if } t \ge 0 \end{cases} \qquad (A, B \ge 0).$$

If X is a random variable with density f and distribution function F, show that $E\rho(X - v)$ is minimized for any v satisfying $F(v) = B/(A + B)$.

1.11 In Example 1.16, find the MRE estimator of ξ when the loss function is given by Problem 1.10.

1.12 Show that an estimator $\delta(X)$ of $g(\theta)$ is risk-unbiased with respect to the loss function of Problem 1.10 if $F_\theta[g(\theta)] = B/(A+B)$, where F_θ is the cdf of $\delta(X)$ under θ.

1.13 Suppose $X_1, \ldots, X_m$ and $Y_1, \ldots, Y_n$ have joint density $f(x_1 - \xi, \ldots, x_m - \xi; y_1 - \eta, \ldots, y_n - \eta)$ and consider the problem of estimating $\Delta = \eta - \xi$. Explain why it is desirable for the loss function $L(\xi, \eta; d)$ to be of the form $\rho(d - \Delta)$ and for an estimator δ of Δ to satisfy $\delta(\mathbf{x} + a, \mathbf{y} + b) = \delta(\mathbf{x}, \mathbf{y}) + (b - a)$.

1.14 Under the assumptions of the preceding problem, prove the equivalents of Theorems 1.4–1.17 and Corollaries 1.11–1.14 for estimators satisfying the restriction.

1.15 In Problem 1.13, determine the totality of estimators satisfying the restriction when $m = n = 1$.

1.16 In Problem 1.13, suppose the X's and Y's are independently normally distributed with known variances σ^2 and τ^2. Find conditions on ρ under which the MRE estimator is $\bar{Y} - \bar{X}$.

1.17 In Problem 1.13, suppose the X's and Y's are independently distributed as $E(\xi, 1)$ and $E(\eta, t)$, respectively, and that $m = n$. Find conditions on ρ under which the MRE estimator of Δ is $Y_{(1)} - X_{(1)}$.

1.18 In Problem 1.13, suppose that $\mathbf{X}$ and $\mathbf{Y}$ are independent and that the loss function is squared error. If $\hat{\xi}$ and $\hat{\eta}$ are the MRE estimators of ξ and η, respectively, the MRE estimator of Δ is $\hat{\eta} - \hat{\xi}$.

1.19 Suppose the X's and Y's are distributed as in Problem 1.17 but with $m \neq n$. Determine the MRE estimator of Δ when the loss is squared error.

1.20 For any density f of $\mathbf{X} = (X_1, \ldots, X_n)$, the probability of the set $A = \{\mathbf{x} : 0 < \int_{-\infty}^{\infty} f(\mathbf{x} - u)\, du < \infty\}$ is 1. [*Hint*: With probability 1, the integral in question is equal to the marginal density of $\mathbf{Y} = (Y_1, \ldots, Y_{n-1})$ where $Y_i = X_i - X_n$, and $P[0 < g(\mathbf{Y}) < \infty] = 1$ holds for any probability density g.]

1.21 Under the assumptions of Theorem 1.10, if there exists an equivariant estimator δ_0 of ξ with finite expected squared error, show that

(a) $E_0(|X_n| \mid \mathbf{Y}) < \infty$ with probability 1;

(b) the set $B = \{\mathbf{x} : \int |u| f(\mathbf{x} - u)\, du < \infty\}$ has probability 1.

[*Hint*: (a) $E|\delta_0| < \infty$ implies $E(|\delta_0| \mid \mathbf{Y}) < \infty$ with probability 1 and hence $E[\delta_0 - v(\mathbf{Y})| \mid \mathbf{Y}] < \infty$ with probability 1 for any $v(\mathbf{Y})$. (b) $P(B) = 1$ if and only if $E(|X_n| \mid \mathbf{Y}) < \infty$ with probability 1.]

1.22 Let δ_0 be location equivariant and let $\mathcal{U}$ be the class of all functions u satisfying (1.20) and such that $u(X)$ is an unbiased estimator of zero. Then, δ_0 is MRE if and only if $\text{cov}[\delta_0, u(X)] = 0$ for all $u \in \mathcal{U}$.[2] (Note the analogy with Theorem 2.1.7.)

Section 2

2.1 Show that the class $G(\mathcal{C})$ is a group.

2.2 In Example 2.2(ii), show that the transformations $\mathbf{x}' = -\mathbf{x}$ together with the identity transformation form a group.

2.3 Let $\{gX, g \in G\}$ be a group of transformations that leave the model (2.1) invariant. If the distributions $P_\theta, \theta \in \Omega$ are distinct, show that the induced transformations $\bar{g}$ are $1:1$ transformations of Ω. [*Hint*: To show that $\bar{g}\theta_1 = \bar{g}\theta_2$ implies $\theta_1 = \theta_2$, use the fact that $P_{\theta_1}(A) = P_{\theta_2}(A)$ for all A implies $\theta_1 = \theta_2$.]

[2] Communicated by P. Bickel.

2.4 Under the assumptions of Problem 2.3, show that

(a) the transformations $\bar{g}$ satisfy $\overline{g_2 g_1} = \bar{g}_2 \cdot \bar{g}_1$ and $(\bar{g})^{-1} = (\overline{g^{-1}})$;

(b) the transformations $\bar{g}$ corresponding to $g \in G$ form a group.

(c) establish (2.3) and (2.4).

2.5 Show that a loss function satisfies (2.9) if and only if it is of the form (2.10).

2.6 (a) The transformations g^* defined by (2.12) satisfy $(g_2 g_1)^* = g_2^* \cdot g_1^*$ and $(g^*)^{-1} = (g^{-1})^*$.

(b) If G is a group leaving (2.1) invariant and $G^* = \{g^*, g \in G\}$, then G^* is a group.

2.7 Let X be distributed as $N(\xi, \sigma^2)$, $-\infty < \xi < \infty, 0 < \sigma$, and let $h(\xi, \sigma) = \sigma^2$. The problem is invariant under the transformations $x' = ax + c; 0 < a, -\infty < c < \infty$. Show that the only equivariant estimator is $\delta(X) \equiv 0$.

2.8 Show that:

(a) If (2.11) holds, the transformations g^* defined by (2.12) are 1 : 1 from $\mathcal{H}$ onto itself.

(b) If $L(\theta, d) = L(\theta, d')$ for all θ implies $d = d'$, then g^* defined by (2.14) is unique, and is a 1 : 1 transformation from $\mathcal{D}$ onto itself.

2.9 If θ is the true temperature in degrees Celsius, then $\theta' = \bar{g}\theta = \theta + 273$ is the true temperature in degrees Kelvin. Given an observation X, in degrees Celsius:

(a) Show that an estimator $\delta(X)$ is functionally equivariant if it satisfies $\delta(x) + a = \delta(x + a)$ for all a.

(b) Suppose our estimator is $\delta(x) = (ax + b\theta_0)/(a + b)$, where x is the observed temperature in degrees Celsius, θ_0 is a prior guess at the temperature, and a and b are constants. Show that for a constant K, $\delta(x + K) \neq \delta(x) + K$, so δ does not satisfy the principle of functional equivariance.

(c) Show that the estimators of part (b) will not satisfy the principle of formal invariance.

2.10 To illustrate the difference between functional equivariance and formal invariance, consider the following.

To estimate the amount of electric power obtainable from a stream, one could use the estimate

$$\delta(x) = c \min\{100, x - 20\}$$

where x = stream flow in m^3/sec, 100 m^3/sec is the capacity of the pipe leading to the turbine, and 20 m^3/sec is the flow reduction necessary to avoid harming the trout. The constant c, in kilowatts /m^3/sec converts the flow to a kilowatt estimate.

(a) If measurements were, instead, made in liters and watts, so $g(x) = 1000x$ and $\bar{g}(\theta) = 1000\theta$, show that functional equivariance leads to the estimate

$$\bar{g}(\delta(x)) = c \min\{10^5, g(x) - 20,000\}.$$

(b) The principle of formal invariance leads to the estimate $\delta(g(x))$. Show that this estimator is not a reasonable estimate of wattage.

(Communicated by L. LeCam.)

2.11 In an invariant probability model, write $X = (T, W)$, where T is sufficient for θ, and W is ancillary.

(a) If the group operation is transitive, show that any invariant statistic must be ancillary.

(b) What can you say about the invariance of an ancillary statistic?

2.12 In an invariant estimation problem, write $X = (T, W)$ where T is sufficient for θ, and W is ancillary. If the group of transformations is transitive, show:

(a) The best equivariant estimator δ^* is the solution to $\min_d E_\theta[L(\theta, d(x))|W = w]$.

(b) If e is the identity element of the group ($g^{-1}g = e$), then $\delta^* = \delta^*(t, w)$ can be found by solving, for each w, $\min_d E_e\{L[e, d(T, w)]|W = w\}$.

2.13 For the situation of Example 2.11:

(a) Show that the class of transformations is a group.

(b) Show that equivariant estimators must satisfy $\delta(n - x) = 1 - \delta(x)$.

(c) Show that, using an invariant loss, the risk of an equivariant estimator is symmetric about $p = 1/2$.

2.14 For the situation of Example 2.12:

(a) Show that the class of transformations is a group.

(b) Show that estimators of the form $\varphi(\bar{x}/s^2)s^2$, where $\bar{x} = 1/n\Sigma x_i$ and $s^2 = \Sigma(x_i - \bar{x})^2$ are equivariant, where φ is an arbitrary function.

(c) Show that, using an invariant loss function, the risk of an equivariant estimator is a function only of $\tau = \mu/\sigma$.

2.15 Prove Corollary 2.13.

2.16 (a) If g is the transformation (2.20), determine $\bar{g}$.

(b) In Example 2.12, show that (2.22) is not only sufficient for (2.14) but also necessary.

2.17 (a) In Example 2.12, determine the smallest group G containing both G_1 and G_2.

(b) Show that the only estimator that is invariant under G is $\delta(\mathbf{X}, \mathbf{Y}) \equiv 0$.

2.18 If $\delta(X)$ is an equivariant estimator of $h(\theta)$ under a group G, then so is $g^*\delta(X)$ with g^* defined by (2.12) and (2.13), provided G^* is commutative.

2.19 Show that:

(a) In Example 2.14(i), X is not risk-unbiased.

(b) The group of transformations $ax + c$ of the real line ($0 < a, -\infty < c < \infty$) is not commutative.

2.20 In Example 2.14, determine the totality of equivariant estimators of Δ under the smallest group G containing G_1 and G_2.

2.21 Let θ be real-valued and h strictly increasing, so that (2.11) is vacuously satisfied. If $L(\theta, d)$ is the loss resulting from estimating θ by d, suppose that the loss resulting from estimating $\theta' = h(\theta)$ by $d' = h(d)$ is $M(\theta', d') = L[\theta, h^{-1}(d')]$. Show that:

(a) If the problem of estimating θ with loss function L is invariant under G, then so is the problem of estimating $h(\theta)$ with loss function M.

(b) If δ is equivariant under G for estimating θ with loss function L, show that $h[\delta(X)]$ is equivariant for estimating $h(\theta)$ with loss function M.

(c) If δ is MRE for θ with L, then $h[\delta(X)]$ is MRE for $h(\theta)$ with M.

2.22 If $\delta(\mathbf{X})$ is MRE for estimating ξ in Example 2.2(i) with loss function $\rho(d - \xi)$, state an optimum property of $e^{\delta(\mathbf{X})}$ as an estimator of e^ξ.

2.23 Let X_{ij}, $j = 1, \ldots, n_i$, $i = 1, \ldots, s$, and W be distributed according to a density of the form

$$\left[\prod_{i=1}^{s} f_i(\mathbf{x}_i - \xi_i)\right] h(w)$$

where $\mathbf{x}_i - \xi_i = (x_{i1} - \xi_i, \ldots, x_{in_i} - \xi_i)$, and consider the problem of estimating $\theta = \Sigma c_i \xi_i$ with loss function $L(\xi_i, \ldots, \xi_s; d) = \rho(d - \theta)$. Show that:

(a) This problem remains invariant under the transformations

$$X'_{ij} = X_{ij} + a_i, \quad \xi'_i = \xi_i + a_i, \quad \theta' = \theta + \Sigma a_i c_i,$$
$$d' = d + \Sigma a_i c_i.$$

(b) An estimator δ of θ is equivariant under these transformations if

$$\delta(\mathbf{x}_1 + a_1, \ldots, \mathbf{x}_s + a_s, w) = \delta(\mathbf{x}_1, \ldots, \mathbf{x}_s, w) + \Sigma a_i c_i.$$

2.24 Generalize Theorem 1.4 to the situation of Problem 2.23.

2.25 If δ_0 is any equivariant estimator of θ in Problem 2.23, and if $\mathbf{y}_i = (x_{i1} - x_{in_i}, x_{i2} - x_{in_i}, \ldots, x_{in_i - 1} - x_{in_i})$, show that the most general equivariant estimator of θ is of the form

$$\delta(\mathbf{x}_1, \ldots, \mathbf{x}_s, w) = \delta_0(\mathbf{x}_1, \ldots, \mathbf{x}_s, w) - v(\mathbf{y}_1, \ldots, \mathbf{y}_s, w).$$

2.26 (a) Generalize Theorem 1.10 and Corollary 1.12 to the situation of Problems 2.23 and 2.25. (b) Show that the MRE estimators of (a) can be chosen to be independent of W.

2.27 Suppose that the variables X_{ij} in Problem 2.23 are independently distributed as $N(\xi_i, \sigma^2)$, σ is known. Show that:

(a) The MRE estimator of θ is then $\Sigma c_i \bar{X}_i - v^*$, where $\bar{X}_i = (X_{i1} + \cdots + X_{in_i})/n_i$, and where v^* minimizes (1.24) with $X = \Sigma c_i \bar{X}_i$.

(b) If ρ is convex and even, the MRE estimator of θ is $\Sigma c_i \bar{X}_i$.

(c) The results of (a) and (b) remain valid when σ is unknown and the distribution of W depends on σ (but not the ξ's).

2.28 Show that the transformation of Example 2.11 and the identity transformation are the only transformations leaving the family of binomial distributions invariant.

Section 3

3.1 (a) A loss function L satisfies (3.4) if and only if it satisfies (3.5) for some γ.

(b) The sample standard deviation, the mean deviation, the range, and the MLE of τ all satisfy (3.7) with $r = 1$.

3.2 Show that if $\delta(\mathbf{X})$ is scale invariant, so is $\delta^*(\mathbf{X})$ defined to be $\delta(\mathbf{X})$ if $\delta(\mathbf{X}) \geq 0$ and $= 0$ otherwise, and the risk of δ^* is no larger than that of δ for any loss function (3.5) for which $\gamma(v)$ is nonincreasing for $v \leq 0$.

3.3 Show that the bias of any equivariant estimator of τ^r in (3.1) is proportional to τ^r.

3.4 A necessary and sufficient condition for δ to satisfy (3.7) is that it is of the form $\delta = \delta_0/u$ with δ_0 and u satisfying (3.7) and (3.9), respectively.

3.5 The function ρ of Corollary 3.4 with γ defined in Example 3.5 is strictly convex for $p \geq 1$.

3.6 Let X be a positive random variable. Show that:

(a) If $EX^2 < \infty$, then the value of c that minimizes $E(X/c - 1)^2$ is $c = EX^2/EX$.

(b) If Y has the gamma distribution with $\Gamma(\alpha, 1)$, then the value of w minimizing $E[(Y/w) - 1]^2$ is $w = \alpha + 1$.

3.7 Let X be a positive random variable.

(a) If $EX < \infty$, then the value of c that minimizes $E|X/c - 1|$ is a solution to $EXI(X \le c) = EXI(X \ge c)$, which is known as a *scale median*.

(b) Let Y have a χ^2-distribution with f degrees for freedom. Then, the minimizing value is $w = f + 2$. [*Hint*: (b) Example 1.5.9.]

3.8 Under the assumptions of Problem 3.7(a), the set of scale medians of X is an interval. If $f(x) > 0$ for all $x > 0$, the scale median of X is unique.

3.9 Determine the scale median of X when the distribution of X is (a) $U(0, \theta)$ and (b) $E(0, b)$.

3.10 Under the assumptions of Theorem 3.3:

(a) Show that the MRE estimator under the loss (3.13) is given by (3.14).

(b) Show that the MRE estimator under the loss (3.15) is given by (3.11), where $w^*(\mathbf{z})$ is any scale median of $\delta_0(\mathbf{x})$ under the distribution of $\mathbf{X}|\mathbf{Z}$.

[*Hint*: Problem 3.7.]

3.11 Let $X_1, \ldots, X_n$ be iid according to the uniform distribution $u(0, \theta)$.

(a) Show that the complete sufficient statistic $X_{(n)}$ is independent of Z [given by Equation (3.8)].

(b) For the loss function (3.13) with $r = 1$, the MRE estimator of θ is $X_{(n)}/w$, with $w = (n+1)/(n+2)$.

(c) For the loss function (3.15) with $r = 1$, the MRE estimator of θ is $[2^{1/(n+1)}]\, X_{(n)}$.

3.12 Show that the MRE estimators of Problem 3.11, parts (b) and (c), are risk-unbiased, but not mean-unbiased.

3.13 In Example 3.7, find the MRE estimator of var(X_1) when the loss function is (a) (3.13) and (b) (3.15) with $r = 2$.

3.14 Let $X_1, \ldots, X_n$ be iid according to the exponential distribution $E(0, \tau)$. Determine the MRE estimator of τ for the loss functions (a) (3.13) and (b) (3.15) with $r = 1$.

3.15 In the preceding problem, find the MRE estimator of var(X_1) when the loss function is (3.13) with $r = 2$.

3.16 Prove formula (3.19).

3.17 Let $X_1, \ldots, X_n$ be iid each with density $(2/\tau)[1 - (x/\tau)]$, $0 < x < \tau$. Determine the MRE estimator (3.19) of τ^r when (a) $n = 2$, (b) $n = 3$, and (c) $n = 4$.

3.18 In the preceding problem, find var(X_1) and its MRE estimator for $n = 2, 3, 4$ when the loss function is (3.13) with $r = 2$.

3.19 (a) Show that the loss function L_s of (3.20) is convex and invariant under scale transformations.

(b) Prove Corollary 3.8.

(c) Show that for the situation of Example 3.7, if the loss function is L_s, then the UMVU estimator is also the MRE.

3.20 Let $X_1, \ldots, X_n$ be iid from the distribution $N(\theta, \theta^2)$.

(a) Show that this probability model is closed under scale transformations.

(b) Show that the MLE is equivariant.

[The MRE estimator is obtainable from Theorem 3.3, but does not have a simple form. See Eaton 1989, Robert 1991, 1994a for more details. Gleser and Healy (1976) consider a similar problem using squared error loss.]

3.21 (a) If δ_0 satisfies (3.7) and $c\delta_0$ satisfies (3.22), show that $c\delta_0$ cannot be unbiased in the sense of satisfying $E(c\delta_0) \equiv \tau^r$.

(b) Prove the statement made in Example 3.10.

3.22 Verify the estimator δ^* of Example 3.12.

3.23 If G is a group, a subset G_0 of G is a *subgroup* of G if G_0 is a group under the group operation of G.

(a) Show that the scale group (3.32) is a subgroup of the location-scale group (3.24)

(b) Show that any equivariant estimator of τ^r that is equivariant under (3.24) is also equivariant under (3.32); hence, in a problem that is equivariant under (3.32), the best scale equivariant estimator is at least as good as the best location-scale equivariant estimator.

(c) Explain why, in general, if $\mathcal{G}_0$ is a subgroup of $\mathcal{G}$, one can expect equivariance under $\mathcal{G}_0$ to produce better estimators than equivariance under $\mathcal{G}$.

3.24 For the situation of Example 3.13:

(a) Show that an estimator is equivariant if and only if it can be written in the form $\varphi(\bar{x}/s)s^2$.

(b) Show that the risk of an equivariant estimator is a function only of ξ/τ.

3.25 If $X_1, \ldots, X_n$ are iid according to $E(\xi, \tau)$, determine the MRE estimator of τ for the loss functions (a) (3.13) and (b) (3.15) with $r = 1$ and the MRE estimator of ξ for the loss function (3.43).

3.26 Show that δ satisfies (3.35) if and only if it satisfies (3.40) and (3.41).

3.27 Determine the bias of the estimator $\delta^*(\mathbf{X})$ of Example 3.18.

3.28 Lele (1993) uses invariance in the study of *mophometrics*, the quantitative analysis of biological forms. In the analysis of a biological object, one measures data $\mathbf{X}$ on k specific points called *landmarks*, where each landmark is typically two- or three-dimensional . Here we will assume that the landmark is two-dimensional (as is a picture), so $\mathbf{X}$ is a $k \times 2$ matrix. A model for $\mathbf{X}$ is

$$\mathbf{X} = (M + \mathbf{Y})\Gamma + \mathbf{t}$$

where $M_{k\times 2}$ is the mean form of the object, $\mathbf{t}$ is a fixed translation vector, and Γ is a 2×2 matrix that rotates the vector $\mathbf{X}$. The random variable $\mathbf{Y}_{k\times 2}$ is a *matrix normal* random variable, that is, each column of $\mathbf{Y}$ is distributed as $N(0, \Sigma_k)$, a k-variate normal random variable, and each row is distributed as $N(0, \Sigma_d)$, a bivariate normal random variable.

(a) Show that $\mathbf{X}$ is a matrix normal random variable with columns distributed as $N_k(M\Gamma_j, \Sigma_k)$ and rows distributed as $N_2(M_i\Gamma, \Gamma'\Sigma_d\Gamma)$, where Γ_j is the jth column of Γ and M_i is the ith row of M.

(b) For estimation of the shape of a biological form, the parameters of interest are M, Σ_k and Σ_d, with $\mathbf{t}$ and Γ being nuisance parameters. Show that, even if there were no nuisance parameters, Σ_k or Σ_d is not identifiable.

(c) It is usually assumed that the (1, 1) element of either Σ_k or Σ_d is equal to 1. Show that this makes the model identifiable.

(d) The form of a biological object is considered an inherent property of the form (a baby has the same form as an adult) and should not be affected by rotations, reflections, or translations. This is summarized by the transformation

$$\mathbf{X}' = \mathbf{X}P + b$$

where P is a 2×2 orthogonal matrix ($P'P = I$) and b is a $k \times 1$ vector. (See Note 9.3 for a similar group.) Suppose we observe n landmarks $\mathbf{X}_1, \cdots, \mathbf{X}_n$. Define the Euclidean distance between two matrices A and B to be $D(A, B) = \sum_{ij}(a_{ij} - b_{ij})^2$, and let the $n \times n$ matrix F have (i, j)th element $f_{ij} = D(\mathbf{X}_i, \mathbf{X}_j)$. Show that F is invariant under this group, that is $F(\mathbf{X}') = F(\mathbf{X})$. (Lele (1993) notes that F is, in fact, maximal invariant.)

3.29 In (9.1), show that the group $\mathbf{X}' = A\mathbf{X} + b$ induces the group $\mu' = A\mu + b$, $\Sigma' = A\Sigma A'$.

3.30 For the situation of Note 9.3, consider the equivariant estimation of μ.

(a) Show that an invariant loss is of the form $L(\mu, \Sigma, \delta) = L((\mu - \delta)'\Sigma^{-1}(\mu - \delta))$.

(b) The equivariant estimators are of the form $\bar{X} + c$, with $c = 0$ yielding the MRE estimator.

3.31 For $\mathbf{X}_1, \ldots, \mathbf{X}_n$ iid as $N_p(\mu, \Sigma)$, the cross-products matrix S is defined by

$$S = \{S_{ij}\} = \sum_{k=1}^{n}(x_{i_k} - \bar{x}_i)(x_{j_k} - \bar{x}_j)$$

where $\bar{x}_i = (1/n)\sum_{k=1}^{n} x_{i_k}$. Show that, for $\Sigma = I$,

(a) $E_I[\text{tr}S] = E_I \sum_{i=1}^{p} \sum_{k=1}^{n}(X_{i_k} - \bar{X}_i)(X_{i_k} - \bar{X}_i) = p(n-1)$,

(b) $E_I[\text{tr}S^2] = E_I \sum_{i=1}^{p} \sum_{j=1}^{p}\{\sum_{k=1}^{n}(X_{i_k} - \bar{X}_i)(X_{j_k} - \bar{X}_j)\}^2 = (n-1)(np - p - 1)$.

[These are straightforward, although somewhat tedious, calculations involving the chi-squared distribution. Alternatively, one can use the fact that S has a Wishart distribution (see, for example, Anderson 1984), and use the properties of that distribution.]

3.32 For the situation of Note 9.3:

(a) Show that equivariant estimators of Σ are of the form cS, where S is the cross-products matrix and c is a constant.

(b) Show that $E_I\{\text{tr}[(cS - I)'(cS - I)]\}$ is minimized by $c = E_I\text{tr}S/E_I\text{tr}S^2$.

[*Hint*: For part (a), use a generalization of Theorem 3.3; see the argument leading to (3.29), and Example 3.11.]

3.33 For the estimation of Σ in Note 9.3:

(a) Show that the loss function in (9.2) is invariant.

(b) Show that Stein's loss $L(\delta, \Sigma) = \text{tr}(\delta\Sigma^{-1}) - \log|\delta\Sigma^{-1}| - p$, where $|A|$ is the determinant of A, is an invariant loss with MRE estimator S/n.

(c) Show that a loss $L(\delta, \Sigma)$ is an invariant loss if and only if it can be written as a function of the eigenvalues of $\delta\Sigma^{-1}$.

[The univariate version of Stein's loss was seen in (3.20) and Example 3.9. Stein (1956b) and James and Stein (1961) used the multivariate version of the loss. See also Dey and Srinivasan 1985, and Dey et al. 1987.]

3.34 Let $X_1, \ldots, X_m$ and $Y_1, \ldots, Y_n$ have joint density

$$\frac{1}{\sigma^m \tau^n} f\left(\frac{x_1}{\sigma}, \ldots, \frac{x_m}{\sigma}; \frac{y_1}{\tau}, \ldots, \frac{y_n}{\tau}\right),$$

and consider the problem of estimating $\theta = (\tau/\sigma)^r$ with loss function $L(\sigma, \tau; d) = \gamma(d/\theta)$. This problem remains invariant under the transformations $X_i' = aX_i, Y_j' = bY_j$, $\sigma' = a\sigma, \tau' = b\tau$, and $d' = (b/a)^r d$ $(a, b > 0)$, and an estimator δ is equivariant under these transformations if $\delta(a\mathbf{x}, b\mathbf{y}) = (b/a)^r \delta(\mathbf{x}, \mathbf{y})$. Generalize Theorems 3.1 and 3.3, Corollary 3.4, and (3.19) to the present situation.

3.35 Under the assumptions of the preceding problem and with loss function $(d-\theta)^2/\theta^2$, determine the MRE estimator of θ in the following situations:

(a) $m = n = 1$ and X and Y are independently distributed as $\Gamma(\alpha, \sigma^2)$ and $\Gamma(\beta, \tau^2)$, respectively (α, β known).

(b) $X_1, \ldots, X_m$ and $Y_1, \ldots, Y_n$ are independently distributed as $N(0, \sigma^2)$ and $N(0, \tau^2)$, respectively.

(c) $X_1, \ldots, X_m$ and $Y_1, \ldots, Y_n$ are independently distributed as $U(0, \sigma)$ and $U(0, \tau)$, respectively.

3.36 Generalize the results of Problem 3.34 to the case that the joint density of $\mathbf{X}$ and $\mathbf{Y}$ is

$$\frac{1}{\sigma^m \tau^n} f\left(\frac{x_1 - \xi}{\sigma}, \ldots, \frac{x_m - \xi}{\sigma}; \frac{y_1 - \eta}{\tau}, \ldots, \frac{y_n - \eta}{\tau}\right).$$

3.37 Obtain the MRE estimator of $\theta = (\tau/\sigma)^r$ with the loss function of Problem 3.35 when the density of Problem 3.36 specializes to

$$\frac{1}{\sigma^m \tau^n} \Pi_i f\left(\frac{x_i - \xi}{\sigma}\right) \Pi_j f\left(\frac{y_j - \eta}{\tau}\right)$$

and f is (a) normal, (b) exponential, or (c) uniform.

3.38 In the model of Problem 3.37 with $\tau = \sigma$, discuss the equivariant estimation of $\Delta = \eta - \xi$ with loss function $(d - \Delta)^2/\sigma^2$ and obtain explicit results for the three distributions of that problem.

3.39 Suppose in Problem 3.37 that an MRE estimator δ^* of $\Delta = \eta - \xi$ under the transformations $X_i' = a + bX_i$ and $Y_j' = a + bY_j$, $b > 0$, exists when the ratio $\tau/\sigma = c$ is known and that δ^* is independent of c. Show that δ^* is MRE also when σ and τ are completely unknown despite the fact that the induced group of transformations of the parameter space is not transitive.

3.40 Let $f(t) = \frac{1}{\pi}\frac{1}{1+t^2}$ be the Cauchy density, and consider the location-scale family

$$\mathcal{F} = \left\{\frac{1}{\sigma} f\left(\frac{x-\mu}{\sigma}\right), -\infty < \mu < \infty, 0 < \sigma < \infty\right\}.$$

(a) Show that this probability model is invariant under the transformation $x' = 1/x$.

(b) If $\mu' = \mu/(\mu^2 + \sigma^2)$ and $\sigma' = \sigma/(\mu^2 + \sigma^2)$, show that $P_{\mu,r}(X \in A) = P_{\mu',\sigma'}(X' \in A)$; that is, if X has the Cauchy density with location parameter μ and scale parameter σ, then X' has the Cauchy density with location parameter $\mu/(\mu^2 + \sigma^2)$ and scale parameter $\sigma/(\mu^2 + \sigma^2)$.

(c) For $\theta = \mu \pm i\sigma$, where $i = \sqrt{-1}$, show that the problem is invariant under the transformation $x \to \frac{ax+b}{cx+d}$ and $\theta \to \frac{a\theta+b}{cx+d}$, where $ad - bc \neq 0$.

[See McCullaugh (1992) for a full development of this model, where it is suggested that the complex plane provides a more appropriate parameter space.]

3.41 Let (X_i, Y_i), $i = 1, \ldots, n$, be distributed as independent bivariate normal random variables with mean $(\mu, 0)$ and covariance matrix

$$\begin{pmatrix} \sigma_{11} & \sigma_{12} \\ \sigma_{21} & \sigma_{22} \end{pmatrix}.$$

(a) Show that the probability model is invariant under the transformations

$$(x', y') = (a + bx, by),$$
$$(\mu', \sigma'_{11}, \sigma'_{12}, \sigma'_{22}) = (a + b\mu, b^2\sigma_{11}, b^2\sigma_{12}, b^2\sigma_{22}).$$

(b) Using the loss function $L(\mu, d) = (\mu - d)^2/\sigma_{11}$, show that this is an invariant estimation problem, and equivariant estimators must be of the form $\delta = \bar{x} + \psi(u_1, u_2, u_3)\bar{y}$, where $u_1 = \Sigma(x_i - \bar{x})^2/\bar{y}^2$, $u_2 = \Sigma(y_i - \bar{y})^2/\bar{y}^2$, and $u_3 = \Sigma(x_i - \bar{x})(y_i - \bar{y})/\bar{y}^2$.

(c) Show that if δ has a finite second moment, then it is unbiased for estimating μ. Its risk function is a function of σ_{11}/σ_{22} and σ_{12}/σ_{22}.

(d) If the ratio σ_{12}/σ_{22} is known, show that $\bar{X} - (\sigma_{12}/\sigma_{22})\bar{Y}$ is the MRE estimator of μ.

[This problem illustrates the technique of *covariance adjustment*. See Berry, 1987.]

3.42 Suppose we let $X_1, \ldots, X_n$ be a sample from an exponential distribution $f(x|\mu, \sigma) = (1/\sigma)e^{-(x-\mu)/\sigma}\ I(x \geq \mu)$. The exponential distribution is useful in reliability theory, and a parameter of interest is often a quantile, that is, a parameter of the form $\mu + b\sigma$, where b is known. Show that, under quadratic loss, the MRE estimator of $\mu + b\sigma$ is $\delta_0 = x_{(1)} + (b - 1/n)(\bar{x} - x_{(1)})$, where $x_{(1)} = \min_i x_i$.

[Rukhin and Strawderman (1982) show that δ_0 is inadmissible, and exhibit a class of improved estimators.]

Section 4

4.1 (a) Suppose $X_i : N(\xi_i, \sigma^2)$ with $\xi_i = \alpha + \beta t_i$. If the first column of the matrix C leading to the canonical form (4.7) is $(1/\sqrt{n}, \ldots, 1/\sqrt{n})'$, find the second column of C.

(b) If $X_i : N(\xi_i, \sigma^2)$ with $\xi_i, = \alpha + \beta t_i + \gamma t_i^2$, and the first two columns of C are those of (a), find the third column under the simplifying assumptions $\Sigma t_i = 0$, $\Sigma t_i^2 = 1$. [*Note*: The orthogonal polynomials that are progressively built up in this way are frequently used to simplify regression analysis.]

4.2 Write out explicit expressions for the transformations (4.10) when Π_Ω is given by (a) $\xi_i = \alpha + \beta t_i$ and (b) $\xi_i = \alpha + \beta t_i + \gamma t_i^2$.

4.3 Use Problem 3.10 to prove (iii) of Theorem 4.3.

4.4 (a) In Example 4.7, determine $\hat{\alpha}$, $\hat{\beta}$, and hence $\hat{\xi}_i$ by minimizing $\Sigma(X_i - \alpha - \beta t_i)^2$.

(b) Verify the expressions (4.12) for α and β, and the corresponding expressions for $\hat{\alpha}$ and $\hat{\beta}$.

4.5 In Example 4.2, find the UMVU estimators of α, β, γ, and σ^2 when $\Sigma t_i = 0$ and $\Sigma t_i^2 = 1$.

4.6 Let X_{ij} be independent $N(\xi_{ij}, \sigma^2)$ with $\xi_{ij} = \alpha_i + \beta t_{ij}$. Find the UMVU estimators of the α_i and β.

4.7 (a) In Example 4.9, show that the vectors of the coefficients in the $\hat{\alpha}_i$ are not orthogonal to the vector of the coefficients of $\hat{\mu}$.

(b) Show that the conclusion of (a) is reversed if $\hat{\alpha}_i$ and $\hat{\mu}$ are replaced by $\hat{\hat{\alpha}}_i$ and $\hat{\hat{\mu}}$.

4.8 In Example 4.9, find the UMVU estimator of μ when the α_i are known to be zero and compare it with $\hat{\mu}$.

4.9 The coefficient vectors of the X_{ijk} given by (4.32) for $\hat{\mu}$, $\hat{\alpha}_i$, and $\hat{\beta}_j$ are orthogonal to the coefficient vectors for the $\hat{\gamma}_{ij}$ given by (4.33).

4.10 In the model defined by (4.26) and (4.27), determine the UMVU estimators of α_i, β_j, and σ^2 under the assumption that the γ_{ij} are known to zero.

4.11 (a) In Example 4.11, show that

$$\Sigma\Sigma\Sigma(X_{ijk} - \mu - \alpha_i - \beta_j - \gamma_{ij})^2 = S^2 + S_\mu^2 + S_\alpha^2 + S_\beta^2 + S_\gamma^2$$

where $S^2 = \Sigma\Sigma\Sigma(X_{ijk} - X_{ij\cdot})^2$, $S_\mu^2 = IJm(X_{\cdots} - \mu)^2$, $S_\alpha^2 = Jm\Sigma(X_{1\cdot\cdot} - X_{\cdots} - \alpha_i)^2$, and S_β^2, $S^2\gamma$ are defined analogously.

(b) Use the decomposition of (a) to show that the least squares estimators of $\mu, \alpha_i, \ldots$ are given by (4.32) and (4.33).

(c) Show that the *error sum of squares* S^2 is equal to $\Sigma\Sigma\Sigma(X_{ijk} - \hat{\xi}_{ij})^2$ and hence in the canonical form to $\Sigma_{j=s+1}^n Y_j^2$.

4.12 (a) Show how the decomposition in Problem 4.11(a) must be modified when it is known that the γ_{ij} are zero.

(b) Use the decomposition of (a) to solve Problem 4.10.

4.13 Let X_{ijk} ($i = 1, \ldots, I$, $j = 1, \ldots, J$, $k = 1, \ldots, K$) be $N(\xi_{ijk}, \sigma^2)$ with

$$\xi_{ijk} = \mu + \alpha_i + \beta_j + \gamma_k$$

where $\Sigma\alpha_i = \Sigma\beta_j = \Sigma\gamma_k = 0$. Express μ, α_i, β_j, and γ_k in terms of the ξ's and find their UMVU estimators. Viewed as a special case of (4.4), what is the value of s?

4.14 Extend the results of the preceding problem to the model

$$\xi_{ijk} = \mu + \alpha_i + \beta_j + \gamma_k + \delta_{ij} + \varepsilon_{ik} + \lambda_{jk}$$

where

$$\sum_i \delta_{ij} = \sum_j \delta_{ij} = \sum_i \varepsilon_{ik} = \sum_k \varepsilon_{ik} = \sum_j \lambda_{jk} = \sum_k \lambda_{jk} = 0.$$

4.15 In the preceding problem, if it is known that the λ's are zero, determine whether the UMVU estimators of the remaining parameters remain unchanged.

4.16 (a) Show that under assumptions (4.35), if $\xi = \theta A$, then the least squares estimate of θ is $\mathbf{x}A(AA')^{-1}$.

(b) If (X, A) is multivariate normal with all parameters unknown, show that the least squares estimator of part (a) is a function of the complete sufficient statistic and, hence, prove part (a) of Theorem 4.14.

4.17 A generalization of the order statistics, to vectors, is given by the following definition.

Definition 8.1 The c_j*-order statistics* of a sample of vectors are the vectors arranged in increasing order according to their jth components.

Let $\mathbf{X}_i$, $i = 1, \ldots, n$, be an iid sample of $p \times 1$ vectors, and let $X = (X_1, \ldots, X_n)$ be a $p \times n$ matrix.

(a) If the distribution of $\mathbf{X}_i$ is completely unknown, show that, for any j, $j = 1, \ldots, p$, the c_j-order statistics of $(X_1, \ldots, X_n)$ are complete sufficient. (That is, the vectors $X_1, \ldots, X_n$ are ordered according to their jth coordinate.)

(b) Let $Y_{1\times n}$ be a random variable with unknown distribution (possibly different from $\mathbf{X}_i$). Form the $(p-1)\times n$ matrix $\begin{pmatrix}\mathbf{x}\\ \mathbf{y}\end{pmatrix}$, and for any $j = 1, \ldots, p$, calculate the c_j-order statistics based on the columns of $\begin{pmatrix}\mathbf{x}\\ \mathbf{y}\end{pmatrix}$. Show that these c_j-order statistics are sufficient.

[Hint: See Problem 1.6.33, and also TSH2, Chapter 4, Problem 12.]

(c) Use parts (a) and (b) to prove Theorem 4.14(b).

[Hint: Part (b) implies that only a symmetric function of (X, A) need be considered, and part (a) implies that an unconditionally unbiased estimator must also be conditionally unbiased. Theorem 4.12 then applies.]

4.18 The proof of Theorem 4.14(c) is based on two results. Establish that:

(a) For large values of θ, the unconditional variance of a linear unbiased estimator will be greater than that of the least squares estimator.

(b) For $\theta = 0$, the variance of $XA(AA')^{-1}$ is greater than that of $XA[E(AA')]^{-1}$. [You may use the fact that $E(AA')^{-1} - [E(AA')]^{-1}$ is a positive definite matrix (Marshall and Olkin 1979; Shaffer 1991). This is a multivariate extension of Jensen's inequality.]

(c) Parts (a) and (b) imply that no best linear unbiased estimator of $\Sigma\gamma_i\xi_i$ exists if EAA' is known.

4.19 (a) Under the assumptions of Example 4.15, find the variance of $\Sigma\lambda_i S_i^2$.

(b) Show that the variance of (a) is minimized by the values stated in the example.

4.20 In the linear model (4.4), a function $\Sigma c_i\xi_i$ with $\Sigma c_i = 0$ is called a *contrast*. Show that a linear function $\Sigma d_i\xi_i$ is a contrast if and only if it is translation invariant, that is, satisfies $\Sigma d_i(\xi_i + a) = \Sigma d_i\xi_i$ for all a, and hence if and only if it is a function of the differences $\xi_i - \xi_j$.

4.21 Determine which of the following are contrasts:

(a) The regression coefficients α, β, or γ of (4.2).

(b) The parameters μ, α_i, β_j, or γ_{ij} of (4.27).

(c) The parameters μ or α_i of (4.23) and (4.24).

Section 5

5.1 In Example 5.1:

(a) Show that the joint density of the Z_{ij} is given by (5.2).

(b) Obtain the joint multivariate normal density of the X_{ij} directly by evaluating their covariance matrix and then inverting it.

[*Hint*: The covariance matrix of $X_{11}, \ldots, X_{1n}; \ldots; X_{s1}, \ldots, X_{sn}$ has the form

$$\Sigma = \begin{pmatrix} \Sigma_1 & 0 & \ldots & 0 \\ 0 & \Sigma_2 & \ldots & 0 \\ \vdots & & \ddots & \vdots \\ 0 & 0 & \ldots & \Sigma_s \end{pmatrix}$$

where each Σ_i is an $n \times n$ matrix with a value a_i for all diagonal elements and a value b_i for all off-diagonal elements. For the inversion of Σ_i, see the next problem.]

5.2 Let $A = (a_{ij})$ be a nonsingular $n \times n$ matrix with $a_{ii} = a$ and $a_{ij} = b$ for all $i \neq j$. Determine the elements of A^{-1}. [*Hint*: Assume that $A^{-1} = (c_{ij})$ with $c_{ii} = c$ and $c_{ij} = d$ for all $i \neq j$, calculate c and d as the solutions of the two linear equations $\Sigma a_{1j}c_{j1} = 1$ and $\Sigma a_{1j}c_{j2} = 0$, and check the product AC.]

5.3 Verify the UMVU estimator of σ_A^2/σ^2 given in Example 5.1.

5.4 Obtain the joint density of the X_{ij} in Example 5.1 in the unbalanced case in which $j = 1, \ldots, n_i$, with the n_i not all equal, and determine a minimal set of sufficient statistics (which depends on the number of distinct values of n_i).

5.5 In the balanced one-way layout of Example 5.1, determine $\lim P(\hat{\sigma}_A^2 < 0)$ as $n \to \infty$ for $\sigma_A^2/\sigma^2 = 0, 0.2, 0.5, 1$, and $s = 3, 4, 5, 6$. [*Hint*: The limit of the probability can be expressed as a probability for a χ_{s-1}^2 variable.]

5.6 In the preceding problem, calculate values of $P(\hat{\sigma}_A^2 < 0)$ for finite n. When would you expect negative estimates to be a problem? [The probability $P(\hat{\sigma}_A^2 < 0)$, which involves an F random variable, can also be expressed using the incomplete beta function, whose values are readily available through either extensive tables or computer packages. Searle et al. (1992, Section 3.5d) look at this problem in some detail.]

5.7 The following problem shows that in Examples 5.1–5.3 every unbiased estimator of the variance components (except σ^2) takes on negative values. (For some related results, see Pukelsheim 1981.)

Let X have distribution $P \in \mathcal{P}$ and suppose that T is a complete sufficient statistic for $\mathcal{P}$. If $g(P)$ is any U-estimable function defined over $\mathcal{P}$ and its UMVU estimator $\eta(T)$ takes on negative values with probability > 0, then show that this is true of every unbiased estimator of $g(P)$. [*Hint*: For any unbiased estimator δ, recall that $E(\delta|T) = \eta(T)$.]

5.8 Modify the car illustration of Example 5.1 so that it illustrates (5.5).

5.9 In Example 5.2, define a linear transformation of the X_{ijk} leading to the joint distribution of the Z_{ijk} stated in connection with (5.6), and verify the complete sufficient statistics (5.7).

5.10 In Example 5.2, obtain the UMVU estimators of the variance components σ_A^2, σ_B^2, and σ^2 when $\sigma_C^2 = 0$, and compare them to those obtained without this assumption.

5.11 For the X_{ijk} given in (5.8), determine a transformation taking them to variables Z_{ijk} with the distribution stated in Example 5.3.

5.12 In Example 5.3, obtain the UMVU estimators of the variance components σ_A^2, σ_B^2, and σ^2.

5.13 In Example 5.3, obtain the UMVU estimators of σ_A^2 and σ^2 when $\sigma_B^2 = 0$ so that the B terms in (5.8) drop out, and compare them with those of Problem 5.12.

5.14 In Example 5.4:

(a) Give a transformation taking the variables X_{ijk} into the W_{ijk} with density (5.11).

(b) Obtain the UMVU estimators of $\mu, \alpha_i, \sigma_B^2$, and σ^2.

5.15 A general class of models containing linear models of Types I and II, and mixed models as special cases assumes that the $1 \times n$ observation vector $\mathbf{X}$ is normally distributed with mean θA as in (4.13) and with covariance matrix $\Sigma_{i=1}^m \gamma_i V_i$ where the γ's are the components of variance and the V_i's are known symmetric positive semidefinite $n \times n$ matrices. Show that the following models are of this type and in each case specify the γ's and V's: (a) (5.1); (b) (5.5); (c) (5.5) without the terms C_{ij}; (d) (5.8); (e) (5.10).

5.16 Consider a nested three-way layout with

$$X_{ijkl} = \mu + a_i + b_{ij} + c_{ijk} + U_{ijkl}$$

$(i = 1, \ldots, I; j = 1, \ldots, J; k = 1, \ldots, K; l = 1, \ldots, n)$ in the versions

(a) $a_i = \alpha_i, b_{ij} = \beta_{ij}, c_{ijk} = \gamma_{ijk}$;
(b) $a_i = \alpha_i, b_{ij} = \beta_{ij}, c_{ijk} = C_{ijk}$;
(c) $a_i = \alpha_i, b_{ij} = B_{ij}, c_{ijk} = C_{ijk}$;
(d) $a_i = A_i, b_{ij} = B_{ij}, c_{ijk} = C_{ijk}$;

where the α's, β's, and γ's are unknown constants defined uniquely by the usual conventions, and the A's, B's, C's, and U's are unobservable random variables, independently normally distributed with means zero and with variances $\sigma_A^2, \sigma_B^2, \sigma_C^2$ and σ^2.
In each case, transform the X_{ijkl} to independent variables Z_{ijkl} and obtain the UMVU estimators of the unknown parameters.

5.17 For the situation of Example 5.5, relax the assumption of normality to only assume that A_i and U_{ij} have zero means and finite second moments. Show that among all linear estimators (of the form $\sum c_{ij}x_{ij}$, c_{ij} known), the UMVU estimator of $\mu + \alpha_i$ (the best linear predictor) is given by (5.14).
[This is a Gauss-Markov theorem for prediction in mixed models. See Harville (1976) for generalizations.]

Section 6

6.1 In Example 6.1, show that $\gamma_{ij} = 0$ for all i, j is equivalent to $p_{ij} = p_{i+}p_{+j}$. [*Hint*: $\gamma_{ij} = \xi_{ij} - \xi_{i\cdot} - \xi_{\cdot j} + \xi_{\cdot\cdot} = 0$ implies $p_{ij} = a_ib_j$ and hence $p_{i+} = ca_i$ and $p_{+j} = b_j/c$ for suitable a_i, b_j, and $c > 0$.]

6.2 In Example 6.2, show that the conditional independence of A, B given C is equivalent to $\alpha_{ijk}^{ABC} = \alpha_{ij}^{AB} = 0$ for all i, j, and k.

6.3 In Example 6.1, show that the conditional distribution of the vectors $(n_{i1}, \ldots, n_{iJ})$ given the values of n_{i+} $(i = 1, \ldots, I)$ is that of I independent vectors with multinomial distribution $M(p_{1|i}, \ldots, p_{J|i}; n_{i+})$ where $p_{j|i} = p_{ij}/p_{i+}$.

6.4 Show that the distribution of the preceding problem also arises in Example 6.1 when the n subjects, rather than being drawn from the population at large, are randomly drawn: n_{1+} from Category $A_1, \ldots, n_{I+}$ from Category A_I.

6.5 An application of log linear models in genetics is through the *Hardy-Weinberg* model of mating. If a parent population contains alleles A, a with frequencies p and $1 - p$, then standard random mating assumptions will result in offspring with genotypes AA, Aa, and aa with frequencies $\theta_1 = p^2, \theta_2 = 2p(1-p)$, and $\theta_3 = (1-p)^2$.

(a) Give the full multinomial model for this situation, and show how the Hardy-Weinberg model is a non-full-rank submodel.

(b) For a sample $X_1, \ldots, X_n$ of n offspring, find the minimal sufficient statistic.

[See Brown (1986a) for a more detailed development of this model.]

6.6 A city has been divided into I major districts and the ith district into J_i subdistricts, all of which have populations of roughly equal size. From the police records for a given year, a random sample of n robberies is obtained. Write the joint multinomial distribution of the numbers n_{ij} of robberies in subdistrict (i, j) for this nested two-way layout as $e^{\Sigma\Sigma n_{ij}\xi_{ij}}$ with $\xi_{ij} = \mu + \alpha_i + \beta_{ij}$ where $\Sigma_i\alpha_i = \Sigma_j\beta_{ij} = 0$, and show that the assumption $\beta_{ij} = 0$ for all i, j is equivalent to the assumption that $p_{ij} = p_{i+}/J_i$ for all i, j.

6.7 Instead of a sample of fixed size n in the preceding problem, suppose the observations consist of all robberies taking place within a given time period, so that n is the value taken on by a random variable N. Suppose that N has a Poisson distribution with unknown expectation λ and that the conditional distribution of the n_{ij} given $N = n$ is the distribution assumed for the n_{ij} in the preceding problem. Find the UMVU estimator of λp_{ij} and show that no unbiased estimator p_{ij} exists. [*Hint*: See the following problem.]

6.8 Let N be an integer-valued random variable with distribution $P_\theta(N = n) = P_\theta(n)$, $n = 0, \ldots,$ for which N is complete. Given $N = n$, let X have the binomial distribution $b(p, n)$ for $n > 0$, with p unknown, and let $X = 0$ when $n = 0$. For the observations (N, X):

(a) Show that (N, X) is complete.

(b) Determine the UMVU estimator of $pE_\theta(N)$.

(c) Show that no unbiased estimator of any function $g(p)$ exists if $P_\theta(0) > 0$ for some θ.

(d) Determine the UMVU estimator of p if $P_\theta(0)$ for all θ.

Section 7

7.1 (a) Consider a population $\{a_1, \ldots, a_N\}$ with the parameter space defined by the restriction $a_1 + \cdots + a_N = A$ (known). A simple random sample of size n is drawn in order to estimate τ^2. Assuming the labels to have been discarded, show that $Y_{(1)}, \ldots, Y_{(n)}$ are not complete.

(b) Show that Theorem 7.1 need not remain valid when the parameter space is of the form $V_1 \times V_2 \times \cdots \times V_N$. [*Hint*: Let $N = 2$, $n = 1$, $V_1 = \{1, 2\}$, $V_2 = \{3, 4\}$.]

7.2 If $Y_1, \ldots, Y_n$ are the sample values obtained in a simple random sample of size n from the finite population (7.2), then (a) $E(Y_i) = \bar{a}$, (b) $\text{var}(Y_i) = \tau^2$, and (c) $\text{cov}(Y_i, Y_j) = -\tau^2/(N-1)$.

7.3 Verify equations (a) (7.6), (b) (7.8), and (c) (7.13).

7.4 For the situation of Example 7.4:

(a) Show that $E\bar{Y}_{\nu-1} = E[\frac{1}{\nu-1}\sum_1^{\nu-1} Y_i] = \bar{a}$.

(b) Show that $[\frac{1}{\nu-1} - \frac{1}{N}]\frac{1}{\nu-2}\sum_1^{\nu-1}(Y_i - \bar{Y}_{\nu-1})^2$ is an unbiased estimator of $\text{var}(\bar{Y}_{\nu-1})$.

[Pathak (1976) proved (a) by first showing that $EY_1 = \bar{a}$, and then that $EY_1|T_0 = \bar{Y}_{\nu-1}$. To avoid trivialities, Pathak also assumes that $C_i + C_j < Q$ for all i, j, so that at least three observations are taken.]

7.5 Random variables $X_1, \ldots, X_n$ are *exchangeable* if any permutation of $X_1, \ldots, X_n$ has the same distribution.

(a) If $X_1, \ldots, X_n$ are iid, distributed as Bernoulli (p), show that given $\sum_1^n X_i = t$, $X_1, \ldots, X_n$ are exchangeable (but not independent).

(b) For the situation of Example 7.4, show that given $T = \{(C_1, X_1), \ldots, (C_\nu, X_\nu)\}$, the $\nu - 1$ preterminal observations are exchangeable.

The idea of exchangeability is due to deFinetti (1974), who proved a theorem that characterizes the distribution of exchangeable random variables as mixtures of iid random variables. Exchangeable random variables play a large role in Bayesian statistics; see Bernardo and Smith 1994 (Sections 4.2 and 4.3).

7.6 For the situation of Example 7.4, assuming that (a) and (b) hold:

(a) Show that $\hat{a}$ of (7.9) is UMVUE for $\bar{a}$.

(b) Defining $S^2 = \sum_{i=1}^{\nu}(Y_i - \bar{Y})/(\nu - 1)$, show that

$$\hat{\sigma}^2 = S^2 - \frac{MS_{[\nu]}\frac{\nu}{\nu-1} - S^2}{\nu - 2}$$

is UMVUE for τ^2 of (7.7), where $MS_{[\nu]}$ is the variance of the observations in the set (7.10).

[Kremers (1986) uses conditional expectation arguments (Rao-Blackwellization), and completeness, to establish these results. He also assumes that at least n_0 observations are taken. To avoid trivialities, we can assume $n_0 \geq 3$.]

7.7 In simple random sampling, with labels discarded, show that a necessary condition for $h(a_1, \ldots, a_N)$ to be U-estimable is that h is symmetric in its N arguments.

7.8 Prove Theorem 7.7.

7.9 Show that the approximate variance (7.16) for stratified sampling with $n_i = nN_i/N$ (proportional allocation) is never greater than the corresponding approximate variance τ^2/n for simple random sampling with the same total sample size.

7.10 Let V_p be the exact variance (7.15) and V_r the corresponding variance for simple random sampling given by (7.6) with $n = \Sigma n_i$, $N = \Sigma N_i$, $n_i/n = N_i/N$ and $\tau^2 = \Sigma\Sigma(a_{ij} - a_{..})^2/N$.

(a) Show that $V_r - V_p = \frac{N-n}{n(N-1)N}\left[\Sigma N_i(a_{i.} - a_{..})^2 - \frac{1}{N}\Sigma\frac{N-N_i}{N_i-1}N_i\tau_i^2\right]$.

(b) Give an example in which $V_r < V_p$.

7.11 The approximate variance (7.16) for stratified sampling with a total sample size $n = n_1 + \cdots + n_s$ is minimized when n_i is proportional to $N_i\tau_i$.

7.12 For sampling designs where the inclusion probabilities $\pi_i = \sum_{s:i\in s} P(s)$ of including the ith sample value Y_i is known, a frequently used estimator of the population total is the Horvitz-Thompson (1952) estimator $\delta_{HT} = \sum_i Y_i/\pi_i$.

(a) Show that δ_{HT} is an unbiased estimator of the population total.

(b) The variance of δ_{HT} is given by

$$\text{var}(\delta_{HT}) = \sum_i Y_i^2\left[\frac{1}{\pi_i} - 1\right] + \sum_{i\neq j} Y_i Y_j\left[\frac{\pi_{ij}}{\pi_i\pi_j} - 1\right],$$

where π_{ij} are the *second-order inclusion probabilities* $\pi_{ij} = \sum_{s:i,j\in s} P(s)$.

Note that it is necessary to know the labels in order to calculate δ_{HT}, thus Theorem 7.5 precludes any overall optimality properties. See Hedayat and Sinha 1991 (Chapters 2 and 3) for a thorough treatment of δ_{HT}.

7.13 Suppose that an auxiliary variable is available for each element of the population (7.2) so that $\theta = \{(1, a_1, b_1), \ldots, (N, a_N, b_N)\}$. If $Y_1, \ldots, Y_n$ and $Z_1, \ldots, Z_n$ denote the values of a and b observed in a simple random sample of size n, and $\bar{Y}$ and $\bar{Z}$ denote their averages, then

$$\text{cov}(\bar{Y}, \bar{Z}) = E(\bar{Y} - \bar{a})(\bar{Z} - \bar{b}) = \frac{N-n}{nN(N-1)}\Sigma(a_i - \bar{a})(b_i - \bar{b}).$$

7.14 Under the assumptions of Problem 7.13, if $B = b_1 + \cdots + b_N$ is known, an alternative unbiased estimator $\bar{a}$ is

$$\left(\frac{1}{n}\sum_{i=1}^{n}\frac{Y_i}{Z_i}\right)\bar{b} + \frac{n(N-1)}{(n-1)N}\left[\bar{Y} - \left(\frac{1}{n}\sum_{i=1}^{n}\frac{Y_i}{Z_i}\right)\bar{Z}\right].$$

[*Hint*: Use the facts that $E(Y_1/Z_1) = (1/N)\Sigma(a_i/b_i)$ and that by the preceding problem

$$E\left[\frac{1}{n-1}\Sigma\frac{Y_i}{Z_i}(Z_i - \bar{Z})\right] = \left[\frac{1}{N-1}\Sigma\frac{a_i}{b_i}(b_i - \bar{b})\right].]$$

7.15 In connection with cluster sampling, consider a set W of vectors $(a_1, \ldots, a_M)$ and the totality G of transformations taking $(a_1, \ldots, a_M)$ into $(a'_1, \ldots, a'_M)$ such that $(a'_1, \ldots, a'_M) \in W$ and $\Sigma a'_i = \Sigma a_i$. Give examples of W such that for any real number a_1 there exist $a_2, \ldots, a_M$ with $(a_1, \ldots, a_M) \in W$ and such that

(a) G consists of the identity transformation only;

(b) G consists of the identity and one other element;

(c) G is transitive over W.

7.16 For cluster sampling with unequal cluster sizes M_i, Problem 7.14 provides an alternative estimator of $\bar{a}$, with M_i in place of b_i. Show that this estimator reduces to $\bar{Y}$ if $b_1 = \cdots = b_N$ and hence when the M_i are equal.

7.17 Show that (7.17) holds if and only if δ depends only on X'', defined by (7.18).

9 Notes

9.1 History

The theory of equivariant estimation of location and scale parameters is due to Pitman (1939), and the first general discussions of equivariant estimation were provided by Peisakoff (1950) and Kiefer (1957). The concept of risk-unbiasedness (but not the term) and its relationship to equivariance were given in Lehmann (1951).

The linear models of Section 3.4 and Theorem 4.12 are due to Gauss. The history of both is discussed in Seal (1967); see also Stigler 1981. The generalization to exponential linear models was introduced by Dempster (1971) and Nelder and Wedderburn (1972).

The notions of *Functional Equivariance* and *Formal Invariance*, discussed in Section 3.2, have been discussed by other authors sometimes using different names. Functional Equivariance is called the *Principle of Rational Invariance* by Berger (1985, Section 6.1), *Measurement Invariance* by Casella and Berger (1990, Section 7.2.4) and *Parameter Invariance* by Dawid (1983). Schervish (1995, Section 6.2.2) argues that this principle is really only a reparametrization of the problem, and has nothing to do with invariance. This is almost in agreement with the principle of functional equivariance, however, it is still the case that when reparameterizing one must be careful to properly reparameterize the estimator, density, and loss function, which is part of the prescription of an invariant problem. This type of invariance is commonly illustrated by the example that if δ measures temperature in degrees Celsius, then $(9/5)\delta + 32$ should be used to measure temperature in degrees Fahrenheit (see Problems 2.9 and 2.10).

What we have called *Formal Invariance* was also called by that name in Casella and Berger (1990), but was called the *Invariance Principle* by Berger (1985) and *Context Invariance* by Dawid (1983).

9.2 Subgroups

The idea of improving an MRE estimator by imposing equivariance only under a subgroup was used by Stein (1964), Brown (1968), and Brewster and Zidek (1974) to find improved estimators of a normal variance. Stein's 1964 proof is also discussed in detail by Maatta and Casella (1990), who give a history of decision-theoretic variance estimation. The proof of Stein (1964) contains key ideas that were further developed by Brown (1968), and led to Brewster and Zidek (1974) finding the best equivariant estimator of the form (2.33). [See Problem 2.14.]

9.3 General Linear Group

The *general linear group* (also called the *full linear group*) is an example of a group that can be thought of as a multivariate extension of the location-scale group. Let $X_1, \ldots, X_n$ be iid according to a p-variate normal distribution $N_p(\mu, \Sigma)$, and define $\mathbf{X}$ as the $p \times n$ matrix $(X_1, \ldots, X_n)$ and $\bar{\mathbf{X}}$ as the $n \times 1$ vector $(\bar{X}_1, \ldots, \bar{X}_n)$. Consider the group of transformations

$$\mathbf{X}' = A\mathbf{X} + b$$

$$\mu' = A\mu + b, \quad \Sigma' = A\Sigma A', \tag{9.1}$$

where A is a $p \times p$ nonsingular matrix and b is a $p \times 1$ vector. [The group of real $p \times p$ nonsingular matrices, with matrix multiplication as the group operation is called the *general linear group*, denoted $\mathcal{G}l_p$ (see Eaton 1989 for a further development). The group (9.1) adds a location component.]

Consider now the estimation of Σ. (The estimation of μ is left to Problem 3.30.) An invariant loss function, analogous to squared error loss, is of the form

$$L(\Sigma, \delta) = \text{tr}[\Sigma^{-1}(\delta - \Sigma)\Sigma^{-1}(\delta - \Sigma)] = \text{tr}[\Sigma^{-1/2}\delta\Sigma^{-1/2} - I]^2, \tag{9.2}$$

where tr[·] is the trace of a matrix (see Eaton 1989, Example 6.2, or Olkin and Selliah 1977). It can be shown that equivariant estimators are of the form cS, where $S = (\mathbf{X} - \mathbf{1}\bar{\mathbf{X}}')(\mathbf{X} - \mathbf{1}\bar{\mathbf{X}}')'$ with $\mathbf{1}$ a $p \times 1$ vector of 1's and c a constant, is the *cross-products matrix* (Problem 3.31). Since the group is transitive, the MRE estimator is given by the value of c that minimizes

$$E_I L(I, cS) = E_I \text{tr}(cS - I)'(cS - I), \tag{9.3}$$

that is, the risk with $\Sigma = I$. Since

$$E_I \text{tr}(cS - I)'(cS - I) = c^2 E_I \text{tr} S^2 - 2cE_I \text{tr} S + p,$$

the minimizing c is given by $c = E_I \text{tr} S / E_I \text{tr} S^2$. Note that, for $p = 1$, this reduces to the best equivariant estimator of quadratic loss in the scalar case. Other equivariant losses, such as Stein's loss (3.20), can be handled in a similar manner. See Problems 3.29-3.33 for details.

9.4 Finite Populations

Estimation in finite populations has, until recently, been developed largely outside the mainstream of statistics. The books by Cassel, Särndal, and Wretman (1977) and Särndal, Swenson, and Wretman (1992) constitute important efforts at a systematic presentation of this topic within the framework of theoretical statistics. The first steps in this direction were taken by Neyman (1934) and by Blackwell and Girshick (1954). The need to consider the labels as part of the data was first emphasized by Godambe (1955). Theorem 7.1 is due to Watson (1964) and Royall (1968), and Theorem 7.5 to Basu (1971).

CHAPTER 4

Average Risk Optimality

1 Introduction

So far, we have been concerned with finding estimators which minimize the risk $R(\theta, \delta)$ at every value of θ. This was possible only by restricting the class of estimators to be considered by an impartiality requirement such as unbiasedness or equivariance. We shall now drop such restrictions, admitting all estimators into competition, but shall then have to be satisfied with a weaker optimality property than uniformly minimum risk. We shall look for estimators that make the risk function $R(\theta, \delta)$ small in some overall sense. Two such optimality properties will be considered: minimizing the (weighted) average risk for some suitable non-negative weight function and minimizing the maximum risk. The second (minimax) approach will be taken up in Chapter 5; the present chapter is concerned with the first of these approaches, the problem of minimizing

$$r(\Lambda, \delta) = \int R(\theta, \delta) d\Lambda(\theta) \tag{1.1}$$

where we shall assume that the weights represented by Λ add up to 1, that is,

$$\int d\Lambda(\theta) = 1, \tag{1.2}$$

so that Λ is a probability distribution. An estimator δ minimizing (1.1) is called a *Bayes estimator* with respect to Λ.

The problem of determining such Bayes estimators arises in a number of different contexts.

(i) As Mathematical Tools

Bayes estimators play a central role in Wald's decision theory. It is one of the main results of this theory that in any given statistical problem, attention can be restricted to Bayes solutions and suitable limits of Bayes solutions; given any other procedure δ, there exists a procedure δ' in this class such that $R(\theta, \delta') \leq R(\theta, \delta)$ for all values of θ. (In view of this result, it is not surprising that Bayes estimators provide a tool for solving minimax problems, as will be seen in the next chapter.)

(ii) As a Way of Utilizing Past Experience

It is frequently reasonable to treat the parameter θ of a statistical problem as the realization of a random variable Θ with known distribution rather than as an unknown constant. Suppose, for example, that we wish to estimate the probability of a penny showing heads when spun on a flat surface. So far, we would have considered n spins of the penny as a set of n binomial trials with an unknown probability p of showing heads. Suppose, however, that we have had considerable experience with spinning pennies, experience perhaps which has provided us with approximate values of p for a large number of similar pennies. If we believe this experience to be relevant to the present penny, it might be reasonable to represent this past knowledge as a probability distribution for p, the approximate shape of which is suggested by the earlier data.

This is not as unlike the modeling we have done in the earlier sections as it may seem at first sight. When assuming that the random variables representing the outcomes of our experiments have normal, Poisson, exponential distributions, and so on, we also draw on past experience. Furthermore, we also realize that these models are in no sense exact but, at best, represent reasonable approximations. There is the difference that in earlier models we have assumed only the shape of the distribution to be known but not the values of the parameters, whereas now we extend our model to include a specification of the prior distribution. However, this is a difference in degree rather than in kind and may be quite reasonable if the past experience is sufficiently extensive.

A difficulty, of course, is the assumption that past experience is relevant to the present case. Perhaps the mint has recently changed its manufacturing process, and the present coin, although it looks like the earlier ones, has totally different spinning properties. Similar kinds of judgment are required also for the models considered earlier. In addition, the conclusions derived from statistical procedures are typically applied not only to the present situation or population but also to those in the future, and extrastatistical judgment is again required in deciding how far such extrapolation is justified.

The choice of the prior distribution Λ is typically made like that of the distributions P_θ by combining experience with convenience. When we make the assumption that the amount of rainfall has a gamma distribution, we probably do not do so because we really believe this to be the case but because the gamma family is a two-parameter family which seems to fit such data reasonably well and which is mathematically very convenient. Analogously, we can obtain a prior distribution by starting with a flexible family that is mathematically easy to handle and selecting a member from this family which approximates our past experience. Such an approach, in which the model incorporates a prior distribution for θ to reflect past experience, is useful in fields in which a large amount of past experience is available. It can be brought to bear, for example, in many applications in agriculture, education, business, and medicine.

There are important differences between the modeling of the distributions P_θ and that of Λ. First, we typically have a number of observations from P_θ and can use these to check the assumption of the form of the distribution. Such a check of Λ is not possible on the basis of one experiment because the value of θ under study represents only a single observation from this distribution. A second difference concerns the meaning of a replication of the experiment. In the models preceding this section, the replication would consist of drawing another set of observations from P_θ with the same value of θ. In the model of the present section, we would replicate the experiment by first drawing another value, θ', of Θ from Λ and then a set of observations from P'_θ. It might be argued that sampling of the θ values (choice of penny, for example) may be even more haphazard and less well controlled than the choice of subjects for an experiment of a study, which assumes these subjects to be a random sample from the population of interest. However, it could also be argued that the assumption of a fixed value of θ is often unrealistic. As we will see, the Bayesian approaches of robust and hierarchical analysis attempt to address these problems.

(iii) As a Description of a State of Mind

A formally similar approach is adopted by the so-called Bayesian school, which interprets Λ as expressing the subjective feeling about the likelihood of different θ values. In the presence of a large amount of previous experience, the chosen Λ would often be close to that made under (ii), but the subjective approach can be applied even when little or no prior knowledge is available. In the latter case, for example, the prior distribution Λ then models the state of ignorance about θ. The subjective Bayesian uses the observations X to modify prior beliefs. After $X = x$ has been observed, the belief about θ is expressed by the posterior (i.e., conditional) distribution of Θ given x.

Detailed discussions of this approach, which we shall not pursue here, can be found, for example, in books by Savage (1954), Lindley (1965), de Finetti (1970, 1974), Box and Tiao (1973), Novick and Jackson (1974), Berger (1985), Bernardo and Smith (1994), Robert (1994a) and Gelman et al. (1995).

A note on notation: In Bayesian (as in frequentist) arguments, it is important to keep track of which variables are being conditioned on. Thus, the density of X will be denoted by $X \sim f(x|\theta)$. Prior distributions will typically be denoted by Π or Λ with their density functions being $\pi(\theta|\lambda)$ or $\gamma(\lambda)$,where λ is another parameter (sometimes called a *hyperparameter*). From these distributions we often calculate conditional distributions such as that of θ given x and λ, or λ given x (called *posterior distributions*). These typically have densities, denoted by $\pi(\theta|x, \lambda)$ or $\gamma(\lambda|x)$. We will also be interested in marginal distributions such as $m(x|\lambda)$. To illustrate, $\pi(\theta|x, \lambda) = f(x|\theta)\pi(\theta|\lambda)/m(x|\lambda)$, where $m(x|\lambda) = \int f(x|\theta)\pi(\theta|\lambda)\, d\theta$.

It is convenient to use boldface to denote vectors, for example, $\mathbf{x} = (x_1, \ldots, x_n)$, so we can write $f(\mathbf{x}|\theta)$ for the sample density $f(x_1, \ldots, x_n|\theta)$.

The determination of a Bayes estimator is, in principle, quite simple. First, consider the situation before any observations are taken. Then, Θ has distribution Λ and the Bayes estimator of $g(\Theta)$ is any number d minimizing $EL(\Theta, d)$. Once the data have been obtained and are given by the observed value x of X, the prior distribution Λ of Θ is replaced by the posterior, that is, conditional, distribution of Θ given x and the Bayes estimator is any number $\delta(x)$ minimizing the posterior risk $E\{L[\Theta, \delta(x)]|x\}$. The following is a precise statement of this result, where, as usual, measurability considerations, are ignored.

Theorem 1.1 *Let Θ have distribution Λ, and given $\Theta = \theta$, let X have distribution P_θ. Suppose, in addition, the following assumptions hold for the problem of estimating $g(\Theta)$ with non-negative loss function $L(\theta, d)$.*

(a) There exists an estimator δ_0 with finite risk.

(b) For almost all x, there exists a value $\delta_\Lambda(x)$ minimizing

$$E\{L[\Theta, \delta(x)]|X = x\}. \tag{1.3}$$

Then, $\delta_\Lambda(X)$ is a Bayes estimator.

Proof. Let δ be any estimator with finite risk. Then, (1.3) is finite a.e. since L is non-negative. Hence,

$$E\{L[\Theta, \delta(x)]|X = x\} \geq E\{L[\Theta, \delta_\Lambda(x)]|X = x\} \quad \text{a.e.,}$$

and the result follows by taking the expectation of both sides. □

[For a discussion of some measurability aspects and more detail when $L(\theta, d) = \rho(d - \theta)$, see DeGroot and Rao 1963. Brown and Purves (1973) provide a general treatment.]

Corollary 1.2 *Suppose the assumptions of Theorem 1.1 hold.*

(a) If $L(\theta, d) = [d - g(\theta)]^2$, then

$$\delta_\Lambda(x) = E[g(\Theta)|x] \tag{1.4}$$

and, more generally, if

$$L(\theta, d) = w(\theta)[d - g(\theta)]^2, \tag{1.5}$$

then

$$\delta_\Lambda(x) = \frac{\int w(\theta)g(\theta)d\Lambda(\theta|x)}{\int w(\theta)d\Lambda(\theta|x)} = \frac{E[w(\Theta)g(\Theta)|x]}{E[w(\Theta)|x]}. \tag{1.6}$$

(b) If $L(\theta, d) = |d - g(\theta)|$, then $\delta_\Lambda(x)$ is any median of the conditional distribution of Θ given x.

(c) If

$$L(\theta, d) = \begin{cases} 0 \text{ when } |d - \theta| \leq c \\ 1 \text{ when } |d - \theta| > c, \end{cases} \tag{1.7}$$

then $\delta_\Lambda(x)$ is the midpoint of the interval I of length $2c$ which maximizes $P[\Theta \in I|x]$.

Proof. To prove part (i), note that by Theorem 1.1, the Bayes estimator is obtained by minimizing

$$E\{[g(\Theta) - \delta(x)]^2|x\}. \tag{1.8}$$

By assumption (a) of Theorem 1.1, there exists $\delta_0(x)$ for which (1.8) is finite for almost all values of x, and it then follows from Example 1.7.17 that (1.8) is minimized by (1.4).

The proofs of the other parts are completely analogous. □

Example 1.3 Poisson. The parameter θ of a Poisson(θ) distribution is both the mean and the variance of the distribution. Although squared error loss $L_0(\theta, \delta) = (\theta - \delta)^2$ is often preferred for the estimation of a mean, some type of scaled squared error loss, for example, $L_k(\theta, \delta) = (\theta - \delta)^2/\theta^k$, may be more appropriate for the estimation of a variance.

If $X_1, \ldots, X_n$ are iid Poisson(θ), and θ has the gamma(a, b) prior distribution, then the posterior distribution is

$$\pi(\theta|\bar{x}) = \text{Gamma}\left(a + n\bar{x}, \frac{b}{1+nb}\right)$$

and the Bayes estimator under L_k is given by (see Problem 1.1)

$$\delta^k(\bar{x}) = \frac{E(\theta^{1-k}|\bar{x})}{E(\theta^{-k}|\bar{x})} = \frac{b}{1+nb}(n\bar{x} + a - k)$$

for $a - k > 0$. Thus, the choice of loss function can have a large effect on the resulting Bayes estimator. ||

It is frequently important to know whether a Bayes solution is unique. The following are sufficient conditions for this to be the case.

Corollary 1.4 *If the loss function $L(\theta, d)$ is squared error, or more generally, if it is strictly convex in d, a Bayes solution δ_Λ is unique (a.e. $\mathcal{P}$), where $\mathcal{P}$ is the class of distributions P_θ, provided*

(a) the average risk of δ_Λ with respect to Λ is finite, and

(b) if Q is the marginal distribution of X given by

$$Q(A) = \int P_\theta(X \in A)d\Lambda(\theta),$$

then a.e. Q implies a.e. $\mathcal{P}$.

Proof. For squared error, if follows from Corollary 1.2 that any Bayes estimator $\delta_\Lambda(x)$ with finite risk must satisfy (1.4) except on a set N of x values with $Q(N) = 0$. For general strictly convex loss functions, the result follows by the same argument from Problem 1.7.26. □

As an example of a case in which condition (b) does not hold, let X have the binomial distribution $b(p, n)$, $0 \le p \le 1$, and suppose that Λ assigns probability 1/2 to each of the values $p = 0$ and $p = 1$. Then, any estimator $\delta(X)$ of p with $\delta(0) = 0$ and $\delta(n) = 1$ is Bayes.

On the other hand, condition (b) is satisfied when the parameter space is an open set which is the support of Λ and if the probability $P_\theta(X \in A)$ is continuous in θ for any A. To see this, note that $Q(N) = 0$ implies $P_\theta(N) = 0$ (a.e. Λ) by (1.2.23). If there exists θ_0 with $P_{\theta_0}(N) > 0$, there exists a neighborhood ω of θ_0 in which $P_\theta(N) > 0$. By the support assumption, $P_\Lambda(\omega) > 0$ and this contradicts the assumption that $P_\theta(N) = 0$ (a.e. Λ).

Three different aspects of the performance of a Bayes estimator, or of any other estimator δ, may be of interest in the present model. These are (a) the Bayes risk (1.1); (b) the risk function $R(\theta, \delta)$ of Section 1.1 [Equation (1.1.10)] [this is the frequentist risk, which is now the conditional risk of $\delta(X)$ given θ]; and (c) the posterior risk given x which is defined by (1.3).

For the determination of the Bayes estimator the relevant criterion is, of course, (a). However, consideration of (b), the conditional risk given θ, as a function of θ provides an important safeguard against an inappropriate choice of Λ (Berger 1985, Section 4.7.5). Finally, consideration of (c) is of interest primarily to the Bayesian. From the Bayesian point of view, the posterior distribution of Θ given x summarizes the investigator's belief about θ in the light of the observation, and hence the posterior risk is the only measure of risk of accuracy that is of interest.

The possibility of evaluating the risk function (b) of δ_Λ suggests still another use of Bayes estimators.

(iv) As a General Method for Generating Reasonable Estimators

Postulating some plausible distributions Λ provides a method for generating interesting estimators which can then be studied in the conventional way. A difficulty with this approach is, of course, the choice of Λ. Methodologies have been developed to deal with this difficulty which sometimes incorporate frequentist measures to assess the choice of Λ. These methods tend to first select not a single prior distribution but a family of priors, often indexed by a parameter (a so-called *hyperparameter*). The family should be chosen so as to balance appropriateness, flexibility, and mathematical convenience. From it, a plausible member is selected to obtain an estimator for consideration. The following are some examples of these approaches, which will be discussed in Sections 4.4 and 4.5.

- *Empirical Bayes*. The parameters of the prior distribution are themselves estimated from the data.
- *Hierarchical Bayes*. The parameters of the prior distribution are, in turn, modeled by another distribution, sometimes called a hyperprior distribution.
- *Robust Bayes*. The performance of an estimator is evaluated for each member of the prior class, with the goal of finding an estimator that performs well (is *robust*) for the entire class.

Another possibility leading to a particular choice of Λ corresponds to the third interpretation (iii), in which the state of mind can be described as "ignorance." One would then select for Λ a *noninformative* prior which tries (in the spirit of invariance) to treat all parameter values equitably. Such an approach was developed by Jeffreys (1939, 1948, 1961), who, on the basis of invariance considerations, suggests as noninformative prior for θ a density that is proportional to $\sqrt{|I(\theta)|}$, where $|I(\theta)|$ is the determinant of the information matrix. A good account of this approach with many applications is given by Berger (1985), Robert (1994a), and Bernardo and Smith (1994). Note 9.6 has a further discussion.

Example 1.5 Binomial. Suppose that X has the binomial distribution $b(p, n)$. A two-parameter family of prior distributions for p which is flexible and for which the calculation of the conditional distribution is particularly simple is the family of beta distributions $B(a, b)$. These densities can take on a variety of shapes (see

Problem 1.2) and we note for later reference that the expectation and variance of a random variable p with density $B(a, b)$ are (Problem 1.5.19).

$$E(p) = \frac{a}{a+b} \quad \text{and} \quad \text{var}(p) = \frac{ab}{(a+b)^2(a+b+1)}. \tag{1.9}$$

To determine the Bayes estimator of a given estimand $g(p)$, let us first obtain the conditional distribution (posterior distribution) of p given x. The joint density of X and p is

$$\binom{n}{x}\frac{\Gamma(a+b)}{\Gamma(a)\Gamma(b)}p^{x+a-1}(1-p)^{n-x+b-1}.$$

The conditional density of p given x is obtained by dividing by the marginal of x, which is a function of x alone (Problem 2.1). Thus, the conditional density of p given x has the form

$$C(a, b, x)p^{x+a-1}(1-p)^{n-x+b-1}. \tag{1.10}$$

Again, this is recognized to be a beta distribution, with parameters

$$a' = a + x, \quad b' = b + n - x. \tag{1.11}$$

Let us now determine the Bayes estimator of $g(p) = p$ when the loss function is squared error. By (1.4), this is

$$\delta_\Lambda(x) = E(p|x) = \frac{a'}{a'+b'} = \frac{a+x}{a+b+n}. \tag{1.12}$$

It is interesting to compare this Bayes estimator with the usual estimator X/n. Before any observations are taken, the estimator from the Bayesian approach is the expectation of the prior: $a/(a+b)$. Once X has been observed, the standard non-Bayesian (for example, UMVU) estimator is X/n. The estimator $\delta_\Lambda(X) = (a+X)/(a+b+n)$ lies between these two. In fact,

$$\frac{a+X}{a+b+n} = \left(\frac{a+b}{a+b+n}\right)\frac{a}{a+b} + \left(\frac{n}{a+b+n}\right)\frac{X}{n} \tag{1.13}$$

is a weighted average of $a/(a+b)$, the estimator of p before any observations are taken, and X/n, the estimator without consideration of a prior.

The estimator (1.13) can be considered as a modification of the standard estimator X/n in the light of the prior information about p expressed by (1.9) or as a modification of the prior estimator $a/(a+b)$ in the light of the observation X. From this point of view, it is interesting to notice what happens as a and $b \to \infty$, with the ratio b/a being kept fixed. Then, the estimator (1.12) tends in probability to $a/(a+b)$, that is, the prior information is so overwhelming that it essentially determines the estimator. The explanation is, of course, that in this case the beta distribution $B(a, b)$ concentrates all its mass essentially at $a/(a+b)$ [the variance in (1.9) tends toward 0], so that the value of p is taken to be essentially known and is not influenced by X. ("Don't confuse me with the facts!")

On the other hand, if a and b are fixed, but $n \to \infty$, it is seen from (1.12) that δ_Λ essentially coincides with X/n. This is the case in which the information provided by X overwhelms the initial information contained in the prior distribution.

The UMVU estimator X/n corresponds to the case $a = b = 0$. However, $B(0, 0)$ is no longer a probability distribution since $\int_0^1 (1/p(1-p))dp = \infty$. Even with such an *improper* distribution (that is, a distribution with infinite mass), it is possible formally to calculate a posterior distribution given x. This possibility will be considered in Example 2.8. ||

This may be a good time to discuss a question facing the reader of this book. Throughout, the theory is illustrated with examples which are either completely formal (that is, without any context) or stated in terms of some vaguely described situation in which such an example might arise. In either case, what is assumed is a model and, in the present section, a prior distribution. Where do these assumptions come from, and how should they be interpreted? "Let X have a binomial distribution $b(p, n)$ and let p be distributed according to a beta distribution $B(a, b)$." Why binomial and why beta?

The assumptions underlying the binomial distribution are (i) independence of the n trials and (ii) constancy of the success probability p throughout the series. While in practice it is rare for either of these two assumptions to hold exactly - consecutive trials typically exhibit some dependence and success probabilities tend to change over time (as in Example 1.8.5) - they are often reasonable approximations and may serve as identifications in a wide variety of situations arising in the real world. Similarly, to a reasonable degree, approximate normality may often be satisfied according to some version of the central limit theorem, or from past experience.

Let us next turn to the assumption of a beta prior for p. This leads to an estimator which, due to its simplicity, is highly prized for a variety of reasons. But simplicity of the solution is of little use if the problem is based on assumptions which bear no resemblance to reality.

Subjective Bayesians, even though perhaps unable to state their prior precisely, will typically have an idea of its shape: It may be bimodal, unimodal (symmetric or skewed), or it may be L- or U-shaped. In the first of these cases, a beta prior would be inappropriate since no beta distribution has more than one mode. However, by proper choice of the parameters a and b, a beta distribution can accommodate itself to each of the other possibilities mentioned (Problem 1.2), and thus can represent a considerable variety of prior shapes.

The modeling of subjective priors discussed in the preceding paragraph correspond to the third of the four interpretations of the Bayes formalism mentioned at the beginning of the section. A very different approach is suggested by the fourth interpretation, where formal priors are used simply as a method of generating a reasonable estimator. A standard choice in this case is to treat all parameter values equally (which corresponds to a subjective prior modeling ignorance). In the nineteenth century, the preferred choice for this purpose in the binomial case was the uniform distribution for p over $(0, 1)$, which is the beta distribution with $a = b = 1$. As an alternative, the Jeffreys prior corresponding to $a = b = 1/2$ (see the discussion preceding Example 1.5) has the advantage of being invariant under change of parameters (Schervisch 1995, Section 2.3.4). The prior density in this case is proportional to $[p(1-p)]^{-1/2}$, which is U-shaped. It is difficult to imagine many real situations in which an investigator believes that it is equally likely for the unknown p to be close to either 0 or 1. In this case, the fourth interpretation would therefore lead to very different priors from those of the third interpretation.

2 First Examples

In constructing Bayes estimators, as functions of the posterior density, some choices are made (such as the choice of prior and loss function). These choices will ultimately affect the properties of the estimators, including not only risk performance (such as bias and admissibility) but also more fundamental considerations (such as sufficiency). In this section, we look at a number of examples to illustrate these points.

Example 2.1 Sequential binomial sampling. Consider a sequence of binomial trials with a stopping rule as in Section 3.3. Let X, Y, and N denote, respectively, the number of successes, the number of failures, and the total number of trials at the moment sampling stops. The probability of any sample path is then $p^x(1-p)^y$ and we shall again suppose that p has the prior distribution $B(a, b)$. What now is the posterior distribution of p given X and Y (or equivalently X and $N = X + Y$)? The calculation in Example 1.5 shows that, as in the fixed sample size case, it is the beta distribution with parameters a' and b' given by (1.11), so that, in particular, the Bayes estimator of p is given by (1.12) *regardless* of the stopping rule. ||

Of course, there are stopping rules which even affect Bayesian inference (for example, "stop when the posterior probability of an event is greater than .9"). However, if the stopping rule is a function only of the data, then the Bayes inference will be independent of it. These so-called *proper stopping rules*, and other aspects of inference under stopping rules, are discussed in detail by Berger and Wolpert (1988, Section 4.2). See also Problem 2.2 for another illustration.

Thus, Example 2.1 illustrates a quite general feature of Bayesian inference: The posterior distribution does not depend on the sampling rule but only on the likelihood of the observed results.

Example 2.2 Normal mean. Let $X_1, \ldots, X_n$ be iid as $N(\theta, \sigma^2)$, with σ known, and let the estimand be θ. As a prior distribution for Θ, we shall assume the normal distribution $N(\mu, b^2)$. The joint density of Θ and $\mathbf{X} = (X_1, \ldots, X_n)$ is then proportional to

$$f(\mathbf{x}, \theta) = \exp\left[-\frac{1}{2\sigma^2}\sum_{i=1}^{n}(x_i - \theta)^2\right]\exp\left[-\frac{1}{2b^2}(\theta - \mu)^2\right]. \tag{2.1}$$

To obtain the posterior distribution of $\Theta|\mathbf{x}$, the joint density is divided by the marginal density of $\mathbf{X}$, so that the posterior distribution has the form $C(\mathbf{x})f(\mathbf{x}|\theta)$. If $C(\mathbf{x})$ is used generically to denote any function of $\mathbf{x}$ not involving θ, the posterior density of $\Theta|\mathbf{x}$ is

$$\begin{aligned} &C(\mathbf{x})e^{-(1/2)\theta^2[n/\sigma^2+1/b^2]+\theta[n\bar{x}/\sigma^2+\mu/b^2]} \\ &= C(\mathbf{x})\exp\left\{\left[-\frac{1}{2}\left(\frac{n}{\sigma^2}+\frac{1}{b^2}\right)\right]\left[\theta^2 - 2\theta\frac{n\bar{x}/\sigma^2+\mu/b^2}{n/\sigma^2+1/b^2}\right]\right\}. \end{aligned}$$

This is recognized to be the normal density with mean

$$E(\Theta|\mathbf{x}) = \frac{n\bar{x}/\sigma^2 + \mu/b^2}{n/\sigma^2 + 1/b^2} \tag{2.2}$$

and variance

$$\operatorname{var}(\Theta|\mathbf{x}) = \frac{1}{n/\sigma^2 + 1/b^2}. \tag{2.3}$$

When the loss is squared error, the Bayes estimator of θ is given by (2.2) and can be rewritten as

$$\delta_\Lambda(\mathbf{x}) = \left(\frac{n/\sigma^2}{n/\sigma^2 + 1/b^2}\right)\bar{x} + \left(\frac{1/b^2}{n/\sigma^2 + 1/b^2}\right)\mu, \tag{2.4}$$

and by Corollary 1.7.19, this result remains true for any loss function $\rho(d - \theta)$ for which ρ is convex and even. This shows δ_Λ to be a weighted average of the standard estimator $\bar{X}$, and the mean μ of the prior distribution, which is the Bayes estimator before any observations are taken. As $n \to \infty$ with μ and b fixed, $\delta_\Lambda(X)$ becomes essentially the estimator $\bar{X}$, and $\delta_\Lambda(X) \to \theta$ in probability. As $b \to 0$, $\delta_\Lambda(X) \to \mu$ in probability, as is to be expected when the prior becomes more and more concentrated about μ. As $b \to \infty$, $\delta_\Lambda(X)$ essentially coincides with $\bar{X}$, which again is intuitively reasonable. These results are analogous to those in the binomial case. See Problem 2.3. ||

It was seen above that $\bar{X}$ is the limit of the Bayes estimators as $b \to \infty$. As $b \to \infty$, the prior density tends to Lebesgue measure. Since the Fisher information $I(\theta)$ of a location parameter is constant, this is actually the Jeffrey's prior mentioned under (iv) earlier in the section. It is easy to check that the posterior distribution calculated from this improper prior is a proper distribution as soon as an observation has been taken. This is not surprising; since X is normally distributed about θ with variance 1, even a single observation provides a good idea of the position of θ.

As in the binomial case, the question arises whether $\bar{X}$ is the Bayes solution also with respect to a proper prior Λ. This question is answered for both cases by the following theorem.

Theorem 2.3 *Let Θ have a distribution Λ, and let P_θ denote the conditional distribution of X given θ. Consider the estimation of $g(\theta)$ when the loss function is squared error. Then, no unbiased estimator $\delta(X)$ can be a Bayes solution unless*

$$E[\delta(X) - g(\Theta)]^2 = 0, \tag{2.5}$$

where the expectation is taken with respect to variation in both X and Θ.

Proof. Suppose $\delta(X)$ is a Bayes estimator and is unbiased for estimating $g(\theta)$. Since $\delta(X)$ is Bayes and the loss is squared error,

$$\delta(X) = E[g(\Theta)|X],$$

with probability 1. Since $\delta(X)$ is unbiased,

$$E[\delta(X)|\theta] = g(\theta) \quad \text{for all } \theta.$$

Conditioning on X and using (1.6.2) leads to

$$E[g(\Theta)\delta(X)] = E\{\delta(X)E[g(\Theta)|X]\} = E[\delta^2(X)].$$

Conditioning instead on Θ, we find

$$E[g(\Theta)\delta(X)] = E\{g(\Theta)E[\delta(X)|\Theta]\} = E[g^2(\Theta)].$$

It follows that

$$E[\delta(X) - g(\Theta)]^2 = E[\delta^2(X)] + E[g^2(\Theta)] - 2E[\delta(X)g(\Theta)] = 0,$$

as was to be proved. □

Let us now apply this result to the case that $\delta(x)$ is the sample mean.

Example 2.4 Sample means. If X_i, $i = 1, \ldots, n$, are iid with $E(X_i) = \theta$ and var $X_i = \sigma^2$ (independent of θ), then the risk of $\bar{X}$ (given θ) is

$$R(\theta, \bar{X}) = E(\bar{X} - \theta)^2 = \sigma^2/n.$$

For any proper prior distribution on Θ,

$$E(\bar{X} - \Theta)^2 = \sigma^2/n \neq 0,$$

so (2.5) cannot be satisfied and, from Theorem 2.3, $\bar{X}$ is not a Bayes estimator.

This argument will apply to any distribution for which the variance of $\bar{X}$ is independent of θ, such as the $N(\theta, \sigma^2)$ distribution in Example 2.2. However, if the variance is a function of θ, the situation is different.

If var $X_i = v(\theta)$, then (2.5) will hold only if

$$\int v(\theta) d\Lambda(\theta) = 0 \tag{2.6}$$

for some proper prior Λ. If $v(\theta) > 0$ (a.e. Λ), then (2.6) cannot hold. For example, if $X_1, \cdots, X_n$ are iid Bernoulli(p) random variables, then the risk function of the sample mean $\delta(\Sigma X_i) = \Sigma X_i/n$ is

$$E\,(\delta(\Sigma X_i) - p)^2 = \frac{p(1-p)}{n},$$

and the left side of (2.5) is therefore

$$\frac{1}{n}\int_0^1 p(1-p)\,d\Lambda(p).$$

The integral is zero if and only if Λ assigns probability 1 to the set $\{0, 1\}$. For such a distribution, Λ,

$$\delta_\Lambda(0) = 0 \quad \text{and} \quad \delta_\Lambda(n) = 1,$$

and any estimator satisfying this condition is a Bayes estimator for such a Λ. Hence, in particular, X/n is a Bayes estimator. Of course, if Λ is true, then the values $X = 1, 2, \ldots, n-1$ are never observed. Thus, X/n is Bayes only in a rather trivial sense. ‖

Extensions and discussion of other consequences of Theorem 2.3 can be found in Bickel and Blackwell (1967), Noorbaloochi and Meeden (1983), and Bickel and Mallows (1988). See Problem 2.4.

The beta and normal prior distributions in the binomial and normal cases are the so-called *conjugate* families of prior distributions. These are frequently defined as distributions with densities proportional to the density of P_θ. It has been pointed out by Diaconis and Ylvisaker (1979) that this definition is ambiguous; they show that in the above examples and, more generally, in the case of exponential families, conjugate priors can be characterized by the fact that the resulting Bayes estimators are linear in X. They also extend the weighted-average representation (1.13) of the Bayes estimator to general exponential families. For one parameter exponential families, MacEachern (1993) gives an alternate characterization of conjugate priors based on the requirement that the posterior mean lies "in between" the prior mean and sample mean.

As another example of the use of conjugate priors, consider the estimation of a normal variance.

Example 2.5 Normal variance, known mean. Let $X_1, \ldots, X_n$ be iid according to $N(0, \sigma^2)$, so that the joint density of the X_i's is $C\tau^r e^{-\tau\Sigma x_i^2}$, where $\tau = 1/2\sigma^2$ and $r = n/2$. As conjugate prior for τ, we take the gamma density $\Gamma(g, 1/\alpha)$ noting that, by (1.5.43),

$$(2.7) \qquad E(\tau) = \frac{g}{\alpha}, \quad E(\tau^2) = \frac{g(g+1)}{\alpha^2},$$

$$E\left(\frac{1}{\tau}\right) = \frac{\alpha}{g-1}, \ g > 1, \quad E\left(\frac{1}{\tau^2}\right) = \frac{\alpha^2}{(g-1)(g-2)}, \ g > 2.$$

Writing $y = \Sigma x_i^2$, we see that the posterior density of τ given the x_i's is

$$C(y)\tau^{r+g-1}e^{-\tau(\alpha+y)},$$

which is $\Gamma[r+g, 1/(\alpha+y)]$. If the loss is squared error, the Bayes estimator of $2\sigma^2 = 1/\tau$ is the posterior expectation of $1/\tau$, which by (2.7) is $(\alpha+y)/(r+g-1)$, $r+g > 1$. The Bayes estimator of $\sigma^2 = 1/2\tau$ is therefore

$$(2.8) \qquad \frac{\alpha+Y}{n+2g-2}, \quad n+2g > 2.$$

In the present situation, we might instead prefer to work with the scale invariant loss function

$$(2.9) \qquad \frac{(d-\sigma^2)^2}{\sigma^4},$$

which leads to the Bayes estimator (Problem 2.6)

$$(2.10) \qquad \frac{E(1/\sigma^2)}{E(1/\sigma^4)} = \frac{E(\tau)}{2E(\tau^2)},$$

and hence by (2.7) after some simplification to

$$(2.11) \qquad \frac{\alpha+Y}{n+2g+2}.$$

Since the Fisher information for σ is proportional to $1/\sigma^2$ (Table 2.5.1), the Jeffreys prior density in the present case is proportional to the improper density $1/\sigma$, which induces for τ the density $(1/\tau)\,d\tau$. This corresponds to the limiting case $\alpha = 0$, $g = 0$, and hence by (2.8) and (2.11) to the Bayes estimators $Y/(n-2)$ and $Y/(n+2)$ for squared error and loss function (2.9), respectively. The first of these has uniformly larger risk than the second which is MRE. ||

We next consider two examples involving more than one parameter.

Example 2.6 Normal variance, unknown mean. Suppose that we let $X_1, \ldots, X_n$ be iid as $N(\theta, \sigma^2)$ and consider the Bayes estimation of θ and σ^2 when the prior assigns to $\tau = 1/2\sigma^2$ the distribution $\Gamma(g, 1/\alpha)$ as in Example 2.5 and takes θ to be independent of τ with (for the sake of simplicity) the uniform improper prior $d\theta$ corresponding to $b = \infty$ in Example 2.2. Then, the joint posterior density of (θ, τ) is proportional to

$$\tau^{r+g-1} e^{-\tau[\alpha+z+n(\bar{x}-\theta)^2]} \tag{2.12}$$

where $z = \Sigma(x_i - \bar{x})^2$ and $r = n/2$. By integrating out θ, it is seen that the posterior distribution of τ is $\Gamma[r + g - 1/2, 1/(\alpha + z)]$ (Problem 1.12). In particular, for $\alpha = g = 0$, the Bayes estimator of $\sigma^2 = 1/2\tau$ is $Z/(n-3)$ and $Z/(n+1)$ for squared error and loss function (2.9), respectively. To see that the Bayes estimator of θ is $\bar{X}$ regardless of the values of α and g, it is enough to notice that the posterior density of θ is symmetric about $\bar{X}$ (Problem 2.9; see also Problem 2.10). ||

A problem for which the theories of Chapters 2 and 3 do not lead to a satisfactory solution is that of components of variance. The following example treats the simplest case from the present point of view.

Example 2.7 Random effects one-way layout. In the model (3.5.1), suppose for the sake of simplicity that μ and Z_{11} have been eliminated either by invariance or by assigning to μ the uniform prior on $(-\infty, \infty)$. In either case, this restricts the problem to the remaining Z's with joint density proportional to

$$\frac{1}{\sigma^{s(n-1)}(\sigma^2 + n\sigma_A^2)^{(s-1)/2}} \exp\left[-\frac{1}{2(\sigma^2 + n\sigma_A^2)} \sum_{i=2}^{s} z_{i1}^2 - \frac{1}{2\sigma^2} \sum_{i=1}^{s} \sum_{j=2}^{n} z_{ij}^2\right]. \tag{2.13}$$

The most natural noninformative prior postulates σ and σ_A to be independent with improper densities $1/\sigma$ and $1/\sigma_A$, respectively. Unfortunately, however, in this case, the posterior distribution of (σ, σ_A) continues to be improper, so that the calculation of a posterior expectation is meaningless (Problem 2.12).

Instead, let us consider the Jeffreys prior Λ which has the improper density $(1/\sigma)(1/\tau)$ but with $\tau^2 = \sigma^2 + n\sigma_A^2$ so that the density is zero for $\tau < \sigma$. (For a discussion of the appropriateness of this and related priors see Hill, Stone and Springer 1965, Tiao and Tan 1965, Box and Tiao 1973, Hobert 1993, and Hobert and Casella 1996.) The posterior distribution is then proper (Problem 2.11). The resulting Bayes estimator δ_Λ of σ_A^2 is obtained by Klotz, Milton, and Zacks (1969), who compare it with the more traditional estimators discussed in Example 5.5. Since the risk of δ_Λ is quite unsatisfactory, Portnoy (1971) replaces squared error by the scale invariant loss function $(d - \sigma_A^2)^2/(\sigma^2 + n\sigma_A^2)^2$, and shows the resulting estimator to be

$$\delta'_\Lambda = \frac{1}{2n}\left[\frac{S_A^2}{a} - \frac{S^2}{c-a-1} + \frac{c-1}{ca(c-a-1)} \cdot \frac{S_A^2 + S^2}{F(R)}\right] \tag{2.14}$$

where $c = \frac{1}{2}(sn+1)$, $a = \frac{1}{2}(s+3)$, $R = S^2/(S_A^2 + S^2)$, and

$$F(R) = \int_0^1 \frac{v^a}{[R + v(1-R)]^{c+1}}\,dv.$$

Portnoy's risk calculations suggest that δ'_A is a satisfactory estimator of σ_A^2 for his loss function or equivalently for squared error loss. The estimation of σ^2 is analogous. ||

Let us next examine the connection between Bayes estimation, sufficiency, and the likelihood function. Recall that if $(X_1, X_2, \ldots, X_n)$ has density $f(x_1, \ldots, x_n|\theta)$, the likelihood function is defined by $L(\theta|\mathbf{x}) = L(\theta|x_1, \ldots, x_n) = f(x_1, \ldots, x_n|\theta)$. If we observe $T = \mathbf{t}$, where T is sufficient for θ, then

$$f(x_1, \ldots, x_n|\theta) = L(\theta|\mathbf{x}) = g(\mathbf{t}|\theta)h(\mathbf{x}),$$

where the function $h(\cdot)$ does not depend on θ. For any prior distribution $\pi(\theta)$, the posterior distribution is then

$$\pi(\theta|\mathbf{x}) = \frac{f(x_1, \ldots, x_n|\theta)\pi(\theta)}{\int f(x_1, \ldots, x_n|\theta')\pi(\theta')\,d\theta'}$$

$$(2.15) \qquad = \frac{L(\theta|\mathbf{x})\pi(\theta)}{\int L(\theta'|\mathbf{x})\pi(\theta')\,d\theta'} = \frac{g(\mathbf{t}|\theta)\pi(\theta)}{\int g(\mathbf{t}|\theta')\pi(\theta')\,d\theta'}$$

so $\pi(\theta|\mathbf{x}) = \pi(\theta|\mathbf{t})$, that is, $\pi(\theta|\mathbf{x})$ depends on $\mathbf{x}$ only through $\mathbf{t}$, and the posterior distribution of θ is the same whether we compute it on the basis of $\mathbf{x}$ or of $\mathbf{t}$. As an illustration, in Example 2.2, rather than starting with (2.1), we could use the fact that the sufficient statistic is $\bar{X} \sim N(\theta, \sigma^2/n)$ and, starting from

$$f(\bar{x}|\theta) \propto e^{-\frac{n}{2\sigma^2}(\bar{x}-\theta)^2} e^{-\frac{1}{2b^2}(\theta-\mu)^2},$$

arrive at the same posterior distribution for θ as before. Thus, Bayesian measures that are computed from posterior distributions are functions of the data only through the likelihood function and, hence, are functions of a minimal sufficient statistic.

Bayes estimators were defined in (1.1) with respect to a proper distribution Λ. It is useful to extend this definition to the case that Λ is a measure satisfying

$$(2.16) \qquad \int d\Lambda(\theta) = \infty,$$

a so-called improper prior. It may then still be the case that (1.3) is finite for each x, so the Bayes estimator can formally be defined.

Example 2.8 Improper prior Bayes. For the situation of Example 1.5, where $X \sim b(p, n)$, the Bayes estimator under a beta(a, b) prior is given by (1.12). For $a = b = 0$, this estimator is x/n, the sample mean, but the prior density, $\pi(p)$, is proportional to $\pi(p) \propto p^{-1}(1-p)^{-1}$, and hence is improper. The posterior distribution in this case is

$$(2.17) \qquad \frac{\binom{n}{x}p^{x-1}(1-p)^{n-x-1}}{\int_0^1 \binom{n}{x}p^{x-1}(1-p)^{n-x-1}dp} = \frac{\Gamma(n)}{\Gamma(x)\Gamma(n-x)}p^{x-1}(1-p)^{n-x-1}$$

which is a proper posterior distribution if $1 \le x \le n-1$ with x/n the posterior mean. When $x = 0$ or $x = n$, the posterior density (2.17) is no longer proper. However, for any estimator $\delta(x)$ that satisfies $\delta(0) = 0$ and $\delta(n) = 1$, the posterior expected loss (1.3) is finite and minimized at $\delta(x) = x/n$ (see Problem 2.16 and Example 2.4). Thus, even though the resulting posterior distribution is not proper for all values of x, $\delta(x) = x/n$ can be considered a Bayes estimator. ||

This example suggests the following definition.

Definition 2.9 An estimator $\delta^\pi(x)$ is a *generalized Bayes estimator* with respect to a measure $\pi(\theta)$ (even if it is not a proper probability distribution) if the posterior expected loss, $E\{L(\Theta, \delta(X))|X = x\}$, is minimized at $\delta = \delta^\pi$ for all x.

As we will see, generalized Bayes estimators play an important part in point estimation optimality, since they often may be optimal under both Bayesian and frequentist criteria.

There is one other useful variant of a Bayes estimator, a *limit of Bayes estimators*.

Definition 2.10 A nonrandomized[1] estimator $\delta(x)$ is a *limit of Bayes estimators* if there exists a sequence of proper priors π_ν and Bayes estimators δ^{π_ν} such that $\delta^{\pi_\nu}(x) \to \delta(x)$ a.e. [with respect to the density $f(x|\theta)$] as $\nu \to \infty$.

Example 2.11 Limit of Bayes estimators. In Example 2.8, it was seen that the binomial estimator X/n is Bayes with respect to an improper prior. We shall now show that it is also a limit of Bayes estimators. This follows since

$$\lim_{\substack{a \to 0 \\ b \to 0}} \frac{a+x}{a+b+n} = \frac{x}{n} \tag{2.18}$$

and the beta(a, b) prior is proper if $a > 0$, $b > 0$. ||

From a Bayesian view, estimators that are limits of Bayes estimators are somewhat more desirable than generalized Bayes estimators. This is because, by construction, a limit of Bayes estimators must be close to a proper Bayes estimator. In contrast, a generalized Bayes estimator may not be close to any proper Bayes estimator (see Problem 2.15).

3 Single-Prior Bayes

As discussed at the end of Section 1, the prior distribution is typically selected from a flexible family of prior densities indexed by one or more parameters. Instead of denoting the prior by Λ, as was done in Section 1, we shall now denote its density by $\pi(\theta|\gamma)$, where the parameter γ can be real- or vector-valued. (Hence, we are implicitly assuming that the prior π is absolutely continuous with respect to a dominating measure $\mu(\theta)$, which, unless specified, is taken to be Lebesgue measure.)

[1] For randomized estimators the convergence can only be in distribution. See Ferguson 1967 (Section 1.8) or Brown 1986a (Appendix).

We can then write a Bayes model in a general form as

$$X|\theta \sim f(x|\theta), \quad \Theta|\gamma \sim \pi(\theta|\gamma). \tag{3.1}$$

Thus, conditionally on θ, X has sampling density $f(x|\theta)$, and conditionally on γ, Θ has prior density $\pi(\theta|\gamma)$. From this model, we calculate the posterior distribution, $\pi(\theta|x, \gamma)$, from which all Bayesian answers would come. The exact manner in which we deal with the parameter γ or, more generally, the prior distribution $\pi(\theta|\gamma)$ will lead us to different types of Bayes analyses. In this section we assume that the functional form of the prior, and the value of γ, is known so we have one completely specified prior. (To emphasize that point, we will sometimes write $\gamma = \gamma_0$.)

Given a loss function $L(\theta, d)$, we then look for the estimator that minimizes

$$\int L(\theta, d(x))\pi(\theta|x, \gamma_0)\, d\theta, \tag{3.2}$$

where $\pi(\theta|x, \gamma_0) = f(x|\theta)\pi(\theta|\gamma_0)/\int f(x|\theta)\pi(\theta|\gamma_0)\, d\theta$.

The calculation of single-prior Bayes estimators has already been illustrated in Section 2. Here is another example.

Example 3.1 Scale uniform. For estimation in the model

$$X_i|\theta \sim \mathcal{U}(0, \theta), \quad i = 1, \ldots, n, \qquad \frac{1}{\theta}|a, b \sim \text{Gamma}(a, b), \quad a, b \text{ known}, \tag{3.3}$$

sufficiency allows us to work only with the density of $Y = \max_i X_i$, which is given by $g(y|\theta) = ny^{n-1}/\theta^n, 0 < y < \theta$. We then calculate the single-prior Bayes estimator of θ under squared error loss. By (4.1.4), this is the posterior mean, given by

$$E(\Theta|y, a, b) = \frac{\int_y^\infty \theta \frac{1}{\theta^{n+a+1}} e^{-1/\theta b}\, d\theta}{\int_y^\infty \frac{1}{\theta^{n+a+1}} e^{-1/\theta b}\, d\theta}. \tag{3.4}$$

Although the ratio of integrals is not expressible in any simple form, calculation is not difficult. See Problem 3.1 for details. ||

In general, the Bayes estimator under squared error loss is given by

$$E(\Theta|x) = \frac{\int \theta f(x|\theta)\pi(\theta)\, d\theta}{\int f(x|\theta)\pi(\theta)\, d\theta} \tag{3.5}$$

where $X \sim f(x|\theta)$ is the observed random variable and $\Theta \sim \pi(\theta)$ is the parameter of interest. While there is a certain appeal about expression (3.5), it can be difficult to work with. It is therefore important to find conditions under which it can be simplified. Such simplification is useful for two somewhat related purposes.

(i) **Implementation**

If a Bayes solution is deemed appropriate, and we want to implement it, we must be able to calculate (3.5). Thus, we need reasonably straightforward, and general, methods of evaluating these integrals.

(ii) **Performance**

By construction, a Bayes estimator minimizes the posterior expected loss and, hence, the Bayes risk. Often, however, we are interested in its performance, and perhaps optimality under other measures. For example, we might examine its mean squared error (or, more generally, its risk function) in looking for admissible or minimax estimators. We also might examine Bayesian measures using other priors, in an investigation of Bayesian robustness.

These latter considerations tend to lead us to look for either manageable expressions for or accurate approximations to the integrals in (3.5). On the other hand, the considerations in (i) are more numerical (or computational) in nature, leading us to algorithms that ease the computational burden. However, even this path can involve statistical considerations, and often gives us insight into the performance of our estimators.

A simplification of (3.5) is possible when dealing with independent prior distributions. If $X_i \sim f(x|\theta_i)$, $i = 1, \cdots, n$, are independent, and the prior is $\pi(\theta_1, \cdots, \theta_n) = \prod_i \pi(\theta_i)$, then the posterior mean of θ_i satisfies

$$E(\theta_i|x_1, \ldots, x_n) = E(\theta_i|x_i), \tag{3.6}$$

that is, the Bayes estimator of θ_i only depends on the data through x_i. Although the simplification provided by (3.6) may prove useful, at this level of generality it is impossible to go further.

However, for exponential families, evaluation of (3.5) is sometimes possible through alternate representations of Bayes estimators. Suppose the distribution of $\mathbf{X} = (X_1, \ldots, X_n)$ is given by the multiparameter exponential family (see (1.5.2)), that is,

$$p_{\eta}(\mathbf{x}) = \exp\left\{\sum_{i=1}^{s} \eta_i T_i(\mathbf{x}) - A(\eta)\right\} h(\mathbf{x}). \tag{3.7}$$

Then, we can express the Bayes estimator as a function of partial derivatives with respect to $\mathbf{x}$. The following theorem presents a general formula for the needed posterior expectation.

Theorem 3.2 *If* $\mathbf{X}$ *has density (3.7), and* η *has prior density* $\pi(\eta)$*, then for* $j = 1, \ldots, n$,

$$E\left(\sum_{i=1}^{s} \eta_i \frac{\partial T_i(\mathbf{x})}{\partial x_j} | \mathbf{x}\right) = \frac{\partial}{\partial x_j} \log m(\mathbf{x}) - \frac{\partial}{\partial x_j} \log h(\mathbf{x}), \tag{3.8}$$

where $m(\mathbf{x}) = \int p_{\eta}(\mathbf{x})\pi(\eta)\, d\eta$ *is the marginal distribution of* X*. Alternatively, the posterior expectation can be expressed in matrix form as*

$$E(\mathcal{T}\eta) = \nabla \log m(\mathbf{x}) - \nabla \log h(\mathbf{x}), \tag{3.9}$$

where $\mathcal{T} = \{\partial T_i/\partial x_j\}$.

Proof. Noting that $\partial \exp\{\sum \eta_i T_i\}/\partial x_j = \sum_i \eta_i(\partial T_i/\partial x_j)\exp\{\sum \eta_i T_i\}$, we can write

$$E\left(\sum \eta_i \frac{\partial T_i(\mathbf{x})}{\partial x_j} | \mathbf{x}\right) = \frac{1}{m(\mathbf{x})} \int \sum_i \left[\eta_i \frac{\partial T_i}{\partial x_j}\right] e^{\Sigma \eta_i T_i - A(\eta)} h(\mathbf{x})\pi(\eta)\, d\eta$$

$$
\begin{aligned}
&= \frac{1}{m(\mathbf{x})} \int \left[\frac{\partial}{\partial x_j} e^{\Sigma_i \eta_i T_i} \right] e^{-A(\boldsymbol{\eta})} h(\mathbf{x}) \pi(\eta)\, d\eta \\
&= \frac{1}{m(\mathbf{x})} \int \left[\left(\frac{\partial}{\partial x_j} e^{\Sigma_i \eta_i T_i} h(\mathbf{x}) \right) \right. \\
&\qquad \left. - \left(\frac{\partial}{\partial x_j} h(\mathbf{x}) \right) e^{\Sigma_i \eta_i T_i} \right] e^{-A(\boldsymbol{\eta})} \pi(\eta)\, d\eta \qquad (3.10) \\
&= \frac{1}{m(\mathbf{x})} \frac{\partial}{\partial x_j} \int \left(e^{\Sigma \eta_i T_i - A(\boldsymbol{\eta})} h(\mathbf{x}) \right) \pi(\eta)\, d\eta \\
&\qquad - \frac{\frac{\partial}{\partial x_j} h(\mathbf{x})}{h(\mathbf{x})} \frac{1}{m(\mathbf{x})} \int e^{\Sigma \eta_i T_i - A(\boldsymbol{\eta})} h(\mathbf{x}) \pi(\eta)\, d\eta \\
&= \frac{\frac{\partial}{\partial x_j} m(\mathbf{x})}{m(\mathbf{x})} - \frac{\frac{\partial}{\partial x_j} h(\mathbf{x})}{h(\mathbf{x})}
\end{aligned}
$$

where, in the third equality, we have used the fact that

$$\left[\frac{\partial T_i}{\partial x_j} \right] e^{\Sigma_i \eta_i T_i} h(\mathbf{x}) = \frac{\partial}{\partial x_j} \left(e^{\Sigma_i \eta_i T_i} h(\mathbf{x}) \right) - e^{\Sigma_i \eta_i T_i} \left[\frac{\partial h(\mathbf{x})}{\partial x_j} \right].$$

In the fourth equality, we have interchanged the order of integration and differentiation (justified by Theorem 1.5.8), and used the definition of $m(\mathbf{x})$. Finally, using logarithms, $E\left(\sum \eta_i \frac{\partial T_i(\mathbf{x})}{\partial x_j} | \mathbf{x} \right)$ can be written as (3.8). □

Although it may appear that this theorem merely shifts calculation from one integral [the posterior of (3.5)] to another [the marginal $m(\mathbf{x})$ of (3.8)], this shift brings advantages which will be seen throughout the remainder of this section (and beyond). These advantages stem from the facts that the calculation of the derivatives of $\log m(\mathbf{x})$ is often feasible and that, with the estimator expressed as (3.8), risk calculations may be simplified. Theorem 3.2 simplifies further when $T_i = X_i$.

Corollary 3.3 *If* $\mathbf{X} = (X_1, \ldots, X_p)$ *has the density*

$$p_{\boldsymbol{\eta}}(\mathbf{x}) = e^{\Sigma_{i=1}^p \eta_i x_i - A(\boldsymbol{\eta})} h(\mathbf{x}) \tag{3.11}$$

and η *has prior density* $\pi(\eta)$, *the Bayes estimator of* η *under the loss* $L(\eta, \delta) = \Sigma(\eta_i - \delta_i)^2$ *is given by*

$$E(\eta_i | \mathbf{x}) = \frac{\partial}{\partial x_i} \log m(\mathbf{x}) - \frac{\partial}{\partial x_i} \log h(\mathbf{x}). \tag{3.12}$$

Proof. Problem 3.3. □

Example 3.4 Multiple normal model. For

$$
\begin{aligned}
X_i | \theta_i &\sim N(\theta_i, \sigma^2), \quad i = 1, \ldots, p, \text{ independent}, \\
\Theta_i &\sim N(\mu, \tau^2), \quad i = 1, \ldots, p, \text{ independent},
\end{aligned}
$$

where σ^2, τ^2, and μ are known, $\eta_i = \theta_i / \sigma^2$ and the Bayes estimator of θ_i is

$$E(\Theta_i | x) = \sigma^2 E(\eta_i | x) = \sigma^2 \left[\frac{\partial}{\partial x_i} \log m(\mathbf{x}) - \frac{\partial}{\partial x_i} \log h(\mathbf{x}) \right]$$

$$= \frac{\tau^2}{\sigma^2+\tau^2} x_i + \frac{\sigma^2}{\sigma^2+\tau^2}\mu,$$

since

$$\begin{aligned}\frac{\partial}{\partial x_i}\log m(\mathbf{x}) &= \frac{\partial}{\partial x_i}\log\left(e^{\frac{1}{2(\sigma^2+\tau^2)}\sum_i (x_i-\mu)^2}\right)\\ &= \frac{-(x_i-\mu)}{\sigma^2+\tau^2}\end{aligned}$$

and

$$\frac{\partial}{\partial x_i}\log h(\mathbf{x}) = \frac{\partial}{\partial x_i}\log\left(e^{-\frac{1}{2}\sum x_i^2/\sigma^2}\right) = -\frac{x_i}{\sigma^2}.$$

||

An application of the representation (3.12) is to the comparison of the risk of the Bayes estimator with the risk of the best unbiased estimator.

Theorem 3.5 *Under the assumptions of Corollary 3.3, the risk of the Bayes estimator (3.12), under the sum of squared error loss, is*

$$\begin{aligned}R[\eta, E(\eta|\mathbf{X})] &= R[\eta, -\nabla\log h(\mathbf{X})]\\ &\quad + \sum_{i=1}^{p} E\left\{2\frac{\partial^2}{\partial X_i^2}\log m(\mathbf{X}) + \left(\frac{\partial}{\partial X_i}\log m(\mathbf{X})\right)^2\right\}.\end{aligned} \tag{3.13}$$

Proof. By an application of Stein's identity (Lemma 1.5.15; see Problem 3.4), it is straightforward to establish that for the situation of Corollary 3.3.

$$E_\eta\left[-\frac{\partial}{\partial X_i}\log h(\mathbf{X})\right] = \int\left[-\frac{\partial}{\partial x_i}\log h(\mathbf{x})\right] p_\eta(\mathbf{x})d\mathbf{x} = \eta_i.$$

Hence, if we write $\nabla\log h(\mathbf{x}) = \{\partial/\partial x_i \log h(\mathbf{x})\}$,

$$-E_\eta\nabla\log h(\mathbf{X}) = \boldsymbol{\eta}. \tag{3.14}$$

Thus, $-\nabla\log h(\mathbf{X})$ is an unbiased estimator of η with risk

$$\begin{aligned}R[\eta, -\nabla\log h(\mathbf{X})] &= E_\eta\sum_{i=1}^{p}\left[\eta_i + \frac{\partial}{\partial X_i}\log h(\mathbf{X})\right]^2\\ &= E_\eta|\eta + \nabla\log h(\mathbf{X})|^2,\end{aligned} \tag{3.15}$$

which can also be further evaluated using Stein's identity (see Problem 3.4). Returning to (3.12), the risk of the Bayes estimator is given by

$$\begin{aligned}R[\eta, E(\eta|\mathbf{X})] &= \sum_{i=1}^{p}[\eta_i - E(\eta_i|\mathbf{X})]^2\\ &= \sum_{i=1}^{p}\left[\eta_i - \left(\frac{\partial}{\partial X_i}\log m(\mathbf{X}) - \frac{\partial}{\partial X_i}\log h(\mathbf{X})\right)\right]^2\\ &= R[\eta, -\nabla\log h(\mathbf{X})]\\ &\quad -2\sum_{i=1}^{p}E\left[(\eta_i + \frac{\partial}{\partial X_i}\log h(\mathbf{X}))\frac{\partial}{\partial X_i}\log m(\mathbf{X})\right]\end{aligned}$$

$$+\sum_{i=1}^{p} E\left[\frac{\partial}{\partial X_i} \log m(\mathbf{X})\right]^2. \tag{3.16}$$

An application of Stein's identity to the middle term leads to

$$E_\eta\left[\left(\eta_i + \frac{\partial}{\partial X_i}\log h(\mathbf{X})\right)\frac{\partial}{\partial X_i}\log m(\mathbf{X})\right] = -E_\eta\left[\frac{\partial^2}{\partial X_i^2}\log m(\mathbf{X})\right],$$

which establishes (3.13). □

From (3.13), we see that if the second term is negative, then the Bayes estimator of η will have smaller risk than the unbiased estimator $-\nabla \log h(\mathbf{X})$, (which is best unbiased if the family is complete). We will exploit this representation (3.13) in Chapter 5, but now just give a simple example.

Example 3.6 Continuation of Example 3.4. To evaluate the risk of the Bayes estimator, we also calculate

$$\frac{\partial^2}{\partial x_i^2}\log m(x) = -\frac{1}{\sigma^2+\tau^2},$$

and hence, from (3.13),

$$\begin{aligned} R[\eta, E(\eta|\mathbf{X})] &= R[\eta, -\nabla \log h(\mathbf{X})] \\ &\quad - \frac{2p}{\sigma^2+\tau^2} + \sum_i E_\eta\left(\frac{X_i-\mu}{\sigma^2+\tau^2}\right)^2. \end{aligned} \tag{3.17}$$

The best unbiased estimator of $\eta_i = \theta_i/\sigma^2$ is

$$-\frac{\partial}{\partial X_i}\log h(\mathbf{X}) = \frac{X_i}{\sigma^2}$$

with risk $R(\eta, -\nabla \log h(\mathbf{X})) = p/\sigma^2$. If $\eta_i = \mu$ for each i, then the Bayes estimator has smaller risk, whereas the Bayes estimator has infinite risk as $|\eta_i - \mu| \to \infty$ for any i (Problem 3.6). ||

We close this section by noting that in exponential families there is a general expression for the conjugate prior distribution and that use of this conjugate prior results in a simple expression for the posterior mean. For the density

$$p_\eta(x) = e^{\eta x - A(\eta)}h(x), \quad -\infty < x < \infty, \tag{3.18}$$

the conjugate prior family is

$$\pi(\eta|k,\mu) = c(k,\mu)e^{k\eta\mu - kA(\eta)}, \tag{3.19}$$

where μ can be thought of as a prior mean and k is proportional to a prior variance (see Problem 3.9).

If $X_1, \ldots, X_n$ is a sample from $p_\eta(x)$ of (3.18), the posterior distribution resulting from (3.19) is

$$\begin{aligned} \pi(\eta|\mathbf{x},k,\mu) &\propto \left[e^{n\eta\bar{x} - nA(\eta)}\right]\left[e^{k\eta\mu - kA(\eta)}\right] \\ &= e^{\eta(n\bar{x}+k\mu) - (n+k)A(\eta)} \end{aligned} \tag{3.20}$$

which is in the same form as (3.19) with $\mu' = (n\bar{x} + k\mu)/(n+k)$ and $k' = n + k$. Thus, using Problem 3.9,

$$E(A'(\eta)|\mathbf{x}, k, \mu) = \frac{n\bar{x} + k\mu}{n+k}. \tag{3.21}$$

As $EX|\eta = A'(\eta)$, we see that the posterior mean is a convex combination of the sample and prior means.

Example 3.7 Conjugate gamma. Let $X_1, \ldots, X_n$ be iid as Gamma(a, b), where a is known. This is in the exponential family form with $\eta = -1/b$ and $A(\eta) = -a\log(-\eta)$. If we use a conjugate prior distribution (3.19) for b for which

$$\begin{aligned} E(A'(\eta)|\mathbf{x}) &= E\left(-\frac{a}{\eta}|\bar{x}\right) \\ &= \frac{n\bar{x} + k\mu}{n+k}. \end{aligned}$$

The resulting Bayes estimator under squared error loss is

$$E(b|\mathbf{x}) = \frac{1}{a}\left[\frac{n\bar{x} + k\mu}{n+k}\right]. \tag{3.22}$$

This is the Bayes estimator based on an inverted gamma prior for b (see Problem 3.10). ||

Using the conjugate prior (3.19) will not generally lead to simplifications in (3.9) and is, therefore, not helpful in obtaining expressions for estimators of the natural parameter: However, there is often more interest in estimating the mean parameter rather than the natural parameter.

4 Equivariant Bayes

Definition 3.2.4 specified what is meant by an estimation problem being invariant under a transformation g of the sample space and the induced transformations $\bar{g}$ and g^* of the parameter and decision spaces, respectively. In such a situation, when considering Bayes estimation, it is natural to select a prior distribution which is also invariant.

Recall that a group family is a family of distributions which is invariant under a group G of transformations for which $\bar{G}$ is transitive over the parameter space. We shall say that a prior distribution Λ for θ is *invariant* with respect to $\bar{G}$ if the distribution of $\bar{g}\theta$ is also Λ for all $\bar{g} \in \bar{G}$; that is, if for all $\bar{g} \in \bar{G}$ and all measurable B

$$P_\Lambda(\bar{g}\theta \in B) = P_\Lambda(\theta \in B) \tag{4.1}$$

or, equivalently,

$$\Lambda(\bar{g}^{-1}B) = \Lambda(B). \tag{4.2}$$

Suppose now that such a Λ exists and that the Bayes solution δ_Λ with respect to it is unique. By (4.1), any δ then satisfies

$$\int R(\theta, \delta)\, d\Lambda(\theta) = \int R(\bar{g}\theta, \delta)\, d\Lambda(\theta). \tag{4.3}$$

Now

$$R(\bar{g}\theta, \delta) = E_{\bar{g}\theta}\{L[\bar{g}\theta, \delta(X)]\} = E_\theta\{L[\bar{g}\theta, \delta(gX)]\}$$
$$(4.4) \qquad = E_\theta\{L[\theta, g^{*-1}\delta(gX)]\}.$$

Here, the second equality follows from (3.2.4) (invariance of the model) and the third from (3.2.14) (invariance of the loss). On substituting this last expression into the right side of (4.3), we see that if $\delta_\Lambda(x)$ minimizes (4.3), so does the estimator $g^{*-1}\delta_\Lambda(gx)$. Hence, if the Bayes estimator is unique, the two must coincide. By (3.2.17), this appears to prove δ_Λ to be equivariant. However, at this point, a technical difficulty arises. Uniqueness can be asserted only up to null sets, that is, sets N with $P_\theta(N) = 0$ for all θ. Moreover, the set N may depend on g. An estimator δ satisfying

$$(4.5) \qquad \delta(x) = g^{*-1}\delta(gx) \quad \text{for all } x \notin N_g$$

where $P_\theta(N_g) = 0$ for all θ is said to be *almost equivariant*. We have therefore proved the following result.

Theorem 4.1 *Suppose that an estimation problem is invariant under a group and that there exists a distribution Λ over Ω such that (4.2) holds for all (measurable) subsets B of Ω and all $g \in G$. Then, if the Bayes estimator δ_Λ is unique, it is almost equivariant.*

Example 4.2 Equivariant binomial. Suppose we are interested in estimating p under squared error loss, where $X \sim$ binomial(n, p). A common group of transformations which leaves the problem invariant is

$$gX = n - X,$$
$$\bar{g}p = 1 - p.$$

For a prior Λ to satisfy (4.1), we must have

$$(4.6) \qquad P_\Lambda(\bar{g}p \le t) = P_\Lambda(p \le t) \quad \text{for all } t.$$

If Λ has density $\gamma(p)$, then (4.1) implies

$$(4.7) \qquad \int_0^t \gamma(p)\,dp = \int_0^t \gamma(1-p)\,dp \quad \text{for all } t,$$

which, upon differentiating, requires $\gamma(t) = \gamma(1-t)$ for all t and, hence, that $\gamma(t)$ must be symmetric about $t = 1/2$. It then follows that, for example, a Bayes rule under a symmetric beta prior is equivariant. See Problem 4.1.

‖

The existence of a proper invariant prior distribution is rather special. More often, the invariant measure for θ will be improper (if it exists at all), and the situation is then more complicated. In particular, (i) the integral (4.3) may not be finite, and the argument leading to Theorem 4.1 is thus no longer valid and (ii) it becomes necessary to distinguish between left- and right-invariant measures Λ. These complications require a level of group-theoretic sophistication that we do not assume. However, for the case of location-scale, we can develop the theory in sufficient detail. [For a more comprehensive treatment of invariant Haar measures, see Berger 1985 (Section 6.6), Robert 1994a (Section 7.4), or Schervisch 1995 (Section 6.2). A general development of the theory of group invariance, and its application to statistics, is given by Eaton (1989) and Wijsman (1990).]

To discuss invariant prior distributions, or more generally invariant measures, over the parameter space, we begin by considering invariant measures over groups. (See Section 1.4 for some of the basics.) Let G be a group and $\mathcal{L}$ be a σ-field of measurable subsets of G, and for a set B in $\mathcal{L}$, let

$$Bh = \{gh : g \in B\}$$

and

$$gB = \{gh : h \in B\}.$$

Then, a measure Λ over $(G, \mathcal{L})$ is right-invariant, a *right invariant Haar measure* if

$$\Lambda(Bh) = \Lambda(B) \quad \text{for all } B \in \mathcal{L},\ h \in G \tag{4.8}$$

and a *left-invariant Haar measure* if

$$\Lambda(gB) = \Lambda(B) \quad \text{for all } B \in \mathcal{L},\ g \in G. \tag{4.9}$$

In our examples, measures satisfying (4.8) or (4.9) exist, have densities, and are unique up to multiplication by a positive constant. We will now look at some location-scale examples.

Example 4.3 Location group. For $\mathbf{x} = (x_1, \ldots, x_n)$ in a Euclidean sample space, consider the transformations

$$g\mathbf{x} = (x_1 + g, \ldots, x_n + g), \qquad -\infty < g < \infty, \tag{4.10}$$

with the composition (group operation)

$$g \circ h = g + h, \tag{4.11}$$

which was already discussed in Sections 1 and 2. Here, G is the set of real numbers g, and for $\mathcal{L}$, we can take the Borel sets. The sets Bh and gB are

$$Bh = \{g + h : g \in B\} \qquad \text{and} \qquad gB = \{g + h : h \in B\}$$

and satisfy $Bg = gB$ since

$$g \circ h = h \circ g. \tag{4.12}$$

When (4.12) holds, the group operation is said to be commutative; groups with this property are called *Abelian*. For an Abelian group, if a measure is right invariant, it is also left invariant and vice versa, and will then be called *invariant*. In the present case, Lebesgue measure is invariant since it assigns the same measure to a set B on the line as to the set obtained by translating B by any fixed account g or h to the right or left. (Abelian groups are a special case of *unimodular* groups, the type of group for which the left- and right-invariant measures agree. See Wijsman (1990, Chapter 7) for details. ‖

There is a difference between transformations acting on parameters or on elements of a group. In the first case, we know what we mean by $\bar{g}\theta$, but $\theta\bar{g}$ makes no sense. On the other hand, for group elements, multiplication is possible both on the left and on the right.

For this reason, a prior distribution Λ (a measure over the parameter space) satisfying (4.1), which only requires left-invariance, is said to be invariant. However, for measures over a group one must distinguish between left- and right-invariance and call such measures invariant only if they are both left and right invariant.

Example 4.4 Scale group. For $\mathbf{x} = (x_1, \ldots, x_n)$ consider the transformations

$$g\mathbf{x} = (gx_1, \cdots, gx_n), \qquad 0 < g < \infty,$$

that is, multiplication of each coordinate by the same positive number g, with the composition

$$g \circ h = g \times h.$$

The sets Bh and gB are obtained by multiplying each element of B by h on the right and g on the left, respectively. Since $gh = hg$, the group is Abelian and the concepts of left- and right-invariance coincide. An invariant measure is given by the density

$$\frac{1}{g}dg. \tag{4.13}$$

To see this, note that

$$\Lambda(B) = \int_B \frac{1}{g}dg = \int_{Bh} \frac{h}{g'} \cdot \frac{dg}{dg'}dg' = \int_{Bh} \frac{1}{g'}dg' = \Lambda(Bh),$$

where the first equality follows by making the change of variables $g' = gh$. ||

Example 4.5 Location-scale group. As a last and somewhat more complicated example, consider the group of transformations

$$g\mathbf{x} = (ax_1 + b, \ldots, ax_n + b), \quad 0 < a < \infty, \quad -\infty < b < \infty.$$

If $g = (a, b)$ and $h = (c, d)$, we have

$$h\mathbf{x} = c\mathbf{x} + d$$

and

$$gh\mathbf{x} = a(c\mathbf{x} + d) + b = ac\mathbf{x} + (ad + b).$$

So, the composition rule is

$$(a, b) \circ (c, d) = (ac, ad + b). \tag{4.14}$$

Since

$$(c, d) \circ (a, b) = (ac, cb + d),$$

it is seen that the group operation is not commutative, and we shall therefore have to distinguish between left and right Haar measures. We shall now show that these are given, respectively, by the densities

$$\frac{1}{a^2}dadb \quad \text{and} \quad \frac{1}{a}dadb. \tag{4.15}$$

To show left-invariance of the first of these, note that the transformation $g(h) = g \circ h$ takes

$$h = (c, d) \quad \text{into} \quad (c' = ac, d' = ad + b). \tag{4.16}$$

If Λ has density $dadb/a^2$, we have

$$\Lambda(B) = \Lambda\{(ac, ad + b) : (c, d) \in B\} \tag{4.17}$$

and, hence,

$$\Lambda(B) = \int_B \frac{1}{c^2} \mathrm{d}c\mathrm{d}d = \int_{gB} \frac{a^2}{(c)^2} \frac{\partial(c, d)}{\partial(c', d')} \mathrm{d}c'\mathrm{d}d', \tag{4.18}$$

where

$$\frac{\partial(c', d')}{\partial(c, d)} = \begin{vmatrix} a & 0 \\ 0 & a \end{vmatrix} = a^2$$

is the Jacobian of the transformation (4.16). The right side of (4.18) therefore reduces to

$$\int_{gB} \frac{1}{(c')^2} \mathrm{d}c'\mathrm{d}d' = \Lambda(gB)$$

and thus proves (4.9).

To prove the right-invariance of the density $dadb/a$, consider the transformation $h(g) = g \circ h$ taking

$$g = (a, b) \quad \text{into} \quad (a' = ac, b' = ad + b). \tag{4.19}$$

We then have

$$\Lambda(B) = \int_B \frac{1}{a} \mathrm{d}a\mathrm{d}b = \int_{Bh} \frac{c}{a'} \frac{\partial(a, b)}{\partial(a', b')} \mathrm{d}a'\mathrm{d}b'. \tag{4.20}$$

The Jacobian of the transformation (4.19) is

$$\frac{\partial(a', b')}{\partial(a, b)} = \begin{vmatrix} c & d \\ 0 & 1 \end{vmatrix} = c,$$

which shows that the right side of (4.20) is equal to $\Lambda(Bh)$. ‖

We introduced invariant measures over groups as a tool for defining measures over the parameter space Ω that, in some sense, share these invariance properties. For this purpose, consider a measure Λ over a transitive group $\bar{G}$ that leaves Ω invariant. Then, Λ induces a measure Λ' by the relation

$$\Lambda'(\omega) = \Lambda\{\bar{g} \in \bar{G} : \bar{g}\theta_0 \in \omega\}, \tag{4.21}$$

where ω is any subset of Ω, and θ_0 any given point of Ω. A disadvantage of this definition is the fact that the resulting measure Λ' will typically depend on θ_0, so that it is not uniquely defined by this construction. However, this difficulty disappears when Λ is right invariant.

Lemma 4.6 *If $\bar{G}$ is transitive over Ω, and Λ is a right-invariant measure over $\bar{G}$, then Λ' defined by (4.21) is independent of θ_0.*

Proof. If θ_1 is any other point of Ω, there exists (by the assumption of transitivity) an element $\bar{h}$ of $\bar{G}$ such that $\theta_1 = h\theta_0$. Let

$$\Lambda''(\omega) = \Lambda\{\bar{g} \in \bar{G} : \bar{g}\theta_1 \in \omega\}$$

and let B be the subset of $\bar{G}$ given by

$$B = \{\bar{g} : \bar{g}\theta_0 \in \omega\}.$$

Then,

$$\{\bar{g} : \bar{g}\bar{h} \in B\} = \{\bar{g}\bar{h}^{-1} : \bar{g} \in B\} = B\bar{h}^{-1}$$

and

$$\begin{aligned}\Lambda''(\omega) &= \Lambda\{\bar{g} : \bar{g}\bar{h}\theta_0 \in \omega\} = \Lambda\{\bar{g}\bar{h}^{-1} : \bar{g}\theta_0 \in \omega\} \\ &= \Lambda(B\bar{h}^{-1}) = \Lambda(B) = \Lambda'(\omega),\end{aligned}$$

where the next to last equation follows from the fact that Λ is right invariant. □

Example 4.7 Continuation of Example 4.3. The group G of Example 4.3 given by (4.10) and (4.11) and (1.2) of Section 1 induces on $\Omega = \{\eta : -\infty < \eta < \infty\}$ the transformation

$$\bar{g}\eta = \eta + \bar{g}$$

and, as we saw in Example 4.3, Lebesgue measure Λ is both right and left invariant over $\bar{G}$. For any point η_0 and any subset ω of Ω, we find

$$\Lambda'(\omega) = \Lambda\{\bar{g} \in \bar{G} : \eta_0 + \bar{g} \in \omega\} = \Lambda\{\bar{g} \in \bar{G} : \bar{g} \in \omega - \eta_0\},$$

where $\omega - \eta_0$ denotes the set ω translated by an amount η_0. Since Lebesgue measure of $\omega - \eta_0$ is the same as that of ω, it follows that Λ' is Lebesgue measure over Ω regardless of the choice of η_0.

Let us now determine the Bayes estimates for this prior measure for η when the loss function is squared error. By (1.6), the Bayes estimator of η is then (Problem 4.2)

$$\delta(x) = \frac{\int u f(x_1 - u, \ldots, x_n - u)\, du}{\int f(x_1 - u, \ldots, x_n - u)\, du}. \tag{4.22}$$

This is the Pitman estimator (3.1.28) of Chapter 3, which in Theorem 1.20 of that chapter was seen to be the MRE estimator of η. ||

Example 4.8 Continuation of Example 4.4. The scale group G given in Example 4.4 and by (3.2) of Section 3.3 induces on $\Omega = \{\tau : 0 < \tau < \infty\}$ the transformations

$$\bar{g} \circ \tau = g \times \tau \tag{4.23}$$

and, as we saw in Example 4.4, the measure Λ with density $\frac{1}{g} dg$ is both left and right invariant over $\bar{G}$. For any point τ_0 and any subset ω of Ω, we find

$$\Lambda'(\omega) = \Lambda\{\bar{g} \in \bar{G} : g\tau_0 \in \omega\} = \Lambda\{\bar{g} \in \bar{G} : g \in \omega/\tau_0\},$$

where ω/τ_0 denotes the set of values in ω each divided by τ_0. The change of variables $g' = \tau_0 g$ shows that

$$\int_{\omega/\tau_0} \frac{1}{g} dg = \int_\omega \frac{\tau_0}{g'} \frac{dg}{dg'} dg' = \int_\omega \frac{1}{g'} dg',$$

and, hence, that

$$\Lambda'(\omega) = \int_\omega \frac{d\tau}{\tau}.$$

Let us now determine the Bayes estimator of τ for this prior distribution when the loss function is

$$L(\tau, d) = \frac{(d-\tau)^2}{\tau^2}.$$

This turns out to be the Pitman estimator (3.19) of Section 3.3 with $r = 1$,

$$\delta(\mathbf{x}) = \frac{\int_0^\infty v^n f(vx_1, \ldots, vx_n)\,dv}{\int_0^\infty v^{n+1} f(vx_1, \ldots, vx_n)\,dv}, \tag{4.24}$$

which is also MRE (Problems 3.3.17 and 4.3). ||

Example 4.9 Continuation of Example 4.5. The location-scale family of distributions (3.23) of Section 3.3 remains invariant under the transformations

$$g\mathbf{x} : x_i' = a + bx_i, \qquad -\infty < a < \infty, \qquad 0 < b,$$

which induce in the parameter space

$$\Omega = \{(\eta, \tau) : -\infty < \eta < \infty, \quad 0 < \tau\}$$

the transformations

$$\eta' = a\eta + b, \quad \tau' = b\tau. \tag{4.25}$$

It was seen in Example 4.5 that the left and right Haar measures Λ_1 and Λ_2 over the group $G = \{g = (a, b) : -\infty < a < \infty, \quad 0 < b\}$ with group operation (4.14) are given by the densities

$$\frac{1}{a^2} da\,db \quad \text{and} \quad \frac{1}{a} da\,db, \tag{4.26}$$

respectively.

Let us now determine the corresponding measures over Ω induced by (4.19). If we describe the elements $\bar{g}$ in this group by (a, b), then for any measure Λ over $\bar{G}$ and any parameter point (η_0, τ_0), the induced measure Λ' over Ω is given by

$$\Lambda'(\omega) = \Lambda\{(a, b) : (a\eta_0 + b, b\tau_0) \in \omega\}. \tag{4.27}$$

Since a measure over the Borel sets is determined by its values over open intervals, it is enough to calculate $\Lambda'(\omega)$ for

$$\omega : \eta_1 < \eta < \eta_2, \qquad \tau_1 < \tau < \tau_2. \tag{4.28}$$

If, furthermore, we assume that Λ has a density λ

$$\begin{aligned}\Lambda'(\omega) &= \Lambda\{(a, b) : \eta_1 - a\eta_0 < b < \eta_2 - a\eta_0, \frac{\tau_1}{\tau_0} < a < \frac{\tau_2}{\tau_0}\} \\ &= \int_{\tau_1/\tau_0}^{\tau_2/\tau_0} \left[\int_{\eta_1 - a\eta_0}^{\eta_2 - a\eta_0} \lambda(a, b)\,db\right] da.\end{aligned}$$

In this integral, let us now change variables from (a, b) to

$$a' = a\tau_0, \qquad b' = b + a\eta_0.$$

The Jacobian of the transformation is

$$\left|\frac{\partial(a', b')}{\partial(a, b)}\right| = \begin{vmatrix} \tau_0 & \eta_0 \\ 0 & 1 \end{vmatrix} = \tau_0$$

and we, therefore, find

$$\Lambda'(\omega) = \int_{\tau_1}^{\tau_2}\int_{\eta_1}^{\eta_2} \lambda'(a, b) da db = \int_{\tau_1}^{\tau_2}\int_{\eta_1}^{\eta_2} \lambda(a, b)\frac{1}{\tau_0} da' db',$$

with $a = a'/\tau_0$ and $b = b' - a\eta_0 = b' - a'/\tau_0$. We can therefore take

$$\lambda'(a', b') = \frac{1}{\tau_0}\lambda\left[\frac{a'}{\tau_0}, \quad b' - \frac{a'}{\tau_0}\right].$$

Now consider the following two cases:

(i) For $\lambda(a, b) = \frac{1}{a}$,

$$\lambda'(a', b') = \frac{1}{\tau_0}\frac{\tau_0}{a'} = \frac{1}{a'}; \tag{4.29}$$

(ii) for $\lambda(a, b) = \frac{1}{a^2}$,

$$\lambda'(a', b') = \frac{1}{\tau_0}\frac{\tau_0^2}{a'^2} = \frac{1}{a'^2}. \tag{4.30}$$

The Bayes estimators of η and τ corresponding to (4.29) are (Problem 4.4)

$$\hat{\eta} = \frac{\int_{-\infty}^{\infty}\int_0^{\infty} \frac{u}{v^{n+3}} f\left(\frac{x_1-u}{v}, \ldots, \frac{x_n-u}{v}\right) dv\, du}{\int_{-\infty}^{\infty}\int_0^{\infty} \frac{1}{v^{n+3}} f\left(\frac{x_1-u}{v}, \ldots, \frac{x_n-u}{v}\right) dv\, du} \tag{4.31}$$

and

$$\hat{\tau} = \frac{\int_{-\infty}^{\infty}\int_0^{\infty} \frac{1}{v^{n+2}} f\left(\frac{x_1-u}{v}, \ldots, \frac{x_n-u}{v}\right) dv\, du}{\int_{-\infty}^{\infty}\int_0^{\infty} \frac{1}{v^{n+3}} f\left(\frac{x_1-u}{v}, \ldots, \frac{x_n-u}{v}\right) dv\, du}. \tag{4.32}$$

These turn out to be the MRE estimators of η and τ under the loss functions $(d - \eta)^2/\tau^2$ and $(d - \tau)^2/\tau^2$, respectively. ||

The treatment of these three examples extends to a number of other important cases (see, for example, Problems 4.6 and 4.7) and suggests that the Bayes estimator with respect to the measure induced by right Haar measure over $\bar{G}$ is equivariant and is, in fact, MRE. For conditions under which these conclusions are valid, see Berger 1985 (Section 6.6), Robert 1994a (Section 7.4), or Schervish 1995 (Section 6.2); a special case is treated in Section 5.4. It is also worth noting that if a Haar measure Λ over a group G is finite, that is, $\Lambda(G) < \infty$, then left and right Haar measures coincide.

At the beginning of the section, we defined invariance of a prior distribution by (4.1) and (4.2), and the same equations define invariance of a measure over Ω even if it is improper. We shall now consider whether the measure Λ' induced over Ω by left- or right-invariant Haar measure is invariant in this sense.

Example 4.10 Invariance of induced measures. We look at the location-scale groups and consider invariance of the induced measures.

(i) *Location Group*. We saw in Example 4.7 that left and right Haar measures Λ coincide in this case and that Λ' is Lebesgue measure which clearly satisfies (4.2)

(ii) *Scale Group*. Again, left and right Haar measure Λ coincide, and by Example 4.4, $\Lambda'(\omega) = \int_\omega \frac{d\tau}{\tau}$. Since $\bar{g}^{-1}(\omega) = \omega/\bar{g}$, as in Example 4.4,

$$\Lambda'[\bar{g}^{-1}(\omega)] = \int_{\omega/\bar{g}} \frac{d\tau}{\tau} = \int_\omega \frac{1}{\bar{g}} d\bar{g} = \Lambda(\omega)$$

so that Λ' is invariant.

(iii) *Location-Scale Group*. Here, the densities induced by the left- and right-invariant Haar measures are given by (4.30) and (4.29), respectively. Calculations similar to those of Example 4.9 show that the former is invariant but the latter is not ||

The general situation is described in the following result.

Theorem 4.11 *Under the assumptions of Theorem 4.1, the measure Λ' over Ω induced by a measure Λ over $\bar{G}$ is invariant provided Λ is left invariant.*

Proof. For any ω and $\theta_0 \in \Omega$, let $B = \{\bar{h} \in \bar{G} : \bar{h}\theta_0 \in \omega\}$, so that $\Lambda'(\omega) = \Lambda(B)$. Then, $\bar{g}B = \{\bar{g}\bar{h} : \bar{h}\theta_0 \in \omega\}$ and

$$\begin{aligned}\Lambda'(\bar{g}\omega) &= \Lambda(\bar{h} : \bar{h}\theta_0 \in \bar{g}\omega) = \Lambda(\bar{h} : \bar{g}^{-1}\bar{h}\theta_0 \in \omega)\\ &= \Lambda(\bar{g}\bar{h} : \bar{h}\theta_0 \in \omega) = \Lambda(\bar{g}B).\end{aligned}$$

Thus, $\Lambda(\bar{g}\omega) = \Lambda'(\omega)$ if and only if $\Lambda(gB) = \Lambda(B)$, and it follows that Λ' is invariant if and only if Λ is left invariant. □

Note that this result does not contradict the remark made after Example 4.9 to the effect that the Bayes estimator under the prior measure induced by right-invariant Haar measure is equivariant. A Bayes estimator can be equivariant under a prior measure Λ even if Λ is not invariant (see Problem 4.8).

When there are no groups leaving the given family of distributions invariant, no Haar measure is available to serve as a noninformative prior. In such situations, transformations that utilize some (perhaps arbitrary) structure of the parameter space may sometimes be used to deduce a form for a "noninformative" prior (Villegas 1990). A discussion of these approaches is given by Berger 1985 (Section 3.3); see also Bernardo and Smith 1994 (Section 5.6.2).

5 Hierarchical Bayes

In a hierarchical Bayes model, rather than specifying the prior distribution as a single function, we specify it in a hierarchy. Thus, we place another level on the model (3.1), and write

$$\begin{aligned}&X|\theta \sim f(x|\theta),\\ (5.1)\qquad &\Theta|\gamma \sim \pi(\theta|\gamma),\\ &\Gamma \sim \psi(\gamma),\end{aligned}$$

where we assume that $\psi(\cdot)$ is known and not dependent on any other unknown *hyperparameters* (as parameters of a prior are sometimes called). Note that we can continue this hierarchical modeling and add more stages to the model, but this is not often done in practice. The class of models (5.1) appears to be more general than the class (3.1) since in (3.1), γ has a fixed value, but in (5.1), it is permitted to have an arbitrary probability distribution. However, this appearance is deceptive. Since $\pi(\theta|\gamma)$ in (3.1) can be any fixed distribution, we can, in particular, take for it $\pi(\theta) = \int \pi(\theta|\gamma)\psi(\gamma)d\gamma$, which reduces the hierarchical model (5.1) to the single-prior model (3.1). However, there is a conceptual and practical advantage to the hierarchical model, in that it allows us to model relatively complicated situations using a series of simpler steps; that is, both $\pi(\theta|\gamma)$ and $\psi(\gamma)$ may be of a simple form (even conjugate), but $\pi(\theta)$ may be more complex. Moreover, there is often a computational advantage to hierarchical modeling. We will illustrate both of these points in this section.

It is also interesting to note that this process can be reversed. Starting from the single-prior model (3.1), we can look for a decomposition of the prior $\pi(\theta)$ of the form $\pi(\theta) = \int \pi(\theta|\gamma)\psi(\gamma)d\gamma$ and thus create the hierarchy (5.1). Such modeling, known as *hidden Markov models*, *hidden mixtures*, or *deconvolution*, has proved very useful (Churchill 1989, Robert 1994a (Section 9.3), Robert and Casella 1998).

Given a loss function $L(\theta, d)$, we would then determine the estimator that minimizes

$$\int L(\theta, d(x))\pi(\theta|x)\,d\theta \tag{5.2}$$

where $\pi(\theta|x) = \int f(x|\theta)\pi(\theta|\gamma)\psi(\gamma)\,d\gamma / \iint f(x|\theta)\pi(\theta|\gamma)\psi(\gamma)\,d\theta\,d\gamma$. Note also that

$$\pi(\theta|x) = \int \pi(\theta|x, \gamma)\pi(\gamma|x)\,d\gamma \tag{5.3}$$

where $\pi(\gamma|x)$ is the posterior distribution of Γ, unconditional on θ. We may then write (5.2) as

$$\int L(\theta, d(x))\pi(\theta|x)\,d\theta = \int \left[\int L(\theta, d(x))\pi(\theta|x, \gamma)\,d\theta\right] \pi(\gamma|x)\,d\gamma, \tag{5.4}$$

which shows that the hierarchical Bayes estimator can be thought of as a mixture of single-prior Bayes estimators. (See Problems 5.1 and 5.2.)

Hierarchical models allow easier modeling of prior distributions with "flatter" tails, which can lead to Bayes estimators with more desirable frequentist properties. This latter end is often achieved by taking $\psi(\cdot)$ to be improper (see, for example, Berger and Robert 1990, or Berger and Strawderman 1996).

Example 5.1 Conjugate normal hierarchy. Starting with the normal distribution and modeling, each stage with a conjugate prior yields the hierarchy

$$\begin{aligned}
&X_i|\theta \sim N(\theta, \sigma^2), \sigma^2 \text{ known}, \quad i = 1, \ldots, n,\\
&\theta|\tau \sim N(0, \tau^2)\\
&\frac{1}{\tau^2} \sim \text{Gamma}(a, b), \quad a, b \text{ known.}
\end{aligned} \tag{5.5}$$

The hierarchical Bayes estimator of θ under squared error loss is

$$
\begin{aligned}
E(\Theta|\mathbf{x}) &= \iint \theta\pi(\theta|\mathbf{x}, \tau^2)\,d\theta\,\pi(\tau^2|\mathbf{x})\,d\tau^2 \\
&= \int \frac{n\tau^2\bar{x}}{n\tau^2+\sigma^2}\pi(\tau^2|\mathbf{x})\,d\tau^2 \qquad (5.6)\\
&= E[E(\Theta|\mathbf{x}, \tau^2)],
\end{aligned}
$$

which is the expectation of the single-prior Bayes estimator using the density $\pi(\tau^2|\mathbf{x})$. (See Problem 5.3 for the form of the posterior distributions.) Although there is no explicit form for $E(\Theta|\mathbf{x})$, calculation is not particularly difficult. ||

It is interesting to note that even though at each stage of the model (5.5) a conjugate prior was used, the resulting Bayes estimator is not from a conjugate prior (the prior $\pi(\theta|a, b) = \int \pi(\theta|\tau)\psi(\tau|a, b)d\tau$ is not conjugate) and is not expressible in a simple form. Such an occurrence is somewhat commonplace in hierarchical Bayes analysis and leads to more reliance on numerical methods.

Example 5.2 Conjugate normal hierarchy, continued. As a special case of the model (5.5), consider the model

$$
\begin{aligned}
&X_i|\theta \sim N(\theta, \sigma^2), \quad i = 1, \ldots, p, \quad \text{independent}, \\
&\Theta|\tau^2 \sim N(0, \tau^2) \qquad (5.7)\\
&\frac{1}{\tau^2} \sim \text{Gamma}\left(\frac{\nu}{2}, \frac{2}{\nu}\right).
\end{aligned}
$$

This leads to a Student's t-prior distribution on Θ, and a posterior mean

$$
E[\Theta|\bar{x}] = \frac{\int_{-\infty}^{\infty}\theta(1+\theta^2/\nu)^{-\frac{\nu+1}{2}}e^{-p/2\sigma^2(\theta-\bar{x})^2}d\theta}{\int_{-\infty}^{\infty}(1+\theta^2/\nu)^{-\frac{\nu+1}{2}}e^{-p/2\sigma^2(\theta-\bar{x})^2}d\theta}, \qquad (5.8)
$$

which is not expressible in a simple form. Numerical evaluation of (5.8) is simple, so calculation of this hierarchical Bayes estimator in practice poses no problem. However, evaluation of the mean squared error or Bayes risk of (5.8) presents a more substantial task. ||

In the preceding example, the hierarchical Bayes estimator was expressible as a ratio of integrals which easily yielded to either direct calculation or simple approximation. There are other cases, however, in which a straightforward hierarchical model can lead to very difficult problems in evaluation of a Bayes estimator.

Example 5.3 Beta-binomial hierarchy. A generalization of the standard beta-binomial hierarchy is

$$
\begin{aligned}
&X|p \sim \text{binomial}\,(p, n), \\
&p|a, b \sim \text{beta}\,(a, b), \qquad (5.9)\\
&(a, b) \sim \psi(a, b),
\end{aligned}
$$

leading to the posterior mean

$$E(p|x) = \frac{\int_0^1 p^{x+1}(1-p)^{n-x}\pi(p)\,dp}{\int_0^1 p^x(1-p)^{n-x}\pi(p)\,dp}, \tag{5.10}$$

where

$$\pi(p) = \iint \frac{\Gamma(a+b)}{\Gamma(a)\Gamma(b)} p^{a-1}(1-p)^{b-1}\psi(a,b)\,da\,db. \tag{5.11}$$

For almost any choice of $\psi(a, b)$, calculation of (5.11), and hence (5.10), is quite difficult. Indeed, there could be difficulty with numerical integration, simulation, and approximation. Moreover, if $\psi(a, b)$ is chosen to be improper, as is typical in such hierarchies, the propriety of $\pi(p|x)$ is not easy to verify (and often does not obtain). George et al. (1993) provide algorithms for calculating expressions such as (5.10), and Hobert (1994) establishes conditions for the propriety of some resulting posterior distributions. ||

To overcome the difficulties in computing hierarchical Bayes estimators, we need to establish either easy-to-use formulas or good approximations, in order to further investigate their risk optimality. The approximation issue will be addressed in the next section. In the remainder of the present section, we consider the evaluation of (3.5) using theory based on Markov chain limiting behavior (see Note 9.4 for a brief discussion). Although this theory does not result in a simple expression for the Bayes estimators in general, it usually allows us to write expressions such as (5.6) as a limit of simple estimators. (Technically, these computations are not approximations, as they are exact in the limit. However, since they involve only a finite number of computations, we think of them as approximations, but realize that any order of precision can be achieved.) The resulting techniques, collectively known as *Markov chain Monte Carlo* (MCMC) techniques (see Tanner 1996, Gilks et al. 1996, or Robert and Casella 1998) can greatly facilitate calculation of a hierarchical Bayes estimator. One of the most popular of these methods is known as the *Gibbs sampler* [brought to statistical prominence by Gelfand and Smith (1990)], which we now illustrate.

Starting with the hierarchy (5.1), suppose we are interested in calculating the posterior distribution $\pi(\theta|\mathbf{x})$ (or $E(\Theta|\mathbf{x})$, or some other feature of the posterior distribution). From (5.1) we calculate the *full conditionals*

$$\begin{aligned} \Theta|\mathbf{x}, \gamma &\sim \pi(\theta|\mathbf{x}, \gamma), \\ \Gamma|\mathbf{x}, \theta &\sim \pi(\gamma|\mathbf{x}, \theta), \end{aligned} \tag{5.12}$$

which are the posterior distributions of each parameter conditional on all others. If, for $i = 1, 2, \ldots, M$, random variables are generated according to

$$\begin{aligned} \Theta_i|\mathbf{x}, \gamma_{i-1} &\sim \pi(\theta|\mathbf{x}, \gamma_{i-1}), \\ \Gamma_i|\mathbf{x}, \theta_i &\sim \pi(\gamma|\mathbf{x}, \theta_i), \end{aligned} \tag{5.13}$$

this defines a Markov chain (Θ_i, Γ_i). It follows from the theory of such chains (see Note 9.4) that there exist distributions $\pi(\theta|\mathbf{x})$ and $\pi(\gamma|\mathbf{x})$ such that

$$\Theta_i \xrightarrow{\mathcal{L}} \Theta \sim \pi(\theta|\mathbf{x}), \tag{5.14}$$

$$\Gamma_i \xrightarrow{\mathcal{L}} \Gamma \sim \pi(\gamma|\mathbf{x})$$

as $i \to \infty$, and

$$\frac{1}{M}\sum_{i=1}^{M} h(\Theta_i) \to E(h(\Theta)|\mathbf{x}) = \int h(\theta)\pi(\theta|\mathbf{x})\,d\theta \tag{5.15}$$

as $M \to \infty$. (A full development of this theory is given in Meyn and Tweedie (1993). See also Resnick 1992 for an introduction to Markov chains and Robert 1994a for more applications to Bayesian calculation.)

It follows from (5.15) that for Θ_i generated according to (5.13), we have

$$\frac{1}{M}\sum_{i=1}^{M} \Theta_i \to E(\Theta|\mathbf{x}), \tag{5.16}$$

the hierarchical Bayes estimator. (Problems 5.8 - 5.11 develop some of the more practical aspects of this theory.)

Example 5.4 Poisson hierarchy with Gibbs sampling. As an example of a Poisson hierarchy (see also Example 6.6), consider

$$\begin{aligned} X|\lambda &\sim \text{Poisson}(\lambda) \\ \Lambda|b &\sim \text{Gamma}(a, b), \quad a \text{ known} \\ \frac{1}{b} &\sim \text{Gamma}(k, \tau), \end{aligned} \tag{5.17}$$

leading to the full conditionals

$$\begin{aligned} \Lambda|x, b &\sim \text{Gamma}\left(a + x, \frac{b}{1+b}\right) \\ \frac{1}{b}|x, \lambda &\sim \text{Gamma}\left(a + k, \frac{\tau}{1+\lambda\tau}\right). \end{aligned} \tag{5.18}$$

Recall that in this hierarchy, $\pi(\lambda|\mathbf{x})$ is not expressible in a simple form. However, if we simulate from (5.18), we obtain a sequence $\{\Lambda_i\}$ satisfying

$$\frac{1}{M}\sum_{i=1}^{M} h(\Lambda_i) \to \int h(\lambda)\pi(\lambda|x)\,d\lambda = E[h(\Lambda|x)]. \tag{5.19}$$

Alternatively, we could use a $\{b_i\}$ sequence and calculate

$$\frac{1}{M}\sum_{i=1}^{M} \pi(\lambda|x, b_i) \to \int \pi(\lambda|x, b)\pi(b|x)\,db = \pi(\lambda|x). \tag{5.20}$$

‖

The Gibbs sampler actually yields two methods of calculating the same quantity. For example, from the hierarchy (5.1), using the full conditionals of (5.12) and the iterations in (5.13), we could estimate $E(h(\Theta)|x)$ by

$$(i) \quad \frac{1}{M}\sum_{i=1}^{M} h(\Theta_i) \to \int h(\theta)\pi(\theta|\mathbf{x})\,d\theta = E(h(\Theta)|x)$$

(5.21) or by

$$(ii) \quad \frac{1}{M}\sum_{i=1}^{M} Eh(\Theta|x,\Gamma_i) \to \int E(h(\Theta)|x,\gamma)\pi(\gamma|x)\,d\gamma = E(h(\Theta)|x).$$

Implementation of the Gibbs sampler is most effective when the full conditionals are easy to work with, and in such cases, it is often possible to calculate $E(h(\Theta)|x, \Gamma_i)$ in a simple form, so (5.21)(ii) is a viable option. To see that it is superior to (5.21)(i), write

$$E(h(\Theta)|x) = E[E(h(\Theta)|x,\gamma)]$$

and apply the Rao-Blackwell theorem (see Problem 5.12).

Example 5.5 Gibbs point estimation. To calculate the hierarchical Bayes estimator of λ in Example 5.4, we use

$$\frac{1}{M}\sum_{i=1}^{M} E(\Lambda|x, b_i) = \frac{1}{M}\sum_{i=1}^{M} \frac{b_i}{1+b_i}(a+x)$$

rather than $(1/M)\sum_{i=1}^{M} \Lambda_i$. Analogously, the posterior density $\pi(\lambda|x)$ can be calculated by

$$\begin{aligned}\hat{\pi}(\lambda|x) &= \frac{1}{M}\sum_{i=1}^{M} \pi(\lambda|x, b_i)\\ &= \frac{\lambda^{a+x-1}}{M\Gamma(a+x)}\sum_{i=1}^{M}\left(\frac{1+b_i}{b_i}\right)^{a+x} e^{-\lambda\frac{(1+b_i)}{b_i}}.\end{aligned}$$

‖

The actual implementation of the Gibbs sampler relies on Monte Carlo techniques to simulate random variables from the distributions in (5.13). Very efficient algorithms for such simulations are available, and Robert (1994a, Appendix B) catalogs a number of them. There are also full developments in Devroye (1985) and Ripley (1987). (See Problems 5.14 and 5.15.)

For many problems, the simulation step is straightforward to implement on a computer so we can take M as large as we like. This makes it possible for the approximations to have any desired precision, with the only limiting factor being computer time. (In this sense we are doing exact calculations.) Many applications of these techniques are given in Tanner (1996).

As a last example, consider the calculation of the hierarchical Bayes estimator of Example 5.2.

Example 5.6 Normal hierarchy. From (5.5), we have the set of full conditionals

$$(5.22) \qquad \begin{aligned}\theta|\bar{x},\tau^2 &\sim N\left(\frac{\tau^2}{\tau^2+n\sigma^2}\bar{x}, \frac{\sigma^2\tau^2}{\sigma^2+n\tau^2}\right)\\ \frac{1}{\tau^2}|\bar{x},\theta &\sim \text{Gamma}\left(a+\frac{1}{2}, \left(\frac{\theta^2}{2}+\frac{1}{b}\right)^{-1}\right).\end{aligned}$$

Note that, conditional on θ, τ^2 is independent of $\bar{x}$. Both of these conditional distributions are easy to simulate from, and we thus use the Gibbs sampler to generate a chain $(\Theta_i, \tau_i^2), i = 1, \ldots, M$, from (5.22). This yields the approximation

$$\begin{aligned} \tilde{E}(\Theta|\bar{x}) &= \frac{1}{M}\sum_{i=1}^{M} E(\Theta|\bar{x}, \tau_i^2) \\ &= \frac{1}{M}\sum_{i=1}^{M} \frac{\tau_i^2}{\tau_i^2 + n\sigma^2}\bar{x} \qquad (5.23) \\ &\to E(\Theta|\bar{x}) \end{aligned}$$

as $M \to \infty$. ||

As mentioned before, one of the purposes of specifying a model in a hierarchy is to make it possible to model more complicated phenomena in a sequence of less complicated steps. In addition, the ordering in the hierarchy allows us both to order the importance of the parameters and to incorporate some of our uncertainty about the prior specification.

To be precise, in the model

$$\begin{aligned} X|\theta &\sim f(x|\theta), \\ \Theta|\lambda &\sim \pi(\theta|\lambda), \qquad (5.24) \\ \Lambda &\sim \psi(\lambda), \end{aligned}$$

we tend to be more exacting in our specification of $\pi(\theta|\lambda)$, and less so in our specification of $\psi(\lambda)$. Indeed, in many cases, $\psi(\lambda)$ is taken to be "flat" or "non-informative" (for example, $\psi(\lambda)$ = Lebesgue measure). In practice, this leads to heavier-tailed prior distributions $\pi(\theta)$, with the resulting Bayes estimators being more robust (Berger and Robert 1990, Fourdrinier et al 1996; see also Example 5.6.7.).

One way of studying the effect that the stages of the hierarchy (5.24) have on each other is to examine, for each parameter, the information contained in its posterior distribution relative to its prior distribution. In effect, this measures how much the data can tell us about the parameter, with respect to the prior distribution.

To measure this information, we can use Kullback-Leibler information (recall Example 1.7.7), which also is known by the longer, and more appropriate name, Kullback-Leibler *information for discrimination* between two densities. For densities f and g, it is defined by

$$K[f, g] = \int \log\left[\frac{f(t)}{g(t)}\right] f(t)\,dt. \qquad (5.25)$$

The interpretation is that as $K[f, g]$ gets larger, it becomes easier to discriminate between the densities f and g; that is, there is more information for discrimination. From the model (5.24), we can assess the information between the data and the parameter by calculating $K[\pi(\theta|x), \pi(\theta)]$, where

$$\pi(\theta) = \int \pi(\theta|\lambda)\psi(\lambda)\,d\lambda, \qquad (5.26)$$

$$\pi(\theta|x) = \frac{f(x|\theta)\pi(\theta)}{\int f(x|\theta)\pi(\theta)\,d\theta} = \frac{f(x|\theta)\pi(\theta)}{m(x)}.$$

By comparison, the information between the data and the hyperparameter is measured by $K[\pi(\lambda|x), \psi(\lambda)]$, where

$$\pi(\lambda|x) = \frac{\int f(x|\theta)\pi(\theta|\lambda)\psi(\lambda)\,d\theta}{m(x)} = \frac{m(x|\lambda)\psi(\lambda)}{m(x)}. \tag{5.27}$$

An important result about the two measures of information for (5.26) and (5.27) is contained in the following theorem.

Theorem 5.7 *For the model (5.24),*

$$K[\pi(\lambda|x), \psi(\lambda)] < K[\pi(\theta|x), \pi(\theta)]. \tag{5.28}$$

From (5.28), we see that the distribution of the data has less effect on hyperpriors than priors, or, turning things around, the posterior distribution of a hyperparameter is less affected by changes in the prior than the posterior distribution of a parameter. This provides justification of the belief that parameters that are deeper in the hierarchy have less effect on inference.

Proof of Theorem 5.7. By definition,

$$\begin{aligned} K[\pi(\lambda|x), \psi(\lambda)] &= \int_\Lambda \pi(\lambda|x)\log\left(\frac{\pi(\lambda|x)}{\psi(\lambda)}\right)d\lambda \\ &= \int_\Lambda \left(\frac{\pi(\lambda|x)}{\psi(\lambda)}\right)\log\left(\frac{\pi(\lambda|x)}{\psi(\lambda)}\right)\psi(\lambda)\,d\lambda. \end{aligned} \tag{5.29}$$

Now, note that

$$\frac{\pi(\lambda|x)}{\psi(\lambda)} = \int_\Omega \left(\frac{f(x|\theta)}{m(x)}\right)\pi(\theta|\lambda)\,d\theta, \tag{5.30}$$

or, more succinctly, $\pi(\lambda|x)/\psi(\lambda) = E[f(x|\theta)/m(x)]$, where the expectation is taken with respect to $\pi(\theta|\lambda)$. We now apply Jensen's inequality to (5.29), using the fact that the function $x\log x$ is convex if $x > 0$, which leads to

$$\begin{aligned} \left(\frac{\pi(\lambda|x)}{\psi(\lambda)}\right)\log\left(\frac{\pi(\lambda|x)}{\psi(\lambda)}\right) &= \left(E\left[\frac{f(x|\theta)}{m(x)}\right]\right)\log\left(E\left[\frac{f(x|\theta)}{m(x)}\right]\right) \\ &\le E\left[\left(\frac{f(x|\theta)}{m(x)}\right)\log\left(\frac{f(x|\theta)}{m(x)}\right)\right] \\ &= \int_\Omega \left[\left(\frac{f(x|\theta)}{m(x)}\right)\log\left(\frac{f(x|\theta)}{m(x)}\right)\right]\pi(\theta|\lambda)\,d\theta. \end{aligned} \tag{5.31}$$

Substituting back into (5.29), we have

$$\begin{aligned} &K[\pi(\lambda|x), \psi(\lambda)] \\ &\quad\le \int_\Lambda\int_\Omega \left(\frac{f(x|\theta)}{m(x)}\right)\log\left(\frac{f(x|\theta)}{m(x)}\right)\pi(\theta|\lambda)\psi(\lambda)\,d\theta\,d\lambda. \end{aligned} \tag{5.32}$$

We now (of course) interchange the order of integration and notice that

$$\int_\Lambda \frac{f(x|\theta)}{m(x)}\pi(\theta|\lambda)\psi(\lambda)\,d\lambda = \pi(\theta|x). \tag{5.33}$$

Substitution into (5.32), together with the fact that $\frac{f(x|\theta)}{m(x)} = \frac{\pi(\theta|x)}{\pi(\theta)}$ yields (5.28). □

Thus, hierarchical modeling allows us to be less concerned about the exact form of $\psi(\lambda)$. This frees the modeler to choose a $\psi(\lambda)$ to yield other good properties without unduly compromising the Bayesian interpretations of the model. For example, as we will see in Chapter 5, $\psi(\lambda)$ can be chosen to yield hierarchical Bayes estimators with reasonable frequentist performance.

A full development of information measures and hierarchical models is given by Goel and DeGroot (1979, 1981); see also Problems 5.16–5.19.

Theorem 5.7 shows how information acts within the levels of a hierarchy, but does not address the, perhaps, more basic question of assessing the information provided by a prior distribution in a particular model. Information measures, such as $K[f, g]$, can also be the basis of answering this latter question. If $X \sim f(x|\theta)$ and $\Theta \sim \pi(\theta)$, then prior distributions that have a large effect on $\pi(\theta|x)$ should produce small values of $K[\pi(\theta|x), \pi(\theta)]$, since the prior and posterior distributions will be close together. Alternatively, prior distributions that have a small effect on $\pi(\theta|x)$ should produce large values of $K[\pi(\theta|x), \pi(\theta)]$, as the posterior will mainly reflect the sampling density. Thus, we may seek to find a prior $\pi(\theta)$ that produces the maximum value of $K[\pi(\theta|x), \pi(\theta)]$. We can consider such a prior to have the least influence on $f(x|\theta)$ and, hence, to be a default, or noninformative, prior.

The above is an informal description of the approach to the construction of a *reference prior*, initiated by Bernardo (1979) and further developed and formalized by Berger and Bernardo (1989, 1992). [See also Robert 1994a, Section 3.4]. This theory is quite involved, but approximations due to Clarke and Barron (1990, 1994) and Clarke and Wasserman (1993) shed some interesting light on the problem. First, we cannot directly use $K[\pi(\theta|x), \pi(\theta)]$ to derive a prior distribution, because it is a function of x. We, thus, consider its expected value with respect to the marginal distribution of X, the *Shannon information*

(5.34)
$$S(\pi) = \int K[\pi(\theta|x), \pi(\theta)] m_\pi(x)\, dx,$$

where $m_\pi(x) = \int f(x|\theta)\pi(\theta)\, d\theta$ is the marginal distribution. The reference prior is the distribution that maximizes $S(\pi)$.

The following theorem is due to Clarke and Barron (1990).

Theorem 5.8 *Let $X_1, \ldots, X_n$ be an iid sample from $f(x|\theta)$, and let $S_n(\pi)$ denote the Shannon information of the sample. Then, as $n \to \infty$,*

(5.35)
$$S_n(\pi) = \frac{k}{2} \log \frac{n}{2\pi e} + \int \pi(\theta) \log \frac{|I(\theta)|^{1/2}}{\pi(\theta)}\, d\theta + o(1)$$

where k is the dimension of θ and $I(\theta)$ is the Fisher information

$$I(\theta) = -E\left[\frac{\partial^2}{\partial\theta^2} \log f(X|\theta)\right].$$

As the integral in the expansion (5.35) is the only term involving the prior $\pi(\theta)$, maximizing that integral will maximize the expansion. Provided that $|I(\theta)|^{1/2}$ is integrable, Corollary 1.7.6 shows that $\pi(\theta) = |I(\theta)|^{1/2}$ is the appropriate choice. This is the Jeffreys prior, which was discussed in Section 4.1.

Example 5.9 Binomial reference prior. For $X_1, \ldots, X_n$ iid as Bernoulli(θ), we have

$$I(\theta) = -E\left[\frac{\partial^2}{\partial\theta^2} \log f(\mathbf{X}|\theta)\right] = \frac{n}{\theta(1-\theta)}, \tag{5.36}$$

which yields the Jeffreys prior $\pi(\theta) \propto [\theta(1-\theta)]^{-1/2}$. This is also the prior that maximizes the integral in $S_n(\pi)$ and, in that sense, imparts the least information on $f(\mathbf{x}|\theta)$. A formal reference prior derivation also shows that the Jeffreys prior is the reference prior. ||

In problems where there are no nuisance parameters, the Jeffreys and reference priors agree, even when they are improper. In fact, the reference prior approach was developed to deal with the nuisance parameter problem, as the Fisher information approach gave no clear-cut guidelines as to how to proceed in that case. Reference prior derivations for nuisance parameter problems are given by Berger and Bernardo (1989, 1992a, 1992b) and Polson and Wasserman (1990). See also Clarke and Wasserman (1993) for an expansion similar to (5.35) that is valid in the nuisance parameter case.

6 Empirical Bayes

Another generalization of single-prior Bayes estimation, empirical Bayes estimation, falls outside of the formal Bayesian paradigm. However, it has proven to be an effective technique of constructing estimators that perform well under both Bayesian and frequentist criteria. One reason for this, as we will see, is that empirical Bayes estimators tend to be more robust against misspecification of the prior distribution.

The starting point is again the model (3.1), but we now treat γ as an unknown parameter of the model, which also needs to be estimated. Thus, we now have two parameters to estimate, necessitating at least two observations. We begin with the Bayes model

$$\begin{aligned} X_i|\theta &\sim f(x|\theta), \quad i = 1, \ldots, p, \\ \Theta|\gamma &\sim \pi(\theta|\gamma). \end{aligned} \tag{6.1}$$

and calculate the marginal distribution of $\mathbf{X}$, with density

$$m(\mathbf{x}|\gamma) = \int \prod f(x_i|\theta)\pi(\theta|\gamma)\, d\theta. \tag{6.2}$$

Based on $m(\mathbf{x}|\gamma)$, we obtain an estimate, $\hat{\gamma}(\mathbf{x})$, of γ. It is most common to take $\hat{\gamma}(\mathbf{x})$ to be the MLE of γ, but this is not essential. We now substitute $\hat{\gamma}(\mathbf{x})$ for γ in $\pi(\theta|\gamma)$ and determine the estimator that minimizes the empirical posterior loss

$$\int L(\theta, \delta(\mathbf{x}))\pi(\theta|\mathbf{x}, \hat{\gamma}(\mathbf{x}))\, d\theta. \tag{6.3}$$

This minimizing estimator is the empirical Bayes estimator.

An alternative definition is obtained by substituting $\hat{\gamma}(\mathbf{x})$ for γ in the Bayes estimator. Although, mathematically, this is equivalent to the definition given here (see Problem 6.1), it is statistically more satisfying to define the empirical Bayes estimator as minimizing the empirical posterior loss (6.3).

Example 6.1 Normal empirical Bayes. To calculate an empirical Bayes estimator for the model (5.7) of Example 5.2, rather than integrate over the prior for τ^2, we estimate τ^2. We determine the marginal distribution of $\mathbf{X}$ (see Problem 6.4),

$$
\begin{aligned}
(6.4)\qquad m(\mathbf{x}|\tau^2) &= \int \prod_{i=1}^{n} f(x_i|\theta)\pi(\theta|\tau^2)\,d\theta \\
&= \frac{1}{(2\pi\sigma^2)^{n/2}} e^{-\frac{1}{2\sigma^2}\Sigma(x_i-\bar{x})^2} \frac{1}{(2\pi\tau^2)^{1/2}} \\
&\quad \times \int_{-\infty}^{\infty} e^{-\frac{n}{2\sigma^2}(\bar{x}-\theta)^2} e^{-\frac{1}{2}\frac{\theta^2}{\tau^2}}\,d\theta \\
&= \frac{1}{(2\pi)^{n/2}} \frac{1}{\sigma^n} \left(\frac{\sigma^2}{\sigma^2+n\tau^2}\right)^{1/2} e^{-\frac{1}{2}\left[\frac{\Sigma(x_i-\bar{x})^2}{\sigma^2}+\frac{n\bar{x}^2}{\sigma^2+n\tau^2}\right]}.
\end{aligned}
$$

(Note the similarity to the density (2.13) in the one-way random effects model.) From this density, we can now estimate τ^2 using maximum likelihood (or some other estimation method). Recalling that we are assuming σ^2 is known, we find the MLE of $\sigma^2 + n\tau^2$ given by $\widehat{\sigma^2 + n\tau^2} = \max\{\sigma^2, n\bar{x}^2\}$. Substituting into the single-prior Bayes estimator, we obtain the empirical Bayes estimator

$$
\begin{aligned}
(6.5)\qquad E(\Theta|\bar{x}, \hat{\tau}) &= \left(1 - \frac{\sigma^2}{\widehat{\sigma^2 + n\tau^2}}\right)\bar{x} \\
&= \left(1 - \frac{\sigma^2}{\max\{\sigma^2, n\bar{x}^2\}}\right)\bar{x}.
\end{aligned}
$$

‖

It is tempting to ask whether the empirical Bayes estimator is ever a Bayes estimator; that is, can we consider $\pi(\theta|x, \hat{\gamma}(x))$ to be a "legitimate" posterior density, in that it be derived from a real prior distribution? The answer is yes, but the prior distribution that leads to such a posterior may sometimes not be proper (see Problem 6.2).

We next consider an example that illustrates the type of situation where empirical Bayes estimation is particularly useful.

Example 6.2 Empirical Bayes binomial. Empirical Bayes estimation is best suited to situations in which there are many problems that can be modeled simultaneously in a common way. For example, suppose that there are K different groups of patients, where each group has n patients. Each group is given a different treatment for the same illness, and in the kth group, we count X_k, $k = 1, \ldots, K$, the number of successful treatments out of n. Since the groups receive different treatments, we expect different success rates; however, since we are treating the same illness, these rates should be somewhat related to each other. These considerations suggest the hierarchy

$$X_k \sim \text{binomial}(p_k, n), \qquad p_k \sim \text{beta}(a, b), \quad k = 1, \ldots, K, \tag{6.6}$$

where the K groups are tied together by the common prior distribution. As in Example 1.5, the single-prior Bayes estimator of p_k under squared error loss is

$$\delta^\pi(x_k) = E(p_k|x_k, a, b) = \frac{a + x_k}{a + b + n}. \tag{6.7}$$

In Example 1.5, a and b are assumed known and all calculations are straightforward. In the empirical Bayes model, however, we consider these hyperparameters unknown and estimate them. To construct an empirical Bayes estimator, we first calculate the marginal distribution

$$\begin{aligned} m(\mathbf{x}|a, b) &= \int_0^1 \cdots \int_0^1 \prod_{k=1}^K \binom{n}{x_k} p_k^{x_k}(1 - p_k)^{n-x_k} \\ &\quad \times \frac{\Gamma(a+b)}{\Gamma(a)\Gamma(b)} p_k^{a-1}(1 - p_k)^{b-1} dp_k \\ &= \prod_{k=1}^K \binom{n}{x_k} \frac{\Gamma(a+b)\Gamma(a+x_k)\Gamma(n-x_k+b)}{\Gamma(a)\Gamma(b)\Gamma(a+b+n)}, \end{aligned} \tag{6.8}$$

a product of beta-binomial distributions. We now proceed with maximum likelihood estimation of a and b based on (6.8). Although the MLEs $\hat{a}$ and $\hat{b}$ are not expressible in closed form, we can calculate them numerically and construct the empirical Bayes estimator

$$\delta^{\hat{\pi}}(x_k) = E(p_k|x_k, \hat{a}, \hat{b}) = \frac{\hat{a} + x_k}{\hat{a} + \hat{b} + n}. \tag{6.9}$$

The Bayes risk of $E(p_k|x_k, \hat{a}, \hat{b})$ is only slightly higher than that of the Bayes estimator (6.7), and is given in Table 6.1. For comparison, we also include the Bayes risk of the unbiased estimator $\mathbf{x}/n$. The first three rows correspond to a prior mean of 1/2, with decreasing prior variance. Notice how the risk of the empirical Bayes estimator is between that of the Bayes estimator and that of X/n. ‖

As Example 6.2 illustrates, and as we will see later in this chapter (Section 7), the Bayes risk performance of the empirical Bayes estimator is often "robust"; that is, its Bayes risk is reasonably close to that of the Bayes estimator no matter what values the hyperparameters attain.

We next turn to the case of exponential families, and find that a number of the expressions developed in Section 3 are useful in evaluating empirical Bayes estimators. In particular, we find an interesting representation for the risk under squared error loss.

Table 6.1. *Bayes Risks for the Bayes, Empirical Bayes, and Unbiased Estimators of Example 6.2, where* $K = 10$ *and* $n = 20$

Prior Parameters		Bayes Risk		
a	b	δ^π of (6.7)	$\delta^{\hat{\pi}}$ of (6.9)	$\mathbf{x}/n$
2	2	.0833	.0850	.1000
6	6	.0721	.0726	.1154
20	20	.0407	.0407	.1220
3	1	.0625	.0641	.0750
9	3	.0541	.0565	.0865
30	10	.0305	.0326	.0915

For the situation of Corollary 3.3, using a prior $\pi(\eta|\lambda)$, where λ is a hyperparameter, the Bayes estimator of (3.12) becomes

$$E(\eta_i|\mathbf{x}, \lambda) = \frac{\partial}{\partial x_i} \log m(\mathbf{x}|\lambda) - \frac{\partial}{\partial x_i} \log h(\mathbf{x}) \tag{6.10}$$

where $m(\mathbf{x}|\lambda) = \int p_\eta(\mathbf{x})\pi(\eta|\lambda)\, d\eta$ is the marginal distribution. Simply substituting an estimate of λ, $\hat{\lambda}(\mathbf{x})$ into (6.10) yields the empirical Bayes estimator

$$E(\eta_i|\mathbf{x}, \hat{\lambda}) = \frac{\partial}{\partial x_i} \log m(\mathbf{x}|\lambda)\Big|_{\lambda=\hat{\lambda}(\mathbf{x})} - \frac{\partial}{\partial x_i} \log h(\mathbf{x}). \tag{6.11}$$

If $\hat{\lambda}$ is, in fact, the MLE of λ based on $m(\mathbf{x}|\lambda)$, then the empirical Bayes estimator has an alternate representation.

Theorem 6.3 *For the situation of Corollary 3.3, with prior distribution* $\pi(\eta|\lambda)$, *suppose* $\hat{\lambda}(\mathbf{x})$ *is the MLE of* λ *based on* $m(\mathbf{x}|\lambda)$. *Then, the empirical Bayes estimator is*

$$E(\eta_i|\mathbf{x}, \hat{\lambda}) = \frac{\partial}{\partial x_i} \log m(\mathbf{x}|\hat{\lambda}(\mathbf{x})) - \frac{\partial}{\partial x_i} \log h(\mathbf{x}). \tag{6.12}$$

Proof. Recall from calculus that if $f(\cdot, \cdot)$ and $g(\cdot)$ are differentiable functions, then

$$\frac{d}{dx} f(x, g(x)) = g'(x) \frac{\partial}{\partial y} f(x, y)\Big|_{y=g(x)} + \frac{\partial}{\partial x} f(x, y)\Big|_{y=g(x)}. \tag{6.13}$$

Applying this to $m(\mathbf{x}|\hat{\lambda}(\mathbf{x}))$ shows that

$$\begin{aligned} \frac{\partial}{\partial x_i} \log m(\mathbf{x}|\hat{\lambda}(\mathbf{x})) &= \frac{\partial}{\partial x_i}\hat{\lambda}(\mathbf{x}) \frac{\partial}{\partial \lambda} \log m(\mathbf{x}|\lambda)\Big|_{\lambda=\hat{\lambda}(\mathbf{x})} \\ &\quad + \frac{\partial}{\partial x_i} \log m(\mathbf{x}|\lambda)\Big|_{\lambda=\hat{\lambda}(\mathbf{x})} \\ &= \frac{\partial}{\partial x_i} \log m(\mathbf{x}|\lambda)\Big|_{\lambda=\hat{\lambda}(\mathbf{x})} \end{aligned}$$

because $(\partial/\partial\lambda) \log m(\mathbf{x}|\lambda)$ is zero at $\lambda = \hat{\lambda}(\mathbf{x})$. Hence, the empirical Bayes estimator is equal to (6.12). □

Thus, for estimating the natural parameter of an exponential family, the empirical Bayes estimator (using the marginal MLE) can be expressed in the same form as a *formal* Bayes estimator. Here we use the adjective *formal* to signify a mathematical equivalence, as the function $m(\mathbf{x}|\hat{\lambda}(\mathbf{x}))$ may not correspond to a proper marginal density. See Bock (1988) for some interesting results and variations on these estimators.

Example 6.4 Normal empirical Bayes, μ unknown. Consider the estimation of θ_i in the model of Example 3.4,

$$X_i|\theta_i \sim N(\theta_i, \sigma^2), \quad i = 1, \ldots, p, \text{ independent,} \tag{6.14}$$

$$\Theta_i \sim N(\mu, \tau^2), \quad i = 1, \ldots, p, \text{ independent,} \tag{6.15}$$

where μ is unknown. We can use Theorem 6.3 to calculate the empirical Bayes estimator, giving

$$E(\Theta_i|\mathbf{x}, \hat{\mu}) = \sigma^2 \left[\frac{\partial}{\partial x_i} \log m(\mathbf{x}|\hat{\mu}) - \frac{\partial}{\partial x_i} h(\mathbf{x})\right]$$

where $\hat{\mu}$ is the MLE of μ from

$$m(\mathbf{x}|\mu) = \frac{1}{[2\pi(\sigma^2+\tau^2)]^{p/2}} e^{-\frac{1}{2(\sigma^2+\tau^2)}\Sigma(x_i-\mu)^2}.$$

Hence, $\hat{\mu} = \bar{x}$ and

$$\frac{\partial}{\partial x_i} \log m(\mathbf{x}|\hat{\mu}) = \frac{\partial}{\partial x_i}\left[\frac{-1}{2(\sigma^2+\tau^2)}\Sigma(x_i - \bar{x})^2\right].$$

This yields the empirical Bayes estimator

$$E(\Theta_i|\mathbf{x}, \hat{\mu}) = \frac{\tau^2}{\sigma^2+\tau^2} x_i + \frac{\sigma^2}{\sigma^2+\tau^2}\bar{x},$$

which is the Bayes estimator under the prior $\pi(\theta|\bar{x})$.

An advantage of the form (6.12) is that it allows us to represent the risk of the empirical Bayes estimator in the form specified by (3.13). The risk of the empirical Bayes estimator (6.12) is given by

$$\begin{aligned} R[\eta, E(\eta|\mathbf{X}, \hat{\lambda}(\mathbf{X}))] &= R[\eta, -\nabla \log h(\mathbf{X})] \\ &\quad + \sum_{i=1}^{p} E_\eta \left\{2\frac{\partial^2}{\partial X_i^2} \log m(\mathbf{X}|\,\hat{\lambda}(\mathbf{X})] \right. \\ &\quad \left. + \left(\frac{\partial}{\partial X_i} \log m[\mathbf{X}|\hat{\lambda}(\mathbf{X})]\right)^2\right\}. \end{aligned} \tag{6.16}$$

Using the MLE $\hat{\mu}(\bar{x}) = \bar{x}$, differentiating the log of $m(\mathbf{x}|\hat{\mu}(\mathbf{x}))$, and substituting into (6.12) shows (Problem 6.10) that

$$\begin{aligned} R[\eta, E\{\eta|\mathbf{X}, \hat{\mu}(\mathbf{X})\}] &= p/\sigma^2 \\ &\quad - \frac{2(p-1)^2}{p(\sigma^2+\tau^2)} + \frac{p-1}{p(\sigma^2+\tau^2)^2} \sum_{i=1}^{p} E_\eta(X_i - \bar{X})^2. \end{aligned}$$

Table 6.2. *Values of the Hierarchical Bayes (HB)(5.8) and Empirical Bayes (EB) Estimate (6.5).*

	Value of $\bar{x}$			
ν	.5	2	5	10
2	.27	1.22	4.36	9.69
10	.26	1.07	3.34	8.89
30	.25	1.02	2.79	7.30
∞	.25	1.00	2.50	5.00
EB	0	1.50	4.80	9.90

‖

As mentioned at the beginning of this section, empirical Bayes estimators can also be useful as approximations to hierarchical Bayes estimators. Since we often have simpler expressions for the empirical Bayes estimator, if its behavior is close to that of the hierarchical Bayes estimator, it becomes a reasonable substitute (see, for example, Kass and Steffey 1989).

Example 6.5 Hierarchical Bayes approximation. Both Examples 5.2 and 6.1 consider the same model, where in Example 5.2 the hierarchical Bayes estimator (5.8) averages over the hyperparameter, and in Example 6.1 the empirical Bayes estimator (6.5) estimates the hyperparameter. A small numerical comparison in Table 6.2 suggests that the empirical Bayes estimator is a reasonable, but not exceptional, approximation to the hierarchical Bayes estimator.

The approximation, of hierarchical Bayes by empirical Bayes, is best for small values of ν [defined in (5.7)] and deteriorates as $\nu \to \infty$. At $\nu = \infty$, the hierarchical Bayes estimator becomes a Bayes estimator under a $N(0, 1)$ prior (see Problem 6.11). Notice that, even though (6.5) provides us with a simple expression for an estimator, it still requires some work to evaluate the mean squared error, or Bayes risk, of (6.5). However, it is important to do so to obtain an overall picture of the performance of the estimator (Problem 6.12). ‖

Although the (admittedly naive) approximation in Example 6.5 is not very accurate, there are other situations where the empirical Bayes estimator, or slight modifications thereof, can provide a good approximation to the hierarchical Bayes estimator. We now look at some of these situations.

For the general hierarchical model (5.1), the Bayes estimator under squared error loss is

$$E(\Theta|\mathbf{x}) = \int \theta\pi(\theta|\mathbf{x})\, d\theta \tag{6.17}$$

which can be written

$$E(\Theta|\mathbf{x}) = \int\int \theta\pi(\theta|\mathbf{x}, \gamma)\pi(\gamma|\mathbf{x})\, d\gamma\, d\theta \tag{6.18}$$

$$= \int E(\Theta|\mathbf{x}, \gamma)\pi(\gamma|\mathbf{x})\, d\gamma$$

where

$$\pi(\gamma|\mathbf{x}) = \frac{m(\mathbf{x}|\gamma)\psi(\gamma)}{m(\mathbf{x})} \tag{6.19}$$

with

$$m(\mathbf{x}|\gamma) = \int f(\mathbf{x}|\theta)\pi(\theta|\gamma)\, d\theta, \tag{6.20}$$

$$m(\mathbf{x}) = \int m(\mathbf{x}|\gamma)\psi(\gamma)\, d\gamma.$$

Now, suppose that $\pi(\gamma|\mathbf{x})$ is quite peaked around its mode, $\hat{\gamma}_\pi$. We might then consider approximating $E(\Theta|\mathbf{x})$ by $E(\Theta|\mathbf{x}, \hat{\gamma}_\pi)$. Moreover, if $\psi(\gamma)$ is relatively flat, as compared to $m(\mathbf{x}|\gamma)$, we would expect $\pi(\gamma|\mathbf{x}) \propto m(\mathbf{x}|\gamma)$ and $\hat{\gamma}_\pi \approx \hat{\gamma}$, the marginal MLE. In such a case, $E(\Theta|\mathbf{x}, \hat{\gamma}_\pi)$ would be close to the empirical Bayes estimator $E(\Theta|\mathbf{x}, \hat{\gamma})$, and hence the empirical Bayes estimator is a good approximation to the hierarchical Bayes estimator (Equation 6.17).

Example 6.6 Poisson hierarchy. Although we might expect the empirical Bayes and hierarchical Bayes estimators to be close if the hyperparameter has a flat-tailed prior, they will, generally, not be equal unless that prior is improper. Consider the model

$$\begin{aligned} &X_i \sim \text{Poisson}\,(\lambda_i), \quad i = 1, \ldots, p, \text{ independent}, \\ &\lambda_i \sim \text{Gamma}(a, b), \quad i = 1, \ldots, p, \text{ independent}, \; a \text{ known}. \end{aligned} \tag{6.21}$$

The marginal distribution of X_i is

$$\begin{aligned} m(x_i|b) &= \int_0^\infty \frac{e^{-\lambda_i}\lambda_i^{x_i}}{x_i!} \frac{1}{\Gamma(a)b^a} \lambda_i^{a-1} e^{-\lambda_i/b}\, d\lambda_i \\ &= \frac{\Gamma(x_i + a)}{x_i!\Gamma(a)} \frac{1}{b^a} \left(1 + \frac{1}{b}\right)^{-(x_i+a)} \\ &= \binom{x_i + a - 1}{a - 1} \left(\frac{b}{b+1}\right)^{x_i} \left(\frac{1}{b+1}\right)^a, \end{aligned}$$

a negative binomial distribution. Thus,

$$m(\mathbf{x}|b) = \left[\prod_{i=1}^p \binom{x_i + a - 1}{a - 1}\right] \left(\frac{b}{b+1}\right)^{\Sigma x_i} \left(\frac{1}{b+1}\right)^{pa} \tag{6.22}$$

and the marginal MLE of b is $\hat{b} = \bar{x}/a$. From (6.21), the Bayes estimator is

$$E(\lambda_i|x_i, b) = \frac{b}{b+1}(a + x_i) \tag{6.23}$$

and, hence, the empirical Bayes estimator is

$$E(\lambda_i|x_i, \hat{b}) = \frac{\bar{x}}{\bar{x} + a}(a + x_i). \tag{6.24}$$

If we add a prior $\psi(b)$ to the hierarchy (6.21), the hierarchical Bayes estimator can be written

$$E(\lambda_i|\mathbf{x}) = \int E(\lambda_i|\mathbf{x}, b)\pi(b|\mathbf{x})\,db \tag{6.25}$$

where

$$\pi(b|\mathbf{x}) = \frac{\left(\frac{b}{b+1}\right)^{p\bar{x}}\left(\frac{1}{b+1}\right)^{pa}\psi(b)}{\int_0^\infty \left(\frac{b}{b+1}\right)^{p\bar{x}}\left(\frac{1}{b+1}\right)^{pa}\psi(b)\,db}. \tag{6.26}$$

From examination of the hierarchy (6.21), a choice of $\psi(b)$ might be an inverted gamma, as this would be conjugate for λ_i. However, these priors will not lead to a simple expression for $E(\lambda_i|\bar{x})$ (although they may lead to good estimators). In general, however, we are less concerned that the hyperprior reflect reality (which is a concern for the prior), since the hyperprior tends to have less influence on our ultimate inference (Theorem 5.7). Thus, we will often base the choice of the hyperprior on convenience.

Let us, therefore, choose as prior for b an F-distribution,

$$\psi(b) \propto \frac{b^{\alpha-1}}{(1+b)^{\alpha+\beta}} \tag{6.27}$$

which is equivalent to putting a beta(α, β) prior on $b/(1+b)$. The denominator of $\pi(b|\mathbf{x})$ in (6.26) is

$$\begin{aligned} &\int_0^\infty \left(\frac{b}{b+1}\right)^{p\bar{x}}\left(\frac{1}{b+1}\right)^{pa}\frac{b^{\alpha-1}}{(1+b)^{\alpha+\beta}}\,db \\ &= \int_0^1 t^{p\bar{x}+\alpha-1}(1-t)^{pa+\beta-1}\,dt \qquad \left(t = \frac{b}{1+b}\right) \\ &= \frac{\Gamma(p\bar{x}+\alpha)\Gamma(pa+\beta)}{\Gamma(p\bar{x}+pa+\alpha+\beta)}, \end{aligned} \tag{6.28}$$

and (6.23), (6.26), and (6.28) lead to the hierarchical Bayes estimator

$$\begin{aligned} E(\lambda_i|\mathbf{x}) &= \int E(\lambda_i|\mathbf{x}, b)\pi(b|\mathbf{x})\,db \\ &= \frac{\Gamma(p\bar{x}+pa+\alpha+\beta)}{\Gamma(p\bar{x}+\alpha)\Gamma(pa+\beta)}(a+x_i) \\ &\quad \times \int_0^\infty \left(\frac{b}{b+1}\right)^{p\bar{x}+1}\left(\frac{1}{b+1}\right)^{pa}\frac{b^{\alpha-1}}{(1+b)^{\alpha+\beta}}\,db \\ &= \left[\frac{\Gamma(p\bar{x}+pa+\alpha+\beta)}{\Gamma(p\bar{x}+\alpha)\Gamma(pa+\beta)}\right]\left[\frac{\Gamma(p\bar{x}+\alpha+1)\Gamma(pa+\beta)}{\Gamma(p\bar{x}+pa+\alpha+\beta+1)}\right](a+x_i) \\ &= \left[\frac{p\bar{x}+\alpha}{p\bar{x}+pa+\alpha+\beta}\right](a+x_i). \end{aligned} \tag{6.29}$$

The hierarchical Bayes estimator will therefore be equal to the empirical Bayes estimator when $\alpha = \beta = 0$. This makes $\psi(b) \propto (1/b)$ an improper prior. However, the calculation of $\pi(b|\mathbf{x})$ from (6.26) and $E(\lambda_i|\mathbf{x})$ from (6.29) will still be valid. (This model was considered by Deely and Lindley (1981), who termed it *Bayes Empirical Bayes.*)

To further see how the empirical Bayes estimator is an approximation to the hierarchical Bayes estimator, write

$$\begin{aligned}\frac{p\bar{x}+\alpha}{p\bar{x}+pa+\alpha+\beta} = \frac{\bar{x}}{\bar{x}+a} - \frac{\beta}{p(\bar{x}+a)} \\ +\frac{pa\alpha+pa\beta+\alpha\beta}{p^2(\bar{x}+a)^2} \\ -\frac{2a\alpha\beta}{p^2(\bar{x}+a)^3}+\cdots.\end{aligned}$$

This shows that the empirical Bayes estimator is the leading term in a Taylor series expansion of the hierarchical Bayes estimator, and we can write

$$E(\lambda_i|\mathbf{x}) = E(\lambda_i|x_i,\hat{b}) + O\left(\frac{1}{p}\right). \tag{6.30}$$

Estimators of the form (6.29) are similar to those developed by Clevenson and Zidek (1975) for estimation of Poisson means. The Clevenson-Zidek estimators, which have $a = 0$ in (6.29), are minimax estimators of λ (see Section 5.7). ||

If interest centers on obtaining an approximation to a hierarchical Bayes estimator, a more direct route would be to look for an accurate approximation to the integral in (6.17). When such an approximation coincides with the empirical Bayes estimator, we can safely consider the empirical Bayes estimator as an approximate hierarchical Bayes estimator.

Example 6.7 Continuation of Example 5.2. In Example 5.2, the hierarchical Bayes estimator (5.8) was approximated by the empirical Bayes estimator (6.5). If, instead, we seek a direct approximation to (5.8), we might start with the Taylor expansion of $(1+\theta^2/\nu)^{-(\nu+1)/2}$ around $\bar{x}$

$$\begin{aligned}\frac{1}{(1+\theta^2/\nu)^{(\nu+1)/2}} = \frac{1}{(1+\bar{x}^2/\nu)^{(\nu+1)/2}} \\ -\frac{\nu+1}{\nu}\frac{\bar{x}}{(1+\bar{x}^2/\nu)^{(\nu+3)/2}}(\theta-\bar{x}) + O[(\theta-\bar{x})^{-2}],\end{aligned} \tag{6.31}$$

and using this in the numerator and denominator of (5.5.8) yields the approximation (Problem 6.15)

$$\hat{E}(\Theta|\mathbf{x}) = \left(1-\frac{(\nu+1)\sigma^2}{p(\nu+\bar{x}^2)}\right)\bar{x} + O\left(\frac{1}{p^{3/2}}\right). \tag{6.32}$$

Notice that the approximation is equal to the empirical Bayes estimator if $\nu = 0$, an extremely flat prior! The approximation (6.32) is better than the empirical Bayes estimator for large values of ν, but worse for small values of ν. ||

The approximation (6.32) is a special case of a *Laplace approximation* (Tierney and Kadane 1986). The idea behind the approximation is to carry out a Taylor series expansion of the integrand around an MLE, which can be summarized as

$$\int b(\lambda)e^{-nh(\lambda)}d\lambda \doteq b(\hat{\lambda})\sqrt{\frac{2\pi}{nh''(\hat{\lambda})}}e^{-nh(\hat{\lambda})}. \tag{6.33}$$

Here, $h(\hat{\lambda})$ is the unique minimum of $h(\lambda)$; that is, $\hat{\lambda}$ is the MLE based on a likelihood proportional to $e^{-nh(\lambda)}$. (See Problem 6.17 for details.) In applying (6.33) to a representation like (6.18), we obtain

$$\begin{aligned} E(\Theta|\mathbf{x}) &= \int E(\Theta|\mathbf{x}, \lambda)\pi(\lambda|\mathbf{x})\,d\lambda \\ &= \int E(\Theta|\mathbf{x}, \lambda)e^{n \log \pi(\lambda|\mathbf{x})^{1/n}}\,d\lambda \\ &\doteq \left[\frac{\sqrt{2\pi}\,\pi(\hat{\lambda}|\mathbf{x})}{-\frac{\partial^2}{\partial\lambda^2}\log \pi(\lambda|\mathbf{x})|_{\lambda=\hat{\lambda}}}\right] E(\Theta|\mathbf{x}, \hat{\lambda}) \end{aligned} \tag{6.34}$$

where $\hat{\lambda}$ is the mode of $\pi(\lambda|\mathbf{x})$. Thus, $E(\Theta|x, \hat{\lambda})$ in (6.34) will be the empirical Bayes estimator if $\pi(\lambda|\mathbf{x}) \propto m(\mathbf{x}|\lambda)$, that is, if $\psi(\lambda) = 1$. Moreover, the expression in square brackets in (6.34) is equal to 1 if $\pi(\lambda|\mathbf{x})$ is normal with mean $\hat{\lambda}$ and variance equal to the inverse of the observed Fisher information (see Problem 6.17).

Both the hierarchical and empirical Bayes approach are generalizations of single-prior Bayes analysis. In each case, we generalize the single prior to a class of priors. Hierarchical Bayes then averages over this class, whereas empirical Bayes chooses a representative member. Moreover, we have considered the functional forms of the prior distribution to be known; that is, even though Θ and γ are unknown, $\pi(\theta|\gamma)$ and $\psi(\gamma)$ are known.

Another generalization of single-prior Bayes analysis is *robust Bayes* analysis, where the class of priors is treated differently. Rather than summarize over the class, we allow the prior distribution to vary through it, and examine the behavior of the Bayes procedures as the prior varies. Moreover, the assumption of knowledge of the functional form is relaxed. Typically, a hierarchy like (3.1) is used, and a class of distributions for $\pi(\cdot|\cdot)$ is specified. For example, a popular class of prior distributions for Θ is given by an *ε-contamination class*

$$\Pi = \{\pi(\theta|\lambda) : \pi(\theta|\lambda) = (1-\varepsilon)\pi_0(\theta|\lambda) + \varepsilon q(\theta), q \in Q\} \tag{6.35}$$

where $\pi_0(\theta|\lambda)$ is a specified prior (sometimes called the *root prior*) and q is any distribution in a class Q. [Here, Q is sometimes taken to be the class of all distributions, but more restrictive classes can often provide estimators and posterior distributions with desirable properties. See, for example, Berger and Berliner 1986. Also, Mattner (1994) showed that for densities specified in the form of ε-contamination classes, the order statistics are complete. See Note 1.10.5.)

Using (6.35), we then proceed in a formal Bayesian way, and derive estimators based on minimizing posterior expected loss resulting from a prior $\pi \in \Pi$, say π^*. The resulting estimator, say δ^{π^*}, is evaluated using measures that range over all $\pi \in \Pi$, to assess the robustness of δ^{π^*} against misspecification of the prior.

For example, one might consider robustness using the posterior expected loss, or robustness using the Bayes risk. In this latter case, we might look at (Berger 1985, Section 4.7.5)

$$\sup_{\pi \in \Pi} r(\pi, \delta) \tag{6.36}$$

and, perhaps, choose an estimator δ that minimizes this quantity. If the loss is squared error, then for any estimator δ, we can write (Problem 6.2)

$$r(\pi, \delta) = r(\pi, \delta^\pi) + E(\delta - \delta^\pi)^2, \tag{6.37}$$

where δ^π is the Bayes estimator under π. From (6.37), we see that a robust Bayes estimator is one that is "close" to the Bayes estimators for all $\pi \in \Pi$. An ultimate goal of robust Bayes analysis is to find a prior $\pi^* \in \Pi$ for which δ^{π^*} can be considered to be robust.

Example 6.8 Continuation of Example 3.1. To obtain a robust Bayes estimator of θ, consider the class of priors

$$\Pi = \{\pi : \pi(\theta) = (1-\varepsilon)\pi_0(\theta|\tau_0) + \varepsilon q(\theta)\} \tag{6.38}$$

where $\pi_0 = N(\theta, \tau_0^2)$, τ_0 is specified, and $q(\theta) = \int \pi(\theta|\tau^2)\pi(\tau^2|a, b)d\tau^2$, as in Problem 6.3(a). The posterior density corresponding to a distribution $\pi \in \Pi$ is given by

$$\pi(\theta|\mathbf{x}) = \lambda(\mathbf{x})\pi_0(\theta|\bar{x}, \tau_0) + (1 - \lambda(\mathbf{x}))q(\theta|\mathbf{x}, a, b) \tag{6.39}$$

where $\lambda(\mathbf{x})$ is given by

$$\lambda(\mathbf{x}) = \frac{(1-\varepsilon)m_{\pi_0}(\bar{x}|\tau_0)}{(1-\varepsilon)m_{\pi_0}(\bar{x}|\tau_0) + \varepsilon m_q(\mathbf{x}|a, b)} \tag{6.40}$$

(see Problem 5.3). Using (6.39) and (6.40), the Bayes estimator for θ under squared error loss is

$$E(\Theta|\mathbf{x}, \tau_0, a, b) = \lambda(x)E(\Theta|\bar{x}, \tau_0) + (1 - \lambda(x))E(\Theta|\mathbf{x}, a, b), \tag{6.41}$$

a convex combination of the single-prior and hierarchical Bayes estimators, with the weights dependent on the marginal distribution. A robust Bayes analysis would proceed to evaluate the behavior (i.e., *robustness*) of this estimator as π ranges though Π. ‖

7 Risk Comparisons

In this concluding section, we look, in somewhat more detail, at the Bayes risk performance of some Bayes, empirical Bayes, and hierarchical Bayes estimators. We will also examine these risks under different prior assumptions, in the spirit of robust Bayes analysis.

Example 7.1 The James-Stein estimator. Let $\mathbf{X}$ have a p-variate normal distribution with mean $\boldsymbol{\theta}$ and covariance matrix $\sigma^2 I$, where σ^2 is known; $\mathbf{X} \sim N_p(\boldsymbol{\theta}, \sigma^2 I)$. We want to estimate $\boldsymbol{\theta}$ under sum-of-squared-errors loss

$$L[\boldsymbol{\theta}, \delta(\mathbf{x})] = |\boldsymbol{\theta} - \delta(\mathbf{x})|^2 = \sum_{i=1}^{p}(\theta_i - \delta_i(\mathbf{x}))^2,$$

using a prior distribution $\Theta \sim N(\mathbf{0}, \tau^2 I)$, where τ^2 is assumed to be known.

The Bayes estimator of $\boldsymbol{\theta}$ is $\delta^\tau(\mathbf{x}) = [\tau^2/(\sigma^2 + \tau^2)]\mathbf{x}$, $\mathbf{x}$ being the vector of componentwise Bayes estimates. It is straightforward to calculate its Bayes risk

$$r(\tau, \delta^\tau) = \frac{p\sigma^2\tau^2}{\sigma^2 + \tau^2}. \tag{7.1}$$

An empirical Bayes approach to this problem would replace τ^2 with an estimate from the marginal distribution of $\mathbf{x}$,

$$m(\mathbf{x}|\tau^2) = \frac{1}{[2\pi(\sigma^2 + \tau^2)]^{p/2}} e^{-\frac{1}{2(\sigma^2+\tau^2)}\Sigma x_i^2}. \tag{7.2}$$

Although, for the most part, we have used maximum likelihood to estimate the hyperparameters in empirical Bayes estimators, unbiased estimation provides an alternative. Using the unbiased estimator of $\tau^2/(\sigma^2 + \tau^2)$, the empirical Bayes estimator is (Problem 7.1)

$$\delta^{JS}(\mathbf{x}) = \left(1 - \frac{(p-2)\sigma^2}{|\mathbf{x}|^2}\right)\mathbf{x}, \tag{7.3}$$

the James-Stein estimator. ||

This estimator was discovered by Stein (1956b) and later shown by James and Stein (1961) to have a smaller mean squared error than the maximum likelihood estimator $\mathbf{X}$ for all $\boldsymbol{\theta}$. Its empirical Bayes derivation can be found in Efron and Morris (1972a).

Since the James-Stein estimator (or any empirical Bayes estimator) cannot attain as small a Bayes risk as the Bayes estimator, it is of interest to see how much larger its Bayes risk $r(\tau, \delta^{JS})$ will be. This, in effect, tells us the penalty we are paying for estimating τ^2.

As a first step, we must calculate $r(\tau, \delta^{JS})$, which is made easier by first obtaining an unbiased estimate of the risk $R(\boldsymbol{\theta}, \delta^{JS})$. The integration over $\boldsymbol{\theta}$ then becomes simple, since the integrand becomes constant in $\boldsymbol{\theta}$.

Recall Theorem 3.5, which gave an expression for the risk of a Bayes estimator of the form (3.3.12). In the normal case, we can apply the theorem to a fairly wide class of estimators to get an unbiased estimator of the risk.

Corollary 7.2 *Let $\mathbf{X} \sim N_p(\boldsymbol{\theta}, \sigma^2 I)$, and let the estimator δ be of the form*

$$\delta(\mathbf{x}) = \mathbf{x} - g(\mathbf{x}),$$

where $g(\mathbf{x}) = \{g_i(\mathbf{x})\}$ is differentiable. If $E_\theta|(\partial/\partial X_i)g_i(\mathbf{X})| < \infty$ for $i = 1, \dots, p$, then

$$\begin{aligned} R(\boldsymbol{\theta}, \delta) &= E_\theta|\boldsymbol{\theta} - \delta(\mathbf{X})|^2 \\ &= p\sigma^2 + E_\theta|g(\mathbf{X})|^2 - 2\sigma^2 \sum_{i=1}^p E_\theta \frac{\partial}{\partial X_i} g_i(\mathbf{X}). \end{aligned} \tag{7.4}$$

Hence,

$$\hat{R}(\delta(\mathbf{x})) = p\sigma^2 + |g(\mathbf{x})|^2 - 2\sigma^2 \sum_{i=1}^p \frac{\partial}{\partial x_i} g_i(\mathbf{x}) \tag{7.5}$$

is an unbiased estimator of the risk $R(\theta, \delta)$.

Proof. In the notation of Theorem 3.5, in the normal case $-\partial/\partial x_i \log h(\mathbf{x}) = x_i/\sigma^2$, and the result now follows by identifying $g(\mathbf{x})$ with $\nabla \log m(\mathbf{x})$, and some calculation. See Problem 7.2. □

For the James-Stein estimator (7.3), we have $g(\mathbf{x}) = (p-2)\sigma^2\mathbf{x}/|\mathbf{x}|^2$; hence,

$$
\begin{aligned}
R(\theta, \delta^{JS}) &= p\sigma^2 + E_\theta\left[\frac{(p-2)^2\sigma^4}{|\mathbf{X}|^2}\right] \\
&\quad -2\sigma^2\sum_{i=1}^{p} E_\theta\left[\frac{\partial}{\partial X_i}\frac{(p-2)\sigma^2 X_i}{|\mathbf{X}|^2}\right] \\
&= p\sigma^2 + (p-2)^2\sigma^4 E_\theta\frac{1}{|\mathbf{X}|^2} \qquad (7.6)\\
&\quad -2(p-2)\sigma^4\sum_{i=1}^{p} E_\theta\left[\frac{|\mathbf{X}|^2 - 2X_i^2}{|\mathbf{X}|^4}\right] \\
&= p\sigma^2 - (p-2)^2\sigma^4 E_\theta\frac{1}{|\mathbf{X}|^2},
\end{aligned}
$$

so $\hat{R}(\delta^{JS}(\mathbf{x})) = p\sigma^2 - (p-2)^2\sigma^4/|\mathbf{x}|^2$.

Example 7.3 Bayes risk of the James-Stein estimator. Under the model of Example 7.1, the Bayes risk of δ^{JS} is

$$
\begin{aligned}
r(\tau, \delta^{JS}) &= \int_\Omega R(\theta, \delta^{JS})\pi(\theta)\,d\theta \\
&= \int_\Omega\int_{\mathcal{X}}\left[p\sigma^2 - \frac{(p-2)^2\sigma^4}{|\mathbf{x}|^2}\right] f(\mathbf{x}|\theta)\pi(\theta)\,d\mathbf{x}\,d\theta \\
&= \int_{\mathcal{X}}\left\{\int_\Omega\left[p\sigma^2 - \frac{(p-2)^2\sigma^4}{|\mathbf{x}|^2}\right]\pi(\theta|\mathbf{x})\,d\theta\right\} m(\mathbf{x})\,d\mathbf{x},
\end{aligned}
$$

where we have used (7.6), and changed the order of integration. Since the integrand is independent of θ, the inner integral is trivially equal to 1, and

$$r(\tau, \delta^{JS}) = p\sigma^2 - (p-2)^2\sigma^4 E\frac{1}{|\mathbf{X}|^2}. \qquad (7.7)$$

Here, the expected value is over the marginal distribution of $\mathbf{X}$ (in contrast to (7.6), where the expectation is over the conditional distribution of $\mathbf{X}|\theta$).

Since, marginally, $E\frac{p-2}{|\mathbf{X}|^2} = \frac{1}{\sigma^2+\tau^2}$, we have

$$
\begin{aligned}
r(\tau, \delta^{JS}) &= p\sigma^2 - \frac{(p-2)\sigma^4}{\sigma^2+\tau^2} \\
&= \frac{p\sigma^2\tau^2}{\sigma^2+\tau^2} + \frac{2\sigma^4}{\sigma^2+\tau^2} \qquad (7.8)\\
&= r(\tau, \delta^\tau) + \frac{2\sigma^4}{\sigma^2+\tau^2}.
\end{aligned}
$$

Here, the second term represents the increase in Bayes risk that arises from estimating τ^2. ||

It is remarkable that δ^{JS} has a reasonable Bayes risk for any value of τ^2, although the latter is unknown to the experimenter. This establishes a degree of Bayesian robustness of the empirical Bayes estimator. Of course, the increase in risk is a function of σ^2 and can be quite large if σ^2 is large. Perhaps a more interesting comparison is obtained by looking at the relative increase in risk

$$\frac{r(\tau,\delta^{JS}) - r(\tau,\delta^{\tau})}{r(\tau,\delta^{\tau})} = \frac{2}{p}\frac{\sigma^2}{\tau^2}.$$

We see that the increase is a decreasing function of the ratio of the sample-to-prior variance and goes to 0 as $\sigma^2/\tau^2 \to 0$. Thus, the risk of the empirical Bayes estimator approaches that of the Bayes estimator as the sampling information gets infinitely better than the prior information.

Example 7.4 Bayesian robustness of the James-Stein estimator. To further explore the robustness of the James-Stein estimator, consider what happens to the Bayes risk if the prior used to calculate the Bayes estimator is different from the prior used to evaluate the Bayes risk (a classic concern of robust Bayesians).

For the model in Example 7.1, suppose we specify a value of τ, say τ_0. The Bayes estimator , δ^{τ_0}, is given by $\delta^{\tau_0}(x_i) = [\tau_0^2/(\tau_0^2+\sigma^2)]x_i$. When evaluating the Bayes risk, suppose we let the prior variance take on any value τ^2, not necessarily equal to τ_0^2. Then, the Bayes risk of δ^{τ_0} is (Problem 7.4)

$$r(\tau,\delta^{\tau_0}) = p\sigma^2\left(\frac{\tau_0^2}{\tau_0^2+\sigma^2}\right)^2 + p\tau^2\left(\frac{\sigma^2}{\tau_0^2+\sigma^2}\right)^2, \tag{7.9}$$

which is equal to the single-prior Bayes risk (7.1) when $\tau_0 = \tau$. However, as $\tau^2 \to \infty$, $r(\tau,\delta^{\tau_0}) \to \infty$, whereas $r(\tau,\delta^{\tau}) \to p\sigma^2$.

In contrast, the Bayes risk of δ^{JS}, given in (7.8), is valid for all τ with $r(\tau,\delta^{JS}) \to p\sigma^2$ as $\tau^2 \to \infty$. Thus, the Bayes risk of δ^{JS} remains finite for any prior in the class, demonstrating robustness. ||

In constructing an empirical Bayes estimator in Example 7.1, the use of unbiased estimation of the hyperparameters led to the James-Stein estimator. If, instead, we had used maximum likelihood, the resulting empirical Bayes estimator would have been (Problem 7.1)

$$\delta^{+}(\mathbf{x}) = \left(1 - \frac{p\sigma^2}{|\mathbf{x}|^2}\right)^{+}\mathbf{x}, \tag{7.10}$$

where $(a)^+ = \max\{0, a\}$. Such estimators are known as positive-part Stein estimators.

A problem with the empirical Bayes estimator (7.3) is that when $|\mathbf{x}|^2$ is small (less than $(p-2)\sigma^2$), the estimator has the "wrong sign"; that is, the signs of the components of δ^{JS} will be opposite those of the Bayes estimator δ^{τ}. This does not happen with the estimator (7.10), and as a result, estimators like (7.10) tend to have improved Bayes risk performance.

Estimators such as (7.3) and (7.10) are called *shrinkage estimators*, since they tend to shrink the estimator $\mathbf{X}$ toward 0, the shrinkage target. Actually, of the two, only (7.10) completely succeeds in this effort since the shrinkage factor $1-(p-$

$2)\sigma^2/|\mathbf{x}|^2$ may take on negative (and even very large negative) values. Nevertheless, the terminology is used to cover also (7.3). The following theorem is due to Efron and Morris (1973a).

Theorem 7.5 *Let* $\mathbf{X} \sim N_p(\boldsymbol{\theta}, \sigma^2 I)$ *and* $\boldsymbol{\theta} \sim N_p(0, \tau^2 I)$, *with loss function* $L(\boldsymbol{\theta}, \delta) = |\boldsymbol{\theta} - \delta|^2$. *If* $\delta(\mathbf{x})$ *is an estimator of the form*

$$\delta(\mathbf{x}) = [1 - B(\mathbf{x})]\mathbf{x}$$

and if

$$\delta^+(\mathbf{x}) = [1 - B(\mathbf{x})]^+\mathbf{x},$$

then $r(\tau, \delta) \geq r(\tau, \delta^+)$, *with strict inequality if* $P_\theta(\delta(\mathbf{X}) \neq \delta^+(\mathbf{X})) > 0$.

Proof. For any estimator $\delta(\mathbf{x})$, the posterior expected loss is given by

$$E[L(\boldsymbol{\theta}, \delta(\mathbf{x}))|\mathbf{x}] = \int_\Omega \sum_{i=1}^p (\theta_i - \delta_i(\mathbf{x}))^2 \pi(\boldsymbol{\theta}|\mathbf{x})\, d\boldsymbol{\theta}$$

$$(7.11) \qquad = \int_\Omega \sum_{i=1}^p \left[(\theta_i - E(\theta_i|\mathbf{x}))^2 + (E(\theta_i|\mathbf{x}) - \delta_i(\mathbf{x}))^2\right] \times \pi(\boldsymbol{\theta}|\mathbf{x})\, d\boldsymbol{\theta}$$

where we have added $\pm E(\theta_i|\mathbf{x})$ and expanded the square, noting that the cross-term is zero. Equation (7.11) can then be written as

$$(7.12) \qquad E[L(\boldsymbol{\theta}, \delta(\mathbf{x}))|\mathbf{x}] = \sum_{i=1}^p \text{var}(\theta_i|\mathbf{x}) + \sum_{i=1}^p [E(\theta_i|\mathbf{x}) - \delta_i(\mathbf{x})]^2.$$

As the first term in (7.12) does not depend on the particular estimator, the difference in posterior expected loss between δ and δ^+ is

$$(7.13) \quad E[L(\boldsymbol{\theta}, \delta(\mathbf{x})|\mathbf{x}] - E[L(\boldsymbol{\theta}, \delta^+(\mathbf{x}))|\mathbf{x}] = \sum_{i=1}^p \left\{[E(\theta_i|\mathbf{x}) - \delta_i(\mathbf{x})]^2 - [E(\theta_i|\mathbf{x})]^2\right\} I(|B(\mathbf{x})| > 1)$$

since the estimators are identical when $|B(\mathbf{x})| \leq 1$. However, since $E(\theta_i|\mathbf{x}) = \tau^2/(\sigma^2 + \tau^2)x_i$, it follows that when $|B(\mathbf{x})| > 1$,

$$\left[\frac{\tau^2}{\sigma^2 + \tau^2}x_i - \delta_i(\mathbf{x})\right]^2 > \left[\frac{\tau^2}{\sigma^2 + \tau^2}x_i\right]^2.$$

Thus, (7.13) is positive for all $\mathbf{x}$, and the result follows by taking expectations. □

In view of results like Theorem 7.5 and other risk results in Chapter 5 (see Theorem 5.5.4), the positive-part Stein estimator

$$(7.14) \qquad \delta^+(\mathbf{x}) = \left(1 - \frac{(p-2)\sigma^2}{|\mathbf{x}|^2}\right)^+ \mathbf{x}$$

is preferred to the ordinary James-Stein estimator (7.3). Moreover, Theorem 7.5 generalizes to the entire exponential family (Problem 7.8). It also supports the use of maximum likelihood estimation in empirical Bayes constructions.

The good Bayes risk performance of empirical Bayes estimators is not restricted to the normal case, nor to squared error loss. We next look at Bayes and empirical Bayes estimation in the Poisson case.

Example 7.6 Poisson Bayes and empirical Bayes estimation. Recall the Poisson model of Example 6.6:

$$(7.15)\qquad \begin{aligned} &X_i \sim \text{Poisson}(\lambda_i), \quad i = 1, \dots, p, \ \text{independent}, \\ &\lambda_i \sim \text{Gamma}(a, b). \end{aligned}$$

For estimation of λ_i under the loss

$$(7.16)\qquad L_k(\lambda, \delta) = \sum_{i=1}^{p} \frac{1}{\lambda_i^k}(\lambda_i - \delta_i)^2,$$

the Bayes estimator (see Example 1.3) is

$$(7.17)\qquad \delta_i^k(\mathbf{x}) = \frac{b}{b+1}(x_i + a - k).$$

The posterior expected loss of $\delta_i^k(\mathbf{x}) = \delta_i^k(x_i)$ is

$$(7.18)\qquad E\left[\frac{1}{\lambda_i^k}[\lambda_i - \delta_i^k(x_i)]^2 | x_i\right] = \frac{1}{\Gamma(a+x_i)\left(\frac{b}{b+1}\right)^{a+x_i}} \times \int_0^\infty (\lambda_i - \delta_i)^2 \lambda_i^{a+x_i-k-1} e^{-\frac{\lambda_i(b+1)}{b}}\, d\lambda_i,$$

since the posterior distribution of $\lambda_i | x_i$ is Gamma$(a + x_i, \frac{b}{b+1})$. Evaluating the integral in (7.18) gives

$$(7.19)\qquad E\left[\frac{1}{\lambda_i^k}[\lambda_i - \delta_i^k(x_i)]^2 | x_i\right] = \frac{\Gamma(a+x_i-k)}{\Gamma(a+x_i)}\left(\frac{b}{b+1}\right)^{2-k}(a + x_i - k).$$

To evaluate the Bayes risk, $r(k, \delta^k)$, we next sum (7.19) with respect to the marginal distribution of X_i, which is Negative Binomial$(a, \frac{1}{b+1})$. For $k = 0$ and $k = 1$, we have

$$r(0, \delta^0) = p\frac{ab^2}{b+1} \quad \text{and} \quad r(1, \delta^1) = p\frac{b}{b+1}.$$

See Problems 7.10 and 7.11 for details.

For the model (7.15) with loss function $L_k(\lambda, \delta)$ of (7.16), an empirical Bayes estimator can be derived (similar to (6.6.24); see Example 6.6) as

$$(7.20)\qquad \delta_{k_i}^{EB}(\mathbf{x}) = \frac{\bar{x}}{\bar{x}+a}(x_i + a - k).$$

We shall now consider the risk of the estimator δ^{EB}. For the loss function (7.16), we can actually evaluate the risk of a more general estimator than δ^{EB}. The coordinatewise posterior expected loss of an estimator of the form $\delta_i^\varphi = \varphi(\bar{x})(x_i + a - k)$

is

$$E\left[\frac{1}{\lambda_i^k}[\lambda_i - \delta_i^\varphi]^2|\mathbf{x}\right]$$

$$(7.21) \quad = \frac{1}{\Gamma(a+x_i)\left(\frac{b}{b+1}\right)^{a+x_i}}\int_0^\infty [\lambda_i - \delta_i^\varphi(\mathbf{x})]^2\lambda_i^{a+x_i-k-1}e^{-\lambda_i\frac{b+1}{b}}\,d\lambda_i$$

$$= \frac{\Gamma(a+x_i-k)}{\Gamma(a+x_i)}\left(\frac{b}{b+1}\right)^{-k} E[(\lambda_i - \delta_i^\varphi(\mathbf{x}))^2|\mathbf{x})]$$

where the expectation is over the random variable λ_i with distribution Gamma($a + x_i - k$, $\frac{b}{b+1}$). Using the same technique as in the proof of Theorem 7.5 [see (7.12)], we add $\pm\delta_i^k(x_i) = \frac{b}{b+1}(a + x_i - k)$ in (7.21) to get

$$E\left[\frac{1}{\lambda_i^k}[\lambda_i - \delta_i^\varphi]^2|\mathbf{x}\right]$$

$$(7.22) \quad = \frac{\Gamma(a+x_i-k)}{\Gamma(a+x_i)}\left(\frac{b}{b+1}\right)^{2-k}(a+x_i-k)$$

$$+\frac{\Gamma(a+x_i-k)}{\Gamma(a+x_i)}\left(\frac{b}{b+1}\right)^{-k}\left(\frac{b}{b+1} - \varphi(\bar{x})\right)^2(a+x_i-k)^2.$$

The first term in (7.22) is the posterior expected loss of the Bayes estimator, and the second term reflects the penalty for estimating b. Evaluation of the Bayes risk, which involves summing over x_i, is somewhat involved (see Problem 7.11). Instead, Table 7.1 provides a few numerical comparisons. Specifically, it shows the Bayes risks for the Bayes (δ^k), empirical Bayes (δ^{EB}), and unbiased estimators (X) of Example 6.6, based on observing p independent Poisson variables, for the loss function (7.16) with $k = 1$. The gamma parameters are chosen so that the prior mean equals 10 and the prior variances are 5 ($a = 20, b = .5$), 10 ($a = 10$, $b = 1$), and 25 ($a = 4, b = 2.5$). It is seen that the empirical Bayes estimator attains a reasonable Bayes risk reduction over that of X, and in some cases, comes quite close to the optimum. ||

As a final example of Bayes risk performance, we turn now to the analysis of variance. Here, we shall consider only the one-way layout (Examples 4.1, 4.6 and 4.9) in detail. Other situations and generalizations are illustrated in the problems (Problems 7.17 and 7.18).

Example 7.7 Empirical Bayes analysis of variance. In the one-way layout (considered earlier in Example 3.4.9 from the point of view of equivariance), we have

$$(7.23) \quad X_{ij} \sim N(\xi_i, \sigma^2), \quad j = 1, \ldots, n_i; \quad i = 1, \ldots, s,$$
$$\xi_i = \mu + \alpha_i, \quad i = 1, \ldots, s$$

where we assume that $\Sigma\alpha_i = 0$ to ensure the identifiability of parameters. With this restriction, the parameterization in terms of μ and α_i is equivalent to that in terms of ξ_i, with the latter parameterization (the so-called *cell means model*; see Searle 1987) being computationally more friendly. As interest often lies in estimation of,

Table 7.1. *Comparisons of Some Bayes Risks for Model (7.15)*

	$p = 5$		
Prior var.	δ^k of (7.17)	δ^{EB} of (7.20)	X
5	1.67	2.28	5.00
10	2.50	2.99	5.00
25	3.57	3.84	5.00
	$p = 20$		
Prior var.	δ^π of (7.17)	δ^{EB} of (7.20)	X
5	6.67	7.31	20.00
10	10.00	10.51	20.00
25	14.29	14.52	20.00

and testing hypotheses about, the differences of the α_is, which are equivalent to the differences of the ξ_i's, we will use the ξ_i version of the model. We will also specialize to the *balanced* case where all n_i's are equal. The more general case requires some (often much) extra effort. (See Problems 7.16 and 7.19).

As an illustration, consider an experiment to assess the effect of linseed oil meal on the digestibility of food by steers. The measurements are a digestibility coefficient, and there are five treatments, representing different amounts of linseed oil meal added to the feed (approximately 1, 2, 3, 4, and 5 kg/animal/day; see Hsu 1982 for more details.) The variable X_{ij} of (7.23) is the jth digestibility measurement in the ith treatment group, where ξ_i is the true coefficient of digestibility of that group. Perhaps the most common hypothesis about the ξ_i's is

$$H_0 : \xi_1 = \xi_2 = \cdots = \xi_s = \mu, \quad \mu \text{ unknown.} \tag{7.24}$$

This specifies that the means are equal and, hence, the treatment groups are equivalent in that they each result in the same (unknown) mean level of digestibility. This hypothesis can be thought of as specifying a submodel where all of the ξ's are equal, which suggests expanding (7.23) into the hierarchical model

$$\begin{aligned} X_{ij}|\xi_i &\sim N(\xi_i, \sigma^2), \; j = 1, \ldots, n, \; i = 1, \ldots, s, \quad \text{independent,} \\ \xi_i|\mu &\sim N(\mu, \tau^2), \; i = 1, \ldots, s, \quad \text{independent.} \end{aligned} \tag{7.25}$$

The model (7.25) is obtained from (7.24) by allowing some variation around the prior mean, μ, in the form of a normal distribution.

In analogy to (4.2.4), the Bayes estimator of ξ_i is

$$\delta^B(\bar{x}_i) = \frac{\sigma^2}{\sigma^2 + n\tau^2}\mu + \frac{n\tau^2}{\sigma^2 + n\tau^2}\bar{x}_i. \tag{7.26}$$

Calculation of an empirical Bayes estimator is straightforward. Since the marginal distribution of $\bar{X}_i$ is

$$\bar{X}_i \sim N\left(\mu, \frac{\sigma^2}{n} + \tau^2\right), \quad i = 1, \ldots, s,$$

the MLE of μ is $\bar{x} = \sum_i \sum_j x_{ij}/ns$ and the resulting empirical Bayes estimator is

$$\delta_i^{EB} = \frac{\sigma^2}{\sigma^2 + n\tau^2}\bar{x} + \frac{n\tau^2}{\sigma^2 + n\tau^2}\bar{x}_i. \tag{7.27}$$

Note that δ^{EB} is a linear combination of $\bar{X}_i$, the UMVU estimator under the full model, and $\bar{X}$, the UMVU estimator under the submodel that specifies $\xi_1 = \cdots = \xi_s$.

If we drop the assumption that τ^2 is known, we can estimate $(\sigma^2 + n\tau^2)^{-1}$ by the unbiased estimator $(s-3)/\Sigma(\bar{x}_i - \bar{x})^2$ and obtain the empirical Bayes estimator

$$\delta_i^L = \bar{x} + \left(1 - \frac{(s-3)\sigma^2}{\Sigma(\bar{x}_i - \bar{x})^2}\right)(\bar{x}_i - \bar{x}), \tag{7.28}$$

which was first derived by Lindley (1962) and examined in detail by Efron and Morris (1972a 1972b, 1973a, 1973b).

Calculation of the Bayes risk of δ^L proceeds as in Example 7.3, and leads to

$$\begin{aligned} r(\xi, \delta^L) &= s\frac{\sigma^2}{n} - (s-3)^2\left(\frac{\sigma^2}{n}\right)^2 E\left[\sum_{i=1}^{s}(\bar{X}_i - \bar{X})^2\right]^{-1} \\ &= r(\xi, \delta^B) + \frac{3(\sigma^2/n)^2}{\sigma^2/n + \tau^2} \end{aligned} \tag{7.29}$$

where $\sum_{i=1}^{s}(\bar{X}_i - \bar{X})^2 \sim (\sigma^2/n + \tau^2)\chi^2_{s-1}$ and $r(\xi, \delta^B)$ is the risk of the Bayes estimator (7.26). See Problem 7.14 for details.

If we compare (7.29) to (7.8), we see that the Bayes risk performance of δ^L, where we have estimated the value of μ, is similar to that of δ^{JS}, where we assume that the value of μ is known. The difference is that δ^L pays an extra penalty for estimating the point that is the shrinkage target. For δ^{JS}, the target is assumed known and taken to be 0, while δ^L estimates it by $\bar{X}$. The penalty for this is that the factor in the term added to the Bayes risk is increased from 2 in (7.8), where $k = 1$ to 3. In general, if we shrink to a k-dimensional subspace, this factor is $2 + k$. ∥

More general submodels can also be incorporated in empirical Bayes analyses, and in many cases, the resulting estimators retain good Bayes risk performance.

Example 7.8 Analysis of variance with regression submodel. Another common hypothesis (or submodel) in the analysis of variance is that of a linear trend in the means, which was considered earlier in Example 3.4.7 and can be written as the null hypothesis

$$H_0 : \xi_i = \alpha + \beta t_i, \; i = 1, \ldots, s, \quad \alpha \text{ and } \beta \text{ unknown}, \quad t_i \text{ known}.$$

For the situation of Example 7.7, this hypothesis would assert that the effect of the quantity of linseed oil meal on digestibility is linear. (We know that as the quantity of linseed oil meal increases, the coefficient of digestibility decreases. But we do not know if this relationship is linear.) In analogy with (7.25), we can translate the hypothesis into the hierarchical model

$$\text{(7.30)} \qquad \begin{aligned} X_{ij}|\xi_i &\sim N(\xi_i, \sigma^2), \quad j = 1, \ldots, n, \quad i = 1, \ldots, s, \\ \xi_i|\alpha, \beta &\sim N(\alpha + \beta t_i, \tau^2), \quad i = 1, \ldots, s. \end{aligned}$$

Again, the hypothesis models the prior mean of the ξ_i's, and we allow variation around this prior mean in the form of a normal distribution. Using squared error loss, the Bayes estimator of ξ_i is

$$\text{(7.31)} \qquad \delta_i^B = \frac{\sigma^2}{\sigma^2 + n\tau^2}(\alpha + \beta t_i) + \frac{n\tau^2}{\sigma^2 + n\tau^2}\bar{X}_i.$$

For an empirical Bayes estimator, we calculate the marginal distribution of $\bar{X}_i$.

$$\bar{X}_i \sim N(\alpha + \beta t_i, \sigma^2 + n\tau^2), \quad i = 1, \ldots, s,$$

and estimate α and β by

$$\hat{\alpha} = \bar{X} - \hat{\beta}\bar{t}, \quad \hat{\beta} = \frac{\Sigma(\bar{X}_i - \bar{X})(t_i - \bar{t})}{\Sigma(t_i - \bar{t})^2},$$

the UMVU estimators of α and β (Section 3.4). This yields the empirical Bayes estimator

$$\text{(7.32)} \qquad \delta_i^{EB_1} = \frac{\sigma^2}{\sigma^2 + n\tau^2}(\hat{\alpha} + \hat{\beta} t_i) + \frac{n\tau^2}{\sigma^2 + n\tau^2}\bar{X}_i.$$

If τ^2 is unknown we can, in analogy to Example 7.7 use the fact that, marginally, $E[\Sigma(\bar{X}_i - \hat{\alpha} + \hat{\beta} t_i)^2]^{-1} = (s-4)/(\sigma^2/n + \tau^2)$ to construct the estimator

$$\text{(7.33)} \qquad \delta_i^{EB_2} = \hat{\alpha} + \hat{\beta} t_i + \left(1 - \frac{(s-4)\sigma^2}{\Sigma(\bar{X}_i - \hat{\alpha} - \hat{\beta} t_i)^2}\right)(\bar{X}_i - \hat{\alpha} - \hat{\beta} t_i)$$

with Bayes risk

$$\text{(7.34)} \qquad \begin{aligned} r(\tau, \delta^{EB_2}) &= s\frac{\sigma^2}{n} - (s-4)^2\left(\frac{\sigma^2}{n}\right)^2 E\left[\sum_{i=1}^{s}(\bar{X}_i - \hat{\alpha} - \hat{\beta} t_i)^2\right] \\ &= r(\xi, \delta^B) + \frac{4(\sigma^2/n)^2}{\sigma^2/n + \tau^2} \end{aligned}$$

where $r(\xi, \delta^B)$ is the risk of the Bayes estimator δ^B of (7.31). See Problem 7.14 for details.

Notice that here we shrunk the estimator toward a two-dimensional submodel, and the factor in the second term of the Bayes risk is 4 $(2 + k)$. We also note that for δ^{EB_2}, as well as δ^{JS} and δ^L, the Bayes risk approaches that of δ^π as $n \to \infty$. ||

In both Examples 7.7 and 7.8, empirical Bayes estimators provide a means for attaining reasonable Bayes risk performance if σ^2/τ^2 is not too large, yet do not require full specification of a prior distribution. An obvious limitation of these results, however, is the dimension of the submodel. The ordinary James-Stein estimator shrinks toward the point 0 [see Equation (7.3)], or any specified point

(Problem 7.6), and hence toward a submodel (subspace) of dimension zero. In the analysis of variance, Example 7.7, the subspace of the submodel has dimension 1, $\{(\xi_1, \ldots, \xi_s) : \xi_i = \mu, i = 1, \ldots, s\}$, and in Example 7.8, it has dimension 2, $\{(\xi_1, \ldots, \xi_s) : \{i = \alpha + \beta t_i, i = 1, \ldots, s\}$. In general, the empirical Bayes strategies developed here will only work if the dimension of the submodel, r, is at least two fewer than that of the full model, s; that is, $s - r > 2$. This is a technical requirement, as the marginal distribution of interest is χ^2_{s-r}, and estimation is problematic if $s - r \leq 2$. The reason for this difficulty is the need to calculate the expectation $E(1/\chi^2_{s-r})$, which is infinite if $s - r \leq 2$. (See Problem 7.6; also see Problem 6.12 for an attempt at empirical Bayes if $s - r \leq 2$.)

In light of Theorem 7.5, we can improve the empirical Bayes estimators of Examples 7.7 and 7.8 by using their positive-part version. Moreover, Problem 7.8 shows that such an improvement will hold throughout the entire exponential family. Thus, the strategy of taking a positive part should always be employed in these cases of empirical Bayes estimation.

Finally, we note that Examples 7.7 and 7.8 can be greatly generalized. One can handle unequal n_i, unequal variances, full covariance matrices, general linear submodels, and more. In some cases, the algebra can become somewhat overwhelming, and details about performance of the estimators may become obscured. We examine a number of these cases in Problems 7.16–7.18.

8 Problems

Section 1

1.1 Verify the expressions for $\pi(\lambda|\bar{x})$ and $\delta^k(\bar{x})$ in Example 1.3.

1.2 Give examples of pairs of values (a, b) for which the beta density $B(a, b)$ is (a) decreasing, (b) increasing, (c) increasing for $p < p_0$ and decreasing for $p > p_0$, and (d) decreasing for $p < p_0$ and increasing for $p > p_0$.

1.3 In Example 1.5, if p has the improper prior density $\frac{1}{p(1-p)}$, show that the posterior density of p given x is proper, provided $0 < x < n$.

1.4 In Example 1.5, find the Jeffreys prior for p and the associated Bayes estimator δ_Λ.

1.5 For the estimator δ_Λ of Problem 1.4,

(a) calculate the bias and maximum bias;

(b) calculate the expected squared error and compare it with that of the UMVU estimator.

1.6 In Example 1.5, find the Bayes estimator δ of $p(1 - p)$ when p has the prior $B(a, b)$.

1.7 For the situation of Example 1.5, the UMVU estimator of $p(1 - p)$ is $\delta' = [x(x - 1)]/[n(n - 1)]$ (see Example 2.3.1 and Problem 2.3.1).

(a) Compare the estimator δ of Problem 1.6 with the UMVU estimator δ'.

(b) Compare the expected squared error of the estimator of $p(1 - p)$ for the Jeffreys prior in Example 1.5 with that of δ'.

1.8 In analogy with Problem 1.2, determine the possible shapes of the gamma density $\Gamma(g, 1/\alpha), \alpha, g > 0$.

1.9 Let $X_1, \ldots, X_n$ be iid according to the Poisson distribution $P(\lambda)$ and let λ have a gamma distribution $\Gamma(g, \alpha)$.

(a) For squared error loss, show that the Bayes estimator $\delta_{\alpha,g}$ of λ has a representation analogous to (1.1.13).

(b) What happens to $\delta_{\alpha,g}$ as (i) $n \to \infty$, (ii) $\alpha \to \infty$, $g \to 0$, or both?

1.10 For the situation of the preceding problem, solve the two parts corresponding to Problem 1.5(a) and (b).

1.11 In Problem 1.9, if λ has the improper prior density $d\lambda/\lambda$ (corresponding to $\alpha = g = 0$), under what circumstances is the posterior distribution proper?

1.12 Solve the problems analogous to Problems 1.9 and 1.10 when the observations consist of a single random variable X having a negative binomial distribution $Nb(p, m)$, p has the beta prior $B(a, b)$, and the estimand is (a) p and (b) $1/p$.

Section 2

2.1 Referring to Example 1.5, suppose that X has the binomial distribution $b(p, n)$ and the family of prior distributions for p is the family of beta distributions $B(a, b)$.

(a) Show that the marginal distribution of X is the *beta-binomial* distribution with mass function

$$\binom{n}{x} \frac{\Gamma(a+b)}{\Gamma(a)\Gamma(b)} \frac{\Gamma(x+a)\Gamma(n-x+b)}{\Gamma(n+a+b)}.$$

(b) Show that the mean and variance of the beta-binomial is given by

$$EX = \frac{na}{a+b} \quad \text{and} \quad \text{var } X = n\left(\frac{a}{a+b}\right)\left(\frac{b}{a+b}\right)\left(\frac{a+b+n}{a+b+1}\right).$$

[*Hint*: For part (b), the identities $EX = E[E(X|p)]$ and $\text{var } X = \text{var}[E(X|p)] + E[\text{var}(X|p)]$ are helpful.]

2.2 For the situation of Example 2.1, Lindley and Phillips (1976) give a detailed account of the effect of stopping rules, which we can illustrate as follows. Let X be the number of successes in n Bernoulli trials with success probability p.

(a) Suppose that the number of Bernoulli trials performed is a prespecifed number n, so that we have the binomial sampling model, $P(X = x) = \binom{n}{x} p^x(1-p)^{n-x}$, $x - 0, 1, \ldots, n$. Calculate the Bayes risk of the Bayes estimator (1.1.12) and the UMVU estimator of p.

(b) Suppose that the number of Bernoulli trials performed is a random variable N. The value $N = n$ was obtained when a prespecified number, x, of successes was observed so that we have the negative binomial sample model, $P(N = n) = \binom{n-1}{x-1} p^x(1-p)^{n-x}$, $n = x$. Calculate the Bayes risk of the Bayes estimator and the UMVU estimator of p.

(c) Calculate the mean squared errors of all three estimators under each model. If it is unknown which sampling mechanism generated the data, which estimator do you prefer overall?

2.3 Show that the estimator (2.2.4) tends in probability (a) to θ as $n \to \infty$, (b) to μ as $b \to 0$, and (c) to θ as $b \to \infty$.

2.4 Bickel and Mallows (1988) further investigate the relationship between unbiasedness and Bayes, specifying conditions under which these properties cannot hold simultaneously. In addition, they show that if a prior distribution is improper, then a posterior mean can be unbiased. Let $X \sim \frac{1}{\theta}f(x/\theta), x > 0$, where $\int_0^\infty tf(t)dt = 1$, and let $\pi(\theta) = \frac{1}{\theta^2}d\theta$, $\theta > 0$.

(a) Show that $E(X|\theta) = \theta$, so X is unbiased.

(b) Show that $\pi(\theta|x) = \frac{x^2}{\theta^3}f(x/\theta)$ is a proper density.

(c) Show that $E(\theta|x) = x$, and hence the posterior mean, is unbiased.

2.5 DasGupta (1994) presents an identity relating the Bayes risk to bias, which illustrates that a small bias can help achieve a small Bayes risk. Let $X \sim f(x|\theta)$ and $\theta \sim \pi(\theta)$. The Bayes estimator under squared error loss is $\delta^\pi = E(\theta|x)$. Show that the Bayes risk of δ^π can be written

$$r(\pi, \delta^\pi) = \int_\Theta \int_\mathcal{X} [\theta - \delta^\pi(x)]^2 f(x|\theta)\pi(\theta)dx d\theta = \int_\Theta \theta b(\theta)\pi(\theta)d\theta$$

where $b(\theta) = E[\delta^\pi(X)|\theta] - \theta$ is the bias of δ^π.

2.6 Verify the estimator (2.2.10).

2.7 In Example 2.6, verify that the posterior distribution of τ is $\Gamma(r + g - 1/2, 1/(\alpha + z))$.

2.8 In Example 2.6 with $\alpha = g = 0$, show that the posterior distribution given the X's of $\sqrt{n}(\theta - \bar{X})/\sqrt{Z/(n-1)}$ is Student's t-distribution with $n - 1$ degrees of freedom.

2.9 In Example 2.6, show that the posterior distribution of θ is symmetric about $\bar{x}$ when the joint prior of θ and σ is of the form $h(\sigma)d\sigma\, d\theta$, where h is an arbitrary probability density on $(0, \infty)$.

2.10 Rukhin (1978) investigates the situation when the Bayes estimator is the same for every loss function in a certain set of loss functions, calling such estimators *universal Bayes estimators*. For the case of Example 2.6, using the prior of the form of Problem 2.9, show that $\bar{X}$ is the Bayes estimator under every even loss function.

2.11 Let X and Y be independently distributed according to distributions P_ξ and Q_η, respectively. Suppose that ξ and η are real-valued and independent according to some prior distributions Λ and Λ'. If, with squared error loss, δ_Λ is the Bayes estimator of ξ on the basis of X, and $\delta'_{\Lambda'}$ is that of η on the basis of Y,

(a) show that $\delta'_{\Lambda'} - \delta_\Lambda$ is the Bayes estimator of $\eta - \xi$ on the basis of (X, Y);

(b) if $\eta > 0$ and $\delta^*_{\Lambda'}$ is the Bayes estimator of $1/\eta$ on the basis of Y, show that $\delta_\Lambda \cdot \delta^*_{\Lambda'}$ is the Bayes estimator of ξ/η on the basis of (X, Y).

2.12 For the density (2.2.13) and improper prior $(d\sigma/\sigma)\cdot(d\sigma_A/\sigma_A)$, show that the posterior distribution of (σ, σ_A) continues to be improper.

2.13 (a) In Example 2.7, obtain the Jeffreys prior distribution of (σ, τ).

(b) Show that for the prior of part (a), the posterior distribution of (σ, τ) is proper.

2.14 Verify the Bayes estimator (2.2.14).

2.15 Let $X \sim N(\theta, 1)$ and $L(\theta, \delta) = (\theta - \delta)^2$.

(a) Show that X is the limit of the Bayes estimators δ^{π_n}, where π_n is $N(0, 1)$. Hence, X is both generalized Bayes and a limit of Bayes estimators.

(b) For the prior measure $\pi(\theta) = e^{a\theta}, a > 0$, show that the generalized Bayes estimator is $X + a$.

(c) For $a > 0$, show that there is no sequence of proper priors for which $\delta^{\pi_n} \to X + a$.

This example is due to Farrell; see Kiefer 1966. Heath and Sudderth (1989), building on the work of Stone (1976), showed that inferences from this model are *incoherent*, and established when generalized Bayes estimators will lead to *coherent* (that is, noncontradictory) inferences. Their work is connected to the theory of "approximable by proper priors," developed by Stein (1965) and Stone (1965, 1970, 1976), which shows when generalized Bayes estimators can be looked upon as Bayes estimators.

2.16 (a) For the situation of Example 2.8, verify that $\delta(x) = x/n$ is a generalized Bayes estimator.

(b) If $X \sim N(0, 1)$ and $L(\theta, \delta) = (\theta - \delta)^2$, show that X is generalized Bayes under the improper prior $\pi(\theta) = 1$.

Section 3

3.1 For the situation of Example 3.1:

(a) Verify that the Bayes estimator will only depend on the data through $Y = \max_i X_i$.

(b) Show that $E(\Theta|y, a, b)$ can be expressed as

$$E(\Theta|y, a, b) = \frac{1}{b(n+a-1)} \frac{P(\chi^2_{2(n+a-1)} < 2/by)}{P(\chi^2_{2(n+a)} < 2/by)}$$

where χ^2_ν is a chi-squared random variable with ν degrees of freedom. (In this form, the estimator is particularly easy to calculate, as many computer packages will have the chi-squared distribution built in.)

3.2 Let $X_1, \ldots, X_n$ be iid from Gamma(a, b) where a is known.

(a) Verify that the conjugate prior for the natural parameter $\eta = -1/b$ is equivalent to an inverted gamma prior on b.

(b) Using the prior in part (a), find the Bayes estimator under the losses (i) $L(b, \delta) = (b - \delta)^2$ and (ii) $L(b, \delta) = (1 - \delta/b)^2$.

(c) Express the estimator in part (b)(i) in the form (3.3.9). Can the same be done for the estimator in part (b)(ii)?

3.3 (a) Prove Corollary 3.3.

(b) Verify the calculation of the Bayes estimator in Example 3.4.

3.4 Using Stein's identity (Lemma 1.5.15), show that if $X_i \sim p_{\eta_i}(x)$ of (3.3.7), then

$$E_\eta(-\nabla \log h(\mathbf{X})) = \eta,$$

$$R(\eta, -\nabla \log h(\mathbf{X})) = \sum_{i=1}^{p} E_\eta \left[-\frac{\partial^2}{\partial X_i^2} \log h(\mathbf{X}) \right].$$

3.5 (a) If $X_i \sim$ Gamma(a, b), $i = 1, \ldots, p$, independent with a known, calculate $-\nabla \log h(\mathbf{x})$ and its expected value.

(b) Apply the results of part (a) to the situation where $X_i \sim N(0, \sigma_i^2)$, $i = 1, \ldots, p$, independent. Does it lead to an unbiased estimator of σ_i^2?

[*Note*: For part (b), squared error loss on the natural parameter $1/\sigma^2$ leads to the loss $L(\sigma^2, \delta) = (\sigma^2\delta - 1)^2/\sigma^4$ for estimation of σ^2.]

(c) If

$$X_i \sim \frac{\tan(a_i\pi)}{\pi} x^{a_i}(1-x)^{-1}, \; 0 < x < 1, \; i = 1, \ldots, p, \text{ independent},$$

evaluate $-\nabla \log h(\mathbf{X})$ and show that it is an unbiased estimator of $\mathbf{a} = (a_1, \ldots, a_p)$.

3.6 For the situation of Example 3.6:

(a) Show that if δ is a Bayes estimator of θ, then $\delta' = \delta/\sigma^2$ is a Bayes estimator of η, and hence $R(\theta, \delta) = \sigma^4 R(\eta, \delta')$.

(b) Show that the risk of the Bayes estimator of η is given by

$$\frac{p\tau^4}{\sigma^2(\sigma^2+\tau^2)^2} + \left(\frac{\sigma^2}{\sigma^2+\tau^2}\right)^2 \sum a_i^2,$$

where $a_i = \eta_i - \mu/\sigma^2$.

(c) If $\Sigma a_i^2 = k$, a fixed constant, then the minimum risk is attained at $\eta_i = \mu/\sigma^2 + \sqrt{k/p}$.

3.7 If $\mathbf{X}$ has the distribution $p_\theta(\mathbf{x})$ of (1.5.1) show that, similar to Theorem 3.2, $E(\mathcal{T}\eta(\theta)) = \nabla \log m_\pi(\mathbf{x}) - \nabla \log h(\mathbf{x})$.

3.8 (a) Use Stein's identity (Lemma 1.5.15) to show that if $X_i \sim p_{\eta_i}(x)$ of (3.3.18), then

$$E_\eta(-\nabla \log h(\mathbf{X})) = \sum_i \eta_i E_\eta \frac{\partial}{\partial X_j} T_i(\mathbf{X}).$$

(b) If X_i are iid from a gamma distribution Gamma(a, b), where the shape parameter a is known, use part (a) to find an unbiased estimator of $1/b$.

(c) If the X_i are iid from a beta(a, b) distribution, can the identity in part (a) be used to obtain an unbiased estimator of a when b is known, or an unbiased estimator of b when a is known?

3.9 For the natural exponential family $p_\eta(x)$ of (3.3.7) and the conjugate prior $\pi(\eta|k, \mu)$ of (3.3.19) establish that:

(a) $E(X) = A'(\eta)$ and var $X = A''(\eta)$, where the expectation is with respect to the sampling density $p_\eta(x)$.

(b) $EA'(\eta) = \mu$ and var$[A(\eta)] = (1/k)EA''(\eta)$, where the expectation is with respect to the prior distribution.

[The results in part (b) enable us to think of μ as a prior mean and k as a prior sample size.]

3.10 For each of the following situations, write the density in the form (3.7), and identify the natural parameter. Obtain the Bayes estimator of $A'(\eta)$ using squared loss and the conjugate prior. Express your answer in terms of the original parameters. (a) $X \sim$ binomial(p, n), (b) $X \sim$ Poisson(λ), and (c) $X \sim$ Gamma(a, b), a known.

3.11 For the situation of Problem 3.9, if $X_1, \ldots, X_n$ are iid as $p_\eta(x)$ and the prior is the conjugate $\pi(\eta|k, \mu)$, then the posterior distribution is $\pi(\eta|k+n, \frac{k\mu+n\bar{x}}{k+n})$.

3.12 If $X_1, \ldots, X_n$ are iid from a one-parameter exponential family, the Bayes estimator of the mean, under squared error loss using a conjugate prior, is of the form $a\bar{X} + b$ for constants a and b.

(a) If $EX_i = \mu$ and var $X_i = \sigma^2$, then no matter what the distribution of the X_i's, the mean squared error is

$$E[(a\bar{X} + b) - \mu]^2 = a^2 \text{ var } \bar{X} + [(a-1)\mu + b]^2.$$

(b) If μ is unbounded, then no estimator of the form $a\bar{X} + b$ can have finite mean squared error for $a \neq 1$.

(c) Can a conjugate-prior Bayes estimator in an exponential family have finite mean squared error?

[This problem shows why conjugate-prior Bayes estimators are considered "non-robust."]

Section 4

4.1 For the situation of Example 4.2:

(a) Show that the Bayes rule under a beta(α, α) prior is equivariant.

(b) Show that the Bayes rule under any prior that is symmetric about $1/2$ is equivariant.

4.2 The Bayes estimator of η in Example 4.7 is given by (4.22).

4.3 The Bayes estimator of τ in Example 4.5 is given by (4.22).

4.4 The Bayes estimators of η and τ in Example 4.9 are given by (4.31) and (4.32). (Recall Corollary 1.2.)

4.5 For each of the following situations, find a group G that leaves the model invariant and determine left- and right-invariant measures over G. The joint density of $\mathbf{X} = (X_1, \ldots, X_n)$ and $\mathbf{Y} = (Y_1, \ldots, Y_n)$ and the estimand are

(a) $f(\mathbf{x} - \eta, \mathbf{y} - \zeta)$, estimand $\eta - \zeta$;

(b) $f\left(\frac{\mathbf{x}-\eta}{\sigma}, \frac{\mathbf{y}-\zeta}{\tau}\right)$, estimand τ/σ;

(c) $f\left(\frac{\mathbf{x}-\eta}{\tau}, /, \frac{\mathbf{y}-\zeta}{\tau}\right)$, τ unknown; estimand $\eta - \zeta$.

4.6 For each of the situations of Problem 4.5, determine the MRE estimator if the loss is squared error with a scaling that makes it invariant.

4.7 For each of the situations of Problem 4.5:

(a) Determine the measure over Ω induced by the right-invariant Haar measure over $\bar{G}$;

(b) Determine the Bayes estimator with respect to the measure found in part (a), and show that it coincides with the MRE estimator.

4.8 In Example 4.9, show that the estimator

$$\hat{\tau}(\mathbf{x}) = \frac{\int\int \frac{1}{v^r} f\left(\frac{x_1-u}{v}, \ldots, \frac{x_n-u}{v}\right) dv du}{\int\int \frac{1}{v^{r+1}} f\left(\frac{x_1-u}{v}, \ldots, \frac{x_n-u}{v}\right) dv du}$$

is equivariant under scale changes; that is, it satisfies $\hat{\tau}(c\mathbf{x}) = c\hat{\tau}(\mathbf{x})$ for all values of r for which the integrals in $\hat{\tau}(\mathbf{x})$ exist.

4.9 If Λ is a left-invariant measure over G, show that Λ^* defined by $\Lambda^*(B) = \Lambda(B^{-1})$ is right invariant, where $B^{-1} = \{g^{-1} : g \in B\}$.
[*Hint*: Express $\Lambda^*(Bg)$ and $\Lambda^*(B)$ in terms of Λ.]

4.10 There is a correspondence between Haar measures and Jeffreys priors in the location and scale cases.

(a) Show that in the location parameter case, the Jeffreys prior is equal to the invariant Haar measure.

(b) Show that in the scale parameter case, the Jeffreys prior is equal to the invariant Haar measure.

(c) Show that in the location-scale case, the Jeffreys prior is equal to the left invariant Haar measure.

[Part c) is a source of some concern because, as mentioned in Section 4.4 (see the discussion following Example 4.9), the best-equivariant rule is Bayes against the right-invariant Haar measure (if it exists).]

4.11 For the model (3.3.23), find a measure ν in the (ξ, τ) plane which remains invariant under the transformations (3.3.24).

The next three problems contain a more formal development of left- and right-invariant Haar measures.

4.12 A measure Λ over a group G is said to be right invariant if it satisfies $\Lambda(Bg) = \Lambda(B)$ and left invariant if it satisfies $\Lambda(gB) = \Lambda(B)$. Note that if G is commutative, the two definitions agree.

(a) If the elements $g \in G$ are real numbers $(-\infty < g < \infty)$ and group composition is $g_2 \cdot g_1 = g_1 + g_2$, the measure ν defined by $\nu(B) = \int_B dx$ (i.e., Lebesgue measure) is both left and right invariant.

(b) If the elements $g \in G$ are the positive real numbers, and composition of g_2 and g_1 is multiplication of the two numbers, the measure ν defined by $\nu(B) = \int_B (1/y)\, dy$ is both left and right invariant.

4.13 If the elements $g \in G$ are pairs of real numbers (a, b), $b > 0$, corresponding to the transformations $gx = a + bx$, group composition by (1.4.8) is

$$(a_2, b_2) \cdot (a_1, b_1) = (a_2 + a_1 b_2, b_1 b_2).$$

Of the measures defined by

$$\nu(B) = \iint_B \frac{1}{y}\, dx\, dy \quad \text{and} \quad \nu(B) = \iint_B \frac{1}{y^2}\, dx\, dy,$$

the first is right but not left invariant, and the second is left but not right invariant.

4.14 The four densities defining the measures ν of Problem 4.12 and 4.13 $(dx, (1/y)dy, (1/y)dxdy, (1/y^2)dxdy)$ are the only densities (up to multiplicative constants) for which ν has the stated invariance properties in the situations of these problems.

[*Hint*: In each case, consider the equation

$$\int_B \pi(\theta)\, d\theta = \int_{gB} \pi(\theta)\, d\theta.$$

In the right integral, make the transformation to the new variable or variables $\theta' = g^{-1}\theta$. If J is the Jacobian of this transformation, it follows that

$$\int_B [\pi(\theta) - J\pi(g\theta)]\, d\theta = 0 \quad \text{for all } B$$

and, hence, that $\pi(\theta) = J\pi(g\theta)$ for all θ except in a null set N_g. The proof of Theorem 4 in Chapter 6 of TSH2 shows that N_g can be chosen independent of g. This proves in Problem 4.12(a) that for all $\theta \notin N$, $\pi(\theta) = \pi(\theta + c)$, and hence that $\pi(c) =$ constant a.e. The other three cases can be treated analogously.]

Section 5

5.1 For the model (3.3.1), let $\pi(\theta|x, \lambda)$ be a single-prior Bayes posterior and $\pi(\theta|x)$ be a hierarchical Bayes posterior. Show that $\pi(\theta|x) = \int \pi(\theta|x, \lambda) \cdot \pi(\lambda|x)\, d\lambda$, where $\pi(\lambda|x) = \int f(x|\theta)\pi(\theta|\lambda)\gamma(\lambda)\, d\theta / \iint f(x|\theta)\pi(\theta|\lambda)\gamma(\lambda)\, d\theta\, d\lambda$.

5.2 For the situation of Problem 5.1, show that:

(a) $E(\theta|x) = E[E(\theta|x, \lambda)]$;

(b) $\text{var}(\theta|x) = E[\text{var}(\theta|x, \lambda)] + \text{var}[E(\theta|x, \lambda)]$;

and hence that $\pi(\theta|x)$ will tend to have a larger variance than $\pi(\theta|x, \lambda_0)$.

5.3 For the model (3.3.3), show that:

(a) The marginal prior of θ, unconditional on τ^2, is given by

$$\pi(\theta) = \frac{\Gamma(a+\frac{1}{2})}{\sqrt{2\pi}\,\Gamma(a)b^a} \frac{1}{\left(\frac{1}{b}+\frac{\theta^2}{2}\right)^{a+1/2}},$$

which for $a = \nu/2$ and $b = 2/\nu$ is Student's t-distribution with ν degrees of freedom.

(b) The marginal posterior of τ^2 is given by

$$\pi(\tau^2|\bar{x}) = \frac{\left[\frac{\sigma^2\tau^2}{\sigma^2+\tau^2}\right]^{1/2} e^{-\frac{1}{2}\frac{\bar{x}^2}{\sigma^2+\tau^2}} \frac{1}{(\tau^2)^{a+3/2}} e^{-1/b\tau^2}}{\int_0^\infty \left[\frac{\sigma^2\tau^2}{\sigma^2+\tau^2}\right]^{1/2} e^{-frac12\frac{\bar{x}^2}{\sigma^2+\tau^2}} \frac{1}{(\tau^2)^{a+3/2}} e^{-1/b\tau^2} dt^2}.$$

5.4 Albert and Gupta (1985) investigate theory and applications of the hierarchical model

$$X_i|\theta_i \sim b(\theta_i, n), \quad i = 1, \ldots, p, \quad \text{independent},$$
$$\theta_i|\eta \sim \text{beta}[k\eta, k(1-\eta)], \quad k \text{ known},$$
$$\eta \sim \text{Uniform}(0, 1).$$

(a) Show that

$$E(\theta_i|\mathbf{x}) = \left(\frac{n}{n+k}\right)\left(\frac{x_i}{n}\right) + \left(\frac{k}{n+k}\right) E(\eta|\mathbf{x}),$$

$$\text{var}(\theta_i|\mathbf{x}) = \frac{k^2}{(n+k)(n+k+1)} \text{var}(\eta|\mathbf{x}).$$

[Note that $E(\eta|\mathbf{x})$ and $\text{var}(\eta|\mathbf{x})$ are not expressible in a simple form.]

(b) Unconditionally on η, the θ_i's have conditional covariance

$$\text{cov}(\theta_i, \theta_j|\mathbf{x}) = \left(\frac{k}{n+k}\right)^2 \text{var}(\eta|\mathbf{x}), \quad i \neq j.$$

(c) Ignoring the prior distribution of η, show how to construct an empirical Bayes estimator of θ_i. (Again, this is not expressible in a simple form.)

[Albert and Gupta (1985) actually consider a more general model than given here, and show how to approximate the Bayes solution. They apply their model to a problem of nonresponse in mail surveys.]

5.5 (a) Analogous to Problem 1.7.9, establish that for any random variable X, Y, and Z,

$$\text{cov}(X, Y) = E[\text{cov}(X, Y)|Z] + \text{cov}[E(X|Z), E(Y|Z)].$$

(b) For the hierarchy

$$X_i|\theta_i \sim f(x|\theta_i), \quad i = 1, \ldots, p, \quad \text{independent},$$
$$\Theta_i|\lambda \sim \pi(\theta_i|\lambda), \quad i = 1, \ldots, p, \quad \text{independent},$$
$$\Lambda \sim \gamma(\lambda),$$

show that $\text{cov}(\Theta_i, \Theta_j|\mathbf{x}) = \text{cov}[E(\Theta_i|\mathbf{x}, \lambda), E(\Theta_j|\mathbf{x}, \lambda)]$.

(c) If $E(\Theta_i|\mathbf{x}, \lambda) = g(x_i) + h(\lambda)$, $i = 1, \ldots, p$, where $g(\cdot)$ and $h(\cdot)$ are known, then

$$\operatorname{cov}(\Theta_i, \Theta_j|\mathbf{x}) = \operatorname{var}[E(\Theta_i|\mathbf{x}, \lambda)].$$

[Part (c) points to what can be considered a limitation in the applicability of some hierarchical models, that they imply a positive correlation structure in the posterior distribution.]

5.6 The one-way random effects model of Example 2.7 (see also Examples 3.5.1 and 3.5.5) can be written as the hierarchical model

$$X_{ij}|\mu, \alpha_i \sim N(\mu + \alpha_i, \sigma^2), \quad j = 1, \ldots, n, \quad i = 1, \ldots, s,$$
$$\alpha_i \sim N(0, \sigma_A^2), \quad i = 1, \ldots, s.$$

If, in addition, we specify that $\mu \sim \text{Uniform}(-\infty, \infty)$, show that the Bayes estimator of $\mu + \alpha_i$ under squared error loss is given by (3.5.13), the UMVU predictor of $\mu + \alpha_i$.

5.7 Referring to Example 6.6:

(a) Using the prior distribution for $\gamma(b)$ given in (5.6.27), show that the mode of the posterior distribution $\pi(b|\mathbf{x})$ is $\hat{b} = (p\bar{x} + \alpha - 1)/(pa + \beta - 1)$, and hence the empirical Bayes estimator based on this $\hat{b}$ does not equal the hierarchical Bayes estimator (5.6.29).

(b) Show that if we estimate $b/(b+1)$ using its posterior expectation $E[b/(b+1)|\mathbf{x}]$, then the resulting empirical Bayes estimator is equal to the hierarchical Bayes estimator.

5.8 The method of Monte Carlo integration allows the calculation of (possibly complicated) integrals by using (possibly simple) generations of random variables.

(a) To calculate $\int h(x) f_X(x)\,dx$, generate a sample $X_1, \ldots, X_m$, iid, from $f_X(x)$. Then, $1/m \sum_{i=1}^m h(x_i) \to \int h(x) f_X(x)\,dx$ as $m \to \infty$.

(b) If it is difficult to generate random variable from $f_X(x)$, then generate pairs of random variables

$$Y_i \sim f_Y(y),$$
$$X_i \sim f_{X|Y}(x|y_i).$$

Then, $1/m \sum_{i=1}^m h(x_i) \to \int h(x) f_X(x)\,dx$ as $m \to \infty$.
[Show that if X is generated according to $Y \sim f_Y(y)$ and $X \sim f_{X|Y}(x|Y)$, then $P(X \le a) = \int_{-\infty}^a f_x(x)\,dx$.]

(c) If it is difficult to generate as in part (b), then generate

$$X_{m_i} \sim f_{X|Y}(x|Y_{m_{i-1}}),$$
$$Y_{m_i} \sim f_{Y|X}(y|X_{m_i}).$$

for $i = 1, \ldots, K$ and $m = 1, \ldots, M$.

Show that:

(i) for each m, $\{X_{m_i}\}$ is a Markov chain. If it is also an ergodic Markov chain $X_{m_i} \xrightarrow{\mathcal{L}} X$, as $i \to \infty$, where X has the stationary distribution of the chain.

(ii) If the stationary distribution of the chain is $f_X(x)$, then

$$\frac{1}{M} \sum_{m=1}^{M} h(x_{m_k}) \to \int h(x) f_X(x)\,dx$$

as $K, M \to \infty$.

[This is the basic theory behind the Gibbs sampler. For each k, we have generated independent random variables X_{m_k}, $m = 1, \ldots, M$, where X_{m_k} is distributed according to $f_{X|Y}(x|y_{m_{k-}})$. It is also the case that for each m and large k, X_{m_k} is approximately distributed according to $f_X(x)$, although the variables are not now independent. The advantages and disadvantages of these computational schemes (*one-long-chain* vs. *many-short-chains*) are debated in Gelman and Rubin 1992; see also Geyer and Thompson 1992 and Smith and Roberts 1992. The prevailing consensus leans toward one long chain.]

5.9 To understand the convergence of the Gibbs sampler, let $(X, Y) \sim f(x, y)$, and define

$$k(x, x') = \int f_{X|Y}(x|y) f_{Y|X}(y|x')\, dy.$$

(a) Show that the function $h^*(\cdot)$ that solves $h^*(x) = \int k(x, x')h^*(x')\,dx'$ is $h^*(x) = f_X(x)$, the marginal distribution of X.

(b) Write down the analogous integral equation that is solved by $f_Y(y)$.

(c) Define a sequence of functions recursively by $h_{i+1}(x) = \int k(x, x')h_i(x')\,dx'_l$ where $h_0(x)$ is arbitrary but satisfies $\sup_x \left|\frac{h_0(x)}{h^*(x)}\right| < \infty$. Show that

$$\int |h_{i+1}(x) - h^*(x)|dx < \int |h_i(x) - h^*(x)|\,dx$$

and, hence, $h_i(x)$ converges to $h^*(x)$.

[The method of part (c) is called *successive substitution*. When there are two variables in the Gibbs sampler, it is equivalent to *data augmentation* (Tanner and Wong 1987). Even if the variables are vector-valued, the above results establish convergence. If the original vector of variables contains more than two variables, then a more general version of this argument is needed (Gelfand and Smith 1990).]

5.10 A direct Monte Carlo implementation of substitution sampling is provided by the *data augmentation* algorithm (Tanner and Wong 1987). If we define

$$h_{i+1}(x) = \int\left[\int f_{X|Y}(x|y) f_{Y|X}(y|x')\,dy\right] h_i(x')\,dx',$$

then from Problem 5.9, $h_i(x) \to f_x(x)$ as $i \to \infty$.

(a) To calculate h_{i+1} using Monte Carlo integration:

(i) Generate $X'_j \sim h_i(x')$, $j = 1, \ldots, J$.
(ii) Generate, for each x'_j, $Y_{jk} \sim f_{Y|X}(y|x'_j)$, $k = 1, \ldots, K$.
(iii) Calculate $\hat{h}_{i+1}(x) = \frac{1}{J}\sum_{j=1}^{J} \frac{1}{K}\sum_{k=1}^{K} f_{X|Y}(x|y_{jk})$.

Then, $\hat{h}_{i+1}(x) \to h_{i+1}(x)$ as $J, K \to \infty$, and hence the data augmentation algorithm converges.

(b) To implement (a)(i), we must be able to generate a random variable from a mixture distribution. Show that if $f_Y(y) = \sum_{i=1}^{n} a_i g_i(y)$, $\Sigma a_i = 1$, then the algorithm

(i) Select g_i with probability a_i
(ii) Generate $Y \sim g_i$

produces a random variable with distribution f_Y. Hence, show how to implement step (a)(i) by generating random variables from $f_{X|Y}$. Tanner and Wong (1987) note that this algorithm will work even if $J = 1$, which yields the approximation

$\hat{h}_{i+1}(x) = \frac{1}{K}\sum_{k=1}^{K} f_{X|Y}(x|y_k)$, identical to the Gibbs sampler. The data augmentation algorithm can also be seen as an application of the process of *multiple imputation* (Rubin 1976, 1987, Little and Rubin 1987).

5.11 Successive substitution sampling can be implemented via the Gibbs sampler in the following way. From Problem 5.8(c), we want to calculate

$$h_M = \frac{1}{M}\sum_{m=1}^{M} k(x|x_{m_k}) = \frac{1}{M}\sum_{m=1}^{M}\int f_{X|Y}(x|y) f_{X|Y}(y|x_{m_k})\,dy.$$

(a) Show that $h_M(x) \to f_X(x)$ as $M \to \infty$.

(b) Given x_{m_k}, a Monte Carlo approximation to $h_M(x)$ is

$$\hat{h}_M(x) = \frac{1}{M}\sum_{m=1}^{M}\frac{1}{J}\sum_{j=1}^{J} f_{X|Y}(x|y_{k_j})$$

where $Y_{k_j} \sim f_{Y|X}(y|x_{m_k})$ and $\hat{h}_M(x) \to h_M(x)$ as $J \to \infty$.

(c) Hence, as $M, J \to \infty$, $\hat{h}_M(x) \to f_X(x)$.

[This is the Gibbs sampler, which is usually implemented with $J = 1$.]

5.12 For the situation of Example 5.6, show that

(a)
$$E\left(\frac{1}{M}\sum_{i=1}^{M}\Theta_i\right) = E\left(\frac{1}{M}\sum_{i=1}^{M}E(\Theta|\mathbf{x}, \tau_i)\right),$$

(b)
$$\text{var}\left(\frac{1}{M}\sum_{i=1}^{M}\Theta_i\right) \geq \text{var}\left(\frac{1}{M}\sum_{i=1}^{M}E(\Theta|\mathbf{x}, \tau_i)\right).$$

(c) Discuss when equality might hold in (b). Can you give an example?

5.13 Show that for the hierarchy (5.5.1), the posterior distributions $\pi(\theta|\mathbf{x})$ and $\pi(\lambda|\mathbf{x})$ satisfy

$$\pi(\theta|\mathbf{x}) = \int\left[\int \pi(\theta|\mathbf{x}, \lambda)\pi(\lambda|\mathbf{x}, \theta')\,d\lambda\right]\pi(\theta'|\mathbf{x})\,d\theta',$$

$$\pi(\lambda|\mathbf{x}) = \int\left[\int \pi(\lambda|\mathbf{x}, \theta)\pi(\theta|\mathbf{x}, \lambda')\,d\theta\right]\pi(\lambda'|\mathbf{x})\,d\lambda',$$

and, hence, are stationary points of the Markov chains in (5.5.13).

5.14 Starting from a uniform random variable $U \sim$ Uniform(0, 1), it is possible to construct many random variables through transformations.

(a) Show that $-\log U \sim \exp(1)$.

(b) Show that $-\sum_{i=1}^{n}\log U_i \sim$ Gamma(n, 1), where $U_1, \ldots, U_n$ are iid as $U(0, 1)$.

(c) Let $X \sim$ Exp(a, b). Write X as a function of U.

(d) Let $X \sim$ Gamma(n, β), n an integer. Write X as a function of $U_1, \ldots, U_n$, iid as $U(0, 1)$.

5.15 Starting with a $U(0, 1)$ random variable, the transformations of Problem 5.14 will not get us normal random variables, or gamma random variables with noninteger shape parameters. One way of doing this is to use the *Accept-Reject Algorithm* (Ripley 1987, Section 3.2), an algorithm for simulating $X \sim f(x)$:

(i) Generate $Y \sim g(y)$, $U \sim U(0, 1)$, independent.
(ii) Calculate $\rho(Y) = \frac{1}{M}\frac{f(Y)}{g(Y)}$ where $M = \sup_t f(t)/g(t)$.
(iii) If $U < \rho(Y)$, set $X = Y$, otherwise return to i).

(a) Show that the algorithm will generate $X \sim f(x)$.
(b) Starting with $Y \sim \exp(1)$, show how to generate $X \sim N(0, 1)$.
(c) Show how to generate a gamma random variable with a noninteger shape parameter.

5.16 Consider the normal hierarchical model

$$\begin{aligned} X|\theta_1 &\sim n(\theta_1, \sigma_1^2), \\ \theta_1|\theta_2 &\sim n(\theta_2|\sigma_2^2), \\ &\vdots \\ \theta_{k-1}|\theta_k &\sim n(\theta_k, \sigma_k^2) \end{aligned}$$

where $\sigma_i^2, i = 1, \ldots, k$, are known.

(a) Show that the posterior distribution of θ_i $(1 \le i \le k-1)$ is

$$\pi(\theta_i|x, \theta_k) = N(\alpha_i x + (1 - \alpha_i)\theta_k, \tau_i^2)$$

where $\tau_i^2 = (\Sigma_1^i \sigma_j^2)(\Sigma_{i+1}^k \sigma_j^2)/\Sigma_1^k \sigma_j^2$ and $\alpha_i = \tau_i^2/\Sigma_1^i \sigma_j^2$.

(b) Find an expression for the Kullback-Leibler information$K[\pi(\theta_i|x, \theta_k), \pi(\theta_i|\theta_k)]$ and show that it is a decreasing function of i.

5.17 The original proof of Theorem 5.7 (Goel and DeGroot 1981) used ***Rényi's entropy function*** (Rényi 1961)

$$R_\alpha(f, g) = \frac{1}{\alpha - 1} \log \int f^\alpha(x) g^{1-\alpha}(x)\, d\mu(x),$$

where f and g are densities, μ is a dominating measure, and α is a constant, $\alpha \neq 1$.

(a) Show that $R_\alpha(f, g)$ satisfies $R_\alpha(f, g) > 0$ and $R_\alpha(f, f) = 0$.
(b) Show that Theorem 5.7 holds if $R_\alpha(f, g)$ is used instead of $K[f, g]$.
(c) Show that $\lim_{\alpha \to 1} R_\alpha(f, g) = K[f, g]$, and provide another proof of Theorem 5.7.

5.18 The Kullback-Leibler information, $K[f, g]$ (5.5.25), is not symmetric in f and g, and a modification, called the ***divergence***, remedies this. Define $J[f, g]$, the divergence between f and g, to be $J[f, g] = K[f, g] + K[g, f]$. Show that, analogous to Theorem 5.7, $J[\pi(\lambda|x), \gamma(\lambda)] < J[\pi(\theta|x), \pi(\theta)]$.

5.19 Goel and DeGroot (1981) define a Bayesian analog of Fisher information [see (2.5.10)] as

$$\mathcal{I}[\pi(\theta|x)] = \int_\Omega \left[\frac{\frac{\partial}{\partial x}\pi(\theta|x)}{\pi(\theta|x)} \right]^2 d\theta,$$

the information that x has about the posterior distribution. As in Theorem 5.7, show that $\mathcal{I}[\pi(\lambda|x)] < \mathcal{I}[\pi(\theta|x)]$, again showing that the influence of λ is less than that of θ.

5.20 Each of m spores has a probability τ of germinating. Of the r spores that germinate, each has probability ω of bending in a particular direction. If s bends in the particular direction, a probability model to describe this process is the ***bivariate binomial***, with mass function

$$f(r, s|\tau, \omega, m) = \binom{m}{r} \tau^r (1 - \tau)^{m-r} \binom{r}{s} \omega^s (1 - \omega)^{r-s}.$$

(a) Show that the Jeffreys prior is $\pi_J(\tau, \omega) = (1-\tau)^{-1/2}\omega^{-1/2}(1-\omega)^{-1/2}$.

(b) If τ is considered a nuisance parameter, the reference prior is

$$\pi_R(\tau, \omega) = \tau^{-1/2}(1-\tau)^{-1/2}\omega^{-1/2}(1-\omega)^{-1/2}.$$

Compare the posterior means $E(\omega|r, s, m)$ under both the Jeffreys and reference priors. Is one more appropriate?

(c) What is the effect of the different priors on the posterior variance?

[Priors for the bivariate binomial have been considered by Crowder and Sweeting (1989), Polson and Wasserman (1990), and Clark and Wasserman (1993), who propose a reference/Jeffreys trade-off prior.]

5.21 Let $\mathcal{F} = \{f(x|\theta); \theta \in \Omega\}$ be a family of probability densities. The Kullback-Leibler information for discrimination between two densities in $\mathcal{F}$ can be written

$$\psi(\theta_1, \theta_2) = \int f(x|\theta_1) \log\left[\frac{f(x|\theta_1)}{f(x|\theta_2)}\right] dx.$$

Recall that the gradient of ψ is $\nabla\psi = \{(\partial/\partial\theta_i)\psi\}$ and the Hessian is $\nabla\nabla\psi = \{(\partial^2/\partial\theta_i\partial\theta_j)\psi\}$.

(a) If integration and differentiation can be interchanged, show that

$$\nabla\psi(\theta, \theta) = 0 \quad \text{and} \quad \det[\nabla\nabla\psi(\theta, \theta)] = I(\theta),$$

where $I(\theta)$ is the Fisher information of $f(x|\theta)$.

(b) George and McCulloch (1993) argue that choosing $\pi(\theta) = (\det[\nabla\nabla\psi(\theta, \theta)])^{1/2}$ is an appealing least informative choice of priors. What justification can you give for this?

Section 6

6.1 For the model (3.3.1), show that $\delta^\lambda(x)|_{\lambda=\hat{\lambda}} = \delta^{\hat{\lambda}(x)}$, where the Bayes estimator $\delta^\lambda(x)$ minimizes $\int L[\theta, d(x)]\pi(\theta|x, \lambda)\,d\theta$ and the empirical Bayes estimator $\delta^{\hat{\lambda}(x)}$ minimizes $\int L[\theta, d(x)]\pi(\theta|x, \hat{\lambda})\,d\theta$.

6.2 This problem will investigate conditions under which an empirical Bayes estimator is a Bayes estimator. Expression (6.6.3) is a true posterior expected loss if $\pi(\theta|\mathbf{x}, \hat{\lambda}(\mathbf{x}))$ is a true posterior.

From the hierarchy

$$X|\theta \sim f(x|\theta),$$
$$\Theta|\lambda \sim \pi(\theta|\lambda),$$

define the joint distribution of $\mathbf{X}$ and Θ to be $(\mathbf{X}, \theta) \sim g(\mathbf{x}, \theta) = f(\mathbf{x}|\theta)\pi(\theta|\hat{\lambda}(\mathbf{x}))$, where $\pi(\theta|\hat{\lambda}(\mathbf{x}))$ is obtained by substituting $\hat{\lambda}(\mathbf{x})$ for λ in $\pi(\theta|\lambda)$.

(a) Show that, for this joint density, the formal Bayes estimator is equivalent to the empirical Bayes estimator from the hierarchical model.

(b) If $f(\cdot|\theta)$ and $\pi(\cdot|\lambda)$ are proper densities, then $\int g(\mathbf{x}, \theta)\,d\theta < \infty$. However, $\int\int g(\mathbf{x}, \theta)\,d\mathbf{x}d\theta$ need not be finite.

6.3 For the model (6.3.1), the Bayes estimator $\delta^\lambda(x)$ minimizes $\int L(\theta, d(x)) \times \pi(\theta|x, \lambda)\,d\theta$ and the empirical Bayes estimator, $\delta^{\hat{\lambda}}(x)$, minimizes $\int L(\theta, d(x))\pi(\theta|x, \hat{\lambda}(x))\,d\theta$. Show that $\delta^\lambda(x)|_{\lambda=\hat{\lambda}(x)} = \delta^{\hat{\lambda}}(x)$.

6.4 For the situation of Example 6.1:

(a) Show that

$$\int_{-\infty}^{\infty} e^{-n/2\sigma^2(\bar{x}-\theta)^2} e^{(-1/2)\theta^2/\tau^2} d\theta = \sqrt{2\pi}\left(\frac{\sigma^2\tau^2}{\sigma^2+\tau^2}\right)^{1/2} e^{(-n/2)\bar{x}^2/\sigma^2+n\tau^2}$$

and, hence, establish (6.6.4).

(b) Verify that the marginal MLE of $\sigma^2 + n\tau^2$ is $n\bar{x}^2$ and that the empirical Bayes estimator is given by (6.6.5).

6.5 Referring to Example 6.2:

(a) Show that the Bayes risk, $r(\pi, \delta^\pi)$, of the Bayes estimator (6.6.7) is given by

$$r(\pi, \delta^\pi) = kE[\text{var}(p_k|x_k)] = \frac{kab}{(a+b)(a+b+1)(a+b+n)}.$$

(b) Show that the Bayes risk of the unbiased estimator $\mathbf{X}/n = (X_1/n, \ldots, X_k/n)$ is given by

$$r(\pi, \mathbf{X}/n) = \frac{kab}{n(a+b+1)(a+b)}.$$

6.6 Extend Theorem 6.3 to the case of Theorem 3.2; that is, if $\mathbf{X}$ has density (3.3.7) and η has prior density $\pi(\eta|\gamma)$, then the empirical Bayes estimator is

$$E\left(\sum \eta_i \frac{\partial T_i(\mathbf{x})}{\partial x_j}|\mathbf{x}, \hat{\gamma}(\mathbf{x})\right) = \frac{\partial}{\partial x_j}\log m(\mathbf{x}|\hat{\gamma}(\mathbf{x})) - \frac{\partial}{\partial x_j}\log h(\mathbf{x}),$$

where $m(\mathbf{x}|\gamma)$ is the marginal distribution of $\mathbf{X}$ and $\hat{\gamma}(\mathbf{x})$ is the marginal MLE of γ.

6.7 (a) For $p_\eta(\mathbf{x})$ of (1.5.2), show that for any prior distribution $\pi(\eta|\lambda)$ that is dependent on a hyperparameter λ, the empirical Bayes estimator is given by

$$E\left[\sum_{i=1}^{s} \eta_i \frac{\partial}{\partial x_j} T_i(\mathbf{x})|\mathbf{x}, \hat{\lambda}\right] = \frac{\partial}{\partial x_j}\log m_\pi(\mathbf{x}|\hat{\lambda}(\mathbf{x})) - \frac{\partial}{\partial x_j}\log h(\mathbf{x}).$$

where $m_\pi(\mathbf{x}) = \int p_\theta(\mathbf{x})\pi(\theta)\,d\theta$.

(b) If $\mathbf{X}$ has the distribution $p_\theta(\mathbf{x})$ of (1.5.1), show that a similar formulas holds, that is,

$$E(\mathcal{T}\eta(\theta)|\hat{\lambda}) = \nabla \log m_\pi(\mathbf{x}|\hat{\lambda}) - \nabla \log h(\mathbf{x}),$$

where $\mathcal{T} = \{\partial T_i/\partial x_j\}$ is the Jacobian of $\mathcal{T}$and ∇a is the gradient vector of a, that is, $\nabla a = \{\partial a/\partial x_i\}$.

6.8 For each of the following situations, write the empirical Bayes estimator of the natural parameter (under squared error loss) in the form (6.6.12), using the marginal likelihood estimator of the hyperparameter λ. Evaluate the expressions as far as possible.

(a) $X_i \sim N(0, \sigma_i^2)$, $i = 1, \ldots, p$, independent; $1/\sigma_i^2 \sim$ Exponential(λ).

(b) $X_i \sim N(\theta_i, 1)$, $i = 1, \ldots, p$, independent, $\theta_i \sim DE(0, \lambda)$.

6.9 Strawderman (1992) shows that the James-Stein estimator can be viewed as an empirical Bayes estimator in an arbitrary location family. Let $\mathbf{X}_{p\times 1} \sim f(x-\theta)$, with $E\mathbf{X} = \theta$ and var $\mathbf{X} = \sigma^2 I$. Let the prior be $\theta \sim f^{*n}$, the n-fold convolution of f with itself. [The *convolution* of f with itself is $f^{*2}(x) = \int f(x-y)f(y)dy$. The *n-fold convolution* is $f^{*n}(x) = \int f^{*(n-1)}(x-y)f(y)dy$.] Equivalently, let $U_i \sim f, i = 0, \cdots, n$, iid, $\theta = \sum_1^n U_i$, and $\mathbf{X} = U_0 + \theta$.

(a) Show that the Bayes rule against squared error loss is $\frac{n}{n+1}\mathbf{x}$. Note that n is a prior parameter.

(b) Show that $|\mathbf{X}|^2/(p\sigma^2)$ is an unbiased estimator of $n+1$, and hence that an empirical Bayes estimator of θ is given by $\delta^{EB} = [1 - (p\sigma^2/|\mathbf{x}|^2)]\mathbf{x}$.

6.10 Show for the hierarchy of Example 3.4, where σ^2 and τ^2 are known but μ is unknown, that:

(a) The empirical Bayes estimator of θ_i, based on the marginal MLE of θ_i, is $\frac{\tau^2}{\sigma^2+\tau^2}X_i + \frac{\sigma^2}{\sigma^2+\tau^2}\bar{X}$.

(b) The Bayes risk, under sum-of-squared-errors loss, of the empirical Bayes estimator from part (a) is

$$p\sigma^2 - \frac{2(p-1)^2\sigma^4}{p(\sigma^2+\tau^2)} + (p-1)\left(\frac{\sigma^2}{\sigma^2+\tau^2}\right)^2 \sum_{i=1}^{p} E(X_i - \bar{X})^2.$$

(c) The minimum risk of the empirical Bayes estimator is attained when all θ_is are equal. [*Hint*: Show that $\sum_{i=1}^{p} E[(X_i - \bar{X})^2] = \sum_{i=1}^{p}(\theta_i - \bar{\theta})^2 + (p-1)\sigma^2$.]

6.11 For $E(\Theta|\mathbf{x})$ of (4.5.8), show that as $\nu \to \infty$, $E(\Theta|x) \to [p/(p+\sigma^2)]\bar{x}$, the Bayes estimator under a $N(0, 1)$ prior.

6.12 (a) Show that the empirical Bayes $\delta^{EB}(\bar{x}) = (1 - \sigma^2/\max\{\sigma^2, p\bar{x}^2\})\bar{x}$ of (6.6.5) has bounded mean squared error.

(b) Show that a variation of $\delta^{EB}(\bar{x})$, of part (a), $\delta^{v}(\bar{x}) = [1 - \sigma^2/(v + p\bar{x}^2)]\bar{x}$, also has bounded mean squared error.

(c) For $\sigma^2 = \tau^2 = 1$, plot the risk functions of the estimators of parts (a) and (b).

[Thompson (1968a, 1968b) investigated the mean squared error properties of estimators like those in part (b). Although such estimators have smaller mean squared error than $\bar{x}$ for small values of θ, they always have larger mean squared error for larger values of θ.]

6.13 (a) For the hierarchy (4.5.7), with $\sigma^2 = 1$ and $p = 10$, evaluate the Bayes risk $r(\pi, \delta^\pi)$ of the Bayes estimator (4.5.8) for $\nu = 2, 5$, and 10.

(b) Calculate the Bayes risk of the estimator δ^v of Problem 6.12(b). Find a value of v that yields a good approximation to the risk of the hierarchical Bayes estimator. Compare it to the Bayes risk of the empirical Bayes estimator of Problem 6.12(a).

6.14 Referring to Example 6.6, show that the empirical Bayes estimator is also a hierarchical Bayes estimator using the prior $\gamma(b) = 1/b$.

6.15 The Taylor series approximation to the estimator (5.5.8) is carried out in a number of steps. Show that:

(a) Using a first-order Taylor expansion around the point $\bar{x}$, we have

$$\frac{1}{(1+\theta^2/\nu)^{(\nu+1)/2}} = \frac{1}{(1+\bar{x}^2/\nu)^{(\nu+1)/2}} - \frac{\nu+1}{\nu}\frac{\bar{x}}{(1+\bar{x}^2/\nu)^{(\nu+3)/2}}(\theta - \bar{x}) + R(\theta - \bar{x})$$

where the remainder, $R(\theta - \bar{x})$, satisfies $R(\theta - \bar{x})/(\theta - \bar{x})^2 \to 0$ as $\theta \to \bar{x}$.

(b) The remainder in part (a) also satisfies

$$\int_{-\infty}^{\infty} R(\theta - \bar{x}) e^{-\frac{p}{2\sigma^2}(\theta - \bar{x})^2} d\theta = O(1/p^{3/2}).$$

(c) The numerator and denominator of (5.5.8) can be written

$$\int_{-\infty}^{\infty} \frac{1}{(1+\theta^2/\nu)^{(\nu+1)/2}} e^{-\frac{p}{2\sigma^2}(\theta-\bar{x})^2} d\theta = \frac{\sqrt{2\pi\sigma^2/p}}{(1+\bar{x}^2/\nu)^{(\nu+1)/2}} + O\left(\frac{1}{p^{3/2}}\right)$$

and

$$\int_{-\infty}^{\infty} \frac{\theta}{(1+\theta^2/\nu)^{(\nu+1)/2}} e^{-\frac{p}{2\sigma^2}(\theta-\bar{x})^2} d\theta$$

$$= \frac{\sqrt{2\pi\sigma^2/p}}{(1+\bar{x}^2/\nu)^{(\nu+1)/2}} \left[1 - \frac{(\nu+1)/\nu}{(1+\bar{x}^2/\nu)}\right] \bar{x} + O\left(\frac{1}{p^{3/2}}\right),$$

which yields (5.6.32).

6.16 For the situation of Example 6.7:

(a) Calculate the values of the approximation (4.6.32) for the values of Table 6.2. Are there situations where the estimator (4.6.32) is clearly preferred over the empirical Bayes estimator (4.6.5) as an approximation to the hierarchical Bayes estimator (4.5.8)?

(b) Extend the argument of Problem 6.15 to calculate the next term in the expansion and, hence, obtain a more accurate approximation to the hierarchical Bayes estimator (4.5.8). For the values of Table 6.2, is this new approximation to (4.5.8) preferable to (4.6.5) and (4.6.32)?

6.17 (a) Show that if $b(\cdot)$ has a bounded second derivative, then

$$\int b(\lambda) e^{-nh(\lambda)} d\lambda = b(\hat{\lambda}) \sqrt{\frac{2\pi}{nh''(\hat{\lambda})}} e^{-nh(\hat{\lambda})} + O\left(\frac{1}{n^{3/2}}\right)$$

where $h(\hat{\lambda})$ is the unique minimum of $h(\lambda)$, $h''(\lambda) \neq 0$, and $nh(\hat{\lambda}) \to$ constant as $n \to \infty$.

[*Hint*: Expand both $b(\cdot)$ and $h(\cdot)$ in Taylor series around $\hat{\lambda}$, up to second-order terms. Then, do the term-by-term integration.]

This is the *Laplace approximation* for an integral. For refinements and other developments of this approximation in Bayesian inference, see Tierney and Kadane 1986, Tierney, Kass, and Kadane 1989, and Robert 1994a (Section 9.2.3).

(b) For the hierarchical model (5.5.1), the posterior mean can be approximated by

$$E(\Theta|x) = e^{-nh(\hat{\lambda})} \left[\frac{2\pi}{nh''(\hat{\lambda})}\right]^{1/2} E(\Theta|x, \hat{\lambda}) + O\left(\frac{1}{n^{3/2}}\right)$$

where $h = \frac{1}{n} \log \pi(\lambda|x)$ and $\hat{\lambda}$ is the mode of $\pi(\lambda|x)$, the posterior distribution of λ.

(c) If $\pi(\lambda|x)$ is the normal distribution with mean $\hat{\lambda}$ and variance $\sigma^2 = [-(\partial^2/\partial\lambda^2) \times \log \pi(\lambda|x)|_{\lambda=\hat{\lambda}}]^{-1}$, then $E(\Theta|x) = E(\Theta|x, \hat{\lambda}) + O(1/n^{3/2})$.

(d) Show that the situation in part (c) arises from the hierarchy

$$X_i|\theta_i \sim N(\theta_i, \sigma^2),$$
$$\theta_i|\lambda \sim N(\lambda, \tau^2),$$
$$\lambda \sim \text{Uniform}(-\infty, \infty).$$

6.18 (a) Apply the Laplace approximation (5.6.33) to obtain an approximation to the hierarchical Bayes estimator of Example 6.6.

(b) Compare the approximation from part (a) with the empirical Bayes estimator (5.6.24). Which is a better approximation to the hierarchical Bayes estimator?

6.19 Apply the Laplace approximation (5.6.33) to the hierarchy of Example 6.7 and show that the resulting approximation to the hierarchical Bayes estimator is given by (5.6.32).

6.20 (a) Verify (6.6.37), that under squared error loss

$$r(\pi, \delta) = r(\pi, \delta^\pi) + E(\delta - \delta^\pi)^2.$$

(b) For $X \sim$ binomial(p, n), $L(p, \delta) = (p - \delta)^2$, and $\pi = \{\pi : \pi = \text{beta}(a, b), a > 0, b > 0\}$, determine whether $\hat{p} = x/n$ or $\delta^0 = (a_0 + x)/(a_0 + b_0 + n)$ is more robust, according to (6.6.37).

(c) Is there an estimator of the form $(c + x)/(c + d + n)$ that you would consider more robust, in the sense of (6.6.37), than either estimator in part (b)?

[In part (b), for fixed n and (a_0, b_0), calculate the Bayes risk of $\hat{p}$ and δ^0 for a number of (a, b) pairs.]

6.21 (a) Establish (6.6.39) and (6.6.40) for the class of priors given by (6.6.38).

(b) Show that the Bayes estimator based on $\pi(\theta) \in \pi$ in (6.6.38), under squared error loss, is given by (6.6.41).

Section 7

7.1 For the situation of Example 7.1:

(a) The empirical Bayes estimator of $\boldsymbol{\theta}$, using an unbiased estimate of $\tau^2/(\sigma^2 + \tau^2)$, is

$$\delta^{\text{EB}} = \left(1 - \frac{(p-2)\sigma^2}{|\mathbf{x}|^2}\right)\mathbf{x},$$

the James-Stein estimator.

(b) The empirical Bayes estimator of $\boldsymbol{\theta}$, using the marginal MLE of $\tau^2/(\sigma^2 + \tau^2)$, is

$$\delta^{\text{EB}} = \left(1 - \frac{p\sigma^2}{|\mathbf{x}|^2}\right)^+ \mathbf{x},$$

which resembles the positive-part Stein estimator.

7.2 Establish Corollary 7.2. Be sure to verify that the conditions on $g(\mathbf{x})$ are sufficient to allow the integration-by-parts argument. [Stein (1973, 1981) develops these representations in the normal case.]

7.3 The derivation of an unbiased estimator of the risk (Corollary 7.2) can be extended to a more general model in the exponential family, the model of Corollary 3.3, where $\mathbf{X} = X_1, \ldots, X_p$ has the density

$$p_{\boldsymbol{\eta}}(\mathbf{x}) = e^{\sum_{i=1}^p \eta_i x_i - A(\boldsymbol{\eta})} h(\mathbf{x}).$$

(a) The Bayes estimator of $\boldsymbol{\eta}$, under squared error loss, is

$$E(\eta_i|\mathbf{x}) = \frac{\partial}{\partial x_i} \log m(\mathbf{x}) - \frac{\partial}{\partial x_i} \log h(\mathbf{x}).$$

Show that the risk of $E(\boldsymbol{\eta}|\underline{X})]$ has unbiased estimator

$$\sum_{i=1}^p \left[\frac{\partial^2}{\partial x_i^2}(\log h(\mathbf{x}) - 2\log m(\mathbf{x})) + \left(\frac{\partial}{\partial x_i}\log m(\mathbf{x})\right)^2\right].$$

[*Hint*: Theorem 3.5 and Problem 3.4.]

(b) Show that the risk of the empirical Bayes estimator

$$E(\eta_i|\mathbf{x}, \hat{\lambda}) = \frac{\partial}{\partial x_i} \log m(\mathbf{x}|\hat{\lambda}(\mathbf{x})) - \frac{\partial}{\partial x_i} \log h(\mathbf{x}).$$

of Theorem 6.3 has unbiased estimator

$$\sum_{i=1}^{p} \left[\frac{\partial^2}{\partial x_i^2} \left(\log h(\mathbf{x}) - 2 \log m(\mathbf{x}|\hat{\lambda}(\mathbf{x})) \right) + \left(\frac{\partial}{\partial x_i} \log m(\mathbf{x}|\hat{\lambda}(\mathbf{x})) \right)^2 \right].$$

(c) Use the results of part (b) to derive an unbiased estimator of the risk of the positive-part Stein estimator of (7.7.10).

7.4 Verify (7.7.9), the expression for the Bayes risk of δ^{τ_0}. (Problem 3.12 may be helpful.)

7.5 A general version of the empirical Bayes estimator (7.7.3) is given by

$$\delta^c(\mathbf{x}) = \left(1 - \frac{c\sigma^2}{|\mathbf{x}|^2}\right)\mathbf{x},$$

where c is a positive constant.

(a) Use Corollary 7.2 to verify that

$$E_\theta|\theta - \delta^c(\mathbf{X})|^2 = p\sigma^2 + c\sigma^4[c - 2(p-2)]E_\theta \frac{1}{|\mathbf{X}|^2}.$$

(b) Show that the Bayes risk, under $\Theta \sim N_p(0, \tau^2 I)$, is given by

$$r(\pi, \delta^c) = \sigma^2 \left[p + \frac{c\sigma^2}{\sigma^2 + \tau^2} \left(\frac{c}{p-2} - 2 \right) \right]$$

and is minimized by choosing $c = p - 2$.

7.6 For the model

$$\mathbf{X}|\theta \sim N_p(\theta, \sigma^2 I),$$
$$\theta|\tau^2 \sim N_p(\mu, \tau^2 I):$$

Show that:

(a) The empirical Bayes estimator, using an unbiased estimator of $\tau^2/(\sigma^2 + \tau^2)$, is the Stein estimator

$$\delta_i^{\text{JS}}(\mathbf{x}) = \mu_i + \left(1 - \frac{(p-2)\sigma^2}{\Sigma(x_i - \mu_i)^2}\right)(x_i - \mu_i).$$

(b) If $p \geq 3$, the Bayes risk, under squared error loss, of δ^{JS} is $r(\tau, \delta^{\text{JS}}) = r(\tau, \delta^\tau) + 2\sigma^4/(\sigma^2 + \tau^2)$, where $r(\tau, \delta^\tau)$ is the Bayes risk of the Bayes estimator.

(c) If $p < 3$, the Bayes risk of δ^{JS} is infinite. [*Hint*: Show that if $Y \sim \chi_m^2$, $E(1/Y) < \infty \iff m < 3$].

7.7 For the model

$$\mathbf{X}|\theta \sim N_p(\theta, \sigma^2 I),$$
$$\theta|\tau^2 \sim N(\mu, \tau^2 I)$$

the Bayes risk of the ordinary Stein estimator

$$\delta_i(\mathbf{x}) = \mu_i + \left(1 - \frac{(p-2)\sigma^2}{\Sigma(x_i - \mu_i)^2}\right)(x_i - \mu_i)$$

is uniformly larger than its positive-part version

$$\delta_i^+(\mathbf{x}) = \mu_i + \left(1 - \frac{(p-2)\sigma^2}{\Sigma(x_i - \mu_i)^2}\right)^+ (x_i - \mu_i).$$

7.8 Theorem 7.5 holds in greater generality than just the normal distribution. Suppose $\mathbf{X}$ is distributed according to the multivariate version of the exponential family $p_\eta(x)$ of (33.7),

$$p_\eta(\mathbf{x}) = e^{\eta'\mathbf{x} - A(\eta)} h(\mathbf{x}), \quad -\infty < x_i < \infty,$$

and a multivariate conjugate prior distribution [generalizing (3.19)] is used.

(a) Show that $E(\mathbf{X}|\eta) = \nabla A(\eta)$.

(b) If $\mu = 0$ in the prior distribution (see 3.19), show that $r(\tau, \delta) \geq r(\tau, \delta^+)$, where $\delta(\mathbf{x}) = [1 - B(\mathbf{x})]\mathbf{x}$ and $\delta^+(\mathbf{x}) = [1 - B(\mathbf{x})]^+\mathbf{x}$.

(c) If $\mu \neq 0$, the estimator $\delta(\mathbf{x})$ would be modified to $\mu + \delta(\mathbf{x} - \mu)$. Establish a result similar to part (b) for this estimator.

[*Hint*: For part (b), the proof of Theorem 7.5, modified to use the Bayes estimator $E(\nabla A(\eta)|\mathbf{x}, k, \mu)$ as in (3.21), will work.]

7.9 (a) For the model (7.7.15), show that the marginal distribution of X_i is negative binomial$(a, 1/b + 1)$; that is,

$$P(X_i = x) = \binom{a+x-1}{x} \left(\frac{b}{b+1}\right)^x \left(\frac{1}{b+1}\right)^a$$

with $EX_i = ab$ and var $X_i = ab(b+1)$.

(b) If $X_1, \ldots, X_m$ are iid according to the negative binomial distribution in part (a), show that the conditional distribution of $X_j | \sum_1^m X_i$ is the *negative hypergeometric* distribution, given by

$$P\left(X_j = x | \sum_1^m X_i = t\right) = \frac{\binom{a+x-1}{x}\binom{(m-1)a+t-x-1}{t-x}}{\binom{ma+t-1}{t}}$$

with $EX_j = t/m$ and var $X_j = (m-1)t(ma+t)/m^2(ma+1)$.

7.10 For the situation of Example 7.6:

(a) Show that the Bayes estimator under the loss $L_k(\lambda, \delta)$ of (7.7.16) is given by (7.7.17).

(b) Verify (7.7.19) and (7.7.20).

(c) Evaluate the Bayes risks $r(0, \delta^1)$ and $r(1, \delta^0)$. Which estimator, δ^0 or δ^1, is more robust?

7.11 For the situation of Example 7.6, evaluate the Bayes risk of the empirical Bayes estimator (7.7.20) for $k = 0$ and 1. What values of the unknown hyperparameter b are least and which are most favorable to the empirical Bayes estimator?

[*Hint*: Using the posterior expected loss (7.7.22) and Problem 7.9(b), the Bayes risk can be expressed as an expectation of a function of ΣX_i only. Further simplification seems unlikely.]

7.12 Consider a hierarchical Bayes estimator for the Poisson model (7.7.15) with loss (7.7.16). Using the distribution (5.6.27) for the hyperparameter b, show that the Bayes estimator is

$$\left(\frac{p\bar{x}+\alpha-k}{p\bar{x}+pa+\alpha+\beta-k}\right)(a+x_i-k).$$

[*Hint*: Show that the Bayes estimator is $E(\lambda^{1-k}|\mathbf{x})/E(\lambda^{-k}|\mathbf{x})$ and that

$$E(\lambda^r|\mathbf{x}) = \frac{\Gamma(p\bar{x}+pa+\alpha+\beta)\Gamma(p\bar{x}+\alpha+r)}{\Gamma(p\bar{x}+\alpha)\Gamma(p\bar{x}+pa+\alpha+\beta+r)}\frac{\Gamma(a+x_i+r)}{\Gamma(a+x_i)}.\Bigg]$$

7.13 Prove the following: Two matrix results that are useful in calculating estimators from multivariate hierarchical models are

(a) For any vector a of the form $a = (I - \frac{1}{s}J)b$, $\mathbf{1}'a = \Sigma a_i = 0$.

(b) If B is an idempotent matrix (that is, $B^2 = I$) and a is a scalar, then

$$(I+aB)^{-1} = I - \frac{a}{1+a}B.$$

7.14 For the situation of Example 7.7:

(a) Show how to derive the empirical Bayes estimator δ^L of (7.7.28).

(b) Verify the Bayes risk of δ^L of (7.7.29).

For the situation of Example 7.8:

(c) Show how to derive the empirical Bayes estimator δ^{EB_2} of (7.7.33).

(d) Verify the Bayes risk of δ^{EB_2}, (7.7.34).

7.15 The empirical Bayes estimator (7.7.27) can also be derived as a hierarchical Bayes estimator. Consider the hierarchical model

$$\begin{aligned} X_{ij}|\xi_i &\sim N(\xi_i, \sigma^2), \quad j = 1,\ldots,n, \quad i = 1,\ldots,s, \\ \xi_i|\mu &\sim N(\mu, \tau^2), \quad i = 1,\ldots,s, \\ \mu &\sim \text{Uniform}(-\infty,\infty) \end{aligned}$$

where σ^2 and τ^2 are known.

(a) Show that the Bayes estimator, with respect to squared error loss, is

$$E(\xi_i|\mathbf{x}) = \frac{\sigma^2}{\sigma^2+n\tau^2}E(\mu|\mathbf{x}) + \frac{n\tau^2}{\sigma^2+n\tau^2}\bar{x}_i$$

where $E(\mu|\mathbf{x})$ is the posterior mean of μ.

(b) Establish that $E(\mu|\mathbf{x}) = \bar{x} = \Sigma x_{ij}/ns$. [This can be done by evaluating the expectation directly, or by showing that the posterior distribution of $\xi_i|\mathbf{x}$ is

$$\xi_i|\mathbf{x} \sim N\left[\frac{\sigma^2}{\sigma^2+n\tau^2}\bar{x} + \frac{n\tau^2}{\sigma^2+n\tau^2}\bar{x}_i, \frac{\sigma^2}{\sigma^2+n\tau^2}\left(n\tau^2+\frac{\sigma^2}{s}\right)\right].$$

Note that the ξ_i's are not independent a posteriori. In fact,

$$\boldsymbol{\xi}|\mathbf{x} \sim N_s\left(\frac{n\tau^2}{\sigma^2+n\tau^2}M, \frac{n\sigma^2\tau^2}{\sigma^2+n\tau^2}M\right),$$

where $M = I + (\sigma^2/n\tau^2)J$].

(c) Show that the empirical Bayes estimator (7.7.32) can also be derived as a hierarchical Bayes estimator, by appending the specification $(\alpha, \beta) \sim \text{Uniform}(\Re^2)$ [that is, $\pi(\alpha, \beta) = d\alpha\, d\beta$, $-\infty < \alpha, \beta < \infty$] to the hierarchy (7.7.30).

7.16 Generalization of model (7.7.23) to the case of unequal n_i is, perhaps, not as straightforward as one might expect. Consider the generalization

$$X_{ij}|\xi_i \sim N(\xi_i, \sigma^2), \quad j = 1, \ldots, n_i, \quad i = 1, \ldots, s,$$
$$\xi_i|\mu \sim N(\mu, \tau_i^2), \quad i = 1, \ldots, s.$$

We also make the assumption that $\tau_i^2 = \tau^2/n_i$. Show that:

(a) The above model is equivalent to

$$\mathbf{Y} \sim N_s(\boldsymbol{\lambda}, \sigma^2 I),$$
$$\boldsymbol{\lambda} \sim N_s(\mathbf{Z}\mu, \tau^2 I)$$

where $Y_i = \sqrt{n_i}\bar{X}_i$, $\lambda_i = \sqrt{n_i}\xi_i$ and $\mathbf{z} = (\sqrt{n_1}, \ldots, \sqrt{n_s})'$.

(b) The Bayes estimator of ξ_i, using squared error loss, is

$$\frac{\sigma^2}{\sigma^2 + \tau^2}\mu + \frac{\tau^2}{\sigma^2 + \tau^2}\bar{x}_i.$$

(c) The marginal distribution of Y_i is $Y_i \sim N_s(\mathbf{Z}\mu, (\sigma^2 + \tau^2)I)$, and an empirical Bayes estimator of ξ is

$$\delta_i^{EB} = \bar{x} + \left(1 - \frac{(s-3)\sigma^2}{\Sigma n_i(\bar{x}_i - \bar{x})^2}\right)(\bar{x}_i - \bar{x})$$

where $\bar{x}_i = \Sigma_j x_{ij}/n_i$ and $\bar{x} = \Sigma_i n_i \bar{x}_i / \Sigma_i n_i$.

[Without the assumption that $\tau_i^2 = \tau^2/n_i$, one cannot get a simple empirical Bayes estimator. If $\tau_i^2 = \tau^2$, the likelihood estimation can be used to get an estimate of τ^2 to be used in the empirical Bayes estimator. This is discussed by Morris (1983a).]

7.17 (Empirical Bayes estimation in a general case). A general version of the hierarchical models of Examples 7.7 and 7.8 is

$$\mathbf{X}|\boldsymbol{\xi} \sim N_s(\boldsymbol{\xi}, \sigma^2 I),$$
$$\boldsymbol{\xi}|\boldsymbol{\beta} \sim N_s(\mathbf{Z}\boldsymbol{\beta}, \tau^2 I)$$

where σ^2 and $\mathbf{Z}_{s\times r}$, of rank r, are known and τ^2 and $\boldsymbol{\beta}_{r\times 1}$ are unknown. Under this model show that:

(a) The Bayes estimator of ξ, under squared error loss, is

$$E(\boldsymbol{\xi}|\mathbf{x}, \boldsymbol{\beta}) = \frac{\sigma^2}{\sigma^2 + \tau^2} z\boldsymbol{\beta} + \frac{\tau^2}{\sigma^2 + \tau^2}\mathbf{x}.$$

(b) Marginally, the distribution of $\mathbf{X}|\boldsymbol{\beta}$ is $\mathbf{X}|\boldsymbol{\beta} \sim N_s(\mathbf{Z}\boldsymbol{\beta}, (\sigma^2 + \tau^2)I)$.

(c) Under the marginal distribution in part (b),

$$E[(\mathbf{Z}'\mathbf{Z})^{-1}\mathbf{Z}'\mathbf{x}] = E\hat{\boldsymbol{\beta}} = \boldsymbol{\beta},$$
$$E\left[\frac{s - r - 2}{|\mathbf{X} - \mathbf{Z}\hat{\boldsymbol{\beta}}|^2}\right] = \frac{1}{\sigma^2 + \tau^2},$$

and, hence, an empirical Bayes estimator of ξ is

$$\delta^{EB} = \mathbf{Z}\hat{\boldsymbol{\beta}} + \left(1 - \frac{(s - r - 2)\sigma^2}{|\mathbf{x} - \mathbf{Z}\hat{\boldsymbol{\beta}}|^2}\right)(\mathbf{x} - \mathbf{Z}\hat{\boldsymbol{\beta}}).$$

(d) The Bayes risk of δ^{EB} is $r(\tau, \delta^\tau) + (r+2)\sigma^4/(\sigma^2+\tau^2)$, where $r(\tau, \delta^\tau)$ is the risk of the Bayes estimator.

7.18 (Hierarchical Bayes estimation in a general case.) In a manner similar to the previous problem, we can derive hierarchical Bayes estimators for the model

$$\begin{aligned} \mathbf{X}|\xi &\sim N_s(\xi, \sigma^2 I), \\ \xi|\beta &\sim N_s(\mathbf{Z}\beta, \tau^2 I), \\ \beta &\sim \text{Uniform}(\Re^r) \end{aligned}$$

where σ^2 and $\mathbf{Z}_{s\times r}$, of rank r, are known and τ^2 is unknown.

(a) The prior distribution of ξ, unconditional on β, is proportional to

$$\pi(\xi) = \int_{\Re^r} \pi(\xi|\beta)\, d\beta \propto e^{-\frac{1}{2}\frac{\xi'(I-H)\xi}{\tau^2}},$$

where $H = \mathbf{Z}(\mathbf{Z}'\mathbf{Z})^{-1}\mathbf{Z}'$ projects from $\Re^s$ to $\Re^r$.
[*Hint*: Establish that

$$\begin{aligned}(\xi - \mathbf{Z}\beta)'(\xi - \mathbf{Z}\beta) &= \xi'(I-H)\xi \\ &\quad + [\beta - (\mathbf{Z}'\mathbf{Z})^{-1}\mathbf{Z}'\xi]'\mathbf{Z}'\mathbf{Z}[\beta - (\mathbf{Z}'\mathbf{Z})^{-1}\mathbf{Z}'\xi]\end{aligned}$$

to perform the integration on β.]

(b) Show that

$$\xi|\mathbf{x} \sim N_s\left(\frac{\tau^2}{\sigma^2+\tau^2}M, \frac{\sigma^2\tau^2}{\sigma^2+\tau^2}M\right)$$

where $M = I + (\sigma^2/\tau^2)H$, and hence that the Bayes estimator is given by

$$\frac{\sigma^2}{\sigma^2+\tau^2}H\mathbf{x} + \frac{\tau^2}{\sigma^2+\tau^2}\mathbf{x},$$

where $\mathbf{Z}\hat{\beta} = H\mathbf{x}$.
[*Hint*: Establish that

$$\begin{aligned}&\frac{1}{\tau^2}\xi'(I-H)\xi + \frac{1}{\sigma^2}(\mathbf{x}-\xi)'(\mathbf{x}-\xi) \\ &= \frac{\sigma^2+\tau^2}{\sigma^2\tau^2}\left[\left(\xi - \frac{\tau^2}{\sigma^2+\tau^2}M\mathbf{x}\right)' M^{-1}\left(\xi - \frac{\tau^2}{\sigma^2+\tau^2}M\mathbf{x}\right)\right] \\ &\quad + \frac{1}{\sigma^2+\tau^2}\mathbf{x}'(I-H)\mathbf{x}\end{aligned}$$

where $M^{-1} = I - \frac{\sigma^2}{\sigma^2+\tau^2}H$.]

(c) Marginally, $\mathbf{X}'(I-H)\mathbf{X} \sim (\sigma^2+\tau^2)\chi^2_{s-r}$. This leads us to the empirical Bayes estimator

$$H\mathbf{x} + \left(1 - \frac{(s-r-2)\sigma^2}{\mathbf{x}'(I-H)\mathbf{x}}\right)(\mathbf{x} - H\mathbf{x})$$

which is equal to the empirical Bayes estimator of Problem 7.17(c).

[The model in this and the previous problem can be substantially generalized. For example, both $\sigma^2 I$ and $\tau^2 I$ can be replaced by full, positive definite matrices. At the cost of an increase in the complexity of the matrix calculations and the loss of simple answers, hierarchical and empirical Bayes estimators can be computed. The covariances, either scalar or matrix, can also be unknown, and inverted gamma (or inverted Wishart) prior distributions can be accommodated. Calculations can be implemented via the Gibbs sampler.

Note that these generalizations encompass the "unequal n_i" case (see Problem 7.16), but there are no simple solutions for this case. Many of these estimators also possess a minimax property, which will be discussed in Chapter 5.]

7.19 As noted by Morris (1983a), an analysis of variance-type hierarchical model, with unequal n_i, will yield closed-form empirical Bayes estimators if the prior variances are proportional to the sampling variances. Show that, for the model

$$X_{ij}|\xi_i \sim N(\xi_i, \sigma^2), \quad j = 1, \ldots, n_i, \quad i = 1, \ldots, s,$$
$$\boldsymbol{\xi}|\boldsymbol{\beta} \sim N_s(\mathbf{Z}\boldsymbol{\beta}, \tau^2 D^{-1})$$

where σ^2 and $\mathbf{Z}_{s\times r}$, of full rank r, are known, τ^2 is unknown, and $D = \text{diag}(n_1, \ldots, n_s)$, an empirical Bayes estimator is given by

$$\delta^{\text{EB}} = \mathbf{Z}\hat{\boldsymbol{\beta}} + \left(1 - \frac{(s-r-2)\sigma^2}{(\bar{\mathbf{x}} - \mathbf{Z}\hat{\boldsymbol{\beta}})' D (\bar{\mathbf{x}} - \mathbf{Z}\hat{\boldsymbol{\beta}})}\right)(\bar{\mathbf{x}} - \mathbf{Z}\hat{\boldsymbol{\beta}})$$

with $\bar{\mathbf{x}}_i = \Sigma_j x_{ij}/n_i$, $\bar{\mathbf{x}} = \{\bar{x}_i\}$, and $\hat{\boldsymbol{\beta}} = (\mathbf{Z}'D\mathbf{Z})^{-1}\mathbf{Z}'D\bar{\mathbf{x}}$.

7.20 An entertaining (and unjustifiable) result which abuses a hierarchical Bayes calculation yields the following derivation of the James-Stein estimator. Let $X \sim N_p(\boldsymbol{\theta}, I)$ and $\boldsymbol{\theta}|\tau^2 \sim N_p(0, \tau^2 I)$.

(a) Verify that conditional on τ^2, the posterior and marginal distributions are given by

$$\pi(\boldsymbol{\theta}|\mathbf{x}, \tau^2) = N_p\left(\frac{\tau^2}{\tau^2+1}\mathbf{x}, \frac{\tau^2}{\tau^2+1}I\right),$$
$$m(\mathbf{x}|\tau^2) = N_p[0, (\tau^2+1)I].$$

(b) Show that, taking $\pi(\tau^2) = 1$, $-1 < \tau^2 < \infty$, we have

$$\iint_{\Re^p} \boldsymbol{\theta}\pi(\boldsymbol{\theta}|\mathbf{x}, \tau^2) m(\mathbf{x}|\tau^2)\, d\boldsymbol{\theta}\, d\tau^2$$
$$= \frac{\mathbf{x}}{(2\pi)^{p/2}(|\mathbf{x}|^2)^{p/2-1}}\left[\Gamma\left(\frac{p-2}{2}\right)2^{(p-2)/2} - \frac{\Gamma(p/2)2^{p/2}}{|\mathbf{x}|^2}\right]$$

and

$$\iint_{\Re^p} \pi(\boldsymbol{\theta}|\mathbf{x}, \tau^2) m(\mathbf{x}|\tau^2) d\boldsymbol{\theta}\, d\tau^2$$
$$= \frac{1}{(2\pi)^{p/2}(|\mathbf{x}|^2)^{p/2-1}}\Gamma\left(\frac{p-2}{2}\right)2^{(p-2)/2}$$

and hence

$$E(\boldsymbol{\theta}|\mathbf{x}) = \left(1 - \frac{p-2}{|\mathbf{x}|^2}\right)\mathbf{x}.$$

(c) Explain some implications of the result in part (b) and, why it cannot be true. [Try to reconcile it with (3.3.12).]

(d) Why are the calculations in part (b) unjustified?

9 Notes

9.1 History

Following the basic paper by Bayes (published posthumously in 1763), Laplace initiated a widespread use of Bayes procedures, particularly with noninformative priors (for example, in his paper of 1774 and the fundamental book of 1820; see Stigler 1983, 1986). However, Laplace also employed non-Bayesian methods, without always making a clear distinction. A systematic theory of statistical inference based on noninformative (locally invariant) priors, generalizing and refining Laplace's approach, was developed by Jeffreys in his book on probability theory (1st edition 1939, 3rd edition 1961). A corresponding subjective theory owes its modern impetus to the work of deFinetti (for example, 1937, 1970) and that of L. J. Savage, particularly in his book on the *Foundations of Statistics* (1954). The idea of selecting an appropriate prior from the conjugate family was put forward by Raiffa and Schlaifer (1961). Interest in Bayes procedures (although not from a Bayesian point of view) also received support from Wald's result (for example, 1950) that all admissible procedures are either Bayes or limiting Bayes (see Section 5.8).

Bayesian attitudes and approaches are continually developing, with some of the most influential work done by Good (1965), DeGroot(1970), Zellner (1971), deFinetti (1974), Box and Tiao (1973), Berger (1985), and Bernardo and Smith (1994). An account of criticisms of the Bayesian approach can be found in Rothenberg (1977), and Berger (1985, Section 4.12). Robert (1994a, Chapter 10) provides a defense of "The Bayesian Choice."

9.2 Modeling

A general Bayesian treatment of linear models is given by Lindley and Smith (1972); the linear mixed model is given a Bayesian treatment in Searle et al. (1992, Chapter 9); sampling from a finite population is discussed from a Bayesian point of view by Ericson (1969) (see also Godambe 1982); a Bayesian approach to contingency tables is developed by Lindley (1964), Good (1965), and Bloch and Watson (1967) (see also Bishop, Fienberg, and Holland 1975 and Leonard 1972). The theory of Bayes estimation in exponential families is given a detailed development by Bernardo and Smith (1994). The fact that the resulting posterior expectations are convex combinations of sample and prior means is a characterization of this situation (Diaconis and Ylvisaker 1979, Goel and DeGroot 1980, MacEachern 1993).

Extensions to nonlinear and generalized linear models are given by Eaves (1983) and Albert (1988). In particular, for the generalized linear model, Ibrahim and Laud (1991) and Natarajan and McCulloch (1995) examine conditions for the propriety of posterior densities resulting from improper priors.

9.3 Computing

One reason why interest in Bayesian methods has flourished is because of the great strides in Bayesian computing. The fundamental work of Geman and Geman (1984) (which built on that of Metropolis et al. (1953) and Hastings 1970) influenced Gelfand and Smith (1990) to write a paper that sparked new interest in Bayesian methods, statistical computing, algorithms, and stochastic processes through the use of computing algorithms such as the Gibbs sampler and the Metropolis-Hastings algorithm. Elementary introductions to these topics can be found in Casella and George (1992) and Chib and Greenberg (1995). More detailed and advanced treatments are given in Tierney (1994), Robert (1994b), Gelman et al. (1995), and Tanner (1996).

9.4 The Ergodic Theorem

The general theorem about convergence of (5.15) in a Markov chain is known as the *Ergodic Theorem*; the name was coined by Boltzmann when investigating the behavior of gases (see Dudley 1989, p. 217). A sequence $X_0, X_1, X_2, \ldots$ is called *ergodic* if the limit of $\sum_{i=1}^n X_i/n$ is independent of the initial value of X_0. The ergodic theorem for *stationary sequences*, those for which $(X_{j_1}, \ldots, X_{j_k})$ has the same distribution as $(X_{j_1+r}, \ldots, X_{j_k+r})$ for all $r = 1, 2, \ldots$ is an assertion of the equality of time and space averages and holds in some generality (Dudley 1989, Section 8.4, Billingsley 1995, Section 24).

As the importance of this theorem led it to have wider applicability, the term "ergodic" has come to be applied in many situations and is often associated with Markov chains. In statistical practice, the usefulness of Markov chains for computations and the importance of the limit being independent of the starting values has brought the study of the ergodic behavior of Markov chains into prominence for statisticians. Good entries to the classical theory of Markov chains can be found in Feller (1968), Kemeny and Snell (1976), Resnick (1992), Ross (1985), or the more advanced treatment by Meyn and Tweedie (1993). In the context of estimation, the papers by Tierney (1994) and Robert (1995) provide detailed introductions to the relevant Markov chain theory. Athreya, Doss, and Sethuraman (1996) rigorously develop limit theorems for Markov chains arising in Gibbs sampling-type situations.

We are mainly concerned with Markov chains $X_0, X_1, \ldots$ that have an *invariant distribution*, F, satisfying $\int_A dF(x) = \int P(X_{n+1} \in A|X_n = x)dF(x)$. The chain is called *irreducible* if all sets with positive probability under the invariant distribution can be reached at some point by the chain. Such an irreducible chain is also *recurrent* (Tierney 1994, Section 3.1). A recurrent chain is one that visits every set infinitely often (i.o.) or, more importantly, a recurrent chain tends not to "drift off" to infinity. Formally, an irreducible Markov chain is *recurrent* if for each A with $\int_A dF(x) > 0$, we have $P(X_k \in A \text{ i.o. } |X_0 = x_0) > 0$ for all x_0, and equal to 1 for almost all x_0 (f). If $P(X_k \in A \text{ i.o. } |X_0 = x_0) = 1$ for all x_0, the chain is called *Harris recurrent*. Finally, if the invariant distribution F has finite mass (as it will in most of the cases we consider here), the chain is *positive recurrent*; otherwise it is *null recurrent*.

The Markov chain is *periodic* if for some integer $m \geq 2$, there exists a collection of disjoint sets $\{A_1, \ldots, A_m\}$ for which $P(X_{k+1} \in A_{j+1}|X_k \in A_j) = 1$ for all $j = 1, \ldots, m-1 \pmod m$. That is, the chain periodically travels through the sets $A_1, \ldots, A_m$. If no such collection of sets exists, the chain is *aperiodic*.

The relationship between these Markov chain properties and their consequences are summarized in the following theorem, based on Theorem 1 of Tierney (1994).

Theorem 9.1 *Suppose that the Markov chain $X_0, X_1, \ldots$ is irreducible with invariant distribution F satisfying $\int dF(x) = 1$. Then, the Markov chain is positive recurrent and F is the unique invariant distribution. If the Markov chain is also aperiodic, then for almost all x_0 (F),*

$$\sup_A |P(X_k \in A|X_0 = x_0) - \int_A dF(x)| \to 0.$$

If the chain is Harris recurrent, the convergence occurs for all x_0.

It is common to call a Markov chain *ergodic* if it is positive Harris recurrent and aperiodic. For such chains, we have the following version of the ergodic theorem.

Theorem 9.2 *Let $X_0, X_1, X_2, \ldots$ be an ergodic Markov chain with invariant distribution F. Then, for any function h with $\int |h(x)|dF(x) < \infty$,*

$$(1/n)\sum_{i=1}^{n} h(X_i) \rightarrow \int h(x)dF(x) \textit{ almost everywhere } (F).$$

9.5 Parametric and Nonparametric Empirical Bayes

The empirical Bayes analysis considered in Section 6 is sometimes referred to as *parametric* empirical Bayes, to distinguish it from the empirical Bayes methodology developed by Robbins (1955), which could be called *nonparametric*. In nonparametric empirical Bayes analysis, no functional form is assumed for the prior distribution, but a nonparametric estimator of the prior is built up and the resulting empirical Bayes mean is calculated. Robbins showed that as the sample size goes to infinity, it is possible to achieve the same Bayes risk as that achieved by the true Bayes estimator. Much research has been done in this area (see, for example, Van Ryzin and Susarla 1977, Susarla 1982, Robbins 1983, and Maritz and Lwin 1989). Due to the nature of this approach, its optimality properties tend to occur in large samples, with the parametric empirical Bayes approach being more suited for estimation in finite-sample problems.

Parametric empirical Bayes methods also have a long history, with major developments evolving in the sequence of papers by Efron and Morris (1971, 1972a, 1972b, 1973a, 1973b, 1975, 1976a, 1976b), where the connection with minimax estimation is explored. The theory and applications of empirical Bayes methods is given by Morris (1983a); a more comprehensive treatment is found in Carlin and Louis (1996). Less technical introductions are given by Casella (1985a, 1992a).

9.6 Robust Bayes

Robust Bayesian methods were effectively coalesced into a practical methodology by Berger (1984). Since then, there has been a great deal of research on this topic. (See, for example, Berger and Berliner 1986, Wasserman 1989, 1990, Sivaganesen and Berger 1989, DasGupta 1991, Lavine 1991a, 1991b, and the review papers by Berger 1990b, 1994 and Wasserman 1994.) The idea of using a class of priors is similar to the *gamma-minimax* approach, first developed by Robbins (1951, 1964) and Good (1952). In this approach, the subject of robustness over the class is usually not an issue, but rather the objective is the construction of an estimator that is minimax over the class (see Problem 5.1.2).

CHAPTER 5

Minimaxity and Admissibility

1 Minimax Estimation

At the beginning of Chapter 4, we introduced two ways in which the risk function $R(\theta, \delta)$ can be minimized in some overall sense: minimizing a weighted-average risk and minimizing the maximum risk. The first of these approaches was the concern of Chapter 4; in the present chapter, we shall consider the second.

Definition 1.1 An estimator δ^M of θ, which minimizes the maximum risk, that is, which satisfies

$$\inf_\delta \sup_\theta R(\theta, \delta) = \sup_\theta R(\theta, \delta^M), \tag{1.1}$$

is called a *minimax* estimator.

The problem of finding the estimator δ^M, which minimizes the maximum risk, is often difficult. Thus, unlike what happened in UMVU, equivariant, and Bayes estimations, we shall not be able to determine minimax estimators for large classes of problems but, rather, will treat problems individually (see Section 5.4).

Example 1.2 A first example. As we will see (Example 2.17), the Bayes estimators of Example 4.1.5, given by (4.1.12), that is,

$$\delta_\wedge(x) = \frac{a+x}{a+b+n}, \tag{1.2}$$

are admissible. Their risk functions are, therefore, incomparable as they all must cross (or coincide). As an illustration, consider the group of three estimators δ^{π_i}, $i = 1, \ldots, 3$, Bayes estimators from beta(1, 3), beta(2, 2) and beta(3, 1) priors, respectively. Based on this construction, each δ^{π_i} will be preferred if it is thought that the true value of the parameter is close to its prior mean (1/4, 1/2, 3/4, respectively). Alternatively, one might choose δ^{π_2} since it can be shown that δ^{π_2} has the smallest maximum risk among the three estimators being considered (see Problem 1.1). Although δ^{π_2} is minimax among these three estimators, it is not minimax overall. See Problems 1.2 and 1.3 for an alternative definition of minimaxity where the class of estimators is restricted. ||

As pointed out in Section 4.1(i), and suggested by Example 1.2, Bayes estimators provide a tool for solving minimax problems. Thus, Bayesian considerations are helpful when choosing an optimal frequentist estimator. Viewed in this light, there is a synthesis of the two approaches. The Bayesian approach provides us with a means of constructing an estimator that has optimal frequentist properties.

This synthesis highlights important features of both the Bayesian and frequentist approaches. The Bayesian paradigm is well suited for the construction of possibly optimal estimators, but is less well suited for their evaluation. The frequentist paradigm is complementary, as it is well suited for risk evaluations, but less well suited for construction. It is important to view these two approaches and hence the contents of Chapters 4 and 5 as complementary rather than adversarial; together they provide a rich set of tools and techniques for the statistician.

If we want to apply this idea to the determination of minimax estimators, we must ask ourselves: For what prior distribution Λ is the Bayes solution δ_Λ likely to be minimax? A minimax procedure, by minimizing the maximum risk, tries to do as well as possible in the worst case. One might, therefore, expect that the minimax estimator would be Bayes for the worst possible distribution. To make this concept precise, let us denote the average risk (Bayes risk) of the Bayes solution δ_Λ by

$$r_\Lambda = r(\Lambda, \delta_\Lambda) = \int R(\theta, \delta_\Lambda)\, d\Lambda(\theta). \tag{1.3}$$

Definition 1.3 A *prior distribution* Λ is *least favorable* if $r_\Lambda \geq r_{\Lambda'}$ for all prior distributions Λ'.

This is the prior distribution which causes the statistician the greatest average loss.

The following theorem provides a simple condition for a Bayes estimator δ_Λ to be minimax.

Theorem 1.4 *Suppose that* Λ *is a distribution on* Θ *such that*

$$r(\Lambda, \delta_\Lambda) = \int R(\theta, \delta_\Lambda)\, d\Lambda(\theta) = \sup_\theta R(\theta, \delta_\Lambda). \tag{1.4}$$

Then:

(i) δ_Λ *is minimax.*

(ii) If δ_Λ *is the unique Bayes solution with respect to* Λ, *it is the unique minimax procedure.*

(iii) Λ *is least favorable.*

Proof.

(i) Let δ be any other procedure. Then,

$$\begin{aligned}\sup_\theta R(\theta, \delta) &\geq \int R(\theta, \delta)\, d\Lambda(\theta)\\ &\geq \int R(\theta, \delta_\Lambda)\, d\Lambda(\theta) = \sup_\theta R(\theta, \delta_\Lambda).\end{aligned}$$

(ii) This follows by replacing $\geq$ by $>$ in the second equality of the proof of (i).

(iii) Let Λ' be some other distribution of θ. Then,

$$\begin{aligned}r_{\Lambda'} &= \int R(\theta, \delta_{\Lambda'})\, d\Lambda'(\theta) \leq \int R(\theta, \delta_\Lambda)\, d\Lambda'(\theta)\\ &\leq \sup_\theta R(\theta, \delta_\Lambda) = r_\Lambda.\end{aligned}$$

□

Condition (1.4) states that the average of $R(\theta, \delta_\Lambda)$ is equal to its maximum. This will be the case when the risk function is constant or, more generally, when Λ assigns probability 1 to the set on which the risk function takes on its maximum value. The following minimax characterizations are variations and simplifications of this requirement.

Corollary 1.5 *If a Bayes solution δ_Λ has constant risk, then it is minimax.*

Proof. If δ_Λ has constant risk, (1.4) clearly holds. □

Corollary 1.6 *Let ω_Λ be the set of parameter points at which the risk function of δ_Λ takes on its maximum, that is,*

$$\omega_\Lambda = \{\theta : R(\theta, \delta_\Lambda) = \sup_{\theta'} R(\theta', \delta_\Lambda)\}. \tag{1.5}$$

Then, δ_Λ is minimax if

$$\Lambda(\omega_\Lambda) = 1. \tag{1.6}$$

This can be rephrased by saying that a sufficient condition for δ_Λ to be minimax is that there exists a set ω such that

$$\Lambda(\omega) = 1$$

and

$$R(\theta, \delta_\Lambda) \text{ attains its maximum at all points of } \omega. \tag{1.7}$$

Example 1.7 Binomial. Suppose that X has the binomial distribution $b(p, n)$ and that we wish to estimate p with squared error loss. To see whether X/n is minimax, note that its risk function $p(1-p)/n$ has a unique maximum at $p = 1/2$. To apply Corollary 1.6, we need to use a prior distribution Λ for p which assigns probability 1 to $p = 1/2$. The corresponding Bayes estimator is $\delta(X) \equiv 1/2$, not X/n. Thus, if X/n is minimax, the approach suggested by Corollary 1.6 does not work in the present case. It is, in fact, easy to see that X/n is not minimax (Problem 1.9).

To determine a minimax estimator by the method of Theorem 1.4, let us utilize the result of Example 4.1.5 and try a beta distribution for Λ. If Λ is $B(a, b)$, the Bayes estimator is given by (4.1.12) and its risk function is

$$\frac{1}{(a+b+n)^2}\{np(1-p) + [a(1-p) - bp]^2\}. \tag{1.8}$$

Corollary 1.5 suggests seeing whether there exist values a and b for which the risk function (1.8) is constant. Setting the coefficients of p^2 and p in (1.8) equal to zero shows that (1.8) is constant if and only if

$$(a+b)^2 = n \quad \text{and} \quad 2a(a+b) = n. \tag{1.9}$$

Since a and b are positive, $a + b = \sqrt{n}$ and, hence,

$$a = b = \frac{1}{2}\sqrt{n}. \tag{1.10}$$

It follows that the estimator

$$\delta = \frac{X + \frac{1}{2}\sqrt{n}}{n + \sqrt{n}} = \frac{X}{n}\frac{\sqrt{n}}{1+\sqrt{n}} + \frac{1}{2}\frac{1}{1+\sqrt{n}} \tag{1.11}$$

is constant risk Bayes and, hence, minimax. Because of the uniqueness of the Bayes estimator (4.1.4), it is seen that (1.11) is the unique minimax estimator of p.

Of course, the estimator (1.11) is biased (Problem 1.10) because X/n is the only unbiased estimator that is a function of X. A comparison of its risk, which is

$$r_n = E(\delta - p)^2 = \frac{1}{4}\frac{1}{(1+\sqrt{n})^2}, \tag{1.12}$$

with the risk function

$$R_n(p) = p(1-p)/n \tag{1.13}$$

of X/n shows that (Problem 1.11) $r_n < R_n(p)$ in an interval $I_n = (1/2 - c_n < p < 1/2 + c_n)$ and $r_n > R_n(p)$ outside I_n. For small values of n, c_n is close to $1/2$, so that the minimax estimator is better (and, in fact, substantially better) for most of the range of p. However, as $n \to \infty$, $c_n \to 0$ and I_n shrinks toward the point $1/2$. Furthermore, $\sup_p R_n(p)/r_n = R_n(1/2)/r_n \to 1$, so that even at $p = 1/2$, where the comparison is least favorable to X/n, the improvement achieved by the minimax estimator is negligible. Thus, for large and even moderate n, X/n is the better of the two estimators. In the limit as $n \to \infty$ (although not for any finite n), X/n dominates the minimax estimator. Problems for which such a *subminimax* sequence does not exist are discussed by Ghosh (1964).

The present example illustrates an asymmetry between parts (ii) and (iii) of Theorem 1.4. Part (ii) asserts the uniqueness of the minimax estimator, whereas no such claim is made in part (iii) for the least favorable Λ. In the present case, it follows from (4.1.4) that for any Λ, the Bayes estimator of p is

$$\delta_\Lambda(x) = \frac{\int_0^1 p^{x+1}(1-p)^{n-x}\,d\Lambda(p)}{\int_0^1 p^x(1-p)^{n-x}\,d\Lambda(p)}. \tag{1.14}$$

Expansion of $(1-p)^{n-x}$ in powers of p shows that $\delta_\Lambda(x)$ depends on Λ only through the first $n+1$ moments of Λ. This shows, in particular, that the least favorable distribution is not unique in the present case. Any prior distribution with the same first $n+1$ moments gives the same Bayes solution and, hence, by Theorem 1.4 is least favorable (Problem 1.13).

Viewed as a loss function, squared error may be unrealistic when estimating p since in many situations an error of fixed size seems much more serious for values of p close to 0 or 1 than for values near $1/2$. To take account of this difficulty, let

$$L(p,d) = \frac{(d-p)^2}{p(1-p)}. \tag{1.15}$$

With this loss function, X/n becomes a constant risk estimator and is seen to be a Bayes estimator with respect to the uniform distribution on $(0, 1)$ and hence a minimax estimator. It is interesting to note that with (1.15), the risk function of the estimator (1.11) is unbounded. This indicates how strongly the minimax property can depend on the loss function. ||

When the loss function is convex in d, as was the case in Example 1.7, it follows from Corollary 1.7.9 that attention may be restricted to nonrandomized estimators. The next example shows that this is no longer true when the convexity assumption is dropped.

Example 1.8 Randomized minimax estimator. In the preceding example, suppose that the loss is zero when $|d - p| \leq \alpha$ and is one otherwise, where $\alpha < 1/2(n+1)$. Since any nonrandomized $\delta(X)$ can take on at most $n+1$ distinct values, the maximum risk of any such δ is then equal to 1. To exhibit a randomized estimator with a smaller maximum risk, consider the extreme case in which the estimator of p does not depend on the data at all but is a random variable U, which is uniformly distributed on (0, 1). The resulting risk function is

$$R(p, U) = 1 - P(|U - p| \leq \alpha) \tag{1.16}$$

and it is easily seen that the maximum of (1.16) is $1 - \alpha < 1$ (Problem 1.14). ||

The loss function in this example was chosen to make the calculations easy, but the possibility of reducing the maximum risk through randomization exists also for other nonconvex loss functions. In particular, for the problem of Example 1.7 with loss function $|d - p|^r (0 < r < 1)$, it can be proved that no nonrandomized estimator can be minimax (Hodges and Lehmann 1950).

Example 1.9 Difference of two binomials. Consider the case of two independent variables X and Y with distributions $b(p_1, m)$ and $b(p_2, n)$, respectively, and the problem of estimating $p_2 - p_1$ with squared error loss. We shall now obtain the minimax estimator when $m = n$; no solution is known when $m \neq n$.

The derivation of the estimator in Example 4.1.5 suggests that in the present case, too, the minimax estimator might be a linear estimator $aX + bY + k$ with constant risk. However, it is easy to see (Problem 1.18) that such a minimax estimator does not exist. Still hoping for a linear estimator, we shall therefore try to apply Corollary 1.6. Before doing so, let us simplify the hoped-for solutions by an invariance consideration.

The problem remains invariant under the transformation

$$(X', Y') = (Y, X), \quad (p_1', p_2') = (p_2, p_1), \quad d' = -d, \tag{1.17}$$

and an estimator $\delta(X, Y)$ is equivariant under this transformation provided $\delta(Y, X) = -\delta(X, Y)$ and hence if

$$(a + b)(x + y) + 2k = 0 \quad \text{for all } x, y.$$

This leads to the condition $a + b = k = 0$ and, therefore, to an estimator of the form

$$\delta(X, Y) = c(Y - X). \tag{1.18}$$

As will be seen in Section 5.4 (see Theorem 4.1 and the discussion following it), if a problem remains invariant under a finite group G and if a minimax estimator exists, then there exists an equivariant minimax estimator. In our search for a linear minimax estimator, we may therefore restrict attention to estimators of the form (1.18).

Application of Corollary 1.6 requires determination of the set ω of pairs (p_1, p_2) for which the risk of (1.18) takes on its maximum. The risk of (1.18) is

$$\begin{aligned} R_c(p_1, p_2) &= E[c(Y - X) - (p_2 - p_1)]^2 \\ &= c^2 n(p_1(1 - p_1) + p_2(1 - p_2)) + (cn - 1)^2 (p_2 - p_1)^2. \end{aligned}$$

Taking partial derivatives with respect to p_1 and p_2 and setting the resulting expressions equal to 0 leads to the two equations

$$\begin{aligned} [2(cn-1)^2 - 2c^2 n]p_1 - 2(cn-1)^2 p_2 &= -c^2 n, \\ -2(cn-1)^2 p_1 + [2(cn-1)^2 - 2c^2 n]p_2 &= -c^2 n. \end{aligned} \tag{1.19}$$

Typically, these equations have a unique solution, say (p_1^0, p_2^0), which is the point of maximum risk. Application of Corollary 1.6 would then have Λ assign probability 1 to the point (p_1^0, p_2^0) and the associated Bayes estimator would be $\delta(X, Y) \equiv p_2^0 - p_1^0$, whose risk does not have a maximum at (p_1^0, p_2^0).

This impasse does not occur if the two equations (1.19) are linearly dependent. This will be the case only if

$$c^2 n = 2(cn - 1)^2$$

and hence if

$$c = \frac{\sqrt{2n}}{n\left[\sqrt{2n} \pm 1\right]}. \tag{1.20}$$

Now, a Bayes estimator (4.1.4) does not take on values outside the convex hull of the range of the estimand, which in the present case is $(-1, 1)$. This rules out the minus sign in the denominator of c. Substituting (1.20) with the plus sign into (1.19) reduces these two equations to the single equation

$$p_1 + p_2 = 1. \tag{1.21}$$

The hoped-for minimax estimator is thus

$$\delta(X, Y) = \frac{\sqrt{2n}}{n\left(\sqrt{2n} + 1\right)}(Y - X). \tag{1.22}$$

We have shown (and it is easily verified directly, see Problem 1.19) that in the (p_1, p_2) plane, the risk of this estimator takes on its maximum value at all points of the line segment (1.21), with $0 < p_1 < 1$, which therefore is the conjectured ω of Corollary 1.6.

It remains to show that (1.22) is the Bayes estimator of a prior distribution Λ, which assigns probability 1 to the set (1.21).

Let us now confine attention to this subset and note that $p_1 + p_2 = 1$ implies $p_2 - p_1 = 2p_2 - 1$. The following lemma reduces the problem of estimating $2p_2 - 1$ to that of estimating p_2.

Lemma 1.10 *Let δ be a Bayes (respectively, UMVU, minimax, admissible) estimator of $g(\theta)$ for squared error loss. Then, $a\delta + b$ is Bayes (respectively, UMVU, minimax, admissible) for $ag(\theta) + b$.*

Proof. This follows immediately from the fact that

$$R(ag(\theta) + b, a\delta + b) = a^2 R(g(\theta), \delta).$$

□

For estimating p_2, we have, in the present case, n binomial trials with parameter $p = p_2$ and n binomial trials with parameter $p = p_1 = 1 - p_2$. If we interchange the meanings of "success" and "failure" in the latter n trials, we have $2n$ binomial trials with success probability p_2, resulting in $Y + (n - X)$ successes. According to Example 1.7, the estimator

$$\frac{Y + n - X}{2n} \frac{\sqrt{2n}}{1 + \sqrt{2n}} + \frac{1}{2} \frac{1}{1 + \sqrt{2n}}$$

is unique Bayes for p_2. Applying Lemma 1.10 and collecting terms, we see that the estimator (1.22) is unique Bayes for estimating $p_2 - p_1 = 2p_2 - 1$ on ω. It now follows from the properties of this estimator and Corollary 1.5 that δ is minimax for estimating $p_2 - p_1$. It is interesting that $\delta(X, Y)$ is not the difference of the minimax estimators for p_2 and p_1. This is unlike the behavior of UMVU estimators.

That $\delta(X, Y)$ is the unique Bayes (and hence minimax) estimator for $p_2 - p_1$, even when attention is not restricted to ω, follows from the remark after Corollary 4.1.4. It is only necessary to observe that the subsets of the sample space which have positive probability are the same whether (p_1, p_2) is in ω or not.

The comparison of the minimax estimator (1.22) with the UMVU estimator $(Y - X)/n$ gives results similar to those in the case of a single p. In particular, the UMVU estimator is again much better for large $m = n$ (Problem 1.20). ||

Equation (1.4) implies that a least favorable distribution exists. When such a distribution does not exist, Theorem 1.4 is not applicable. Consider, for example, the problem of estimating the mean θ of a normal distribution with known variance. Since all possible values of θ play a completely symmetrical role, in the sense that none is easier to estimate than any other, it is natural to conjecture that the least favorable distribution is "uniform" on the real line, that is, that the least favorable distribution is Lebesgue measure. This is the Jeffreys prior and, in this case, is not a proper distribution.

There are two ways in which the approach of Theorem 1.4 can be generalized to include such improper priors.

(a) As was seen in Section 4.1, it may turn out that the posterior distribution given $\mathbf{x}$ is a proper distribution. One can then compute the expectation $E[g(\Theta)|\mathbf{x}]$ for this distribution, a *generalized* Bayes estimator, and hope that it is the desired estimator. This approach is discussed, for example, by Sacks (1963), Brown (1971), and Berger and Srinivasan (1978).

(b) Alternatively, one can approximate the improper prior distribution with a sequence of proper distributions; for example, Lebesgue measure by the uniform distributions on $(-N, N)$, $N = 1, 2, \ldots$, and generalize the concept of least favorable distribution to that of least favorable sequence. We shall here follow the second approach.

Definition 1.11 A *sequence of prior distributions* $\{\Lambda_n\}$ is *least favorable* if for every prior distribution Λ we have

$$r_\Lambda \le r = \lim_{n\to\infty} r_{\Lambda_n}, \tag{1.23}$$

where

$$r_{\Lambda_n} = \int R(\theta, \delta_n)\, d\Lambda_n(\theta) \tag{1.24}$$

is the Bayes risk under Λ_n.

Theorem 1.12 *Suppose that* $\{\Lambda_n\}$ *is a sequence of prior distributions with Bayes risks* r_n *satisfying (1.23) and that* δ *is an estimator for which*

$$\sup_\theta R(\theta, \delta) = r. \tag{1.25}$$

Then

(i) δ *is minimax and*

(ii) the sequence $\{\Lambda_n\}$ *is least favorable.*

Proof.

(i) Suppose δ' is any other estimator. Then,

$$\sup_\theta R(\theta, \delta') \ge \int R(\theta, \delta')\, d\Lambda_n(\theta) \ge r_{\Lambda_n},$$

and this holds for every n. Hence,

$$\sup_\theta R(\theta, \delta') \ge \sup_\theta R(\theta, \delta),$$

and δ is minimax.

(ii) If Λ is any distribution, then

$$r_\Lambda = \int R(\theta, \delta_\Lambda)\, d\Lambda(\theta) \le \int R(\theta, \delta)\, d\Lambda(\theta) \le \sup_\theta R(\theta, \delta) = r.$$

This completes the proof. □

This theorem is less satisfactory than Theorem 1.4 in two respects. First, even if the Bayes estimators δ_n are unique, it is not possible to conclude that δ is the unique minimax estimator. The reason for this is that the second inequality in the second line of the proof of (i), which is strict when δ_n is unique Bayes, becomes weak under the limit operation.

The other difficulty is that in order to check condition (1.25), it is necessary to evaluate r and hence the Bayes risk r_{Λ_n}. This evaluation is often easy when the Λ_n are conjugate priors. Alternatively, the following lemma sometimes helps.

Lemma 1.13 *If δ_Λ is the Bayes estimator of $g(\theta)$ with respect to Λ and if*

$$r_\Lambda = E[\delta_\Lambda(\mathbf{X}) - g(\Theta)]^2 \tag{1.26}$$

is its Bayes risk, then

$$r_\Lambda = \int \text{var}[g(\Theta)|\mathbf{x}]\, dP(\mathbf{x}). \tag{1.27}$$

In particular, if the posterior variance of $g(\Theta)|\mathbf{x}$ is independent of $\mathbf{x}$, then

$$r_\Lambda = \text{var}[g(\Theta)|\mathbf{x}]. \tag{1.28}$$

Proof. The right side of (1.26) is equal to

$$\int \left\{E\,[g(\Theta) - \delta_\Lambda(\mathbf{x})]^2\,|\mathbf{x}\right\}\, dP(\mathbf{x})$$

and the result follows from (4.5.2). □

Example 1.14 Normal mean. Let $\mathbf{X} = (X_1, \ldots, X_n)$, with the X_i iid according to $N(\theta, \sigma^2)$. Let the estimand be θ, the loss squared error, and suppose, at first, that σ^2 is known. We shall prove that $\bar{X}$ is minimax by finding a sequence of Bayes estimators δ_n satisfying (1.23) with $r = \sigma^2/n$.

As prior distribution for θ, let us try the conjugate normal distribution $N(\mu, b^2)$. Then, it follows from Example 4.2.2 that the Bayes estimator is

$$\delta_\Lambda(\mathbf{x}) = \frac{n\bar{x}/\sigma^2 + \mu/b^2}{n/\sigma^2 + 1/b^2}. \tag{1.29}$$

The posterior variance is given by (4.2.3) and is independent of $\mathbf{x}$, so that

$$r_\Lambda = \frac{1}{n/\sigma^2 + 1/b^2}. \tag{1.30}$$

As $b \to \infty$, $r_\Lambda \uparrow \sigma^2/n$, and this completes the proof of the fact that $\bar{X}$ is minimax.

Suppose, now, that σ^2 is unknown. It follows from the result just proved that the maximum risk of every estimator will be infinite unless σ^2 is bounded. We shall therefore assume that

$$\sigma^2 \le M. \tag{1.31}$$

Under this restriction, the maximum risk of $\bar{X}$ is

$$\sup_{(\theta,\sigma^2)} E(\bar{X} - \theta)^2 = \frac{M}{n}.$$

That $\bar{X}$ is minimax subject to (1.31), then, is an immediate consequence of Lemma 1.15 below.

It is interesting to note that although the boundedness condition (1.31) was required for the minimax problem to be meaningful, the minimax estimator does not, in fact, depend on the value of M.

An alternative modification, when σ^2 is unknown, is to consider the loss function

$$L(\theta, \delta) = \frac{1}{\sigma^2}(\theta - \delta)^2. \tag{1.32}$$

For this loss function, the risk of $\bar{X}$ is bounded, and $\bar{X}$ is again minimax (Problem 1.21). ‖

We now prove a lemma which is helpful in establishing minimaxity in nonparametric situations.

Lemma 1.15 *Let X be a random quantity with distribution F, and let $g(F)$ be a functional defined over a set $\mathcal{F}_1$ of distributions F. Suppose that δ is a minimax estimator of $g(F)$ when F is restricted to some subset $\mathcal{F}_0$ of $\mathcal{F}_1$. Then, if*

$$\sup_{F\in\mathcal{F}_0} R(F,\delta) = \sup_{F\in\mathcal{F}_1} R(F,\delta), \tag{1.33}$$

δ is minimax also when F is permitted to vary over $\mathcal{F}_1$.

Proof. If an estimator δ' existed with smaller sup risk over $\mathcal{F}_1$ than δ, it would also have smaller sup risk over $\mathcal{F}_0$ and thus contradict the minimax property of δ over $\mathcal{F}_0$. □

Example 1.16 Nonparametric mean. Let $X_1, \ldots, X_n$ be iid with distribution F and finite expectation θ, and consider the problem of estimating θ with squared error loss. If the maximum risk of every estimator of θ is infinite, the minimax problem is meaningless. To rule this out, we shall consider two possible restrictions on F:

(a) Bounded variance,

$$\text{var}_F(X_i) \le M < \infty; \tag{1.34}$$

(b) bounded range,

$$-\infty < a < X_i < b < \infty. \tag{1.35}$$

Under (a), it is easy to see that $\bar{X}$ is minimax by applying Lemma 1.15 with $\mathcal{F}_1$ the family of all distributions F satisfying (1.34), and $\mathcal{F}_0$ the family of normal distributions satisfying (1.34). Then, $\bar{X}$ is minimax for $\mathcal{F}_0$ by Example 1.14. Since (1.33) holds with $\delta = \bar{X}$, it follows that $\bar{X}$ is minimax for $\mathcal{F}_1$. We shall see in the next section that it is, in fact, the unique minimax estimator of θ.

To find a minimax estimator of θ under (b), suppose without loss of generality that $a = 0$ and $b = 1$, and let $\mathcal{F}_1$ denote the class of distributions F with $F(1) - F(0) = 1$. It seems plausible in the present case that a least favorable distribution over $\mathcal{F}_1$ would concentrate on those distributions $F \in \mathcal{F}_1$ which are as spread out as possible, that is, which put all their mass on the points 0 and 1. But these are just binomial distributions with $n = 1$. If this conjecture is correct, the minimax estimator of θ should reduce to (1.11) when all the X_i are 0 or 1, with X in (1.11) given by $X = \Sigma X_i$. This suggests the estimator

$$\delta(X_1, \ldots, X_n) = \frac{\sqrt{n}}{1+\sqrt{n}}\bar{X} + \frac{1}{2}\frac{1}{1+\sqrt{n}}, \tag{1.36}$$

and we shall now prove that (1.36) is, indeed, a minimax estimator of θ.

Let $\mathcal{F}_0$ denote the set of distributions F according to which

$$P(X_i = 0) = 1 - p, \quad P(X_i = 1) = p, \quad 0 < p < 1.$$

Then, it was seen in Example 1.7 that (1.36) is the minimax estimator of $p = E(X_i)$ as F varies over $\mathcal{F}_0$. To prove that (1.36) is minimax with respect to $\mathcal{F}_1$, it is, by Lemma 1.15, enough to prove that the risk function of the estimator (1.36) takes on its maximum over $\mathcal{F}_0$.

Let $R(F, \delta)$ denote the risk of (1.36). Then,

$$R(F, \delta) = E\left[\frac{\sqrt{n}}{1+\sqrt{n}}\bar{X} + \frac{1}{2(1+\sqrt{n})} - \theta\right]^2.$$

By adding and subtracting $[\sqrt{n}/(1+\sqrt{n})]\theta$ inside the square brackets, this is seen to simplify to

$$R(F, \delta) = \frac{1}{(1+\sqrt{n})^2}\left[\operatorname{var}_F(X) + \left(\frac{1}{2} - \theta\right)^2\right]. \tag{1.37}$$

Now,

$$\operatorname{var}_F(X) = E(X-\theta)^2 = E(X^2) - \theta^2 \le E(X) - \theta^2$$

since $0 \le X \le 1$ implies $X^2 \le X$. Thus,

$$\operatorname{var}_F(X) \le \theta - \theta^2. \tag{1.38}$$

Substitution of (1.38) into (1.37) shows, after some simplification, that

$$R(F, \delta) \le \frac{1}{4(1+\sqrt{n})^2}. \tag{1.39}$$

Since the right side of (1.39) is the (constant) risk of δ over $\mathcal{F}_0$, the minimax property of δ follows. ||

Let us next return to the situation, considered at the beginning of Section 3.7, of estimating the mean $\bar{a}$ of a population $\{a_1, \ldots, a_N\}$ from a simple random sample $Y_1, \ldots, Y_n$ drawn from this population. To make the minimax estimation of $\bar{a}$ meaningful, restrictions on the a's are needed. In analogy to (1.34) and (1.35), we shall consider the following cases:

(a) Bounded population variance

$$\frac{1}{N}\Sigma(a_i - \bar{a})^2 \le M; \tag{1.40}$$

(b) Bounded range,

$$0 \le a_i \le 1, \tag{1.41}$$

to which the more general case $a \le a_i \le b$ can always be reduced. The loss function will be squared error, and for the time being, we shall ignore the labels. It will be seen in Section 5.4 that the minimax results remain valid when the labels are included in the data.

Example 1.17 Simple random sampling. We begin with case (b) and consider first the special case in which all the values of a are either 1 or 0, say D equal to

1, $N - D$ equal to 0. The total number X of 1's in the sample is then a sufficient statistic and has the hypergeometric distribution

$$P(X = x) = \binom{D}{x}\binom{N-D}{n-x}\Big/\binom{N}{n} \tag{1.42}$$

where $\max[0, n - (N - D)] \le x \le \min(n, D)$ (Problem 1.28) and where D can take on the values $0, 1, \ldots, N$. The estimand is $\bar{a} = D/N$, and, following the method of Example 4.1.5, one finds that $\alpha X/n + \beta$ with

$$\alpha = \frac{1}{1 + \sqrt{\frac{N-n}{n(N-1)}}}, \quad \beta = \frac{1}{2}(1 - \alpha) \tag{1.43}$$

is a linear estimator with constant risk (Problem 1.29). That (1.43) is minimax is then a consequence of the fact that it is the Bayes estimator of D/N with respect to the prior distribution

$$P(D = d) = \int_0^1 \binom{N}{d} p^d q^{N-d} \frac{\Gamma(a+b)}{\Gamma(a)\Gamma(b)} p^{a-1} q^{b-1} dp, \tag{1.44}$$

where

$$a = b = \frac{\beta}{\alpha/n - 1/N}. \tag{1.45}$$

It is easily checked that as $N \to \infty$, (1.43) $\to$ (1.11) and (1.45) $\to 1/2\sqrt{n}$, as one would expect since the hypergeometric distribution then tends toward the binomial.

The special case just treated plays the same role as a tool for the problem of estimating $\bar{a}$ subject to (1.41) that the binomial case played in Example 1.16. To show that

$$\delta = \alpha\bar{Y} + \beta \tag{1.46}$$

is minimax, it is only necessary to check that

$$E(\delta - \bar{a})^2 = \alpha^2 \operatorname{var}(\bar{Y}) + [\beta + (\alpha - 1)\bar{a}]^2 \tag{1.47}$$

takes on its maximum when all the values of a are 0 or 1, and this is seen as in Example 1.16 (Problem 1.31). Unfortunately, δ shares the poor risk properties of the binomial minimax estimator for all but very small n.

The minimax estimator of $\bar{a}$ subject to (1.40), as might be expected from Example 1.16, is $\bar{Y}$. For a proof of this result, which will not be given here, see Bickel and Lehmann (1981) or Hodges and Lehmann (1981). ||

As was seen in Examples 1.7 and 1.8, minimax estimators can be quite unsatisfactory over a large part of the parameter space. This is perhaps not surprising since, as a Bayes estimator with respect to a least favorable prior, a minimax estimator takes the most pessimistic view possible. This is illustrated by Example 1.7, in which the least favorable prior, $B(a_n, b_n)$ with $a_n = b_n = \sqrt{n}/2$, concentrates nearly its entire attention on the neighborhood of $p = 1/2$ for which accurate estimation of p is most difficult. On the other hand, a Bayes estimator corresponding to a personal prior may expose the investigator to a very high maximum risk, which may well be realized if the prior has badly misjudged the situation. It is possible to avoid the worst consequences of both these approaches through a compromise which permits the use of personal judgment and yet provides adequate protection against unacceptably high risks.

Suppose that M is the maximum risk of the minimax estimator. Then, one may be willing to consider estimators whose maximum risk exceeds M, if the excess is controlled, say, if

$$R(\theta, \delta) \le M(1+\varepsilon) \quad \text{for all } \theta \tag{1.48}$$

where ε is the proportional increase in risk that one is willing to tolerate. A *restricted Bayes estimator* is then obtained by minimizing, subject to (1.48), the average risk (4.1.1) for the prior Λ of one's choice.

Such restricted Bayes estimators are typically quite difficult to calculate. There is, however, one class of situations in which the evaluation is trivial: If the maximum risk of the unrestricted Bayes estimator satisfies (1.48), it, of course, coincides with the restricted Bayes estimator. This possibility is illustrated by the following example.

Example 1.18 Binomial restricted Bayes estimator. In Example 4.1.5, suppose we believe p to be near zero (it may, for instance, be the probability of a rarely occurring disease or accident). As a prior distribution for p, we therefore take $B(1, b)$ with a fairly high value of b. The Bayes estimator (4.11.12) is then $\delta = (X+1)/(n+b+1)$ and its risk is

$$E(\delta - p)^2 = \frac{np(1-p) + [(1-p) - bp]^2}{[n+b+1]^2}. \tag{1.49}$$

At $p = 1$, the risk is $[b/(n+b+1)]^2$, which for fixed n and sufficiently large b can be arbitrarily close to 1, while the constant risk of the minimax estimator is only $1/4(1+\sqrt{n})^2$. On the other hand, for fixed b, an easy calculation shows that (Problem 1.32).

$$4(1+\sqrt{n})^2 \sup R(p, \delta) \to 1 \quad \text{as } n \to \infty.$$

For any given b and $\varepsilon > 0$, δ will therefore satisfy (1.48) for sufficiently large values of n. ||

A quite different, and perhaps more typical, situation is illustrated by the normal case.

Example 1.19 Normal. If in the situation of Example 4.2.2, without loss of generality, we put $\sigma = 1$ and $\mu = 0$, the Bayes estimator (4.2.2) reduces to $c\bar{X}$ with $c = nb^2/(1+nb^2)$. Since its risk function is unbounded for all n, while the minimax risk is $1/n$, no such Bayes estimator can be restricted Bayes.

As a compromise, Efron and Morris (1971) propose an estimator of the form

$$\delta = \begin{cases} \bar{x} + M & \text{if } \bar{x} < -M/(1-c) \\ c\bar{x} & \text{if } |\bar{x}| \le M/(1-c) \\ \bar{x} - M & \text{if } \bar{x} > M/(1-c) \end{cases} \tag{1.50}$$

for $0 \le c \le 1$. The risk of these estimators is bounded (Problem 1.33) with maximum risk tending toward $1/n$ as $M \to 0$. On the other hand, for large M

values, (1.50) is close to the Bayes estimator. Although (1.50) is not the exact optimum solution of the restricted Bayes problem, Efron and Morris (1971) and Marazzi (1980) show it to be close to optimal. ||

2 Admissibility and Minimaxity in Exponential Families

It was seen in Example 2.2.6 that a UMVU estimator δ need not be admissible. If a biased estimator δ' has uniformly smaller risk, the choice between δ and δ' is not clear-cut: One must balance the advantage of unbiasedness against the drawback of larger risk. The situation is, however, different for minimax estimators. If δ' dominates a minimax estimator δ, then δ' is also minimax and, thus, definitely preferred. It is, therefore, particularly important to ascertain whether a proposed minimax estimator is admissible. In the present section, we shall obtain some admissibility results (and in the process, some minimax results) for exponential families, and in the next section, we shall consider the corresponding problem for group families.

To prove inadmissibility of an estimator δ, it is sufficient to produce an estimator δ' which dominates it. An example was given in Lemma 2.2.7. The following is another instance.

Lemma 2.1 *Let the range of the estimand $g(\theta)$ be an interval with end-points a and b, and suppose that the loss function $L(\theta, d)$ is positive when $d \neq g(\theta)$ and zero when $d = g(\theta)$, and that for any fixed θ, $L(\theta, d)$ is increasing as d moves away from $g(\theta)$ in either direction. Then, any estimator δ taking on values outside the closed interval $[a, b]$ with positive probability is inadmissible.*

Proof. δ is dominated by the estimator δ', which is a or b when $\delta < a$ or $> b$, and which otherwise is equal to δ. □

Example 2.2 Randomized response. The following is a survey technique sometimes used when delicate questions are being asked. Suppose, for example, that the purpose of a survey is to estimate the proportion p of students who have ever cheated on an exam. Then, the following strategy may be used. With probability a (known), the student is asked the question "Have you ever cheated on an exam?", and with probability $(1 - a)$, the question "Have you always been honest on exams?" The survey taker does not know which question the student answers, so the answer cannot incriminate the respondent (hence, honesty is encouraged). If a sample of n students is questioned in this way, the number of positive responses is a binomial random variable $X^* \sim b(p^*, n)$ with

$$p^* = ap + (1 - a)(1 - p), \tag{2.1}$$

where p is the probability of cheating, and

$$\min\{a, 1 - a\} < p^* < \max\{a, 1 - a\}. \tag{2.2}$$

For estimating the probability $p = [p^* - (1 - a)]/(1 - 2a)$, the method of moments estimator $\tilde{p} = [\hat{p}^* - (1 - a)]/(1 - 2a)$ is inadmissible by Lemma 2.1. The MLE of p, which is equal to $\tilde{p}$ if it falls in the interval specified in (2.2) and takes on the endpoint values if $\tilde{p}$ is not in the interval, is also inadmissible, although this fact does not follow directly from Lemma 2.1. (Inadmissibility of the MLE of p follows from Moors (1981); see also Hoeffding 1982 and Chaudhuri and Mukerjee 1988). ||

Example 2.3 Variance components. Another application of Lemma 2.1 occurs in the estimation of variance components. In the one-way layout with random effects (see Example 3.5.1 or 4.2.7), let

$$X_{ij} = \mu + A_i + \mu_{ij}, \quad j = 1, \ldots, n_i, \ i = 1, \ldots, s, \tag{2.3}$$

where the variables $A_i \sim N(0, \sigma_A^2)$ and $U_{ij} \sim N(0, \sigma^2)$ are independent. The parameter σ_A^2 has range $[0, \infty)$; hence, any estimator δ taking on negative values is an inadmissible estimator of σ_A^2 (against any loss function for which the risk function exists). The UMVU estimator of σ_A^2 [see (3.5.4)] has this property and hence is inadmissible. ||

A principal method for proving admissibility is the following result.

Theorem 2.4 *Any unique*[1] *Bayes estimator is admissible.*

Proof. If δ is unique Bayes with respect to the prior distribution Λ and is dominated by δ', then

$$\int R(\theta, \delta')\, d\Lambda(\theta) \leq \int R(\theta, \delta)\, d\Lambda(\theta),$$

which contradicts uniqueness. □

An example is provided by the binomial minimax estimator (1.11) of Example 1.7. For the corresponding nonparametric minimax estimator (1.36) of Example 1.16, admissibility was proved by Hjort (1976) who showed that it is the essentially unique minimax estimator with respect to a class of Dirichlet-process priors described by Ferguson (1973).

We shall, in the present section, illustrate a number of ideas and results concerning admissibility on the estimation of the mean and variance of a normal distribution and then indicate some of their generalizations. Unless stated otherwise, the loss function will be assumed to be squared error.

Example 2.5 Admissibility of linear estimators. Let $X_1, \ldots, X_n$ be independent, each distributed according to a $N(\theta, \sigma^2)$, with σ^2 known. In the preceding section, $\bar{X}$ was seen to be minimax for estimating θ. Is it admissible? Instead of attacking this question directly, we shall consider the admissibility of an arbitrary linear function $a\bar{X} + b$.

From Example 2.2, it follows that the unique Bayes estimator with respect to the normal prior for θ with mean μ and variance τ^2 is

$$\frac{n\tau^2}{\sigma^2 + n\tau^2}\bar{X} + \frac{\sigma^2}{\sigma^2 + n\tau^2}\mu \tag{2.4}$$

and that the associated Bayes risk is finite (Problem 2.2). It follows that $a\bar{X} + b$ is unique Bayes and hence admissible whenever

[1] Uniqueness here means that any two Bayes estimators differ only on a set N with $P_\theta(N) = 0$ for all θ.

$$0 < a < 1. \tag{2.5}$$

||

To see what can be said about other values of a, we shall now prove an inadmissibility result for linear estimators, which is quite general and in particular does not require the assumption of normality.

Theorem 2.6 *Let X be a random variable with mean θ and variance σ^2. Then, $aX + b$ is an inadmissible estimator of θ under squared error loss whenever*

(i) $a > 1$, or

(ii) $a < 0$, or

(iii) $a = 1$ and $b \neq 0$.

Proof. The risk of $aX + b$ is

$$\rho(a, b) = E(aX + b - \theta)^2 = a^2\sigma^2 + [(a-1)\theta + b]^2. \tag{2.6}$$

(i) If $a > 1$, then

$$\rho(a, b) \geq a^2\sigma^2 > \sigma^2 = \rho(1, 0)$$

so that $aX + b$ is dominated by X.

(ii) If $a < 0$, then $(a - 1)^2 > 1$ and hence

$$\rho(a, b) \geq [(a-1)\theta + b]^2 = (a-1)^2\left[\theta + \frac{b}{a-1}\right]^2$$
$$> \left(\theta + \frac{b}{a-1}\right)^2 = \rho\left(0, -\frac{b}{a-1}\right).$$

Thus, $aX + b$ is dominated by the constant estimator $\delta \equiv -b/(a-1)$.

(iii) In this case, $aX + b = X + b$ is dominated by X (see Lemma 2.2.7). □

Example 2.7 Continuation of Example 2.5. Combining the results of Example 2.5 and Theorem 2.6, we see that the estimator $a\bar{X} + b$ is admissible in the strip $0 < a < 1$ in the (a, b) plane, that it is inadmissible to the left $(a < 0)$ and to the right $(a > 1)$.

The left boundary $a = 0$ corresponds to the constant estimators $\delta = b$ which are admissible since $\delta = b$ is the only estimator with zero risk at $\theta = b$. Finally, the right boundary $a = 1$ is inadmissible by (iii) of Theorem 2.6, with the possible exception of the point $a = 1, b = 0$. ||

We have thus settled the admissibility of $a\bar{X} + b$ for all cases except $\bar{X}$ itself, which was the estimator of primary interest. In the next example, we shall prove that $\bar{X}$ is indeed admissible.

Example 2.8 Admissibility of $\bar{X}$. The admissibility of $\bar{X}$ for estimating the mean of a normal distribution is not only of great interest in itself but can also be regarded as the starting point of many other admissibility investigations. For this reason, we shall now give two proofs of this fact—they represent two principal methods for proving admissibility and are seen particularly clearly in this example because of its great simplicity.

First Proof of Admissibility (the Limiting Bayes Method). Suppose that $\bar{X}$ is not admissible, and without loss of generality, assume that $\sigma = 1$. Then, there exists δ^* such that

$$R(\theta, \delta^*) \leq \frac{1}{n} \quad \text{for all } \theta,$$

$$R(\theta, \delta^*) < \frac{1}{n} \quad \text{for at least some } \theta.$$

Now, $R(\theta, \delta)$ is a continuous function of θ for every δ so that there exists $\varepsilon > 0$ and $\theta_0 < \theta_1$ such that

$$R(\theta, \delta^*) < \frac{1}{n} - \varepsilon \quad \text{for all } \theta_0 < \theta < \theta_1.$$

Let r_τ^* be the average risk of δ^* with respect to the prior distribution $\Lambda_\tau = N(0, \tau^2)$, and let r_τ be the Bayes risk, that is, the average risk of the Bayes solution with respect to Λ_τ. Then, by (1.30) with $\sigma = 1$ and τ in place of b,

$$\frac{\frac{1}{n} - r_\tau^*}{\frac{1}{n} - r_\tau} = \frac{\frac{1}{\sqrt{2\pi}\tau} \int_{-\infty}^{\infty} \left[\frac{1}{n} - R(\theta, \delta^*)\right] e^{-\theta^2/2\tau^2}\, d\theta}{\frac{1}{n} - \frac{\tau^2}{1+n\tau^2}}$$

$$\geq \frac{n(1+n\tau^2)\varepsilon}{\tau\sqrt{2\pi}} \int_{\theta_0}^{\theta_1} e^{-\theta^2/2\tau^2}\, d\theta.$$

The integrand converges monotonically to 1 as $\tau \to \infty$. By the Lebesgue monotone convergence theorem (TSH2, Theorem 2.2.1), the integral therefore converges to $\theta_1 - \theta_0$, and, hence, as $\tau^2 \to \infty$,

$$\frac{1/n - r_\tau^*}{1/n - r_\tau} \to \infty.$$

Thus, there exists τ_0 such that $r_{\tau_0}^* < r_{\tau_0}$, which contradicts the fact that r_{τ_0} is the Bayes risk for Λ_{τ_0}. This completes the proof.

A more general version of this approach, known as *Blyth's method*, will be given in Theorem 7.13.

Second Proof of Admissibility (the Information Inequality Method). Another useful tool for establishing admissibility is based on the information inequality and solutions to a differential inequality, a method due to Hodges and Lehmann (1951).

It follows from the information inequality (2.5.33) and the fact that

$$R(\theta, \delta) = E(\delta - \theta)^2 = \text{var}_\theta(\delta) + b^2(\theta),$$

where $b(\theta)$ is the bias of δ, that

$$R(\theta, \delta) \geq \frac{[1 + b'(\theta)]^2}{nI(\theta)} + b^2(\theta), \tag{2.7}$$

where the first term on the right is the information inequality variance bound for estimators with expected value $\theta + b(\theta)$. Note that, in the present case with $\sigma^2 = 1$, $I(\theta) = 1$ from Table 2.5.1.

Suppose, now, that δ is any estimator satisfying

(2.8) $$R(\theta, \delta) \leq \frac{1}{n} \quad \text{for all } \theta$$

and hence

(2.9) $$\frac{[1 + b'(\theta)]^2}{n} + b^2(\theta) \leq R(\theta, \delta) \leq \frac{1}{n} \quad \text{for all } \theta.$$

We shall then show that (2.9) implies

(2.10) $$b(\theta) \equiv 0,$$

that is, that δ is unbiased.

(i) Since $|b(\theta)| \leq 1/\sqrt{n}$, the function b is bounded.

(ii) From the fact that

$$1 + 2b'(\theta) + [b'(\theta)]^2 \leq 1,$$

it follows that $b'(\theta) \leq 0$, so that b is nonincreasing.

(iii) We shall show, next, that there exists a sequence of values θ_i tending to ∞ and such that $b'(\theta_i) \to 0$. Suppose that $b'(\theta)$ were bounded away from 0 as $\theta \to \infty$, say $b'(\theta) \leq -\varepsilon$ for all $\theta > \theta_0$. Then $b(\theta)$ cannot be bounded as $\theta \to \infty$, which contradicts (i).

(iv) Analogously, it is seen that there exists a sequence of values $\theta_i \to -\infty$ and such that $b'(\theta_i) \to 0$ (Problem 2.3).

Inequality (2.9) together with (iii) and (iv) shows that $b(\theta) \to 0$ as $\theta \to \pm\infty$, and (2.10) now follows from (ii).

Since (2.10) implies that $b(\theta) = b'(\theta) = 0$ for all θ, it implies by (2.7) that

$$R(\theta, \delta) \geq \frac{1}{n} \quad \text{for all } \theta$$

and hence that

$$R(\theta, \delta) \equiv \frac{1}{n}.$$

This proves that $\bar{X}$ is admissible and minimax. That it is, in fact, the only minimax estimator is an immediate consequence of Theorem 1.7.10.

For another application of this second method of proof, see Problem 2.7. ||

Admissibility (hence, minimaxity) of $\bar{X}$ holds not only for squared error loss but for large classes of loss functions $L(\theta, d) = \rho(d - \theta)$. In particular, it holds if $\rho(t)$ is nondecreasing as t moves away from 0 in either direction and satisfies the growth condition

$$\int |t|\rho(2|t|)\phi(t)\,dt < \infty,$$

with the only exceptions being the loss functions

$$\rho(0) = a, \quad \rho(t) = b \quad \text{for} \quad |t| \neq 0, \quad a < b.$$

This result[2] follows from Brown (1966, Theorem 2.1.1); it is also proved under somewhat stronger conditions in Hájek (1972).

[2] Communicated by L. Brown.

Example 2.9 Truncated normal mean. In Example 2.8, suppose it is known that $\theta > \theta_0$. Then, it follows from Lemma 2.1 that $\bar{X}$ is no longer admissible. However, assuming that $\sigma^2 = 1$ and using the method of the second proof of Example 2.8, it is easy to show that $\bar{X}$ continues to be minimax. If it were not, there would exist an estimator δ and an $\varepsilon > 0$ such that

$$R(\theta, \delta) \leq \frac{1}{n} - \varepsilon \quad \text{for all } \theta > \theta_0$$

and hence

$$\frac{[1 + b'(\theta)]^2}{n} + b^2(\theta) \leq \frac{1}{n} - \varepsilon \quad \text{for all } \theta > \theta_0.$$

As a consequence, $b(\theta)$ would be bounded and satisfy $b'(\theta) \leq -\varepsilon n/2$ for all $\theta > \theta_0$, and these two statements are contradictory.

This example provides an instance in which the minimax estimator is not unique and the constant risk estimator $\bar{X}$ is inadmissible. A uniformly better estimator which a fortiori is also minimax is $\max(\theta_0, \bar{X})$, but it, too, is inadmissible [see Sacks (1963), in which a characterization of all admissible estimators is given]. Admissible minimax estimators in this case were found by Katz (1961) and Sacks (1963); see also Gupta and Rohatgi 1980.

If θ is further restricted to satisfy $a \leq \theta \leq b$, $\bar{X}$ is not only inadmissible but also no longer minimax. If $\bar{X}$ were minimax, the same would be true of its improvement, the MLE

$$\delta^*(X) = \begin{cases} a & \text{if } \bar{X} < a \\ \bar{X} & \text{if } a \leq \bar{X} \leq b \\ b & \text{if } \bar{X} > b, \end{cases}$$

so that

$$\sup_{a \leq \theta \leq b} R(\theta, \delta^*) = \sup_{a \leq \theta \leq b} R(\theta, \bar{X}) = \frac{1}{n}.$$

However, $R(\theta, \delta^*) < R(\theta, \bar{X}) = 1/n$ for all $a \leq \theta \leq b$. Furthermore, $R(\theta, \delta^*)$ is a continuous function of θ and hence takes on its maximum at some point $a \leq \theta_0 \leq b$. Thus,

$$\sup_{a \leq \theta \leq b} R(\theta, \delta^*) = R(\theta_0, \delta^*) < \frac{1}{n},$$

which provides a contradiction.

It follows from Wald's general decision theory (see Section 5.8) that in the present situation, there exists a probability distribution Λ over $[a, b]$ which satisfies (1.4) and (1.6). We shall now prove that the associated set ω_Λ of (1.5) consists of a finite number of points. Suppose the contrary were true. Then, ω_Λ contains an infinite sequence of points with a limit point. Since $R(\theta, \delta_\Lambda)$ is constant over these points and since it is an analytic function of θ, it follows that $R(\theta, \delta_\Lambda)$ is constant, not only in $[a, b]$ but for all θ. Example 2.8 then shows that $\delta_\Lambda = \bar{X}$, which is in contradiction to the fact that $\bar{X}$ is not minimax for the present problem.

To simplify matters, and without losing generality [Problem 2.9(a)], we can take $a = -m$ and $b = m$, and, thus, consider θ to be restricted to the interval $[-m, m]$. To determine a minimax estimator, let us consider the form of a least favorable prior. Since Λ is concentrated on a finite number of points, it is reasonable to suspect

Figure 2.1. *Risk functions of bounded mean estimators for $m = 1.056742$, $n = 1$.*

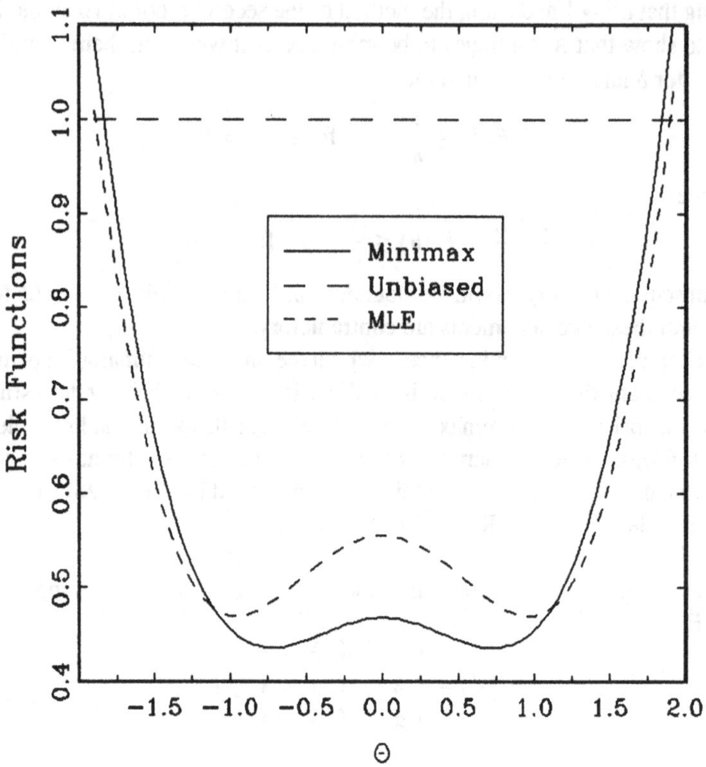

that these points would be placed at a distances neither too close together nor too far apart, where "close" is relative to the standard deviation of the density of X. (If the points are either much closer together or much further apart, then the prior might be giving us information.) One might therefore conjecture that the number of points in ω_Λ increases with m, and for small m, look at the Bayes estimator for the two-point prior Λ that puts mass $1/2$ at $\pm m$.

The Bayes estimator, against squared error loss, is [Problem 2.9(b)]

$$\delta^\Lambda(\bar{x}) = m \tanh(mn\bar{x}) \tag{2.11}$$

where tanh(·) is the *hyperbolic tangent* function. For $m \leq 1.05/\sqrt{n}$, Corollary 1.6 can be used to show that δ is minimax and provides a substantial risk decrease over $\bar{x}$. Moreover, for $m < 1/\sqrt{n}$, δ also dominates the MLE δ^* [Problem 2.9(c)]. This is illustrated in Figure 2.1, where we have taken m to be the largest value for which (2.11) is minimax. Note that the risk of δ^Λ is equal at $\theta = 0$ and $\theta = m$.

As m increases, so does the number of points in ω_Λ. The range of values of m, for which the associated Bayes estimators is minimax, was established by Casella and Strawderman (1981) for 2- and 3-point priors and Kempthorne (1988a, 1988b) for 4-point priors. Some interesting results concerning Λ and δ^Λ, for large m, are given by Bickel (1981). An alternative estimator, the Bayes estimator against a uniform prior on $[-m, m]$, was studied by Gatsonis et al. (1987) and shown to perform reasonably when compared to δ^{Λ} and to dominate δ^* for $|\theta| \le m/\sqrt{n}$. Many of these results were discovered independently by Zinzius (1981, 1982), who derived minimax estimators for θ restricted to the interval $[0, c]$, where c is known and small. ||

Example 2.10 Linear minimax risk. Suppose that in Example 2.9 we decide to restrict attention to linear estimators $\delta^{(a,b)} = a\bar{X} + b$ because of their simplicity.

With $\sigma^2 = 1$, from the proof of Theorem 2.6 [see also Problem 4.3.12(a)],

$$R(\theta, a\bar{X} + b) = a^2 \operatorname{var}\bar{X} + [(a-1)\theta + b]^2$$
$$= a^2/n + [(a-1)\theta + b]^2,$$

and from Theorem 2.6, we only need consider $0 \le a \le 1$. It is straightforward to establish (Problem 2.10) that

$$\max_{\theta \in [-m,m]} R(\theta, a\bar{X} + b) = \max\{R(-m, a\bar{X} + b), R(m, a\bar{X} + b)\}$$

and that $\delta^* = a^*\bar{X}$, with $a^* = m^2/(\frac{1}{n} + m^2)$ is minimax among linear estimators.

Donoho et al. (1990) provide bounds on the ratio of the linear minimax risk to the minimax risk. They show that, surprisingly, this ratio is approximately 1.25 and, hence, that the linear minimax estimators may sometimes be reasonable substitutes for the full minimax estimators. ||

Example 2.11 Linear model. Consider the general linear model of Section 3.4 and suppose we wish to estimate some linear function of the ξ's. Without loss of generality, we can assume that the model is expressed in the canonical form (4.8) so that $Y_1, \ldots, Y_n$ are independent, normal, with common variance σ^2, and $E(Y_i) = \eta_i$ $(i = 1, \ldots, s)$; $E(Y_{s+1}) = \cdots = E(Y_n) = 0$. The estimand can be taken to be η_1. If $Y_2, \ldots, Y_n$ were not present, it would follow from Example 2.8 that Y_1 is admissible for estimating η_1. It is obvious from the Rao-Blackwell theorem (Theorem 1.7.8) that the presence of $Y_{s+1}, \ldots, Y_n$ cannot affect this result. The following lemma shows that, as one would expect, the same is true for $Y_2, \ldots, Y_s$. ||

Lemma 2.12 *Let X and Y be independent (possibly vector-valued) with distributions F_ξ and G_η, respectively, where ξ and η vary independently. Then, if $\delta(X)$ is admissible for estimating ξ when Y is not present, it continues to be so in the presence of Y.*

Proof. Suppose, to the contrary, that there exists an estimator $T(X, Y)$ satisfying

$$R(\xi, \eta; T) \le R(\xi; \delta) \quad \text{for all } \xi, \eta,$$
$$R(\xi_0, \eta_0; T) < R(\xi_0; \delta) \quad \text{for some } \xi_0, \eta_0.$$

Consider the case in which it is known that $\eta = \eta_0$. Then, $\delta(X)$ is admissible on the basis of X and Y (Problem 2.11). On the other hand,

$$R(\xi, \eta_0; T) \le R(\xi; \delta) \quad \text{for all } \xi,$$
$$R(\xi_0, \eta_0; T) < R(\xi_0; \delta) \quad \text{for some } \xi_0,$$

and this is a contradiction. □

The examples so far have been concerned with normal means. Let us now turn to the estimation of a normal variance.

Example 2.13 Normal variance. Under the assumptions of Example 4.2.5, let us consider the admissibility, using squared error loss, of linear estimators $aY + b$ of $1/\tau = 2\sigma^2$. The Bayes solutions

$$\frac{\alpha + Y}{r + g - 1}, \tag{2.12}$$

derived there for the prior distributions $\Gamma(g, 1/\alpha)$, appear to prove admissibility of $aY + b$ with

$$0 < a < \frac{1}{r - 1}, \quad 0 < b. \tag{2.13}$$

In particular, this includes the estimators $(1/r)Y + b$ for any $b > 0$. On the other hand, it follows from (2.7) that $E(Y) = r/\tau$, so that $(1/r)Y$ is an unbiased estimator of $1/\tau$, and hence from Lemma 2.2.7, that $(1/r)Y + b$ is inadmissible for any $b > 0$. What went wrong?

Conditions (i) and (ii) of Corollary 4.1.4 indicate two ways in which the uniqueness (hence, admissibility) of a Bayes estimator may be violated. The second of these clearly does not apply here since the gamma prior assigns positive density to all values $\tau > 0$. This leaves the first possibility as the only visible suspect. Let us, therefore, consider the Bayes risk of the estimator (2.12).

Given τ, we find [by adding and subtracting the expectation of $Y/(g + r - 1)$], that

$$E\left(\frac{Y + \alpha}{g + r - 1} - \frac{1}{\tau}\right)^2 = \frac{1}{(g + r - 1)^2}\left[\frac{1}{\tau^2} - \left(\alpha - \frac{g - 1}{\tau}\right)^2\right].$$

The Bayes risk will therefore be finite if and only if $E(1/\tau^2) < \infty$, where the expectation is taken with respect to the prior and, hence, if and only if $g > 2$. Applying this condition to (2.12), we see that admissibility has not been proved for the region (2.13), as seemed the case originally, but only for the smaller region

$$0 < a < \frac{1}{r + 1}, \quad 0 < b. \tag{2.14}$$

In fact, it is not difficult to prove inadmissibility for all $a > 1/(r + 1)$ (Problem 2.12), whereas for $a < 0$ and for $b < 0$, it, of course, follows from Lemma 2.1.

The left boundary $a = 0$ of the strip (2.14) is admissible as it was in Example 2.5; the bottom boundary $b = 0$ was seen to be inadmissible for any positive $a \neq 1/(r + 1)$ in Example 2.2.6. This leaves in doubt only the point $a = b = 0$, which is inadmissible (Problem 2.13), and the right boundary, corresponding to the estimators

$$\frac{1}{r + 1}Y + b, \quad 0 \leq b < \infty. \tag{2.15}$$

Admissibility of (2.15) for $b = 0$ was first proved by Karlin (1958), who considered the case of general one-parameter exponential families. His proof was extended to other values of b by Ping (1964) and Gupta (1966). We shall follow Ping's proof, which uses the second method of Example 2.3, whereas Karlin (1958) and Stone (1967) employed the first method. ||

Let X have probability density

$$p_\theta(x) = \beta(\theta)e^{\theta T(x)} \quad (\theta, T \text{ real-valued}) \tag{2.16}$$

with respect to μ and let Ω be the natural parameter space. Then, Ω is an interval, with endpoints, say, $\underline{\theta}$ and $\bar{\theta}$ $(-\infty \le \underline{\theta} \le \bar{\theta} \le \infty)$ (see Section 1.5). For estimating $E_\theta(T)$, the estimator $aT + b$ is inadmissible if $a < 0$ or $a > 1$ and is a constant for $a = 0$. To state Karlin's sufficient condition in the remaining cases, it is convenient to write the estimator as

$$\delta_{\lambda,\gamma}(x) = \frac{1}{1+\lambda}T + \frac{\gamma\lambda}{1+\lambda}, \tag{2.17}$$

with $0 \le \lambda < \infty$ corresponding to $0 < a \le 1$.

Theorem 2.14 (Karlin's Theorem) *Under the above assumptions, a sufficient condition for the admissibility of the estimator (2.17) for estimating $g(\theta) = E_\theta(T)$ with squared error loss is that the integral of $e^{-\gamma\lambda\theta}[\beta(\theta)]^{-\lambda}$ diverges at $\underline{\theta}$ and $\bar{\theta}$; that is, that for some (and hence for all) $\underline{\theta} < \theta_0 < \bar{\theta}$, the two integrals*

$$\int_{\theta_0}^{\theta^*} \frac{e^{-\gamma\lambda\theta}}{[\beta(\theta)]^\lambda}\,d\theta \quad \text{and} \quad \int_{\theta^*}^{\theta_0} \frac{e^{-\gamma\lambda\theta}}{[\beta(\theta)]^\lambda}\,d\theta \tag{2.18}$$

tend to infinity as θ^* tends to $\bar{\theta}$ and $\underline{\theta}$, respectively.

Proof. It is seen from (1.5.14) and (1.5.15) that

$$g(\theta) = E_\theta(T) = \frac{-\beta'(\theta)}{\beta(\theta)} \tag{2.19}$$

and

$$g'(\theta) = \text{var}_\theta(T) = I(\theta), \tag{2.20}$$

where $I(\theta)$ is the Fisher information defined in (2.5.10). For any estimator $\delta(X)$, we have

$$\begin{aligned} E_\theta[\delta(X) - g(\theta)]^2 &= \text{var}_\theta[\delta(X)] + b^2(\theta) \\ &\ge \frac{[g'(\theta) + b'(\theta)]^2}{I(\theta)} + b^2(\theta) \quad \text{[information inequality]} \\ &= \frac{[I(\theta) + b'(\theta)]^2}{I(\theta)} + b^2(\theta) \end{aligned} \tag{2.21}$$

where $b(\theta) = E_\theta[\delta(X)] - g(\theta)$ is the bias of $\delta(x)$. If $\delta = \delta_{\lambda,\gamma}$ of (2.17), then its bias is $b_{\lambda,\gamma}(\theta) = \frac{\lambda}{1+\lambda}[\gamma - g(\theta)]$ with $b'(\theta) = -\frac{\lambda}{1+\lambda}g'(\theta)$ and

$$\begin{aligned} E_\theta[\delta_{\lambda,\gamma}(X) - g(\theta)]^2 &= E_\theta\left[\frac{T+\gamma\lambda}{1+\lambda} - g(\theta)\right]^2 \\ &= \frac{I(\theta)}{(1+\lambda)^2} + \frac{\lambda^2[g(\theta) - \gamma]^2}{(1+\lambda)^2} \end{aligned} \tag{2.22}$$

$$= \frac{[I(\theta) + b'_{\lambda,\gamma}(\theta)]^2}{I(\theta)} + b^2_{\lambda,\gamma}(\theta).$$

Thus, for estimators (2.17), the information inequality risk bound is an equality. Now, suppose that δ_0 satisfies

$$(2.23) \qquad E_\theta[\delta_{\lambda,\gamma}(X) - g(\theta)]^2 \geq E_\theta[\delta_0(X) - g(\theta)]^2 \quad \text{for all } \theta.$$

Denote the bias of δ_0 by $b_0(\theta)$, apply inequality (2.21) to the right side of (2.23), and apply Equation (2.22) to the left side of (2.23) to get

$$(2.24) \qquad b^2_{\lambda,\gamma}(\theta) + \frac{[I(\theta) + b'_{\lambda,\gamma}(\theta)]^2}{I(\theta)} \geq b_0^2(\theta) + \frac{[I(\theta) + b'_0(\theta)]^2}{I(\theta)}.$$

If $h(\theta) = b_0(\theta) - b_{\lambda,\gamma}(\theta)$, (2.24) reduces to

$$(2.25) \qquad h^2(\theta) - \frac{2\lambda}{1+\lambda} h(\theta)[g(\theta) - \gamma] + \frac{2}{1+\lambda} h'(\theta) + \frac{[h'(\theta)]^2}{I(\theta)} \leq 0,$$

which implies

$$(2.26) \qquad h^2(\theta) - \frac{2\lambda}{1+\lambda} h(\theta)[g(\theta) - \gamma] + \frac{2}{1+\lambda} h'(\theta) \leq 0.$$

Finally, let

$$\kappa(\theta) = h(\theta)\beta^\lambda(\theta)e^{\gamma\lambda\theta}.$$

Differentiation of $\kappa(\theta)$ and use of (2.19) reduces (2.26) to (Problem 2.7)

$$(2.27) \qquad \kappa^2(\theta)\beta^{-\lambda}(\theta)e^{-\gamma\lambda\theta} + \frac{2}{1+\lambda}\kappa'(\theta) \leq 0.$$

We shall now show that (2.27) with $\lambda \geq 0$ implies that $\kappa(\theta) \geq 0$ for all θ. Suppose to the contrary that $\kappa(\theta_0) < 0$ for some θ_0. Then, $\kappa(\theta) < 0$ for all $\theta \geq \theta_0$ since $\kappa'(\theta) < 0$, and for $\theta > \theta_0$, we can write (2.27) as

$$\frac{d}{d\theta}\left[\frac{1}{\kappa(\theta)}\right] \geq \frac{1+\lambda}{2}\beta^{-\lambda}(\theta)e^{-\gamma\lambda\theta}.$$

Integrating both sides from θ_0 to θ^* leads to

$$\frac{1}{\kappa(\theta)} - \frac{1}{\kappa(\theta_0)} \geq \frac{1+\lambda}{2}\int_{\theta_0}^{\theta^*} \beta^{-\lambda}(\theta)e^{-\gamma\lambda\theta}\,d\theta.$$

As $\theta^* \to \bar{\theta}$, the right side tends to infinity, and this provides a contradiction since the left side is $< -1/\kappa(\theta_0)$.

Similarly, $\kappa(\theta) \leq 0$ for all θ. It follows that $\kappa(\theta)$ and, hence, $h(\theta)$ is zero for all θ. This shows that for all θ equality holds in (2.25), (2.24), and, thus, (2.23). This proves the admissibility of (2.17). □

Under some additional restrictions, it is shown by Diaconis and Ylvisaker (1979) that when the sufficient condition of Theorem 2.14 holds, $aX + b$ is Bayes with respect to a proper prior distribution (a member of the conjugate family) and has finite risk. This, of course, implies that it is admissible.

Karlin (1958) conjectured that the sufficient condition of Theorem 2.14 is also necessary for the admissibility of (2.17). Despite further work on this problem (Morton and Raghavachari 1966, Stone 1967, Joshi 1969a), this conjecture has not yet been settled. See Brown (1986a) for further discussion.

Let us now see whether Theorem 2.14 settles the admissibility of $Y/(r+1)$, which was left open in Example 2.13.

Example 2.15 Continuation of Example 2.13. The density of Example 4.2.5 is of the form (2.16) with

$$\theta = -r\tau, \quad \beta(\theta) = \left(\frac{-\theta}{r}\right)^r, \quad \frac{Y}{r} = T(X), \quad \underline{\theta} = -\infty, \quad \bar{\theta} = 0.$$

Here, the parameterization is chosen so that

$$E_\theta[T(X)] = \frac{1}{\tau}$$

coincides with the estimand of Example 2.13. An estimator

$$\frac{1}{1+\lambda}\frac{Y}{r} + \frac{\gamma\lambda}{1+\lambda} \tag{2.28}$$

is therefore admissible, provided the integrals

$$\int_{-\infty}^{-c} e^{-\gamma\lambda\theta}\left(\frac{-\theta}{r}\right)^{-r\gamma} d\theta = C\int_c^\infty e^{\gamma\lambda\theta}\theta^{-r\lambda}\,d\theta$$

and

$$\int_0^c e^{\gamma\lambda\theta}\theta^{-r\lambda}\,d\theta$$

are both infinite.

The conditions for the first integral to be infinite are that either

$$\gamma = 0 \quad \text{and} \quad r\lambda \le 1, \quad \text{or} \quad \gamma\lambda > 0.$$

For the second integral, the factor $e^{\gamma\lambda\theta}$ plays no role, and the condition is simply

$$r\lambda \ge 1.$$

Combining these conditions, we see that the estimator (2.28) is admissible if either

(a) $$\gamma = 0 \quad \text{and} \quad \lambda = \frac{1}{r}$$

or

(b) $$\lambda \ge \frac{1}{r} \quad \text{and} \quad \gamma > 0 \text{ (since } r > 0).$$

If we put $a = 1/(1+\lambda)r$ and $b = \gamma\lambda/(1+\lambda)$, it follows that $aY + b$ is admissible if either

(a') $$b = 0 \quad \text{and} \quad a = \frac{1}{1+r}$$

or

(b') $$b > 0 \quad \text{and} \quad 0 < a \le \frac{1}{1+r}.$$

The first of these results settles the one case that was left in doubt in Example 2.13; the second confirms the admissibility of the interior of the strip (2.14), which had already been established in that example. The admissibility of $Y/(r+1)$ for estimating $1/\tau = 2\sigma^2$ means that

$$\frac{1}{2r+2}Y = \frac{1}{n+2}Y \tag{2.29}$$

is admissible for estimating σ^2. The estimator (2.29) is the MRE estimator for σ^2 found in Section 3.3 (Example 3.3.7). ||

Example 2.16 Normal variance, unknown mean. Admissibility of the estimator $\Sigma X_i^2/(n+2)$ when the X's are from $N(0, \sigma^2)$ naturally raises the corresponding question for

$$\Sigma(X_i - \bar{X})^2/(n+1), \tag{2.30}$$

the MRE estimator of σ^2 when the X's are from $N(\xi, \sigma^2)$ with ξ unknown (Example 3.3.11). The surprising answer, due to Stein (1964), is that (2.30) is not admissible (see Examples 3.3.13 and 5.2.15). An estimator with uniformly smaller risk is

$$\delta^s = \min\left(\frac{\Sigma(X_i - \bar{X})^2}{n+1}, \frac{\Sigma X_i^2}{n+2}\right). \tag{2.31}$$

The estimator (2.30) is MRE under the location-scale group, that is, among estimators that satisfy

$$\delta(ax_1 + b, \ldots, ax_n + b) = a\delta(x_1, \ldots, x_n). \tag{2.32}$$

To search for a better estimator of σ^2 than (2.30), consider the larger class of estimators that are only scale invariant. These are the estimators of σ^2 that satisfy (2.32) with $b = 0$, and are of the form

$$\delta(\bar{x}, s) = \varphi(\bar{x}/s)s^2. \tag{2.33}$$

The estimator δ^s is of this form.

As a motivation of δ^s, suppose that it is thought a priori likely but by no means certain that $\xi = 0$. One might then wish to test the hypothesis $H : \xi = 0$ by the usual t-test. If

$$\frac{|\sqrt{n}\bar{X}|}{\sqrt{\Sigma\left(X_i - \bar{X}\right)^2/(n-1)}} < c, \tag{2.34}$$

one would accept H and correspondingly estimate σ^2 by $\Sigma X_i^2/(n+2)$; in the contrary case, H would be rejected and σ^2 estimated by (2.30). For the value $c = \sqrt{(n-1)/(n+1)}$, it is easily checked that (2.34) is equivalent to

$$\frac{1}{n+2}\Sigma X_i^2 < \frac{1}{n+1}\Sigma(X_i - \bar{X})^2, \tag{2.35}$$

and the resulting estimator then reduces to (2.31).

While (2.30) is inadmissible, it is clear that no substantial improvement is possible, since $\Sigma(X_i - \bar{X})^2/\sigma^2$ has the same distribution as $\Sigma(X_i - \xi)^2/\sigma^2$ with n replaced by $n-1$ so that ignorance of σ^2 can be compensated for by one additional observation. Rukhin (1987) shows that the maximum relative risk improvement is approximately 4%. ||

Let us now return to Theorem 2.14 and apply it to the binomial case as another illustration.

Example 2.17 Binomial. Let X have the binomial distribution $b(p, n)$, which we shall write as

$$P(X = x) = \binom{n}{x}(1-p)^n e^{(x/n)n\log(p/(1-p))}. \tag{2.36}$$

Putting $\theta = n\log(p/(1-p))$, we have

$$\beta(\theta) = (1-p)^n = [1 + e^{\theta/n}]^{-n}$$

and

$$g(\theta) = E_\theta\left(\frac{X}{n}\right) = p = \frac{e^{\theta/n}}{1+e^{\theta/n}}.$$

Furthermore, as p ranges from 0 to 1, θ ranges from $\underline{\theta} = -\infty$ to $\bar{\theta} = +\infty$.

The integral in question is then

$$\int e^{-\gamma\lambda\theta}(1 + e^{\theta/n})^{\lambda n}\, d\theta \tag{2.37}$$

and the estimator $X/[n(1+\lambda)] + \gamma\lambda/(1+\lambda)$ is admissible, provided this integral diverges at both $-\infty$ and $+\infty$. If $\lambda < 0$, the integrand is $\leq e^{-\gamma\lambda\theta}$ and the integral cannot diverge at both limits, whereas for $\lambda = 0$, the integral does diverge at both limits. Suppose, therefore, that $\lambda > 0$. Near infinity, the dominating term (which is also a lower bound) is

$$\int e^{-\gamma\lambda\theta + \lambda\theta}\, d\theta,$$

which diverges provided $\gamma \leq 1$. At the other end, we have

$$\int_{-\infty}^{-c} e^{-\gamma\lambda\theta}(1 + e^{\theta/n})^{\lambda n}\, d\theta = \int_c^\infty e^{\gamma\lambda\theta}\left(1 + \frac{1}{e^{\theta/n}}\right)^{\lambda n} d\theta.$$

The factor in parentheses does not affect the convergence or divergence of this integral, which therefore diverges if and only if $\gamma\lambda \geq 0$. The integral will therefore diverge at both limits, provided

$$\lambda > 0 \quad\text{and}\quad 0 \leq \gamma \leq 1, \quad\text{or}\quad \lambda = 0. \tag{2.38}$$

With $a = 1/(1+\lambda)$ and $b = \gamma\lambda/(1+\lambda)$, this condition is seen to be equivalent (Problem 2.7) to

$$0 < a \leq 1, \quad 0 \leq b, \quad a + b \leq 1. \tag{2.39}$$

The estimator, of course, is also admissible when $a = 0$ and $0 \leq b \leq 1$, and it is easy to see that it is inadmissible for the remaining values of a and b (Problem 2.8). The region of admissibility is, therefore, the closed triangle $\{(a, b) : a \geq 0, b \geq 0, a + b \leq 1\}$. ||

Theorem 2.14 provides a simple condition for the admissibility of T as an estimator of $E_\theta(T)$.

Corollary 2.18 *If the natural parameter space of (2.16) is the whole real line so that $\underline{\theta} = -\infty, \bar{\theta} = \infty$, then T is admissible for estimating $E_\theta(T)$ with squared error loss.*

Proof. With $\lambda = 0$ and $\gamma = 1$, the two integrals (2.18) clearly tend toward infinity as $\theta \to \pm\infty$. □

The condition of this corollary is satisfied by the normal (variance known), binomial, and Poisson distributions, but not in the gamma or negative binomial case (Problem 2.25).

The starting point of this section was the question of admissibility of some minimax estimators. In the opposite direction, it is sometimes possible to use the admissibility of an estimator to prove that it is minimax.

Lemma 2.19 *If an estimator has constant risk and is admissible, it is minimax.*

Proof. If it were not, another estimator would have smaller maximum risk and, hence, uniformly smaller risk. □

This lemma together with Corollary 2.18 yields the following minimax result.

Corollary 2.20 *Under the assumptions of Corollary 2.18, T is the unique minimax estimator of $g(\theta) = E_\theta(T)$ for the loss function $[d - g(\theta)]^2/\text{var}_\theta(T)$.*

Proof. For this loss function, T is a constant risk estimator which is admissible by Corollary 2.18 and unique by Theorem 1.7.10. □

A companion to Lemma 2.19 allows us to deduce admissibility from unique minimaxity.

Lemma 2.21 *If an estimator is unique minimax, it is admissible.*

Proof. If it were not admissible, another estimator would dominate it in risk and, hence, would be minimax. □

Example 2.22 Binomial admissible minimax estimator. If X has the binomial distribution $b(p, n)$, then, by Corollary 2.20, X/n is the unique minimax estimator of p for the loss function $(d - p)^2/pq$ (which was seen in Example 1.7). By Lemma 2.21, X/n is admissible for this loss function. ||

The estimation of a normal variance with unknown mean provided a surprising example of a reasonable estimator which is inadmissible. We shall conclude this section with an example of a totally unreasonable estimator that is admissible.

Example 2.23 Two binomials. Let X and Y be independent binomial random variables with distributions $b(p, m)$ and $b(\pi, n)$, respectively. It was shown by Makani (1977) that a necessary and sufficient condition for

$$a\frac{X}{m} + b\frac{Y}{n} + c \tag{2.40}$$

to be admissible for estimating p with squared error loss is that either

$$\begin{aligned} &0 \le a < 1, \quad 0 \le c \le 1, \quad 0 \le a + c \le 1, \\ &0 \le b + c \le 1, \quad 0 \le a + b + c \le 1, \end{aligned} \tag{2.41}$$

or

(2.42) $$a = 1 \quad \text{and} \quad b = c = 0.$$

We shall now prove the sufficiency part, which is the result of interest; for necessity, see Problem 2.21.

Suppose there exists another estimator $\delta(X, Y)$ with risk uniformly at least as small as that of (2.40), so that

$$E\left(a\frac{X}{m} + b\frac{Y}{n} + c - p\right)^2 \geq E[\delta(X, Y) - p]^2 \quad \text{for all} \quad p.$$

Then

(2.43) $$\sum_{x=0}^{m}\sum_{k=0}^{n}\left(a\frac{x}{m} + b\frac{k}{n} + c - p\right)^2 P(X = x, Y = k)$$
$$\geq \sum_{x=0}^{m}\sum_{k=0}^{n}[\delta(x, k) - p]^2 P(X = x, Y = k).$$

Letting $\pi \to 0$, this leads to

$$\sum_{x=0}^{m}\left(a\frac{x}{m} + c - p\right)^2 P(X = x) \geq \sum_{x=0}^{m}[\delta(x, 0) - p]^2 P(X = x)$$

for all p. However, $a(X/m) + c$ is admissible by Example 2.17; hence $\delta(x, 0) = a(x/m) + c$ for all $x = 0, 1, \ldots, m$.

The terms in (2.43) with $k = 0$, therefore, cancel. The remaining terms contain a common factor π which can also be canceled and one can now proceed as before. Continuing in this way by induction over k, one finds at the $(k + 1)$st stage that

$$\sum_{x=0}^{m}\left(a\frac{x}{m} + b\frac{k}{n} + c - p\right)^2 P(X = x) \geq \sum_{x=0}^{m}[\delta(x, k) - p]^2 P(X = x)$$

for all p. However, $aX/m + bk/n + c$ is admissible by Example 2.17 since

$$a + b\frac{k}{n} + c \leq 1$$

and, hence,

$$\delta(x, k) = a\frac{x}{m} + b\frac{k}{n} + c \quad \text{for all } x.$$

This shows that (2.43) implies

$$\delta(x, y) = a\frac{x}{m} + b\frac{y}{n} + c \quad \text{for all } x \text{ and } y$$

and, hence, that (2.40) is admissible.

Putting $a = 0$ in (2.40), we see that estimates of the form $b(Y/n) + c$ $(0 \leq c \leq 1, 0 \leq b + c \leq 1)$ are admissible for estimating p despite the fact that only the distribution of X depends on p and that X and Y are independent. This paradoxical result suggests that admissibility is an extremely weak property. While it is somewhat embarrassing for an estimator to be inadmissible, the fact that it is admissible in no way guarantees that it is a good or even halfway reasonable estimator. ||

The result of Example 2.23 is not isolated. An exactly analogous result holds in the Poisson case (Problem 2.22) and a very similar one due to Brown for normal distributions (see Example 7.2); that an exactly analogous example is not possible in the normal case follows from Cohen (1965a).

3 Admissibility and Minimaxity in Group Families

The two preceding sections dealt with minimax estimators and their admissibility in exponential families. Let us now consider the corresponding problems for group families. As was seen in Section 3.2, in these families there typically exists an MRE estimator δ_0 for any invariant loss function, and it is a constant risk estimator. If δ_0 is also a Bayes estimator, it is minimax by Corollary 1.5 and admissible if it is unique Bayes.

Recall Theorem 4.4.1, where it was shown that a Bayes estimator under an invariant prior is (almost) equivariant. It follows that under the assumptions of that theorem, there exists an almost equivariant estimator which is admissible. Furthermore, it turns out that under very weak additional assumptions, given any almost equivariant estimator δ, there exists an equivariant estimator δ' which differs from δ only on a fixed null set N. The existence of such a δ' is obvious in the simplest case, that of a finite group. We shall not prove it here for more general groups (a precise statement and proof can be found in TSH2, Section 6.5, Theorem 4). Since δ and δ' then have the same risk function, this establishes the existence of an equivariant estimator that is admissible.

Theorem 4.4.1 does not require $\bar{G}$ to be transitive over Ω. If we add the assumption of transitivity, we get a stronger result.

Theorem 3.1 *Under the conditions of Theorem 4.4.1, if $\bar{G}$ is transitive over Ω, then the MRE estimator is admissible and minimax.*

The crucial assumption in this approach is the existence of an invariant prior distribution. The following example illustrates the rather trivial case in which the group is finite.

Example 3.2 Finite group. Let $X_1, \ldots, X_n$ be iid according to the normal distribution $N(\xi, 1)$. Then, the problem of estimating ξ with squared error loss remains invariant under the two-element group G, which consists of the identity transformation e and the transformation

$$g(x_1, \ldots, x_n) = (-x_1, \ldots, -x_n); \quad \bar{g}\xi = -\xi; \quad g^*d = -d.$$

In the present case, any distribution Λ for ξ which is symmetric with respect to the origin is invariant. Under the conditions of Theorem 4.4.1, it follows that for any such Λ, there is a version of the Bayes solution which is equivariant, that is, which satisfies $\delta(-x_1, \ldots, -x_n) = -\delta(x_1, \ldots x_n)$. The group $\bar{G}$ in this case is, of course, not transitive over Ω. ||

As an example in which G is not finite, we shall consider the following version of the location problem on the circle.

Example 3.3 Circular location family. Let $U_1, \ldots, U_n$ be iid on $(0, 2\pi)$ according to a distribution F with density f. We shall interpret these variables as n points chosen at random on the unit circle according to F. Suppose that each point is translated on the circle by an amount θ $(0 \le \theta < 2\pi)$ (i.e., the new positions are those obtained by rotating the circle by an amount, θ). When a value $U_i + \theta$ exceeds 2π, it is, of course, replaced by $U_i + \theta - 2\pi$. The resulting values are the observations $X_1, \ldots, X_n$. It is then easily seen (Problem 3.2) that the density of X_i is

$$\begin{aligned} &f(x_i - \theta + 2\pi) \quad \text{when} \quad 0 < x_i < \theta, \\ &f(x_i - \theta) \quad \text{when} \quad \theta < x_i < 2\pi. \end{aligned} \tag{3.1}$$

This can also be written as

$$f(x_i - \theta)I(\theta < x_i) + f(x_i - \theta + 2\pi)I(x_i < \theta) \tag{3.2}$$

where $I(a < b)$ is 1 when $a < b$, and 0 otherwise.

If we straighten the circle to a straight-line segment of length 2π, we can also represent this family of distributions in the following form. Select n points at random on $(0, 2\pi)$ according to F. Cut the line segment at an arbitrary point θ $(0 < \theta < 2\pi)$. Place the upper segment so that its endpoints are $(0, 2\pi - \theta)$ and the lower segment so that its endpoints are $(2\pi - \theta, 2\pi)$, and denote the coordinates of the n points in their new positions by $X_1, \ldots, X_n$. Then, the density of X_i is given by (3.1).

As an illustration of how such a family of distributions might arise, suppose that in a study of gestation in rats, n rats are impregnated by artificial insemination at a given time, say at midnight on day zero. The observations are the n times $Y_1, \ldots, Y_n$ to birth, recorded as the number of days plus a fractional day. It is assumed that the Y's are iid according to $G(y - \eta)$ where G is known and η is an unknown location parameter. A scientist who is interested in the time of day at which births occur abstracts from the data the fractional parts $X_i' = Y_i - [Y_i]$. The variables $X_i = 2\pi X_i'$ have a distribution of the form (3.1) where θ is 2π times the fractional part of η.

Let us now return to (3.2) and consider the problem of estimating θ. The model as originally formulated remains invariant under rotations of the circle. To represent these transformations formally, consider for any real number a the unique number a^*, $0 \le a^* < 2\pi$, for which $a = 2\kappa\pi + a^*$ (κ an integer). Then, the group G of rotations can be represented by

$$x_i' = (x_i + c)^*, \quad \theta' = (\theta + c)^*, \quad d' = (d + c)^*.$$

A loss function $L(\theta, d)$ remains invariant under G if and only if it is of the form $L(\theta, d) = \rho[(d - \theta)^*]$ (Problem 3.3.). Typically, one would want it to depend only on $(d - \theta)^{**} = \min\{(d - \theta)^*, (2\pi - (d - \theta))^*\}$, which is the difference between d and θ along the smaller of the two arcs connecting them. Thus, the loss might be $((d - \theta)^{**})^2$ or $|(d - \theta)^{**}|$. It is important to notice that neither of these is convex (Problem 3.4).

The group G is transitive over Ω and an invariant distribution for θ is the uniform distribution over $(0, 2\pi)$. By applying an obvious extension of the construction

(3.20) or (3.21) below, one obtains an admissible equivariant (and, hence, constant risk) Bayes estimator, which a fortiori is also minimax. If the loss function is not convex in d, only the extension of (3.20) is available and the equivariant Bayes procedure may be randomized. ||

Let us next turn to the question of the admissibility and minimaxity of MRE estimators which are Bayes solutions with respect to improper priors. We begin with the location parameter case.

Example 3.4 Location family on the line. Suppose that $\mathbf{X} = (X_1, \ldots, X_n)$ has density

$$f(\mathbf{x} - \theta) = f(x_1 - \theta, \ldots, x_n - \theta), \tag{3.3}$$

and let G and $\bar{G}$ be the groups of translations $x_i' = x_i + a$ and $\theta' = \theta + a$. The parameter space Ω is the real line, and from Example 4.4.3, the invariant measure on Ω is the measure ν which to any interval I assigns its length, that is, Lebesgue measure.

Since the measure ν is improper, we proceed as in Section 4.3 and look for a generalized Bayes estimator. The posterior density of θ given $\mathbf{x}$ is given by

$$\frac{f(\mathbf{x} - \theta)}{\int f(\mathbf{x} - \theta)\,d\theta}. \tag{3.4}$$

This quantity is non-negative and its integral, with respect to θ, is equal to 1. It therefore defines a proper distribution for θ, and by Section 4.3, the generalized Bayes estimator of θ, with loss function L, is obtained by minimizing the posterior expected loss

$$\int L[\theta, \delta(\mathbf{x})] f(\mathbf{x} - \theta)\,d\theta \Big/ \int f(\mathbf{x} - \theta)\,d\theta. \tag{3.5}$$

For the case that L is squared error, the minimizing value of $\delta(\mathbf{x})$ is the expectation of θ under (3.4), which was seen to be the Pitman estimator (3.1.28) in Example 4.4.7. The agreement of the estimator minimizing (3.5) with that obtained in Section 3.1 of course holds also for all other invariant loss functions.

Up to this point, the development here is completely analogous to that of Example 3.3. However, since ν is not a probability distribution, Theorem 4.4.1 is not applicable and we cannot conclude that the Pitman estimator is admissible or even minimax. ||

The minimax character of the Pitman estimator was established in the normal case in Example 1.14 by the use of a least favorable sequence of prior distributions. We shall now consider the minimax and admissibility properties of MRE estimators more generally in group families, beginning with the case of a general location family.

Theorem 3.5 *Suppose* $\mathbf{X} = (X_1, \ldots, X_n)$ *is distributed according to the density (3.3) and that the Pitman estimator* δ^* *given by (1.28) has finite variance. Then,* δ^* *is minimax for squared error loss.*

Proof. As in Example 1.14, we shall utilize Theorem 1.12, and for this purpose, we require a least favorable sequence of prior distributions. In view of the discussion at the beginning of Example 3.4, one would expect a sequence of priors that approximates Lebesgue measure to be suitable. The sequence of normal distributions with variance tending toward infinity used in Example 1.14 was of this kind. Here, it will be more convenient to use instead a sequence of uniform densities.

$$\pi_T(u) = \begin{cases} 1/2T & \text{if } |u| < T \\ 0 & \text{otherwise,} \end{cases} \tag{3.6}$$

with T tending to infinity. If δ_T is the Bayes estimator with respect to (3.6) and r_T its Bayes risk, the minimax character of δ^* will follow if it can be shown that r_T tends to the constant risk $r^* = E_0\delta^{*2}(\mathbf{X})$ of δ^* as $T \to \infty$. Since $r_T \le r^*$ for all T, it is enough to show

$$\liminf r_T \ge r^*. \tag{3.7}$$

We begin by establishing the lower bound for r_T

$$r_T \ge (1-\varepsilon) \inf_{\substack{a \le -\varepsilon T \\ b \ge \varepsilon T}} E_0\delta^2_{a,b}(\mathbf{X}), \tag{3.8}$$

where ε is any number between 0 and 1, and $\delta_{a,b}$ is the Bayes estimator with respect to the uniform prior on (a, b) so that, in particular, $\delta_T = \delta_{-T,T}$. Then, for any c (Problem 3.7),

$$\delta_{a,b}(\mathbf{x}+c) = \delta_{a-c,b-c}(\mathbf{x}) + c \tag{3.9}$$

and hence

$$E_\theta[\delta_{-T,T}(\mathbf{X}) - \theta]^2 = E_0[\delta_{-T-\theta,T-\theta}(\mathbf{X})]^2.$$

It follows that for any $0 < \varepsilon < 1$,

$$\begin{aligned} r_T &= \frac{1}{2T}\int_{-T}^{T} E_0[\delta_{-T-\theta,T-\theta}(\mathbf{X})]^2\,d\theta \\ &\ge (1-\varepsilon) \inf_{|\theta| \le (1-\varepsilon)T} E_0[\delta_{-T-\theta,T-\theta}(\mathbf{X})]^2. \end{aligned}$$

Since $-T - \theta \le -\varepsilon T$ and $T - \theta \ge \varepsilon T$ when $|\theta| \le (1-\varepsilon)T$, this implies (3.8).

Next, we show that

$$\liminf_{T\to\infty} r_T \ge E_0\left[\liminf_{\substack{a\to-\infty \\ b\to\infty}} \delta^2_{a,b}(\mathbf{X})\right] \tag{3.10}$$

where the lim inf on the right side is defined as the smallest limit point of all sequences $\delta^2_{a_n,b_n}(\mathbf{X})$ with $a_n \to -\infty$ and $b_n \to \infty$. To see this, note that for any function h of two real arguments, one has (Problem 3.8).

$$\liminf_{T\to\infty}\left[\inf_{\substack{a\le -T \\ b\ge T}} h(a,b)\right] = \liminf_{\substack{a\to-\infty \\ b\to\infty}} h(a,b). \tag{3.11}$$

Taking the lim inf of both sides of (3.8), and using (3.11) and Fatou's Lemma (Lemma 1.2.6) proves (3.10).

We shall, finally, show that as $a \to -\infty$ and $b \to \infty$,

$$\delta_{a,b}(\mathbf{X}) \to \delta^*(\mathbf{X}) \quad \text{with probability 1.} \tag{3.12}$$

From this, it follows that the right side of (3.10) is r^*, which will complete the proof. The limit (3.12) is seen from the fact that

$$\delta_{a,b}(\mathbf{x}) = \int_a^b uf(\mathbf{x}-u)\,du \Big/ \int_a^b f(\mathbf{x}-u)\,du$$

and that, by Problems 3.1.20, and 3.1.21 the set of points $\mathbf{x}$ for which

$$0 < \int_{-\infty}^{\infty} f(\mathbf{x}-u)\,du < \infty \quad \text{and} \quad \int_{-\infty}^{\infty} |u| f(\mathbf{x}-u)\,du < \infty$$

has probability 1. □

Theorem 3.5 is due to Girshick and Savage (1951), who proved it somewhat more generally without assuming a probability density and under the sole assumption that there exists an estimator (not necessarily equivariant) with finite risk. The streamlined proof given here is due to Peter Bickel.

Of course, one would like to know whether the constant risk minimax estimator δ^* is admissible. This question was essentially settled by Stein (1959). We state without proof the following special case of his result.

Theorem 3.6 *If $X_1, \ldots, X_n$ are independently distributed with common probability density $f(x-\theta)$, and if there exists an equivariant estimator δ_0 of θ for which $E_0|\delta_0(\mathbf{X})|^3 < \infty$, then the Pitman estimator δ^* is admissible under squared error loss.*

It was shown by Perng (1970) that this admissibility result need not hold when the third-moment condition is dropped.

In Example 3.4, we have, so far, restricted attention to squared error loss. Admissibility of the MRE estimator has been proved for large classes of loss functions by Farrell (1964), Brown (1966), and Brown and Fox (1974b). A key assumption is the uniqueness of the MRE estimator. An early counterexample when that assumption does not hold was given by Blackwell (1951). A general inadmissibility result in the case of nonuniqueness is due to Farrell (1964).

Examples 3.3 and 3.4 involved a single parameter θ. That an MRE estimator of θ may be inadmissible in the presence of nuisance parameters, when the corresponding estimator of θ with known values of the nuisance parameters is admissible, is illustrated by the estimator (2.30). Other examples of this type have been studied by Brown (1968), Zidek (1973), and Berger (1976bc), among others. An important illustration of the inadmissibility of the MRE estimator of a vector-valued parameter constitutes the principal subject of the next two sections.

Even when the best equivariant estimator is not admissible, it may still be—and frequently is—minimax. Conditions for an MRE estimator to be minimax are given by Kiefer (1957) or Robert (1994a, Section 7.5). (See Note 9.3.) The general treatment of admissibility and minimaxity of MRE estimators is beyond the scope of this book. However, roughly speaking, MRE estimators will typically not be admissible except in the simplest situations, but they have a much better chance of being minimax.

The difference can be seen by comparing Example 1.14 and the proof of Theorem 3.5 with the first admissibility proof of Example 2.8. If there exists an invariant measure over the parameter space of the group family (or equivalently over the group, see Section 3.2) which can be suitably approximated by a sequence of probability distributions, one may hope that the corresponding Bayes estimators will tend to the MRE estimator and Theorem 3.5 will become applicable. In comparison, the corresponding proof in Example 2.8 is much more delicate because it depends on the rate of convergence of the risks (this is well illustrated by the attempted admissibility proof at the beginning of the next section).

As a contrast to Theorem 3.5, we shall now give some examples in which the MRE estimator is not minimax.

Example 3.7 MRE not minimax. Consider once more the estimation of Δ in Example 3.2.12 with loss 1 when $|d-\Delta|/\Delta > 1/2$, and 0 otherwise. The problem remains invariant under the group G of transformations

$$X_1' = a_1X_1 + a_2X_2, \quad Y_1' = c(a_1Y_1 + a_2Y_2),$$
$$X_2' = b_1X_1 + b_2X_2, \quad Y_2' = c(b_1Y_1 + b_2Y_2)$$

with $a_1b_2 \neq a_2b_1$ and $c > 0$. The only equivariant estimator is $\delta(\mathbf{x}, \mathbf{y}) \equiv 0$ and its risk is 1 for all values of Δ. On the other hand, the risk of the estimator $k^*Y_2^2/X_2^2$ obtained in Example 3.2.12 is clearly less than 1. ‖

Example 3.8 A random walk.[3] Consider a walk in the plane. The walker at each step goes one unit either right, left, up, or down and these possibilities will be denoted by a, a^-, b, and b^-, respectively. Such a walk can be represented by a finite "path" such as

$$bba^-b^-a^-a^-a^-a^-.$$

In reporting a path, we shall, however, cancel any pair of successive steps which reverse each other, such as a^-a or bb^-. The resulting set of all finite paths constitutes the parameter space Ω. A typical element of Ω will be denoted by

$$\theta = \pi_1 \cdots \pi_m,$$

its length by $l(\theta) = m$. Being a parameter, θ (as well as m) is assumed to be unknown. What is observed is the path X obtained from θ by adding one more step, which is taken in one of the four possible directions at random, that is, with probability $1/4$ each. If this last step is π_{m+1}, we have

$$X = \begin{cases} \theta\pi_{m+1} & \text{if } \pi_m \text{ and } \pi_{m+1} \text{ do not cancel each other,} \\ \pi_1 \cdots \pi_{m-1} & \text{otherwise.} \end{cases}$$

A special case occurs if θ or X, after cancellation, reduce to a path of length 0; this happens, for example, if $\theta = a^-$ and the random step leading to X is a. The resulting path will then be denoted by e.

The problem is to estimate θ, having observed $X = x$; the loss will be 1 if the estimated path $\delta(x)$ is $\neq \theta$, and 0 if $\delta(x) = \theta$.

If we observe X to be

$$x = \pi_1 \cdots \pi_k,$$

[3] A more formal description of this example is given in TSH2 [Chapter 1, Problem 11(ii)]. See also Diaconis (1988) for a general treatment of random walks on groups.

the natural estimate is

$$\delta_0(x) = \pi_1 \cdots \pi_{k-1}.$$

An exception occurs when $x = e$. In that case, which can arise only when $l(\theta) = 1$, let us arbitrarily put $\delta_0(e) = a$. The estimator defined in this way clearly satisfies

$$R(\theta, \delta_0) \leq \frac{1}{4} \quad \text{for all } \theta.$$

Now, consider the transformations that modify the paths θ, x, and $\delta(x)$ by having each preceded by an initial segment $\pi_{-r} \cdots \pi_{-1}$ on the left, so that, for example, $\theta = \pi_1 \cdots \pi_m$ is transformed into

$$\bar{g}\theta = \pi_{-r} \cdots \pi_{-1}\pi_1 \cdots \pi_m$$

where, of course, some cancellations may occur. The group G is obtained by considering the addition in this manner of all possible initial path segments. Equivariance of an estimator δ under this group is expressed by the condition

$$\delta(\pi_{-r} \cdots \pi_{-1}x) = \pi_{-r} \cdots \pi_{-1}\delta(x) \tag{3.13}$$

for all x and all $\pi_{-r} \cdots \pi_{-1}$, $r = 1, 2 \ldots$. This implies, in particular, that

$$\delta(\pi_{-r} \cdots \pi_{-1}) = \pi_{-r} \cdots \pi_{-1}\delta(e), \tag{3.14}$$

and this condition is sufficient as well as necessary for δ to be equivariant because (3.14) implies that

$$\pi_{-r} \cdots \pi_{-1}\delta(x) = \pi_{-r} \cdots \pi_{-1}x\delta(e) = \delta(\pi_{-r} \cdots \pi_{-1}x).$$

Since $\bar{G}$ is clearly transitive over Ω, the risk function of any equivariant estimator is constant. Let us now determine the MRE estimator. Suppose that $\delta(e) = \pi_{10} \cdots \pi_{k0}$, so that by (3.14),

$$\delta(x) = x\pi_{10} \cdots \pi_{k0}.$$

The only possibility of $\delta(x)$ being equal to θ occurs when π_{10} cancels the last element of x. The best choice for k is clearly $k = 1$, and the choice of π_{10} (fixed or random) is then immaterial; in any case, the probability of cancellation with the last element of X is $1/4$, so that the risk of the MRE estimator (which is not unique) is $3/4$. Comparison with δ_0 shows that a best equivariant estimator in this case is not only not admissible but not even minimax. ||

The following example, in which the MRE estimator is again not minimax but where G is simply the group of translations on the real line, is due to Blackwell and Girshick (1954).

Example 3.9 Discrete location family. Let $X = U + \theta$ where U takes on the values $1, 2, \ldots$ with probabilities

$$P(U = k) = p_k.$$

We observe x and wish to estimate θ with loss function

$$L(\theta, d) = \begin{cases} d - \theta & \text{if } d > \theta \\ 0 & \text{if } d \leq \theta. \end{cases} \tag{3.15}$$

The problem remains invariant under arbitrary translation of X, θ, and d by the same amount. It follows from Section 3.1 that the only equivariant estimators are those of the form $X - c$. The risk of such an estimator, which is constant, is given by

$$\sum_{k>c}(k-c)p_k. \tag{3.16}$$

If the p_k tend to 0 sufficiently slowly, an equivariant estimator will have infinite risk. This is the case, for example, when

$$p_k = \frac{1}{k(k+1)} \tag{3.17}$$

(Problem 3.11). The reason is that there is a relatively large probability of substantially overestimating θ for which there is a heavy penalty. This suggests a deliberate policy of grossly underestimating θ, for which, by (3.15), there is no penalty. One possible such estimator (which, of course, is not equivariant) is

$$\delta(x) = x - M|x|, \quad M > 1, \tag{3.18}$$

and it is not hard to show that its maximum risk is finite (Problem 3.12). ||

The ideas of the present section have relevance beyond the transitive case for which they were discussed so far. If $\bar{G}$ is not transitive, we can no longer ask whether the uniform minimum risk equivariant (UMRE) estimator is minimax since a UMRE estimator will then typically not exist. Instead, we can ask whether there exists a minimax estimator which is equivariant. Similarly, the question of the admissibility of the UMRE estimator can be rephrased by asking whether an estimator which is admissible among equivariant estimators is also admissible within the class of all estimators.

The conditions for affirmative answers to these two questions are essentially the same as in the transitive case. In particular, the answer to both questions is affirmative when G is finite. A proof along the lines of Theorem 4.1 is possible but not very convenient because it would require a characterization of all admissible (within the class of equivariant estimators) equivariant estimators as Bayes solutions with respect to invariant prior distributions. Instead, we shall utilize the fact that for every estimator δ, there exists an equivariant estimator whose average risk (to be defined below) is no worse than that of δ.

Let the elements of the finite group G be $g_1, \ldots, g_N$ and consider the estimators

$$\delta_i(x) = g_i^{*-1}\delta(g_i x). \tag{3.19}$$

When δ is equivariant, of course, $\delta_i(x) = \delta(x)$ for all i. Consider the randomized estimator δ^* for which

$$\delta^*(x) = \delta_i(x) \quad \text{with probability } 1/N \text{ for each } i = 1, \ldots, N, \tag{3.20}$$

and assuming the set $\mathcal{D}$ of possible decisions to be convex, the estimator

$$\delta^{**}(x) = \frac{1}{N}\sum_{i=1}^{N}\delta_i(x) \tag{3.21}$$

which, for given x, is the expected value of $\delta^*(x)$. Then, $\delta^{**}(x)$ is equivariant, and so is $\delta^*(x)$ in the sense that $g^{*-1}\delta^*(gx)$ again is equal to $\delta_i(x)$ with probability $1/N$ for each i (Problem 3.13). For these two estimators, it is easy to prove that (Problem 3.14):

(i) for any loss function L,

$$R(\theta, \delta^*) = \frac{1}{N}\Sigma R(\bar{g}_i\theta, \delta) \tag{3.22}$$

and

(ii) for any loss function $L(\theta, d)$ which is convex in d,

$$R(\theta, \delta^{**}) \leq \frac{1}{N}\Sigma R(\bar{g}_i\theta, \delta). \tag{3.23}$$

From (3.22) and (3.23), it follows immediately that

$$\sup R(\theta, \delta^*) \leq \sup R(\theta, \delta) \quad \text{and} \quad \sup R(\theta, \delta^{**}) \leq \sup R(\theta, \delta),$$

which proves the existence of an equivariant minimax estimator provided a minimax estimator exists.

Suppose, next, that δ_0 is admissible among all equivariant estimators. If δ_0 is not admissible within the class of all estimators, it is dominated by some δ. Let δ^* and δ^{**} be as above. Then, (3.22) and (3.23) imply that δ^* and δ^{**} dominate δ_0, which is a contradiction.

Of the two constructions, δ^{**} has the advantage of not requiring randomization, whereas δ^* has the advantage of greater generality since it does not require L to be convex. Both constructions easily generalize to groups that admit an invariant measure which is finite (Problems 4.4.12–4.4.14). Further exploration of the relationship of equivariance to admissibility and the minimax property leads to the Hunt-Stein theorem (see Notes 9.3).

4 Simultaneous Estimation

So far, we have been concerned with the estimation of a single real-valued parameter $g(\theta)$. However, one may wish to estimate several parameters simultaneously, for example, several physiological constants of a patient, several quality characteristics of an industrial or agricultural product, or several dimensions of musical ability. One is then dealing with a vector-valued estimand

$$g(\boldsymbol{\theta}) = [g_1(\boldsymbol{\theta}), \ldots, g_r(\boldsymbol{\theta})]$$

and a vector-valued estimator

$$\delta = (\delta_1, \ldots, \delta_r).$$

A natural generalization of squared error as a measure of accuracy is

$$\Sigma[\delta_i - g_i(\boldsymbol{\theta})]^2, \tag{4.1}$$

a sum of squared error losses, which we shall often simply call squared error loss. More generally, we shall consider loss functions $L(\boldsymbol{\theta}, \delta)$ where $\delta = (\delta_1, \ldots, \delta_r)$, and then denote the risk of an estimator δ by

(4.2) $$R(\boldsymbol{\theta}, \delta) = E_{\theta} L[\boldsymbol{\theta}, \delta(\mathbf{X})].$$

Another generalization of expected squared error loss is the matrix $\mathcal{R}(\boldsymbol{\theta}, \delta)$ whose (i, j)th element is

(4.3) $$E\{[\delta_i(\mathbf{X}) - g_i(\boldsymbol{\theta})][\delta_j(\mathbf{X}) - g_j(\boldsymbol{\theta})]\}.$$

We shall say that δ is *more concentrated* about $g(\boldsymbol{\theta})$ than δ' if

(4.4) $$\mathcal{R}(\boldsymbol{\theta}, \delta') - \mathcal{R}(\boldsymbol{\theta}, \delta)$$

is positive semidefinite (but not identically zero). This definition differs from that based on (4.2) by providing only a partial ordering of estimators, since (4.4) may be neither positive nor negative semidefinite.

Lemma 4.1

(i) δ is more concentrated about $g(\boldsymbol{\theta})$ than δ' if and only if

(4.5) $$E\{\Sigma k_i[\delta_i(\mathbf{X}) - g_i(\boldsymbol{\theta})]\}^2 \le E\{\Sigma k_i[\delta_i'(\mathbf{X}) - g_i(\boldsymbol{\theta})]\}^2$$

for all constants $k_1, \ldots, k_r$.

(ii) In particular, if δ is more concentrated about $g(\boldsymbol{\theta})$ than δ', then

(4.6) $$E[\delta_i(\mathbf{X}) - g_i(\boldsymbol{\theta})]^2 \le E[\delta_i'(\mathbf{X}) - g_i(\boldsymbol{\theta})]^2 \quad \textit{for all } i.$$

(iii) If $R(\boldsymbol{\theta}, \delta) \le R(\boldsymbol{\theta}, \delta')$ for all convex loss functions, then δ is more concentrated about $g(\boldsymbol{\theta})$ than δ'.

Proof.

(i) If $E\{\Sigma k_i[\delta_i(\mathbf{X}) - g_i(\boldsymbol{\theta})]\}^2$ is expressed as a quadratic form in the k_i, its matrix is $\mathcal{R}(\boldsymbol{\theta}, \delta)$.

(ii) This is a special case of (i).

(iii) This follows from the fact that $\{\Sigma k_i[d_i - g_i(\boldsymbol{\theta})]\}^2$ is a convex function of $d = (d_1, \ldots, d_r)$.

□

Let us now consider the extension of some of the earlier theory to the case of simultaneous estimation of several parameters.

(1) *The Rao-Blackwell theorem* (Theorem 1.7.8). The proof of this theorem shows that its results remain valid when δ and g are vector-valued. In particular, for any convex loss function, the risk of any estimator is reduced by taking its expectation given a sufficient statistic. It follows that for such loss functions, one can dispense with randomized estimators. Also, Lemma 4.1 shows that an estimator δ is always less concentrated about $g(\boldsymbol{\theta})$ than the expectation of $\delta(\mathbf{X})$, given a sufficient statistic.

(2) ***Unbiased estimation.*** In the vector-valued case, an estimator δ of $g(\boldsymbol{\theta})$ is said to be unbiased if

(4.7) $$E_{\theta}[\delta_i(\mathbf{X})] = g_i(\boldsymbol{\theta}) \quad \text{for all } i \text{ and } \boldsymbol{\theta}.$$

For unbiased estimators, the concentration matrix $\mathcal{R}$ defined by (4.3) is just the covariance matrix of δ.

From the Rao-Blackwell theorem, it follows, as in Theorem 2.1.11 for the case $r = 1$, that if L is convex and if a complete sufficient statistic T exists, then any U-estimable g has a unique unbiased estimator depending only on T. This estimator uniformly minimizes the risk among all unbiased estimators and, thus, is also more concentrated about $g(\boldsymbol{\theta})$ than any other unbiased estimator.

(3) *Equivariant estimation.* The definitions and concepts of Section 3.2 apply without changes. They are illustrated by the following example, which will be considered in more detail later in the section.

Example 4.2 Several normal means. Let $\mathbf{X} = (X_1, \ldots, X_r)$, with the X_i independently distributed as $N(\theta_i, 1)$, and consider the problem of estimating the vector mean $\boldsymbol{\theta} = (\theta_1, \ldots, \theta_r)$ with squared error loss. This problem remains invariant under the group G_1 of translations

$$\begin{aligned} g\mathbf{X} &= (X_1 + a_1, \ldots, X_r + a_r), \\ \bar{g}\boldsymbol{\theta} &= (\theta_1 + a_1, \ldots, \theta_r + a_r), \\ g^*d &= (d_1 + a_1, \ldots, d_r + a_r). \end{aligned} \tag{4.8}$$

The only equivariant estimators are those of the form

$$\delta(\mathbf{X}) = (X_1 + c_1, \ldots, X_r + c_r) \tag{4.9}$$

and an easy generalization of Example 3.1.16 shows that $\mathbf{X}$ is the MRE estimator of $\boldsymbol{\theta}$.

The problem also remains invariant under the group G_2 of orthogonal transformations

$$g\mathbf{X} = \mathbf{X}\Gamma, \quad \bar{g}\boldsymbol{\theta} = \boldsymbol{\theta}\Gamma, \quad g^*d = d\Gamma \tag{4.10}$$

where Γ is an orthogonal $r \times r$ matrix. An estimator δ is equivariant if and only if it is of the form (Problem 4.1)

$$\delta(\mathbf{X}) = u(\mathbf{X}) \cdot \mathbf{X}, \tag{4.11}$$

where $u(\mathbf{X})$ is any scalar satisfying

$$u(\mathbf{X}\Gamma) = u(\mathbf{X}) \quad \text{for all orthogonal } \Gamma \text{ and all } \mathbf{X} \tag{4.12}$$

and, hence, is an arbitrary function of ΣX_i^2 (Problem 4.2). The group $\bar{G}$ defined by (4.10) is not transitive over the parameter space, and a UMRE estimator of $\boldsymbol{\theta}$, therefore, cannot be expected. ||

(4) *Bayes estimators.* The following result frequently makes it possible to reduce Bayes estimation of a vector-valued estimand to that of its components.

Lemma 4.3 *Suppose that $\delta_i^*(\mathbf{X})$ is the Bayes estimator of $g_i(\boldsymbol{\theta})$ when $\boldsymbol{\theta}$ has the prior distribution Λ and the loss is squared error. Then, $\delta^* = (\delta_1^*, \ldots, \delta_r^*)$ is more concentrated about $g(\boldsymbol{\theta})$ in the Bayes sense that it minimizes*

$$E[\Sigma k_i(\delta_i(\mathbf{X}) - g_i(\boldsymbol{\theta}))]^2 = E[\Sigma k_i\delta_i(\mathbf{X}) - \Sigma k_i g_i(\boldsymbol{\theta})]^2 \tag{4.13}$$

for all k_i, where the expectation is taken over both θ and **X**.

Proof. The result follows from the fact that the estimator $\Sigma k_i \delta_i(\mathbf{X})$ minimizing (4.13) is

$$E[\Sigma k_i g_i(\theta)|\mathbf{X}] = \Sigma k_i E[g(\theta_i)|\mathbf{X}] = \Sigma k_i \delta_i^*(\mathbf{X}).$$

□

Example 4.4 Multinomial Bayes. Let $\mathbf{X} = (X_0, \ldots, X_s)$ have the multinomial distribution $M(n; p_0, \ldots, p_s)$, and consider the Bayes estimation of the vector $\mathbf{p} = (p_0, \ldots, p_s)$ when the prior distribution of $\mathbf{p}$ is the Dirichlet distribution Λ with density

$$\text{(4.14)} \quad \frac{\Gamma(a_0, \ldots, a_s)}{\Gamma(a_0) \ldots \Gamma(a_s)} p_0^{a_0-1} \cdots p_s^{a_s-1} \quad (a_i > 0,\ 0 \le p_i \le 1,\ \Sigma p_i = 1).$$

The Bayes estimator of p_i for squared error loss is (Problem 4.3)

$$\text{(4.15)} \qquad \delta_i(X) = \frac{a_i + X_i}{\Sigma a_j + n},$$

and by Lemma 4.3, the estimator $[\delta_0(\mathbf{X}), \ldots, \delta_s(\mathbf{X})]$ is then most concentrated in the Bayes sense. As a check, note that $\Sigma \delta_i(\mathbf{X}) = 1$ as, of course, it must since Λ assigns probability 1 to $\Sigma p_i = 1$. ||

(5) *Minimax estimators.* In generalization of the binomial minimax problem treated in Example 1.7, let us now determine the minimax estimator of $(p_0, \ldots, p_s)$ for the multinomial model of Example 4.4.

Example 4.5 Multinomial minimax. Suppose the loss function is squared error. In light of Example 1.7, one might guess that a least favorable distribution is the Dirichlet distribution (4.14) with $a_0 = \cdots = a_s = a$. The Bayes estimator (4.15) reduces to

$$\text{(4.16)} \qquad \delta_i(\mathbf{X}) = \frac{a + X_i}{(s+1)a + n}.$$

The estimator $\delta(\mathbf{X})$ with components (4.16) has constant risk over the support of (4.14), provided $a = \sqrt{n}/(s+1)$, and for this value of a, $\delta(\mathbf{X})$ is therefore minimax by Corollary 1.5. [Various versions of this problem are discussed by Steinhaus (1957), Trybula (1958), Rutkowska (1977), and Olkin and Sobel (1979).] ||

Example 4.6 Independent experiments. Suppose the components X_i of $\mathbf{X} = (X_1, \ldots, X_r)$ are independently distributed according to distributions P_{θ_i}, where the θ_i vary independently over Ω_i, so that the parameter space for $\boldsymbol{\theta} = (\theta_1, \ldots, \theta_r)$ is $\Omega = \Omega_1 \times \cdots \times \Omega_r$. Suppose, further, that for the ith component problem of estimating θ_i with squared error loss, Λ_i is least favorable for θ_i, and the minimax estimator δ_i is the Bayes solution with respect to Λ_i, satisfying condition (1.5) with $\omega_i = \omega_{\Lambda_i}$. Then, $\delta = (\delta_1, \ldots, \delta_r)$ is minimax for estimating $\boldsymbol{\theta}$ with squared error loss. This follows from the facts that (i) δ is a Bayes estimator with respect to the prior distribution Λ for $\boldsymbol{\theta}$, according to which the components θ_i are independently distributed with distribution Λ_i, (ii) $\Lambda(\omega) = 1$ where $\omega = \omega_1 \times \cdots \times \omega_r$, and (iii) the set of points $\boldsymbol{\theta}$ at which $R(\boldsymbol{\theta}, \delta)$ attains its maximum is exactly ω.

The analogous result holds if the component minimax estimators δ_i are not Bayes solutions with respect to least favorable priors but have been obtained through a least favorable sequence by Theorem 1.12. As an example, suppose that X_i $(i = 1, \ldots, r)$ are independently distributed as $N(\theta_i, 1)$. Then, it follows that $(X_1, \ldots, X_r)$ is minimax for estimating $(\theta_1, \ldots, \theta_r)$ with squared error loss. ||

The extensions so far have brought no great surprises. The results for general r were fairly straightforward generalizations of those for $r = 1$. This will no longer always be the case for the last topic to be considered.

(6) *Admissibility*. The multinomial minimax estimator (4.16) was seen to be a unique Bayes estimator and, hence, is admissible. To investigate the admissibility of the minimax estimator **X** for the case of r normal means considered at the end of Example 4.6, one might try the argument suggested following Theorem 4.1. It was seen in Example 4.2 that the problem under consideration remains invariant under the group G_1 of translations and the group G_2 of orthogonal transformations, given by (4.8) and (4.10), respectively. Of these, G_1 is transitive; if there existed an invariant probability distribution over G_1, the remark following Theorem 4.1 would lead to an admissible estimator, hopefully **X**. However, the measures $c\nu$, where ν is Lebesgue measure, are the only invariant measures (Problem 4.14) and they are not finite. Let us instead consider G_2. An invariant probability distribution over G_2 does exist (TSH2, Example 6 of Chapter 9). However, the approach now fails because $\bar{G}_2$ is not transitive. Equivariant estimators do not necessarily have constant risk and, in fact, in the present case, a UMRE estimator does not exist (Strawderman 1971).

Since neither of these two attempts works, let us try the limiting Bayes method (Example 2.8, first proof) instead, which was successful in the case $r = 1$. For the sake of convenience, we shall take the loss to be the average squared error,

$$L(\boldsymbol{\theta}, d) = \frac{1}{r}\Sigma(d_i - \theta_i)^2. \tag{4.17}$$

If **X** is not admissible, there exists an estimator δ^*, a number $\varepsilon > 0$, and intervals $(\theta_{i0}, \theta_{i1})$ such that

$$R(\boldsymbol{\theta}, \delta^*) \begin{cases} \le 1 & \text{for all } \boldsymbol{\theta} \\ < 1 - \varepsilon & \text{for } \boldsymbol{\theta} \text{ satisfying } \theta_{i0} < \theta_i < \theta_{i1} \text{ for all } i. \end{cases}$$

A computation analogous to that of Example 2.8 now shows that

$$\frac{1 - r_\tau^*}{1 - r_\tau} \ge \frac{\varepsilon(1 + \tau^2)}{(\sqrt{2\pi}\tau)^r} \int_{\theta_{10}}^{\theta_{11}} \cdots \int_{\theta_{r0}}^{\theta_{r1}} \exp(-\Sigma\theta_i^2/2\tau^2) d\theta_1 \cdots d\theta_r. \tag{4.18}$$

Unfortunately, the factor preceding the integral no longer tends to infinity when $r > 1$, and so this proof breaks down too.

It was shown by Stein (1956b) that **X** is, in fact, no longer admissible when $r \ge 3$ although admissibility continues to hold for $r = 2$. (A limiting Bayes proof will work for $r = 2$, although not with normal priors. See Problem 4.5). For $r \ge 3$, there are many different estimators whose risk is uniformly less than that of **X**.

To produce an improved estimator, Stein (1956b) gave a "large r and $|\boldsymbol{\theta}|$" argument based on the observation that with high probability, the true $\boldsymbol{\theta}$ is in the sphere $\{\boldsymbol{\theta} : |\boldsymbol{\theta}|^2 \leq |\mathbf{x}|^2 - r\}$. Since the usual estimator $\mathbf{X}$ is approximately the same size as $\boldsymbol{\theta}$, it will almost certainly be outside of this sphere. Thus, we should cut down the estimator $\mathbf{X}$ to bring it inside the sphere. Stein argues that $\mathbf{X}$ should be cut down by a factor of $(|\mathbf{X}|^2 - r)/|\mathbf{X}|^2 = 1 - r/|\mathbf{X}|^2$, and as a more general form, he considers the class of estimators

$$\delta(\mathbf{x}) = [1 - h(|\mathbf{x}|^2)]\mathbf{x}, \tag{4.19}$$

with particular emphasis on the special case

$$\delta(\mathbf{x}) = \left(1 - \frac{r}{|\mathbf{x}|^2}\right)\mathbf{x}. \tag{4.20}$$

See Problem 4.6 for details.

Later, James and Stein (1961) established the complete dominance of (4.20) over $\mathbf{X}$, and (4.20) remains the basic underlying form of almost all improved estimators. In particular, the appearance of the squared term in the shrinkage factor is essential for optimality (Brown 1971; Berger 1976a; Berger 1985, Section 8.9.4).

Since Stein (1956b) and James and Stein (1961), the proof of domination of the estimator (4.20) over the maximum likelihood estimator, $\mathbf{X}$, has undergone many modifications and updates. More recent proofs are based on the representation of Corollary 4.7.2 and can be made to apply to cases other than the normal. We defer treatment of this topic until Section 5.6. At present, we only make some remarks about the estimator (4.20) and the following modifications due to James and Stein (1961). Let

$$\delta_i = \mu_i + \left(1 - \frac{r-2}{|\mathbf{x} - \boldsymbol{\mu}|^2}\right)(x_i - \mu_i) \tag{4.21}$$

where $\boldsymbol{\mu} = (\mu_1, \ldots, \mu_r)$ are given numbers and

$$|\mathbf{x} - \boldsymbol{\mu}| = \left[\Sigma(x_i - \mu_i)^2\right]^{1/2}. \tag{4.22}$$

A motivation for the general structure of the estimator (4.21) can be obtained by using arguments similar to the empirical Bayes arguments in Examples 4.7.7 and 4.7.8 (see also Problems 4.7.6 and 4.7.7). Suppose, a priori, it was thought likely, though not certain, that $\theta_i = \mu_i$ $(i = 1, \ldots, r)$. Then, it might be reasonable first to test

$$H : \theta_1 = \mu_1, \ldots, \theta_r = \mu_r$$

and to estimate $\boldsymbol{\theta}$ by $\boldsymbol{\mu}$ when H is accepted and by $\mathbf{X}$ otherwise. The best acceptance region has the form $|\mathbf{x} - \boldsymbol{\mu}| \leq C$ so that the estimator becomes

$$\delta = \begin{cases} \boldsymbol{\mu} & \text{if } |\mathbf{x} - \boldsymbol{\mu}| \leq C \\ \mathbf{x} & \text{if } |\mathbf{x} - \boldsymbol{\mu}| > C. \end{cases} \tag{4.23}$$

A smoother approach is provided by an estimator with components of the form

$$\delta_i = \psi(|\mathbf{x} - \boldsymbol{\mu}|)x_i + [1 - \psi(|\mathbf{x} - \boldsymbol{\mu}|)]\mu_i \tag{4.24}$$

where ψ, instead of being two-valued as in (4.23), is a function increasing continuously with $\psi(0) = 0$ to $\psi(\infty) = 1$. The estimator (4.21) is of the form (4.24) (although with $\psi(0) = -\infty$), but the argument given above provides no explanation of the particular choice for ψ. We note, however, that many hierarchical Bayes estimators (such as given in Example 5.2) will result in estimators of this form. We will return to this question in Section 5.6.

For the case of unknown σ, the estimator corresponding to (4.23) has been investigated by Sclove, Morris, and Radhakrishnan (1972). They show that it does not provide a uniform improvement over $\mathbf{X}$ and that its risk is uniformly greater than that of the corresponding James-Stein estimator. Although these so-called *pretest estimators* tend not to be optimal, they have been the subject of considerable research (see, for example, Sen and Saleh 1985, 1987).

Unlike $\mathbf{X}$, the estimator δ is, of course, biased. An aspect that in some circumstances is disconcerting is the fact that the estimator of θ_i depends not only on X_i but also on the other (independent) X's. Do we save enough in risk to make up for these drawbacks? To answer this, we take a closer look at the risk function.

Under the loss (4.17), it will be shown in Theorem 5.1 that the risk function of the estimator (4.21) can be written as

$$R(\boldsymbol{\theta}, \delta) = 1 - \frac{r-2}{r} E_{\theta}\left(\frac{r-2}{|\mathbf{X}-\boldsymbol{\mu}|^2}\right). \tag{4.25}$$

Thus, δ has uniformly smaller risk than the constant estimator $\mathbf{X}$ when $r \geq 3$, and, in particular, δ is then minimax by Example 4.6. More detailed information can be obtained from the fact that $|\mathbf{X}-\boldsymbol{\mu}|^2$ has a noncentral χ^2-distribution with noncentrality parameter $\lambda = \Sigma(\theta_i - \mu_i)^2$ and that, therefore, the risk function (4.25) is an increasing function of λ. (See TSH2 Chapter 3, Lemma 2 and Chapter 7, Problem 4 for details). The risk function tends to 1 as $\lambda \to \infty$, and takes on its minimum value at $\lambda = 0$. For this value, $|\mathbf{X}-\boldsymbol{\mu}|^2$ has a χ^2-distribution with r degrees of freedom, and it follows from Example 2.1 that (Problem 4.7)

$$E\left(\frac{1}{|\mathbf{X}-\boldsymbol{\mu}|^2}\right) = \frac{1}{r-2}$$

and hence $R(\boldsymbol{\mu}, \delta) = 2/r$. Particularly for large values of r, the savings over the risk of $\mathbf{X}$ (which is equal to 1 for all $\boldsymbol{\theta}$) can therefore be substantial. (See Bondar 1987 for further discussion.)

We thus have the surprising result that $\mathbf{X}$ is not only inadmissible when $r \geq 3$ but that even substantial risk savings are possible. This is the case not only for squared error loss but also for a wide variety of loss functions which in a suitable way combine the losses resulting from the r component problems. In particular, Brown (1966, Theorem 3.1.1) proves that $\mathbf{X}$ is inadmissible for $r \geq 3$ when $L(\boldsymbol{\theta}, d) = \rho(d - \boldsymbol{\theta})$, where ρ is a convex function satisfying, in addition to some mild conditions, the requirement that the $r \times r$ matrix $\mathcal{R}$ with the (i, j)th element

$$E_0\left[X_i \frac{\partial}{\partial X_j} \rho(\mathbf{X})\right] \tag{4.26}$$

is nonsingular. Here, the derivative in (4.26) is replaced by zero whenever it does not exist.

Example 4.7 A variety of loss functions. Consider the following loss functions $\rho_1, \ldots, \rho_4$:

$$\rho_1(\mathbf{t}) = \Sigma v_i t_i^2 \qquad (\text{all } v_i > 0);$$
$$\rho_2(\mathbf{t}) = \max_i t_i^2;$$
$$\rho_3(\mathbf{t}) = t_1^2,$$
$$\rho_4(\mathbf{t}) = \left(\frac{1}{r}\Sigma t_i\right)^2.$$

All four are convex, and $\mathcal{R}$ is nonsingular for ρ_1 and ρ_2 but singular for ρ_3 and ρ_4 (Problem 4.8). For $r \geq 3$, it follows from Brown's theorem that **X** is inadmissible for ρ_1 and ρ_2. On the other hand, it is admissible for ρ_3 and ρ_4 (Problem 4.10). ||

Other ways in which the admissibility of **X** depends on the loss function are indicated by the following example (Brown, 1980b) in which $L(\boldsymbol{\theta}, d)$ is not of the form $\rho(d - \boldsymbol{\theta})$.

Example 4.8 Admissibility of X. Let X_i $(i = 1, \ldots, r)$ be independently distributed as $N(\theta_i, 1)$ and consider the estimation of $\boldsymbol{\theta}$ with loss function

$$L(\boldsymbol{\theta}, d) = \sum_{i=1}^{r} \frac{v(\theta_i)}{\Sigma v(\theta_j)}(\theta_i - d_i)^2. \tag{4.27}$$

Then the following results hold:

(i) When $v(t) = e^{kt}$ $(k \neq 0)$, **X** is inadmissible if and only if $r \geq 2$.

(ii) When $v(t) = (1 + t^2)^{k/2}$,

(a) **X** is admissible for $k < 1$, $1 \leq r < (2 - k)/(1 - k)$ and for $k \geq 1$, all r;

(b) **X** is inadmissible for $k < 1$, $r > (2 - k)/(1 - k)$.

Parts (i) and (ii(b)) will not be proved here. For the proof of (ii(a)), see Problem 4.11. ||

In the formulations considered so far, the loss function in some way combines the losses resulting from the different component problems. Suppose, however, that the problems of estimating $\theta_1, \ldots, \theta_r$ are quite unrelated and that it is important to control the error on each of them. It might then be of interest to minimize

$$\max_i \left[\sup_{\theta_i} E(\delta_i - \theta_i)^2\right]. \tag{4.28}$$

It is easy to see that **X** is the unique estimator minimizing (4.28) and is admissible from this point of view. This follows from the fact that X_i is the unique estimator for which

$$\sup_{\theta_i} E(\delta_i - \theta_i)^2 \leq 1.$$

[On the other hand, it follows from Example 4.7 that **X** is inadmissible for $r \geq 3$ when $L(\boldsymbol{\theta}, d) = \max_i (d_i - \theta_i)^2$.]

The performance measure (4.28) is not a risk function in the sense defined in Chapter 1 because it is not the expected value of some loss but the maximum of a number of such expectations. An interesting way of looking at such a criterion was proposed by Brown (1975) [see also Bock 1975, Shinozaki 1980, 1984]. Brown considers a family $\mathcal{L}$ of loss functions L, with the thought that it is not clear which of these loss functions will be most appropriate. (It may not be clear how the data will be used, or they may be destined for multiple uses. In this connection, see also Rao 1977.) If

$$R_L(\boldsymbol{\theta}, \delta) = E_\theta L[\boldsymbol{\theta}, \delta(\mathbf{X})], \tag{4.29}$$

Brown defines δ to be admissible with respect to the class $\mathcal{L}$ if there exists no δ' such that

$$R_L(\boldsymbol{\theta}, \delta') \le R_L(\boldsymbol{\theta}, \delta) \quad \text{for all} \quad L \in \mathcal{L} \text{ and all } \boldsymbol{\theta}$$

with strict inequality holding for at least one $L = L_0$ and $\boldsymbol{\theta} = \boldsymbol{\theta}_0$.

The argument following (4.28) shows that $\mathbf{X}$ is admissible when $\mathcal{L}$ contains the r loss functions $L_i(\boldsymbol{\theta}, d) = (d_i - \theta_i)^2$, $i = 1, \ldots, r$, and hence, in particular, when $\mathcal{L}$ is the class of all loss functions

$$\sum_{i=1}^{r} c_i(\delta_i - \theta_i)^2, \quad 0 \le c_i < \infty. \tag{4.30}$$

On the other hand, Brown shows that if the ratios of the weights c_i to each other are bounded,

$$c_i/c_j < K, \quad i, j = 1, \ldots, r, \tag{4.31}$$

then no matter how large K, the estimator $\mathbf{X}$ is inadmissible with respect to the class $\mathcal{L}$ of loss functions (4.30) satisfying (4.31). Similar results persist in even more general settings, such as when $\mathcal{L}$ is not restricted to squared error loss. See Hwang 1985, Brown and Hwang 1989, and Problem 4.14.

The above considerations make it clear that the choice between $\mathbf{X}$ and competitors such as (4.21) must depend on the circumstances. (In this connection, see also Robinson 1979a, 1979b). A more detailed discussion of some of these issues will be given in the next section.

5 Shrinkage Estimators in the Normal Case

The simultaneous consideration of a number of similar estimation problems involving independent variables and parameters (X_i, θ_i) often occurs in repetitive situations in which it may be reasonable to view the $\boldsymbol{\theta}$'s themselves as random variables. This leads to the Bayesian approach of Examples 4.7.1, 4.7.7, and 4.7.8. In the simplest normal case, we assume, as in Example 4.7.1, that the X_i are independent normal with mean θ_i and variance σ^2, and that the θ_i's are also normal, say with mean $\boldsymbol{\xi}$ and variance A (previously denoted by τ^2), that is, $\mathbf{X} \sim N_r(\boldsymbol{\theta}, \sigma^2 I)$ and $\boldsymbol{\theta} \sim N_r(\boldsymbol{\xi}, AI)$. This model has some similarity with the Model II version of the one-way layout considered in Section 3.5. There, however, interest centered on the variances σ^2 and A, while we now wish to estimate the θ_i's.

To simplify the problem, we shall begin by assuming that σ and ξ are known, say $\sigma = 1$ and $\xi = 0$, so that only A and θ are unknown. The empirical Bayes arguments of Example 4.7.1 led to the estimator

$$\delta_i^{\hat{\pi}} = (1 - \hat{B})x_i, \tag{5.1}$$

where $\hat{B} = (r-2)/\Sigma x_i^2$, that is, the James-Stein estimator (4.21) with $\mu = 0$. We now prove that, as previously claimed, the risk function of (5.1) is given by (4.25). However, we shall do so for the more general estimator (5.1) with $\hat{B} = c(r-2)/\Sigma x_i^2$ where c is a positive constant. (The value $c = 1$ minimizes both the Bayes risk (Problem 4.7.5) and the frequentist risk among estimators of this form.)

Theorem 5.1 *Let X_i, $i = 1, \ldots, r$ $(r > 2)$, be independent, with distributions $N(\theta_i, 1)$ and let the estimator δ_c of $\boldsymbol{\theta}$ be given by*

$$\delta_c(\mathbf{x}) = \left(1 - c\frac{r-2}{|\mathbf{x}|^2}\right)\mathbf{x}, \quad |\mathbf{x}|^2 = \Sigma x_j^2. \tag{5.2}$$

Then, the risk function of δ_c, with loss function (4.17), is

$$R(\boldsymbol{\theta}, \delta_c) = 1 - \frac{(r-2)^2}{r} E_{\boldsymbol{\theta}}\left[\frac{c(2-c)}{|\mathbf{X}|^2}\right]. \tag{5.3}$$

Proof. From Corollary 4.7.2, using the loss function (4.17), the risk of δ_c is

$$R(\boldsymbol{\theta}, \delta_c) = 1 + \frac{1}{r}E_{\boldsymbol{\theta}}|g(\mathbf{X})|^2 - \frac{2}{r}\sum_{i=1}^r E_{\boldsymbol{\theta}}\frac{\partial}{\partial X_i}g_i(\mathbf{X}) \tag{5.4}$$

where $g_i(\mathbf{x}) = c(r-2)x_i/|\mathbf{x}|^2$ and $|g(\mathbf{x})|^2 = c^2(r-2)^2/|\mathbf{x}|^2$. Differentiation shows

$$\frac{\partial}{\partial x_i}g_i(\mathbf{x}) = \frac{c(r-2)}{|\mathbf{x}|^4}[|\mathbf{x}|^2 - 2x_i^2] \tag{5.5}$$

and hence

$$\begin{aligned}\sum_{i=1}^r \frac{\partial}{\partial x_i}g_i(\mathbf{x}) &= \frac{c(r-2)}{|\mathbf{x}|^4}\sum_{i=1}^r[|\mathbf{x}|^2 - 2x_i^2] \\ &= \frac{c(r-2)}{|\mathbf{x}|^2}(r-2),\end{aligned}$$

and substitution into (5.4) gives

$$\begin{aligned}R(\boldsymbol{\theta}, \delta_c) &= 1 + \frac{1}{r}E_{\boldsymbol{\theta}}\left[\frac{c^2(r-2)^2}{|\mathbf{X}|^2}\right] - \frac{2}{r}E_{\boldsymbol{\theta}}\left[\frac{c(r-2)^2}{|\mathbf{X}|^2}\right] \\ &= 1 - \frac{(r-2)^2}{r}E_{\boldsymbol{\theta}}\left[\frac{c(2-c)}{|\mathbf{X}|^2}\right].\end{aligned}$$

□

Note that

$$E_{\boldsymbol{\theta}}\left[\frac{1}{|\mathbf{X}|^2}\right] \le E_0\left[\frac{1}{|\mathbf{X}|^2}\right], \tag{5.6}$$

so $R(\boldsymbol{\theta}, \delta_c) < \infty$ only if the latter expectation is finite, which occurs when $r \geq 3$ (see Problem 5.2).

From the expression (5.3) for the risk, we immediately get the following results.

Corollary 5.2 *The estimator δ_c defined by (5.2) dominates $\mathbf{X}$ ($\delta_c = \mathbf{X}$ when $c = 0$), provided $0 < c < 2$ and $r \geq 3$.*

Proof. For these values, $c(2-c) > 0$ and, hence, $R(\boldsymbol{\theta}, \delta_c) < 1$ for all $\boldsymbol{\theta}$. Note that $R(\boldsymbol{\theta}, \delta_c) = R(\boldsymbol{\theta}, X)$ for $c = 2$. □

Corollary 5.3 *The James-Stein estimator δ, which equals δ_c with $c = 1$, dominates all estimators δ_c with $c \neq 1$.*

Proof. The factor $c(2-c)$ takes on its maximum value 1 if and only if $c = 1$. □

For $c = 1$, formula (5.3) verifies the risk formula (4.25). Since the James-Stein estimator dominates all estimators δ_c with $c \neq 1$, one might hope that it is admissible. However, unfortunately, this is not the case, as is shown by the following theorem, which strengthens Theorem 4.7.5 by extending the comparison from the average (Bayes) risk to the risk function.

Theorem 5.4 *Let δ be any estimator of the form (5.1) with $\hat{B}$ any strictly decreasing function of the x_i's and suppose that*

$$P_\theta(\hat{B} > 1) > 0. \tag{5.7}$$

Then,

$$R(\boldsymbol{\theta}, \hat{\delta}) < R(\boldsymbol{\theta}, \delta)$$

where

$$\hat{\delta}_i = \max[(1-\hat{B}), 0]x_i. \tag{5.8}$$

Proof. By (4.17),

$$R(\boldsymbol{\theta}, \delta) - R(\boldsymbol{\theta}, \hat{\delta}) = \frac{1}{r}\sum\left[E_\theta\left(\delta_i^2 - \hat{\delta}_i^2\right) - 2\theta_i E_\theta\left(\delta_i - \hat{\delta}_i\right)\right].$$

To show that the expression in brackets is always > 0, calculate the expectations by first conditioning on $\hat{B}$. For any value $\hat{B} \leq 1$, we have $\delta_i = \hat{\delta}_i$, so it is enough to show that the right side is positive when conditioned on any value $\hat{B} = b > 1$. Since in that case $\hat{\delta}_i = 0$, it is finally enough to show that for any $b > 1$,

$$\theta_i E_\theta[\delta_i | \hat{B} = b] = \theta_i(1-b)E_\theta(X_i | \hat{B} = b) \leq 0$$

and hence that $\theta_i E_\theta(X_i | \hat{B} = b) \geq 0$. Now $\hat{B} = b$ is equivalent to $|\mathbf{X}|^2 = c$ for some c and hence to $X_1^2 = c - (X_2^2 + \cdots + X_r^2)$. Conditioning further on $X_2, \ldots, X_r$, we find that

$$\begin{aligned} E_\theta(\theta_1 X_1 | |\mathbf{X}|^2 = c, x_2, \ldots, x_r) &= \theta_1 E_\theta(X_1 | X_1^2 = y^2) \\ &= \frac{\theta_1 y(e^{\theta_1 y} - e^{-\theta_1 y})}{e^{\theta_1 y} + e^{-\theta_1 y}} \end{aligned}$$

where $y = \sqrt{c - (x_2^2 + \cdots + x_r^2)}$. This is an increasing function of $|\theta_1 y|$, which is zero when $\theta_1 y = 0$, and this completes the proof. □

Theorem 5.4 shows in particular that the James-Stein estimator (δ_c with $c = 1$) is dominated by another minimax estimator,

$$\delta_i^+ = \left(1 - \frac{r-2}{|\mathbf{x}|^2}\right)^+ x_i \tag{5.9}$$

where $(\cdot)^+$ indicates that the quantity in parentheses is replaced by 0 whenever it is negative. We shall call

$$(\cdot)^+ = \max[(\cdot), 0]$$

the positive part of $(\cdot)$. The risk functions of the ordinary and positive-part Stein estimators are shown in Figure 5.1.

Figure 5.1. *Risk functions of the ordinary and positive-part Stein estimators, for r=4.*

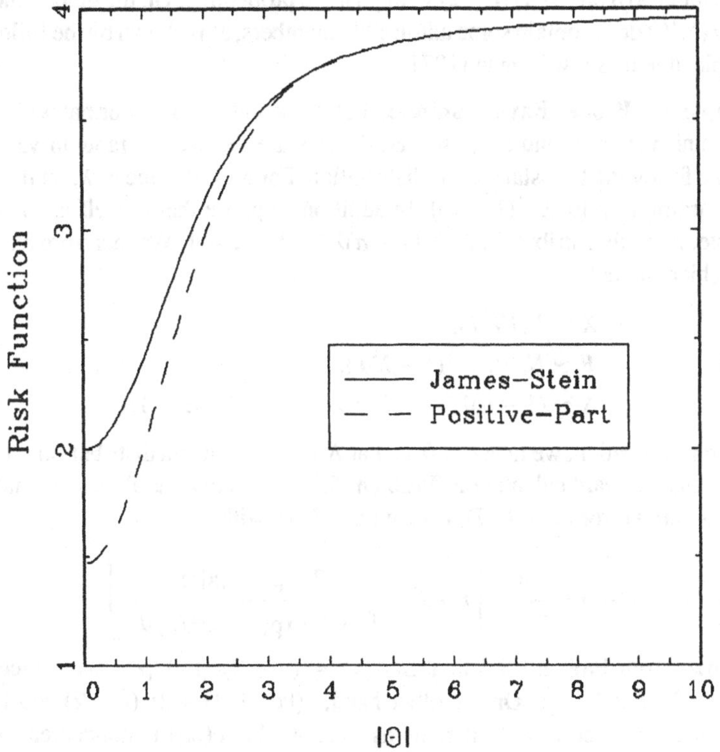

Unfortunately, it can be shown that even δ^+ is inadmissible because it is not smooth enough to be either Bayes or limiting Bayes, as we will see in Section 5.7. However, as suggested by Efron and Morris (1973a, Section 5), the positive-part estimator δ^+ is difficult to dominate, and it took another twenty years until a dominating estimator was found by Shao and Strawderman (1994). There exist, in fact, many admissible minimax estimators, but they are of a more general form than (5.1) or (5.9). To obtain such an estimator, we state the following generalization of Corollary 5.2, due to Baranchik (1970) (see also Strawderman 1971, and Efron and Morris 1976a). The proof is left to Problem 5.4.

Theorem 5.5 *For* $\mathbf{X} \sim N_r(\boldsymbol{\theta}, I)$, $r \geq 3$, *and loss* $L(\boldsymbol{\theta}, d) = \frac{1}{r}\Sigma(d_i - \theta_i)^2$, *an estimator of the form*

$$\delta_i = \left[1 - c(|\mathbf{x}|)\frac{r-2}{|\mathbf{x}|^2}\right]x_i \tag{5.10}$$

is minimax provided

(i) $0 \leq c(\cdot) \leq 2$ *and*

(ii) the function c is nondecreasing.

It is interesting to note how very different the situation for $r \geq 3$ is from the one-dimensional problem discussed in Sections 5.2 and 5.3. There, minimax estimators were unique (although recall Example 2.9); here, they constitute a rich collection. It follows from Theorem 5.4 that the estimators (5.10) are inadmissible whenever $c(|\mathbf{x}|)/|\mathbf{x}|^2 > 1/(r-2)$ with positive probability. On the other hand, the family (5.10) does contain some admissible members, as is shown by the following example, due to Strawderman (1971).

Example 5.6 Proper Bayes minimax. Let X_i be independent normal with mean θ_i and unit variance, and suppose that the θ_i's are themselves random variables with the following two-stage prior distribution. For a fixed value of λ, let the θ_i be iid according to $N[0, \lambda^{-1}(1-\lambda)]$. In addition, suppose that λ itself is a random variable, Λ, with distribution $\Lambda \sim (1-a)\lambda^{-a}, 0 \leq a < 1$. We therefore have the hierarchical model

$$\begin{aligned}
&\mathbf{X} \sim N_r(\boldsymbol{\theta}, I),\\
&\boldsymbol{\theta} \sim N_r(0, \lambda^{-1}(1-\lambda)I),\\
&\Lambda \sim (1-a)\lambda^{-a}, \quad 0 < \lambda < 1, \quad 0 < a < 1.
\end{aligned} \tag{5.11}$$

Here, for illustration, we take $a = 0$ so that Λ has the uniform distribution $U(0, 1)$.

A straightforward calculation (Problem 5.5) shows that the Bayes estimator δ, under squared error loss (4.17), is given by (5.10) with

$$c(|\mathbf{x}|) = \frac{1}{r-2}\left[r + 2 - \frac{2\exp(-\frac{1}{2}|\mathbf{x}|^2)}{\int_0^1 \lambda^{r/2}\exp(-\lambda|\mathbf{x}|^2/2)\,d\lambda}\right]. \tag{5.12}$$

It follows from Problem 5.4 that $E(\Lambda|\mathbf{x}) = (r-2)c(|\mathbf{x}|)/|\mathbf{x}|^2$ and hence that $c(|\mathbf{x}|) \geq 0$ since $\lambda < 1$. On the other hand, $c(|\mathbf{x}|) \leq (r+2)/(r-2)$ and hence $c(|\mathbf{x}|) \leq 2$ provided $r \geq 6$. It remains to show that $c(|\mathbf{x}|)$ is nondecreasing or, equivalently, that

$$\int_0^1 \lambda^{r/2}\exp\left[\frac{1}{2}|\mathbf{x}|^2(1-\lambda)\right]d\lambda$$

is nondecreasing in $|\mathbf{x}|$. This is obvious since $0 < \lambda < 1$.

Thus, the estimator (5.10) with $c(|\mathbf{x}|)$ given by (5.12) is a proper Bayes (and admissible) minimax estimator for $r \geq 6$. ‖

Although neither the James-Stein estimator nor its positive-part version of (5.9) are admissible, it appears that no substantial improvements over the latter are possible (see Example 7.3). We shall now turn to some generalizations of these estimators, where we no longer require equal variances or equal weights in the loss function.

We first look at the case where the covariance matrix is no longer $\sigma^2 I$, but may be any positive definite matrix Σ. Conditions for minimaxity of the estimator

$$\delta(\mathbf{x}) = \left(1 - \frac{c(|\mathbf{x}|^2)}{|\mathbf{x}|^2}\right)\mathbf{x} \tag{5.13}$$

will now involve this covariance matrix.

Theorem 5.7 *For* $\mathbf{X} \sim N_r(\boldsymbol{\theta}, \Sigma)$ *with* Σ *known, an estimator of the form (5.13) is minimax against the loss* $L(\boldsymbol{\theta}, \delta) = |\boldsymbol{\theta} - \delta|^2$, *provided*

(i) $0 \le c(|\mathbf{x}|^2) \le 2[\text{tr}(\Sigma)/\lambda_{\max}(\Sigma)] - 4$,

(ii) the function $c(\cdot)$ *is nondecreasing,*

where $\text{tr}(\Sigma)$ *denotes the trace of the matrix* Σ *and* $\lambda_{\max}(\Sigma)$ *denotes its largest eigenvalue.*

Note that the covariance matrix must satisfy $\text{tr}(\Sigma)/\lambda_{\max}(\Sigma) > 2$ for δ to be different from X. If $\Sigma = I$, $\text{tr}(\Sigma)/\lambda_{\max}(\Sigma) = r$, so this is the dimensionality restriction in another guise. Bock (1975) (see also Brown 1975) shows that X is unique minimax among *spherically symmetric estimators* if $\text{tr}(\Sigma)/\lambda_{\max}(\Sigma) < 2$. (An estimator is said to be *spherically symmetric* if it is equivariant under orthogonal transformations. Such estimators were characterized by (4.11), to which (5.13) is equivalent.)

When the bound on $c(\cdot)$ is displayed in terms of the covariance matrix, we get some idea of the types of problems in which we can expect improvement from shrinkage estimators. The condition $\text{tr}(\Sigma) > 2\lambda_{\max}(\Sigma)$ will be satisfied when the eigenvalues of Σ are not too different (see Problem 5.10). If this condition is not satisfied, then estimators which allow different coordinatewise shrinkage are needed to obtain minimaxity (see Notes 9.6).

Proof of Theorem 5.7. The risk of $\delta(\mathbf{x})$ is

$$\begin{aligned} R(\boldsymbol{\theta}, \delta) &= E_\theta\left[(\boldsymbol{\theta} - \delta(\mathbf{X}))'(\boldsymbol{\theta} - \delta(\mathbf{X}))\right] \\ &= E_\theta\left[(\boldsymbol{\theta} - \mathbf{X})'(\boldsymbol{\theta} - \mathbf{X})\right] \\ &\quad -2E_\theta\left[\frac{c(|\mathbf{X}|^2)}{|\mathbf{X}|^2}\mathbf{X}'(\boldsymbol{\theta} - \mathbf{X})\right] + E_\theta\left[\frac{c^2(|\mathbf{X}|^2)}{|\mathbf{X}|^2}\right] \end{aligned} \tag{5.14}$$

where $E_\theta(\boldsymbol{\theta} - \mathbf{X})'(\boldsymbol{\theta} - \mathbf{X}) = \text{tr}(\Sigma)$, the trace of the matrix Σ, is the minimax risk (Problem 5.8). We can now apply integration by parts (see Problem 5.9) to write

$$\begin{aligned} R(\boldsymbol{\theta}, \delta) &= \text{tr}(\Sigma) + E_\theta\left\{\frac{c(|\mathbf{X}|^2)}{|\mathbf{X}|^2}\left[(c(|\mathbf{X}|^2) + 4)\frac{\mathbf{X}'\Sigma\mathbf{X}}{|\mathbf{X}|^2} - 2\,\text{tr}\,\Sigma\right]\right\} \\ &\quad -4E_\theta\frac{c'(|\mathbf{X}|^2)}{|\mathbf{X}|^2}\mathbf{X}'\Sigma\mathbf{X}. \end{aligned} \tag{5.15}$$

Since $c'(\cdot) \ge 0$, an upper bound on $R(\boldsymbol{\theta}, \delta)$ results by dropping the last term. We then note [see Equation (2.6.5)]

$$\frac{\mathbf{x}'\Sigma\mathbf{x}}{|\mathbf{x}|^2} = \frac{\mathbf{x}'\Sigma\mathbf{x}}{\mathbf{x}'\mathbf{x}} \le \lambda_{\max}(\Sigma) \tag{5.16}$$

and the result follows. □

Theorem 5.5 can also be extended by considering more general loss functions in which the coordinates may have different weights.

Example 5.8 Loss functions in the one-way layout. In the one-way layout we observe

$$X_{ij} \sim N(\xi_i, \sigma_i^2), \quad j = 1, \ldots, n_i, \quad i = 1, \ldots, r, \tag{5.17}$$

where the usual estimator of ξ_i is $\bar{X}_i = \sum_j X_{ij}/n_i$. If the assumptions of Theorem 5.7 are satisfied, then the estimator $\bar{\mathbf{X}} = (\bar{X}_1, \ldots, \bar{X}_r)$ can be improved. If the ξ_i's represent mean responses for different treatments, for example, crop yields from different fertilization treatments or mean responses from different drug therapies, it may be unrealistic to penalize the estimation of each coordinate by the same amount. In particular, if one drug is uncommonly expensive or if a fertilization treatment is quite difficult to apply, this could be reflected in the loss function. ||

The situation described in Example 5.8 can be generalized to

$$\begin{aligned} &\mathbf{X} \sim N(\boldsymbol{\theta}, \Sigma), \\ &L(\boldsymbol{\theta}, \delta) = (\boldsymbol{\theta} - \delta)'Q(\boldsymbol{\theta} - \delta), \end{aligned} \tag{5.18}$$

where both Σ and Q are positive definite matrices. We again ask under what conditions the estimator (5.13) is a minimax estimator of $\boldsymbol{\theta}$.

Before answering this question, we first note that, without loss of generality, we can consider one of Σ or Q to be the identity (see Problem 5.11). Hence, we take $\Sigma = I$ in the following theorem, whose proof is left to Problem 5.12; see also Problem 5.13 for a more general result.

Theorem 5.9 *Let $\mathbf{X} \sim N(\boldsymbol{\theta}, I)$. An estimator of the form (5.13) is minimax against the loss $L(\boldsymbol{\theta}, \delta) = (\boldsymbol{\theta} - \delta)'Q(\boldsymbol{\theta} - \delta)$, provided*

(i) $0 \le c(|\mathbf{x}|^2) \le 2[\text{tr}(Q)/\lambda_{\max}(Q)] - 4$,

(ii) the function $c(\cdot)$ is nondecreasing.

Theorem 5.9 can also be viewed as a robustness result, since we have shown δ to be minimax against any Q, which provides an upper bound for $c(|\mathbf{x}|^2)$ in (i). This is in the same spirit as the results of Brown (1975), mentioned in Section 5.5. (See Problem 5.14.)

Thus far, we have been mainly concerned with one form of shrinkage estimator, the estimator (5.13). We shall now obtain a more general class of minimax estimators by writing δ as $\delta(\mathbf{x}) = \mathbf{x} - g(\mathbf{x})$ and utilizing the resulting expression (5.4) for the risk of δ. As first noted by Stein, (5.4) can be combined with the identities derived in Section 4.3 (for the Bayes estimator in an exponential family) to obtain a set of sufficient conditions for minimaxity in terms of the condition of *superharmonicity* of the marginal distribution (see Section 1.7).

In particular, for the case of $\mathbf{X} \sim N(\boldsymbol{\theta}, I)$, we can, by Corollary 3.3, write a Bayes estimator of $\boldsymbol{\theta}$ as

$$E(\theta_i|\mathbf{x}) = \frac{\partial}{\partial x_i}\log m(\mathbf{x}) - \frac{\partial}{\partial x_i}\log h(\mathbf{x}) \tag{5.19}$$

with $-\partial \log h(\mathbf{x})/\partial x_i = x_i$ (Problem 5.17) so that the Bayes estimators are of the form

$$\delta(\mathbf{x}) = \mathbf{x} + \nabla \log m(\mathbf{x}). \tag{5.20}$$

Theorem 5.10 *If $\mathbf{X} \sim N_r(\boldsymbol{\theta}, I)$, the risk, under squared error loss, of the estimator (5.20) is given by*

$$\begin{aligned} R(\boldsymbol{\theta}, \delta) &= 1 + \frac{4}{r}E_\theta\left\{\sum_{i=1}^{r}\frac{\frac{\partial^2}{\partial x_i^2}\sqrt{m(\mathbf{X})}}{\sqrt{m(\mathbf{X})}}\right\} \\ &= 1 + \frac{4}{r}E_\theta\frac{\nabla^2\sqrt{m(\mathbf{X})}}{\sqrt{m(\mathbf{X})}} \end{aligned} \tag{5.21}$$

where $\nabla^2 f = \sum\{(\partial^2/\partial x_i^2) f\}$ is the Laplacian of f.

Proof. The ith component of the estimator (5.20) is

$$\begin{aligned} \delta_i(\mathbf{x}) &= x_i + (\nabla \log m(\mathbf{x}))_i \\ &= x_i + \frac{\partial}{\partial x_i}\log m(\mathbf{x}) = x_i + \frac{m_i'(\mathbf{x})}{m(\mathbf{x})} \end{aligned}$$

where, for simplicity of notation, we write $m_i'(\mathbf{x}) = (\partial/\partial x_i)m(\mathbf{x})$. In the risk identity (5.4), set $g_i(\mathbf{x}) = -m_i'(\mathbf{x})/m(\mathbf{x})$ to obtain

$$\begin{aligned} R(\boldsymbol{\theta}, \delta) &= 1 + \frac{1}{r}E_\theta\left\{\sum_{i=1}^{r}\left[\frac{m_i'(\mathbf{X})}{m(\mathbf{X})}\right]^2\right\} + \frac{2}{r}E_\theta\left\{\sum_{i=1}^{r}\left[\frac{\partial}{\partial x_i}\frac{m_i'(\mathbf{X})}{m(\mathbf{X})}\right]\right\} \\ &= 1 + \frac{1}{r}\sum_{i=1}^{r}E_\theta\left\{2\frac{m_i''(\mathbf{X})}{m(\mathbf{X})} - \left[\frac{m_i'(\mathbf{X})}{m(\mathbf{X})}\right]^2\right\} \end{aligned} \tag{5.22}$$

where $m_i''(\mathbf{x}) = (\partial^2/\partial x_i^2)m(\mathbf{x})$, and the second expression follows from straightforward differentiation and gathering of terms. Finally, notice the differentiation identity

$$\frac{\partial^2}{\partial x_i^2}[g(\mathbf{x})]^{1/2} = \frac{g_i''(\mathbf{x})}{2[g(\mathbf{x})]^{1/2}} - \frac{[g_i'(\mathbf{x})]^2}{4[g(\mathbf{x})]^{3/2}}. \tag{5.23}$$

Using (5.23), we can rewrite the risk (5.22) in the form (5.21). □

The form of the risk function (5.21) shows that the estimator (5.20) is minimax (provided all expectations are finite) if $\sum(\partial^2/\partial x_i^2)[m(\mathbf{x})]^{1/2} < 0$, and hence by Theorem 1.7.24 if $[m(\mathbf{x})]^{1/2}$ is superharmonic. We, therefore, have established the following class of minimax estimators.

Corollary 5.11 *Under the conditions of Theorem 5.10, if $E_\theta\{\nabla^2\sqrt{m(\mathbf{X})}/\sqrt{m(\mathbf{X})}\} < \infty$ and $[m(\mathbf{x})]^{1/2}$ is a superharmonic function, then $\delta(\mathbf{x}) = \mathbf{x} + \log \nabla m(\mathbf{x})$ is minimax.*

A useful consequence of Corollary 5.11 follows from the fact that superharmonicity of $m(\mathbf{x})$ implies that of $[m(\mathbf{x})]^{1/2}$ (Problem 1.7.16) and is often easier to verify.

Corollary 5.12 *Under the conditions of Theorem 5.10, if $E_\theta|\nabla m(\mathbf{X})/ m(\mathbf{X})|^2 < \infty$, $E_\theta|\nabla^2 m(\mathbf{X})/m(\mathbf{X})| < \infty$, and $m(\mathbf{x})$ is a superharmonic function, then $\delta(\mathbf{x}) = x + \log \nabla m(\mathbf{x})$ is minimax.*

Proof. From the second expression in (5.22), we see that

$$R(\theta, \delta) \le 1 + \frac{2}{r}\sum_{i=1}^{r} E_\theta \frac{m_i''(\mathbf{X})}{m(\mathbf{X})}, \tag{5.24}$$

which is ≤ 1 if $m(\mathbf{x})$ is superharmonic. □

Example 5.13 Superharmonicity of the marginal. For the model in (5.11) of Example 5.6, we have

$$m(\mathbf{x}) \propto \int_0^1 \lambda^{(r/2)-a} e^{\frac{-\lambda}{2}|\mathbf{x}|^2}\, d\lambda$$

and

$$\begin{aligned}\sum_{i=1}^{r} \frac{\partial^2}{\partial x_i^2} m(\mathbf{x}) &\propto \int_0^1 \lambda^{(r/2)-a+1}[\lambda|\mathbf{x}|^2 - r] e^{\frac{-\lambda}{2}|\mathbf{x}|^2}\, d\lambda \\ &= \frac{1}{(|\mathbf{x}|^2)^{(r/2)-a+2}} \int_0^{|\mathbf{x}|^2} t^{(r/2)-a+1}[t - r]e^{-t/2}\, dt.\end{aligned} \tag{5.25}$$

Thus, a sufficient condition for $m(\mathbf{x})$ to be superharmonic is

$$\int_0^{|\mathbf{x}|^2} t^{(r/2)-a+1}[t - r]e^{-t/2}\, dt \le 0. \tag{5.26}$$

From Problem 5.18, we have

$$\begin{aligned}\int_0^{|\mathbf{x}|^2} t^{(r/2)-a+1}[t - r]e^{-t/2}\, dt &\le \int_0^{\infty} t^{(r/2)-a+1}[t - r]e^{-t/2}\, dt \\ &= \Gamma\left(\frac{r}{2} - a + 2\right) 2^{(r/2)-a+2}[-2a + 4],\end{aligned} \tag{5.27}$$

so $m(\mathbf{x})$ is superharmonic if $-2a + 4 \le 0$ or $a \ge 2$. For the choice $a = 2$, the Strawderman estimator is (Problem 5.5)

$$\delta(\mathbf{x}) = \left(1 - \frac{r-2}{|\mathbf{x}|^2}\left[\frac{P(\chi_r^2 \le |\mathbf{x}|^2)}{P(\chi_{r-2}^2 \le |\mathbf{x}|^2)}\right]\right)\mathbf{x},$$

which resembles the positive-part Stein estimator. Note that this estimator is not a proper Bayes estimator, as the prior distribution is improper. ‖

There are many other characterizations of superharmonic functions that lead to different versions of minimax theorems. A most useful one, noticed by Stein (1981), is the following.

Corollary 5.14 *Under the conditions of Theorem 5.10 and Corollary 5.11, if the prior $\pi(\boldsymbol{\theta})$ is superharmonic, then $\delta(\mathbf{x}) = \mathbf{x} + \nabla \log m(\mathbf{x})$ is minimax.*

Proof. The marginal density can be written as

$$m(\mathbf{x}) = \int \phi_r(\mathbf{x} - \boldsymbol{\theta})\pi(\boldsymbol{\theta})\, d\boldsymbol{\theta}, \tag{5.28}$$

where $\phi_r(\mathbf{x} - \boldsymbol{\theta})$ is the r-variate normal density. From Problem 1.7.16, the superharmonicity of $m(\mathbf{x})$ follows. □

Example 5.15 Superharmonic prior. The hierarchical Bayes estimators of Faith (1978) (Problem 5.7) are based on the multivariate t prior

$$\pi(\boldsymbol{\theta}) \propto \left(\frac{2}{b} + |\boldsymbol{\theta}|^2\right)^{-(a+r/2)}. \tag{5.29}$$

It is straightforward to verify that this prior is superharmonic if $a \le -1$, allowing a simple verification of minimaxity of an estimator that can only be expressed as an integral.

The superharmonic condition, although sometimes difficult to verify, has often proved helpful in not only establishing minimaxity but also in understanding what types of prior distributions may lead to minimax Bayes estimators. See Note 9.7 for further discussion. ||

We close this section with an examination of componentwise risk. For $X_i \sim N(\theta_i, 1)$, independent, and risk function

$$\bar{R}(\boldsymbol{\theta}, \delta) = \frac{1}{r}\Sigma R(\theta_i, \delta_i) \tag{5.30}$$

with $R(\theta_i, \delta_i) = E(\delta_i - \theta_i)^2$, it was seen in Section 5.3 that it is not possible to find a δ_i for which $R(\theta_i, \delta_i)$ is uniformly better than $R(\theta_i, X_i) = 1$.

Thus, the improvement in the average risk can be achieved only though increasing some of the component risks, and it becomes of interest to consider the maximum possible component risk

$$\max_i \sup_{\theta_i} R(\theta_i, \delta_i). \tag{5.31}$$

For given $\lambda = \Sigma\theta_j^2$, it can be shown (Baranchik 1964) that (5.31) attains its maximum when all but one of the θ_i's are zero, say $\theta_2 = \cdots = \theta_r = 0$, $\theta_1 = \sqrt{\lambda}$, and that this maximum risk $\rho_r(\lambda)$ as a function of λ increases from a minimum of $2/r$ at $\lambda = 0$ to a maximum and then decreases and tends to 1 as $\lambda \to \infty$; see Figure 6.2.

The values of $\max_\lambda \rho_r(\lambda)$ and the value λ_r at which the maximum is attained, shown in Table 5.1, are given by Efron and Morris (1972a). The table suggests that shrinkage estimators will typically not be appropriate when the component problems concern different clients. No one wants his or her blood test subjected to the possibility of large errors in order to improve a laboratory's average performance.

To get a feeling for the behavior of the James-Stein estimator (4.21) with $\mu = 0$, in a situation in which most of the θ_i's are at or near zero (representing the standard

Figure 5.2. *Maximum component risk $\rho_r(\lambda)$ of the ordinary James-Stein estimator, and the componentwise risk of X, the UMVU estimator, for r=4.*

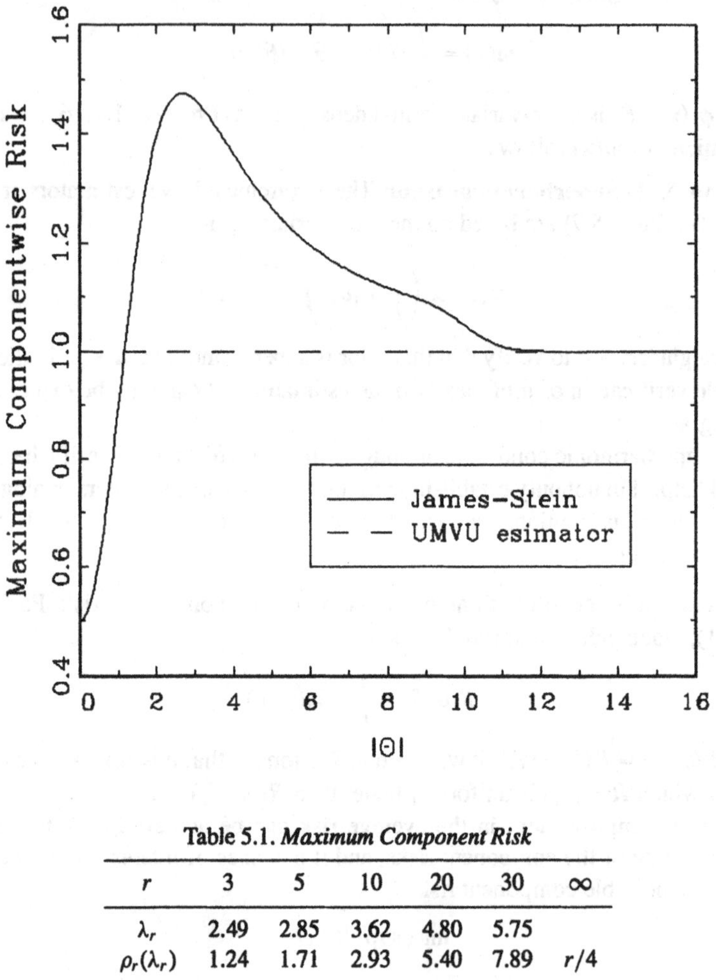

Table 5.1. *Maximum Component Risk*

r	3	5	10	20	30	∞
λ_r	2.49	2.85	3.62	4.80	5.75	
$\rho_r(\lambda_r)$	1.24	1.71	2.93	5.40	7.89	$r/4$

or normal situation of no effect) but a few relatively large θ_i's are present, consider the 20-component model

$$X_i \sim N(\theta_i, 1), \quad i = 1, \ldots, 20,$$

where the vector $\theta = (\theta_1, \ldots, \theta_{20})$ is taken to have one of three configurations:

(a) $\theta_1 = \cdots = \theta_{19} = 0, \quad \theta_{20} = 2, 3, 4,$

(b) $\theta_1 = \cdots = \theta_{18} = 0, \quad \theta_{19} = i, \quad \theta_{20} = j, \quad 2 \le i \le j \le 4,$

(c) $\theta_1 = \cdots = \theta_{17} = 0, \quad \theta_{18} = i, \quad \theta_{19} = j,$
$\theta_{20} = k, \quad 2 \le i \le j \le k \le 4.$

The resulting shrinkage factor, $1 - (r-2)/|\mathbf{x}|^2$, by which the observation is multiplied to obtain the estimators δ_i of θ_i, has expected value

$$E\left(1 - \frac{r-2}{|\mathbf{X}|^2}\right) = 1 - (r-2)E\frac{1}{\chi_r^2(\lambda)} = 1 - (r-2)\sum_{k=0}^{\infty}\frac{e^{-\lambda/2}(\lambda/2)^k}{(r+2k-2)k!} \tag{5.32}$$

where $\lambda = |\boldsymbol{\theta}|^2$ (see Problem 5.23). Its values are given in Table 5.2.

Table 5.2. *Expected Value of the Shrinkage Factor*

	(a) θ_{20}			(b) $\theta_{19}\theta_{20}$					
θ's $\neq 0$	2	3	4	22	23	33	24	34	44
Factor	.17	.37	.46	.29	.40	.49	.51	.57	.63
λ	4	9	16	8	13	18	20	25	32

(c) $\theta_{18}\theta_{19}\theta_{20}$

θ's $\neq 0$	222	223	224	233	234	244	333	334	344	444
Factor	.38	.47	.56	.54	.61	.66	.59	.64	.69	.78
λ	12	17	24	22	29	36	27	34	41	64

To see the effect of the shrinkage explicitly, suppose, for example, that the observation X_{20} corresponding to $\theta_{20} = 2$ turned out to be 2.71. The modified estimate ranges from $2.71 \times .17 = .46$ (when $\theta_1 = \cdots = \theta_{19} = 0$) to $2.71 \times .66 = 1.79$ (when $\theta_1 = \cdots = \theta_{17} = 0, \theta_{18} = \theta_{19} = 4$).

What is seen here can be summarized roughly as follows:

(i) If all the θ_i's are at or fairly close to zero, then the James-Stein estimator will reduce the $\mathbf{X}$'s very substantially in absolute value and thereby typically will greatly improve the accuracy of the estimated values.

(ii) If there are some very large θ_i's or a substantial number of moderate ones, the factor by which the $\mathbf{X}$'s are multiplied will not be very far from 1, and the modification will not have a great effect.

Neither of these situations causes much of a problem: In (ii), the modification presents an unnecessary but not particularly harmful complication; in (i), it is clearly very beneficial. The danger arises in the following intermediate situation.

(iii) Most of the θ_i's are close to zero, but there are a few moderately large θ_i's (of the order of two to four standard deviations, say). These represent the cases in which something is going on, about which we will usually want to know. However, in these cases, the estimated values are heavily shrunk toward the norm, with the resulting risk of their being found "innocent by association."

If one is interested in minimizing the average risk (5.30) but is concerned about the possibility of large component risks, a compromise is possible along the lines of restricted Bayes estimation mentioned in Section 5.2. One can impose an upper bound on the maximum component risk, say 10% or 25% above the minimax risk of 1 (when $\sigma = 1$). Subject to this restriction, one can then try to minimize the average risk, for example, in the sense of obtaining a Bayes or empirical Bayes solution. An approximation to such an approach has been developed by Efron and Morris (1971, 1972a), Berger (1982a, 1988b), Bickel (1983, 1984), and Kempthorne (1988a, 1988b). See Example 6.7 for an illustration.

The results discussed in this and the preceding section for the simultaneous estimation of normal means have been extended, particularly to various members of the exponential family and to general location parameter families, with and without nuisance parameters. The next section contains a number of illustrations.

6 Extensions

The estimators of the previous section have all been constructed for the case of the estimation of $\boldsymbol{\theta}$ based on observing $\mathbf{X} \sim N_r(\boldsymbol{\theta}, I)$. The applicability of shrinkage estimation, now often referred to as Stein estimation, goes far beyond this case. In this section, through examples, we will try to illustrate some of the wide ranging applicability of the "Stein effect," that is, the ability to improve individual estimates by using ensemble information.

Also, the shrinkage estimators previously considered were designed to obtain the greatest risk improvement in a specified region of the parameter space. For example, the maximum risk improvement of (5.10) occurs at $\theta_1 = \theta_2 = \cdots = \theta_r = 0$, while that of (4.21) occurs at $\theta_1 = \mu_1, \theta_2 = \mu_2, \ldots, \theta_r = \mu_r$. In the next three examples, we look at modifications of Stein estimators that shrink toward adaptively chosen targets, that is, targets selected by the data. By doing so, it is hoped that a maximal risk improvement will be realized.

Although we only touch upon the topic of selecting a shrinkage target, the literature is vast. See Note 9.7 for some references.

The first two examples examine estimators that we have seen before, in the context of empirical Bayes analysis of variance and regression (Examples 4.7.7 and 4.7.8). These estimators shrink toward subspaces of the parameter space rather than specified points. Moreover, we can allow the data to help choose the specific shrinkage target. We now establish minimaxity of such estimators.

Example 6.1 Shrinking toward a common mean. In problems where it is thought there is some similarity between components, a reasonable choice of shrinkage target may be the linear subspace where all the components are equal. This was illustrated in Example 4.7.7, where the estimator (4.7.28) shrunk the coordinates toward an estimated common mean value rather than a specified point.

For the average squared error loss $L(\boldsymbol{\theta}, \delta) = \frac{1}{r}|\boldsymbol{\theta} - \delta|^2$, the estimator (4.7.28) with coordinates

$$\delta_i^L(\mathbf{x}) = \bar{x} + \left(1 - \frac{c(r-3)}{\sum_j (x_j - \bar{x})^2}\right)(x_i - \bar{x}), \tag{6.1}$$

has risk given by

$$R(\boldsymbol{\theta}, \delta^L) = 1 + \frac{(r-3)^2}{r} E_\theta \left[\frac{c(c-2)}{\sum_j (X_j - \bar{X})^2} \right]. \tag{6.2}$$

Hence, δ^L is minimax if $r \geq 4$ and $c \leq 2$. The minimum risk is attained at $\boldsymbol{\theta}$ values that satisfy $\sum(\theta_i - \bar{\theta})^2 = 0$, that is, where $\theta_1 = \theta_2 = \cdots = \theta_r$. Moreover, the best value of c is $c = 1$, which results in a minimum risk of $3/r$. This is greater than the minimum of $2/r$ (for the case of a known value of θ) but is attained on a larger set. See Problems 6.1 and 6.2. ||

Example 6.2 Shrinking toward a linear subspace. The estimator δ^L given by (6.1) shrinks toward the subspace of the parameter space defined by

$$\mathcal{L} = \{\boldsymbol{\theta} : \theta_1 = \theta_2 = \cdots = \theta_r\} = \left\{ \boldsymbol{\theta} : \frac{1}{r} J\boldsymbol{\theta} = \boldsymbol{\theta} \right\} \tag{6.3}$$

where J is a matrix of 1's, $J = \mathbf{1}\mathbf{1}'$.

Another useful submodel, which is a generalization of (6.3), is

$$\theta_i = \alpha + \beta t_i \tag{6.4}$$

where the t_i's are known but α and β are unknown. This corresponds to the (sub)model of a linear trend in the means (see Example 4.7.8). If we define

$$T = \begin{pmatrix} 1 1 \cdots 1 \\ t_1 t_2 \cdots t_r \end{pmatrix}', \tag{6.5}$$

then the θ_i's satisfying (6.4) constitute the subspace

$$\mathcal{L} = \left\{ \boldsymbol{\theta} : T^* \boldsymbol{\theta} = \boldsymbol{\theta} \right\}, \tag{6.6}$$

where $T^* = T(T'T)^{-1}T'$ is the matrix that projects any vector $\boldsymbol{\theta}$ into the subspace. (Such projection matrices are symmetric and idempotent, that is, they satisfy $(T^*)^2 = I$.)

The models (6.3) and (6.6) suggest what the more general situation would look like when the target is a linear subspace defined by

$$\mathcal{L}_k = \{\boldsymbol{\theta} : K\boldsymbol{\theta} = \boldsymbol{\theta}, \quad K \text{ idempotent of rank } s\}. \tag{6.7}$$

If we shrink toward the MLE of $\boldsymbol{\theta} \in \mathcal{L}_k$, which is given by $\hat{\boldsymbol{\theta}}_k = K\mathbf{x}$, the resulting Stein estimator is

$$\delta^k(\mathbf{x}) = \hat{\boldsymbol{\theta}}_k + \left(1 - \frac{r-s-2}{|\mathbf{x} - \hat{\boldsymbol{\theta}}_k|^2} \right)(\mathbf{x} - \hat{\boldsymbol{\theta}}_k) \tag{6.8}$$

and is minimax provided $r - s > 2$. (See Problem 6.3.) More general linear restrictions are possible: one can take $\mathcal{L} = \{\boldsymbol{\theta} : H\boldsymbol{\theta} = \boldsymbol{m}\}$ where $H_{s\times r}$ and $\boldsymbol{m}_{s\times 1}$ are specified (see Casella and Hwang, 1987). ||

Example 6.3 Combining biased and unbiased estimators. Green and Strawderman (1991) show how the Stein effect can be used to combine biased and unbiased estimators, and they apply their results to a problem in forestry.

An important attribute of a forest stand (a homogeneous group of trees) is the *basal area per acre*, **B**, defined as the sum of the cross-sectional areas 4.5 feet above the ground of all trees. Regression models exist that predict log **B** as a function of stand age, number of trees per acre, and the average height of the dominant trees in the stand. The average prediction, **Y**, from the regression is a biased estimator of **B**. Green and Strawderman investigated how to combine this estimator with **X**, the sample mean basal area from a small sample of trees, to obtain an improved estimator of **B**. They formulated the problem in the following way.

Suppose $\mathbf{X} \sim N_r(\boldsymbol{\theta}, \sigma^2 I)$ and $\mathbf{Y} \sim N_r(\boldsymbol{\theta} + \boldsymbol{\xi}, \tau^2 I)$, independent, where σ^2 and τ^2 are known, and the loss function is $L(\boldsymbol{\theta}, \delta) = |\boldsymbol{\theta} - \delta|^2 / r\sigma^2$. Thus, $\boldsymbol{\xi}$ is an unknown nuisance parameter. The estimator

$$\delta^c(\mathbf{x}, \mathbf{y}) = \mathbf{y} + \left(1 - \frac{c(r-2)\sigma^2}{|\mathbf{x} - \mathbf{y}|^2}\right)(\mathbf{x} - \mathbf{y}) \tag{6.9}$$

is a minimax estimator of $\boldsymbol{\theta}$ if $0 \le c \le 2$, which follows from noting that

$$R(\boldsymbol{\theta}, \delta^c) = 1 - \sigma^2 \frac{(r-2)^2}{r} E \frac{c(2-c)}{|\mathbf{X} - \mathbf{Y}|^2} \tag{6.10}$$

and that the minimax risk is 1. If $\boldsymbol{\xi} = 0$ and, hence, **Y** is also an unbiased estimator of $\boldsymbol{\theta}$, then the optimal linear combined estimator

$$\delta^{\text{comb}}(\mathbf{x}, \mathbf{y}) = \frac{\tau^2 \mathbf{x} + \sigma^2 \mathbf{y}}{\sigma^2 + \tau^2} \tag{6.11}$$

dominates $\delta^1(\mathbf{x}, \mathbf{y})$ in risk. However, the risk of δ^{comb} becomes unbounded as $|\boldsymbol{\xi}| \to \infty$, whereas that of δ^1 is bounded by 1. (See Problems 6.5 and 6.6.) ||

The next example looks at the important case of unknown variance.

Example 6.4 Unknown variance. The James-Stein estimator (4.21) which shrinks **X** toward a given point $\boldsymbol{\mu}$ was obtained under the assumption that $\mathbf{X} \sim N(\boldsymbol{\theta}, I)$. We shall now generalize this estimator to the situations, first, that the common variance of the X_i has a known value σ^2 and then that σ^2 is unknown.

In the first case, the problem can be reduced to that with unit variance by considering the variables X_i/σ and estimating the means θ_i/σ and then multiplying the estimator of θ_i/σ by σ to obtain an estimator of θ_i. This argument leads to replacing (4.21) by

$$\delta_i = \mu_i + \left(1 - \frac{r-2}{|\mathbf{x} - \boldsymbol{\mu}|^2/\sigma^2}\right)(x_i - \mu_i), \tag{6.12}$$

where $|\mathbf{x} - \boldsymbol{\mu}|^2 = \Sigma(x_i - \mu_i)^2$, with risk function (see Problem 6.7)

$$R(\boldsymbol{\theta}, \delta) = \sigma^2 \left[1 - \frac{r-2}{r} E_\theta \left(\frac{r-2}{|\mathbf{X} - \boldsymbol{\mu}|^2/\sigma^2}\right)\right]. \tag{6.13}$$

Suppose now that σ^2 is unknown. We shall then suppose that there exists a random variable S^2, independent of **X** and such that $S^2/\sigma^2 \sim \chi^2_\nu$, and we shall in (6.12) replace σ^2 by $\hat{\sigma}^2 = kS^2$, where k is a positive constant. The estimator is then modified to

$$(6.14) \qquad \delta_i = \mu_i + \left(1 - \frac{r-2}{|\mathbf{x} - \mu|^2/\hat{\sigma}^2}\right)(x_i - \mu_i)$$
$$= \mu_i + \left(1 - \frac{r-2}{|\mathbf{x} - \mu|^2/\sigma^2} \cdot \frac{\hat{\sigma}^2}{\sigma^2}\right)(x_i - \mu_i).$$

The conditional risk of δ given $\hat{\sigma}$ is given by (5.4) with $|\mathbf{x} - \mu|^2$ replaced by $|\mathbf{x} - \mu|^2/\sigma^2$ and $c = \hat{\sigma}^2/\sigma^2$. Because of the independence of S^2 and $|\mathbf{x} - \mu|^2$, we thus have

$$(6.15) \qquad R(\boldsymbol{\theta}, \delta) = 1 - \frac{(r-2)^2}{r} E_\theta \left[\frac{\sigma^2}{|\mathbf{X} - \mu|^2}\right] E\left[2k\frac{S^2}{\sigma^2} - k^2\left(\frac{S^2}{\sigma^2}\right)^2\right].$$

Now, $E(S^2/\sigma^2) = \nu$ and $E(S^2/\sigma^2)^2 = \nu(\nu+2)$, so that the second expectation is

$$(6.16) \qquad E\left[2k\frac{S^2}{\sigma^2} - k^2\left(\frac{S^2}{\sigma^2}\right)^2\right] = 2k\nu - k^2\nu(\nu+2).$$

This is positive (making the estimator minimax) if $k \le 2/(\nu+2)$, and (6.16) is maximized at $k = 1/(\nu+2)$ where its value is $\nu/(\nu+2)$.

The best choice of k in $\hat{\sigma}^2$ thus leads to using the estimator

$$(6.17) \qquad \hat{\sigma}^2 = S^2/(\nu+2)$$

and the risk of the resulting estimator is

$$(6.18) \qquad R(\boldsymbol{\theta}, \delta) = 1 - \frac{\nu}{\nu+2}\frac{(r-2)^2}{r} E_\theta \frac{\sigma^2}{|\mathbf{X} - \mu|^2}.$$

The improvement in risk over $\mathbf{X}$ is thus reduced from that of (4.25) by a factor of $\nu/(\nu+2)$. (See Problems 6.7–6.11.) ||

For distributions other than the normal, Strawderman (1974) determined minimax Stein estimators for the following situation.

Example 6.5 Mixture of normals. Suppose that, given σ, the vector $\mathbf{X}$ is distributed as $N(\boldsymbol{\theta}, \sigma^2 I)$, and that σ is a random variable with distribution G, so that the density of $\mathbf{X}$ is

$$(6.19) \qquad f(|\mathbf{x} - \boldsymbol{\theta}|) = \frac{1}{(2\pi)^{r/2}} \int_0^\infty e^{-(1/2\sigma^2)|\mathbf{x}-\boldsymbol{\theta}|^2} \sigma^{-r}\, dG(\sigma),$$

a scale mixture of normals, including, in particular, the multivariate Student's t-distribution. Since $E(|\mathbf{X} - \boldsymbol{\theta}|^2 \mid \sigma) = r\sigma^2$, it follows that with loss function $L(\boldsymbol{\theta}, \delta) = |\boldsymbol{\theta} - \delta|^2/r$, the risk of the estimator $\mathbf{X}$ is $E(\sigma^2)$. On the other hand, the risk of the estimator

$$(6.20) \qquad \delta(\mathbf{x}) = \left(1 - \frac{c}{|\mathbf{x}|^2}\right)\mathbf{x}$$

is given by

$$(6.21) \qquad E_\theta L[\boldsymbol{\theta}, \delta(\mathbf{X})] = E_\sigma E_{\theta|\sigma} L[\boldsymbol{\theta}, \delta(\mathbf{X})].$$

Calculations similar to those in the proofs of Theorems 4.7.2 and 5.1 show that

$$E_{\theta|\sigma} L[\theta, \delta(\mathbf{X})] = \sigma^2 - \frac{1}{r}\left[2c(r-2) - \frac{c^2}{\sigma^2}\right] E_{\theta|\sigma}\left(\frac{\sigma^2}{|\mathbf{X}|^2}\right)$$

and, hence,

$$R(\theta, \delta) = E_\sigma \sigma^2 - \frac{1}{r} E_\sigma \left\{\left[2c(r-2) - \frac{c^2}{\sigma^2}\right]\left[E_{\theta|\sigma}\left(\frac{\sigma^2}{|\mathbf{X}|^2}\right)\right]\right\}. \tag{6.22}$$

An upper bound on the risk can be obtained from the following lemma, whose proof is left to Problem 6.13.

Lemma 6.6 *Let Y be a random variable, and $g(y)$ and $h(y)$ any functions for which $E[g(Y)]$, $E[(h(Y)]$, and $E(g(Y)h(Y)]$ exist. Then:*

(a) If one of the functions $g(\cdot)$ and $h(\cdot)$ is nonincreasing and the other is nondecreasing,

$$E[g(Y)h(Y)] \le E[g(Y)]E[h(Y)].$$

(b) If both functions are either nondecreasing or nonincreasing,

$$E[g(Y)h(Y)] \ge E[g(Y)]E[h(Y)].$$

Returning to the risk function (6.22), we see that $[2c(r-2)-c^2/\sigma^2]$ is an increasing function of σ^2, and $E_{\theta|\sigma}(\sigma^2/|\mathbf{X}|^2)$ is also an increasing function of σ^2. (This latter statement follows from the fact that, given σ^2, $|\mathbf{X}|^2/\sigma^2$ has a noncentral χ^2-distribution with noncentrality parameter $|\theta|^2/\sigma^2$, and that, therefore, as was pointed out following (4.25), the expectation is increasing in σ^2.)

Therefore, by Lemma 6.6,

$$E_\sigma\left\{\left[2c(r-2) - \frac{c^2}{\sigma^2}\right] E_{\theta|\sigma}\left(\frac{\sigma^2}{|\mathbf{X}|^2}\right)\right\}$$
$$\ge E_\sigma\left[2c(r-2) - \frac{c^2}{\sigma^2}\right] E_\sigma \frac{\sigma^2}{|\mathbf{X}|^2}$$

Hence, $\delta(\mathbf{x})$ will dominate $\mathbf{x}$ if

$$2c(r-2) - c^2 E_\sigma \frac{1}{\sigma^2} \ge 0$$

or

$$0 < c < 2(r-2)/E_\sigma \frac{1}{\sigma^2} = 2/E_0|\mathbf{X}|^{-2},$$

where $E_0|\mathbf{X}|^{-2}$ is the expectation when $\theta = 0$ (see Problem 6.12).

If $f(|\mathbf{x} - \theta|)$ is the normal density $N(\theta, I)$, then $E_0|\mathbf{X}|^{-2} = (r-2)^{-1}$, and we are back to a familiar condition. The interesting fact is that, for a wide class of scale mixtures of normals, $E_0|\mathbf{X}|^{-2} > (r-2)^{-1}$. This holds, for example, if $1/\sigma^2 \sim \chi^2_\nu/\nu$ so $f(|\mathbf{x} - \theta|)$ is multivariate Student's t. This implies a type of robustness of the estimator (6.20); that is, for $0 \le c \le 2(r-2)$, $\delta(\mathbf{X})$ dominates

X under a multivariate t-distribution and hence retains its minimax property (see Problem 6.12). ‖

Bayes estimators minimize the average risk under the prior, but the maximum of their risk functions can be large and even infinite. On the other hand, minimax estimators often have relatively large Bayes risk under many priors. The following example, due to Berger (1982a) , shows how it is sometimes possible to construct estimators having good Bayes risk properties (with respect to a given prior), while at the same time being minimax. The resulting estimator is a compromise between a Bayes estimator and a Stein estimator.

Example 6.7 Bayesian robustness. For $\mathbf{X} \sim N_r(\boldsymbol{\theta}, \sigma^2 I)$ and $\boldsymbol{\theta} \sim \pi = N_r(0, \tau^2 I)$, the Bayes estimator against squared error loss is

$$\delta^\pi(\mathbf{x}) = \frac{\tau^2}{\sigma^2 + \tau^2}\mathbf{x} \tag{6.23}$$

with Bayes risk $r(\pi, \delta^\pi) = r\sigma^2\tau^2/(\sigma^2 + \tau^2)$. However, δ^π is not minimax and, in fact, has unbounded risk (Problem 4.3.12). The Stein estimator

$$\delta^c(\mathbf{x}) = \left(1 - c\frac{(r-2)\sigma^2}{|\mathbf{x}|^2}\right)\mathbf{x} \tag{6.24}$$

is minimax if $0 \le c \le 2$, but its Bayes risk

$$r(\pi, \delta^c) = r(\pi, \delta^\pi) + \frac{\sigma^4}{\sigma^2 + \tau^2}[r + c(c-2)(r-2)],$$

at the best value $c = 1$, is $r(\pi, \delta') = r(\pi, \delta^\pi) + 2\sigma^4/(\sigma^2 + \tau^2)$.

To construct a minimax estimator with small Bayes risk, consider the compromise estimator

$$\delta^R(\mathbf{x}) = \begin{cases} \delta^\pi(\mathbf{x}) & \text{if } |\mathbf{x}|^2 < c(r-2)(\sigma^2 + \tau^2) \\ \delta^c(\mathbf{x}) & \text{if } |\mathbf{x}|^2 \ge c(r-2)(\sigma^2 + \tau^2). \end{cases} \tag{6.25}$$

This estimator is minimax if $0 \le c \le 2$ (Problem 6.14). If $|\mathbf{x}|^2 > c(r-2)(\sigma^2+\tau^2)$, the data do not support the prior specification and we, therefore, put $\delta^R = \delta^c$; if $|\mathbf{x}|^2 < c(r-2)(\sigma^2 + \tau^2)$, we tend to believe that the data support the prior specification since $\frac{|\mathbf{x}|^2}{\sigma^2+\tau^2} \sim \chi^2_r$, and we are, therefore, willing to gamble on π and put $\delta^R = \delta^\pi$.

The Bayes risk of δ^R is

$$\begin{aligned} r(\pi, \delta^R) = {} & E|\boldsymbol{\theta} - \delta^\pi(\mathbf{X})|^2 I\left(|\mathbf{X}|^2 < c(r-2)(\sigma^2 + \tau^2)\right) \\ & + E|\boldsymbol{\theta} - \delta^c(\mathbf{X})|^2 I\left(|\mathbf{X}|^2 \ge c(r-2)(\sigma^2 + \tau^2)\right), \end{aligned} \tag{6.26}$$

where the expectation is over the joint distribution of **X** and $\boldsymbol{\theta}$. Adding $\pm\delta^\pi$ to the second term in (6.26) yields

$$\begin{aligned} r(\pi, \delta^R) &= E|\boldsymbol{\theta} - \delta^\pi(\mathbf{X})|^2 \\ & \quad + E|\delta^\pi(\mathbf{X}) - \delta^c(\mathbf{X})|^2 I\left(|\mathbf{X}|^2 \ge c(r-2)(\sigma^2 + \tau^2)\right) \\ &= r(\pi, \delta^\pi) + E|\delta^\pi(\mathbf{X}) - \delta^c(\mathbf{X})|^2 I\left(|\mathbf{X}|^2 \ge c(r-2)(\sigma^2 + \tau^2)\right). \end{aligned} \tag{6.27}$$

Here, we have used the fact that, marginally, $|\mathbf{X}|^2/(\sigma^2+\tau^2) \sim \chi_r^2$. We can write (6.27) as (see Problem 6.14)

$$r(\pi, \delta^R) = r(\pi, \delta^\pi)$$

$$+\frac{1}{r-2}\frac{\sigma^4}{\sigma^2+\tau^2}E[Y - c(r-2)]^2 I[Y > c(r-2)], \tag{6.28}$$

where $Y \sim \chi_{r-2}^2$. An upper bound on (6.28) is obtained by dropping the indicator function, which gives

$$\begin{aligned} r(\pi, \delta^R) &\le r(\pi, \delta^\pi) + \frac{1}{r-2}\frac{\sigma^4}{\sigma^2+\tau^2}E[Y - c(r-2)]^2 \\ &= r(\pi, \delta^\pi) + [r + c(c-2)(r-2)]\frac{\sigma^4}{\sigma^2+\tau^2} \qquad (6.29)\\ &= r(\pi, \delta^c). \end{aligned}$$

This shows that δ^R has smaller Bayes risk than δ^c while remaining minimax.

Since $E(Y-a)^2 I(Y > a)$ is a decreasing function of a (Problem 6.14), the value $c = 2$ minimizes (6.28) and therefore, among the estimators (6.25), determines the minimax estimator with minimum Bayes risk. However, for $c = 2$, δ^c has the same (constant) risk as X, so we are trading optimal Bayes risk for minimal frequentist risk improvement over $\mathbf{X}$, the constant risk minimax estimator. Thus, it may be better to choose $c = 1$, which gives optimal frequentist risk performance and still provides good Bayes risk reduction over δ'. Table 6.1 shows the relative Bayes savings

$$r^* = \frac{r(\pi, \delta^c) - r(\pi, \delta^\pi)}{r(\pi, \delta^c)}$$

for $c = 1$.

Table 6.1. *Values of r^*, the Relative Bayes Risk Savings of δ^R over δ^c, with $c = 1$*

r	3	4	5	7	10	20
r^*	.801	.736	.699	.660	.629	.587

For other approaches to this "compromise" decision problem, see Bickel (1983, 1984), Kempthorne (1988a, 1988b), and DasGupta and Rubin (1988). ‖

Thus far, we have considered only continuous distributions, but the Stein effect continues to hold also in discrete families. Minimax proofs in discrete families have developed along two different lines. The first method, due to Clevenson and Zidek (1975), is illustrated by the following result.

Theorem 6.8 *Let $X_i \sim$ Poisson(λ_i), $i = 1, \ldots, r$, $r \ge 2$, be independent, and let the loss be given by*

$$L(\boldsymbol{\lambda}, \delta) = \sum_{i=1}^r (\lambda_i - \delta_i)^2/\lambda_i. \tag{6.30}$$

The estimator

(6.31) $$\delta_{cz}(\mathbf{x}) = \left(1 - \frac{c(\Sigma x_i)}{\Sigma x_i + b}\right)\mathbf{x}$$

is minimax if

(i) $c(\cdot)$ *is nondecreasing,*

(ii) $0 \le c(\cdot) \le 2(r-1)$,

(iii) $b \ge r - 1$.

Recall (Corollary 2.20) that the usual minimax estimator here is $\mathbf{X}$, with constant risk r. Note also that, in contrast to the normal-squared-error-loss case, by (ii) there exist positive values of c for which δ_{cz} is minimax provided $r \ge 2$.

Proof. If $Z = \Sigma X_i$, the risk of δ_{cz} can be written as

$$R(\boldsymbol{\lambda}, \delta_{cz}) = E\left[\sum_{i=1}^{r} \frac{1}{\lambda_i}\left(\lambda_i - X_i + \frac{c(Z)X_i}{Z+b}\right)^2\right]$$

(6.32) $$= r + 2E\left\{\frac{c(Z)}{Z+b}\sum_{i=1}^{r} X_i(\lambda_i - X_i)\right\}$$

$$+E\left\{\frac{c^2(Z)}{(Z+b)^2}\sum_{i=1}^{r} X_i^2\right\}.$$

Let us first evaluate the expectations conditional on Z. The distribution of $X_i|Z$ is multinomial with $E(X_i|Z) = Z(\lambda_i/\Lambda)$ and $\text{var}(X_i|Z) = Z(\lambda_i/\Lambda)(1 - \lambda_i/\Lambda)$, where $\Lambda = \Sigma\lambda_i$. Hence,

(6.33) $$E\left[\sum_{i=1}^{r} X_i(\lambda_i - X_i)|Z\right] = \frac{Z}{\Lambda}[\Lambda - (Z + r - 1)]$$

$$E\left[\sum_{i=1}^{r} X_i^2|Z\right] = \frac{Z}{\Lambda}(Z + r - 1),$$

and, so, after some rearrangement of terms,

$$R(\boldsymbol{\lambda}, \delta_{cz}) = r + E\left\{\frac{c(Z)Z}{\Lambda(Z+b)}\left[2(\Lambda - Z) - 2(r-1) + c(Z)\frac{Z+r-1}{Z+b}\right]\right\}.$$

(6.34)

Now, if $b \ge r-1$, $z + r - 1 < z + b$, and $c(z) < 2(r-1)$, we have

$$-2(r-1) + c(z)\frac{z+r-1}{z+b} \le -2(r-1) + c(z) \le 0,$$

so the risk of δ_{cz} is bounded above by

$$R(\boldsymbol{\lambda}, \delta_{cz}) \le r + 2E\left[\left(\frac{c(Z)Z}{\Lambda(Z+b)}\right)(\Lambda - Z)\right].$$

But this last expectation is the product of an increasing and a decreasing function of z; hence, by Lemma 6.6,

(6.35) $$E\left(\frac{c(Z)Z}{\Lambda(Z+b)}\right)(\Lambda - Z) \le E\frac{c(Z)Z}{\Lambda(Z+b)}E(\Lambda - z) = 0,$$

since $Z \sim \text{Poisson}(\Lambda)$. Hence, $R(\boldsymbol{\lambda}, \delta_{cz}) \le r$ and δ_{cz} is minimax. □

If we recall Example 4.6.6 [in particular, Equation (4.6.29)], we see a similarity between δ_{cz} and the hierarchical Bayes estimators derived there. It is interesting to note that δ_{cz} is also a Bayes estimator (Clevenson and Zidek 1975; see Problem 6.15).

The above method of proof, which relies on being able to evaluate the conditional distribution of $X_i|\Sigma X_i$ and the marginal distribution of $\sum X_i$, works for other discrete families, in particular the negative binomial and the binomial (where n is the parameter to be estimated). (See Problem 6.16.) However, there exists a more powerful method (similar to that of Stein's lemma) which is based on the following lemma due to Hudson (1978) and Hwang (1982a). The proof is left to Problem 6.17.

Lemma 6.9 *Let X_i, $i = 1, \ldots, r$, be independent with probabilities*

$$p_i(x|\theta_i) = c_i(\theta_i)h_i(x)\theta_i^x, \quad x = 0, 1, \ldots, \tag{6.36}$$

that is, $p_i(x|\theta_i)$ is in the exponential family. Then, for any real-valued function $g(\mathbf{x})$ with $E_\theta|g(\mathbf{X})| < \infty$, and any number m for which $g(\mathbf{x}) = 0$ when $x + i < m$,

$$E_\theta \theta_i^m g(\mathbf{X}) = E_\theta \left\{ g(\mathbf{X} - m\mathbf{e}_i) \frac{h_i(X_i - m)}{h_i(X_i)} \right\} \tag{6.37}$$

where $\mathbf{e}_i$ is the unit vector with ith coordinate equal to 1 and the rest equal to 0.

The principal application of Lemma 6.9 is to find an unbiased estimator of the risk of estimators of the form $\mathbf{X} + g(\mathbf{X})$, analogous to that of Corollary 4.7.2.

Theorem 6.10 *Let $X_1, \ldots, X_r$ be independently distributed according to (6.36), and let $\delta^0(\mathbf{x}) = \{h_i(x_i - 1)/h_i(x_i)\}$ [the estimator whose ith coordinate is $h_i(x_i - 1)/h_i(x_i)$] be the UMVU estimator of $\boldsymbol{\theta}$. For the loss function*

$$L_m(\boldsymbol{\theta}, \delta) = \sum_{i=1}^r \theta_i^{m_i}(\theta_i - \delta_i)^2, \tag{6.38}$$

where $\mathbf{m} = (m_1, \ldots, m_r)$ are known numbers, the risk of the estimator $\delta(\mathbf{x}) = \delta^0(\mathbf{x}) + g(\mathbf{x})$ is given by

$$R(\boldsymbol{\theta}, \delta) = R(\boldsymbol{\theta}, \delta^0) + E_\theta \mathcal{D}(\mathbf{x}) \tag{6.39}$$

with

$$\begin{aligned} \mathcal{D}(\mathbf{x}) = \sum_{i=1}^r \Bigg\{ & \frac{2h_i(x_i - m_i - 1)}{h_i(x_i)}[g_i(\mathbf{x} - m_i\mathbf{e}_i - \mathbf{e}_i) - g_i(\mathbf{x} - m_i\mathbf{e}_i)] \\ & + \frac{h_i(x_i - m_i)}{h_i(x_i)} g_i^2(\mathbf{x} - m_i\mathbf{e}_i) \Bigg\}. \end{aligned} \tag{6.40}$$

Proof. Write

$$R(\boldsymbol{\theta}, \delta) = R(\boldsymbol{\theta}, \delta^0) - 2E_\theta \left\{ \sum_{i=1}^r \theta_i^m g_i(\mathbf{X}) \left(\theta_i - \frac{h_i(X_i - 1)}{h_i(X_i)} \right) \right\}$$

$$+E_\theta \left\{ \sum_{i=1}^{r} \theta_i^m g_i^2(\mathbf{X}) \right\}.$$

and apply Lemma 6.9 (see Problem 6.17.) □

Hwang (1982a) established general conditions on $g(\mathbf{x})$ for which $\mathcal{D}(\mathbf{x}) \leq 0$, leading to improved estimators of $\boldsymbol{\theta}$ (see also Ghosh et al. 1983, Tsui 1979a, 1979b, and Hwang 1982b). We will only look at some examples.

Example 6.11 Improved estimation for independent Poissons. The Clevenson-Zidek estimator (6.31) dominates **X** (and is minimax) for the loss $L_{-1}(\theta, \delta)$ of (6.38); however, Theorem 6.8 does not cover squared error loss, $L_0(\boldsymbol{\theta}, \delta)$. For this loss, if $X_i \sim \text{Poisson}(\lambda_i)$, independent, the risk of an estimator $\delta(\mathbf{x}) = \mathbf{x} + g(\mathbf{x})$ is given by (6.39) with $\delta^0 = \mathbf{x}$ and

$$\mathcal{D}(\mathbf{x}) = \sum_{i=1}^{r} \left\{ 2x_i[g_i(\mathbf{x}) - g_i(\mathbf{x} - \mathbf{e}_i)] + g_i^2(\mathbf{x}) \right\}. \tag{6.41}$$

The estimator with

$$g_i(\mathbf{x}) = \frac{c(\mathbf{x})k(x_i)}{\sum_{j=1}^{r} k(x_j)k(x_j+1)}, \quad k(x) = \sum_{l=1}^{x} \frac{1}{l}, \tag{6.42}$$

and $c(\mathbf{x})$ nondecreasing in each coordinate with

$$0 \leq c(\mathbf{x}) \leq 2[\#(x_i's > 1) - 2]$$

satisfies $\mathcal{D}(\mathbf{x}) \leq 0$ and hence dominates **x** under L_0. (The notation $\#(a_is > b)$ denotes the number of a_i s that are greater than b.)

For the loss function $L_{-1}(\boldsymbol{\theta}, \delta)$, the situation is somewhat easier, and the estimator $\mathbf{x} + g(\mathbf{x})$, with

$$g_i(\mathbf{x}) = c(\mathbf{x} - \mathbf{e}_i)\frac{x_i}{\sum_{j=1}^{r} x_i}, \tag{6.43}$$

where $c(\cdot)$ is nondecreasing with $0 \leq c(\cdot) \leq 2(p-1)$, will satisfy $\mathcal{D}(\mathbf{x}) \leq 0$ and, hence, is minimax for L_{-1}. Note that (6.43) includes the Clevenson-Zidek estimator as a special case. (See Problem 6.18.) ‖

As might be expected, these improved estimators, which shrink toward 0, perform best and give the greatest risk improvement, when the θ_i's are close to zero and, more generally, when they are close together. Numerical studies (Clevenson and Zidek 1975, Tsui, 1979a, 1979b, Hudson and Tsui, 1981) quantify this improvement, which can be substantial. Other estimators, which shrink toward other targets in the parameter space, can optimize the region of greatest risk reduction (see, for example, Ghosh et al. 1983, Hudson 1985).

Just as the minimaxity of Stein estimators carried over from the normal distribution to mixtures of normals, minimaxity carries over from the Poisson to mixtures of Poissons, for example, the negative binomial distribution (see Example 4.6.6).

Example 6.12 Improved negative binomial estimation. For $X_1, \ldots, X_r$ independent negative binomial random variables with distribution

$$p_i(x|\theta_i) = \binom{t_i + x - 1}{x} \theta_i^x (1-\theta_i)^{t_i}, \quad x = 0, 1, \ldots \tag{6.44}$$

the UMVU estimator of θ_i is $\delta_i^0(x_i) = x_i/(x_i + t_i - 1)$ (where $\delta_i^0(0) = 0$ for $t_i = 1$). Using Theorem 6.10, this estimator can be improved. For example, for the loss $L_{-1}(\boldsymbol{\theta}, \delta)$ of (6.38), the estimator $\delta_0(\mathbf{x}) + g(\mathbf{x})$, with

$$g_i(\mathbf{x}) = \frac{c(\mathbf{x} - \mathbf{e}_i)x_i}{\sum_{j=1}^r (x_j^2 + x_j)} \tag{6.45}$$

and $c(\cdot)$ nondecreasing with $0 \le c \le 2(r-2)/\min_i\{t_i\}$, satisfies $\mathcal{D}(\mathbf{x}) \le 0$ and, hence, has uniformly smaller risk than $\delta^0(\mathbf{x})$. Similar results can be obtained for other loss functions (see Problem 6.19). Surprisingly, however, similar domination results do not hold for the MLE $\hat{\theta}_i = x_i/(x_i + t_i)$. Chow (1990) has shown that the MLE is admissible in all dimensions (see also Example 7.14). ||

Finally, we turn to a situation where the Stein effect fails to yield improved estimators.

Example 6.13 Multivariate binomial. For $X_i \sim b(\theta_i, n_i)$, $i = 1, \ldots, r$, independent, that is, with distribution

$$p_i(x|\theta_i) = \binom{n_i}{x} \theta_i^x (1-\theta_i)^{n_i - x}, \quad x = 0, 1, \ldots, n_i, \tag{6.46}$$

it seems reasonable to expect that estimators of the form $\mathbf{x}+g(\mathbf{x})$ exist that dominate the UMVU estimator $\mathbf{x}$. This expectation is partially based on the fact that (6.46) is a discrete exponential family. However, Johnson (1971) showed that such estimators do not exist in the binomial problem for squared error loss (see Example 7.23 and Problem 7.28).

Theorem 6.14 *If $k_i(\theta_i)$, $i = 1, \ldots, r$, are continuous functions and $\delta_i(x_i)$ is an admissible estimator of $k_i(\theta_i)$ under squared error loss, then $(\delta_1(x_1), \ldots, \delta_r(x_r))$ is an admissible estimator of $(k_1(\theta_1), \ldots, k_r(\theta_r))$ under sum-of-squared-error loss.*

Thus, there is no "Stein effect" in the binomial problem. In particular, as X_i is an admissible estimator of θ_i under squared error loss (Example 2.16), $\mathbf{X}$ is an admissible estimator of $\boldsymbol{\theta}$. ||

It turns out that the absence of the Stein effect is not a property of the binomial distribution, but rather a result of the finiteness of the sample space (Gutmann 1982a; see also Brown 1981). See Note 9.7 for further discussion.

7 Admissibility and Complete Classes

In Section 1.7, we defined the admissibility of an estimator which can be formally stated as follows.

Definition 7.1 An estimator $\delta = \delta(\mathbf{X})$ of $\boldsymbol{\theta}$ is *admissible* [with respect to the loss function $L(\boldsymbol{\theta}, \delta)$] if there exists no estimator δ' that satisfies

$$\text{(7.1)} \qquad \begin{aligned} &\text{(i)} \quad R(\boldsymbol{\theta}, \delta') \le R(\boldsymbol{\theta}, \delta) \quad \text{for all } \boldsymbol{\theta}, \\ &\text{(ii)} \quad R(\boldsymbol{\theta}, \delta') < R(\boldsymbol{\theta}, \delta) \quad \text{for some } \boldsymbol{\theta}, \end{aligned}$$

where $R(\boldsymbol{\theta}, \delta) = E_{\boldsymbol{\theta}} L(\boldsymbol{\theta}, \delta)$. If such an estimator δ' exists, then δ is *inadmissible*. When a pair of estimators δ and δ' satisfy (7.1), δ' is said to *dominate* δ.

Although admissibility is a desirable property, it is a very weak requirement. This is illustrated by Example 2.23, where an admissible estimator was completely unreasonable since it used no information from the relevant distribution. Here is another example (from Makani 1977, who credits it to L.D. Brown).

Example 7.2 Unreasonable admissible estimator. Let X_1 and X_2 be independent random variables, X_i distributed as $N(\theta_i, 1)$, and consider the estimation of θ_1 with loss function $L((\theta_1, \theta_2), \delta) = (\theta_1 - \delta)^2$. Then, $\delta = \text{sign}(X_2)$ is an admissible estimator of θ_1, although its distribution does not depend on θ_1. The result is established by showing that δ cannot be simultaneously beaten at $(\theta_1, \theta_2) = (1, \theta_2)$ and $(-1, \theta_2)$. (See Problem 7.1.) ||

Conversely, there exist inadmissible decision rules that perform quite well.

Example 7.3 The positive-part Stein estimator. For $X \sim N_r(\boldsymbol{\theta}, I)$, the positive-part Stein estimator

$$\delta^+(\mathbf{x}) = \left(1 - \frac{r-2}{|\mathbf{x}|^2}\right)^+ \mathbf{x} \tag{7.2}$$

is a good estimator of $\boldsymbol{\theta}$ under squared error loss, being both difficult to improve upon and difficult to dominate. However, as was pointed out by Baranchik (1964), it is not admissible. (This follows from Theorem 7.17, as δ^+ is not smooth enough to be a generalized Bayes estimator.) Thus, there exists an estimator that uniformly dominates it.

How much better can such a dominating estimator be? Efron and Morris (1973a, Section 5) show that δ^+ is "close" to a Bayes rule (Problem 7.2). Brown (1988; see also Moore and Brook 1978) writing

$$\begin{aligned} R(\boldsymbol{\theta}, \delta^+) &= E_\theta\left[\frac{1}{r}\sum_{i=1}^r (\theta_i - \delta_i(\mathbf{X}))^2\right] \\ &= E_\theta \mathcal{D}_{\delta^+}(\mathbf{X}), \end{aligned} \tag{7.3}$$

where

$$\mathcal{D}_{\delta^+}(\mathbf{x}) = 1 + \frac{m^2(\mathbf{x})|\mathbf{x}|^2}{r} - \frac{2}{r}\{(r-2)m(\mathbf{x}) + 2I[m(\mathbf{x}) = 1]\}$$

with $m(\mathbf{x}) = \min\{1, c(r-2)/|x|^2\}$ (see Corollary 4.7.2), proves that no estimator δ exists for which $\mathcal{D}_\delta(\mathbf{x}) \le \mathcal{D}_{\delta^+}(\mathbf{x})$ for all $\mathbf{x}$. These observations imply that the inadmissible δ^+ behaves similar to a Bayes rule and has a risk that is close to that of an admissible estimator. ||

However, since admissibility generally is a desirable property, it is of interest to determine the totality of admissible estimators.

Definition 7.4 A class of $\mathcal{C}$ of estimators is *complete* if for any δ not in $\mathcal{C}$ there exists an estimator δ' in $\mathcal{C}$ such that (7.1) holds; $\mathcal{C}$ is *essentially complete* if for any δ not in $\mathcal{C}$ there exists an estimator δ' in $\mathcal{C}$ such that (7.1)(i) holds.

It follows from this definition that any estimator outside a complete class is inadmissible. If $\mathcal{C}$ is essentially complete, an estimator δ outside of $\mathcal{C}$ may be admissible, but there will then exist an estimator δ' in $\mathcal{C}$ with the same risk function.

It is therefore reasonable, in the search for an optimal estimator, to restrict attention to a complete or essentially complete class. The following result provides two examples of such classes.

Lemma 7.5

(i) If C is the class of all (including randomized) estimators based on a sufficient statistic, then C is essentially complete.

(ii) If the loss function $L(\theta, d)$ is convex in d, then the class of nonrandomized estimators is complete.

Proof. These results are immediate consequences of Theorem 1.6.1 and Corollary 1.7.9. □

Although a complete class contains all admissible estimators, it may also contain many inadmissible ones. (This is, for example, the case for the two complete classes of Lemma 7.5.) A complete class is most useful if it is as small as possible.

Definition 7.6 A complete class C of estimators is *minimal* complete if no proper subset of C is complete.

Lemma 7.7 *If a minimal complete class C exists, then it is exactly the class of all admissible estimators.*

Proof. It is clear that C contains all admissible rules, so we only need to prove that it cannot contain any inadmissible ones. Let $\delta \in C$ and suppose that δ is inadmissible. Then, there is a $\delta' \in C$ that dominates it, and, hence, the class $C \setminus \{\delta\}$ (C with the estimator δ removed) is a complete class. This contradicts the fact that C is minimal complete. □

Note that Lemma 7.7 requires the existence of a minimal complete class. The following example illustrates the possibility that a minimal complete class may not exist. (For another example, see Blackwell and Girshick 1954, Problem 5.2.1.)

Example 7.8 Nonexistence of a minimal complete class. Let X be normally distributed as $N(\theta, 1)$ and consider the problem of estimating θ with loss function

$$L(\theta, d) = \begin{cases} d-\theta & \text{if } \theta < d \\ 0 & \text{if } \theta \geq d. \end{cases}$$

Then, if $\delta(x) \leq \delta'(x)$ for all x, we have $R(\theta, \delta) \leq R(\theta, \delta')$ with strict inequality if $P_\theta[\delta(X) \leq \delta'(X)] > 0$.

Many complete classes exist in this situation. For example, if δ_0 is any estimator of θ, then the class of all estimators with $\delta(x) < \delta_0(x)$ for some x is complete (Problem 7.4). We shall now show that there exists no minimal complete class. Suppose C is minimal complete and δ_0 is any member of C. Then, some estimator δ_1 dominating δ_0 must also lie in C. If not, there would be no members of C left to dominate such estimators and C would not be complete. On the other hand, if δ_1 dominates δ_0, and δ_1 and δ_0 are both in C, the class C is not minimal since δ_0 could be removed without disturbing completeness. ‖

Despite this example, the minimal complete class typically coincides with the class of admissible estimators, and the search for a minimal complete class is therefore equivalent to the determination of all admissible estimators. The following results are concerned with these two related aspects, admissibility and completeness, and with the relation of both to Bayes estimators.

Theorem 2.4 showed that any unique Bayes estimator is admissible. The following result replaces the uniqueness assumption by some other conditions.

Theorem 7.9 *For a possibly vector-valued parameter* $\boldsymbol{\theta}$, *suppose that* δ^π *is a Bayes estimator having finite Bayes risk with respect to a prior density* π *which is positive for all* $\boldsymbol{\theta}$, *and that the risk function of every estimator* δ *is a continuous function of* $\boldsymbol{\theta}$. *Then,* δ^π *is admissible.*

Proof. If δ^π is not admissible, there exists an estimator δ such that

$$R(\boldsymbol{\theta}, \delta) \le R(\boldsymbol{\theta}, \delta^\pi) \text{ for all } \boldsymbol{\theta}$$

and

$$R(\boldsymbol{\theta}, \delta) < R(\boldsymbol{\theta}, \delta^\pi) \text{ for some } \boldsymbol{\theta}.$$

It then follows from the continuity of the risk functions that $R(\boldsymbol{\theta}, \delta) < R(\boldsymbol{\theta}, \delta^\pi)$ for all $\boldsymbol{\theta}$ in some open subset Ω_0 of the parameter space and hence that

$$\int R(\boldsymbol{\theta}, \delta)\pi(\boldsymbol{\theta})d\boldsymbol{\theta} < \int R(\boldsymbol{\theta}, \delta^\pi)\pi(\boldsymbol{\theta})d\boldsymbol{\theta},$$

which contradicts the definition of δ^π. □

A basic assumption in this theorem is the continuity of all risk functions. The following example provides an important class of situations for which this assumptions holds.

Example 7.10 Exponential families have continuous risks. Suppose that we let $p(x|\eta)$ be the exponential family of (5.2). Then, it follows from Theorem 1.5.8 that for any loss function $L(\eta, \delta)$ for which $R(\eta, \delta) = E_\eta L(\eta, \delta)$ is finite, $R(\eta, \delta)$ is continuous. (See Problem 7.6.) ||

There are many characterizations of problems in which all risk functions are continuous. With assumptions on both the loss function and the density, theorems can be established to assert the continuity of risks. (See Problem 7.7 for a set of conditions involving boundedness of the loss function.) The following theorem, which we present without proof, is based on a set of assumptions that are often satisfied in practice.

Theorem 7.11 *Consider the estimation of* θ *with loss* $L(\theta, \delta)$, *where* $X \sim f(x|\theta)$ *has monotone likelihood ratio and is continuous in* θ *for each* x. *If the loss function* $L(\theta, \delta)$ *satisfies*

(i) $L(\theta, \delta)$ *is continuous in* θ *for each* δ,

(ii) L *is decreasing in* δ *for* $\delta < \theta$ *and increasing in* δ *for* $\delta > \theta$,

(iii) *there exist functions* a *and* b, *which are bounded on all bounded subsets of the parameter space, such that for all* δ

$$L(\theta, \delta) \le a(\theta, \theta')L(\theta', \delta) + b(\theta, \theta'),$$

then the estimators with finite-valued, continuous risk functions $R(\theta, \delta) = E_\theta L(\theta, \delta)$ form a complete class.

Theorems similar to Theorem 7.11 can be found in Ferguson (1967, Section 3.7, Brown 1986a, Berk, Brown, and Cohen 1981, Berger 1985, Section 8.8, or Robert 1994a, Section 6.2.1). Also see Problem 7.9 for another version of this theorem.

The assumptions on the loss are relatively simple to check. In fact, assumptions (i) and (ii) are almost self-evident, whereas (iii) will be satisfied by most interesting loss functions.

Example 7.12 Squared error loss. For $L(\theta, \delta) = (\theta - \delta)^2$, we have

$$(\theta - \delta)^2 = (\theta - \theta' + \theta' - \delta)^2 = (\theta - \theta')^2 + 2(\theta - \theta')(\theta' - \delta) + (\theta' - \delta)^2. \tag{7.4}$$

Now, since $2xy \le x^2 + y^2$,

$$2(\theta - \theta')(\theta' - \delta) \le (\theta - \theta')^2 + (\theta' - \delta)^2$$

and, hence,

$$(\theta - \delta)^2 \le 2(\theta - \theta')^2 + (\theta' - \delta)^2,$$

so condition (iii) is satisfied with $a(\theta, \theta') = 2$ and $b(\theta, \theta') = 2(\theta - \theta')^2$. ||

Since most problems that we will be interested in will satisfy the conditions of Theorem 7.11, we now only need consider estimators with finite-valued continuous risks. Restriction to continuous risk, in turn, allows us to utilize the method of proving admissibility that we previously saw in Example 2.8. (But note that this restriction can be relaxed somewhat; see Gajek 1983.) The following theorem extends the admissibility of Bayes estimators to sequences of Bayes estimators.

Theorem 7.13 (Blyth's Method) *Suppose that the parameter space $\Omega \in \Re^r$ is open, and estimators with continuous risk functions form a complete class. Let δ be an estimator with a continuous risk function, and let $\{\pi_n\}$ be a sequence of (possibly improper) prior measures such that*

(a) $r(\pi_n, \delta) < \infty$ for all n,

(b) for any nonempty open set $\Omega_0 \in \Omega$, there exist constants $B > 0$ and N such that

$$\int_{\Omega_0} \pi_n(\theta)\, d\theta \ge B \quad \textit{for all } n \ge N,$$

(c) $r(\pi_n, \delta) - r(\pi_n, \delta^{\pi_n}) \to 0$ as $n \to \infty$.

Then, δ is an admissible estimator.

Proof. Suppose δ is inadmissible, so that there exists δ' with $R(\theta, \delta') \le R(\theta, \delta)$, with strict inequality for some θ. By the continuity of the risk functions, this implies that there exists a set Ω_0 and $\varepsilon > 0$ such that $R(\theta, \delta) - R(\theta, \delta') > \varepsilon$ for $\theta \in \Omega_0$. Hence, for all $n \ge N$,

$$r(\pi_n, \delta) - r(\pi_n, \delta') > \varepsilon \int_{\Omega_0} \pi_n(\theta)\, d\theta \ge \varepsilon B \tag{7.5}$$

and therefore (c) cannot hold. □

Note that condition (b) prevents the possibility that as $n \to \infty$, all the mass of π_n escapes to ∞. This is similar to the requirement of *tightness* of a family of measures (see Chung 1974, Section 4.4, or Billingsley 1995, Section 25). [It is possible to combine conditions (b) and (c) into one condition involving a ratio (Problem 7.12) which is how Blyth's method was applied in Example 2.8.]

Example 7.14 Admissible negative binomial MLE. As we stated in Example 6.12 (but did not prove), the MLE of a negative binomial success probability is admissible under squared error loss. We can now prove this result using Theorem 7.13.

Let X have the negative binomial distribution

$$p(x|\theta) = \binom{r+x-1}{x} \theta^x (1-\theta)^r, \quad 0 < \theta < 1. \tag{7.6}$$

The ML estimator of θ is $\delta^0(x) = x/(x+r)$.

To use Blyth's method, we need a sequence of priors π for which the Bayes risks $r(\pi, \delta^\pi)$ get close to the Bayes risk of δ^0. When θ has the beta prior $\pi = B(a,b)$, the Bayes estimator is the posterior mean $\delta^\pi = (x+a)/(x+r+a+b)$. Since $\delta^\pi(x) \to \delta^0(x)$ as $a, b \to 0$, it is reasonable to try a sequence of priors $B(a,b)$ with $a, b \to 0$.

It is straightforward to calculate the posterior expected losses

$$\begin{aligned} E\left\{[\delta^\pi(x) - \theta]^2 | x\right\} &= \frac{(x+a)(r+b)}{(x+r+a+b)^2(x+r+a+b+1)} \\ E\left\{[\delta^0(x) - \theta]^2 | x\right\} &= \frac{(bx-ar)^2}{(x+r)^2(x+r+a+b)^2} + E\left\{[\delta^\pi(x) - \theta]^2 | x\right\}, \end{aligned} \tag{7.7}$$

and hence the difference is

$$D(x) = \frac{(bx-ar)^2}{(x+r)^2(x+r+a+b)^2}. \tag{7.8}$$

Before proceeding further, we must check that the priors satisfy condition (b) of Theorem 7.13. [The normalized priors will not, since, for example, the probability of the interval $(\varepsilon, 1-\varepsilon)$ under $B(a,b)$ tends to zero as $a, b \to 0$.] Since we are letting $a, b \to 0$, we only need consider $0 < a, b < 1$. We then have for any $0 < \varepsilon < \varepsilon' < 1$,

$$\int_\varepsilon^{\varepsilon'} \theta^{a-1}(1-\theta)^{b-1}\, d\theta \geq \int_\varepsilon^{\varepsilon'} \theta^{-1}(1-\theta)^{-1}\, d\theta = \log\left(\frac{1-\varepsilon}{\varepsilon}\frac{\varepsilon'}{1-\varepsilon'}\right), \tag{7.9}$$

satisfying condition (b).

To compute the Bayes risk, we next need the marginal distribution of X, which is given by

$$P(X = x) = \frac{\Gamma(r+x)}{\Gamma(x+1)\Gamma(r)} \frac{\Gamma(r+b)\Gamma(x+a)}{\Gamma(r+x+a+b)} \frac{\Gamma(a+b)}{\Gamma(a)\Gamma(b)}, \tag{7.10}$$

the *beta-Pascal* distribution. Hence, the difference in Bayes risks, using the unnormalized priors, is

$$\sum_{x=0}^{\infty} \frac{\Gamma(r+x)}{\Gamma(x+1)\Gamma(r)} \frac{\Gamma(r+b)\Gamma(x+a)}{\Gamma(r+x+a+b)} D(x), \tag{7.11}$$

which we must show goes to 0 as $a, b \to 0$.

Note first that the x=0 term in (7.11) is

$$\frac{\Gamma(r)\Gamma(a+1)}{\Gamma(r+a+b)} \frac{a}{(r+a+b)^2} \to 0$$

as $a \to 0$. Also, for $x \geq 1$ and $a \leq 1$, $\frac{\Gamma(r+x)}{\Gamma(x+1)} \frac{\Gamma(x+a)}{\Gamma(r+x+a+b)} \leq 1$, so it is sufficient to show

$$\sum_{x=1}^{\infty} D(x) \to 0 \text{ as } a, b \to 0.$$

From (7.8), using the facts that

$$\sup_{x \geq 0} \frac{(bx - ar)^2}{(x+r)^2} = \max\{a^2, b^2\} \quad \text{and} \quad \frac{1}{(x+r+a+b)^2} \leq \frac{1}{x^2},$$

we have

$$\sum_{x=1}^{\infty} D(x) \leq \max\{a^2, b^2\} \sum_{x=1}^{\infty} \frac{1}{x^2} \to 0$$

as $a, b \to 0$, establishing the admissibility of the ML estimator of θ. ||

Theorem 7.13 shows that one of the sufficient conditions for an estimator to be admissible is that its Bayes risk is approachable by a sequence of Bayes risks of Bayes estimators. It would be convenient if it were possible to replace the risks by the estimators themselves. That this is not the case can be seen from the fact that the normal sample mean in three or more dimensions is not admissible although it is the limit of Bayes estimators.

However, under certain conditions the converse is true: That every admissible estimator is a limit of Bayes estimators.[4] We present, but do not prove, the following necessary conditions for admissibility. (This is essentially Theorem 4A.12 of Brown (1986a); see his Appendix to Chapter 4 for a detailed proof.)

Theorem 7.15 *Let* $\mathbf{X} \sim f(\mathbf{x}|\theta)$ *be a density relative to a σ-finite measure ν, such that $f(\mathbf{x}|\theta) > 0$ for all $\mathbf{x} \in \mathcal{X}, \theta \in \Omega$. Let the loss function $L(\theta, \delta)$ be continuous, strictly convex in δ for every θ, and satisfy*

$$\lim_{|\delta| \to \infty} L(\theta, \delta) = \infty \quad \textit{for all } \theta \in \Omega.$$

Then, to every admissible procedure $\delta(\mathbf{x})$ there corresponds a sequence π_n of prior distributions with support on a finite set (and hence with finite Bayes risk) for which

$$\delta^{\pi_n}(\mathbf{x}) \to \delta(\mathbf{x}) \quad \text{a.e. } (\nu), \tag{7.12}$$

where δ^{π_n} is the Bayes estimator for π_n.

[4] The remaining material of this section is of a somewhat more advanced nature. It is sketched here to give the reader some idea of these developments and to serve as an introduction to the literature.

As an immediate corollary to Theorem 7.15, we have the following complete class theorem.

Corollary 7.16 *Under the assumptions of Theorem 7.15, the class of all estimators $\delta(\mathbf{x})$ that satisfy (7.12) is complete.*

For exponential families, the assumptions of Theorem 7.15 are trivially satisfied, so limits of Bayes estimators are a complete class. More importantly, if **X** has a density in the r-variate exponential family, and if δ^π is a limit of Bayes estimators δ^{π_n}, then a subsequence of measures $\{\pi_{n'}\}$ can be found such that $\pi_{n'} \to \pi$ and δ^π is generalized Bayes against π. Such a result was originally developed by Sacks (1963) and extended by Brown (1971) and Berger and Srinivasan (1978) to the following theorem.

Theorem 7.17 *Under the assumptions of Theorem 7.15, if the densities of X constitute an r-variate exponential family, then any admissible estimator is a generalized Bayes estimator. Thus, the generalized Bayes estimators form a complete class.*

Further characterizations of generalized Bayes estimators were given by Strawderman and Cohen (1971) and Berger and Srinivasan (1978). See Berger 1985 for more details. Note that it is not the case that all generalized Bayes estimators are admissible. Farrell (1964) gave examples of inadmissible generalized Bayes estimators in location problem, in particular $X \sim N(\theta, 1), \pi(\theta) = e^\theta$. (See also Problem 4.2.15.) Thus, it is of interest to determine conditions under which generalized Bayes estimators are admissible. We do so in the following examples, where we look at a number of characterizations of admissible estimators in specific situations. Although these characterizations have all been derived using the tools (or their generalizations) that have been described here, in some cases the exact derivations are complex.

We begin with a fundamental identity.

Example 7.18 Brown's identity. In order to understand what types of estimators are admissible, it would be helpful if the convergence of risk functions in Blyth's method were more explicitly dependent on the convergence of the estimators. Brown (1971) gave an identity that makes this connection clearer.

Let $\mathbf{X} \sim N_r(\boldsymbol{\theta}, I)$ and $L(\boldsymbol{\theta}, \delta) = |\boldsymbol{\theta} - \delta|^2$, and for a given prior $\pi(\boldsymbol{\theta})$, let $\delta^\pi(\mathbf{x}) = \mathbf{x} + \nabla \log m_\pi(\mathbf{x})$ be the Bayes estimator, where $m_\pi(\mathbf{x}) = \int_\Omega f(\mathbf{x}|\boldsymbol{\theta})\pi(\boldsymbol{\theta})\,d\boldsymbol{\theta}$ is the marginal density. Suppose that $\delta^g(\mathbf{x}) = \mathbf{x} + \nabla \log m_g(\mathbf{x})$ is another estimator. First note that

$$r(\pi, \delta^\pi) - r(\pi, \delta^g) = E\left|\delta^\pi(\mathbf{X}) - \delta^g(\mathbf{X})\right|^2, \tag{7.13}$$

(see Problem 7.16); hence, we have the identity

$$r(\pi, \delta^\pi) - r(\pi, \delta^g) = \int \left|\nabla \log m_\pi(\mathbf{x}) - \nabla \log m_g(\mathbf{x})\right|^2 m_\pi(\mathbf{x})\,d\mathbf{x}. \tag{7.14}$$

We now have the estimator explicitly in the integral, but we must develop (7.14) a bit further to be more useful in helping to decide admissibility. Two paths have been taken. On the first, we note that if we were going to use (7.14) to establish the admissibility of δ^g, we might replace the prior $\pi(\cdot)$ with a sequence $\pi_n(\cdot)$. However, it would be more useful to have the measure of integration not depend on n [since $m_\pi(\mathbf{x})$ would now equal $m_{\pi_n}(\mathbf{x})$]. To this end, write $k_n(\mathbf{x}) = m_{\pi_n}(\mathbf{x})/m_g(\mathbf{x})$, and

$$(7.15)\qquad r(\pi, \delta^\pi) - r(\pi, \delta^g) = \int \left|\nabla \log \left(m_{\pi_n}(\mathbf{x})/m_g(\mathbf{x})\right)\right|^2 m_{\pi_n}(\mathbf{x})\, d\mathbf{x}$$
$$= \int \frac{|\nabla k_n(\mathbf{x})|^2}{k_n(\mathbf{x})} m_g(\mathbf{x})\, d\mathbf{x},$$

where the second equality follows from differentiation (see Problem 7.16), and now the integration measure does not depend on n. Thus, if we could apply Lebesgue's dominated convergence, then $\frac{|\nabla k_n(\mathbf{x})|^2}{k_n(\mathbf{x})} \to 0$ would imply the admissibility of δ^g. This is the path taken by Brown (1971), who established a relationship between (7.15) and the behavior of a diffusion process in r dimensions, and then gave necessary and sufficient conditions for the admissibility of δ^g. For example, the admissibility of the sample mean in one and two dimensions is linked to the recurrence of a random walk in one and two dimensions, and the inadmissibility is linked to its transience in three or more dimensions. This is an interesting and fruitful approach, but to pursue it fully requires the development of properties of diffusions, which we will not do here. [Johnstone 1984 (see also Brown and Farrell 1985) developed similar theorems for the Poisson distribution (Problem 7.25), and Eaton (1992) investigated another related stochastic process; the review paper by Rukhin (1995) provides an excellent entry into the mathematics of this literature.]

Another path, developed in Brown and Hwang (1982), starts with the estimator δ^g and constructs a sequence $g_n \to g$ that leads to a simplified condition for the convergence of (7.14) to zero, and uses Blyth's method to establish admissibility. Although they prove their theorem for exponential families, we shall only state it here for the normal distribution. (See Problem 7.19 for a more general statement.) ||

Theorem 7.19 *Let* $\mathbf{X} \sim N_r(\boldsymbol{\theta}, I)$ *and* $L(\boldsymbol{\theta}, \delta) = |\boldsymbol{\theta} - \delta|^2$. *Let* $\delta^g(\mathbf{x}) = \mathbf{x} + \nabla \log m_g(\mathbf{x})$ *where* $m_g(\mathbf{x}) = \int_\Omega f(\mathbf{x}|\boldsymbol{\theta})g(\boldsymbol{\theta})\,d\boldsymbol{\theta}$. *Assume that* $g(\cdot)$ *satisfies*

(a) $$\int_{\{\boldsymbol{\theta}:|\boldsymbol{\theta}|>1\}} \frac{g(\boldsymbol{\theta})}{|\boldsymbol{\theta}|^2 \max\{\log|\boldsymbol{\theta}|, \log 2\}^2}\, d\boldsymbol{\theta} < \infty,$$

(b) $$\int_\Omega \frac{|\nabla g(\boldsymbol{\theta})|^2}{g(\boldsymbol{\theta})}\, d\boldsymbol{\theta} < \infty,$$

(c) $\sup\{R(\boldsymbol{\theta}, \delta^g) : \boldsymbol{\theta} \in K\} < \infty$ *for all compact sets* $K \in \Omega$.

Then, $\delta^g(\mathbf{x})$ *is admissible.*

Proof. The proof follows from (7.14) by taking the sequence of priors $g_n(\boldsymbol{\theta}) \to g(\boldsymbol{\theta})$, where $g_n(\boldsymbol{\theta}) = h_n^2(\boldsymbol{\theta})g(\boldsymbol{\theta})$ and

$$(7.16)\qquad h_n(\boldsymbol{\theta}) = \begin{cases} 1 & \text{if } |\boldsymbol{\theta}| \le 1 \\ 1 - \frac{\log(|\boldsymbol{\theta}|)}{\log(n)} & \text{if } 1 < |\boldsymbol{\theta}| \le n \\ 0 & \text{if } |\boldsymbol{\theta}| > n \end{cases}$$

for $n = 2, 3, \ldots$. See Problem 7.18 for details. □

Example 7.20 Multivariate normal mean. The conditions of Theorem 7.19 relate to the tails of the prior, which are crucial in determining whether the integral are finite. Priors with polynomial tails, that is, priors of the form $g(\boldsymbol{\theta}) = 1/|\boldsymbol{\theta}|^k$, have received a lot of attention. Perhaps the reason for this is that using a Laplace approximation (4.6.33), we can write

$$\begin{aligned}
\delta^g(\mathbf{x}) &= \mathbf{x} + \nabla \log m_g(\mathbf{x}) \\
&= \mathbf{x} + \frac{\nabla m_g(\mathbf{x})}{m_g(\mathbf{x})} \\
&\approx \mathbf{x} + \frac{\nabla g(\mathbf{x})}{g(\mathbf{x})} \qquad (7.17) \\
&= \left(1 - \frac{k}{|\mathbf{x}|^2}\right)\mathbf{x}.
\end{aligned}$$

See Problem 7.20 for details.

Now what can we say about the admissibility of δ^g? For $g(\boldsymbol{\theta}) = 1/|\boldsymbol{\theta}|^k$, condition (a) of Theorem 7.19 becomes, upon transforming to polar coordinates,

$$\begin{aligned}
&\int_{\{\boldsymbol{\theta}:|\boldsymbol{\theta}|>1\}} \frac{g(\boldsymbol{\theta})}{|\boldsymbol{\theta}|^2 \max\{\log|\boldsymbol{\theta}|, \log 2\}^2}\, d\boldsymbol{\theta} \qquad (7.18) \\
&= \int_0^{2\pi} \sin^{r-2}\boldsymbol{\beta}\, d\boldsymbol{\beta} \int_1^\infty \frac{1}{t^{k+2}\max\{\log(t), \log 2\}^2} t^{r-1}\, dt
\end{aligned}$$

where $t = |\boldsymbol{\theta}|$ and $\boldsymbol{\beta}$ is a vector of direction cosines. The integral over $\boldsymbol{\beta}$ is finite, and if we ignore the log term, a sufficient condition for this integral to be finite is

$$\int_1^\infty \frac{1}{t^{k-r+3}}\, d\,t < \infty, \qquad (7.19)$$

which is satisfied if $k > r - 2$. If we keep the log term and work a little harder, condition (a) can be verified for $k \geq r - 2$ (see Problem 7.22). ||

Example 7.21 Continuation of Example 7.20. The characterization of admissible estimators by Brown (1971) goes beyond that of Theorem 7.19, as he was able to establish both necessary and sufficient conditions. Here is an example of these results.

Using a spherically symmetric prior (see, for example, Corollary 4.3.3), all generalized Bayes estimators are of the form

$$\delta^\pi(\mathbf{x}) = \mathbf{x} + \nabla \log m(\mathbf{x}) = (1 - h(|\mathbf{x}|))\mathbf{x}. \qquad (7.20)$$

The estimator δ^π is

(a) *inadmissible* if there exists $\varepsilon > 0$ and $M < \infty$ such that

$$0 \leq h(|\mathbf{x}|) < \frac{r-2-\varepsilon}{|\mathbf{x}|^2} \quad \text{for } |\mathbf{x}| > M,$$

(b) *admissible* if $h(|\mathbf{x}|)|\mathbf{x}|$ is bounded and there exists $M < \infty$ such that

$$1 \geq h(|\mathbf{x}|) \geq \frac{r-2}{|\mathbf{x}|^2} \quad \text{for } |\mathbf{x}| > M.$$

It is interesting how the factor $r-2$ appears again and supports its choice as the optimal constant in the James-Stein estimator (even though that estimator is not generalized Bayes, and hence inadmissible). Bounds such as these are called *semitail upper bounds* by Hwang (1982b), who further developed their applicability.

For Strawderman's estimator (see Problem 5.5), we have

$$h(|\mathbf{x}|) = \frac{r-2a+2}{|\mathbf{x}|^2} \frac{P(\chi^2_{r-2a+4} \le |\mathbf{x}|^2)}{P(\chi^2_{r-2a+2} \le |\mathbf{x}|^2)} \tag{7.21}$$

and it is admissible (Problem 7.23) as long as $r-2a+2 \ge r-2$, or $r \ge 1$ and $a \le 2$. ‖

Now that we have a reasonably complete picture of the types of estimators that are admissible estimators of a normal mean, it is interesting to see how the admissibility conditions fit in with minimaxity conditions. To do so requires the development of some general necessary conditions for minimaxity. This was first done by Berger (1976a), who derived conditions for an estimator to be *tail minimax*.

Example 7.22 Tail minimaxity. Let $\mathbf{X} \sim N_r(\boldsymbol{\theta}, I)$ and $L(\boldsymbol{\theta}, \delta) = |\boldsymbol{\theta} - \delta|^2$. Since the estimator $\mathbf{X}$ is minimax with constant risk $R(\boldsymbol{\theta}, \mathbf{X})$, another estimator δ is *tail minimax* if there exists $M > 0$ such that $R(\boldsymbol{\theta}, \delta) \le R(\boldsymbol{\theta}, \mathbf{X})$ for all $|\boldsymbol{\theta}| > M$. (Berger investigated tail minimaxity for much more general situations than are considered here, including non-normal distributions and nonquadratic loss.) Since tail minimaxity is a necessary condition for minimaxity, it can help us see which admissible estimators have the possibility of also being minimax. An interesting characterization of $h(|\mathbf{x}|)$ of (7.20) is obtained if admissibility is considered together with tail minimaxity.

Using a risk representation similar to (5.4), the risk of $\delta(\mathbf{x}) = [1 - h(|\mathbf{x}|)]\mathbf{x}$ is

$$R(\boldsymbol{\theta}, \delta) = r + E_{\boldsymbol{\theta}}\left[|\mathbf{X}|^2 h^2(|\mathbf{X}|) - 2rh(|\mathbf{X}|) - 4|\mathbf{X}|^2 h'(|\mathbf{X}|)\right]. \tag{7.22}$$

If we now use a Laplace approximation on the expectation in (7.22), we have

$$\begin{aligned} &E_{\boldsymbol{\theta}}\left[|\mathbf{X}|^2 h^2(|\mathbf{X}|) - 2rh(|\mathbf{X}|) - 4|\mathbf{X}|^2 h'(|\mathbf{X}|)\right] \\ &\approx |\boldsymbol{\theta}|^2 h^2(|\boldsymbol{\theta}|) - 2rh(|\boldsymbol{\theta}|) - 4|\boldsymbol{\theta}|^2 h'(|\boldsymbol{\theta}|) \\ &= \mathcal{B}(\boldsymbol{\theta}). \end{aligned} \tag{7.23}$$

By carefully working with the error terms in the Laplace approximation, Berger showed that the error of approximation was $o(|\boldsymbol{\theta}|^{-2})$, that is,

$$R(\boldsymbol{\theta}, \delta) = r + \mathcal{B}(\boldsymbol{\theta}) + o(|\boldsymbol{\theta}|^{-2}). \tag{7.24}$$

In order to ensure that the estimator is tail minimax, we must be able to ensure that $\mathcal{B}(\boldsymbol{\theta}) + o(|\boldsymbol{\theta}|^{-2}) < 0$ for sufficiently large $|\boldsymbol{\theta}|$. This would occur if, for some $\varepsilon > 0$, $|\boldsymbol{\theta}|^{-2}\mathcal{B}(\boldsymbol{\theta}) \le -\varepsilon$ for sufficiently large $|\boldsymbol{\theta}|$, that is,

$$|\boldsymbol{\theta}|^2 h^2(|\boldsymbol{\theta}|) - 2rh(|\boldsymbol{\theta}|) - 4|\boldsymbol{\theta}|^2 h'(|\boldsymbol{\theta}|) \le \frac{-\varepsilon}{|\boldsymbol{\theta}|^2} \tag{7.25}$$

for sufficiently large $|\boldsymbol{\theta}|$.

Now, for $\delta(\mathbf{x}) = [1 - h(|\mathbf{x}|)]\mathbf{x}$ to be admissible, we must have $h(|\mathbf{x}|) \geq (r-2)/|\mathbf{x}|^2$. Since $|\mathbf{x}|h(|\mathbf{x}|)$ must be bounded, this suggests that, for large $|\mathbf{x}|$, we could have $h(|\mathbf{x}|) \approx k/|\mathbf{x}|^{2\alpha}$, for some α, $1/2 \leq \alpha \leq 1$. We now show that for $\delta(\mathbf{x})$ to be minimax, it is necessary that $\alpha = 1$.

For $h(|\mathbf{x}|) = k/|\mathbf{x}|^{2\alpha}$, (7.25) is equal to

$$\text{(7.26)} \qquad \frac{k}{|\theta|^{2\alpha}}\left[\frac{k}{|\theta|^{2\alpha-2}} - 2(r-2\alpha)\right] < \frac{-\varepsilon}{|\theta|^2} \quad \text{for } |\mathbf{x}| > M$$

which, for $r \geq 3$, cannot be satisfied if $1/2 \leq \alpha < 1$. Thus, the only possible admissible minimax estimators are those for which $h(|\mathbf{x}|) \approx k/|\mathbf{x}|^2$, with $r-2 \leq k \leq 2(r-2)$. ‖

Theorem 7.17 can be adapted to apply to discrete distributions (the assumption of a density can be replaced by a probability mass function), and an interesting case is the binomial distribution. It turns out that the fact that the sample space is finite has a strong influence on the form of the admissible estimators. We first look at the following characterization of admissible estimators, due to Johnson (1971).

Example 7.23 Binomial estimation. For the problem of estimating $h(p)$, where $h(\cdot)$ is a continuous real-valued function on $[0, 1]$, $X \sim \text{binomial}(p, n)$, and $L(h(p), \delta) = (h(p) - \delta)^2$, a minimal complete class is given by

$$\text{(7.27)} \qquad \delta^\pi(x) = \begin{cases} h(0) & \text{if } x \leq r \\ \dfrac{\int_0^1 h(p)p^{x-r-1}(1-p)^{s-x-1}d\pi(p)}{\int_0^1 p^{x-r-1}(1-p)^{s-x-1}d\pi(p)} & \text{if } r+1 \leq x < s-1 \\ h(1) & \text{if } x \geq s, \end{cases}$$

where p has the prior distribution

$$\text{(7.28)} \qquad p \sim k(p)d\pi(p)$$

with

$$k(p) = \frac{P(r+1 \leq X \leq s-1|p)}{p^{r+1}(1-p)^{n-s+1}},$$

r and s are integers, $-1 \leq r < s \leq n+1$, and π is a probability measure with $\pi(\{0\} \cup \{1\}) < 1$, that is, π does not put all of its mass on the endpoints of the parameter space.

To see that δ^π is admissible, let δ' be another estimator of $h(p)$ that satisfies $R(p, \delta^\pi) \geq R(p, \delta')$. We will assume that $s \geq 0$, $r \leq n$, and $r+1 < s$ (as the cases $r = -1$, $s = n+1$, and $r+1 = s$ are straightforward). Also, if $\delta'(x) = h(0)$ for $x \leq r'$, and $\delta'(x) = h(1)$ for $x \geq s'$, then it follows that $r' \geq r$ and $s' \leq s$. Define

$$R_{r,s}(p, \delta) = \sum_{x=r+1}^{s-1} \binom{n}{x} [h(p) - \delta(x)]^2 p^{x-r-1}(1-p)^{s-x-1}k(p)^{-1}.$$

Now, $R(p, \delta') \leq R(p, \delta^\pi)$ for all $p \in [0, 1]$ if and only if $R_{r,s}(p, \delta') \leq R_{r,s}(p, \delta^\pi)$ for all $p \in [0, 1]$. However, for the prior (7.28), $\int_0^1 R_{r,s}(p, \delta) \times k(p)d\pi(p)$ is uniquely minimized by $[\delta^\pi(r+1), \ldots, \delta^\pi(s-1)]$, which establishes the admissibility of δ^π. The converse assertion, that any admissible estimator is of the form (7.27), follows from Theorem 7.17. (See Problem 7.27.)

For $h(p) = p$, we can take $r = 0$, $s = n$, and $\pi(p) = \text{Beta}(a, b)$. The resulting estimators are of the form $\alpha\frac{x}{n} + (1-\alpha)\frac{a}{a+b}$, so we obtain conditions on admissibility of linear estimators. In particular, we see that x/n is an admissible estimator. If $h(p) = p(1-p)$, we find an admissible estimator of the variance to be $\frac{n}{n+1}(x/n)(1-x/n)$. (See Problem 7.26.)

Brown (1981, 1988) has generalized Johnson's results and characterizes a minimal complete class for estimation in a wide class of problems with finite sample spaces. ‖

Johnson (1971) was further able to establish the somewhat surprising result that if δ_1 is an admissible estimator of $h(p_1)$ in the binomial $b(p_1, n_1)$ problem and δ_2 is an admissible estimator of $h(p_2)$ in the binomial $b(p_2, n_2)$ problem, then $[\delta_1, \delta_2]$ is an admissible estimator of $[h(p_1), h(p_2)]$ if the loss function is the sum of the losses. This result can be extended to higher dimensions, and thus there is no Stein effect in the binomial problem. The following example gives conditions under which this can be expected.

Example 7.24 When there is no Stein effect. For $i = 1, 2$, let $X_i \sim f_i(x|\theta_i)$ and suppose that $\delta_i^*(x_i)$ is a unique Bayes (hence, admissible) estimator of θ_i under the loss $L(\theta_i, \delta)$, where L satisfies $L(a, a) = 0$ and $L(a, a') > 0$, $a \neq a'$, and all risk functions are continuous. Suppose there is a value θ^* such that if $\theta_2 = \theta^*$,

(i) $X_2 = x^*$ with probability 1,

(ii) $\delta_2^*(x^*) = \theta^*$,

then $(\delta_1^*(x_1), \delta_2^*(x_2))$ is admissible for (θ_1, θ_2) under the loss $\sum_i L(\theta_i, \delta)$; that is, there is no Stein effect.

To see why this is so, let $\delta' = (\delta_1'(x_1, x_2), \delta_2'(x_1, x_2))$ be a competitor. At the parameter value $(\theta_1, \theta_2) = (\theta_1, \theta^*)$, we have

$$\begin{aligned} R[(\theta_1, \theta^*), \delta'] &= E_{(\theta_1,\theta^*)}L[\theta_1, \delta_1'(X_1, X_2)] + E_{(\theta_1,\theta^*)}L[\theta_2, \delta_2'(X_1, X_2)] \\ &= E_{(\theta_1,\theta^*)}L[\theta_1, \delta_1'(X_1, x^*)] + E_{(\theta_1,\theta^*)}L[\theta^*, \delta_2'(X_1, x^*)] \\ &= E_{\theta_1}L[\theta_1, \delta_1'(X_1, x^*)] + E_{\theta_1}L[\theta^*, \delta_2'(X_1, x^*)], \end{aligned} \tag{7.29}$$

while for $(\delta_1^*(x_1), \delta_2^*(x_2))$,

$$\begin{aligned} R[(\theta_1, \theta^*), \delta^*] &= E_{(\theta_1,\theta^*)}L[\theta_1, \delta_1^*(X_1)] + E_{(\theta_1,\theta^*)}L[\theta_2, \delta_2^*(X_2)] \\ &= E_{\theta_1}L[\theta_1, \delta_1^*(X_1)] + E_{\theta^*}L[\theta^*, \delta_2^*(x^*)] \\ &= E_{\theta_1}L[\theta_1, \delta_1^*(X_1)] \end{aligned} \tag{7.30}$$

as $E_{\theta^*}L[\theta^*, \delta_2^*(x^*)] = 0$.

Since δ_1^* is a unique Bayes estimator of θ_1,

$$E_{\theta_1}L[\theta_1, \delta_1^*(X_1)] < E_{\theta_1}L[\theta_1, \delta_1'(X_1, x^*)] \text{ for some } \theta_1.$$

Since $E_{\theta_1}L[\theta^*, \delta_2'(X_1, x^*)] \geq 0$, it follows that $R[(\theta_1, \theta^*), \delta^*] < R[(\theta_1, \theta^*), \delta']$ for some θ_1, and hence that δ^* is an admissible estimator of (θ_1, θ_2). By induction, the result can be extended to any number of coordinates (see Problem 7.28).

If $X \sim b(\theta, n)$, then we can take $\theta^* = 0$ or 1, and the above result applies. The absence of the Stein effect persists in other situations, such as any problem with a finite sample space (Gutmann 1982a; see also Brown 1981). Gutmann (1982b) also demonstrates a sequential context in which the Stein effect does not hold (see Problem 7.29). ||

Finally, we look at the admissibility of linear estimators. There has always been interest in characterizing admissibility of linear estimators, partly due to the ease of computing and using linear estimators, and also due to a search for a converse to Karlin's theorem (Theorem 2.14) (which gives sufficient conditions for admissibility of linear estimators). Note that we are concerned with the admissibility of linear estimators in the class of all estimators, not just in the class of linear estimators. (This latter question was addressed by La Motte (1982).)

Example 7.25 Admissible linear estimators. Let $X \sim N_r(\boldsymbol{\theta}, I)$, and consider estimation of $\varphi'\boldsymbol{\theta}$, where $\varphi_{r\times 1}$ is a known vector, and $L(\varphi'\boldsymbol{\theta}, \delta) = (\varphi'\boldsymbol{\theta} - \delta)^2$. For $r = 1$, the results of Karlin (1958); see also Meeden and Ghosh (1977), show that ax is admissible if and only if $0 \le a \le \varphi$. This result was generalized by Cohen (1965a) to show that $a'x$ is admissible if and only if a is in the sphere:

$$\{a : (a - \varphi/2)'(a - \varphi/2) \le \varphi'\varphi/4\} \tag{7.31}$$

(see Problem 7.30). Note that the extension to known covariance matrix is straightforward, and (7.31) becomes an ellipse.

For the problem of estimating $\boldsymbol{\theta}$, the linear estimator $C\mathbf{x}$, where C is an $r \times r$ symmetric matrix, is admissible if and only if all of the eigenvalues of C are between 0 and 1, with at most two equal to 1 (Cohen 1966).

Necessary and sufficient conditions for admissibility of linear estimators have also been described for multivariate Poisson estimation (Brown and Farrell, 1985a, 1985b) and for estimation of the scale parameters in the multivariate gamma distribution (Farrell et al., 1989). This latter result also has application to the estimation of variance components in mixed models. ||

8 Problems

Section 1

1.1 For the situation of Example 1.2:

(a) Plot the risk functions of $\delta^{1/4}$, $\delta^{1/2}$, and $\delta^{3/4}$ for $n = 5, 10, 25$.

(b) For each value of n in part (a), find the range of prior values of p for which each estimator is preferred.

(c) If an experimenter has no prior knowledge of p, which of $\delta^{1/4}$, $\delta^{1/2}$, and $\delta^{3/4}$ would you recommend? Justify your choice.

1.2 The principle of *gamma-minimaxity* [first used by Hodges and Lehmann (1952); see also Robbins 1964 and Solomon 1972a, 1972b)] is a Bayes/frequentist synthesis. An estimator δ^* is *gamma-minimax* if

$$\inf_{\delta\in\mathcal{D}} \sup_{\pi\in\Gamma} r(\pi, \delta) = \sup_{\pi\in\Gamma} r(\pi, \delta^*)$$

where Γ is a specified class of priors. Thus, the estimator δ^* minimizes the maximum Bayes risk over those priors in the class Γ. (If Γ = all priors, then δ^* would be minimax.)

(a) Show that if $\Gamma = \{\pi_0\}$, that is, Γ consists of one prior, then the Bayes estimator is Γ minimax.

(b) Show that if Γ = {all priors}, then the minimax estimator is Γ minimax.

(c) Find the Γ-minimax estimator among the three estimators of Example 1.2.

1.3 Classes of priors for Γ-minimax estimation have often been specified using moment restrictions.

(a) For $X \sim b(p, n)$, find the Γ-minimax estimator of p under squared error loss, with

$$\Gamma_\mu = \left\{\pi(p) : \pi(p) = \text{beta}(a, b),\ \mu = \frac{a}{a+b}\right\}$$

where μ is considered fixed and known.

(b) For $X \sim N(\theta, 1)$, find the Γ-minimax estimator of θ under squared error loss, with

$$\Gamma_{\mu,\tau} = \left\{\pi(\theta) : E(\theta) = \mu,\ \text{var}\ \theta = \tau^2\right\}$$

where μ and τ are fixed and known.

[*Hint*: In part (b), show that the Γ-minimax estimator is the Bayes estimator against a normal prior with the specified moments (Jackson et al. 1970; see Chen, Eichenhauer-Herrmann, and Lehn 1990 for a multivariate version). This somewhat nonrobust Γ-minimax estimator is characteristic of estimators derived from moment restrictions and shows why robust Bayesians tend to not use such classes. See Berger 1985, Section 4.7.6 for further discussion.]

1.4 (a) For the random effects model of Example 4.2.7 (see also Example 3.5.1), show that the restricted maximum likelihood (REML) likelihood of σ_A^2 and σ^2 is given by (4.2.13), which can be obtained by integrating the original likelihood against a uniform $(-\infty, \infty)$ prior for μ.

(b) For $n_i = n$ in

$$X_{ij} = \mu + A_i + u_{ij} \quad (j = 1, \ldots, n_i, \quad i = 1, \ldots, s)$$

calculate the expected value of the REML estimate of σ_A^2 and show that it is biased. Compare REML to the unbiased estimator of σ_A^2. Which do you prefer?

(Construction of REML-type marginal likelihoods, where some effects are integrated out against priors, becomes particularly useful in nonlinear and generalized linear models. See, for example, Searle et al. 1992, Section 9.4 and Chapter 10.)

1.5 Establishing the fact that (9.1) holds, so S^2 is conditionally biased, is based on a number of steps, some of which can be involved. Define $\phi(a, \mu, \sigma^2) = (1/\sigma^2)E_{\mu,\sigma^2}[S^2 \mid |\bar{x}|/s < a]$.

(a) Show that $\phi(a, \mu, \sigma^2)$ only depends on μ and σ^2 through μ/σ. Hence, without loss of generality, we can assume $\sigma = 1$.

(b) Use the fact that the density $f(s \mid |\bar{x}|/s < a, \mu)$ has monotone likelihood ratio to establish $\phi(a, \mu, 1) \geq \phi(a, 0, 1)$.

(c) Show that

$$\lim_{a\to\infty} \phi(a, 0, 1) = 1 \quad \text{and} \quad \lim_{a\to 0} \phi(a, 0, 1) = \frac{E_{0,1}S^3}{E_{0,1}S} = \frac{n}{n-1}.$$

(d) Combine parts (a), (b), and (c) to establish (19.1).

The next three problems explore conditional properties of estimators. A detailed development of this theory is found in Robinson (1979a, 1979b), who also explored the relationship between admissibility and conditional properties.

1.6 Suppose that $X \sim f(x|\theta)$, and $T(x)$ is used to estimate $\tau(\theta)$. One might question the worth of $T(x)$ if there were some set $A \in \mathcal{X}$ for which $T(x) > \tau(\theta)$ for $x \in A$ (or if the reverse inequality holds). This leads to the conditional principle of never using an estimator if there exists a set $A \in \mathcal{X}$ for which $E_\theta\{[T(X) - \tau(\theta)]I(X \in A)\} \geq 0 \quad \forall\theta$, with strict inequality for some θ (or if the equivalent statement holds with the inequality reversed). Show that if $T(x)$ is the posterior mean of $\tau(\theta)$ against a proper prior, where both the prior and $f(x|\theta)$ are continuous in θ, then no such A can exist. (If such an A exists, it is called a *semirelevant* set. Elimination of semirelevant sets is an extremely strong requirement. A weaker requirement, elimination of *relevant* sets, seems more appropriate.)

1.7 Show that if there exists a set $A \in \mathcal{X}$ and an $\varepsilon > 0$ for which $E_\theta\{[T(X) - \tau(\theta)]I(X \in A)\} > \varepsilon$, then $T(x)$ is inadmissible for estimating $\tau(\theta)$ under squared error loss. (A set A satisfying the this inequality is an example of a *relevant* set.)

[*Hint*: Consider the estimator $T(x) + \varepsilon I(x \in A)$]

1.8 To see why elimination of semirelevant sets is too strong a requirement, consider the estimation of θ based on observing $X \sim f(x - \theta)$. Show that for any constant a, the Pitman estimator X satisfies

$$E_\theta[(X - \theta)I(X < a)] \leq 0 \quad \forall\theta \quad \text{or} \quad E_\theta[(X - \theta)I(X > a)] \geq 0 \quad \forall\theta,$$

with strict inequality for some θ. Thus, there are semirelevant sets for the Pitman estimator, which is, by most accounts, a fine estimator.

1.9 In Example 1.7, let $\delta^*(X) = X/n$ with probability $1 - \varepsilon$ and $= 1/2$ with probability ε. Determine the risk function of δ^* and show that for $\varepsilon = 1/(n+1)$, its risk is constant and less than $\sup R(p, X/n)$.

1.10 Find the bias of the minimax estimator (1.11) and discuss its direction.

1.11 In Example 1.7,

(a) determine c_n and show that $c_n \to 0$ as $n \to \infty$,

(b) show that $R_n(1/2)/r_n \to 1$ as $n \to \infty$.

1.12 In Example 1.7, graph the risk functions of X/n and the minimax estimator (1.11) for $n = 1, 4, 9, 16$, and indicate the relative positions of the two graphs for large values of n.

1.13 (a) Find two points $0 < p_0 < p_1 < 1$ such that the estimator (1.11) for $n = 1$ is Bayes with respect to a distribution Λ for which $P_\Lambda(p = p_0) + P_\Lambda(p = p_1) = 1$.

(b) For $n = 1$, show that (1.11) is a minimax estimator of p even if it is known that $p_o \leq p \leq p_1$.

(c) In (b), find the values p_0 and p_1 for which $p_1 - p_0$ is as small as possible.

1.14 Evaluate (1.16) and show that its maximum is $1 - \alpha$.

1.15 Let $X = 1$ or 0 with probabilities p and q, respectively, and consider the estimation of p with loss $= 1$ when $|d - p| \geq 1/4$, and 0 otherwise. The most general randomized estimator is $\delta = U$ when $X = 0$, and $\delta = V$ when $X = 1$ where U and V are two random variables with known distributions.

(a) Evaluate the risk function and the maximum risk of δ when U and V are uniform on $(0, 1/2)$ and $(1/2, 1)$, respectively.

(b) Show that the estimator δ of (a) is minimax by considering the three values $p = 0$, 1/2, 1.

[*Hint*: (b) The risk at $p = 0, 1/2, 1$ is, respectively, $P(U > 1/4)$, $1/2[P(U < 1/4) + P(V > 3/4)]$, and $P(V < 3/4)$].

1.16 Show that the problem of Example 1.8 remains invariant under the transformations

$$X' = n - X, \quad p' = 1 - p, \quad d' = 1 - d.$$

This illustrates that randomized equivariant estimators may have to be considered when $\bar{G}$ is not transitive.

1.17 Let r_Λ be given by (1.3). If $r_\Lambda = \infty$ for some Λ, show that any estimator δ has unbounded risk.

1.18 In Example 1.9, show that no linear estimator has constant risk.

1.19 Show that the risk function of (1.22) depends on p_1 and p_2 only through $p_1 + p_2$ and takes on its maximum when $p_1 + p_2 = 1$.

1.20 (a) In Example 1.9, determine the region in the (p_1, p_2) unit square in which (1.22) is better than the UMVU estimator of $p_2 - p_1$ for $m = n = 2, 8, 18$, and 32.

(b) Extend Problems 1.11 and 1.12 to Example 1.9.

1.21 In Example 1.14, show that $\bar{X}$ is minimax for the loss function $(d - \theta)^2/\sigma^2$ without any restrictions on σ.

1.22 (a) Verify (1.37).

(b) Show that equality holds in (1.39) if and only if $P(X_i = 0) + P(X_i = 1) = 1$.

1.23 In Example 1.16(b), show that for any $k > 0$, the estimator

$$\delta = \frac{\sqrt{n}}{1 + \sqrt{n}} \frac{1}{n} \sum_{i=1}^{n} X_i^k + \frac{1}{2(1 + \sqrt{n})}$$

is a Bayes estimator for the prior distribution Λ over $\mathcal{F}_0$ for which (1.36) was shown to be Bayes.

1.24 Let X_i $(i = 1, \ldots, n)$ and Y_j $(j = 1, \ldots, n)$ be independent with distributions F and G, respectively. If $F(1) - F(0) = G(1) - G(0) = 1$ but F and G are otherwise unknown, find a minimax estimator for $E(Y_j) - E(X_i)$ under squared error loss.

1.25 Let X_i $(i = 1, \ldots, n)$ be iid with unknown distribution F. Show that

$$\delta = \frac{\text{No. of } X_i \leq 0}{\sqrt{n}} \cdot \frac{1}{1 + \sqrt{n}} + \frac{1}{2(1 + \sqrt{n})}$$

is minimax for estimating $F(0) = P(X_i \leq 0)$ with squared error loss. [*Hint*: Consider the risk function of δ.]

1.26 Let $X_1, \ldots, X_m$ and $Y_1, \ldots, Y_n$ be independently distributed as $N(\xi, \sigma^2)$ and $N(\eta, \tau^2)$, respectively, and consider the problem of estimating $\Delta = \eta - \xi$ with squared error loss.

(a) If σ and τ are known, $\bar{Y} - \bar{X}$ is minimax.

(b) If σ and τ are restricted by $\sigma^2 \leq A$ and $\tau^2 \leq B$, respectively (A, B known and finite), $\bar{Y} - \bar{X}$ continues to be minimax.

1.27 In the linear model (3.4.4), show that $\Sigma a_i \hat{\xi}_i$ (in the notation of Theorem 3.4.4) is minimax for estimating $\theta = \Sigma a_i \xi_i$ with squared error loss, under the restriction $\sigma^2 \leq M$. [*Hint*: Treat the problem in its canonical form.]

1.28 For the random variable X whose distribution is (1.42), show that x must satisfy the inequalities stated below (1.42).

1.29 Show that the estimator defined by (1.43)

(a) has constant risk,

(b) is Bayes with respect to the prior distribution specified by (1.44) and (1.45).

1.30 Show that for fixed X and n, (1.43) $\to$ (1.11) as $N \to \infty$.

1.31 Show that var($\bar{Y}$) given by (3.7.6) takes on its maximum value subject to (1.41) when all the a's are 0 or 1.

1.32 (a) If $R(p, \delta)$ is given by (1.49), show that $\sup R(p, \delta) \cdot 4(1+\sqrt{n})^2 \to 1$ as $n \to \infty$.

(b) Determine the smallest value of n for which the Bayes estimator of Example 1.18 satisfies (1.48) for $r = 1$ and $b = 5$, 10, and 20.

1.33 (Efron and Morris 1971)

(a) Show that the estimator δ of (1.50) is the estimator that minimizes $|\delta - c\bar{x}|$ subject to the constraint $|\delta - \bar{x}| \le M$. In this sense, it is the estimator that is closest to a Bayes estimator, $c\bar{x}$, while not straying too far from a minimax estimator, $\bar{x}$.

(b) Show that for the situation of Example 1.19, $R(\theta, \delta)$ is bounded for δ of (1.50).

(c) For the situation of Example 1.19, δ of (1.50) satisfies $\sup_\theta R(\theta, \delta) = (1/n) + M^2$.

Section 2

2.1 Lemma 2.1 has been extended by Berger (1990a) to include the case where the estimand need not be restricted to a finite interval, but, instead, attains a maximum or minimum at a finite parameter value.

Lemma 8.1 *Let the estimand $g(\theta)$ be nonconstant with global maximum or minimum at a point $\theta^* \in \Omega$ for which $f(x|\theta^*) > 0$ a.e. (with respect to a dominating measure μ), and let the loss $L(\theta, d)$ satisfy the assumptions of Lemma 2.1. Then, any estimator δ taking values above the maximum of $g(\theta)$, or below the minimum, is inadmissible.*

(a) Show that if θ^* minimizes $g(\theta)$, and if $\hat{g}(x)$ is an unbiased estimator of $g(\theta)$, then there exists $\epsilon > 0$ such that the set $A_\epsilon = \{x \in \mathcal{X} : \hat{g}(x) < g(\theta) - \epsilon\}$ satisfies $P(A_\epsilon) > 0$. A similar conclusion holds if $g(\theta^*)$ is a maximum.

(b) Suppose $g(\theta^*)$ is a minimum. (The case of a maximum is handled similarly.) Show that the estimator

$$\delta(x) = \begin{cases} \hat{g}(x) & \text{if } \hat{g}(x) \ge g(\theta^*) \\ g(\theta^*) & \text{if } \hat{g}(x) < g(\theta^*) \end{cases}$$

satisfies $R(\delta, \theta) - R(\hat{g}(x), \theta) < 0$.

(c) For the situation of Example 2.3, apply Lemma 8.1 to establish the inadmissibility of the UMVU estimator of σ_A^2. Also, explain why the hypotheses of Lemma 8.1 are not satisfied for the estimation of σ^2

2.2 Determine the Bayes risk of the estimator (2.4) when θ has the prior distribution $N(\mu, \tau^2)$.

2.3 Prove part (d) in the second proof of Example 2.8, that there exists a sequence of values $\theta_i \to -\infty$ with $b'(\theta_i) \to 0$.

2.4 Show that an estimator $aX + b$ $(0 \le a \le 1)$ of $E_\theta(X)$ is inadmissible (with squared error loss) under each of the following conditions:

(a) if $E_\theta(X) \ge 0$ for all θ, and $b < 0$;

(b) if $E_\theta(X) \le k$ for all θ, and $ak + b > k$.

[*Hint*: In (b), replace X by $X' = k - X$ and $aX + b$ by $k - (aX + b) = aX' + k - b - ak$, respectively and use (a).]

2.5 Show that an estimator $[1/(1+\lambda)+\varepsilon]X$ of $E_\theta(X)$ is inadmissible (with squared error loss) under each of the following conditions:

(a) if $\text{var}_\theta(X)/E_\theta^2(X) > \lambda > 0$ and $\varepsilon > 0$,

(b) if $\text{var}_\theta(X)/E_\theta^2(X) < \lambda$ and $\varepsilon < 0$.

[*Hint*: (a) Differentiate the risk function of the estimator with respect to ε to show that it decreases as ε decreases (Karlin 1958).]

2.6 Show that if $\text{var}_\theta(X)/E_\theta^2(X) > \lambda > 0$, an estimator $[1/(1+\lambda)+\varepsilon]X+b$ is inadmissible (with squared error loss) under each of the following conditions:

(a) if $E_\theta(X) > 0$ for all θ, $b > 0$ and $\varepsilon > 0$;

(b) if $E_\theta(X) < 0$ for all θ, $b < 0$ and $\varepsilon > 0$ (Gupta 1966).

2.7 Brown (1986a) points out a connection between the information inequality and the unbiased estimator of the risk of Stein-type estimators.

(a) Show that (2.7) implies

$$R(\theta, \delta) \ge \frac{[1 + b'(\theta)]^2}{n} + b^2(\theta) \ge \frac{1}{n} + \frac{2b'(\theta)}{n} + b^2(\theta)$$

and, hence, if $R(\theta, \delta) \le R(\theta, \bar{X})$, then $\frac{2b'(\theta)}{n} + b^2 \le 0$.

(b) Show that a nontrivial solution $b(\theta)$ would lead to an improved estimator $x - g(x)$, for $p = 1$, in Corollary 4.7.2.

2.8 A density function $f(x|\theta)$ is *variation reducing of order* $n + 1$ (VR_{n+1}) if, for any function $g(x)$ with k ($k \le n$) sign changes (ignoring zeros), the expectation $E_\theta g(X) = \int g(x) f(x|\theta)\,dx$ has at most k sign changes. If $E_\theta g(X)$ has exactly k sign changes, they are in the same order.

Show that $f(x|\theta)$ is VR_2 if and only if it has monotone likelihood ratio. (See TSH2, Lemma 2, Section 3.3 for the "if" implication).

Brown et al. (1981) provide a thorough introduction to this topic, including VR characterizations of many families of distributions (the exponential family is VR_∞, as is the χ_ν^2 with ν the parameter, and the noncentral $\chi_\nu^2(\lambda)$ in λ). There is an equivalence between VR_n and TP_n, Karlin's (1968) *total positivity of order n*, in that $VR_n = TP_n$.

2.9 For the situation of Example 2.9, show that:

(a) without loss of generality, the restriction $\theta \in [a, b]$ can be reduced to $\theta \in [-m, m]$, $m > 0$.

(b) If Λ is the prior distribution that puts mass 1/2 on each of the points $\pm m$, then the Bayes estimator against squared error loss is

$$\delta^\Lambda(\bar{x}) = m\frac{e^{mn\bar{x}} - e^{-mn\bar{x}}}{e^{mn\bar{x}} + e^{-mn\bar{x}}} = m\tanh(mn\bar{x}).$$

(c) For $m < 1/\sqrt{n}$,

$$\max_{\theta\in[-m,m]} R(\theta, \delta(\bar{X})) = \max\left\{R(-m, \delta^\Lambda(\bar{X})), R(m, \delta^\Lambda(\bar{X}))\right\}$$

and hence, by Corollary 1.6, δ^Λ is minimax.

[*Hint*: Problem 2.8 can be used to show that the derivative of the risk function can have at most one sign change, from negative to positive, and hence any interior extrema can only be a minimum.]

(d) For $m > 1.05/\sqrt{n}$, $\delta^\wedge$ of part (b) is no longer minimax. Explain why this is so and suggest an alternate estimator in this case.
[*Hint*: Consider $R(0, \delta^\wedge)$].

2.10 For the situation of Example 2.10, show that:

(a) $\max_{\theta \in [-m,m]} R(\theta, a\bar{X} + b) = \max\{R(-m, a\bar{X} + b), R(m, a\bar{X} + b)\}$.

(b) The estimator $a^*\bar{X}$, with $a^* = m^2/(\frac{1}{n} + m^2)$, is the linear minimax estimator for all m with minimax risk a^*/n.

(c) $\bar{X}$ is the linear minimax estimator for $m = \infty$.

2.11 Suppose X has distribution F_ξ and Y has distribution G_η, where ξ and η vary independently. If it is known that $\eta = \eta_0$, then any estimator $\delta(X, Y)$ can be improved upon by

$$\delta^*(x) = E_Y \delta(x, Y) = \int \delta(x, y)\, dG_{\eta_0}(y).$$

[*Hint*: Recall the proof of Theorem 1.6.1.]

2.12 In Example 2.13, prove that the estimator $aY + b$ is inadmissible when $a > 1/(r+1)$.
[*Hint*: Problems 2.4–2.6]

2.13 Let $X_1, \ldots, X_n$ be iid according to a $N(0, \sigma^2)$ density, and let $S^2 = \sum X_i^2$. We are interested in estimating σ^2 under squared error loss using linear estimators $cS^2 + d$, where c and d are constants. Show that:

(a) admissibility of the estimator $aY + b$ in Example 2.13 is equivalent to the admissibility of $cS^2 + d$, for appropriately chosen c and d.

(b) the risk of $cS^2 + d$ is given by $R(cS^2 + d, \sigma^2) = 2nc^2\sigma^2 + [(nc - 1)\sigma^2 + d]^2$

(c) for $d = 0$, $R(cS^2, \sigma^2) < R(0, \sigma^2)$ when $c < 2/(n+2)$, and hence the estimator $aY + b$ in Example 2.13 is inadmissible when $a = b = 0$.

[This exercise illustrates the fact that constants are not necessarily admissible estimators.]

2.14 For the situation of Example 2.15, let $Z = \bar{X}/S$.

(a) Show that the risk, under squared error loss, of $\delta = \varphi(z)s^2$ is minimized by taking

$$\varphi(z) = \varphi^*_{\mu,\sigma}(z) = E(S^2/\sigma^2|z)/E((S^2/\sigma^2)^2|z).$$

(b) Stein (1964) showed that $\varphi^*_{\mu,\sigma}(z) \le \varphi^*_{0,1}(z)$ for every μ, σ. Assuming this is so, deduce that $\varphi_s(Z)S^2$ dominates $[1/(n+1)]S^2$ in squared error loss, where

$$\varphi_s(z) = \min\left\{\varphi^*_{0,1}(z), \frac{1}{n+1}\right\}.$$

(c) Show that $\varphi^*_{0,1}(z) = (1 + z^2)/(n+2)$, and, hence, $\varphi_s(Z)S^2$ is given by (2.31).

(d) The best equivariant estimator of the form $\varphi(Z)S^2$ was derived by Brewster and Zidek (1974) and is given by

$$\varphi_{BZ}(z) = \frac{E(S^2|Z \le z)}{E(S^4|Z \le z)},$$

where the expectation is calculated assuming $\mu = 0$ and $\sigma = 1$. Show that $\varphi_{BZ}(Z)S^2$ is generalized Bayes against the prior

$$\pi(\mu, \sigma) = \frac{1}{\sigma}\int_0^\infty u^{-1/2}(1+u)^{-1}e^{-un\mu^2/\sigma^2}\,du\,d\mu\,d\sigma.$$

[Brewster and Zidek did not originally derive their estimator as a Bayes estimator, but rather first found the estimator and then found the prior. Brown (1968) considered a family of estimators similar to those of Stein (1964), which took different values depending on a cutoff point for z^2. Brewster and Zidek (1974) showed that the number of cutoff points can be arbitrarily large. They constructed a sequence of estimators, with decreasing risks and increasingly dense cutoffs, whose limit was the best equivalent estimator.]

2.15 Show the equivalence of the following relationships: (a) (2.26) and (2.27), (b) (2.34) and (2.35) when $c = \sqrt{(n-1)/(n+1)}$, and (c) (2.38) and (2.39).

2.16 In Example 2.17, show that the estimator $aX/n+b$ is inadmissible for all (a, b) outside the triangle (2.39).

[*Hint*: Problems 2.4–2.6.]

2.17 Prove admissibility of the estimators corresponding to the interior of the triangle (2.39), by applying Theorem 2.4 and using the results of Example 4.1.5.

2.18 Use Theorem 2.14 to provide an alternative proof for the admissibility of the estimator $a\bar{X} + b$ satisfying (2.6), in Example 2.5.

2.19 Determine which estimators $aX+b$ are admissible for estimating $E(X)$ in the following situations, for squared error loss:

(a) X has a Poisson distribution.

(b) X has a negative binomial distribution (Gupta 1966).

2.20 Let X have the Poisson(λ) distribution, and consider the estimation of λ under the loss $(d-\lambda)^2/\lambda$ with the restriction $0 \le \lambda \le m$, where m is known.

(a) Using an argument similar to that of Example 2.9, show that X is not minimax, and a least favorable prior distribution must have a set w_Λ [of (1.5)] consisting of a finite number of points.

(b) Let Λ_a be a prior distribution that puts mass a_i, $i = 1, \ldots, k$, at parameter points b_i, $i = 1, \ldots, k$. Show that the Bayes estimator associated with this prior is

$$\delta^{\Lambda_a}(x) = \frac{1}{E(\lambda^{-1}|x)} = \frac{\sum_{i=1}^k a_i b_i^x e^{-b_i}}{\sum_{i=1}^k a_i b_i^{x-1} e^{-b_i}}.$$

(c) Let m_0 be the solution to $m = e^{-m}$ ($m_0 \approx .57$). Show that for $0 \le \lambda \le m$, $m \le m_0$ a one-point prior ($a_i = 1$, $b_1 = m$) yields the minimax estimator. Calculate the minimax risk and compare it to that of X.

(d) Let m_1 be the first positive zero of $(1+\delta^\Lambda(m))^2 = 2+m^2/2$, where Λ is a two-point prior ($a_1 = a, b_1 = 0; a_2 = 1-a, b_2 = m$). Show that for $0 \le \lambda \le m, m_0 < m \le m_1$, a two-point prior yields the minimax estimator (use Corollary 1.6). Calculate the minimax risk and compare it to that of X.

[As m increases, the situation becomes more complex and exact minimax solutions become intractable. For these cases, linear approximations can be quite satisfactory. See Johnstone and MacGibbon 1992, 1993.]

2.21 Show that the conditions (2.41) and (2.42) of Example 2.22 are not only sufficient but also necessary for admissibility of (2.40).

2.22 Let X and Y be independently distributed according to Poisson distributions with $E(X) = \xi$ and $E(Y) = \eta$, respectively. Show that $aX + bY + c$ is admissible for estimating ξ with squared error loss if and only if either $0 \le a < 1, b \ge 0, c \ge 0$ or $a = 1, b = c = 0$ (Makani 1972).

2.23 Let X be distributed with density $\frac{1}{2}\beta(\theta)e^{\theta x}e^{-|x|}$, $|\theta| < 1$.

(a) Show that $\beta(\theta) = 1 - \theta^2$.

(b) Show that $aX + b$ is admissible for estimating $E_\theta(X)$ with squared error loss if and only if $0 \le a \le 1/2$.

[*Hint*: (b) To see necessity, let $\delta = (1/2 + \varepsilon)X + b$ $(0 < \varepsilon \le 1/2)$ and show that δ is dominated by $\delta' = (1 - \frac{1}{2}\alpha + \alpha\varepsilon)X + (b/\alpha)$ for some α with $0 < \alpha < 1/(1/2 - \varepsilon)$.]

2.24 Let X be distributed as $N(\theta, 1)$ and let θ have the improper prior density $\pi(\theta) = e^\theta$ $(-\infty < \theta < \infty)$. For squared error loss, the formal Bayes estimator of θ is $X + 1$, which is neither minimax nor admissible. (See also Problem 2.15.)

Conditions under which the formal Bayes estimator corresponding to an improper prior distribution for θ in Example 3.4 is admissible are given by Zidek (1970).

2.25 Show that the natural parameter space of the family (2.16) is $(-\infty, \infty)$ for the normal (variance known), binomial, and Poisson distribution but not in the gamma or negative binomial case.

Section 3

3.1 Show that Theorem 3.2.7 remains valid for almost equivariant estimators.

3.2 Verify the density (3.1).

3.3 In Example 3.3, show that a loss function remains invariant under G if and only if it is a function of $(d - \theta)^*$.

3.4 In Example 3.3, show that neither of the loss functions $[(d - \theta)^{**}]^2$ or $|(d - \theta)^{**}|$ is convex.

3.5 Let Y be distributed as $G(y - \eta)$. If $T = [Y]$ and $X = Y - T$, find the distribution of X and show that it depends on η only through $\eta - [\eta]$.

3.6 (a) If $X_1, \ldots, X_n$ are iid with density $f(x - \theta)$, show that the MRE estimator against squared error loss [the Pitman estimator of (3.1.28)] is the Bayes estimator against right-invariant Haar measure.

(b) If $X_1, \ldots, X_n$ are iid with density $1/\tau f[(x - \mu)/\tau]$, show that:

(i) Under squared error loss, the Pitman estimator of (3.1.28) is the Bayes estimator against right-invariant Haar measure.

(ii) Under the loss (3.3.17), the Pitman estimator of (3.3.19) is the Bayes estimator against right-invariant Haar measure.

3.7 Prove formula (3.9).

3.8 Prove (3.11).

[*Hint*: In the term on the left side, lim inf can be replaced by lim. Let the left side of (3.11) be A and the right side B, and let $A_N = \inf h(a, b)$, where the inf is taken over $a \le -N, b \ge N, N = 1, 2, \ldots$, so that $A_N \to A$. There exist (a_N, b_N) such that $|h(a_N, b_N) - A_N| \le 1/N$. Then, $h(a_N, b_N) \to A$ and $A \ge B$.]

3.9 In Example 3.8, let $h(\theta)$ be the length of the path θ after cancellation. Show that h does not satisfy conditions (3.2.11).

3.10 Discuss Example 3.8 for the case that the random walk instead of being in the plane is (a) on the line and (b) in three-space.

3.11 (a) Show that the probabilities (3.17) add up to 1.

(b) With p_k given by (3.17), show that the risk (3.16) is infinite.

[*Hint*: (a) $1/k(k+1) = (1/k) - 1/(k+1)$.]

3.12 Show that the risk $R(\theta, \delta)$ of (3.18) is finite.

[*Hint*: $R(\theta,\delta) < \Sigma_{k>M|k+\theta|} 1/(k+1) \le \Sigma_{c<k<d} 1/(k+1) < \int_c^{d+1} dx/x$, where $c = M|\theta|/(M+1)$ and $d = M|\theta|/(M-1)$. The reason for the second inequality is that values of k outside (c, d) make no contribution to the sum.]

3.13 Show that the two estimators δ^* and δ^{**}, defined by (3.20) and (3.21), respectively, are equivariant.

3.14 Prove the relations (3.22) and (3.23).

3.15 Let the distribution of X depend on parameters θ and ϑ, let the risk function of an estimator $\delta = \delta(x)$ of θ be $R(\theta, \vartheta; \delta)$, and let $r(\theta, \delta) = \int R(\theta, \vartheta; \delta)\, dP(\vartheta)$ for some distribution P. If δ_0 minimizes $\sup_\theta r(\theta, \delta)$ and satisfies $\sup_\theta r(\theta, \delta_0) = \sup_{\theta,\vartheta} R(\theta, \vartheta; \delta_0)$, show that δ_0 minimizes $\sup_{\theta,\vartheta} R(\theta, \vartheta; \delta)$.

Section 4

4.1 In Example 4.2, show that an estimator δ is equivariant if and only if it satisfies (4.11) and (4.12).

4.2 Show that a function μ satisfies (4.12) if and only if it depends only on ΣX_i^2.

4.3 Verify the Bayes estimator (4.15).

4.4 Let X_i be independent with binomial distribution $b(p_i, n_i), i = 1, \ldots, r$. For estimating $p = (p_1, \ldots, p_r)$ with average squared error loss (4.17), find the minimax estimator of p, and determine whether it is admissible.

4.5 Establishing the admissibility of the normal mean in two dimensions is quite difficult, made so by the fact that the conjugate priors fail in the limiting Bayes method. Let

$$X \sim N_2(\theta, I) \quad \text{and} \quad L(\theta, \delta) = |\theta - \delta|^2.$$

The conjugate priors are $\theta \sim N_2(0, \tau^2 I), \tau^2 > 0$.

(a) For this sequence of priors, verify that the limiting Bayes argument, as in Example 2.8, results in inequality (4.18), which does not establish admissibility.

(b) Stein (in James and Stein 1961), proposed the sequence of priors that works to prove X is admissible by the limiting Bayes method. A version of these priors, given by Brown and Hwang (1982), is

$$g_n(\theta) = \begin{cases} 1 & \text{if } |\theta| \le 1 \\ 1 - \dfrac{\log|\theta|}{\log n} & \text{if } 1 \le |\theta| \le n \\ 0 & \text{if } |\theta| \ge n \end{cases}$$

for $n = 2, 3, \ldots$. Show that $\delta^{g_n}(x) \to x$ a.e. as $n \to \infty$.

(c) A special case of the very general results of Brown and Hwang (1982) state that for the prior $\pi_n(\theta)^2 g(\theta)$, the limiting Bayes method (Blyth's method) will establish the admissibility of the estimator $\delta^g(x)$ [the generalized Bayes estimator against $g(\theta)$] if

$$\int_{\{\theta:|\theta|>1\}} \frac{g(\theta)\,d\theta}{|\theta|^2[\max\{\log|\theta|, \log 2\}]^2} < \infty.$$

Show that this holds for $g(\theta) = 1$ and that $\delta^g(x) = x$, so x is admissible.

[Stein (1956b) originally established the admissibility of X in two dimensions using an argument based on the information inequality. His proof was complicated by the fact that he needed some additional invariance arguments to establish the result. See Theorem 7.19 and Problem 7.19 for more general statements of the Brown/Hwang result.]

4.6 Let $X_1, X_2, \ldots, X_r$ be independent with $X_i \sim N(\theta_i, 1)$. The following heuristic argument, due to Stein (1956b), suggests that it should be possible, at least for large r and hence large $|\theta|$, to improve on the estimator $\mathbf{X} = (X_1, X_2, \ldots, X_r)$.

(a) Use a Taylor series argument to show

$$|\mathbf{x}|^2 = r + |\theta|^2 + O_p[(r + |\theta|^2)^{1/2}],$$

so, with high probability, the true θ is in the sphere $\{\theta : |\theta|^2 \le |\mathbf{x}|^2 - r\}$. The usual estimator $\mathbf{X}$ is approximately the same size as θ and will almost certainly be outside of this sphere.

(b) Part (a) suggested to Stein an estimator of the form $\delta(\mathbf{x}) = [1 - h(|\mathbf{x}|^2)]\mathbf{x}$. Show that

$$|\theta - \delta(\mathbf{x})|^2 = (1-h)^2|\mathbf{x} - \theta|^2 - 2h(1-h)\theta'(\mathbf{x} - \theta) + h^2|\theta|^2.$$

(c) Establish that $\theta'(\mathbf{x} - \theta)/|\theta| = Z \sim N(0, 1)$, and $|\mathbf{x} - \theta|^2 \approx r$, and, hence,

$$|\theta - \delta(\mathbf{x})|^2 \approx (1-h)^2 r + h^2|\theta|^2 + O_p[(r + |\theta|^2)^{1/2}].$$

(d) Show that the leading term in part (c) is minimized at $h = r/(r + |\theta|^2)$, and since $|\mathbf{x}|^2 \approx r + |\theta|^2$, this leads to the estimator $\delta(\mathbf{x}) = \left(1 - \frac{r}{|\mathbf{x}|^2}\right)\mathbf{x}$ of (4.20).

4.7 If S^2 is distributed as χ^2_r, use (2.2.5) to show that $E(S^{-2}) = 1/(r-2)$.

4.8 In Example 4.7, show that $\mathcal{R}$ is nonsingular for ρ_1 and ρ_2 and singular for ρ_3 and ρ_4.

4.9 Show that the function ρ_2 of Example 4.7 is convex.

4.10 In Example 4.7, show that X is admissible for (a) ρ_3 and (b) ρ_4.
[*Hint*: (a) It is enough to show that X_1 is admissible for estimating θ_1 with loss $(d_1 - \theta_1)^2$. This can be shown by letting $\theta_2, \ldots, \theta_r$ be known. (b) Note that X is admissible minimax for $\theta = (\theta_1, \ldots, \theta_r)$ when $\theta_1 = \cdots = \theta_r$.]

4.11 In Example 4.8, show that $\mathbf{X}$ is admissible under the assumptions (ii)(a).
[*Hint*:

i. If $v(t) > 0$ is such that

$$\int \frac{1}{v(t)} e^{-t^2/2\tau^2}\,dt < \infty,$$

show that there exists a constant $k(\tau)$ for which

$$\lambda_\tau(\theta) = k(\tau)\left[\Sigma v(\theta_j)\right] \exp\left(-\frac{1}{2\tau^2}\Sigma\theta_j^2\right) / \Pi v(\theta_j)$$

is a probability density for $\theta = (\theta_1, \ldots, \theta_r)$.

ii. If the X_i are independent $N(\theta_i, 1)$ and $\boldsymbol{\theta}$ has the prior $\lambda_\tau(\boldsymbol{\theta})$, the Bayes estimator of $\boldsymbol{\theta}$ with loss function (4.27) is $\tau^2 X/(1+\tau^2)$.

iii. To prove X admissible, use (4.18) with $\lambda_\tau(\boldsymbol{\theta})$ instead of a normal prior.]

4.12 Let $\mathcal{L}$ be a family of loss functions and suppose there exists $L_0 \in \mathcal{L}$ and a minimax estimator δ_0 with respect to L_0 such that in the notation of (4.29),

$$\sup_{L,\theta} R_L(\theta, \delta_0) = \sup_{\theta} R_{L_0}(\theta, \delta_0).$$

Then, δ_0 is minimax with respect to $\mathcal{L}$; that is, it minimizes $\sup_{L,\theta} R_L(\theta, \delta)$.

4.13 Assuming (4.25), show that $E = 1 - [(r-2)^2/r|\mathbf{X} - \mu|^2]$ is the unique unbiased estimator of the risk (5..4.25), and that E is inadmissible. [The estimator E is also unbiased for estimation of the loss $L(\boldsymbol{\theta}, \delta)$. See Note 9.5.]

4.14 A natural extension of risk domination under a particular loss is to risk domination under a class of losses. Hwang (1985) defines *universal domination* of δ by δ' if the inequality

$$E_{\boldsymbol{\theta}} L(|\boldsymbol{\theta} - \delta'(\mathbf{X})|) \le E_{\boldsymbol{\theta}} L(|\boldsymbol{\theta} - \delta(\mathbf{X})|) \text{ for all } \boldsymbol{\theta}$$

holds for all loss functions $L(\cdot)$ that are nondecreasing, with at least one loss function producing nonidentical risks.

(a) Show that δ' universally dominates δ if and only if it *stochastically dominates* δ, that is, if and only if

$$P_{\boldsymbol{\theta}}(|\boldsymbol{\theta} - \delta'(\mathbf{X})| > k) \le P_{\boldsymbol{\theta}}(|\boldsymbol{\theta} - \delta(\mathbf{X})| > k)$$

for all k and $\boldsymbol{\theta}$ with strict inequality for some $\boldsymbol{\theta}$.

[*Hint*: For a positive random variable Y, recall that $EY = \int_0^\infty P(Y > t)\,dt$. Alternatively, use the fact that stochastic ordering on random variables induces an ordering on expectations. See Lemma 1, Section 3.3 of TSH2.]

(b) For $X \sim N_r(\boldsymbol{\theta}, I)$, show that the James-Stein estimator $\delta^c(\mathbf{x}) = (1 - c/|\mathbf{x}|^2)\mathbf{x}$ does not universally dominate $\mathbf{x}$. [From (a), it only need be shown that $P_{\boldsymbol{\theta}}(|\boldsymbol{\theta} - \delta^c(\mathbf{X})| > k) > P_{\boldsymbol{\theta}}(|\boldsymbol{\theta} - \mathbf{X}| > k)$ for some $\boldsymbol{\theta}$ and k. Take $\boldsymbol{\theta} = 0$ and find such a k.]

Hwang (1985) and Brown and Hwang (1989) explore many facets of universal domination. Hwang (1985) shows that even δ^+ does not universally dominate X unless the class of loss functions is restricted.

We also note that although the inequality in part (a) may seem reminiscent of the "Pitman closeness" criterion, there is really no relation. The criterion of Pitman closeness suffers from a number of defects not shared by stochastic domination (see Robert et al. 1993).

Section 5

5.1 Show that the estimator δ_c defined by (5.2) with $0 < c = 1 - \Delta < 1$ is dominated by any δ_d with $|d - 1| < \Delta$.

5.2 In the context of Theorem 5.1, show that

$$E_{\boldsymbol{\theta}}\left[\frac{1}{|\mathbf{X}|^2}\right] \le E_0\left[\frac{1}{|\mathbf{X}|^2}\right] < \infty.$$

[*Hint*: The chi-squared distribution has monotone likelihood ratio in the noncentrality parameter.]

5.3 Stigler (1990) presents an interesting explanation of the Stein phenomenon using a regression perspective, and also gives an identity that can be used to prove the minimaxity of the James-Stein estimator. For $\mathbf{X} \sim (N_r\boldsymbol{\theta}, I)$, and $\delta^c(\mathbf{x}) = \left(1 - \frac{c}{|\mathbf{x}|^2}\right)\mathbf{x}$:

(a) Show that

$$E_{\boldsymbol{\theta}}|\boldsymbol{\theta} - \delta^c(\mathbf{X})|^2 = r - 2cE_{\boldsymbol{\theta}}\left[\frac{\mathbf{X}'\boldsymbol{\theta} + (c/2)}{|\mathbf{X}|^2} - 1\right].$$

(b) The expression in square brackets is increasing in c. Prove the minimaxity of δ^c for $0 \le c \le 2(r-2)$ by establishing Stigler's identity

$$E_{\boldsymbol{\theta}}\left[\frac{\mathbf{X}'\boldsymbol{\theta} + r - 2}{|\mathbf{X}|^2}\right] = 1.$$

[*Hint*: Part (b) can be established by transforming to polar coordinates and directly integrating, or by writing $\frac{\mathbf{x}'\boldsymbol{\theta}}{|\mathbf{x}|^2} = \frac{\mathbf{x}'(\boldsymbol{\theta}-\mathbf{x})+|\mathbf{x}|^2}{|\mathbf{x}|^2}$ and using Stein's identity.]

5.4 (a) Prove Theorem 5.5.

(b) Apply Theorem 5.5 to establish conditions for minimaxity of Strawderman's (1971) proper Bayes estimator given by (5.10) and (5.12).

[*Hint*: (a) Use the representation of the risk given in (5.4), with $g(\mathbf{x}) = c(|\mathbf{x}|)(r-2)\mathbf{x}/|\mathbf{x}|^2$. Show that $R(\boldsymbol{\theta}, \delta)$ can be written

$$R(\boldsymbol{\theta}, \delta) = 1 - \frac{(r-2)^2}{r}E_\theta\left[\frac{c(|\mathbf{X}|)(2 - c(|\mathbf{X}|))}{|\mathbf{X}|^2}\right] - \frac{2(r-2)}{r}E_\theta\frac{\Sigma X_i \frac{\partial}{\partial X_i}c(|\mathbf{X}|)}{|\mathbf{X}|^2}$$

and an upper bound on $R(\boldsymbol{\theta}, \delta)$ is obtained by dropping the last term. It is not necessary to assume that $c(\cdot)$ is differentiable everywhere; it can be nondifferentiable on a set of Lebesgue measure zero.]

5.5 For the hierarchical model (5.11) of Strawderman (1971):

(a) Show that the Bayes estimator against squared error loss is given by $E(\boldsymbol{\theta}|\mathbf{x}) = [1 - E(\lambda|\mathbf{x})]\mathbf{x}$ where

$$E(\lambda|\mathbf{x}) = \frac{\int_0^1 \lambda^{r/2-a+1}e^{-1/2\lambda|\mathbf{x}|^2}\,d\lambda}{\int_0^1 \lambda^{r/2-a}e^{-1/2\lambda|\mathbf{x}|^2}\,d\lambda}.$$

(b) Show that $E(\lambda|\mathbf{x})$ has the alternate representations

$$E(\lambda|\mathbf{x}) = \frac{r-2a+2}{|\mathbf{x}|^2}\frac{P(\chi^2_{r-2a+4} \le |\mathbf{x}|^2)}{P(\chi^2_{r-2a+2} \le |\mathbf{x}|^2}, $$

$$E(\lambda|\mathbf{x}) = \frac{r-2a+2}{|\mathbf{x}|^2} - \frac{2e^{-1/2|\mathbf{x}|^2}}{|\mathbf{x}|^2\int_0^1 \lambda^{r/2-a}e^{-1/2\lambda|\mathbf{x}|^2}}\,d\lambda,$$

and hence that $a = 0$ gives the estimator of (5.12).

(c) Show that $|\mathbf{x}|^2E(\lambda|\mathbf{x})$ is increasing in $|\mathbf{x}|^2$ with maximum $r - 2a + 2$. Hence, the Bayes estimator is minimax if $r - 2a + 2 \le 2(r-2)$ or $r \ge 2(3-a)$. For $0 \le a \le 1$, this requires $r \ge 5$.

[Berger (1976b) considers matrix generalizations of this hierarchical model and derives admissible minimax estimators. Proper Bayes minimax estimators only exist if $r \ge 5$ (Strawderman 1971); however, formal Bayes minimax estimators exist for $r = 3$ and 4.]

5.6 Consider a generalization of the Strawderman (1971) hierarchical model of Problem 5.5:

$$\mathbf{X}|\boldsymbol{\theta} \sim N(\boldsymbol{\theta}, I),$$
$$\boldsymbol{\theta}|\lambda \sim N(0, \lambda^{-1}(1-\lambda)I),$$
$$\lambda \sim \pi(\lambda).$$

(a) Show that the Bayes estimator against squared error loss is $[1 - E(\lambda|\mathbf{x})]\mathbf{x}$, where

$$E(\lambda|\mathbf{x}) = \frac{\int_0^1 \lambda^{r/2+1} e^{-1/2\lambda|\mathbf{x}|^2} \pi(\lambda)\, d\lambda}{\int_0^1 \lambda^{r/2} e^{-1/2\lambda|\mathbf{x}|^2} \pi(\lambda)\, d\lambda}.$$

(b) Suppose $\lambda \sim \text{beta}(\alpha, \beta)$, with density

$$\pi(\lambda) = \frac{\Gamma(\alpha+\beta)}{\Gamma(\alpha)\Gamma(\beta)} \lambda^{\alpha-1}(1-\lambda)^{\beta-1}.$$

Show that the Bayes estimator is minimax if $\beta \geq 1$ and $0 \leq a \leq (r-4)/2$.

[*Hint*: Use integration by parts on $E(\lambda|\mathbf{x})$, and apply Theorem 5.5. These estimators were introduced by Faith (1978).]

(c) Let $t = \lambda^{-1}(1-\lambda)$, the prior precision of $\boldsymbol{\theta}$. If $\lambda \sim \text{beta}(\alpha, \beta)$, show that the density of t is proportional to $t^{\alpha-1}/(1+t)^{\alpha+\beta}$, that is, $t \sim F_{2\alpha,2\beta}$, the F-distribution with 2α and 2β degrees of freedom.

[Strawderman's prior of Problem 5.5 corresponds to $\beta = 1$ and $0 < \alpha < 1$. If we take $\alpha = 1/2$ and $\beta = 1$, then $t \sim F_{1,2}$.]

(d) Two interesting limiting cases are $\alpha = 1, \beta = 0$ and $\alpha = 0, \beta = 1$. For each case, show that the resulting prior on t is proper, and comment on the minimaxity of the resulting estimators.

5.7 Faith (1978) considered the hierarchical model

$$\mathbf{X}|\boldsymbol{\theta} \sim N(\boldsymbol{\theta}, I),$$
$$\boldsymbol{\theta}|t \sim N\left(0, \frac{1}{t}I\right),$$
$$t \sim \text{Gamma}(a, b),$$

that is,

$$\pi(t) = \frac{1}{\Gamma(a)b^a} t^{a-1} e^{-t/b}.$$

(a) Show that the marginal prior for $\boldsymbol{\theta}$, unconditional on t, is

$$\pi(\boldsymbol{\theta}) \propto (2/b + |\boldsymbol{\theta}|^2)^{-(a+r/2)},$$

a multivariate Student's t-distribution.

(b) Show that $a \leq -1$ is a sufficient condition for $\sum_i \frac{\partial^2 \pi(\boldsymbol{\theta})}{\partial \theta_i^2} \geq 0$ and, hence, is a sufficient condition for the minimaxity of the Bayes estimator against squared error loss.

(c) Show, more generally, that the Bayes estimator against squared error loss is minimax if $a \leq (r-4)/2$ and $a \leq 1/b + 3$.

(d) What choices of a and b would produce a multivariate Cauchy prior for $\pi(\boldsymbol{\theta})$? Is the resulting Bayes estimator minimax?

5.8 (a) Let $X \sim N(\theta, \Sigma)$ and consider the estimation of θ under the loss $L(\theta, \delta) = (\theta - \delta)'(\theta - \delta)$. Show that $R(\theta, \mathbf{X}) = \operatorname{tr} \Sigma$, the minimax risk. Hence, $\mathbf{X}$ is a minimax estimator.

(b) Let $X \sim N(\theta, I)$ and consider estimation of θ under the loss $L(\theta, \delta) = (\theta - \delta)' Q(\theta - \delta)$, where Q is a known positive definite matrix. Show that $R(\theta, \mathbf{X}) = \operatorname{tr} Q$, the minimax risk. Hence, $\mathbf{X}$ is a minimax estimator.

(c) Show that the calculations in parts (a) and (b) are equivalent.

5.9 In Theorem 5.7, verify

$$E_\theta \frac{c(|\mathbf{X}|^2)}{|\mathbf{X}|^2}\mathbf{X}'(\theta - \mathbf{X}) = E_\theta \left\{ \frac{c(|\mathbf{X}|^2)}{|\mathbf{X}|^2}\operatorname{tr}(\Sigma) - 2\frac{c(|\mathbf{X}|^2)}{|\mathbf{X}|^4}\mathbf{X}'\Sigma\mathbf{X} + 2\frac{c'(|\mathbf{X}|^2)}{|\mathbf{X}|^2}\mathbf{X}'\Sigma\mathbf{X} \right\}.$$

[*Hint*: There are several ways to do this:

(a) Write

$$\begin{aligned} E_\theta \frac{c(|\mathbf{X}|^2)}{|\mathbf{X}|^2}\mathbf{X}'(\theta - \mathbf{X}) &= E_\theta \frac{c(\mathbf{Y}'\mathbf{Y}}{\mathbf{Y}'\mathbf{Y}}\mathbf{Y}'\Sigma(\eta - \mathbf{Y}) \\ &= \sum_i E_\theta \left\{ \frac{c(\mathbf{Y}'\mathbf{Y}}{\mathbf{Y}'\mathbf{Y}} \sum_j Y_j \sigma_{ji}(\eta_i - Y_i) \right\} \end{aligned}$$

where $\Sigma = \{\sigma_{ij}\}$ and $\mathbf{Y} = \Sigma^{-1/2}\mathbf{X} \sim N(\Sigma^{-1/2}\theta, I) = N(\eta, I)$. Now apply Stein's lemma.

(b) Write $\Sigma = PDP'$, where P is an orthogonal matrix ($P'P = I$) and D=diagonal matrix of eigenvalues of Σ, $D = \text{diagonal}\{d_i\}$. Then, establish that

$$E_\theta \frac{c(|\mathbf{X}|^2)}{|\mathbf{X}|^2}\mathbf{X}'(\theta - \mathbf{X}) = \sum_j E_\theta \frac{c(\sum_i d_i Z_i^2)}{\sum_i d_i Z_i^2} d_j Z_j(\eta_j^* - Z_j)$$

where $\mathbf{Z} = P\Sigma^{-1/2}\mathbf{X}$ and $\eta^* = P\Sigma^{-1/2}\theta$. Now apply Stein's lemma.

5.10 In Theorem 5.7, show that condition (i) allows the most shrinkage when $\Sigma = \sigma^2 I$, for some value of σ^2. That is, show that for all $r \times r$ positive definite Σ,

$$\max_\Sigma \frac{\operatorname{tr}\Sigma}{\lambda_{\max}(\Sigma)} = \frac{\operatorname{tr}\sigma^2 I}{\lambda_{\max}(\sigma^2 I)} = r.$$

[*Hint*: Write $\frac{\operatorname{tr}\Sigma}{\lambda_{\max}(\Sigma)} = \sum \lambda_i/\lambda_{\max}$, where the λ_i's are the eigenvalues of Σ.]

5.11 The estimation problem of (5.18),

$$\begin{gathered} \mathbf{X} \sim N(\theta, \Sigma) \\ L(\theta, \delta) = (\theta - \delta)' Q(\theta - \delta), \end{gathered}$$

where both Σ and Q are positive definite matrices, can always be reduced, without loss of generality, to the simpler case,

$$\begin{gathered} \mathbf{Y} \sim N(\eta, I) \\ L(\eta, \delta^*) = (\eta - \delta^*)' D_{q^*}(\eta - \delta^*), \end{gathered}$$

where D_{q^*} is a diagonal matrix with elements $(q_1^*, \ldots, q_r^*)$, using the following argument. Define $R = \Sigma^{1/2}B$, where $\Sigma^{1/2}$ is a symmetric square root of Σ (that is, $\Sigma^{1/2}\Sigma^{1/2} = \Sigma$), and B is the matrix of eigenvectors of $\Sigma^{1/2}Q\Sigma^{1/2}$ (that is, $B'\Sigma^{1/2}Q\Sigma^{1/2}B = D_{q^*}$).

(a) Show that R satisfies

$$R'\Sigma^{-1}R = I, \quad R'QR = D_{q^*}$$

(b) Define $\mathbf{Y} = R^{-1}\mathbf{X}$. Show that $\mathbf{Y} \sim N(\eta, I)$, where $\eta = R^{-1}\theta$.

(c) Show that estimations problems are equivalent if we define $\delta^*(\mathbf{Y}) = R^{-1}\delta(R\mathbf{Y})$.

[*Note*: If Σ has the eigenvalue-eigenvector decomposition $P'\Sigma P = D = \text{diagonal}(d_1, \cdots, d_r)$, then we can define $\Sigma^{1/2} = PD^{1/2}P'$, where $D^{1/2}$ is a diagonal matrix with elements $\sqrt{d_i}$. Since Σ is positive definite, the d_i's are positive.]

5.12 Complete the proof of Theorem 5.9.

(a) Show that the risk of $\delta(\mathbf{x})$ is

$$R(\theta, \delta) = E_\theta\left[(\theta - \mathbf{X})'Q(\theta - \mathbf{X})\right]$$
$$-2E_\theta\left[\frac{c(|\mathbf{X}|^2)}{|\mathbf{X}|^2}\mathbf{X}'Q(\theta - \mathbf{X})\right]$$
$$+E_\theta\left[\frac{c^2(|\mathbf{X}|^2)}{|\mathbf{X}|^4}\mathbf{X}'Q\mathbf{X}\right]$$

where $E_\theta(\theta - \mathbf{X})'Q(\theta - \mathbf{X}) = \text{tr}(Q)$.

(b) Use Stein's lemma to verify

$$E_\theta\frac{c(|\mathbf{X}|^2)}{|\mathbf{X}|^2}\mathbf{X}'Q(\theta - \mathbf{X})$$
$$= E_\theta\left\{\frac{c(|\mathbf{X}|^2)}{|\mathbf{X}|^2}\text{tr}(Q) - 2\frac{c(|\mathbf{X}|^2)}{|\mathbf{X}|^4}\mathbf{X}'Q\mathbf{X} + 2\frac{c'(|\mathbf{X}|^2)}{|\mathbf{X}|^2}\mathbf{X}'Q\mathbf{X}\right\}.$$

Use an argument similar to the one in Theorem 5.7.

[*Hint*: Write

$$E_\theta\frac{c(|\mathbf{X}|^2)}{|\mathbf{X}|^2}\mathbf{X}'Q(\theta - \mathbf{X}) = \sum_i E_\theta\left\{\frac{c(|\mathbf{X}|^2)}{|\mathbf{X}|^2}\sum_j X_j q_{ji}(\theta_i - X_i)\right\}$$

and apply Stein's lemma.]

5.13 Prove the following "generalization" of Theorem 5.9.

Theorem 8.2 *Let $\mathbf{X} \sim N(\theta, \Sigma)$. An estimator of the form (5.13) is minimax against the loss $L(\theta, \delta) = (\theta - \delta)'Q(\theta - \delta)$, provided*

(i) $0 \le c(|\mathbf{x}|^2) \le 2[\text{tr}(Q^*)/\lambda_{\max}(Q^*)] - 4$,

(ii) the function $c(\cdot)$ is nondecreasing,

where $Q^* = \Sigma^{1/2}Q\Sigma^{1/2}$.

5.14 Brown (1975) considered the performance of an estimator against a class of loss functions

$$\mathcal{L}(C) = \left\{L : L(\theta, \delta) = \sum_{i=1}^r c_i(\theta_i - \delta_i)^2; \quad (c_1, \ldots, c_r) \in C\right\}$$

for a specified set C, and proved the following theorem.

Theorem 8.3 *For $\mathbf{X} \sim N_r(\theta, I)$, there exists a spherically symmetric estimator δ, that is, $\delta(\mathbf{x}) = [1 - h(|\mathbf{x}|^2)]\mathbf{x}$, where $h(|\mathbf{x}|^2) \neq 0$, such that $R(\theta, \delta) \le R(\theta, \mathbf{X})$ for all $L \in \mathcal{L}(C)$ if, for all $(c_1, \ldots, c_r) \in C$, the inequality $\sum_{j=1}^r c_i > 2c_k$ holds for $k = 1, \ldots, r$.*

Show that this theorem is equivalent to Theorem 5.9 in that the above inequality is equivalent to part (i) of Theorem 5.9, and the estimator (5.13) is minimax.

[*Hint*: Identify the eigenvalues of Q with $c_1, \ldots, c_r$.]

Bock (1975) also establishes this theorem; see also Shinozaki (1980).

5.15 There are various ways to seemingly generalize Theorems 5.5 and 5.9. However, if both the estimator and loss function are allowed to depend on the covariance and loss matrix, then linear transformations can usually reduce the problem.

Let $X \sim N_r(\theta, \Sigma)$, and let the loss function be $L(\theta, \delta) = (\theta - \delta)' Q(\theta - \delta)$, and consider the following "generalizations" of Theorems 5.5 and 5.9.

(a) $$\delta(\mathbf{x}) = \left(1 - \frac{c(\mathbf{x}'\Sigma^{-1}\mathbf{x})}{\mathbf{x}'\Sigma^{-1}\mathbf{x}}\right)\mathbf{x}, \quad Q = \Sigma^{-1},$$

(b) $$\delta(\mathbf{x}) = \left(1 - \frac{c(\mathbf{x}'Q\mathbf{x})}{\mathbf{x}'Q\mathbf{x}}\right)\mathbf{x}, \quad \Sigma = I \text{ or } \Sigma = Q,$$

(c) $$\delta(\mathbf{x}) = \left(1 - \frac{c(\mathbf{x}'\Sigma^{-1/2}Q\Sigma^{-1/2}\mathbf{x})}{\mathbf{x}'\Sigma^{-1/2}Q\Sigma^{-1/2}\mathbf{x}}\right)\mathbf{x}.$$

In each case, use transformations to reduce the problem to that of Theorem 5.5 or 5.9, and deduce the condition for minimaxity of δ.

[*Hint*: For example, in (a) the transformation $Y = \Sigma^{-1/2}\mathbf{X}$ will show that δ is minimax if $0 < c(\cdot) < 2(r-2)$.]

5.16 A natural extension of the estimator (5.10) is to one that shrinks toward an arbitrary known point $\mu = (\mu_1, \ldots, \mu_r)$,

$$\delta_\mu(\mathbf{x}) = \mu + \left[1 - c(S)\frac{r-2}{|\mathbf{x} - \mu|^2}\right](\mathbf{x} - \mu)$$

where $|\mathbf{x} - \mu|^2 = \Sigma(x_i - \mu_i)^2$.

(a) Show that, under the conditions of Theorem 5.5, δ_μ is minimax.

(b) Show that its positive-part version is a better estimator.

5.17 Let $\mathbf{X} \sim N_r(\theta, I)$. Show that the Bayes estimator of θ, against squared error loss, is given by $\delta(\mathbf{x}) = \mathbf{x} + \nabla \log m(\mathbf{x})$ where $m(\mathbf{x})$ is the marginal density function and $\nabla f = \{\partial/\partial x_i f\}$.

5.18 Verify (5.27).

[*Hint*: Show that, as a function of $|\mathbf{x}|^2$, the only possible interior extremum is a minimum, so the maximum must occur either at $|\mathbf{x}|^2 = 0$ or $|\mathbf{x}|^2 = \infty$.]

5.19 The property of superharmonicity, and its relationship to minimaxity, is not restricted to Bayes estimators. For $X \sim N_r(\theta, I)$, a ***pseudo-Bayes*** estimator (so named, and investigated by Bock, 1988) is an estimator of the form

$$\mathbf{x} + \nabla \log m(\mathbf{x})$$

where $m(\mathbf{x})$ is not necessarily a marginal density.

(a) Show that the positive-part Stein estimator

$$\delta_a^+ = \mu + \left(1 - \frac{a}{|\mathbf{x} - \mu|^2}\right)^+ (\mathbf{x} - \mu)$$

is a pseudo-Bayes estimator with

$$m(\mathbf{x}) = \begin{cases} e^{-(1/2)|\mathbf{x}-\mu|^2} & \text{if } |\mathbf{x} - \mu|^2 < a \\ (|\mathbf{x} - \mu|^2)^{-a/2} & \text{if } |\mathbf{x} - \mu|^2 \ge a. \end{cases}$$

(b) Show that, except at the point of discontinuity, if $a \le r-2$, then $\sum_{i=1}^r \frac{\partial^2}{\partial x_i^2} m(\mathbf{x}) \le 0$, so $m(\mathbf{x})$ is superharmonic.

(c) Show how to modify the proof of Corollary 5.11 to accommodate superharmonic functions $m(\mathbf{x})$ with a finite number of discontinuities of measure zero.

This result is adapted from George (1986a, 1986b), who exploits both pseudo-Bayes and superharmonicity to establish minimaxity of an interesting class of estimators that are further investigated in the next problem.

5.20 For $X|\theta \sim N_r(\theta, I)$, George (1986a, 1986b) looked at *multiple shrinkage* estimators, those that can shrink to a number of different targets. Suppose that $\theta \sim \pi(\theta) = \sum_{j=1}^k \omega_i \pi_i(\theta)$, where the ω_i are known positive weights, $\sum \omega_i = 1$.

(a) Show that the Bayes estimator against $\pi(\theta)$, under squared error loss, is given by $\delta^*(\mathbf{x}) = \mathbf{x} + \nabla \log m^*(\mathbf{x})$ where $m^*(\mathbf{x}) = \sum_{j=1}^k \omega_j m_j(\mathbf{x})$ and

$$m_i(\mathbf{x}) = \int_\Omega \frac{1}{(2\pi)^{p/2}} e^{-(1/2)|\mathbf{x}-\theta|^2} \pi_i(\theta)\, d\theta.$$

(b) Clearly, δ^* is minimax if $m^*(\mathbf{x})$ is superharmonic. Show that $\delta^*(\mathbf{x})$ is minimax if either (i) $m_i(\mathbf{x})$ is superharmonic, $i = 1, \dots, k$, or (ii) $\pi_i(\theta)$ is superharmonic, $i = 1, \dots, k$. [*Hint*: Problem 1.7.16]

(c) The real advantage of δ^* occurs when the components specify different targets. For $\rho_j = \omega_j m_j(x)/m^*(x)$, let $\delta^*(\mathbf{x}) = \sum_{j=1}^k \rho_j \delta_j^+(\mathbf{x})$ where

$$\delta_j^+(\mathbf{x}) = \mu_j + \left(1 - \frac{r-2}{|\mathbf{x}-\mu_j|^2}\right)^+ (\mathbf{x}-\mu_j)$$

and the μ_j's are target vectors. Show that $\delta^*(\mathbf{x})$ is minimax. [*Hint*: Problem 5.19]

[George (1986a, 1986b) investigated many types of multiple targets, including multiple points, subspaces, and clusters and subvectors. The subvector problem was also considered by Berger and Dey (1983a, 1983b). Multiple shrinkage estimators were also investigated by Ki and Tsui (1990) and Withers (1991).]

5.21 Let X_i, Y_j be independent $N(\xi_i, 1)$ and $N(\eta_j, 1)$, respectively ($i = 1, \dots, r$; $j = 1, \dots, s$).

(a) Find an estimator of $(\xi_1, \dots, \xi_r; \eta_1, \dots, \eta_s)$ that would be good near $\xi_i = \cdots = \xi_r = \xi$, $\eta_1 = \cdots = \eta_s = \eta$, with ξ and η unknown, if the variability of the ξ's and η's is about the same.

(b) When the loss function is (4.17), determine the risk function of your estimator.

[*Hint*: Consider the Bayes situation in which $\xi_i \sim N(\xi, A)$ and $\eta_j \sim N(\eta, A)$. See Berger 1982b for further development of such estimators].

5.22 The early proofs of minimaxity of Stein estimators (James and Stein 1961, Baranchik 1970) relied on the representation of a noncentral χ^2-distribution as a Poisson sum of central χ^2 (TSH2, Problem 6.7). In particular, if $\chi_r^2(\lambda)$ is a noncentral χ^2 random variable with noncentrality parameter λ, then

$$E_\lambda h(\chi_r^2(\lambda)) = E[Eh(\chi_{r+2K}^2)|K]$$

where $K \sim \text{Poisson}(\lambda)$ and χ_{r+2k}^2 is a central χ^2 random variable with $r + 2k\, df$. Use this representation, and the properties of the central χ^2-distribution, to establish the following identities for $X \sim N_r(\theta, I)$ and $\lambda = |\theta|^2$.

(a) $E_\theta \frac{\mathbf{x}'\theta}{|\mathbf{x}|^2} = |\theta|^2 E \frac{1}{\chi^2_{r+2}(\lambda)}$.

(b) $(r-2)E\frac{1}{\chi^2_r(\lambda)} + |\theta|^2 E \frac{1}{\chi^2_{r+2}(\lambda)} = 1$.

(c) For $\delta(\mathbf{x}) = (1 - c/|\mathbf{x}|^2)\mathbf{x}$, use the identities (a) and (b) to show that for $L(\theta, \delta) = |\theta - \delta|^2$,

$$R(\theta, \delta) = r + 2c|\theta|^2 E \frac{1}{\chi^2_{r+2}(\lambda)} - 2c + c^2 E \frac{1}{\chi^2_r(\lambda)}$$
$$= r + 2c\left[1 - (r-2)E\frac{1}{\chi^2_r(\lambda)}\right] - 2c + c^2 E \frac{1}{\chi^2_r(\lambda)}$$

and, hence, that $\delta(\mathbf{x})$ is minimax if $0 \le c \le 2(r-2)$.

[See Bock 1975 or Casella 1980 for more identities involving noncentral χ^2 expectations.]

5.23 Let $\chi^2_r(\lambda)$ be a χ^2 random variable with r degrees of freedom and noncentrality parameter λ.

(a) Show that $E\frac{1}{\chi^2_r(\lambda)} = E\left[E\frac{1}{\chi^2_{r+2K}}|K\right] = E\left[\frac{1}{r-2+2K}\right]$, where $K \sim$ Poisson$(\lambda/2)$.

(b) Establish (5.32).

5.24 For the most part, the risk function of a Stein estimator increases as $|\theta|$ moves away from zero (if zero is the shrinkage target). To guarantee that the risk function is monotone increasing in $|\theta|$ (that is, that there are no "dips" in the risk as in Berger's 1976a tail minimax estimators) requires a somewhat stronger assumption on the estimator (Casella 1990). Let $X \sim N_r(\theta, I)$ and $L(\theta, \delta) = |\theta - \delta|^2$, and consider the Stein estimator

$$\delta(\mathbf{x}) = \left(1 - c(|\mathbf{x}|^2)\frac{(r-2)}{|\mathbf{x}|^2}\right)\mathbf{x}.$$

(a) Show that if $0 \le c(\cdot) \le 2$ and $c(\cdot)$ is concave and twice differentiable, then $\delta(\mathbf{x})$ is minimax. [*Hint*: Problem 1.7.7.]

(b) Under the conditions in part (a), the risk function of $\delta(\mathbf{x})$ is nondecreasing in $|\theta|$. [*Hint*: The conditions on $c(\cdot)$, together with the identity

$$(d/d\lambda)E_\lambda[h(\chi^2_p(\lambda))] = E_\lambda\{[\partial/\partial\chi^2_{p+2}(\lambda)]h(\chi^2_{p+2}(\lambda))\},$$

where $\chi^2_p(\lambda)$ is a noncentral χ^2 random variable with p degrees of freedom and noncentrality parameter λ, can be used to show that $(\partial/\partial|\theta|^2)R(\theta, \delta) > 0$.]

5.25 In the spirit of Stein's "large r and $|\theta|$" argument, Casella and Hwang (1982) investigated the limiting risk ratio of $\delta^{JS}(\mathbf{x}) = (1 - (r-2)/|\mathbf{x}|^2)\mathbf{x}$ to that of $\mathbf{x}$. If $X \sim N_r(\theta, I)$ and $L(\theta, \delta) = |\theta - \delta|^2$, they showed

$$\lim_{\substack{r\to\infty \\ \frac{|\theta|^2}{r} \to c}} \frac{R(\theta, \delta^{JS})}{R(\theta, \mathbf{X})} = \frac{c}{c+1}.$$

To establish this limit we can use the following steps.

(a) Show that $\frac{R(\theta, \delta^{JS})}{R(\theta, \mathbf{x})} = 1 - \frac{(r-2)^2}{r} E_\theta \frac{1}{|\mathbf{X}|^2}$.

(b) Show that $\frac{1}{p-2+|\theta|^2} \le E_\theta \frac{1}{|\mathbf{X}|^2} \le \frac{1}{p-2}\left(\frac{p}{p+|\theta|^2}\right)$.

(c) Show that the upper and lower bounds on the risk ratio both have the same limit.

[*Hint*: (b) The upper bound is a consequence of Problem 5.22(b). For the lower bound, show $E_\theta(1/|\mathbf{X})|^2) = E(1/p - 2 + K)$, where $K \sim \text{Poisson}(|\theta|^2)$ and use Jensen's inequality.]

Section 6

6.1 Referring to Example 6.1, this problem will establish the validity of the expression (6.2) for the risk of the estimator δ^L of (6.1), using an argument similar to that in the proof of Theorem 5.7.

(a) Show that

$$R(\theta, \delta^L) = \sum_i E_\theta[\theta_i - \delta_i^L(\mathbf{X})]^2$$
$$= \sum_i E_\theta \left\{ (\theta_i - X_i)^2 + \frac{2c(r-3)}{S}(\theta_i - X_i)(X_i - \bar{\mathbf{X}}) + \frac{[c(r-3)]^2}{S^2}(X_i - \bar{\mathbf{X}})^2 \right\}$$

where $S = \sum_j (X_j - \bar{\mathbf{X}})^2$.

(b) Use integration by parts to show

$$E_\theta \frac{(\theta_i - X_i)(X_i - \bar{\mathbf{X}})}{S} = -E_\theta \frac{\frac{r-1}{r}S + 2(X_i - \bar{\mathbf{X}})^2}{S^2}.$$

[*Hint*: Write the cross-term as $-E_\theta\left[\frac{(X_i - \bar{\mathbf{X}})}{S}\right](X_i - \theta_i)$ and adapt Stein's identity (Lemma 1.5.15).]

(c) Use the results of parts (a) and (b) to establish (6.2).

6.2 In Example 6.1, show that:

(a) The estimator δ^L is minimax if $r \geq 4$ and $c \leq 2$.

(b) The risk of δ^L is infinite if $r \leq 3$

(c) The minimum risk is equal to $3/r$, and is attained at $\theta_1 = \theta_2 = \cdots = \theta$.

(d) The estimator δ^L is dominated in risk by its positive-part version

$$\delta^{L^+} = \bar{x}\mathbf{1} + \left(1 - \frac{c(r-3)}{|\mathbf{x} - \bar{x}\mathbf{1}|^2}\right)^+ (\mathbf{x} - \bar{x}\mathbf{1}).$$

6.3 In Example 6.2:

(a) Show that $k\mathbf{x}$ is the MLE if $\theta \in \mathcal{L}_k$.

(b) Show that $\delta^k(\mathbf{x})$ of (6.8) is minimax under squared error loss.

(c) Verify that θ_i of the form (6.4) satisfy $T(T'T)^{-1}T'\theta = \theta$ for T of (6.5), and construct a minimax estimator that shrinks toward this subspace.

6.4 Consider the problem of estimating the mean based on $X \sim N_r(\theta, I)$, where it is thought that $\theta_i = \sum_{j=1}^s \beta_j t_i^j$ where $(t_i, \ldots, t_r)$ are known, $(\beta_1, \ldots, \beta_s)$ are unknown, and $r - s > 2$.

(a) Find the MLE of θ, say $\hat{\theta}_R$, if θ is assumed to be in the linear subspace

$$\mathcal{L} = \left\{ \theta : \sum_{j=1}^s \beta_j t_i^j = \theta_i, \; i = 1, \ldots, r \right\}.$$

(b) Show that $\mathcal{L}$ can be written in the form (6.7), and find K.

(c) Construct a Stein estimator that shrinks toward the MLE of part (a) and prove that it is minimax.

6.5 For the situation of Example 6.3:

(a) Show that $\delta^c(\mathbf{x}, \mathbf{y})$ is minimax if $0 \le c \le 2$.

(b) Show that if $\xi = 0$, $R(\theta, \delta^1) = 1 - \frac{\sigma^2}{\sigma^2+\tau^2}\frac{r-2}{r}$, $R(\theta, \delta^{\text{comb}}) = 1 - \frac{\sigma^2}{\sigma^2+\tau^2}$, and, hence, $R(\theta, \delta^1) > R(\theta, \delta^{\text{comb}})$.

(c) For $\xi \neq 0$, show that $R(\theta, \delta^{\text{comb}}) = 1 - \frac{\sigma^2}{\sigma^2+\tau^2} + \frac{|\xi|^2\sigma^2}{r(\sigma^2+\tau^2)}$ and hence is unbounded as $|\xi| \to \infty$.

6.6 The Green and Strawderman (1991) estimator $\delta^c(\mathbf{x}, \mathbf{y})$ can be derived as an empirical Bayes estimator.

(a) For $X|\theta \sim N_r(\theta, \sigma^2 I)$, $Y|\theta, \xi \sim N_r(\theta + \xi, \tau^2 I)$, $\xi \sim N(0, \gamma^2 I)$, and $\theta_i \sim$ Uniform$(-\infty, \infty)$, with σ^2 and τ^2 assumed to be known, show how to derive $\delta^{r-2}(\mathbf{x}, \mathbf{y})$ as an empirical Bayes estimator.

(b) Calculate the Bayes estimator, δ^π, against squared error loss.

(c) Compare $r(\pi, \delta^\pi)$ and $r(\pi, \delta^{r-2})$.

[*Hint*: For part (a), Green and Strawderman suggest starting with $\theta \sim N(0, \kappa^2 I)$ and let $\kappa^2 \to \infty$ get the uniform prior.]

6.7 In Example 6.4:

(a) Verify the risk function (6.13).

(b) Verify that for unknown σ^2, the risk function of the estimator (6.14) is given by (6.15).

(c) Show that the minimum risk of the estimator (6.14) is $1 - \frac{\nu}{\nu+2}\frac{r-2}{r}$.

6.8 For the situation of Example 6.4, the analogous modification of the Lindley estimator (6.1) is

$$\delta^L = \bar{x}\mathbf{1} + \left(1 - \frac{r-3}{\Sigma(x_i - \bar{x})^2/\hat{\sigma}^2}\right)(\mathbf{x} - \bar{x}\mathbf{1}),$$

where $\hat{\sigma}^2 = S^2/(\nu + 2)$ and $S^2/\sigma^2 \sim \chi^2_\nu$, independent of $\mathbf{X}$.

(a) Show that $R(\theta, \delta^L) = 1 - \frac{\nu}{\nu+2}\frac{(r-3)^2}{r}E_\theta\frac{\sigma^2}{\Sigma(x_i-\bar{x})^2}$.

(b) Show that both δ^L and δ of (6.14) can be improved by using their positive-part versions.

6.9 The major application of Example 6.4 is to the situation

$$Y_{ij} \sim N(\theta_i, \sigma^2),\ i = 1, \ldots, s,\ j = 1, \ldots, n, \text{ independent}$$

with $\bar{Y}_i = (1/n)\Sigma_j Y_{ij}$ and $\hat{\sigma}^2 = \Sigma_{ij}(Y_{ij} - \bar{Y}_i)^2/s(n-1)$. Show that the estimator

$$\delta_i = \bar{\bar{y}} + \left(1 - c\frac{(s-3)\hat{\sigma}^2}{\Sigma(\bar{y}_i - \bar{\bar{y}})^2}\right)^+(\bar{y}_i - \bar{\bar{y}})$$

is a minimax estimator, where $\bar{\bar{y}} = \Sigma_{ij}y_{ij}/sn$, as long as $0 \le c \le 2$.

[The case of unequal sample sizes n_i is not covered by what we have done so far. See Efron and Morris 1973b, Berger and Bock 1976, and Morris 1983 for approaches to this problem. The case of totally unknown covariance matrix is considered by Berger et al. (1977) and Gleser (1979, 1986).]

6.10 The positive-part Lindley estimator of Problem 6.9 has an interesting interpretation in the one-way analysis of variance, in particular with respect to the usual test performed, that of $H_0 : \theta_1 = \theta_2 = \cdots = \theta_s$. This hypothesis is tested with the statistic

$$F = \frac{\Sigma(\bar{y}_i - \bar{\bar{y}})^2/(s-1)}{\Sigma(y_{ij} - \bar{y}_i)^2/s(n-1)},$$

which, under H_0, has an F-distribution with $s-1$ and $s(n-1)$ degrees of freedom.

(a) Show that the positive-part Lindley estimator can be written as

$$\delta_i = \bar{\bar{y}} + \left(1 - c\frac{s-3}{s-1}\frac{1}{F}\right)^+ (\bar{y}_i - \bar{\bar{y}}).$$

(b) The null hypothesis is rejected if F is large. Show that this corresponds to using the MLE under H_0 if F is small, and a Stein estimator if F is large.

(c) The null hypothesis is rejected at level α if $F > F_{s-1,s(n-1),\alpha}$. For $s = 8$ and $n = 6$:

(i) What is the level of the test that corresponds to choosing $c = 1$, the optimal risk choice?

(ii) What values of c correspond to choosing $\alpha = .05$ or $\alpha = .01$, typical α levels. Are the resulting estimators minimax?

6.11 Prove the following extension of Theorem 5.5 to the case of unknown variance, due to Strawderman (1973).

Theorem 8.4 *Let $X \sim N_r(\theta, \sigma^2 I)$ and let $S^2/\sigma^2 \sim \chi^2_\nu$, independent of X. The estimator*

$$\delta^c(\mathbf{x}) = \left(1 - \frac{c(F, S^2)}{S^2}\frac{r-2}{\nu+2}\right)\mathbf{x},$$

where $F = \Sigma x_i^2/S^2$, is a minimax estimator of θ, provided

(i) for each fixed S^2, $c(\cdot, S^2)$ is nondecreasing,

(ii) for each fixed F, $c(F, \cdot)$ is nonincreasing,

(iii) $0 \le c(\cdot,\cdot) \le 2$.

[Note that, here, the loss function is taken to be scaled by σ^2, $L(\theta, \delta) = |\theta - \delta|^2/\sigma^2$, otherwise the minimax risk is not finite. Strawderman (1973) went on to derive proper Bayes minimax estimators in this case.]

6.12 For the situation of Example 6.5:

(a) Show that $E_\sigma \frac{1}{\sigma^2} = E_0 \frac{r-2}{|\mathbf{X}|^2}$.

(b) If $1/\sigma^2 \sim \chi^2_\nu/\nu$, then $f(|\mathbf{x} - \theta|)$ of (6.19) is the multivariate t-distribution, with ν degrees of freedom and $E_0|\mathbf{X}|^{-2} = (r-2)^{-1}$.

(c) If $1/\sigma^2 \sim Y$, where χ^2_ν/ν is stochastically greater than Y, then $\delta(\mathbf{x})$ of (6.20) is minimax for this mixture as long as $0 \le c \le 2(r-2)$.

6.13 Prove Lemma 6.2.

6.14 For the situation of Example 6.7:

(a) Verify that the estimator (6.25) is minimax if $0 \le c \le 2$. (Theorem 5.5 will apply.)

(b) Referring to (6.27), show that

$$E|\delta^\pi(\mathbf{X}) - \delta^c(\mathbf{X})|^2 I[|\mathbf{X}|^2 \geq c(r-2)(\sigma^2+\tau^2)]$$
$$= \frac{\sigma^4}{\sigma^2+\tau^2} E\frac{1}{Y}[Y - c(r-2)]^2 I[Y \geq c(r-2)]$$

where $Y \sim \chi^2_r$.

(c) If χ^2_ν denotes a chi-squared random variable with ν degrees of freedom, establish the identity

$$Eh(\chi^2_\nu) = \nu E\frac{h(\chi^2_{\nu+2})}{\chi^2_{\nu+2}}$$

to show that

$$r(\pi, \delta^R) = r(\pi, \delta^\pi) + \frac{1}{r-2}\frac{\sigma^4}{\sigma^2+\tau^2} E[Y - c(r-2)]^2 I[Y \geq c(r-2)]$$

where, now, $Y \sim \chi^2_{r-2}$.

(d) Verify (6.29), hence showing that $r(\pi, \delta^R) \leq r(\pi, \delta^c)$.

(e) Show that $E(Y-a)^2 I(Y > a)$ is a decreasing function of a, and hence the maximal Bayes risk improvement, while maintaining minimaxity, is obtained at $c = 2$.

6.15 For $X_i \sim \text{Poisson}(\lambda_i)$ $i = 1, \ldots, r$, independent, and loss function $L(\boldsymbol{\lambda}, \delta) = \Sigma(\lambda_i - \delta_i)^2/\lambda_i$:

(a) For what values of a, α, and β are the estimators of (4.6.29) minimax? Are they also proper Bayes for these values?

(b) Let $\Lambda = \Sigma\lambda_i$ and define $\theta_i = \lambda_i/\Lambda, i = 1, \ldots, r$. For the prior distribution $\pi(\theta, \Lambda) = m(\Lambda)d\Lambda \prod_{i=1}^r d\theta_i$, show that the Bayes estimator is

$$\delta^\pi(\mathbf{x}) = \frac{\psi_\pi(z)}{z+r-1}\mathbf{x},$$

where $z = \Sigma x_i$ and

$$\psi_\pi(z) = \frac{\int \Lambda^z e^{-\Lambda} m(\Lambda)\,d\Lambda}{\int \Lambda^{z-1} e^{-\Lambda} m(\Lambda)\,d\Lambda}.$$

(c) Show that the choice $m(\Lambda) = 1$, yields the estimator $\delta(\mathbf{x}) = [1-(r-1)/(z+r-1)]\mathbf{x}$, which is minimax.

(d) Show that the choice $m(\Lambda) = (1+\Lambda)^{-\beta}$, $1 \leq \beta \leq r-1$ yields an estimator that is proper Bayes minimax for $r > 2$.

(e) The estimator of part (d) is difficult to evaluate. However, for the prior choice

$$m(\Lambda) = \int_0^\infty \frac{t^{-r}e^{-1/t}}{(1+\Lambda t)^\beta}\,dt, \quad 1 \leq \beta \leq r-1,$$

show that the generalized Bayes estimator is

$$\delta^\pi(\mathbf{x}) = \frac{z}{z+\beta+r-1}\mathbf{x},$$

and determine conditions for its minimaxity. Show that it is proper Bayes if $\beta > 1$.

6.16 Let $X_i \sim \text{binomial}(p, n_i), i = 1, \ldots, r$, where n_i are unknown and p is known. The estimation target is $\mathbf{n} = (n_1, \ldots, n_r)$ with loss function

$$L(\mathbf{n}, \delta) = \sum_{i=1}^r \frac{1}{n_i}(n_i - \delta_i)^2.$$

(a) Show that the usual estimator $\mathbf{x}/p$ has constant risk $r(1-p)/p$.

(b) For $r \geq 2$, show that the estimator

$$\delta(\mathbf{x}) = \left(1 - \frac{a}{z+r-1}\right)\frac{\mathbf{x}}{p}$$

dominates $\mathbf{x}/p$ in risk, where $z = \Sigma x_i$ and $0 < a < 2(r-1)(1-p)$.

[*Hint*: Use an argument similar to Theorem 6.8, but here $X_i|Z$ is hypergeometric, with $E(X_i|z) = z\frac{n_i}{N}$ and $\text{var}(X_i|z) = z\frac{n_i}{N}\left(1-\frac{n_i}{N}\right)\frac{N-z}{N-1}$, where $N = \Sigma n_i$.]

(c) Extend the argument from part (b) and find conditions on the function $c(\cdot)$ and constant b so that

$$\delta(\mathbf{x}) = \left(1 - \frac{c(z)}{z+b}\right)\frac{\mathbf{x}}{p}$$

dominates $\mathbf{x}/p$ in risk.

Domination of the usual estimator of $\mathbf{n}$ was looked at by Feldman and Fox (1968), Johnson (1987), and Casella and Strawderman (1994). The problem of $\mathbf{n}$ estimation for the binomial has some interesting practical applications; see Olkin et al. 1981, Carroll and Lombard 1985, Casella 1986. Although we have made the unrealistic assumption that p is known, these results can be adapted to the more practical unknown p case (see Casella and Strawderman 1994 for details).

6.17 (a) Prove Lemma 6.9. [*Hint*: Change variables from $\mathbf{x}$ to $\mathbf{x} - e_i$, and note that h_i must be defined so that $\delta^0(0) = 0$.]

(b) Prove that for $X \sim p_i(x|\theta)$, where $p_i(x|\theta)$ is given by (6.36), $\delta^0(x) = h_i(x-1)/h_i(x)$ is the UMVU estimator of θ (Roy and Mitra 1957).

(c) Prove Theorem 6.10.

6.18 For the situation of Example 6.11:

(a) Establish that $\mathbf{x} + g(\mathbf{x})$, where $g(\mathbf{x})$ is given by (6.42), satisfies $\mathcal{D}(\mathbf{x}) \leq 0$ for the loss $L_0(\boldsymbol{\theta}, \delta)$ of (6.38), and hence dominates $\mathbf{x}$ in risk.

(b) Derive $\mathcal{D}(\mathbf{x})$ for $X_i \sim \text{Poisson}(\lambda_i)$, independent, and loss $L_{-1}(\boldsymbol{\lambda}, \delta)$ of (6.38). Show that $\mathbf{x} + g(\mathbf{x})$, for $g(\mathbf{x})$ given by (6.43), satisfies $\mathcal{D}(\mathbf{x}) \leq 0$ and hence is a minimax estimator of $\boldsymbol{\lambda}$.

6.19 For the situation of Example 6.12:

(a) Show that the estimator $\delta_0(\mathbf{x}) + g(\mathbf{x})$, for $g(\mathbf{x})$ of (6.45) dominates δ^0 in risk under the loss $L_{-1}(\boldsymbol{\theta}, \delta)$ of (6.38) by establishing that $\mathcal{D}(\mathbf{x}) \leq 0$.

(b) For the loss $L_0(\boldsymbol{\theta}, \delta)$ of (6.38), show that the estimator $\delta^0(\mathbf{x}) + g(\mathbf{x})$, where

$$g_i(\mathbf{x}) = \frac{c(\mathbf{x})k_i(x_i)}{\sum_{j=1}^r [k_j^2(x_j) + \left(\frac{1+t_j}{2}\right)k_j(x_j)]},$$

with $k_i(x) = \Sigma_{\ell=1}^x (t_i - 1 + \ell)/\ell$ and $c(\cdot)$ nondecreasing with $0 \leq c(\cdot) \leq 2[(\#x_i s > 1) - 2]$ has $\mathcal{D}(\mathbf{x}) \leq 0$ and hence dominates $\delta^0(\mathbf{x})$ in risk.

6.20 In Example 6.12, we saw improved estimators for the success probability of negative binomial distributions. Similar results hold for estimating the means of the negative binomial distributions, with some added features of interest. Let $X_1, \ldots, X_r$ be independent negative binomial random variables with mass function (6.44), and suppose we want to estimate $\mu = \{\mu_i\}$, where $\mu_i = t_i\theta_i/(1-\theta_i)$, the mean of the ith distribution, using the loss $L(\mu, \delta) = \Sigma(\mu_i - \delta_i)^2/\mu_i$.

(a) Show that the MLE of μ is X, and the risk of an estimator $\delta(\mathbf{x}) = \mathbf{x} + g(\mathbf{x})$ can be written

$$R(\mu, \delta) = R(\mu, \mathbf{X}) + E_\mu[\mathcal{D}_1(\mathbf{X}) + \mathcal{D}_2(\mathbf{X})]$$

where

$$\mathcal{D}_1(\mathbf{x}) = \sum_{i=1}^{r} \left\{ 2[g_i(\mathbf{x}+e_i) - g_i(\mathbf{x})] + \frac{g_i^2(\mathbf{x}+e_i)}{x_i+1} \right\}$$

and

$$\mathcal{D}_2(\mathbf{x}) = \sum_{i=1}^{r} \left\{ 2\frac{x_i}{t_i}[g_i(\mathbf{x}+e_i) - g_i(\mathbf{x})] + \frac{g_i^2(\mathbf{x}+e_i)}{t_i}\left[\frac{x_i}{x_{i+1}} - 1\right] \right\}$$

so that a sufficient condition for domination of the MLE is $\mathcal{D}_1(\mathbf{x}) + \mathcal{D}_2(\mathbf{x}) \le 0$ for all $\mathbf{x}$. [Use Lemma 6.9 in the form $Ef(\mathbf{X})/\theta_i = E\left[\frac{t_i+X_i}{X_i+1} f(\mathbf{X}+e_i)\right]$.]

(b) Show that if X_i are Poisson(θ_i) (instead of negative binomial), then $\mathcal{D}_2(\mathbf{x}) = 0$. Thus, any estimator that dominates the MLE in the negative binomial case also dominates the MLE in the Poisson case.

(c) Show that the Clevenson-Zidek estimator

$$\delta_{cz}(\mathbf{x}) = \left(1 - \frac{c(r-1)}{\Sigma x_i + r - 1}\right)\mathbf{x}$$

satisfies $\mathcal{D}_1(\mathbf{x}) \le 0$ and $\mathcal{D}_2(\mathbf{x}) \le 0$ and, hence, dominates the MLE under both the Poisson and negative binomial model.

This robustness property of Clevenson-Zidek estimators was discovered by Tsui (1984) and holds for more general forms of the estimator. Tsui (1984, 1986) also explores other estimators of Poisson and negative binomial means and their robustness properties.

Section 7

7.1 Establish the claim made in Example 7.2. Let X_1 and X_2 be independent random variables, $X_i \sim N(\theta_i, 1)$, and let $L((\theta_1, \theta_2), \delta) = (\theta_1 - \delta)^2$. Show that $\delta = \text{sign}(X_2)$ is an admissible estimate of θ_1, even though its distribution does not depend on θ_1.

7.2 Efron and Morris (1973a) give the following derivation of the positive-part Stein estimator as a *truncated* Bayes estimator. For $X \sim N_r(\theta, \sigma^2 I)$, $r \ge 3$, and $\theta \sim N(0, \tau^2 I)$, where σ^2 is known and τ^2 is unknown, define $t = \sigma^2/(\sigma^2 + \tau^2)$ and put a prior $h(t)$, $0 < t < 1$ on t.

(a) Show that the Bayes estimator against squared error loss is given by $E(\theta|\mathbf{x}) = [1 - E(t|\mathbf{x})]\mathbf{x}$ where

$$\pi(t|\mathbf{x}) = \frac{t^{r/2}e^{-t|\mathbf{x}|^2/2}h(t)}{\int_0^1 t^{r/2}e^{-t|\mathbf{x}|^2/2}h(t)\,dt}.$$

(b) For estimators of the form $\delta^\tau(\mathbf{x}) = \left(1 - \tau(|\mathbf{x}|^2)\frac{r-2}{|\mathbf{x}|^2}\right)\mathbf{x}$, the estimator that satisfies

(i) $\tau(\cdot)$ is nondecreasing,
(ii) $\tau(\cdot) \le c$,
(iii) δ^τ minimizes the Bayes risk against $h(t)$

has $\tau(|\mathbf{x}|^2) = \tau^*(|\mathbf{x}|^2) = \min\{c, \frac{|\mathbf{x}|^2}{r-2}E(t|\mathbf{x})\}$. (This is a truncated Bayes estimator, and is minimax if $c \le 2$.)

(c) Show that if $h(t)$ puts all of its mass on $t = 1$, then

$$\tau^*(|\mathbf{x}|^2) = \min\left\{c, \frac{|\mathbf{x}|^2}{r-2}\right\}$$

and the resulting truncated Bayes estimator is the positive-part estimator.

7.3 Fill in the details of the proof of Lemma 7.5.

7.4 For the situation of Example 7.8, show that if δ_0 is any estimator of θ, then the class of all estimators with $\delta(x) < \delta_0(x)$ for some x is complete.

7.5 A decision problem is *monotone* (as defined by Karlin and Rubin 1956; see also Brown, Cohen and Strawderman 1976 and Berger 1985, Section 8.4) if the loss function $L(\theta, \delta)$ is, for each θ, minimized at $\delta = \theta$ and is an increasing function of $|\delta - \theta|$. An estimator δ is *monotone* if it is a nondecreasing function of x.

(a) Show that if $L(\theta, \delta)$ is convex, then the monotone estimators form a complete class.

(b) If $\delta(x)$ is not monotone, show that the monotone estimator δ' defined implicitly by

$$P_t(\delta'(X) \le t) = P_t(\delta(X) \le t) \quad \text{for every } t$$

satisfies $R(\theta, \delta') \le R(\theta, \delta)$ for all θ.

(c) If $X \sim N(\theta, 1)$ and $L(\theta, \delta) = (\theta - \delta)^2$, construct a monotone estimator that dominates

$$\delta^a(x) = \begin{cases} -2a - x & \text{if } x < -a \\ x & \text{if } |x| \le a \\ 2a - x & \text{if } x > a. \end{cases}$$

7.6 Show that, in the following estimation problems, all risk functions are continuous.

(a) Estimate θ with $L(\theta, \delta(x)) = [\theta - \delta(x)]^2$, $X \sim N(\theta, 1)$.

(b) Estimate θ with $L(\theta, \delta(\mathbf{x})) = |\theta - \delta(\mathbf{x})|^2$, $X \sim N_r(\theta, I)$.

(c) Estimate λ with $L(\lambda, \delta(\mathbf{x})) = \sum_{i=1}^r \lambda_i^{-m}(\lambda_i - \delta_i(\mathbf{x}))^2$, $X_i \sim \text{Poisson}(\lambda_i)$, independent.

(d) Estimate β with $L(\beta, \delta(\mathbf{x})) = \sum_{i=1}^r \beta_i^{-m}(\beta_i - \delta_i(\mathbf{x}))^2$, $X_i \sim \text{Gamma}(\alpha_i, \beta_i)$, independent, α_i known.

7.7 Prove the following theorem, which gives sufficient conditions for estimators to have continuous risk functions.

Theorem 8.5 (Ferguson 1967, Theorem 3.7.1) *Consider the estimation of θ with loss $L(\theta, \delta)$, where $X \sim f(x|\theta)$. Assume*

(i) the loss function $L(\theta, \delta)$ is bounded and continuous in θ uniformly in δ (so that $\lim_{\theta \to \theta_0} \sup_\delta |L(\theta, \delta) - L(\theta_0, \delta)| = 0$);

(ii) for any bounded function φ, $\int \varphi(x) f(x|\theta) d\mu(x)$ is continuous in θ.

Then, the risk function $R(\theta, \delta) = E_\theta L(\theta, \delta)$ is continuous in θ.

[*Hint*: Show that

$$|R(\theta', \delta) - R(\theta, \delta)| \le \int |L(\theta', \delta(x)) - L(\theta, \delta(x))| f(x|\theta')\, dx + \int L(\theta, \delta(x)) |f(x|\theta') - f(x|\theta)|\, dx,$$

and use (i) and (ii) to make the first integral $< \varepsilon/2$, and (i) and (iii) to make the second integral $< \varepsilon/2$.]

7.8 Referring to Theorem 8.5, show that condition (iii) is satisfied by

(a) the exponential family,

(b) continuous densities in which θ is a one-dimensional location or scale parameter.

7.9 A family of functions $\mathcal{F}$ is *equicontinuous at the point* x_0 if, given $\varepsilon > 0$, there exists δ such that $|f(x) - f(x_0)| < \varepsilon$ for all $|x - x_0| < \delta$ and all $f \in \mathcal{F}$. (The same δ works for all f.) The family is *equicontinuous* if it is equicontinuous at each x_0.

Theorem 8.6 *(Communicated by L. Gajek) Consider estimation of θ with loss $L(\theta, \delta)$, where $X \sim f(x|\theta)$ is continuous in θ for each x. If*

(i) The family $L(\theta, \delta(x))$ is equicontinuous in θ for each δ.

(ii) For all $\theta, \theta' \in \Omega$,

$$\sup_x \frac{f(x|\theta')}{f(x|\theta)} < \infty.$$

Then, any finite-valued risk function $R(\theta, \delta) = E_\theta L(\theta, \delta)$ is continuous in θ and, hence, the estimators with finite, continuous risks form a complete class.

(a) Prove Theorem 8.6.

(b) Give an example of an equicontinuous family of loss functions. [*Hint*: Consider squared error loss with a bounded sample space.]

7.10 Referring to Theorem 7.11, this problem shows that the assumption of continuity of $f(x|\theta)$ in θ cannot be relaxed. Consider the density $f(x|\theta)$ that is $N(\theta, 1)$ if $\theta \le 0$ and $N(\theta + 1, 1)$ if $\theta > 0$.

(a) Show that this density has monotone likelihood ratio, but is not continuous in θ.

(b) Show that there exists a bounded continuous loss function $L(\theta - \delta)$ for which the risk $R(\theta, X)$ is discontinuous.

7.11 For $\mathbf{X} \sim f(\mathbf{x}|\theta)$ and loss function $L(\theta, \delta) = \sum_{i=1}^r \theta_i^m (\theta_i - \delta_i)^2$, show that condition (iii) of Theorem 7.11 holds.

7.12 Prove the following (equivalent) version of Blyth's Method (Theorem 7.13).

Theorem 8.7 *Suppose that the parameter space $\Omega \in \Re^r$ is open, and estimators with continuous risks are a complete class. Let δ be an estimator with a continuous risk function, and let $\{\pi_n\}$ be a sequence of (possibly improper) prior measures such that*

(i) $r(\pi_n, \delta) < \infty$ for all n,

(ii) for any nonempty open set $\Theta_0 \in \Omega$,

$$\frac{r(\pi_n, \delta) - r(\pi_n, \delta^{\pi_n})}{\int_{\Theta_0} \pi_n(\theta)\, d\theta} \to 0 \quad \text{as } n \to \infty.$$

Then, δ is an admissible estimator.

7.13 Fill in some of the gaps in Example 7.14:

(i) Verify the expressions for the posterior expected losses of δ^0 and δ^π in (7.7).

(ii) Show that the normalized beta priors will not satisfy condition (b) of Theorem 7.13, and then verify (7.9).

(iii) Show that the marginal distribution of X is given by (7.10).

(iv) Show that

$$\sum_{x=1}^{\infty} D(x) \le \max\{a^2, b^2\} \sum_{x=1}^{\infty} \frac{1}{x^2} \to 0,$$

and hence that δ^0 is admissible.

7.14 Let $X \sim$ Poisson(λ). Use Blyth's method to show that $\delta^0 = X$ is an admissible estimator of λ under the loss function $L(\lambda, \delta) = (\lambda - \delta)^2$ with the following steps:

(a) Show that the unnormalized gamma priors $\pi_n(\lambda) = \lambda^{a-1}e^{-\lambda/n}$ satisfy condition (b) of Theorem 7.13 by verifying that for any c,

$$\lim_{n\to\infty} \int_0^c \pi_n(\lambda)\, d\lambda = \text{constant}.$$

Also show that the normalized gamma priors will not work.

(b) Show that under the priors $\pi_n(\lambda)$, the Bayes risks of δ^0 and $\delta^{\pi'_n}$, the Bayes estimator, are given by

$$r(\pi'_n, \delta^0) = n^a\Gamma(a),$$
$$r(\pi'_n, \delta^{\pi'_n}) = \frac{n}{n+1} n^a \Gamma(a).$$

(c) The difference in risks is

$$r(\pi'_n, \delta^0) - r(\pi'_n, \delta^{\pi'_n}) = \Gamma(a)n^a\left(1 - \frac{n}{n+1}\right),$$

which, for fixed $a > 0$, goes to infinity as $n \to \infty$ (Too bad!). However, show that if we choose $a = a(n) = 1/\sqrt{n}$, then $\Gamma(a)n^a\left(1 - \frac{n}{n+1}\right) \to 0$ as $n \to \infty$. Thus, the difference in risks goes to zero.

(d) Unfortunately, we must go back and verify condition (b) of Theorem 7.13 for the sequence of priors with $a = 1/\sqrt{n}$, as part (a) no longer applies. Do this, and conclude that $\delta^0(x) = x$ is an admissible estimator of λ.

[*Hint*: For large n, since $t \le c/n$, use Taylor's theorem to write $e^{-t} = 1 - t +$ error, where the error can be ignored.]

(Recall that we have previously considered the admissibility of $\delta^0 = X$ in Corollaries 2.18 and 2.20, where we saw that δ^0 is admissible.)

7.15 Use Blyth's method to establish admissibility in the following situations.

(a) If $X \sim$ Gamma(α, β), α known, then x/α is an admissible estimator of β using the loss function $L(\beta, \delta) = (\beta - \delta)^2/\beta^2$.

(b) If $X \sim$ Negative binomial(k, p), then X is an admissible estimator of $\mu = k(1 - p)/p$ using the loss function $L(\mu, \delta) = (\mu - \delta)^2/(\mu + \frac{1}{k}\mu^2)$.

7.16 (i) Show that, in general, if δ^π is the Bayes estimator under squared error loss, then

$$r(\pi, \delta^\pi) - r(\pi, \delta^g) = E\,|\delta^\pi(\mathbf{X}) - \delta^g(\mathbf{X})|^2,$$

thus establishing (7.13).

(ii) Prove (7.15).

(iii) Use (7.15) to prove the admissibility of X in one dimension.

7.17 The identity (7.14) can be established in another way. For the situation of Example 7.18, show that

$$r(\pi, \delta^g) = r - 2\int [\nabla \log m_\pi(\mathbf{x})][\nabla \log m_g(\mathbf{x})] m_\pi(\mathbf{x})\, d\mathbf{x} + \int |\nabla \log m_g(\mathbf{x})|^2\, m_\pi(\mathbf{x})\, d\mathbf{x},$$

which implies

$$r(\pi, \delta^\pi) = r - \int |\nabla \log m_\pi(\mathbf{x})|^2\, m_\pi(\mathbf{x})\, d\mathbf{x},$$

and hence deduce (7.14).

7.18 This problem will outline the argument needed to prove Theorem 7.19:

(a) Show that $\nabla m_g(\mathbf{x}) = m_{\nabla g}(\mathbf{x})$, that is,

$$\nabla \int g(\boldsymbol{\theta}) e^{-|\mathbf{x}-\boldsymbol{\theta}|^2}\, d\boldsymbol{\theta} = \int [\nabla g(\boldsymbol{\theta})]\, e^{-|\mathbf{x}-\boldsymbol{\theta}|^2}\, d\boldsymbol{\theta}.$$

(b) Using part (a), show that

$$\begin{aligned} r(\pi, \delta^g) - r(g_n, \delta^{g_n}) &= \int \left|\nabla \log m_g(\mathbf{x}) - \nabla \log m_{g_n}(\mathbf{x})\right|^2 m_{g_n}(\mathbf{x})\, d\mathbf{x} \\ &= \int \left|\frac{\nabla m_g(\mathbf{x})}{m_g(\mathbf{x})} - \frac{\nabla m_{g_n}(\mathbf{x})}{m_{g_n}(\mathbf{x})}\right|^2 m_{g_n}(\mathbf{x})\, d\mathbf{x} \\ &\le 2\int \left|\frac{\nabla m_g(\mathbf{x})}{m_g(\mathbf{x})} - \frac{m_{h_n^2 \nabla g}(\mathbf{x})}{m_{g_n}(\mathbf{x})}\right|^2 m_{g_n}(\mathbf{x})\, d\mathbf{x} \\ &\quad + 2\int \left|\frac{m_{g \nabla h_n^2}(\mathbf{x})}{m_{g_n}(\mathbf{x})}\right|^2 m_{g_n}(\mathbf{x})\, d\mathbf{x} \\ &= B_n + A_n. \end{aligned}$$

(c) Show that

$$A_n = 4\int \left|\frac{m_{g h_n \nabla h_n}(\mathbf{x})}{m_{g h_n^2}(\mathbf{x})}\right|^2 m_{g_n}(\mathbf{x})\, d\mathbf{x} \le 4 \int m_{g(\nabla h_n)^2}(\mathbf{x})\, d\mathbf{x}$$

and this last bound $\to 0$ by condition (a).

(d) Show that the integrand of $B_n \to 0$ as $n \to \infty$, and use condition (b) together with the dominated convergence theorem to show $B_n \to 0$, proving the theorem.

7.19 Brown and Hwang (1982) actually prove Theorem 7.19 for the case $f(\mathbf{x}|\boldsymbol{\theta}) = e^{\boldsymbol{\theta}'\mathbf{x} - \psi(\boldsymbol{\theta})}$, where we are interested in estimating $\tau(\boldsymbol{\theta}) = E_{\boldsymbol{\theta}}(\mathbf{X}) = \nabla\psi(\boldsymbol{\theta})$ under the loss $L(\boldsymbol{\theta}, \delta) = |\tau(\boldsymbol{\theta}) - \delta|^2$. Prove Theorem 7.19 for this case. [The proof is similar to that outlined in Problem 7.18.]

7.20 For the situation of Example 7.20:

(a) Using integration by parts, show that

$$\begin{aligned} \frac{\partial}{\partial x_i} \int g(\boldsymbol{\theta}) e^{-|\mathbf{x}-\boldsymbol{\theta}|^2}\, d\boldsymbol{\theta} &= -\int (x_i - \theta_i) g(\boldsymbol{\theta}) e^{-|\mathbf{x}-\boldsymbol{\theta}|^2}\, d\boldsymbol{\theta} \\ &= \int \left[\frac{\partial}{\partial \theta_i} g(\boldsymbol{\theta})\right] e^{-|\mathbf{x}-\boldsymbol{\theta}|^2}\, d\boldsymbol{\theta} \end{aligned}$$

and hence

$$\frac{\nabla m_g(\mathbf{x})}{m_g(\mathbf{x})} = \frac{\int [\nabla g(\boldsymbol{\theta})] e^{-|\mathbf{x}-\boldsymbol{\theta}|^2}\, d\boldsymbol{\theta}}{\int g(\boldsymbol{\theta}) e^{-|\mathbf{x}-\boldsymbol{\theta}|^2}\, d\boldsymbol{\theta}}.$$

(b) Use the Laplace approximation (4.6.33) to show that

$$\frac{\int [\nabla g(\boldsymbol{\theta})] e^{-|\mathbf{x}-\boldsymbol{\theta}|^2}\, d\boldsymbol{\theta}}{\int g(\boldsymbol{\theta}) e^{-|\mathbf{x}-\boldsymbol{\theta}|^2}\, d\boldsymbol{\theta}} \approx \frac{\nabla g(\mathbf{x})}{g(\mathbf{x})},$$

and that

$$\delta^g(\mathbf{x}) \approx \mathbf{x} + \frac{\nabla g(\mathbf{x})}{g(\mathbf{x})}.$$

(c) If $g(\boldsymbol{\theta}) = 1/|\boldsymbol{\theta}|^k$, show that

$$\delta^g(\mathbf{x}) \approx \left(1 - \frac{k}{\mathbf{x}^2}\right)\mathbf{x}.$$

7.21 In Example 7.20, if $g(\boldsymbol{\theta}) = 1/|\boldsymbol{\theta}|^k$ is a proper prior, then δ^g is admissible. For what values of k is this the case?

7.22 Verify that the conditions of Theorem 7.19 are satisfied for $g(\boldsymbol{\theta}) = 1/|\boldsymbol{\theta}|^k$ if (a) $k > r - 2$ and (b) $k = r - 2$.

7.23 Establish conditions for the admissibility of Strawderman's estimator (Example 5.6)

(a) using Theorem 7.19,

(b) using the results of Brown (1971), given in Example 7.21.

(c) Give conditions under which Strawderman's estimator is an admissible minimax estimator.

(See Berger 1975, 1976b for generalizations).

7.24 (a) Verify the Laplace approximation of (7.23).

(b) Show that, for $h(|\mathbf{x}|) = k/|x|^{2\alpha}$, (7.25) can be written as (7.26) and that $\alpha = 1$ is needed for an estimator to be both admissible and minimax.

7.25 Theorem 7.17 also applies to the Poisson($\boldsymbol{\lambda}$) case, where Johnstone (1984) obtained the following characterization of admissible estimators for the loss $L(\boldsymbol{\lambda}, \delta) = \sum_{i=1}^r (\lambda_i - \delta_i)^2/\lambda_i$.

A generalized Bayes estimator of the form $\delta(\mathbf{x}) = [1 - h(\Sigma x_i)]\mathbf{x}$ is

(i) *inadmissible* if there exists $\varepsilon > 0$ and $M < \infty$ such that

$$h(\Sigma x_i) < \frac{r - 1 - \varepsilon}{\Sigma x_i} \quad \text{for } \Sigma x_i > M,$$

(ii) *admissible* if $h(\Sigma x_i)(\Sigma x_i)^{1/2}$ is bounded and there exits $M < \infty$ such that

$$h(\Sigma x_i) \geq \frac{r - 1}{\Sigma x_i} \quad \text{for } \Sigma x_i > M.$$

(a) Use Johnstone's characterization of admissible Poisson estimators (Example 7.22) to find an admissible Clevenson-Zidek estimator (6.31).

(b) Determine conditions under which the estimator is both admissible and minimax.

7.26 For the situation of Example 7.23:

(a) Show that X/n and $(n/n + 1)(X/n)(1 - X/n)$ are admissible for estimating p and $p(1 - p)$, respectively.

(b) Show that $\alpha(X/n) + (1-\alpha)(a/(a+b))$ is an admissible estimator of p, where $\alpha = n/(n+a+b)$. Compare the results here to that of Theorem 2.14 (Karlin's theorem). [Note that the results of Diaconis and Ylvisaker (1979) imply that $\pi(\cdot)$ = uniform are the only priors that give linear Bayes estimators.]

7.27 Fill in the gaps in the proof that estimators δ^π of the form (7.27) are a complete class.

(a) Show that δ^π is admissible when $r = -1$, $s = n+1$, and $r+1 = s$.

(b) For any other estimator $\delta'(x)$ for which $\delta'(x) = h(0)$ for $x \le r'$ and $\delta'(x) = h(1)$ for $x \ge s'$, show that we must have $r' \ge r$ and $s' \le s$.

(c) Show that $R(p, \delta') \le R(p, \delta^\pi)$ for all $p \in [0, 1]$ if and only if $R_{r,s}(p, \delta') \le R_{r,s}(p, \delta^\pi)$ for all $p \in [0, 1]$.

(d) Show that $\int_0^1 R_{r,s}(p, \delta)k(p)d\pi(p)$ is uniquely minimized by $[\delta^\pi(r+1), \ldots, \delta^\pi(s-1)]$, and hence deduce the admissibility of δ^π.

(e) Use Theorem 7.17 to show that any admissible estimator of $h(p)$ is of the form (7.27), and hence that (7.27) is a minimal complete class.

7.28 For $i = 1, 2, \ldots, k$, let $X_i \sim f_i(x|\theta_i)$ and suppose that $\delta_i^*(x_i)$ is a unique Bayes estimator of θ_i under the loss $L_i(\theta_i, \delta)$, where L_i satisfies $L_i(a, a) = 0$ and $L_i(a, a') > 0$, $a \ne a'$. Suppose that for some j, $1 \le j \le k$, there is a value θ^* such that if $\theta_j = \theta^*$,

(i) $X_j = x^*$ with probability 1,

(ii) $\delta_j^*(x^*) = \theta^*$.

Show that $(\delta_1^*(x_1), \delta_2^*(x_2), \ldots, \delta_k^*(x_k))$ is admissible for $(\theta_1, \theta_2, \cdots, \theta_k)$ under the loss $\sum_i L_i(\theta_i, \delta)$; that is, there is no Stein effect.

7.29 Suppose we observe $X_1, X_2, \ldots$ sequentially, where $X_i \sim f_i(x|\theta_i)$. An estimator of $\boldsymbol{\theta}_j = (\theta_1, \theta_2, \ldots, \theta_j)$ is called *nonanticipative* (Gutmann 1982b) if it only depends on $(X_1, X_2, \ldots, X_j)$. That is, we cannot use information that comes later, with indices $> j$. If $\delta_i^*(x_i)$ is an admissible estimator of θ_i, show that it cannot be dominated by a nonanticipative estimator. Thus, this is again a situation in which there is no Stein effect. [*Hint*: It is sufficient to consider $j = 2$. An argument similar to that of Example 7.24 will work.]

7.30 For $X \sim N_r(\boldsymbol{\theta}, I)$, consider estimation of $\varphi'\boldsymbol{\theta}$ where $\varphi_{r\times 1}$ is known, using the estimator $a'X$ with loss function $L(\varphi'\boldsymbol{\theta}, \delta) = (\varphi'\boldsymbol{\theta} - \delta)^2$.

(a) Show that if a lies outside the sphere (7.31), then $a'X$ is inadmissible.

(b) Show that the Bayes estimator of $\varphi'\boldsymbol{\theta}$ against the prior $\boldsymbol{\theta} \sim N(0, V)$ is given by

$$E(\varphi'\boldsymbol{\theta}|\mathbf{x}) = (I + V)^{-1}\varphi\mathbf{x}.$$

(c) Find a covariance matrix V such that $E(\varphi'\boldsymbol{\theta}|\mathbf{x})$ lies inside the sphere (7.31) [V will be of rank one, hence of the form vv' for some $r \times 1$ vector v].

Parts (a)–(c) show that all linear estimators inside the sphere (7.31) are admissible, and those outside are inadmissible. It remains to consider the boundary, which is slightly more involved. See Cohen 1966 for details.

7.31 *Brown's ancillarity paradox*. Let $\mathbf{X} \sim N_r(\mu, I)$, $r > 2$, and consider the estimation of $\mathbf{w}'\mu = \Sigma_{i=1}^r w_i\mu_i$, where $\mathbf{w}$ is a known vector with $\Sigma w_i^2 > 0$, using loss function $L(\mu, d) = (\mathbf{w}'\mu - \mathbf{w}'d)^2$.

(a) Show that the estimator $\mathbf{w}'\mathbf{X}$ is minimax and admissible.

(b) Assume no that **w** is the realized value of a random variable **W**, with distribution independent of **X**, where $\mathbf{V} = E(\mathbf{W}'\mathbf{W})$ is known. Show that the estimator $\mathbf{w}'d^*$, where

$$d^*(\mathbf{x}) = \left(I - \frac{c\mathbf{V}^{-1}}{\mathbf{x}\mathbf{V}^{-1}\mathbf{x}}\right)\mathbf{x},$$

with $0 < c < 2(r-2)$, dominates $\mathbf{w}'\mathbf{X}$ in risk.

[*Hint*: Establish and use the fact that $E[L(\mu, d)] = E[(d-\mu)'\mathbf{V}(d-\mu)]$. This is a special case of results established by Brown (1990a). It is referred to as a *paradox* because the distribution of the ancillary, which should not affect the estimation of μ, has an enormous effect on the properties of the standard estimator. Brown showed how these results affect the properties of coefficient estimates in multiple regression when the assumption of random regressors is made. In that context, the ancillarity paradox also relates to Shaffer's (1991) work on best linear unbiased estimation (see Theorem 3.4.14 and Problems 3.4.16-3.4.18.]

7.32 Efron (1990), in a discussion of Brown's (1990a) ancillarity paradox, proposed an alternate version.

Suppose $\mathbf{X} \sim N_r(\mu, I)$, $r > 2$, and with probability $1/r$, independent of **X**, the value of the random variable $J = j$ is observed, $j = 1, 2, \ldots, r$. The problem is to estimate θ_j using the loss function $L(\theta_j, d) = (\theta_j - d)^2$. Show that, conditional on $J = j$, X_j is a minimax and admissible estimator of θ_j. However, unconditionally, X_j is dominated by the jth coordinate of the James-Stein estimator. This version of the paradox may be somewhat more transparent. It more clearly shows how the presence of the ancillary random variable forces the problem to be considered as a multivariate problem, opening the door for the Stein effect.

9 Notes

9.1 History

Deliberate efforts to develop statistical inference and decision making not based on "inverse probability" (i.e., without assuming prior distributions) were mounted by R.A. Fisher (for example, 1922, 1930, and 1935; see also Lane 1980), by Neyman and Pearson (for example, 1933ab), and by Wald (1950). The latter's general decision theory introduced, as central notions, the minimax principle and least favorable distributions in close parallel to the corresponding concepts of the theory of games. Many of the examples of Section 5.2 were first worked out by Hodges and Lehmann (1950). Admissibility is another basic concept of Wald's decision theory. The admissibility proofs in Example 2.8 are due to Blyth (1951) and Hodges and Lehmann (1951). A general necessary and sufficient condition for admissibility was obtained by Stein (1955). Theorem 2.14 is due to Karlin (1958), and the surprising inadmissibility results of Section 5.5 had their origin in Stein's seminal paper (1956b). The relationship between equivariance and the minimax property was foreshadowed in Wald (1939) and was developed for point estimation by Peisakoff (1950), Girshick and Savage (1951), Blackwell and Girshick (1954), Kudo (1955), and Kiefer (1957).

Characterizations of admissible estimators and complete classes have included techniques such as Blyth's method and the information inequality. The pathbreaking paper of Brown (1971) was influential in shaping the understanding of admissibility problems, and motivated further study of differential inequalities (Brown 1979, 1988) and associated stochastic processes and Markov chains (Brown 1971, Johnstone 1984, Eaton 1992).

9.2 Synthesis

The strengths of combining the Bayesian and frequentist approach are evident in Problem 1.4. The Bayes approach provides a clear methodology for constructing estimators (of which REML is a version), while the frequentist approach provides the methodology for evaluation. There are many other approaches to statistical problems, that is, many other statistical philosophies. For example, there is the fiducial approach (Fisher 1959), structural inference (Fraser 1968, 1979), pivotal inference (Barnard 1985), conditional inference (Fisher 1956, Cox 1958, Buehler 1959, Robinson 1979a, 1979b), likelihood-based conditional inference (Barndorff-Nielsen 1980, 1983, Barndorff-Nielsen and Cox, 1979, 1994), and many more. Moreover, within each philosophy, there are many subdivisions, for example, robust Bayesian, conditional frequentist, and so on. Examination of conditional inference, with both synthesis and review in mind, can also be found in Casella (1987, 1988, 1992b).

An important difference among these different philosophies is the role of conditional and unconditional inference, that is, whether the criterion for evaluation of an estimator is allowed to depend on the data.

Example 9.1 Conditional bias. If $X_1, \ldots, X_n$ are distributed iid as $N(\mu, \sigma^2)$, both unknown, the estimator $S^2 = \frac{1}{n-1}\sum_{i=1}^n (X_i - \bar{X})^2$ is an unbiased estimator of σ^2; that is, the unconditional expectation satisfies $E_{\sigma^2}[S^2] = \sigma^2$, for all values of σ^2. In doing a conditional evaluation, we might ask if there is a set in the sample space [a *reference set* or *recognizable subset* according to Fisher (1959)] on which the conditional expectation is always biased. Robinson (1979b) showed that there exist constants a and $\delta > 0$ such that

$$E_{\sigma^2}[S^2 \mid |\bar{X}|/S < a] > (1+\delta)\sigma^2 \quad \text{for all } \mu, \sigma^2, \tag{9.1}$$

showing that S^2 is conditionally biased. See Problem 1.5 for details.

The importance of a result such as (9.1) is that the experimenter knows whether the recognizable set $\{(x_1, \ldots, x_n) : \bar{x}/s < a\}$ has occurred. If it has, then the claim that S^2 is unbiased may be suspect if the inference is to apply to experiments in which the recognizable set occurs. ||

The study of conditional properties is actually better suited to examination of confidence procedures, which we are not covering here. (However, see TSH2, Chapter 10 for an introduction to conditional inference in testing and, hence, in confidence procedures.) The variance inequality (9.1) has many interesting consequences in interval estimation for normal parameters, both for the mean (Brown 1968, Goutis and Casella 1992) and the variance (Stein 1964, Maata and Casella 1987, Goutis and Casella 1997 and Shorrock 1990).

9.3 The Hunt-Stein Theorem

The relationship between equivariance and minimaxity finds an expression in the Hunt-Stein theorem. Although these authors did not publish their result, it plays an important role in mathematical statistics.

The work of Hunt and Stein took place in the 1940s, but it was not until the landmark paper by Kiefer (1957) that a comprehensive treatment of the topic, and a very general version of the theorem, was given. (See also Kiefer 1966 for an expanded discussion.)

The basis of the theorem is that in invariant statistical problems, if the group satisfies certain assumptions, then the existence of a minimax estimator implies the existence of an equivariant minimax estimator. Intuitively, we expect such a theorem to exist in invariant decision problems for which the group is transitive and a right-invariant Haar measure exists. If the Haar measure were proper, then Theorem 4.1 and Theorem 3.1 (see also the end of Section 5.4) would apply. The question that the Hunt-Stein theorem addresses is whether an improper right-invariant Haar measure can yield a minimax estimator.

The theorem turns out to be not true for all groups, but only for groups possessing certain properties. A survey of these properties, and some interrelationships, was made by Stone and van Randow (1968). A later paper by Bondar and Milnes (1981) reviews and establishes many of the group-theoretic equivalences conjectured by Stone and van Randow. From this survey, two equivalent group-theoretic conditions, which we discuss informally, emerge as the appropriate conditions on the group.

A. *Amenability*. A group is ***amenable*** if there exists a right-invariant mean. That is, if we define the sequence of functionals

$$m_n(f) = \frac{1}{2n}\int_{-n}^{n} f(x)dx,$$

where $f \in \mathcal{L}_\infty$, then there exists a functional $m(\cdot)$ such that for any $f_1, \ldots, f_k \in \mathcal{L}_\infty$ and $\varepsilon > 0$ there is an n_0 such that

$$|m_n(f_i) - m(f_i)| < \varepsilon \text{ for } i = 1, \ldots, k \text{ and all } n > n_0.$$

B. *Approximable by proper priors*, or the existence of a sequence of proper probability distributions that converge to the right-invariant Haar measure. [The concept of approximable by proper priors can be traced to Stein (1965), and was further developed by Stone (1970) and Heath and Sudderth (1989).]

With these conditions, we can state the theorem

Theorem 9.2 (Hunt-Stein) *If the decision problem is invariant with respect to a group G that satisfies condition A (equivalently condition B), then if a minimax estimator exists, an equivariant minimax estimator exists. Conversely, if there exists an equivariant estimator that is minimax among equivariant estimators, it is minimax overall.*

The proof of this theorem has a history almost as rich as the theorem itself. The original published proof of Kiefer (1957) was improved upon by use of a fixed-point theorem. This elegant method is attributed to LeCam and Huber, and is used in the general development of the Hunt-Stein theorem by LeCam (1986, Section 8.6) and Strasser (1985, Section 48). An outline of such a proof is given by Kiefer (1966). Brown (1986b) provides an interesting commentary on Kiefer's 1957 paper, and also sketches Huber's method of proof. Robert (1994a, Section 7.5) gives a particularly readable sketch of the proof. If the group is finite, then the assumptions of the Hunt-Stein theorem are satisfied, and a somewhat less complex proof will work. See Berger (1985, Section 6.7) for the proof for finite groups. In TSH2, Section 9.5, a version of the Hunt-Stein theorem for testing problems is stated and proved under condition B.

Theorem 9.2 reduces the problem to a property of groups, and to apply the theorem we need to identify which groups satisfy the A/B conditions. Bondar and Milnes (1981) provide a nice catalog of groups, and we note that the amenable groups include finite groups, location/scale groups, the ***triangular group*** $T(n)$ of $n \times n$ nonsingular upper triangular matrices, and many permutation groups. "Large" groups, such as those arising in nonparametric problems, are often not amenable. A famous group that is not amenable is the ***general linear group*** GL_n, $n > 2$, of nonsingular $n \times n$ matrices (see Note 3.9.3). See also Examples 3.7-3.9 for MRE estimators that are not minimax.

9.4 Recentered Sets

The topic of set estimation has not been ignored because of its lack of importance, but rather because the subject is so vast that it really needs a separate book-length treatment. (TSH2 covers many aspects of standard set estimation theory.) Here, we will only comment on some of the developments in set estimators that are centered at Stein estimators, so-called *recentered sets*.

The remarkable paper of Stein (1962) gave heuristic arguments that showed why recentered sets of the form

$$C^+ = \{\theta : |\theta - \delta^+(\mathbf{x})| \le c\}$$

would dominate the usual confidence set C

$$C^0 = \{\theta : |\theta - \mathbf{x}| \le c\}$$

in the sense that $P_\theta(\theta \in C^+(\mathbf{X})) > P_\theta(\theta \in C^0(\mathbf{X}))$ for all θ, where $\mathbf{X} \sim N_r(\theta, I), r \ge 3$, and δ^+ is the positive-part Stein estimator. Stein's argument was heuristic, but Brown (1966) and Joshi (1967) proved the inadmissibility of C^0 if $r \ge 3$ (without giving an explicit dominating procedure). Joshi (1969b) also showed that C^0 was admissible if $r \le 2$.

Advances in this problem were made by Olshen (1977), Morris (1977, 1983a), Faith (1976), and Berger (1980), each demonstrating (but not proving) dominance of C^0 by Stein-like set estimators. Analytic dominance of C^0 by C^+ was established by Hwang and Casella (1982, 1984) and, in subsequent papers (Casella and Hwang 1983, 1987), dominance in both coverage probability and volume was achieved (the latter was only demonstrated numerically).

Many other results followed. Generalizations were given by Ki and Tsui (1985) and Shinozaki (1989), and domination results for non-normal distributions by Hwang and Chen (1986), Robert and Casella (1990), and Hwang and Ullah (1994).

All of these improved confidence sets have the property that their coverage probability is uniformly greater than that of C^0, but the infimum of the coverage probability (the *confidence coefficient*) is equal to that of C^0. As this is the value that is usually reported, unless there is a great reduction in volume, the practical advantages of such sets may be minimal. For example, recentered sets such as C^+ will present the same volume and confidence coefficient to an experimenter. Other sets, which attain some volume reduction but maintain the same confidence coefficient as C^0, still are somewhat "wasteful" because they have coverage probabilities higher than that of C^0. However, this deficiency now seems to be overcome. By adapting results of Brown et al. (1995), Tseng and Brown (1997) have constructed an improved confidence set, C^*, with the property that $P_\theta(\theta \in C^*(\mathbf{X})) = P_\theta(\theta \in C^0(\mathbf{X}))$ for every θ, and $\text{vol}(C^*) < \text{vol}(C^0)$, achieving a maximal amount of volume reduction while maintaining the same coverage probability as C^0.

9.5 Estimation of the Loss Function

In the proof of Theorem 5.1, the integration-by-parts technique yielded an unbiased estimate of the risk, that is, a function $\mathcal{D}(\mathbf{x})$ satisfying $E_\theta \mathcal{D}(\mathbf{X}) = E_\theta L(\theta, \delta) = R(\theta, \delta)$. Of course, we could also consider $\mathcal{D}(\mathbf{x})$ as an estimate of $L(\theta, \delta)$, and ask if $\mathcal{D}(\mathbf{x})$ is a reasonable estimator using, perhaps, another loss function such as

$$\mathcal{L}(L, \mathcal{D}) = (L(\theta, \delta) - \mathcal{D}(x))^2.$$

If we think of $L(\theta, \delta)$ as a measure of accuracy of δ, then we are looking for good estimators of this accuracy. Note, however, that this problem is slightly more complex than the ones we have considered, as the "parameter" $L(\theta, \delta(\mathbf{x}))$ is a function of both θ and $\mathbf{x}$.

Loss estimation was first addressed by Rukhin (1988a, 1988b, 1988c) and Johnstone (1988). Rukhin considered "decision-precision" losses, that combined the estimation of θ and $L(\theta, \delta)$ into one loss. Johnstone looked at the multivariate normal problem, and showed that 1 is an inadmissible estimator of $L(\theta, \mathbf{x}) = 1/r \sum_{i=1}^{r}(\theta_i - x_i)^2$ (recall that $E_\theta[L(\theta, \mathbf{X})] = 1$) and showed that estimates of the form $\mathcal{D}(\mathbf{x}) = 1 - (c/r)/|\mathbf{x}|^2$ dominate it, where $0 \le c \le 2(r-4)$. Note that this implies $r \ge 5$. Further advances in loss estimation are found in Lu and Berger (1989a, 1989b) and Fourdrinier and Wells (1995).

The loss estimation problem is closely tied to the problem of set estimation, and actually transforms the set estimation problem into one of point estimation. Suppose $C(\mathbf{x})$ is a set estimator (or *confidence set*) for θ, and we measure the worth of $C(\mathbf{x})$ with the loss function

$$L(\theta, C) = \begin{cases} 0 & \text{if } \theta \in C \\ 1 & \text{if } \theta \notin C. \end{cases}$$

We usually calculate $R(\theta, C) = E_\theta L(\theta, C(\mathbf{X})) = P_\theta(\theta \in C(\mathbf{X}))$, the probability of coverage of C. Moreover, $1 - \alpha = \inf_\theta P_\theta(\theta \in C(\mathbf{X}))$ is usually reported as our confidence in C. However, it is really of interest to estimate $L(\theta, C)$, the actual coverage. We can thus ask how well $1 - \alpha$ estimates this quantity, and if there are estimators $\gamma(\mathbf{x})$ (known as *estimators of accuracy*) that are better. Using the loss function

$$\mathcal{L}(\theta, \gamma) = (L(\theta, C) - \gamma(\mathbf{x}))^2,$$

a number of interesting (and some surprising) results have been obtained. In the multivariate normal problem improved estimates have been found for the accuracy of the usual confidence set (Lu and Berger 1989a, 1989b, Robert and Casella 1994) and for the accuracy of Stein-type confidence sets (George and Casella, 1994). However, Hwang and Brown (1991) have shown that under an additional constraint (that of *frequency validity*), the estimator $1 - \alpha$ is an admissible estimator of the accuracy of the usual confidence set.

Other situations have also been considered. Goutis and Casella (1992) have demonstrated that the accuracy statement of Student's t interval can be uniformly increased, and will still dominate $1 - \alpha$ under squared error loss. Hwang et al. (1992) have looked at accuracy estimation in the context of testing, where complete classes are described and the question of the admissibility of the p value is addressed. More recently, Lindsay and Yi (1996) have shown that, up to second-order terms, the observed Fisher information is the best estimator of the expected Fisher information, which is the variance (or loss) of the MLE. One can think of this result as a decision-theoretic formalization of the work of Efron and Hinkley (1978).

This variant of loss estimation as confidence estimation also has roots in the work of Kiefer (1976, 1977) who considered an alternate approach to the assignment of confidence (see also Brown 1978).

9.6 *Shrinkage and Multicollinearity*

In Sections 5 and 6, we have assumed that the covariance is known, and hence, without loss of generality, that it is the identity. This has led to shrinkage estimators of the form $\delta_i(\mathbf{x}) = (1 - h(\mathbf{x}))x_i$, that is, estimators that shrink every coefficient by the same fraction.

If the original variances are unequal, say $X \sim N_r(\theta, \Sigma)$, then it may be more desirable to shrink some coordinates more than others (Efron and Morris 1973a, 1973b, Morris

1983a). Intuitively, it seems reasonable for the amount of shrinkage to be proportional to the size of the componentwise variance, that is, the greater the variance the greater the shrinkage. This would tend to leave alone the coefficients with small variance, and to shrink the coefficients with higher variance relatively more, bringing the ensemble information more to bear on these coefficients (and improving the variance of the coefficient estimates). This strategy reflects the shrinkage pattern of a Bayes estimator with prior $\theta \sim N(0, \tau^2 I)$.

However, general minimax estimators of Hudson (1974) and Berger (1976a, 1976b) (see also Berger 1985, Section 5.4.3, and Chen 1988), on the contrary, shrink the lower variance coordinates more than the higher variance coordinates. What happens is that the variance/bias trade-off is profitable for coordinates with low variance, but not so for coordinates with high variance, where X_i is minimax anyway.

This minimax shrinkage pattern is directly opposite to what is advocated to relieve the problem of multicollinearity in multiple regression problems. For that problem, work that started from ridge regression (Hoerl and Kennard 1971a, 1971b) advocated shrinkage patterns that are similar to those arising from the $N(0, \tau^2 I)$ prior—and were in the opposite direction of the minimax pattern. There is a large literature on ridge regression, with much emphasis on applications and data analysis, and less on this dichotomy of shrinkage patterns. A review of ridge regression is given by Draper and van Nostrand (1979), and some theoretical properties are investigated by Brown and Zidek (1980), Casella (1980), and Obenchain (1981); see also Oman 1985 for a discussion of appropriate prior distributions. Casella (1985b) attempts to resolve the minimax/multicollinear shrinkage dilemma.

9.7 *Other Minimax Considerations*

The following provides a guide to some additional minimax literature.

(i) ***Bounded mean***

The multiparameter version of Example 2.9 suffers from the additional complication that many different shapes of the bounding set may be of interest. Shapes that have been considered are convex sets (DasGupta, 1985), spheres and rectangles (Berry, 1990), and hyperrectangles (Donoho, et al. 1990). Other versions of this problem that have been investigated include different loss functions (Bischoff and Fieger 1992, Eichenauer-Herrmann and Fieger 1992), gamma-minimax estimation (Vidakovic and DasGupta 1994), other distributions (Eichenauer-Herrmann and Ickstadt 1992), and other restrictions (Fan and Gijbels 1992, Spruill 1986, Feldman 1991). Some other advances in this problem have come from application of bounds on the risk function, often derived using the information inequality (Gajek 1987, 1988, Brown and Gajek 1990, Brown and Low 1991, Gajek and Kaluzka 1995). Truncated mean problems also underlie many deeper problems in estimation, as illustrated by Donoho (1994) and Johnstone (1994).

(ii) ***Selection of shrinkage target***

In Stein estimation, much has been written on the problem of selecting a shrinkage target. Berger (1982a, 1982b) shows how to specify elliptical regions in which maximal risk improvement is obtained, and also shows that desirable Bayesian properties can be maintained. Oman (1982a, 1982b) and Casella and Hwang (1987) describe shrinking toward linear subspaces, and Bock (1982) shows how to shrink toward convex polyhedra. George (1986a, 1986b) constructs estimators that shrink toward multiple targets, using properties of superharmonic functions establish minimaxity of *multiple shrinkage* estimators.

(iii) *A Bayes/minimax compromise*

Bayes estimation subject to a bound on the maximum risk was first considered by Hodges and Lehmann (1952). Although such problems tend to be computationally difficult, the resulting estimators often show good performance on both frequentist and Bayesian measures (Berger 1982a, 1982b, 1985, Section 4.7.7, DasGupta and Bose 1988, Chen 1988, DasGupta and Studden 1989) Some of these properties are related to those of Stein-type shrinkage estimators (Bickel 1983, 1984, Kempthorne 1988a, 1988b) and some are discussed in Section 5.7.

(iv) *Superharmonicity*

The superharmonic condition, although often difficult to verify, has sometimes proved helpful in not only establishing minimaxity, but also in understanding what types of prior distributions may lead to minimax Bayes estimators. Berger and Robert (1990) applied the condition to a family of hierarchical Bayes estimators and Haff and Johnson (1986) generalized it to the estimation of means in exponential families. More recently, Fourdrinier, Strawderman, and Wells (1998) have shown that no superharmonic prior can be proper, and were able to use Corollary 5.11 to establish minimaxity of a class of proper Bayes estimators, in particular, the Bayes estimator using a Cauchy prior. The review article of Brandwein and Strawderman (1990) contains other examples.

(v) *Minimax robustness*

Minimax robustness of Stein estimators, that is, the fact that minimaxity holds over a wide range of densities, has been established for many different spherically symmetric densities. Strawderman (1974) was the first author to exhibit minimax Stein estimators for distributions other than the normal. [The work of Stein (1956b) and Brown (1966) had established the inadmissibility of the best invariant estimator, but explicit improvements had not been given.] Brandwein and Strawderman (1978, 1980) have established minimax results for wide classes of mixture distributions, under both quadratic and concave loss. Elliptical distributions were considered by Srivastava and Bilodeau (1989) and Cellier, Fourdrinier, and Robert (1989), where domination was established for an entire class of distributions.

(vi) *Stein estimation*

Other topics of Stein estimation that have received attention include matrix estimation (Efron and Morris 1976b, Haff 1979, Dey and Srinivasan 1985, Bilodeau and Srivastava 1988, Carter et al. 1990, Konno 1991), regression problems (Zidek 1978, Copas 1983, Casella 1985b, Jennrich and Oman 1986, Gelfand and Dey 1988, Rukhin 1988c, Oman 1991), nonnormal distributions (Bravo and MacGibbon 1988, Chen and Hwang 1988, Srivastava and Bilodeau 1989, Cellier et al. 1989, Ralescu et al. 1992), robust estimation (Liang and Waclawiw 1990, Konno 1991), sequential estimation (Natarajan and Strawderman 1985, Sriram and Bose 1988, Ghosh et al. 1987), and unknown variances (Berger and Bock 1976, Berger et al. 1977, Gleser 1979, 1986, DasGupta and Rubin 1988, Honda 1991, Tan and Gleser 1992).

9.8 Other Admissibility Considerations

The following provides a guide to some additional admissibility literature.

(i) *Establishing admissibility*

Theorems 7.13 and 7.15 (and their generalizations; see, for example, Farrell 1968) represent the major tools for establishing admissibility. The other admissibility result we have seen (Karlin's theorem, Theorem 2.14) can actually be derived using Blyth's method. (See Zidek 1970, Portnoy 1971, and Brown and Hwang 1982 for more general Karlin-type theorems, and Berger 1982b for a partial converse.) Combining these theorems with a thorough investigation of the differential inequality that results from an integration by parts can also lead to some interesting characterizations of the behavior of admissible estimators (Portnory 1975, Berger 1976d, 1976e, 1980a, Brown 1979, 1988). A detailed survey of admissibility is given by Rukhin (1995).

(ii) *Dimension doubling*

Note that for the Poisson case, in contrast to the normal case, the factor $r-1$ tends to appear (instead of $r-2$). This results in the Poisson sample mean being inadmissible in two dimensions. This occurrence was first explained by Brown (1978, Section 2.3), who noted that the Poisson problem in k dimensions is "qualitatively similar" to the location problem in $2k$ dimensions (in terms of a differential inequality derived to establish admissibility). In Johnstone and MacGibbon (1992), this idea of "dimension doubling" also occurs and provides motivation for the transformed version of the Poisson problem that they consider.

(iii) *Finite populations*

Although we did not cover the topic of admissibility in finite population sampling, there are interesting connections between admissibility in multinomial, nonparametric, and finite population problems.

Using results of Meeden and Ghosh (1983) and Cohen and Kuo (1983), Meeden, Ghosh and Vardeman (1985) present a theorem that summarizes the admissibility connection, relating admissibility in a multinomial problem to admissibility in a nonparametric problem.

Stepwise Bayes arguments, which originated with Johnson (1971) [see also Alam 1979, Hsuan 1979, Brown 1981] are useful tools for establishing admissibility in these situations.

CHAPTER 6

Asymptotic Optimality

1 Performance Evaluations in Large Samples

The performance of the estimators developed in the preceding chapters—UMVU, MRE, Bayes, and minimax—is often difficult to evaluate exactly. This difficulty can frequently be overcome by computing power (particularly by simulation). Although such an approach works well on a case-by-case basis, it lacks the ability to provide an overall picture of performance which is needed, for example, to assess robustness and efficiency. We shall consider an alternative approach here: to obtain approximations to or limits of performance measures as the sample size gets large. Some of the probabilistic tools required for this purpose were treated in Section 1.8.

One basic result of that section concerned the consistency of estimators, that is, their convergence in probability to the parameters that they are estimating. For example, if $X_1, X_2, \ldots$ are iid with $E(X_i) = \xi$ and $\text{var}(X_i) = \sigma^2 < \infty$, it was seen in Example 1.8.3 that the sample mean $\bar{X}$ is a consistent estimator of ξ. More detailed information about the large-sample behavior of $\bar{X}$ can be obtained from the central limit theorem (Theorem 1.8.9), which states that

$$\sqrt{n}(\bar{X} - \xi) \xrightarrow{\mathcal{L}} N(0, \sigma^2). \tag{1.1}$$

The limit theorem (1.1) suggests the approximation

$$P\left(\bar{X} \leq \xi + \frac{\Delta}{\sqrt{n}}\right) \approx \Phi\left(\frac{\Delta}{\sqrt{n}}\right) \tag{1.2}$$

where Φ denotes the standard normal distribution function.

Instead of the probabilities (1.2), one may be interested in the expectation, variance, and higher moments of $\bar{X}$ and then find

$$\begin{aligned} &E(\bar{X}) = \xi, \quad E(\bar{X} - \xi)^2 = \frac{\sigma^2}{n}; \\ &E(\bar{X} - \xi)^3 = \frac{1}{n^2}\mu_3, \quad E(\bar{X} - \xi)^4 = \frac{1}{n^3}\mu_4 + \frac{3(n-1)}{n^3}\sigma^4 \end{aligned} \tag{1.3}$$

where $\mu_k = E(X_1 - \xi)^k$. We shall be concerned with the behavior corresponding to (1.1) and (1.3) not only of $\bar{X}$ but also of functions of $\bar{X}$.

As we shall see, performance evaluation of statistics $h(\bar{X}_n)$ based on, respectively, (1.1) and (1.3)—the asymptotic distribution and limiting moment approach—agree often but not always. It is convenient to have both approaches since they tend to be applicable in different circumstances. In the present section, we shall begin with (1.3) and then take up (1.1).

(a) The limiting moment approach

Theorem 1.1 *Let $X_1, \ldots, X_n$ be iid with $E(X_1) = \xi$, $var(X_1) = \sigma^2$, and finite fourth moment, and suppose h is a function of a real variable whose first four derivatives $h'(x)$, $h''(x)$, $h'''(x)$, and $h^{(iv)}(x)$ exist for all $x \in I$, where I is an interval with $P(X_1 \in I) = 1$. Furthermore, suppose that $|h^{(iv)}(x)| \leq M$ for all $x \in I$, for some $M < \infty$. Then,*

$$E[h(\bar{X})] = h(\xi) + \frac{\sigma^2}{2n} h''(\xi) + R_n, \tag{1.4}$$

and if, in addition, the fourth derivative of h^2 is also bounded,

$$\text{var}[h(\bar{X})] = \frac{\sigma^2}{n}[h'(\xi)]^2 + R_n, \tag{1.5}$$

where the remainder R_n in both cases is $O(1/n^2)$, that is, there exist n_0 and $A < \infty$ such that $R_n(\xi) < A/n^2$ for $n > n_0$ and all ξ.

Proof. The reason for the possibility of such a result is the strong set of assumptions concerning h, which permit an expansion of $h(\bar{X}_n)$ about $h(\xi)$ with bounded coefficients. Using the assumptions on the fourth derivative $h^{(iv)}(x)$, we can write

$$\begin{aligned} h(\bar{x}_n) = h(\xi) &+ h'(\xi)(\bar{x}_n - \xi) + \frac{1}{2} h''(\xi)(\bar{x}_n - \xi)^2 \\ &+ \frac{1}{6} h'''(\xi)(\bar{x}_n - \xi)^3 + R(x_n, \xi) \end{aligned} \tag{1.6}$$

where

$$|R(\bar{x}_n, \xi)| \leq \frac{M(\bar{x}_n - \xi)^4}{24}. \tag{1.7}$$

Using (1.3) and taking expectations of both sides of (1.6), we find

$$E[h(\bar{X}_n)] = h(\xi) + \frac{1}{2} h''(\xi) \frac{\sigma^2}{n} + O\left(\frac{1}{n^2}\right). \tag{1.8}$$

Here, the term in $h'(\xi)$ is missing since $E(\bar{X}_n) = \xi$, and the order of the remainder term follows from (1.3) and (1.7).

To obtain an expansion of $\text{var}[h(\bar{X}_n)]$, apply (1.8) to h^2 in place of h, using the fact that

$$[h^2(\xi)]'' = 2\{h(\xi)h''(\xi) + [h'(\xi)]^2\}. \tag{1.9}$$

This yields

$$E[h^2(\bar{X}_n)] = h^2(\xi) + [h(\xi)h''(\xi) + (h'(\xi))^2] \frac{\sigma^2}{n} + O\left(\frac{1}{n^2}\right), \tag{1.10}$$

and it follows from (1.8) that

$$[Eh(\bar{X}_n)]^2 = h^2(\xi) + h(\xi)h''(\xi) \frac{\sigma^2}{n} + O\left(\frac{1}{n^2}\right). \tag{1.11}$$

Taking the difference proves the validity of (1.5). □

Equation (1.4) suggests the following definition.

Definition 1.2 A sequence of estimators δ_n of $h(\xi)$ is *unbiased in the limit* if

$$E[\delta_n] \to h(\xi) \quad \text{as } n \to \infty, \tag{1.12}$$

that is, if the bias of δ_n, $E[\delta_n] - h(\xi)$, tends to 0 as $n \to \infty$.

Whenever (1.4) holds, the estimator $h(\bar{X})$ of $h(\xi)$ is unbiased in the limit.

Example 1.3 Approximate variance of a binomial UMVU estimator. Consider the UMVU estimator $T(n - T)/n(n - 1)$ of pq in Example 2.3.1. Note that $\xi = E(X) = p$ and $\sigma^2 = pq$ and that $\bar{X} = T/n$, and write the estimator as

$$\delta(\bar{X}) = \bar{X}(1 - \bar{X})\frac{n}{n-1}.$$

To obtain an approximation of its variance, let us consider first $h(\bar{X}) = \bar{X}(1 - \bar{X})$. Then, $h'(p) = 1 - 2p = q - p$ and $\text{var}[h(\bar{X})] = (1/n)pq(q-p)^2 + O(1/n^2)$. Also,

$$\left(\frac{n}{n-1}\right)^2 = \frac{1}{(1-1/n)^2} = 1 + \frac{2}{n} + O\left(\frac{1}{n^2}\right).$$

Thus,

$$\begin{aligned}\text{var}\ \delta(\bar{X}) &= \left(\frac{n}{n-1}\right)^2 \text{var}\ h(\bar{X}) \\ &= \left[\frac{pq(q-p)^2}{n} + O\left(\frac{1}{n^2}\right)\right]\left[1 + \frac{2}{n} + O\left(\frac{1}{n^2}\right)\right] \\ &= \frac{pq(q-p)^2}{n} + O\left(\frac{1}{n^2}\right).\end{aligned}$$

The exact variance of $\delta(\bar{X})$ given in Problem 2.3.1(b) shows that the error is $2p^2q^2/n(n-1)$ which is, indeed, of the order $1/n^2$. The maximum absolute error occurs at $p = 1/2$ and is $1/8n(n-1)$. It is a decreasing function of n which, for $n = 10$, equals 1/720. On the other hand, the relative error will tend to be large, unless p is close to 0 or 1 (Problem 1.2; see also Examples 1.8.13 and 1.8.15). ||

In this example, the bounded derivative condition of Theorem 1.1 is satisfied for all polynomials h because $\bar{X}$ is bounded. On the other hand, the condition fails when h is polynomial of degree $k \geq 4$ and the X's are, for example, normally distributed. However, (1.5) continues to hold in these circumstances. To see this, carry out an expansion like (1.6) to the $(k-1)$st power. The kth derivative of h is then a constant M, and instead of (1.7), the remainder will satisfy $R = M(\bar{X} - \delta)^k/k!$. This result then follows from the fact that all moments of the X's of order $\leq k$ exist and from Problem 1.1. This argument proves the following variant of Theorem 1.1.

Theorem 1.4 *In the situation of Theorem 1.1, formulas (1.4) and (1.5) remain valid if for some $k \geq 3$ the function h has k derivatives, the kth derivative is bounded, and the first k moments of the X's exist.*

To cover estimators such as

$$\delta_n(\bar{X}) = \Phi\left(\sqrt{\frac{n}{n-1}}(u - \bar{X})\right) \tag{1.13}$$

of p in Example 2.2.2, in which the function h depends on n, a slight generalization of Theorem 1.1 is required.

Theorem 1.5 *Suppose that the assumptions of Theorem 1.1 hold, and that c_n is a sequence of constants satisfying*

$$c_n = 1 + \frac{a}{n} + O\left(\frac{1}{n^2}\right). \tag{1.14}$$

Then, the variance of

$$\delta_n(\bar{X}) = h(c_n\bar{X}) \tag{1.15}$$

satisfies

$$\text{var}[\delta_n(\bar{X})] = \frac{\sigma^2}{n}[h'(\xi)]^2 + O\left(\frac{1}{n^2}\right). \tag{1.16}$$

The proof is left as an exercise (Problem 1.3).

Example 1.6 Approximate variance of a normal probability estimator. For the estimator $\delta_n(\bar{X})$ given by (1.13), we have

$$c_n = \sqrt{\frac{n}{n-1}} = \left(1 - \frac{1}{n}\right)^{-1/2} = 1 + \frac{1}{2n} + O\left(\frac{1}{n^2}\right)$$

and

$$\delta_n = h(c_n\bar{Y}) = \Phi(-c_n\bar{Y}) \quad \text{where } Y_i = X_i - u.$$

Thus,

$$h'(\xi) = -\phi(\xi),\ h''(\xi) = \xi\phi(\xi) \tag{1.17}$$

and hence from (1.16)

$$\text{var}\,\delta_n(\bar{X}) = \frac{1}{n}\phi^2(u - \xi) + O\left(\frac{1}{n^2}\right). \tag{1.18}$$

Since ξ is unknown, it is of interest to note that to terms of order $1/n$, the maximum variance is $1/2\pi n$.

If the factor $\sqrt{n/(n-1)}$ is neglected and the maximum likelihood estimator $\delta(\bar{X}) = \Phi(u - \bar{X})$ is used instead of δ_n, the variance is unchanged (up to the order $1/n$); however, the estimator is now biased. It follows from (1.8) and (1.17) that

$$E\delta(\bar{X}) = p + \frac{\xi - u}{2n}\phi(u - \xi) + O\left(\frac{1}{n^2}\right)$$

so that the bias is of the order $1/n$. The MLE is therefore unbiased in the limit. ||

The approximations of the accuracy of an estimator indicated by the above theorems may appear somewhat restrictive in that they apply only to functions of sample means. However, this covers all sufficiently smooth estimators based on samples from one-parameter exponential families. For on the basis of such a sample, $\bar{T} = \Sigma T(X_i)/n$ [in the notation of (8.1)] is a sufficient statistic, so that attention can be restricted to estimators that are functions of $\bar{T}$ (and possibly n). Extensions of the approximations of Theorems 1.1 - 1.5 to functions of higher sample moments are given by Cramér (1946a, Sections 27.6 and 28.4) and Serfling (1980, Section 2.2). On the other hand, this type of approximation is not applicable to optimal estimators for distributions whose support depends on the unknown parameter, such as the uniform or exponential distributions. Here, the minimal sufficient statistics and the estimators based on them are governed by different asymptotic laws with different convergence rates. (Problems 1.15–1.18 and Examples 7.11 - 7.14).

The conditions on the function $h(\cdot)$ of Theorem 1.1 are fairly stringent and do not apply, for example, to $h(x) = 1/x$ or $\sqrt{x}$ (unless the X_i are bounded away from zero) and the corresponding fact also limits the applicability of multivariate versions of these theorems (see Problem 1.27). When the assumptions of the theorems are not satisfied, the conclusions may or may not hold, depending on the situation.

Example 1.7 Limiting variance in the exponential distribution. Suppose that $X_1, \ldots, X_n$ are iid from the exponential distribution with density $(1/\theta)e^{-x/\theta}, x > 0, \theta > 0$, so that $EX_i = \theta$ and $\text{var} X_i = \theta^2$. The assumptions of Theorem 1.1 do not hold for $h(x) = \sqrt{x}$, so we cannot use (1.5) to approximate $\text{var}(\sqrt{\bar{X}})$. However, an exact calculation shows that (Problem 1.14)

$$\text{var}\left(\sqrt{\bar{X}}\right) = \left[1 - \frac{1}{n}\left(\frac{\Gamma(n+1/2)}{\Gamma(n)}\right)^2\right]\theta$$

and that $\lim_{n\to\infty} n\, \text{var}(\sqrt{\bar{X}}) = \theta/4 = \theta^2[h'(\theta)]^2$.

Thus, although the assumptions of Theorem 1.1 do not apply, the limit of the approximation (1.5) is correct. For an example in which the conclusions of the theorem do not hold, see Problem 1.13(a). ||

Let us next take up the second approach mentioned at the beginning of the section.

(b) The asymptotic distribution approach

Instead of the behavior of the moments $E[h(\bar{X})]$ and $\text{var}[h(\bar{X})]$, we now consider the probabilistic behavior of $h(\bar{X})$.

Theorem 1.8 *If $X_1, \ldots, X_n$ are iid with expectation ξ, and h is any function which is continuous at ξ, then*

$$h(\bar{X}) \xrightarrow{P} h(\xi) \text{ as } n \to \infty. \tag{1.19}$$

Proof. It was seen in Section 1.8 (Example 8.3) that $\bar{X} \xrightarrow{P} \xi$. The conclusion (1.19) is then a consequence of the following general result. □

Theorem 1.9 *If a sequence of random variables T_n tends to ξ in probability and if h is continuous at ξ, then $h(T_n) \xrightarrow{P} h(\xi)$.*

Proof. To show that $P(|h(T_n) - h(\xi)| < a) \to 1$, it is enough to notice that by the continuity of h, the difference $|h(T_n) - h(\xi)|$ will be arbitrarily small if T_n is close to ξ, and that for any a, the probability that $|T_n - \xi| < a$ tends to 1 as $n \to \infty$. We leave a detailed proof to Problem 1.30 □

Consistency can be viewed as a probabilistic analog of unbiasedness in the limit but, as Theorem 1.8 shows, requires much weaker assumptions on h. Unlike those needed for Theorem 1.1, they are satisfied, for example, when $\xi \neq 0$ and $h(x) = 1/x$ or $h(x) = 1/\sqrt{x}$.

The assumptions of Theorem 1.1 provide sufficient conditions in order that

$$\text{var}[\sqrt{n}\, h(\bar{X})] \to \sigma^2[h'(\xi)]^2 \quad \text{as } n \to \infty. \tag{1.20}$$

On the other hand, it follows from Theorem 8.12 of Section 1.8 that $v^2 = \sigma^2[h'(\xi)]$ is also the variance of the limiting distribution $N(0, v^2)$ of $\sqrt{n}[h(\bar{X}) - h(\xi)]$. This asymptotic normality holds under the following weak assumptions on h.

Theorem 1.10 *Let $X_1, \ldots, X_n$ be iid with $E(X_i) = \xi$ and $\text{var}(X_i) = \sigma^2$. Suppose that*

(a) the function h has a derivative h' with $h'(\xi) \neq 0$,

(b) the constants c_n satisfy $c_n = 1 + a/n + O(1/n^2)$.

Then,

(i) $\sqrt{n}[h(c_n\bar{X}) - h(\xi)]$ has the normal limit distribution with mean zero and variance $\sigma^2[h'(\xi)]^2$;

(ii) if $h'(\xi) = 0$ but $h''(\xi)$ exists and is not 0, then $n[h(c_n\bar{x}) - h(\xi)] \xrightarrow{\mathcal{L}} \frac{1}{2}\sigma^2 h''(\xi)\chi_1^2$.

Proof. Immediate consequence of Theorems 1.8.10 - 1.8.14. □

Example 1.11 Continuation of Example 1.6. For $Y_i = X_i - u$, we have $EY_i = \xi - u$ and var $Y_i = \sigma^2$. The maximum likelihood estimator of $\Phi(u - \xi)$ is given by $\delta'_n = \Phi(-\bar{Y}_n)$ and Theorem 1.10 shows

$$\sqrt{n}[\delta'_n - \Phi(u - \xi)] \xrightarrow{\mathcal{D}} N(0, \phi^2(u - \xi)).$$

The UMVU estimator is $\delta_n = \Phi(-c_n\bar{Y}_n)$ as in (1.13), and again by Theorem 1.10, $\sqrt{n}[\delta_n - \Phi(u - \xi)] \xrightarrow{\mathcal{D}} N(0, \varphi^2(u - \xi))$. ||

Example 1.12 Asymptotic distribution of squared mean estimators. Let $X_1, \ldots, X_n$ be iid $N(\theta, \sigma^2)$, and let the estimand be θ^2. Three estimators of θ^2 (Problems 2.2.1 and 2.2.2) are

$$\delta_{1n} = \bar{X}^2 - \frac{\sigma^2}{n} \quad \text{(UMVU when } \sigma \text{ is known)},$$

$$\delta_{2n} = \bar{X}^2 - \frac{S^2}{n(n-1)} \quad \text{(UMVU when } \sigma \text{ is unknown)}$$

$$\text{where } S^2 = \Sigma(X_i - \bar{X})^2,$$

$$\delta_{3n} = \bar{X}^2 \quad \text{(MLE in either case)}.$$

For each of these three sequences of estimators $\delta(\mathbf{X})$, let us find the limiting distribution of $[\delta(\mathbf{X}) - \theta^2]$ suitably normalized. Now, $\sqrt{n}(\delta_{3n} - \theta) \to N(0, \sigma^2)$ in law by the central limit theorem. Using $h(u) = u^2$ in Theorem 1.8.10, it follows that

$$\sqrt{n}(\bar{X}^2 - \theta^2) \xrightarrow{\mathcal{L}} N(0, 4\sigma^2\theta^2), \tag{1.21}$$

provided $h'(\theta) = 2\theta \neq 0$.

Next, consider δ_{1n}. Since

$$\sqrt{n}\left(\bar{X}^2 - \frac{\sigma^2}{n} - \theta^2\right) = \sqrt{n}(\bar{X}^2 - \theta^2) - \frac{\sigma^2}{\sqrt{n}},$$

it follows from Theorem 1.8.9 that

$$\sqrt{n}(\delta_{1n} - \theta^2) \xrightarrow{\mathcal{L}} N(0, 4\sigma^2\theta^2). \tag{1.22}$$

Finally, consider

$$\sqrt{n}\left(\bar{X}^2 - \frac{S^2}{n(n-1)} - \theta^2\right) = \sqrt{n}(\bar{X}^2 - \theta^2) - \frac{1}{\sqrt{n}}\left(\frac{S^2}{n-1}\right).$$

Now, $S^2/(n-1)$ tends to σ^2 in probability, so $S^2/\sqrt{n}\,(n-1)$ tends to 0 in probability. Thus,

$$\sqrt{n}(\delta_{2n} - \theta^2) \xrightarrow{\mathcal{L}} N(0, 4\sigma^2\theta^2).$$

Hence, when $\theta \neq 0$, all three estimators have the same limit distribution.

There remains the case $\theta = 0$. It is seen from the Taylor expansion [for example, Equation (1.6)] that if $h'(\theta) = 0$, then

$$\sqrt{n}[h(T_n) - h(\theta)] \to 0 \quad \text{in probability.}$$

Thus, in particular, in the present situation, when $\theta = 0$, $\sqrt{n}[\delta(\mathbf{X}) - \theta^2] \to 0$ in probability for all three estimators. When $h'(\theta) = 0$, $\sqrt{n}$ is no longer the appropriate normalizing factor: it tends to infinity too slowly.

Let us therefore apply the second part of Theorem 1.10 to the three estimators δ_{in} $(i = 1, 2, 3)$ when $\theta = 0$. Since $h''(0) = 2$, it follows that for $\delta_{3n} = \bar{X}^2$,

$$n(\bar{X}^2 - \theta^2) = n(\bar{X}^2 - 0^2) \to \frac{1}{2}\sigma^2\,(2\chi_1^2) = \sigma^2\chi_1^2.$$

Actually, since the distribution of $\sqrt{n}\bar{X}$ is $N(0, \sigma^2)$ for each n, the statistic $n\bar{X}^2$ is distributed exactly as $\sigma^2\chi_1^2$, so that no asymptotic argument would have been necessary.

For δ_{1n}, we find

$$n\left(\bar{X}^2 - \frac{\sigma^2}{n} - \theta^2\right) = n\bar{X}^2 - \sigma^2,$$

and the right-hand side tends in law to $\sigma^2(\chi_1^2 - 1)$. In fact, here too, this is the exact rather than just a limit distribution. Finally, consider δ_{2n}. Here,

$$n\left(\bar{X}^2 - \frac{S^2}{n(n-1)} - \theta^2\right) = n\bar{X}^2 - \frac{S^2}{n-1},$$

and since $S^2/(n-1)$ tends in probability to σ^2, the limiting distribution is again $\sigma^2(\chi_1^2 - 1)$.

Although, for $\theta \neq 0$, the three sequences of estimators have the same limit distribution, this is no longer true when $\theta = 0$. In this case, the limit distribution of $n(\delta - \theta^2)$ is $\sigma^2(\chi_1^2 - 1)$ for δ_{1n} and δ_{2n} but $\sigma^2\chi_1^2$ for the MLE δ_{3n}. These two distributions differ only in their location. The distribution of $\sigma^2(\chi_1^2 - 1)$ is centered so that its expectation is zero, while that of $\sigma^2\chi_1^2$ has expectation σ^2. So, although δ_{1n} and δ_{2n} suffer from the disadvantage of taking on negative values with probability > 0, asymptotically they are preferable to the MLE δ_{3n}.

The estimators δ_{1n} and δ_{2n} of Example 1.12 can be thought of as bias-corrected versions of the MLE δ_{3n}. Typically, the MLE $\hat{\theta}_n$ has bias of order $1/n$, say

$$b_n(\theta) = \frac{B(\theta)}{n} + O\left(\frac{1}{n^2}\right).$$

The order of the bias can be reduced by subtracting from $\hat{\theta}_n$ an estimator of the bias based on the MLE. This leads to the bias-corrected ML estimator

$$\hat{\hat{\theta}}_n = \hat{\theta}_n - \frac{B(\hat{\theta}_n)}{n} \tag{1.23}$$

whose bias will be of order $1/n^2$. (For an example, see Problem 1.25; see also Example 7.15.) ||

To compare these bias-correcting approaches, consider the following example.

Example 1.13 A family of estimators. The estimators δ_{1n} and δ_{3n} for θ^2 of Example 1.12 are special cases (with $c = 1$ or 0) of the family of estimators

$$\delta_n^{(c)} = \bar{X}^2 - \frac{c\sigma^2}{n}. \tag{1.24}$$

As in Example 1.12, it follows from Theorems 1.8.10 and 1.8.12 that for $\theta \neq 0$,

$$\sqrt{n}[\delta_n^{(c)} - \theta^2] \xrightarrow{\mathcal{L}} N(0, 4\sigma^2\theta^2), \tag{1.25}$$

so that the asymptotic variance is $4\sigma^2\theta^2$.

If, instead, we apply Theorem 1.1 with $h(\theta) = \theta^2$, we see that

$$E\left(\bar{X}^2\right) = \theta^2 + \frac{\sigma^2}{2n} + O\left(\frac{1}{n^2}\right), \tag{1.26}$$

$$\text{var}\left(\bar{X}^2\right) = \frac{4\sigma^2\theta^2}{n} + O\left(\frac{1}{n^2}\right).$$

Thus, to the first order, the two approaches give the same result.

Since the common value of the asymptotic and limiting variance does not involve c, this first-order approach does not provide a useful comparison of the estimators (1.25) corresponding to different values of c. To obtain such a comparison, we must take the next-order terms into account. This is easy for approach (a), where we only need to take the Taylor expansions (1.10) and (1.11) a step further. In fact, in the present case, it is easy to calculate $\text{var}(\delta_n^{(c)})$ exactly (Problem 1.26).

However, since the estimators $\delta_n^{(c)}$ are biased when $c \neq 1$, they should be compared not in terms of their variances but in terms of the expected squared error, which is (Problem 1.26)

$$E\left(\bar{X}^2 - \frac{c\sigma^2}{n} - \theta^2\right)^2 = \frac{4\sigma^2\theta^2}{n} + \frac{(c^2 - 2c + 3)\sigma^4}{n^2}. \tag{1.27}$$

In terms of this measure, the estimator is better the smaller $c^2 - 2c + 3$ is. This quadratic has its minimum at $c = 1$, and the UMVU estimator $\delta_n^{(1)}$ therefore minimizes the risk (1.27) among all estimators (1.25).

Equality of the asymptotic variance and the limit of the variance does not always hold (Problem 1.38). However, what can be stated quite generally is that the appropriately normalized limit of the variance is greater than or equal to the asymptotic variance. To see this, let us state the following lemma.

Lemma 1.14 *Let Y_n, $n = 1, 2, \ldots$, be a sequence of random variables such that $Y_n \xrightarrow{\mathcal{D}} Y$, where $E(Y) = 0$ and $\text{var}(Y^2) = E(Y^2) = v^2 < \infty$. For a constant A, define $Y_{n_A} = Y_n I(|Y_n| \leq A) + AI(|Y_n| > A)$, the random variable Y_n truncated at A. Then,*

(a) $\lim_{A\to\infty} \lim_{n\to\infty} E(Y_{n_A}^2) = \lim_{A\to\infty} \lim_{n\to\infty} E[\min(Y_n^2, A)] = v^2$,

(b) if $EY_n^2 \to w^2$, then $w^2 \geq v^2$.

Proof. (a) By Theorem 1.8.8,

$$\lim_{n\to\infty} E(Y_{nA}^2) = E[Y^2 I(|Y| \leq A)] + A^2 P(|Y| > A),$$

and as $A \to \infty$, the right side tends to v^2.

(b) It follows from Problem 1.39 that

$$\lim_{A\to\infty} \lim_{n\to\infty} E(Y_{n_A}^2) \leq \lim_{n\to\infty} \lim_{A\to\infty} E(Y_{n_A}^2), \tag{1.28}$$

provided the indicated limit exists. Now, $\lim_{A\to\infty} E(Y_{n_A}^2) = E(Y_n^2)$, so that the right side of (1.28) is w^2, while the left side is v^2 by part (a). □

Suppose now that T_n is a sequence of statistics for which $Y_n = k_n[T_n - E(T_n)]$ tends in law to a random variable Y with zero expectation. Then, the asymptotic variance $v^2 = \text{var}(Y)$ and the limit of the variances $w^2 = \lim E(Y_n^2)$, if it exists, satisfy $v^2 \leq w^2$ as was claimed. (Note that w^2 need not be finite.) Conditions for v^2 and w^2 to coincide are given by Chernoff (1956). For the special case that T_n is a function of a sample mean of iid variables, the two coincide under the assumptions of Theorems 1.1 and 1.8.12. ‖

2 Asymptotic Efficiency

The large-sample approximations of the preceding section not only provide a convenient method for assessing the performance of an estimator and for comparing different estimators, they also permit a new approach to optimality that is less restrictive than the theories of unbiased and equivariant estimation developed in Chapters 2 and 3.

It was seen that estimators[1] of interest typically are consistent as the sample sizes tend to infinity and, suitably normalized, are asymptotically normally distributed about the estimand with a variance $v(\theta)$ (the asymptotic variance), which provides a reasonable measure of the accuracy of the estimator sequence. (In this connection, see Problem 2.1.) Within this class of consistent asymptotically normal estimators, it turns out that under additional restrictions, there exist estimators that uniformly minimize $v(\theta)$. The remainder of this chapter is mainly concerned with the development of fairly explicit methods of obtaining such *asymptotically efficient* estimators.

Before embarking on this program, it may be helpful to note an important difference between the present large-sample approach and the small-sample results here and elsewhere. Both UMVU and MRE estimators tend to be unique (Theorem 1.7.10) and so are at least some of the minimax estimators derived in Chapter 5. On the other hand, it is in the nature of asymptotically optimal solutions not to be unique, since asymptotic results refer to the limiting behavior of sequences, and the same limit is shared by many different sequences. More specifically, if

$$\sqrt{n}[\delta_n - g(\theta)] \xrightarrow{\mathcal{L}} N(0, v)$$

and $\{\delta_n\}$ is asymptotically optimal in the sense of minimizing v, then $\delta_n + R_n$ is also optimal, provided

$$\sqrt{n}\, R_n \to 0 \quad \text{in probability.}$$

As we shall see later, asymptotically equivalent optimal estimators can be obtained from quite different starting points.

The goal of minimizing the asymptotic variance is reasonable only if the estimators under consideration have the same asymptotic expectation. In particular, we shall be concerned with estimators whose asymptotic expectation is the quantity being estimated.

Definition 2.1 If $k_n[\delta_n - g(\theta)] \xrightarrow{\mathcal{L}} H$ for some sequence k_n, the estimator δ_n of $g(\theta)$ is *asymptotically unbiased* if the expectation of H is zero.

Note that the definition of asymptotic unbiasedness is analogous to that of asymptotic variance. Unlike Definition 1.2, it is concerned with properties of the limiting distribution rather than limiting properties of the distribution of the estimator sequence. To see that Definition 2.1 is independent of the normalizing constant, see Problem 2.2.

Under the conditions of Theorem 1.8.12, the estimator $h(T_n)$ is asymptotically unbiased for the parameter $h(\theta)$. The estimator of Theorem 1.1 is unbiased in the limit.

Example 2.2 Large-sample behavior of squared mean estimators. In Example 1.12, all three estimators of θ^2 are asymptotically unbiased and unbiased in the limit. We note that these results continue to hold if the assumption of normality is replaced by that of finite variance. ||

[1] We shall frequently use *estimator* instead of the more accurate but cumbersome term *estimator sequence.*

In Example 1.12, the MLE was found to be asymptotically unbiased for $\theta \neq 0$ but asymptotically biased when the range of the distribution depends on the parameter.

Example 2.3 Asymptotically biased estimator. If $X_1, \ldots, X_n$ are iid as $U(0, \theta)$, the MLE of θ is $X_{(n)}$ and satisfies (Problem 2.6)

$$n(\theta - X_{(n)}) \xrightarrow{\mathcal{L}} E(0, \theta),$$

and, hence, is asymptotically biased, but it is unbiased in the limit.

A similar situation occurs in sampling from an exponential $E(a, b)$ distribution. See Examples 7.11 and 7.12 for further details. ||

The asymptotic analog of a UMVU estimator is an asymptotically unbiased estimator with minimum asymptotic variance. In the theory of such estimators, an important role is played by an asymptotic analog of the information inequality (2.5.31). If $X_1, \ldots, X_n$ are iid according to a density $f_\theta(x)$ (with respect to μ) satisfying suitable regularity conditions, this inequality states that the variance of any unbiased estimator δ of $g(\theta)$ satisfies

$$\text{var}_\theta(\delta) \geq \frac{[g'(\theta)]^2}{nI(\theta)}, \tag{2.1}$$

where $I(\theta)$ is the amount of information in a single observation defined by (2.5.10).

Suppose now that $\delta_n = \delta_n(X_1, \ldots, X_n)$ is asymptotically normal, say that

$$\sqrt{n}[\delta_n - g(\theta)] \xrightarrow{\mathcal{L}} N[0, v(\theta)], \quad v(\theta) > 0. \tag{2.2}$$

Then, it turns out that under some additional restrictions, one also has

$$v(\theta) \geq \frac{[g'(\theta)]^2}{I(\theta)}. \tag{2.3}$$

However, although the lower bound (2.1) is attained only in exceptional circumstances (Section 2.5), there exist sequences $\{\delta_n\}$ that satisfy (2.2) with $v(\theta)$ equal to the lower bound (2.3) subject only to quite general regularity conditions.

Definition 2.4 A sequence $\{\delta_n\} = \{\delta_n(X_1, \ldots, X_n)\}$, satisfying (2.2) with

$$v(\theta) = \frac{[g'(\theta)]^2}{I(\theta)} \tag{2.4}$$

is said to be *asymptotically efficient.*

At first glance, (2.3) might be thought to be a consequence of (2.1). Two differences between the inequalities (2.1) and (2.3) should be noted, however.

(i) The estimator δ in (2.1) is assumed to be unbiased, whereas (2.2) only implies asymptotic unbiasedness and consistency of $\{\delta_n\}$. It does not imply that δ_n is unbiased or even that its bias tends to zero (Problem 2.11).

(ii) The quantity $v(\theta)$ in (2.3) is an asymptotic variance whereas (2.1) refers to the actual variance of δ. It follows from Lemma 1.14 that

$$v(\theta) \leq \liminf[n \,\text{var}_\theta \delta_n] \tag{2.5}$$

but equality need not hold. Thus, (2.3) is a consequence of (2.1), provided

$$\text{(2.6)} \qquad \operatorname{var}\{\sqrt{n}[\delta_n - g(\theta)]\} \to v(\theta)$$

and if δ_n is unbiased, but not necessarily if these requirements do not hold.

For a long time, (2.3) was nevertheless believed to be valid subject only to regularity conditions on the densities f_θ. This belief was exploded by the example (due to Hodges; see Le Cam 1953) given below. Before stating the example, note that in discussing the inequality (2.3) under assumption (2.2), if θ is real-valued and $g(\theta)$ is differentiable, it is enough to consider the case $g(\theta) = \theta$, for which (2.3) reduces to

$$\text{(2.7)} \qquad v(\theta) \geq \frac{1}{I(\theta)}.$$

For if

$$\text{(2.8)} \qquad \sqrt{n}(\delta_n - \theta) \xrightarrow{\mathcal{L}} N[0, v(\theta)]$$

and if g has derivative g', it was seen in Theorem 1.8.12 that

$$\text{(2.9)} \qquad \sqrt{n}[g(\delta_n) - g(\theta)] \xrightarrow{\mathcal{L}} N[0, v(\theta)[g'(\theta)]^2].$$

After the obvious change of notation, this implies (2.3).

Example 2.5 Superefficient estimator. Let $X_1, \ldots, X_n$ be iid according to the normal distribution $N(\theta, 1)$ and let the estimand be θ. It was seen in Table 2.5.1 that in this case, $I(\theta) = 1$ so that (2.7) reduces to $v(\theta) \geq 1$. On the other hand, consider the sequence of estimators,

$$\delta_n = \begin{cases} \bar{X} & \text{if } |\bar{X}| \geq 1/n^{1/4} \\ a\bar{X} & \text{if } |\bar{X}| < 1/n^{1/4}. \end{cases}$$

Then (Problem 2.8),

$$\sqrt{n}(\delta_n - \theta) \xrightarrow{\mathcal{L}} N[0, v(\theta)],$$

where $v(\theta) = 1$ when $\theta \neq 0$ and $v(\theta) = a^2$ when $\theta = 0$. If $a < 1$, inequality (2.3) is therefore violated at $\theta = 0$. ||

This phenomenon is quite general (Problems 2.4 - 2.5). There will typically exist estimators satisfying (2.8) but with $v(\theta)$ violating (2.7) for at least some values of θ, called points of *superefficiency*. However, (2.7) is almost true, for it was shown by Le Cam (1953) that for any sequence δ_n satisfying (2.8), the set S of points of superefficiency has Lebesgue measure zero. The following version of this result, which we shall not prove, is due to Bahadur (1964). The assumptions are somewhat stronger but similar to those of Theorem 2.5.15.

Remark on notation. Recall that we are using X_i and X, and x_i and x for real-valued random variables and the values they take on, respectively, and $\mathbf{X}$ and $\mathbf{x}$ for the vectors $(X_1, \ldots, X_n)$ and $(x_1, \ldots, x_n)$, respectively.

Theorem 2.6 *Let $X_1, \ldots, X_n$ be iid, each with density $f(x|\theta)$ with respect to a σ-finite measure μ, where θ is real-valued, and suppose the following regularity conditions hold.*

(a) The parameter space Ω is an open interval (not necessarily finite).

(b) The distributions P_θ of the X_i have common support, so that the set $A = \{x : f(x|\theta) > 0\}$ is independent of θ.

(c) For every $x \in A$, the density $f(x|\theta)$ is twice differentiable with respect to θ, and the second derivative is continuous in θ.

(d) The integral $\int f(x|\theta)\,d\mu(x)$ can be twice differentiated under the integral sign.

(e) The Fisher information $I(\theta)$ defined by (3.5.10) satisfies $0 < I(\theta) < \infty$.

(f) For any given $\theta_0 \in \Omega$, there exists a positive number c and a function $M(x)$ (both of which may depend on θ_0) such that

$$|\partial^2 \log f(x|\theta)/\partial\theta^2| \le M(x)$$

$$\textit{for all } x \in A, \quad \theta_0 - c < \theta < \theta_0 + c$$

and

$$E_{\theta_0}[M(X)] < \infty.$$

Under these assumptions, if $\delta_n = \delta_n(X_1, \ldots, X_n)$ is any estimator satisfying (2.8), then $v(\theta)$ satisfies (2.7) except on a set of Lebesgue measure zero.

Note that by Lemma 2.5.3, condition (d) ensures that for all $\theta \in \Omega$

(g) $\quad E[\partial \log f(X|\theta)/\partial\theta] = 0$

and

(h) $\quad E[-\partial^2 \log f(X|\theta)/\partial\theta^2] = E[\partial \log f(X|\theta)/\partial\theta]^2 = I(\theta).$

Condition (d) can be replaced by conditions (g) and (h) in the statement of the theorem.

The example makes it clear that no regularity conditions on the densities $f(x|\theta)$ can prevent estimators from violating (2.7). This possibility can be avoided only by placing restrictions on the sequence of estimators also. In view of the information inequality (2.5.31), an obvious sufficient condition is (2.6) [with $g(\theta) = \theta$] together with

(2.10) $$b_n'(\theta) \to 0$$

where $b_n(\theta) = E_\theta(\delta_n) - \theta$ is the bias of δ_n.

If $I(\theta)$ is continuous, as will typically be the case, a more appealing assumption is perhaps that $v(\theta)$ also be continuous. Then, (2.7) clearly cannot be violated at any point since, otherwise, it would be violated in an interval around this point in contradiction to Theorem 2.6. As an alternative, which under mild assumptions on f implies continuity of $v(\theta)$, Rao (1963) and Wolfowitz (1965) require the convergence in (2.2) to be uniform in θ. By working with coverage probabilities rather than asymptotic variance, the latter author also removes the unpleasant assumption that the limit distribution in (2.2) must be normal. An analogous result is proved by Pfanzagl, (1970), who requires the estimators to be asymptotically median unbiased.

The search for restrictions on the sequence $\{\delta_n\}$, which would ensure (2.7) for all values of θ, is motivated in part by the hope of the existence, within the restricted class, of uniformly best estimators for which $v(\theta)$ attains the lower bound. It is further justified by the fact, brought out by Le Cam (1953), Huber (1966), and Hájek (1972), that violation of (2.7) at a point θ_0 entails certain unpleasant properties of the risk of the estimator in the neighborhood of θ_0.

This behavior can be illustrated in the Hodges example.

Example 2.7 Continuation of Example 2.5. The normalized risk function

$$R_n(\theta) = nE(\delta_n - \theta)^2 \tag{2.11}$$

of the Hodges estimator δ_n can be written as

$$R_n(\theta) = 1 - (1 - a^2)\int_{\underline{I}_n}^{\bar{I}_n} (x + \sqrt{n}\,\theta)^2 \phi(x)\,dx$$

$$+2\theta\sqrt{n}(1 - a)\int_{\underline{I}_n}^{\bar{I}_n} (x + \sqrt{n}\,\theta)\phi(x)\,dx$$

where $\bar{I}_n = \sqrt[4]{n} - \sqrt{n}\,\theta$ and $\underline{I}_n = -\sqrt[4]{n} - \sqrt{n}\,\theta$. When the integrals are broken up into their three and two terms, respectively, and the relations

$$\Phi'(x) = \phi(x) \quad \text{and} \quad \phi'(x) = -x\phi(x)$$

are used, $R_n(\theta)$ reduces to

$$R_n(\theta) = \left[1 - (1 - a^2)\int_{\underline{I}_n}^{\bar{I}_n} x^2\phi(x)\,dx\right]$$

$$+n\theta^2(1 - a)^2[\Phi(\bar{I}_n) - \Phi(\underline{I}_n)]$$

$$+2\sqrt{n}\,\theta a(1 - a)[\phi(\bar{I}_n) - \phi(\underline{I}_n)].$$

Consider now the sequence of parameter values $\theta_n = 1/\sqrt[4]{n}$, so that

$$\sqrt{n}\,\theta_n = n^{1/4}, \quad \underline{I}_n = -2n^{1/4}, \quad \bar{I}_n = 0.$$

Then,

$$\sqrt{n}\,\theta_n\phi(\underline{I}_n) \to 0,$$

so that the third term tends to infinity as $n \to \infty$. Since the second term is positive and the first term is bounded, it follows that

$$R_n(\theta_n) \to \infty \quad \text{for} \quad \theta_n = 1/\sqrt[4]{n},$$

and hence, a fortiori, that

$$\sup_\theta R_n(\theta) \to \infty.$$

Let us now compare this result with the fact that (Problem 2.12) for any fixed θ

$$R_n(\theta) \to 1 \quad \text{for} \quad \theta \neq 0, \quad R_n(0) \to a^2.$$

(This shows that in the present case, the limiting risk is equal to the asymptotic variance (see Problem 2.4).) The functions $R_n(\theta)$ are continuous functions of θ with discontinuous limit function

$$L(\theta) = 1 \quad \text{for} \quad \theta \neq 0, \quad L(0) = a^2.$$

However, each of the functions with large values of n rises to a high above the limit value 1, at values of θ tending to the origin with n, and with the value of the

Figure 2.1. *Risk functions $R_n(\theta)$ of the superefficient estimator δ_n of Example 2.5 for a = .5.*

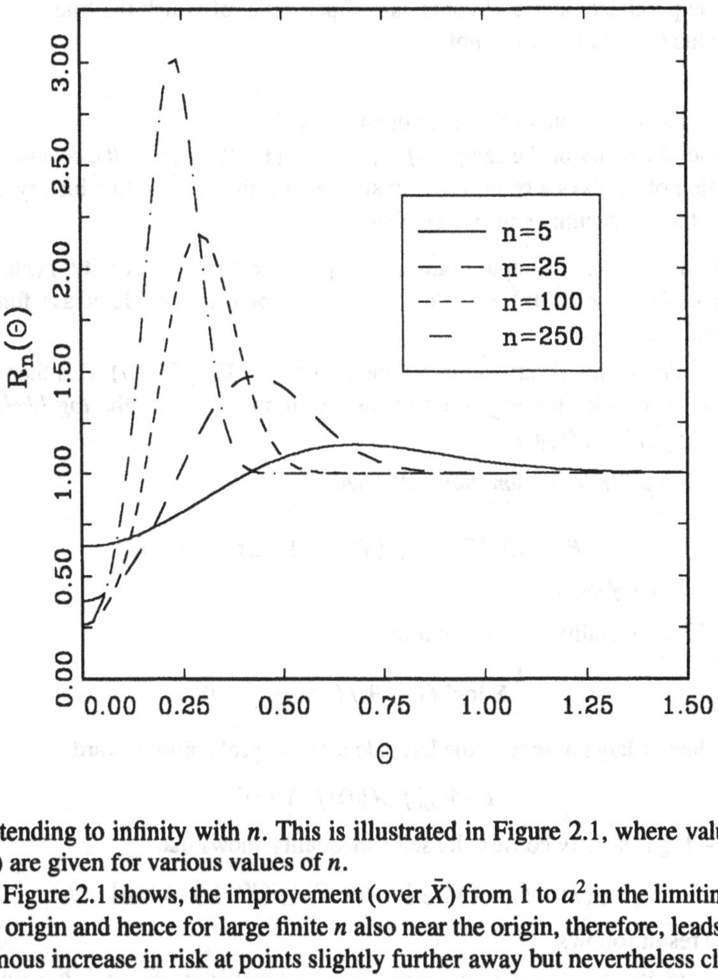

peak tending to infinity with n. This is illustrated in Figure 2.1, where values of $R_n(\theta)$ are given for various values of n.

As Figure 2.1 shows, the improvement (over $\bar{X}$) from 1 to a^2 in the limiting risk at the origin and hence for large finite n also near the origin, therefore, leads to an enormous increase in risk at points slightly further away but nevertheless close to the origin. (In this connection, see Problem 2.14.) ‖

3 Efficient Likelihood Estimation

Under smoothness assumptions similar to those of Theorem 2.6, we shall in the present section prove the existence of asymptotically efficient estimators and provide a method for determining such estimators which, in many cases, leads to an explicit solution.

We begin with the following assumptions:

(A0) The distributions P_θ of the observations are distinct (otherwise, θ cannot be estimated consistently[2]).

(A1) The distributions P_θ have common support.

[2] But see Redner (1981) for a different point of view

(A2) The observations are $\mathbf{X} = (X_1, \ldots, X_n)$, where the X_i are iid with probability density $f(x_i|\theta)$ with respect to μ.

(A3) The parameter space Ω contains an open set ω of which the true parameter value θ_0 is an interior point.

Note: The true value of θ will be denoted by θ_0.

The joint density of the sample, $f(x_1|\theta)\cdots f(x_n|\theta) = \Pi f(x_i|\theta)$, considered as a function of θ, plays a central role in statistical estimation, with a history dating back to the eighteenth century (see Note 10.1).

Definition 3.1 For a sample point $\mathbf{x} = (x_1, \ldots, x_n)$ from a density $f(\mathbf{x}|\theta)$, the *likelihood function* $L(\theta|\mathbf{x}) = f(\mathbf{x}|\theta)$ is the sample density considered as a function of θ for fixed $\mathbf{x}$.

In the case of iid observations, we have $L(\theta|\mathbf{x}) = \Pi_{i=1}^n f(x_i|\theta)$. It is then often easier to work with the logarithm of the likelihood function, the *log likelihood* $l(\theta|\mathbf{x}) = \sum_{i=1}^n \log f(x_i|\theta)$.

Theorem 3.2 *Under assumptions (A0)–(A2),*

$$P_{\theta_0}(L(\theta_0|\mathbf{X}) > L(\theta|\mathbf{X})) \to 1 \quad \textit{as } n \to \infty \tag{3.1}$$

for any fixed $\theta \neq \theta_0$.

Proof. The inequality is equivalent to

$$\frac{1}{n}\Sigma \log[f(X_i|\theta)/f(X_i|\theta_0)] < 0.$$

By the law of large numbers, the left side tends in probability toward

$$E_{\theta_0} \log[f(X|\theta)/f(X|\theta_0)].$$

Since $-\log$ is strictly convex, Jensen's inequality shows that

$$E_{\theta_0} \log[f(X|\theta)/f(X|\theta_0)] < \log E_{\theta_0}[f(X|\theta)/f(X|\theta_0)] = 0,$$

and the result follows. □

By (3.1), the density of $\mathbf{X}$ at the true θ_0 exceeds that at any other fixed θ with high probability when n is large. We do not know θ_0, but we can determine the value $\hat{\theta}$ of θ which maximizes the density of $\mathbf{X}$, that is, which maximizes the likelihood function at the observed $\mathbf{X} = \mathbf{x}$. If this value exists and is unique, it is the *maximum likelihood estimator* (MLE) of θ.[3] The MLE of $g(\theta)$ is defined to be $g(\hat{\theta})$. If g is 1:1 and $\xi = g(\theta)$, this agrees with the definition of $\hat{\xi}$ as the value of ξ that maximizes the likelihood, and the definition is consistent also in the case that g is not 1:1. (In this connection, see Zehna 1966 and Berk 1967b.)

Theorem 3.2 suggests that if the density of $\mathbf{X}$ varies smoothly with θ, the MLE of θ typically should be close to the true value of θ, and hence be a reasonable estimator.

[3] For a more general definition, see Strasser (1985, Sections 64.4 and 84.2) or Scholz (1980, 1985). A discussion of the MLE as a summarizer of the data rather than an estimator is given by Efron (1982a).

Example 3.3 Binomial MLE. Let X have the binomial distribution $b(p, n)$. Then, the MLE of p is obtained by maximizing $\binom{n}{x} p^x q^{n-x}$ and hence is $\hat{p} = X/n$ (Problem 3.1). ||

Example 3.4 Normal MLE. If $X_1, \ldots, X_n$ are iid as $N(\xi, \sigma^2)$, it is convenient to obtain the MLE by maximizing the logarithm of the density, $-n \log \sigma - 1/2\sigma^2 \Sigma(x_i - \xi)^2 - c$. When (ξ, σ) are both unknown, the maximizing values are $\hat{\xi} = \bar{x}$, $\hat{\sigma}^2 = \Sigma(x - \bar{x})^2/n$ (Problem 3.3). ||

As a first question regarding the MLE for iid variables, let us ask whether it is consistent. We begin with the case in which Ω is finite, so that θ can take on only a finite number of values. In this case, a sequence δ_n is consistent if and only if

$$P_\theta(\delta_n = \theta) \to 1 \quad \text{for all } \theta \in \Omega \tag{3.2}$$

(Problem 3.6).

Corollary 3.5 *Under assumptions (A0)–(A2) if Ω is finite, the MLE $\hat{\theta}_n$ exists, it is unique with probability tending to 1, and it is consistent.*

Proof. The result is an immediate consequence of Theorem 3.2 and the fact that if $P(A_{in}) \to 1$ for $i = 1, \ldots, k$, then also $P[A_{1n} \cap \cdots \cap A_{kn}] \to 1$ as $n \to \infty$. □

The proof of Corollary 3.5 breaks down when Ω is not restricted to be finite. That the consistency conclusion itself can break down even if Ω is only countably infinite is shown by the following example due to Bahadur (1958) and Le Cam (1979b, 1990).

Example 3.6 An inconsistent MLE. Let h be a continuous function defined on $(0, 1]$, which is strictly decreasing, with $h(x) \geq 1$ for all $0 < x \leq 1$ and satisfying

$$\int_0^1 h(x)\,dx = \infty. \tag{3.3}$$

Given a constant $0 < c < 1$, let a_k, $k = 0, 1, \ldots$, be a sequence of constants defined inductively as follows: $a_0 = 1$; given $a_0, \ldots, a_{k-1}$, the constant a_k is defined by

$$\int_{a_k}^{a_{k-1}} [h(x) - c]\,dx = 1 - c. \tag{3.4}$$

It is easy to see that there exists a unique value $0 < a_k < a_{k-1}$ satisfying (3.4) (Problem 3.8). Since the sequence $\{a_k\}$ is decreasing, it tends to a limit $a \geq 0$. If a were > 0, the left side of (3.4) would tend to zero which is impossible. Thus, $a_k \to 0$ as $k \to \infty$.

Consider now the sequence of densities

$$f_k(x) = \begin{cases} c & \text{if } x \leq a_k \text{ or } x > a_{k-1} \\ h(x) & \text{if } a_k < x \leq a_{k-1}, \end{cases}$$

and the problem of estimating the parameter k on the basis of independent observations $X_1, \ldots, X_n$ from f_k. We shall show that the MLE exists and that it tends to infinity in probability regardless of the true value k_0 of k and is, therefore, not consistent, provided $h(x) \to \infty$ sufficiently fast as $x \to 0$.

Let us denote the joint density of the X's by

$$p_k(\mathbf{x}) = f_k(x_1)\cdots f_k(x_n).$$

That the MLE exists follows from the fact that $p_k(\mathbf{x}) = c^n < 1$ for any value of k for which the interval $I_k = (a_k, a_{k-1}]$ contains none of the observations, so that the maximizing value of k must be one of the $\leq n$ values for which I_k contains at least one of the x's.

For $n = 1$, the MLE is the value of k for which $X_1 \in I_k$, and for $n = 2$, the MLE is the value of k for which $X_{(1)} \in I_k$. For $n = 3$, it may happen that one observation lies in I_k and two in $I_l (k < l)$, and whether the MLE is k or l then depends on whether $c \cdot h(x_{(1)})$ is greater than or less than $h(x_{(2)})h(x_{(3)})$.

We shall now prove that the MLE $\hat{K}_n$ (which is unique with probability tending to 1) tends to infinity in probability, that is, that

$$P(\hat{K}_n > k) \to 1 \quad \text{for every } k, \tag{3.5}$$

provided h satisfies

$$h(x) \geq e^{1/x^2} \tag{3.6}$$

for all sufficiently small values of x.

To prove (3.5), we will show that for any fixed j,

$$P[p_{K_n^*}(\mathbf{X}) > p_j(\mathbf{X})] \to 1 \quad \text{as } n \to \infty \tag{3.7}$$

where K_n^* is the value of k for which $X_{(1)} \in I_k$. Since $p_{\hat{K}_n}(\mathbf{X}) \geq p_{K_n^*}(\mathbf{X})$, it then follows that for any fixed k,

$$p[\hat{K}_n > k] \geq P[p_{\hat{K}_n}(\mathbf{X}) > p_j(\mathbf{X}) \text{ for } j = 1, \ldots, k] \to 1.$$

To prove (3.7), consider

$$L_{jk} = \log \frac{f_k(x_1)\cdots f_k(x_n)}{f_j(x_1)\cdots f_j(x_n)} = \Sigma_i^{(1)} \log \frac{h(x_i)}{c} - \Sigma_i^{(2)} \log \frac{h(x_i)}{c}$$

where $\Sigma^{(1)}$ and $\Sigma^{(2)}$ extend over all i for which $x_i \in I_k$ and $x_i \in I_j$, respectively. Now $x_i \in I_j$ implies that $h(x_i) < h(a_j)$, so that

$$\Sigma^{(2)} \log[h(x_i)/c] < \nu_{jn} \log[h(a_j)/c]$$

where ν_{jn} is the number of x's in I_j. Similarly, for $k = K_n^*$,

$$\Sigma^{(1)} \log[h(x_i)/c] \geq \log[h(x_{(1)})/c]$$

since $\log[h(x)/c] \geq 0$ for all x. Thus,

$$\frac{1}{n} L_{j,K_n^*} \geq \frac{1}{n} \log \frac{h(x_{(1)})}{c} - \frac{1}{n} \nu_{jn} \log \frac{h(a_j)}{c}.$$

Since ν_{jn}/n tends in probability to $P(X_1 \in I_j) < 1$, it only remains to show that

$$\frac{1}{n} \log h(X_{(1)}) \to \infty \quad \text{in probability.} \tag{3.8}$$

Instead of $X_1, \ldots, X_n$, consider a sample $Y_1, \ldots, Y_n$ from the uniform distribution $U(0, 1/c)$. Then, for any x, $P(Y_i > x) \geq P(X_i > x)$ and hence

$$P[h(Y_{(1)}) > x] \leq P[h(X_{(1)}) > x],$$

and it is therefore enough to prove that $(1/n)\log h(Y_{(1)}) \to \infty$ in probability. If h satisfies (3.6), $(1/n)\log h(Y_{(1)}) \geq 1/nY_{(1)}^2$, and the right side tends to infinity in probability since $nY_{(1)}$ tends to a limit distribution (Problem 2.6). This completes the proof. ‖

For later reference, note that the proof has established not only (2.5) but the fact that for any fixed A (Problem 3.9),

$$P[p_{K_n^*}(\mathbf{X}) > A^n p_j(\mathbf{X})] \to 1. \tag{3.9}$$

The example suggests (and this suggestion will be verified in the next section) that also for densities depending smoothly on a continuously varying parameter θ, the MLE need not be consistent. We shall now show, however, that a slightly weaker conclusion is possible under relatively mild conditions. Throughout the present section, we shall assume θ to be real-valued. The case of several parameters will be taken up in Section 6.5.

In the following, we shall frequently use the shorthand notation $l(\theta)$ for the *log likelihood*

$$l(\theta|\mathbf{x}) = \Sigma \log f(x_i|\theta), \tag{3.10}$$

and $l'(\theta), l''(\theta), \ldots$ for its derivatives with respect to θ.

A way around the difficulty presented by this example was found by Cramér (1946a, 1946b), who replaced the search for a global maximum of the likelihood function with that for a local maximum.

Theorem 3.7 *Let $X_1, \ldots, X_n$ satisfy (A0)–(A3) and suppose that for almost all x, $f(x|\theta)$ is differentiable with respect to θ in ω, with derivative $f'(x|\theta)$. Then, with probability tending to 1 as $n \to \infty$, the likelihood equation*

$$\frac{\partial}{\partial\theta} l(\theta|\mathbf{x}) = 0 \tag{3.11}$$

or, equivalently, the equation

$$l'(\theta|\mathbf{x}) = \Sigma \frac{f'(x_i|\theta)}{f(x_i|\theta)} = 0 \tag{3.12}$$

has a root $\hat{\theta}_n = \hat{\theta}_n(x_1, \ldots, x_n)$ such that $\hat{\theta}_n(X_1, \ldots, X_n)$ tends to the true value θ_0 in probability.

Proof. Let $a > 0$ be small enough so that $(\theta_0 - a, \theta_0 + a) \subset \omega$, and let

$$S_n(a) = \{\mathbf{x} : l(\theta_0|\mathbf{x}) > l(\theta_0 - a|\mathbf{x}) \text{ and } l(\theta_0|\mathbf{x}) > l(\theta_0 + a|\mathbf{x})\}. \tag{3.13}$$

Then for any $\mathbf{x} \in S_n(a)$, there exists a value $\hat{\theta}_n(a) \in (\theta_0 - a, \theta_0 + a)$ at which $l(\theta)$ has a local maximum, so that $l'(\hat{\theta}_n(a)) = 0$. Moreover, by Theorem 3.2, $P_{\theta_0}(S_n(a)) \to 1$ for any such a, and from this it follows that there exists $a_n \downarrow 0$ such that $P(S_n(a_n)) \to 1$ as $n \to \infty$.

Let $\hat{\theta}_n^* = \hat{\theta}_n(a_n)$ for $\mathbf{x} \in S_n(a_n)$ (and equal to an arbitrary constant otherwise). Then $P_{\theta_0}(l_n'(\hat{\theta}_n^*) = 0) \geq P_{\theta_0}(S_n(a_n)) \to 1$, and for any fixed $a > 0$ and n sufficiently large,

$$P_{\theta_0}(|\hat{\theta}_n^* - \theta| < a) \geq P_{\theta_0}(|\hat{\theta}_n^* - \theta| < a_n) \geq P_{\theta_0}(S_n(a_n)) \to 1, \tag{3.14}$$

completing the proof. (We are grateful to Brett Presnell for pointing out an error in the original proof and for providing us with this correction.) □

In connection with this theorem, the following comments should be noted.

1. The proof yields the additional fact that with probability tending to 1, the roots $\hat{\theta}_n(a)$ can be chosen to be local maxima and so, therefore, can the θ_n^* if we let θ_n^* be the closest root corresponding to a maximum.
2. On the other hand, the theorem does not establish the existence of a consistent estimator sequence since, with the true value θ_0 unknown, the data do not tell us which root to choose so as to obtain a consistent sequence. An exception, of course, is the case in which the root is unique.
3. It should also be emphasized that the existence of a root $\hat{\theta}_n$ is not asserted for all $\mathbf{x}$ (or for a given n even for any $\mathbf{x}$). This does not affect consistency, which only requires $\hat{\theta}_n$ to be defined on a set S_n', the probability of which tends to 1 as $n \to \infty$.
4. Although the likelihood equation can have many roots, the consistent sequence of roots generated by Theorem 3.7 is essentially unique. For a more precise statement of this result, which is due to Huzurbazar (1948), see Problem 3.28.
5. Finally, there is a technical question concerning the measurability of the estimator sequence $\hat{\theta}_n(a)$, and hence of the sequence $\hat{\theta}_n^*$. Recall from Section 1.2 that $\hat{\theta}_n(a)$ is measurable function if the set $\{a : \hat{\theta}_n(a) > t\}$ is a measurable set for every t. Since $\hat{\theta}_n(a)$ is defined implicitly, its measurability (and also that of $\hat{\theta}_n^*$) is not immediately obvious. Happily, it turns out that the sequences $\hat{\theta}_n(a)$ and $\hat{\theta}_n^*$ are measurable. (For details, see Problem 3.29.)

Corollary 3.8 *Under the assumptions of Theorem 3.7, if the likelihood equation has a unique root δ_n for each n and all $\mathbf{x}$, then $\{\delta_n\}$ is a consistent sequence of estimators of θ. If, in addition, the parameter space is an open interval $(\underline{\theta}, \bar{\theta})$ (not necessarily finite), then with probability tending to 1, δ_n maximizes the likelihood, that is, δ_n is the MLE, which is therefore consistent.*

Proof. The first statement is obvious. To prove the second, suppose the probability of δ_n being the MLE does not tend to 1. Then, for sufficiently large values of n, the likelihood must tend to a supremum as θ tends toward $\underline{\theta}$ or $\bar{\theta}$ with positive probability. Now with probability tending to 1, δ_n is a local maximum. This contradicts the assumed uniqueness of the root. □

The conclusion of Corollary 3.8 holds, of course, not only when the root of the likelihood equation is unique but also when the probability of multiple roots tends to zero as $n \to \infty$. On the other hand, even when the root is unique, the corollary says nothing about its properties for finite n.

Example 3.9 Minimum likelihood. Let X take on the values 0, 1, 2 with probabilities $6\theta^2 - 4\theta + 1, \theta - 2\theta^2$, and $3\theta - 4\theta^2$ $(0 < \theta < 1/2)$. Then, the likelihood equation has a unique root for all x, which is a minimum for $x = 0$ and a maximum for $x = 1$ and 2 (Problem 3.11). ||

Theorem 3.7 establishes the existence of a consistent root of the likelihood equation. The next theorem asserts that any such sequence is asymptotically normal and efficient.

Theorem 3.10 *Suppose that $X_1, \ldots, X_n$ are iid and satisfy the assumptions of Theorem 2.6, with (c) and (d) replaced by the corresponding assumptions on the third (rather than the second) derivative, that is, by the existence of a third derivative satisfying*

$$\left|\frac{\partial^3}{\partial\theta^3}\log f(x|\theta)\right| \le M(x) \tag{3.15}$$

$$\text{for all } x \in A, \quad \theta_0 - c < \theta < \theta_0 + c$$

with

$$E_{\theta_0}[M(\mathbf{X})] < \infty. \tag{3.16}$$

Then, any consistent sequence $\hat{\theta}_n = \hat{\theta}_n(X_1, \ldots, X_n)$ of roots of the likelihood equation satisfies

$$\sqrt{n}(\hat{\theta}_n - \theta) \xrightarrow{\mathcal{L}} N\left(0, \frac{1}{I(\theta)}\right). \tag{3.17}$$

We shall call such a sequence $\hat{\theta}_n$ an *efficient likelihood estimator* (ELE) of θ. It is typically (but need not be, see Example 4.1) provided by the MLE. Note also that any sequence $\hat{\theta}_n^*$ satisfying (3.19) is *asymptotically efficient* in the sense of Definition 2.4.

Proof of Theorem 3.10. For any fixed $\mathbf{x}$, expand $l'(\hat{\theta}_n)$ about θ_0,

$$l'(\hat{\theta}_n) = l'(\theta_0) + (\hat{\theta}_n - \theta_0)l''(\theta_0) + \frac{1}{2}(\hat{\theta}_n - \theta_0)^2 l'''(\theta_n^*)$$

where θ_n^* lies between θ_0 and $\hat{\theta}_n$. By assumption, the left side is zero, so that

$$\sqrt{n}(\hat{\theta}_n - \theta_0) = \frac{(1/\sqrt{n})\,l'(\theta_0)}{-(1/n)l''(\theta_0) - (1/2n)(\hat{\theta}_n - \theta_0)l'''(\theta_n^*)}$$

where it should be remembered that $l(\theta)$, $l'(\theta)$, and so on are functions of $\mathbf{X}$ as well as θ. We shall show that

$$\frac{1}{\sqrt{n}}l'(\theta_0) \xrightarrow{\mathcal{L}} N[0, I(\theta_0)], \tag{3.18}$$

that

$$-\frac{1}{n}l''(\theta_0) \xrightarrow{P} I(\theta_0) \tag{3.19}$$

and that

$$\frac{1}{n}l'''(\theta_n^*) \quad \text{is bounded in probability.} \tag{3.20}$$

The desired result then follows from Theorem 1.8.10.

Of the above statements, (3.18) follows from the fact that

$$\frac{1}{\sqrt{n}}l'(\theta_0) = \sqrt{n}\,\frac{1}{n}\sum\left[\frac{f'(X_i|\theta_0)}{f(X_i|\theta_0)} - E_{\theta_0}\frac{f'(X_i|\theta_0)}{f(X_i|\theta_0)}\right]$$

since the expectation term is zero, and then from the central limit theorem (CLT) and the definition of $I(\theta)$.

Next, (3.19) follows because

$$-\frac{1}{n}l''(\theta_0) = \frac{1}{n}\sum \frac{f'^2(X_i|\theta_0) - f(X_i|\theta_0)f''(X_i|\theta_0)}{f^2(X_i|\theta_0)},$$

and, by the law of large numbers, this tends in probability to

$$I(\theta_0) - E_{\theta_0}\frac{f''(X_i|\theta_0)}{f(X_i|\theta_0)} = I(\theta_0).$$

Finally, (3.20) is established by noting

$$\frac{1}{n}l'''(\theta) = \frac{1}{n}\sum \frac{\partial^3}{\partial\theta^3}\log f(X_i|\theta)$$

so that by (3.15),

$$\left|\frac{1}{n}l'''(\theta_n^*)\right| < \frac{1}{n}[M(X_1) + \cdots + M(X_n)]$$

with probability tending to 1. The right side tends in probability to $E_{\theta_0}[M(X)]$, and this completes the proof. □

Although the conclusions of Theorem 3.10 are quite far-reaching, the proof is remarkably easy. The reason is that Theorem 3.7 already puts $\hat{\theta}_n$ into the neighborhood of the true value θ_0, so that an expansion about θ_0 essentially linearizes the problem and thereby prepares the way for application of the central limit theorem.

Corollary 3.11 *Under the assumptions of Theorem 3.10, if the likelihood equation has a unique root for all n and* **x**, *and more generally if the probability of multiple roots tends to zero as* $n \to \infty$, *the MLE is asymptotically efficient.*

To establish the assumptions of Theorem 3.10, one must verify the following two conditions that may not be obvious.

(a) That $\int f(x|\theta)\,d\mu(x)$ can be differentiated twice with respect to θ by differentiating under the integral sign.

(b) The third derivative is uniformly bounded by an integrable function [see (3.15)].

Conditions when (a) holds are given in books on calculus (see also Casella and Berger 1990, Section 2.4) although it is often easier simply to calculate the difference quotient and pass to the limit.

Condition (b) is usually easy to check after realizing that it is not necessary for (3.15) to hold for all θ, but that it is enough if there exist $\theta_1 < \theta_0 < \theta_2$ such that (3.15) holds for all $\theta_1 \le \theta \le \theta_2$.

Example 3.12 One-parameter exponential family. Let $X_1, \ldots, X_n$ be iid according to a one-parameter exponential family with density

$$f(x_i|\eta) = e^{\eta T(x_i) - A(\eta)} \tag{3.21}$$

with respect to a σ-finite measure μ, and let the estimand be η. The likelihood equation is

$$\frac{1}{n}\Sigma T(x_i) = A'(\eta), \tag{3.22}$$

which, by (1.5.14), is equivalent to

$$E_\eta[T(X_j)] = \frac{1}{n}\Sigma T(x_i). \tag{3.23}$$

The left side of (3.23) is a strictly increasing function of η since, by (1.5.15),

$$\frac{d}{d\eta}[E_\eta T(X_j)] = \text{var}_\eta T(X_j) > 0.$$

It follows that Equation (3.23) has at most one solution. The conditions of Theorem 3.10 are easily checked in the present case. In particular, condition (a) follows from Theorem 1.5.8 and (b) from the fact that the third derivative of $\log f(x|\eta)$ is independent of x and a continuous function of η. With probability tending to 1, (3.23) therefore has a solution $\hat{\eta}$. This solution is unique, consistent, and asymptotically efficient, so that

$$\sqrt{n}(\hat{\eta} - \eta) \xrightarrow{\mathcal{L}} N(0, \text{var}\, T) \tag{3.24}$$

where $T = T(X_i)$ and the asymptotic variance follows from (2.5.18). ||

Example 3.13 Truncated normal. As an illustration of the preceding example, consider a sample of n observations from a normal distribution $N(\xi, 1)$, truncated at two fixed points $a < b$. The density of a single X is then

$$\frac{1}{\sqrt{2\pi}} \exp\left[-\frac{1}{2}(x-\xi)^2\right] / [\Phi(b-\xi) - \Phi(a-\xi)], \quad a < x < b,$$

which satisfies (3.21) with $\eta = \xi$, $T(x) = x$. An ELE will therefore be the unique solution of $E_\xi(X) = \bar{x}$ if it exists. To see that this equation has a solution for any value $a < \bar{x} < b$, note that as $\xi \to -\infty$ or $+\infty$, X tends in probability to a or b, respectively (Problem 3.12). Since X is bounded, this implies that also $E_\xi(X)$ tends to a or b. Since $E_\xi(X)$ is continuous, the existence of $\hat{\xi}$ follows. ||

For densities that are members of location or scale families, it is fairly straightforward to determine the existence and behavior of the MLE. (See Problems 3.15 –3.19.)

We turn to one last example, which is not covered by Theorem 3.10.

Example 3.14 Double exponential. For the double exponential density $DE(\theta, 1)$ given in Table 1.5.1, it is not true that for all (or almost all) x, $f(x-\theta)$ is differentiable with respect to θ, since for every x there exists a value ($\theta = x$) at which the derivative does not exist. Despite this failure, the MLE (which is the median of the X's) satisfies the conclusion of Theorem 3.10 and is asymptotically normal with variance $1/n$ (see Problem 3.25). This was established by Daniels (1961), who proved a general theorem, not requiring differentiability of the density, that was motivated by this problem. (See Note 10.2.) ||

4 Likelihood Estimation: Multiple Roots

When the likelihood equation has multiple roots, the assumptions of Theorem 3.10 are no longer sufficient to guarantee consistency of the MLE, even when it exists for all n. This is shown by the following example due to Le Cam (1979b, 1990), which is obtained by embedding the sequence $\{f_k\}$ of Example 3.6 in a sufficiently smooth continuous-parameter family.

Example 4.1 Continuation of Example 3.6. For $k \le \theta < k+1$, $k = 1, 2, \ldots,$ let

$$f(x|\theta) = [1 - u(\theta - k)]f_k(x) + u(\theta - k)f_{k+1}(x), \tag{4.1}$$

with f_k defined as in Example 3.6 and u defined on $(-\infty, \infty)$ such that $u(x) = 0$ for $x \le 0$ and $u(x) = 1$ for $x \ge 1$ is strictly increasing on $(0, 1)$ and infinitely differentiable on $(-\infty, \infty)$ (Problem 4.1). Let $X_1, \ldots, X_n$ be iid, each with density $f(x|\theta)$, and let $p(\mathbf{x}|\theta) = \Pi f(x_i|\theta)$.

Since for any given $\mathbf{x}$, the density $p(\mathbf{x}|\theta)$ is bounded and continuous in θ and is equal to c^n for all sufficiently large θ and greater than c^n for some θ, it takes on its maximum for some finite θ, and the MLE $\hat{\theta}_n$ therefore exists.

To see that $\hat{\theta}_n \to \infty$ in probability, note that for $k \le \theta < k+1$,

$$p(\mathbf{x}|\theta) \le \Pi \max[f_k(x_i), f_{k+1}(x_i)] = \bar{p}_k(\mathbf{x}). \tag{4.2}$$

If $\hat{K}_n$ and K_n^* are defined as in Example 3.6, the argument of that example shows that it is enough to prove that for any fixed j,

$$P[p_{K_n^*}(\mathbf{X}) > \bar{p}_j(\mathbf{X})] \to 1 \quad \text{as } n \to \infty, \tag{4.3}$$

where $p_k(\mathbf{x}) = p(\mathbf{x}|k)$. Now

$$\bar{L}_{jk} = \frac{p_k(\mathbf{x})}{\bar{p}_j(\mathbf{x})} = \Sigma^{(1)} \log \frac{h(x_i)}{c} - \Sigma^{(2)} \log \frac{h(x_i)}{c} - \Sigma^{(3)} \log \frac{h(x_i)}{c}$$

where $\Sigma^{(1)}, \Sigma^{(2)}$, and $\Sigma^{(3)}$ extend over all i for which $x_i \in I_k, x_i \in I_j$, and $x_i \in I_{j+1}$, respectively. The argument is now completed as before to show that $\hat{\theta}_n \to \infty$ in probability regardless of the true value of θ and is therefore not consistent.

The example is not yet completely satisfactory since $\partial f(x|\theta)/\partial\theta = 0$ and, hence, $I(\theta) = 0$ for $\theta = 1, 2, \ldots$ [The remaining conditions of Theorem 3.10 are easily checked (Problem 4.2).] To remove this difficulty, define

$$g(x|\theta) = \frac{1}{2}[f(x|\theta) + f(x|\theta + \alpha e^{-\theta^2})], \quad \theta \ge 1, \tag{4.4}$$

for some fixed $\alpha < 1$.

If $X_1, \ldots, X_n$ are iid according to $g(x|\theta)$, we shall now show that the MLE $\hat{\theta}_n$ continues to tend to infinity for any fixed θ. We have, as before

$$P[\hat{\theta}_n > k] \ge P[\Pi g(x_i|K_n^*) > \Pi g(x_i|\theta) \quad \text{for all } \theta \le k]$$
$$\ge P\left\{\frac{1}{2^n}\Pi f(x_i|K_n^*) > \Pi\left(\frac{1}{2}\left[f(x_i|\theta) + f(x_i|\theta + \alpha e^{-\theta^2})\right]\right)\right\}.$$

For $j \le \theta < j+1$, it is seen from (4.1) that $[f(x_i|\theta) + f(x_i|\theta + \alpha e^{-\theta^2})]/2$ is a weighted average of $f_j(x_i)$, $f_{j+1}(x_i)$, and possibly $f_{j+2}(x_i)$. By using $\bar{p}_j(\mathbf{x}) = \Pi \max[f_j(x_i), f_{j+1}(x_i), f_{j+2}(x_i)]$ in place of $p_j(\mathbf{x})$, the proof can now be completed as before. Since the densities $g(x_i|\theta)$ satisfy the conditions of Theorem

3.10 (Problem 4.3), these conditions are therefore not enough to ensure the consistency of the MLE. (For another example, see Ferguson 1982.) ||

Even under the assumptions of Theorem 3.10, one is thus, in the case of multiple roots, still faced with the problem of identifying a consistent sequence of roots. Following are three possible approaches.

(a) In many cases, the maximum likelihood estimator is consistent. Conditions which ensure this were given by, among others, Wald (1949), Wolfowitz (1965), Le Cam (1953, 1955, 1970), Kiefer and Wolfowitz (1956), Kraft and Le Cam (1956), Bahadur (1967), and Perlman (1972). A survey of the literature can be found in Perlman (1983). This material is technically difficult, and even when the conditions are satisfied, the determination of the MLE may present problems (see Barnett 1966). We shall therefore turn to somewhat simpler alternatives.

The following two methods require that some sequence of consistent (but not necessarily efficient) estimators be available. In any given situation, it is usually easy to construct a consistent sequence, as will be illustrated below and in the next section.

(b) Suppose that δ_n is any consistent estimator of θ and that the assumptions of Theorem 3.10 hold. Then, the root $\hat{\theta}_n$ of the likelihood equation closest to δ_n (which exists by the proof of Theorem 3.7) is also consistent, and hence is efficient by Theorem 3.10.

To see this, note that by Theorem 3.10, there exists a consistent sequence of roots, say $\hat{\theta}_n^*$. Since $\hat{\theta}_n^* - \delta_n \to 0$ in probability, so does $\hat{\theta}_n - \delta_n$.

The following approach, which does not require the determination of the closest root and in which the estimators are no longer exact roots of the likelihood equation, is often more convenient.

(c) The usual iterative methods for solving the likelihood equation

$$l'(\theta) = 0 \tag{4.5}$$

are based on replacing the left side by the linear terms of its Taylor expansion about an approximate solution $\tilde{\theta}$. If $\hat{\theta}$ denotes a root of (4.5), this leads to the approximation

$$0 = l'(\hat{\theta}) \doteq l'(\tilde{\theta}) + (\hat{\theta} - \tilde{\theta})l''(\tilde{\theta}), \tag{4.6}$$

and hence to

$$\hat{\theta} = \tilde{\theta} - \frac{l'(\tilde{\theta})}{l''(\tilde{\theta})}. \tag{4.7}$$

The procedure is then iterated by replacing $\tilde{\theta}$ by the value $\tilde{\tilde{\theta}}$ of the right side of (4.7), and so on. This is the *Newton-Raphson* iterative process. (For a discussion of the performance of this procedure, see, for example, Barnett 1966, Stuart and Ord 1991, Section 18.21, or Searle et al. 1992, Section 8.2.)

Here, we are concerned only with the first step and with the performance of the one-step approximation (4.7) as an estimator of θ. The following result gives conditions on $\tilde{\theta}$ under which the resulting sequence of estimators is consistent, asymptotically normal, and efficient. It relies on the sequence of estimators possessing the following property.

Definition 4.2 A sequence of estimators δ_n is *$\sqrt{n}$-consistent* for θ if $\sqrt{n}(\delta_n - \theta)$ is bounded in probability, that is, if $\delta_n - \theta = O_p(1/\sqrt{n})$.

Theorem 4.3 *Suppose that the assumptions of Theorem 3.10 hold and that $\tilde{\theta}_n$ is not only a consistent but a $\sqrt{n}$-consistent*[4] *estimator of θ. Then, the estimator sequence*

$$\delta_n = \tilde{\theta}_n - \frac{l'(\tilde{\theta}_n)}{l''(\tilde{\theta}_n)} \tag{4.8}$$

is asymptotically efficient, that is, it satisfies (3.17) with δ_n in place of $\hat{\theta}_n$.

Proof. As in the proof of Theorem 3.10, expand $l'(\tilde{\theta}_n)$ about θ_0 as

$$l'(\tilde{\theta}_n) = l'(\theta_0) + (\tilde{\theta}_n - \theta_0)l''(\theta_0) + \frac{1}{2}(\tilde{\theta}_n - \theta_0)^2 l'''(\theta_n^*)$$

where θ_n^* lies between θ_0 and $\tilde{\theta}_n$. Substituting this expression into (4.8) and simplifying, we find

$$\sqrt{n}(\delta_n - \theta_0) = \frac{(1/\sqrt{n})l'(\theta_0)}{-(1/n)l''(\tilde{\theta}_n)} + \sqrt{n}(\tilde{\theta}_n - \theta_0) \times \left[1 - \frac{l''(\theta_0)}{l''(\tilde{\theta}_n)} - \frac{1}{2}(\tilde{\theta}_n - \theta_0)\frac{l'''(\theta_n^*)}{l''(\tilde{\theta}_n)}\right]. \tag{4.9}$$

The result now follows from the following facts:

(a) $\dfrac{(1/\sqrt{n})l'(\theta_0)}{-(1/n)l''(\theta_0)} \xrightarrow{\mathcal{L}} N(0, I^{-1}(\theta_0)]$ [(3.18) and (3.19)]

(b) $\sqrt{n}(\hat{\theta}_n - \theta_0) = O_p(1)$ [assumption]

(c) $\dfrac{l'''(\theta_n^*)}{l''(\tilde{\theta}_n)} = O_p(1)$ [(3.19) and (3.20)]

(d) $\dfrac{l''(\tilde{\theta}_n)}{l''(\theta_0)} \to 1$ in probability [see below]

Here, (d) follows from the fact that

$$\frac{1}{n}l''(\tilde{\theta}_n) = \frac{1}{n}l''(\theta_0) + \frac{1}{n}(\tilde{\theta}_n - \theta_0)l'''(\theta_n^{**}), \tag{4.10}$$

for some θ_n^{**} between θ_0 and $\tilde{\theta}_n$. Now (3.19), (3.20), and consistency of $\tilde{\theta}_n$ applied to (4.10) imply (d). In turn, (b)-(d) show that the entire second term in (4.9) converges to zero in probability, and (a) shows that the first term has the correct limit distribution. □

[4] A general method for constructing $\sqrt{n}$-consistent estimators is given by Le Cam (1969, p. 103). See also Bickel et al. (1993).

Corollary 4.4 *Suppose that the assumptions of Theorem 4.3 hold and that the Fisher information $I(\theta)$ is a continuous function of θ. Then, the estimator*

$$\delta_n' = \tilde{\theta}_n + \frac{l'(\tilde{\theta}_n)}{nI(\tilde{\theta}_n)} \tag{4.11}$$

is also asymptotically efficient.

Proof. By (d) in the proof of Theorem 4.3, condition (h) of Theorem 2.6, and the law of large numbers, $-(1/n)l''(\tilde{\theta}_n) \to I(\theta_0)$ in probability. Also, since $I(\theta)$ is continuous, $I(\tilde{\theta}_n) \to I(\theta_0)$ in probability, so that $-(1/n)l''(\tilde{\theta}_n)/I(\tilde{\theta}_n) \to 1$ in probability, and this completes the proof. □

The estimators (4.8) and (4.11) are compared by Stuart (1958), who gives a heuristic argument why (4.11) might be expected to be closer to the ELE than (4.8) and provides a numerical example supporting this argument. See also Efron and Hinkley 1978 and Lindsay and Yi 1996.

Example 4.5 Location parameter. Consider the case of a symmetric location family, with density $f(x - \theta)$, in which the likelihood equation

$$\sum \frac{f'(x_i - \theta)}{f(x_i - \theta)} = 0 \tag{4.12}$$

has multiple roots. [For the Cauchy distribution, for example, it has been shown by Reeds (1985) that if (4.12) has $K + 1$ roots, then as $n \to \infty$, K tends in law to a Poisson distribution with expectation $1/\pi$. The Cauchy case has also been considered by Barnett (1966) and Bai and Fu (1987).] If $\text{var}(X) < \infty$, it follows from the CLT that the sample mean $\bar{X}_n$ is $\sqrt{n}$-consistent and that an asymptotically efficient estimator of θ is therefore provided by (4.8) or (4.11) with $\tilde{\theta}_n = \bar{X}$ as long as $f(x - \theta)$ satisfies the conditions of Theorem 3.10. For distributions such as the Cauchy for which $E(X^2) = \infty$, one can, instead, take for $\tilde{\theta}_n$ the sample median provided $f(0) > 0$; other robust estimators provide still further possibilities (see, for example, Huber 1973, 1981 or Haberman 1989). ‖

Example 4.6 Grouped or censored observations. Suppose that $X_1, \ldots, X_n$ are iid according to a location family with cdf $F(x - \theta)$, with F known and with $0 < F(x) < 1$ for all x, but that it is only observed whether each X_i falls below a, between a and b, or above b where $a < b$ are two given constants. The n observations constitute n trinomial trials with probabilities $p_1 = p_1(\theta) = F(a-\theta)$, $p_2(\theta) = F(b - \theta) - F(a - \theta)$, $p_3(\theta) = 1 - F(b - \theta)$ for the three outcomes. If V denotes the number of observations less than a, then

$$\sqrt{n}\left[\frac{V}{n} - p_1\right] \xrightarrow{\mathcal{L}} N[0, p_1(1 - p_1)] \tag{4.13}$$

and, by Theorem 1.8.12,

$$\tilde{V}_n = a - F^{-1}\left(\frac{V}{n}\right) \tag{4.14}$$

is a $\sqrt{n}$-consistent estimator of θ. Since the estimator is not defined when $V = 0$ or $V = n$, some special definition has to be adopted in these cases whose probability however tends to zero as $n \to \infty$.

If the trinomial distribution for a single trial satisfies the assumptions of Theorem 3.10 as will be the case under mild assumptions on F, the estimator (4.8) is asymptotically efficient (but see the comment following Example 7.15). The approach applies, of course, equally to the case of more than three groups.

A very similar situation arises when the X's are *censored*, say at a fixed point a. For example, they might be lengths of life of light bulbs or patients, with observation discontinued at time a. The observations can then be represented as

$$Y_i = \begin{cases} X_i & \text{if } X_i < a \\ a & \text{if } X_i \geq a. \end{cases} \tag{4.15}$$

Here, the value a of Y_i when $X_i \geq a$ has no significance; it simply indicates that the value of X_i is $\geq a$. The Y's are then iid with density

$$g(y|\theta) = \begin{cases} f(y-\theta) & \text{if } y < a \\ 1 - F(a-\theta) & \text{if } y = a \end{cases} \tag{4.16}$$

with respect to the measure μ which is Lebesgue measure on $(-\infty, a)$ and assigns measure 1 to the point $y = a$.

The estimator (4.14) continues to be $\sqrt{n}$-consistent in the present situation. An alternative starting point is, for example, the best linear combination of the ordered X's less than a (see, for example, Chan 1967). ||

Example 4.7 Mixtures. Let $X_1, \ldots, X_n$ be a sample from a distribution $\theta G + (1-\theta)H$, $0 < \theta < 1$, where G and H are two specified distributions with densities g and h. The log likelihood of a single observation is a concave function of θ, and so therefore is the log likelihood of a sample (Problem 4.5). It follows that the likelihood equation has at most one solution. [The asymptotic performance of the ML estimator is studied by Hill (1963).]

Even when the root is unique, as it is here, Theorem 4.3 provides an alternative, which may be more convenient than the MLE. In the mixture problem, as in many other cases, a $\sqrt{n}$-consistent estimator can be obtained by the *method of moments*, which consists in equating the first k moments of X to the corresponding sample moments, say

$$E_\theta(X_i^r) = \frac{1}{n}\sum_{j=1}^{n} X_j^r, \quad r = 1, \ldots, k, \tag{4.17}$$

where k is the number of unknown parameters. (For further discussion, see, for example, Cramér 1946a, Section 33.1 and Serfling 1980, Section 4.3.1). In the present case, suppose that $E(X_i) = \xi$ or η when X is distributed as G or H where $\eta \neq \xi$ and G and H have finite variance. Since $k = 1$, the method of moments estimates θ as the solution of the equation

$$\xi\theta + \eta(1-\theta) = \bar{X}_n$$

and hence by

$$\tilde{\theta}_n = \frac{\bar{X}_n - \eta}{\xi - n}.$$

[If $\eta = \xi$ but the second moments of X_i under H and G differ, one can, instead, equate $E(X_i^2)$ with $\Sigma X_j^2/n$ (Problem 4.6).] An asymptotically efficient estimator is then provided by (4.8).

Estimation under a mixture distribution provides interesting challenges, and has many application in practice. There is a large literature on mixtures, and entry can be found through the books by Everitt and Hand (1981), Titterington et al. (1985), McLachlan and Basford (1988), and McLachlan (1997). ||

In the context of choosing a $\sqrt{n}$-consistent estimator $\tilde{\theta}_n$ for (4.8), it is of interest to note that in sufficiently regular situations good efficiency of $\tilde{\theta}_n$ is equivalent to high correlation with $\hat{\theta}_n$. This is made precise by the following result, which is concerned only with first-order approximations.

Theorem 4.8 *Suppose $\hat{\theta}_n$ is an ELE estimator and $\tilde{\theta}_n$ a $\sqrt{n}$-consistent estimator, for which the joint distribution of*

$$T_n = \sqrt{n}(\hat{\theta}_n - \theta) \quad \text{and} \quad T_n' = \sqrt{n}(\tilde{\theta}_n - \theta)$$

tends to a bivariate limit distribution H with zero means and covariance matrix $\Sigma = ||\sigma_{ij}||$. Let (T, T') have distribution H and suppose that the means and covariance matrix of (T_n, T_n') tend toward those of (T, T') as $n \to \infty$. Then,

$$\frac{\text{var } T}{\text{var } T'} = \rho^2 \tag{4.18}$$

where $\rho = \sigma_{12}/\sqrt{\sigma_{11}\sigma_{22}}$ is the correlation coefficient of (T, T').

Proof. Consider $\text{var}[(1-\alpha)T_n + \alpha T_n']$ which tends to

$$\text{var}[(1-\alpha)T + \alpha T'] = (1-\alpha)^2\sigma_{11} + 2\alpha(1-\alpha)\sigma_{12} + \alpha^2\sigma_{22}. \tag{4.19}$$

This is non-negative for all values of α and takes on its minimum at $\alpha = 0$ since $\hat{\theta}_n$ is asymptotically efficient. Evaluating the derivative of (4.19) at $\alpha = 0$ shows that we must have $\sigma_{11} = \sigma_{12}$ (Problem 4.7). Thus, $\rho = \sqrt{\sigma_{11}/\sigma_{22}}$, as was to be proved. □

The ratio of the asymptotic variances in (4.18) is a special case of *asymptotic relative efficiency (ARE)*. See Definition 6.6.

In Examples 4.6 and 4.7, we used the method of moments to obtain $\sqrt{n}$-consistent estimators and then applied the one-step estimator (4.8) or (4.11). An alternative approach, when the direct calculation of an ELE is difficult, is the following expectation-maximization (EM) algorithm for obtaining a stationary point of the likelihood.

The idea behind the EM algorithm is to replace one computationally difficult likelihood maximization with a sequence of easier maximizations whose limit is the answer to the original problem. More precisely, let $Y_1, \ldots, Y_n$ be iid with density $g(y|\theta)$, and suppose that the object is to compute the value $\hat{\theta}$ that maximizes $L(\theta|\mathbf{y}) = \prod_{i=1}^n g(y_i|\theta)$. If $L(\theta|\mathbf{y})$ is difficult to work with, we can sometimes augment the data $\mathbf{y} = (y_1, \ldots, y_n)$ and create a new likelihood function $L(\theta|\mathbf{y}, \mathbf{z})$ that has a simpler form.

Example 4.9 Censored data likelihood. Suppose that we observe $Y_1, \ldots, Y_n$, iid, with density (4.16), and we have ordered the observations so that $(y_1, \ldots, y_m)$

are uncensored and $(y_{m+1}, \ldots, y_n)$ are censored (and equal to a). The likelihood function is then

$$(4.20) \qquad L(\theta|\mathbf{y}) = \prod_{i=1}^{n} g(y_i|\theta) = \prod_{i=1}^{m} f(y_i|\theta)\,[1 - F(a-\theta)]^{n-m}.$$

If we had observed the last $n - m$ values, say $\mathbf{z} = (z_{m+1}, \ldots, z_n)$, the likelihood would have had the simpler form

$$L(\theta|\mathbf{y}, \mathbf{z}) = \prod_{i=1}^{m} f(y_i|\theta) \prod_{i=m+1}^{n} f(z_i|\theta).$$

More generally, the EM algorithm is useful when the density of interest, $g(y_i|\theta)$, can be expressed as

$$(4.21) \qquad g(\mathbf{y}|\theta) = \int_Z f(\mathbf{y}, \mathbf{z}|\theta)\, d\mathbf{z},$$

for some simpler function $f(\mathbf{y}, \mathbf{z}|\theta)$. The $\mathbf{z}$ vector merely serves to simplify calculations, and its choice does not affect the value of the estimator. An illustration of a typical construction of the density f is the case of "filling in" missing data, for example, by turning an unbalanced data set into a balanced one.

Example 4.10 EM in a one-way layout. In a one-way layout (Example 3.4.9), suppose there are four treatments with the following data

Treatments			
1	2	3	4
y_{11}	y_{12}	y_{13}	y_{14}
y_{21}	y_{22}	y_{23}	y_{24}
z_1	y_{32}	z_3	y_{34}

where the y_{ij}'s represent the observed data, and the dummy variables z_1 and z_3 represent missing observations. Under the usual assumptions, the Y_{ij}'s are independently normally distributed as $N(\mu + \alpha_i, \sigma^2)$. If we let $\theta = (\mu, \alpha_1, \ldots, \alpha_4, \sigma^2)$ and let n_{ij} denote the number of observations per treatment, the incomplete-data likelihood is given by

$$L(\theta|\mathbf{y}) = g(\mathbf{y}|\theta) = \left(\frac{1}{\sqrt{2\pi\sigma^2}}\right)^{10} e^{\sum_{i=1}^{4}\sum_{j=1}^{n_{ij}}(y_{ij}-\mu-\alpha_i)^2/\sigma^2}$$

while the complete-data likelihood is

$$L(\theta|\mathbf{y}, \mathbf{z}) = f(\mathbf{y}, \mathbf{z}|\theta) = \left(\frac{1}{\sqrt{2\pi\sigma^2}}\right)^{12} e^{\sum_{i=1}^{4}\sum_{j=1}^{3}(y_{ij}-\mu-\alpha_i)^2/\sigma^2},$$

where $y_{31} = z_1$ and $y_{33} = z_3$. By integrating out z_1 and z_3, the original likelihood is recovered.

Although estimation in the original problem (with only the y_{ij}'s) is not difficult, it is easier in the augmented problem. [The computational advantage of the EM algorithm becomes more obvious as we move to higher-order designs, for example, the two-way layout (see Problem 4.14).] ||

The EM algorithm is often useful for obtaining an MLE when, as in Example 4.10, we should like to maximize $L(\theta|\mathbf{y})$, but it would be much easier, if certain additional observations $\mathbf{z}$ were available, to work with the joint density $f(\mathbf{y},\mathbf{z}|\theta) = L(\theta|\mathbf{y},\mathbf{z})$ and the conditional density of $\mathbf{Z}$ given $\mathbf{y}$, that is,

$$(4.22) \qquad L(\theta|\mathbf{y},\mathbf{z}) \quad \text{and} \quad k(\mathbf{z}|\theta,\mathbf{y}) = \frac{f(\mathbf{y},\mathbf{z}|\theta)}{g(\mathbf{y}|\theta)}.$$

These quantities are related by the identity

$$(4.23) \qquad \log L(\theta|\mathbf{y}) = \log L(\theta|\mathbf{y},\mathbf{z}) - \log k(\mathbf{z}|\theta,\mathbf{y}).$$

Since $\mathbf{z}$ is not available, we replace the right side of (4.23) with its expectation, using the conditional distribution of $\mathbf{Z}$ given $\mathbf{y}$. With an initial guess θ_0 (to start the iterations), we define

$$(4.24) \qquad Q(\theta|\theta_0,\mathbf{y}) = \int \log L(\theta|\mathbf{y}k(\mathbf{z}|\theta_0,\mathbf{y})\,d\mathbf{z},$$

$$H(\theta|\theta_0,\mathbf{y}) = \int \log k(\mathbf{z}|\theta,\mathbf{y})|\theta_0,\mathbf{y})\,d\mathbf{z}.$$

As the left side of (4.23) does not depend on $\mathbf{z}$, the expected value of $\log L(\theta|\mathbf{y})$ is then given by

$$(4.25) \qquad L(\theta|\mathbf{y}) = Q(\theta|\theta_0,\mathbf{y}) - H(\theta|\theta_0,\mathbf{y}).$$

Let the value of θ maximizing $Q(\theta|\theta_0,\mathbf{y})$ be $\hat{\theta}_{(1)}$. The process is then repeated with θ_0 in (4.24) and (4.22) replaced by the updated value $\hat{\theta}_{(1)}$, so that (4.24) is replaced by $Q(\theta|\hat{\theta}_{(1)},\mathbf{y})$. In this manner, a sequence of estimators $\hat{\theta}_{(j)}$, $j = 1, 2, \ldots$ is obtained iteratively where $\hat{\theta}_{(j)}$ is defined as the value of θ maximizing $Q(\theta|\hat{\theta}_{(j-1)},\mathbf{y})$, that is,

$$(4.26) \qquad Q(\hat{\theta}_{(j)}|\hat{\theta}_{(j-1)},\mathbf{y}) = \max_{\theta} Q(\theta|\hat{\theta}_{(j-1)},\mathbf{y}).$$

(It is sometimes written $\hat{\theta}_{(j)} = \text{argmax}_\theta Q(\theta|\hat{\theta}_{(j-1)},\mathbf{y})$, that is, $\hat{\theta}_{(j)}$ is the value of the argument θ that maximizes Q.)

The quantities $\log L(\theta|\mathbf{y})$, $\log L(\theta|\mathbf{y},\mathbf{z})$, and $Q(\theta|\theta_0,\mathbf{y})$ are referred to as the *incomplete*, *complete*, and *expected log likelihood*. The term EM for this algorithm stands for *Expectation-Maximization* since the jth step of the iteration consists of the calculating the expectation (4.24), with θ_0 replaced by $\hat{\theta}_{(j-1)}$, and then maximizing it.

The following is a key property of the sequence $\{\hat{\theta}_{(j)}\}$.

Theorem 4.11 *The sequence* $\{\hat{\theta}_{(j)}\}$ *defined by (4.26) satisfies*

$$(4.27) \qquad L(\hat{\theta}_{(j+1)}|\mathbf{y}) \geq L(\hat{\theta}_{(j)}|\mathbf{y}),$$

with equality holding if and only if $Q(\hat{\theta}_{(j+1)}|\hat{\theta}_{(j)},\mathbf{y}) = Q(\hat{\theta}_{(j)}|\hat{\theta}_{(j)},\mathbf{y})$.

Proof. On successive iterations, the difference between the logarithms of the left and right sides of (4.25) is

$$\log L(\hat{\theta}_{(j+1)}|\mathbf{y}) - \log L(\hat{\theta}_{(j)}|\mathbf{y})$$

$$(4.28) \qquad = \left[Q(\hat\theta_{(j+1)}|\hat\theta_{(j)}, \mathbf{y}) - Q(\hat\theta_{(j)}|\hat\theta_{(j)}, \mathbf{y})\right] - \left[H(\hat\theta_{(j+1)}|\hat\theta_{(j)}, \mathbf{y}) - H(\hat\theta_{(j)}|\hat\theta_{(j)}, \mathbf{y})\right].$$

The first expression in (4.28) is non-negative by definition of $\hat\theta_{(j+1)}$. It remains to show that the second term is non-negative, that is,

$$(4.29) \qquad \int \left[\log k(\mathbf{z}|\hat\theta_{(j+1)}, \mathbf{y}) - \log k(\mathbf{z}|\hat\theta_{(j)}, \mathbf{y})\right] k(\mathbf{z}|\hat\theta_{(j)}, \mathbf{y})\, d\mathbf{z} \le 0.$$

Since the difference of the logarithms is the logarithm of the ratio, this integral can be written as

$$(4.30) \qquad \int \log\left[\frac{k(\mathbf{z}|\hat\theta_{(j+1)}, \mathbf{y})}{k(\mathbf{z}|\hat\theta_{(j)}, \mathbf{y})}\right] k(\mathbf{z}|\hat\theta_{(j)}, \mathbf{y})\, d\mathbf{z} \le \log \int k(\mathbf{z}|\hat\theta_{(j+1)}, \mathbf{y}) d\mathbf{z} = 0.$$

The inequality follows from Jensen's inequality (see Example 1.7.7, Inequality (1.7.13), and Problem 4.17), and this completes the proof. □

Although Theorem 4.11 guarantees that the likelihood will increase at each iteration, we still may not be able to conclude that the sequence $\{\hat\theta_{(j)}\}$ converges to a maximum likelihood estimator.

To ensure convergence, we require further conditions on the mapping $\hat\theta_{(j)} \to \hat\theta_{(j+1)}$. These conditions are investigated by Boyles (1983) and Wu (1983); see also Finch et al. 1989. The following theorem is, perhaps, the most easily applicable condition to guarantee convergence to a *stationary point*, which may be a local maximum or saddlepoint.

Theorem 4.12 *If the expected complete-data likelihood $Q(\theta|\theta_0, \mathbf{y})$ is continuous in both θ and θ_0, then all limit points of an EM sequence $\{\hat\theta_{(j)}\}$ are stationary points of $L(\theta|\mathbf{y})$, and $L(\hat\theta_{(j)}|\mathbf{y})$ converges monotonically to $L(\hat\theta|\mathbf{y})$ for some stationary point $\hat\theta$.*

Example 4.13 Continuation of Example 4.9. The situation of Example 4.9 does not quite fit the conditions under which the EM algorithm was described above since the observations $y_{m+1}, \ldots, y_n$ are not missing completely but only partially. (We know that they are $\ge a$.) However, the situation reduces to the earlier one if we just ignore $y_{m+1}, \ldots, y_n$, so that $\mathbf{y}$ now stands for $(y_1, \ldots, y_m)$. To be specific, let the density $f(y|\theta)$ of (4.16) be the $N(\theta, 1)$ density, so that the likelihood function (4.20) is

$$L(\theta|\mathbf{y}) = \frac{1}{(2\pi)^{m/2}} e^{-\frac{1}{2}\sum_{i=1}^m (y_i-\theta)^2}.$$

We replace $y_{m+1}, \ldots, y_n$ with $n-m$ phantom variables $\mathbf{z} = (z_1, \ldots, z_{n-m})$ which are distributed as $n-m$ iid variables from the conditional normal distribution given that they are all $\ge a$; thus, for $z_i \ge a$, $i = 1, \ldots, n-m$,

$$k(\mathbf{z}|\theta, \mathbf{y}) = \frac{1}{(\sqrt{2\pi})^{(n-m)/2}} \frac{\exp\left\{-\frac{1}{2}\sum_{i=1}^{n-m}(z_i-\theta)^2\right\}}{[1-\Phi(a-\theta)]^{n-m}}.$$

At the jth step in the EM sequence, we have

$$Q(\theta|\hat{\theta}_{(j)}, \mathbf{y}) \propto -\frac{1}{2}\sum_{i=1}^{m}(y_i - \theta)^2 - \frac{1}{2}\sum_{i=1}^{n-m}\int_a^\infty (z_i - \theta)^2 k(z|\hat{\theta}_{(j)}, \mathbf{y})\,dz_i,$$

and differentiating with respect to θ yields

$$m(\bar{y} - \theta) + (n - m)\left[E(Z|\hat{\theta}_{(j)}) - \theta\right] = 0$$

or

$$\hat{\theta}_{(j+1)} = \frac{m\bar{y} + (n - m)E(Z|\hat{\theta}_{(j)})}{n}$$

where

$$E(Z|\hat{\theta}_{(j)}) = \int_a^\infty zk(z|\hat{\theta}_{(j)}, \mathbf{y})\,dz = \hat{\theta}_{(j)} + \frac{\phi(a - \hat{\theta}_{(j)})}{1 - \Phi(a - \hat{\theta}_{(j)})}.$$

Thus, the EM sequence is defined by

$$\hat{\theta}_{(j+1)} = \frac{m}{n}\bar{y} + \frac{n - m}{n}\left[\hat{\theta}_{(j)} + \frac{\phi(a - \hat{\theta}_{(j)})}{1 - \Phi(a - \hat{\theta}_{(j)}}\right],$$

which converges to the MLE $\hat{\theta}$ (Problem 4.8). ||

Quite generally, in an exponential family, computations are somewhat simplified because we can write

$$\begin{aligned}
Q(\theta|\hat{\theta}_{(j)}, \mathbf{y}) &= E_{\hat{\theta}_{(j)}}\left[\log L(\theta|\mathbf{y}, \mathbf{Z})|\mathbf{y})\right] \\
&= E_{\hat{\theta}_{(j)}}\left[\log\left(h(\mathbf{y}, \mathbf{Z})\; e^{\sum \eta_i(\theta)T_i - B(\theta)}\right)|\mathbf{y}\right] \\
&= E_{\hat{\theta}_{(j)}}\left[\log h(\mathbf{y}, \mathbf{Z})\right] + \sum \eta_i(\theta)E_{\hat{\theta}_{(j)}}\left[T_i|\mathbf{y}\right] - B(\theta).
\end{aligned}$$

Thus, calculating the complete-data MLE only involves the simpler expectation $E_{\hat{\theta}_{(j)}}\left[T_i|\mathbf{y}\right]$.

The books by Little and Rubin (1987), Tanner (1996), and McLachlan and Krishnan (1997) provide good overviews of the EM literature. Other references include Louis (1982), Laird et al. (1987), Meng and Rubin (1993), Smith and Roberts (1993), and Liu and Rubin (1994).

5 The Multiparameter Case

In the preceding sections, asymptotically efficient estimators were obtained when the distribution depends on a single parameter θ. When extending this theory to probability models involving several parameters $\theta_1, \ldots, \theta_s$, one may be interested either in the simultaneous estimation of these parameters (or certain functions of them) or with the estimation of one of the parameters at a time, the remaining parameters then playing the role of nuisance or incidental parameters. In the present section, we shall primarily take the latter point of view.

Let $X_1, \ldots, X_n$ be iid with a distribution that depends on $\boldsymbol{\theta} = (\theta_1, \ldots, \theta_s)$ and satisfies assumptions (A0)–(A3) of Section 6.3. For the time being, we shall assume

s to be fixed. Suppose we wish to estimate θ_j. Then, it was seen in Section 2.6 that the variance of any unbiased estimator δ_n of θ_j, based on n observations, satisfies the inequality

$$\text{var}(\delta_n) \geq [I(\boldsymbol{\theta})]^{-1}_{jj}/n \tag{5.1}$$

where the numerator on the right side is the jjth element of the inverse of the information matrix $I(\boldsymbol{\theta})$ with elements $I_{jk}(\boldsymbol{\theta})$, $j, k = 1, \ldots, s$, defined by

$$I_{jk}(\boldsymbol{\theta}) = \text{cov}\left[\frac{\partial}{\partial\theta_j}\log f(\mathbf{X}|\boldsymbol{\theta}), \quad \frac{\partial}{\partial\theta_k}\log f(\mathbf{X}|\boldsymbol{\theta})\right]. \tag{5.2}$$

It was further shown by Bahadur (1964) under conditions analogous to those of Theorem 2.6 that for any sequence of estimators δ_n of θ_j satisfying

$$\sqrt{n}\,(\delta_n - \theta_j) \xrightarrow{\mathcal{L}} N(0, v(\boldsymbol{\theta})], \tag{5.3}$$

the asymptotic variance v satisfies

$$v(\boldsymbol{\theta}) \geq [I(\boldsymbol{\theta})]^{-1}_{jj}, \tag{5.4}$$

except on a set of values $\boldsymbol{\theta}$ having measure zero.

We shall now show under assumptions generalizing those of Theorem 3.10 that with probability tending to 1, there exist solutions $\hat{\boldsymbol{\theta}}_n = (\hat{\theta}_{1n}, \ldots, \hat{\theta}_{sn})$ of the likelihood equations

$$\frac{\partial}{\partial\theta_j}[f(x_1|\boldsymbol{\theta})\cdots f(x_n|\boldsymbol{\theta})] = 0, \quad j = 1, \ldots, s, \tag{5.5}$$

or, equivalently,

$$\frac{\partial}{\partial\theta_j}[l(\boldsymbol{\theta})] = 0, \quad j = 1, \ldots, s, \tag{5.6}$$

such that $\hat{\theta}_{jn}$ is consistent for estimating θ_j and asymptotically efficient in the sense of satisfying (5.3) with

$$v(\boldsymbol{\theta}) = [I(\boldsymbol{\theta})]^{-1}_{jj}. \tag{5.7}$$

We state first some assumptions:

(A) There exists an open subset ω of Ω containing the true parameter point $\boldsymbol{\theta}^0$ such that for almost all x, the density $f(\mathbf{x}|\boldsymbol{\theta})$ admits all third derivatives $(\partial^3/\partial\theta_j\partial\theta_k\partial\theta_l)f(\mathbf{x}|\boldsymbol{\theta})$ for all $\boldsymbol{\theta} \in \omega$.

(B) The first and second logarithmic derivatives of f satisfy the equations

$$E_\theta\left[\frac{\partial}{\partial\theta_j}\log f(\mathbf{X}|\boldsymbol{\theta})\right] = 0 \quad \text{for } j = 1, \ldots, s \tag{5.8}$$

and

$$\begin{aligned} I_{jk}(\boldsymbol{\theta}) &= E_\theta\left[\frac{\partial}{\partial\theta_j}\log f(\mathbf{X}|\boldsymbol{\theta}) \cdot \frac{\partial}{\partial\theta_k}\log f(\mathbf{X}|\boldsymbol{\theta})\right] \\ &= E_\theta\left[-\frac{\partial^2}{\partial\theta_j\partial\theta_k}\log f(\mathbf{X}|\boldsymbol{\theta})\right]. \end{aligned} \tag{5.9}$$

Clearly, (5.8) and (5.9) imply (5.2).

(C) Since the $s \times s$ matrix $I(\boldsymbol{\theta})$ is a covariance matrix, it is positive semidefinite. In generalization of condition (v) of Theorem 2.6, we shall assume that the $I_{jk}(\boldsymbol{\theta})$ are finite and that the matrix $I(\boldsymbol{\theta})$ is positive definite for all $\boldsymbol{\theta}$ in ω, and hence that the s statistics

$$\frac{\partial}{\partial\theta_1}\log f(\mathbf{X}|\boldsymbol{\theta}), \ldots, \frac{\partial}{\partial\theta_s}\log f(\mathbf{X}|\boldsymbol{\theta})$$

are affinely independent with probability 1.

(D) Finally, we shall suppose that there exist functions M_{jkl} such that

$$\left|\frac{\partial^3}{\partial\theta_j\partial\theta_k\partial\theta_l}\log f(\mathbf{x}|\boldsymbol{\theta})\right| \le M_{jkl}(\mathbf{x}) \quad \text{for all } \boldsymbol{\theta} \in \omega$$

where

$$m_{jkl} = E_{\boldsymbol{\theta}^0}[M_{jkl}(\mathbf{X})] < \infty \quad \text{for all } j, k, l.$$

Theorem 5.1 *Let $X_1, \ldots, X_n$ be iid, each with a density $f(x|\boldsymbol{\theta})$ (with respect to μ) which satisfies (A0)–(A2) of Section 6.3 and assumptions (A)–(D) above. Then, with probability tending to 1 as $n \to \infty$, there exist solutions $\hat{\boldsymbol{\theta}}_n = \hat{\boldsymbol{\theta}}_n(X_1, \ldots, X_n)$ of the likelihood equations such that*

(a) $\hat{\theta}_{jn}$ is consistent for estimating θ_j,

(b) $\sqrt{n}(\hat{\boldsymbol{\theta}}_n - \boldsymbol{\theta})$ is asymptotically normal with (vector) mean zero and covariance matrix $[I(\boldsymbol{\theta})]^{-1}$, and

(c) $\hat{\theta}_{jn}$ is asymptotically efficient in the sense that

$$\sqrt{n}(\hat{\theta}_{jn} - \theta_j) \xrightarrow{\mathcal{L}} N\{0, [I(\boldsymbol{\theta})]^{-1}_{jj}\}. \tag{5.10}$$

Proof. (a) Existence and Consistency. To prove the existence, with probability tending to 1, of a sequence of solutions of the likelihood equations which is consistent, we shall consider the behavior of the log likelihood $l(\boldsymbol{\theta})$ on the sphere Q_a with center at the true point $\boldsymbol{\theta}^0$ and radius a. We will show that for any sufficiently small a, the probability tends to 1 that

$$l(\boldsymbol{\theta}) < l(\boldsymbol{\theta}^0)$$

at all points $\boldsymbol{\theta}$ on the surface of Q_a, and hence that $l(\boldsymbol{\theta})$ has a local maximum in the interior of Q_a. Since at a local maximum the likelihood equations must be satisfied, it will follow that for any $a > 0$, with probability tending to 1 as $n \to \infty$, the likelihood equations have a solution $\hat{\boldsymbol{\theta}}_n(a)$ within Q_a and the proof can be completed as in the one-dimensional case.

To obtain the needed facts concerning the behavior of the likelihood on Q_a for small a, we expand the log likelihood about the true point $\boldsymbol{\theta}^0$ and divide by n to find

$$\frac{1}{n}l(\boldsymbol{\theta}) - \frac{1}{n}l(\boldsymbol{\theta}^0)$$
$$= \frac{1}{n}\Sigma A_j(\mathbf{x})(\theta_j - \theta_j^0)$$

$$+ \frac{1}{2n}\Sigma\Sigma B_{jk}(\mathbf{x})(\theta_j - \theta_j^0)(\theta_k - \theta_k^0)$$
$$+ \frac{1}{6n}\sum_j\sum_k\sum_l(\theta_j - \theta_j^0)(\theta_k - \theta_k^0)(\theta_l - \theta_l^0)\sum_{i=1}^n \gamma_{jkl}(x_i)M_{jkl}(x_i)$$
$$= S_1 + S_2 + S_3$$

where

$$A_j(\mathbf{x}) = \left.\frac{\partial}{\partial\theta_j}l(\boldsymbol{\theta})\right|_{\theta=\theta^0}$$

and

$$B_{jk}(\mathbf{x}) = \left.\frac{\partial^2}{\partial\theta_j\partial\theta_k}l(\boldsymbol{\theta})\right|_{\theta=\theta^0},$$

and where, by assumption (D),

$$0 \le |\gamma_{jkl}(\mathbf{x})| \le 1.$$

To prove that the maximum of this difference for $\boldsymbol{\theta}$ on Q_a is negative with probability tending to 1 if a is sufficiently small, we will show that with high probability the maximum of S_2 is negative while S_1 and S_3 are small compared to S_2. The basic tools for showing this are the facts that by (5.8), (5.9), and the law of large numbers,

$$\frac{1}{n}A_j(\mathbf{X}) = \frac{1}{n}\left.\frac{\partial}{\partial\theta_j}l(\boldsymbol{\theta})\right|_{\theta=\theta^0} \to 0 \quad \text{in probability} \tag{5.11}$$

and

$$\frac{1}{n}B_{jk}(\mathbf{X}) = \frac{1}{n}\left.\frac{\partial^2}{\partial\theta_j\partial\theta_k}l(\boldsymbol{\theta})\right|_{\theta=\theta^0} \to -I_{jk}(\theta^0) \quad \text{in probability.} \tag{5.12}$$

Let us begin with S_1. On Q_a, we have

$$|S_1| \le \frac{1}{n}a\Sigma|A_j(\mathbf{X})|.$$

For any given a, it follows from (5.11) that $|A_j(\mathbf{X})|/n < a^2$ and hence that $|S_1| < sa^3$ with probability tending to 1. Next, consider

$$\begin{aligned} 2S_2 = {} & \Sigma\Sigma[-I_{jk}(\theta^0)(\theta_j - \theta_j^0)(\theta_k - \theta_k^0)] \\ & +\Sigma\Sigma\left\{\frac{1}{n}B_{jk}(\mathbf{X}) - [-I_{jk}(\theta^0)]\right\}(\theta_j - \theta_j^0)(\theta_k - \theta_k^0). \end{aligned} \tag{5.13}$$

For the second term, it follows from an argument analogous to that for S_1 that its absolute value is less than s^2a^3 with probability tending to 1. The first term is a negative (nonrandom) quadratic form in the variables $(\theta_j - \theta_j^0)$. By an orthogonal transformation, this can be reduced to diagonal form $\Sigma\lambda_i\zeta_i^2$ with Q_a becoming $\Sigma\zeta_i^2 = a^2$. Suppose that the λ's that are negative are numbered so that $\lambda_s \le \lambda_{s-1} \le \cdots \le \lambda_1 < 0$. Then, $\Sigma\lambda_i\zeta_i^2 \le \lambda_i\Sigma\zeta_i^2 = \lambda_1a^2$. Combining the first and second terms, we see that there exist $c > 0$ and $a_0 > 0$ such that for $a < a_0$

$$S_2 < -ca^2$$

with probability tending to 1.

Finally, with probability tending to 1,

$$\left|\frac{1}{n}\Sigma M_{jkl}(X_i)\right| < 2m_{jkl}$$

and hence $|S_3| < ba^3$ on Q_a where

$$b = \frac{s^3}{3}\Sigma\Sigma\Sigma m_{jkl}.$$

Combining the three inequalities, we see that

$$\max(S_1 + S_2 + S_3) < -ca^2 + (b+s)a^3, \tag{5.14}$$

which is less than zero if $a < c/(b+s)$, and this completes the proof of (i).

(b) and (c) Asymptotic Normality and Efficiency. This part of the proof is basically the same as that of Theorem 3.10. However, the single equation derived there from the expansion of $\hat{\theta}_n - \theta_0$ is now replaced by a system of s equations which must be solved for the differences $(\hat{\theta}_{jn} - \theta_j^0)$. This makes the details of the argument somewhat more cumbersome. In preparation, it will be convenient to consider quite generally a set of random linear equations in s unknowns,

$$\sum_{k=1}^{s} A_{jkn}Y_{kn} = T_{jn} \quad (j = 1, \ldots, s). \tag{5.15}$$

□

Lemma 5.2 *Let $(T_{1n}, \ldots, T_{sn})$ be a sequence of random vectors tending weakly to $(T_1, \ldots, T_s)$ and suppose that for each fixed j and k, A_{jkn} is a sequence of random variables tending in probability to constants a_{jk} for which the matrix $A = ||a_{jk}||$ is nonsingular. Let $B = ||b_{jk}|| = A^{-1}$. Then, if the distribution of $(T_1, \ldots, T_s)$ has a density with respect to Lebesgue measure over E_s, the solutions $(Y_{1n}, \ldots, Y_{sn})$ of (5.15) tend in probability to the solutions $(Y_1, \ldots, Y_s)$ of*

$$\sum_{k=1}^{s} a_{jk}Y_k = T_j \quad (j = 1, \ldots, s) \tag{5.16}$$

given by

$$Y_j = \sum_{k=1}^{s} b_{jk}T_k. \tag{5.17}$$

Proof. With probability tending to 1, the matrices $||A_{jkn}||$ are nonsingular, and by Theorem 1.8.19 (Problem 5.1), the elements of the inverse of $||A_{jkn}||$ tend in probability to the elements of B. Therefore, by a slight extension of Theorem 1.8.10, the solutions of (5.15) have the same limit distribution as those of

$$Y_{jn} = \sum_{k=1}^{s} b_{jk}T_{kn}. \tag{5.18}$$

By applying Theorem 1.8.19 to the set S,

$$\Sigma b_{1k}T_k \le y_1, \ldots, \Sigma b_{sk}T_k \le y_s, \tag{5.19}$$

it is only necessary to show that the distribution of $(T_1, \ldots, T_s)$ assigns probability zero to the boundary of (5.19). Since this boundary is contained in the union of the hyperplanes $\Sigma b_{jk} T_k = y_j$, the result follows. □

Proof of Parts (b) and (c) of Theorem 5.1. In the generalization of the proof of Theorem 3.10, expand $\partial l(\boldsymbol{\theta})/\partial\theta_j = l'_j(\boldsymbol{\theta})$ about $\boldsymbol{\theta}^0$ to obtain

$$l'_j(\boldsymbol{\theta}) = l'_j(\boldsymbol{\theta}^0) + \Sigma(\theta_k - \theta_k^0) l''_{jk}(\boldsymbol{\theta}^0) + \frac{1}{2}\Sigma\Sigma(\theta_k - \theta_k^0)(\theta_l - \theta_l^0) l'''_{jkl}(\boldsymbol{\theta}^*) \tag{5.20}$$

where l''_{jk} and l'''_{jkl} denote the indicated second and third derivatives of l and where $\boldsymbol{\theta}^*$ is a point on the line segment connecting $\boldsymbol{\theta}$ and $\boldsymbol{\theta}^0$. In this expansion, replace $\boldsymbol{\theta}$ by a solution $\hat{\boldsymbol{\theta}}_n$ of the likelihood equations, which by part (a) of the theorem can be assumed to exist with probability tending to 1 and to be consistent. The left side of (5.20) is then zero and the resulting equations can be written as

$$\sqrt{n}\,\Sigma(\hat{\theta}_k - \theta_k^0)\left[\frac{1}{n} l''_{jk}(\boldsymbol{\theta}^0) + \frac{1}{2n}\Sigma(\hat{\theta}_l - \theta_l^0) l'''_{jkl}(\boldsymbol{\theta}^*)\right] = -\frac{1}{\sqrt{n}} l'_j(\boldsymbol{\theta}^0). \tag{5.21}$$

These have the form (5.15) with

$$\begin{aligned} Y_{kn} &= \sqrt{n}\,(\hat{\theta}_k - \theta_k^0), \\ A_{jkn} &= \frac{1}{n} l''_{jk}(\boldsymbol{\theta}^0) + \frac{1}{2n}\Sigma(\hat{\theta}_l - \theta_l^0) l'''_{jkl}(\boldsymbol{\theta}^*), \\ T_{jn} &= -\frac{1}{\sqrt{n}} l'_j(\boldsymbol{\theta}^0) = -\sqrt{n}\left[\frac{1}{n}\sum_{i=1}^{n} \frac{\partial}{\partial\theta_j} \log f(X_i|\boldsymbol{\theta})\right]_{\boldsymbol{\theta}=\boldsymbol{\theta}^0}. \end{aligned} \tag{5.22}$$

Since $E_{\boldsymbol{\theta}^0}[(\partial/\partial\theta_j)\log f(X_i|\boldsymbol{\theta})] = 0$, the multivariate central limit theorem (Theorem 1.8.21) shows that $(T_{1n}, \ldots, T_{sn})$ has a multivariate normal limit distribution with mean zero and covariance matrix $I(\boldsymbol{\theta}^0)$.

On the other hand, it is easy to see—again in parallel to the proof of Theorem 3.10—that

$$A_{jkn} \xrightarrow{P} a_{jk} = E[l''_{jk}(\boldsymbol{\theta}^0)] = -I_{jk}(\boldsymbol{\theta}^0). \tag{5.23}$$

The limit distribution of the Y's is therefore that of the solution $(Y_1, \ldots, Y_s)$ of the equations

$$\sum_{k=1}^{s} I_{jk}(\boldsymbol{\theta}^0) Y_k = T_j \tag{5.24}$$

where $T = (T_1, \ldots, T_s)$ is multivariate normal with mean zero and covariance matrix $I(\boldsymbol{\theta}^0)$. It follows that the distribution of Y is that of

$$[I(\boldsymbol{\theta}^0)]^{-1} T,$$

which is a multivariate distribution with zero mean and covariance matrix $[I(\boldsymbol{\theta}^0)]^{-1}$.

This completes the proof of asymptotic normality and efficiency. □

If the likelihood equations have a unique solution $\hat{\theta}_n$, then $\hat{\theta}_n$ is consistent, asymptotically normal, and efficient. It is, however, interesting to note that even if the parameter space is an open interval, it does not follow as in Corollary 3.8 that the MLE exists and hence is consistent (Problem 5.6). Sufficient conditions for existence and uniqueness are given in Mäkeläinen, Schmidt, and Styan (1981).

As in the one-parameter case, if the solution of the likelihood equations is not unique, Theorem 5.1 does not establish the existence of an efficient estimator of $\boldsymbol{\theta}$. However, the methods mentioned in Section 2.5 also work in the present case. In particular, if $\tilde{\boldsymbol{\theta}}_n$ is a consistent sequence of estimators of $\boldsymbol{\theta}$, then the solutions $\hat{\boldsymbol{\theta}}_n$ of the likelihood equations closest to $\tilde{\boldsymbol{\theta}}_n$, for example, in the sense that $\Sigma(\hat{\theta}_{jn} - \tilde{\theta}_{jn})^2$ is smallest, is asymptotically efficient.

More convenient, typically, is the approach of Theorem 4.3, which we now generalize to the multiparameter case.

Theorem 5.3 *Suppose that the assumptions of Theorem 5.1 hold and that $\tilde{\theta}_{jn}$ is a $\sqrt{n}$-consistent estimator of θ_j for $j = 1, \ldots, s$. Let $\{\delta_{kn}, k = 1, \ldots, s\}$ be the solution of the equations*

$$\sum_{k=1}^{s} (\delta_{kn} - \tilde{\theta}_{kn}) l''_{jk}(\tilde{\boldsymbol{\theta}}_n) = -l'_j(\tilde{\boldsymbol{\theta}}_n). \tag{5.25}$$

Then, $\delta_n = (\delta_{1n}, \ldots, \delta_{sn})$ satisfies (5.10) with δ_{jn} in place of $\hat{\theta}_{jn}$ and, thus, is asymptotically efficient.

Proof. The proof is a simple combination of the proofs of Theorem 4.3 and 5.1 and we shall only sketch it. Expanding the right side about $\boldsymbol{\theta}^0$ allows us to rewrite (5.25) as

$$\Sigma_k(\delta_{kn} - \tilde{\theta}_{kn}) l''_{jk}(\tilde{\boldsymbol{\theta}}_n) = -l'_j(\boldsymbol{\theta}^0) - \Sigma_k(\tilde{\theta}_{kn} - \theta_k^0) l''_{jk}(\boldsymbol{\theta}^0) + R_n$$

where

$$R_n = -\frac{1}{2}\Sigma_k \Sigma_l (\tilde{\theta}_{kn} - \theta_k^0)(\tilde{\theta}_{ln} - \theta_l^0) l'''_{jkl}(\boldsymbol{\theta}_n^*)$$

and hence as

$$\sqrt{n}\, \Sigma_k(\delta_{kn} - \theta_k^0) \frac{1}{n} l''_{jk}(\tilde{\boldsymbol{\theta}}_n)$$

$$= -\frac{1}{\sqrt{n}} l'_j(\boldsymbol{\theta}^0) + \sqrt{n}\, \Sigma_k(\tilde{\theta}_{kn} - \theta_k^0) \left[\frac{1}{n} l''_{jk}(\tilde{\boldsymbol{\theta}}_n) - \frac{1}{n} l''_{jk}(\boldsymbol{\theta}^0)\right] + \frac{1}{\sqrt{n}} R_n. \tag{5.26}$$

This has the form (5.15), and it is easy to check (Problem 5.2) that the limits (in probability) of the A_{jkn} are the same a_{jk} as in (5.23) and that the second and third terms on the right side of (5.26) tend toward zero in probability. Thus, the joint distribution of the right side is the same as that of the T_{jn} given by (5.22). If follows that the joint limit distribution of the $\sqrt{n}\,(\delta_{kn} - \theta_k^0)$ is the same as that of the $\sqrt{n}\,(\hat{\theta}_{kn} - \theta_k^0)$ in Theorem 3.2, and this completes the proof. □

The following result generalizes Corollary 4.4 to the multiparameter case.

Corollary 5.4 *Suppose that the assumptions of Theorem 5.3 hold and that the elements $I_{jk}(\boldsymbol{\theta})$ of the information matrix of the* **X***'s are continuous. Then, the* *solutions δ'_{kn} of the equations*

$$n\Sigma(\delta'_{kn} - \tilde{\theta}_{kn})I_{jk}(\tilde{\theta}_n) = l'_j(\tilde{\theta}_n) \tag{5.27}$$

are asymptotically efficient.

The proof is left to Problem 5.5.

6 Applications

Maximum likelihood (together with some of its variants) is the most widely used method of estimation, and a list of its applications would cover practically the whole field of statistics. [For a survey with a comprehensive set of references, see Norden 1972-1973 or Scholz 1985.] In this section, we will discuss a few applications to illustrate some of the issues arising. The discussion, however, is not carried to the practical level, and in particular, the problem of choosing among alternative asymptotically efficient methods is not addressed. Such a choice must be based not only on theoretical considerations but requires empirical evidence on the performance of the estimators at various sample sizes. For any specific example, the relevant literature should be consulted.

Example 6.1 Weibull distribution. Let $X_1, \ldots, X_n$ be iid according to a two-parameter Weibull distribution, whose density it is convenient to write in a parameterization suggested by Cohen (1965b) as

$$\frac{\gamma}{\beta}x^{\gamma-1}e^{-x^\gamma/\beta}, \quad x > 0, \quad \beta > 0, \quad \gamma > 0, \tag{6.1}$$

where γ is a shape parameter and $\beta^{1/\gamma}$ a scale parameter. The likelihood equations, after some simplification, reduce to (Problem 6.1)

$$h(\gamma) = \frac{\Sigma x_i^\gamma \log x_i}{\Sigma x_i^\gamma} - \frac{1}{\gamma} = \frac{1}{n}\Sigma \log x_i \tag{6.2}$$

and

$$\beta = \Sigma x_i^\gamma / n. \tag{6.3}$$

To show that (6.2) has at most one solution, note that $h'(\gamma)$ exceeds the derivative of the first term, which equals (Problem 6.2) $\Sigma a_i^2 p_i - (\Sigma a_i p_i)^2$ with

$$a_i = \log x_i, \quad p_i = e^{\gamma a_i} / \sum_j e^{\gamma a_j}. \tag{6.4}$$

It follows that $h'(\gamma) > 0$ for all $\gamma > 0$. That (6.2) always has a solution follows from (Problem 6.2):

$$-\infty = \lim_{\gamma\to 0} h(\gamma) < \frac{1}{n}\Sigma \log x_i < \log x_{(n)} = \lim_{\gamma\to\infty} h(\gamma). \tag{6.5}$$

This example, therefore, illustrates the simple situation in which the likelihood equations always have a unique solution. ||

Example 6.2 Location-scale families. Let $X_1, \ldots, X_n$ be iid, each with density $(1/a)f[(x-\xi)/a]$. The calculation of an ELE is easy when the likelihood equations have a unique root $(\hat{\xi}, \hat{a})$. It was shown by Barndorff-Nielsen and Blaesild (1980) that sufficient conditions for this to be the case is that $f(x)$ is positive, twice differentiable for all x, and strongly unimodal. Surprisingly, Copas (1975) showed that it is unique also when f is Cauchy, despite the fact that the Cauchy density is not strongly unimodal and that in this case the likelihood equation can have multiple roots when a is known. Ferguson (1978) gave explicit formulas for the Cauchy MLEs for $n = 3$ or 4. See also Haas et al. 1970 and McCullagh 1992.

In the presence of multiple roots, the simplest approach typically is that of Theorem 5.3. The $\sqrt{n}$-consistent estimators of ξ and a required by this theorem are easily obtained in the present case. As was pointed out in Example 4.5, the mean or median of the X's will usually have the desired property for ξ. (When f is asymmetric, this requires that ξ be specified to be some particular location measure such as the mean or median of the distribution of the X_i.) If $E(X_i^4) < \infty$, $\hat{a}_n = \sqrt{\Sigma(X_i - \bar{X})^2/n}$ will be $\sqrt{n}$-consistent for a if the latter is taken to be the population standard deviation. If $E(X_i^4) = \infty$, one can instead, for example, take a suitable multiple of the interquartile range $X_{(k)} - X_{(j)}$, where $k = [3n/4]$ and $j = [n/4]$ (see, for example, Mosteller 1946).

If f satisfies the assumptions of Theorem 5.1, then $[\sqrt{n}(\hat{\xi}_n - \xi), \sqrt{n}(\hat{a}_n - a)]$ have a joint bivariate normal distribution with zero means and covariance matrix $I^{-1}(a) = ||I_{ij}(a)||^{-1}$, which is independent of ξ and where $I(a)$ is given by (2.6.20) and (2.6.21). ||

If the distribution of the X_i depends on $\boldsymbol{\theta} = (\theta_1, \ldots, \theta_s)$, it is interesting to compare the estimation of θ_j when the other parameters are unknown with the situation in which they are known. The mathematical meaning of this distinction is that an estimator is permitted to depend on known parameters but not on unknown ones. Since the class of possible estimators is thus more restricted when the nuisance parameters are unknown, it follows from Theorems 3.10 and 5.1 that the asymptotic variance of an efficient estimator when some of the θ's are unknown can never fall below its value when they are known, so that

$$\frac{1}{I_{jj}(\boldsymbol{\theta})} \leq [I(\boldsymbol{\theta})]^{-1}_{jj}, \tag{6.6}$$

as was already shown in Section 2.6 as (2.6.25). There, it was also proved that equality holds in (6.6) whenever

$$\text{cov}\left[\frac{\partial}{\partial\theta_j}\log f(\mathbf{X}|\boldsymbol{\theta}), \frac{\partial}{\partial\theta_k}\log f(\mathbf{X}|\boldsymbol{\theta})\right] = 0 \quad \text{for all } j \neq k, \tag{6.7}$$

and that this condition, which states that

$$I(\boldsymbol{\theta}) \quad \text{is diagonal}, \tag{6.8}$$

is also necessary for equality. For the location-scale families of Example 6.2, it follows from (2.6.21) that $I_{12} = 0$ whenever f is symmetric about zero but not necessarily otherwise. For symmetric f, there is therefore no loss of asymptotic efficiency in estimating ξ or a when the other parameter is unknown.

Quite generally, if the off-diagonal elements of the information matrix are zero, the parameters are said to be orthogonal. Although it is not always possible to find an entire set of orthogonal parameters, it is always possible to obtain orthogonality between a scalar parameter of interest and the remaining (nuisance) parameters. See Cox and Reid 1987 and Problem 6.5.

As another illustration of efficient likelihood estimation, consider a multiparameter exponential family. Here, UMVU estimators often are satisfactory solutions of the estimation problem. However, the estimand may not be U-estimable and then another approach is needed. In some cases, even when a UMVU estimator exists, the MLE has the advantage of not taking on values outside the range of the estimand.

Example 6.3 Multiparameter exponential families. Let $\mathbf{X} = (X_1, \ldots, X_n)$ be distributed according to an s-parameter exponential family with density (1.5.2) with respect to a σ-finite measure μ, where $\mathbf{x}$ takes the place of x and where it is assumed that $T_1(\mathbf{X}), \ldots, T_s(\mathbf{X})$ are affinely independent with probability 1. Using the fact that

$$\frac{\partial}{\partial \eta_j}[l(\eta)] = -\frac{\partial}{\partial \eta_j}[A(\eta)] + T_j(\mathbf{x}) \tag{6.9}$$

and other properties of the densities (1.5.2), one sees that the conditions of Theorem 5.1 are satisfied when the X's are iid. By (1.5.14), the likelihood equations for the η's reduce to

$$T_j(\mathbf{x}) = E_\eta[T_j(\mathbf{X})]. \tag{6.10}$$

If these equations have a solution, it is unique (and is the MLE) since $l(\eta)$ is a strictly concave function of η. This follows from Theorem 1.7.13 and the fact that, by (1.5.15),

$$-\frac{\partial^2}{\partial \eta_j \partial \eta_k}[l(\eta)] = \frac{\partial^2}{\partial \eta_j \partial \eta_k}[A(\eta)] = \text{cov}[T_j(\mathbf{X}), T_k(\mathbf{X})] \tag{6.11}$$

and that, by assumption, the matrix with entries (6.11) is positive definite.

Sufficient conditions for the existence of a solution of the likelihood equations are given by Crain (1976) and Barndorff-Nielsen (1978, Section 9.3, 9.4), where they are shown to be satisfied for the two-parameter gamma family of Table 1.5.1.

An alternative method for obtaining asymptotically efficient estimators for the parameters of an exponential family is based on the mean-value parameterization (2.6.17). Slightly changing the formulation of the model, consider a sample $(X_1, \ldots, X_n)$ of size n from the family (1.5.2), and let $\bar{T}_j = [T_j(X_1) + \cdots + T_j(X_n)]/n$ and $\theta_j = E(T_j)$. By the CLT, the joint distribution of the $\sqrt{n}(\bar{T}_j - \theta_j)$ is multivariate normal with zero means and covariance matrix $\sigma_{ij} = \text{cov}[T_i(\mathbf{X}), T_j(\mathbf{X})]$. This proves the $\bar{T}_j$ to be asymptotically efficient estimators by (2.6.18). ||

For further discussion of maximum likelihood estimation in exponential families, see Berk 1972b, Sundberg 1974, 1976, Barndorff-Nielsen 1978, Johansen 1979, Brown 1986a, and Note 10.4.

In the next two examples, we shall consider in somewhat more detail the most important case of Example 6.3, the multivariate and, in particular, the bivariate normal distribution.

Example 6.4 Multivariate normal distribution. Suppose we let $(X_{1\nu}, \ldots, X_{p\nu})$, $\nu = 1, \ldots, n$, be a sample from a nonsingular normal distribution with means $E(X_{i\nu}) = \xi_i$ and covariances $\text{cov}(X_{i\nu}, X_{j\nu}) = \sigma_{ij}$. By (1.4.15), the density of the X's is given by

$$|\Xi|^{n/2}(2\pi)^{-pn/2}\exp\left(-\frac{1}{2}\Sigma\Sigma\eta_{jk}S_{jk}\right) \tag{6.12}$$

where

$$S_{jk} = \Sigma_\nu(X_{j\nu} - \xi_j)(X_{k\nu} - \xi_k), \quad j, k = 1, \ldots, p, \tag{6.13}$$

and where $\Xi = ||\eta_{jk}||$ is the inverse of the covariance matrix $||\sigma_{jk}||$.

Consider, first, the case in which the ξ's are known. Then, (6.12) is an exponential family with $T_{jk} = -(1/2)S_{jk}$. If the matrix $||\sigma_{jk}||$ is nonsingular, the T_{jk} are affinely independent with probability 1, so that the result of the preceding example applies. Since $E(S_{jk}) = n\sigma_{jk}$, the likelihood equations (6.10) reduce to $n\sigma_{jk} = S_{jk}$ and thus have the solutions

$$\hat{\sigma}_{jk} = \frac{1}{n}S_{jk}. \tag{6.14}$$

The sample moments and correlations are, therefore, ELEs of the population variances, covariances, and correlation coefficients. Also, the (jk)th element of $||\hat{\sigma}_{jk}||^{-1}$ is an asymptotically efficient estimator of η_{jk}. In addition to being the MLE, $\hat{\sigma}_{jk}$ is the UMVU estimator of σ_{jk} (Example 2.2.4).

If the ξ's are unknown,

$$\hat{\xi}_j = \frac{1}{n}\Sigma_\nu X_{j\nu} = X_j,$$

and $\hat{\sigma}_{jk}$, given by (6.14) but with S_{jk} now defined as

$$S_{jk} = \Sigma_\nu(X_{j\nu} - X_{j\cdot})(X_{k\nu} - X_{k\cdot}), \tag{6.15}$$

continue to be ELEs for ξ_j and σ_{jk} (Problem 6.6).

If ξ is known, the asymptotic distribution of S_{jk} given by (6.13) is immediate from the central limit theorem since S_{jk} is the sum of n iid variables with expectation

$$E(X_{j\nu} - \xi_j)(X_{k\nu} - \xi_k) = \sigma_{jk}$$

and variance

$$E[(X_{j\nu} - \xi_j)^2(X_{k\nu} - \xi_k)^2] - \sigma_{jk}^2.$$

If $j \neq k$, it follows from Problem 1.5.26 that

$$E[(X_{j\nu} - \xi_j)^2(X_{k\nu} - \xi_k)^2] = \sigma_{jj}\sigma_{kk} + 2\sigma_{jk}^2$$

so that

$$\text{var}[(X_{j\nu} - \xi_j)(X_{k\nu} - \xi_k)] = \sigma_{jj}\sigma_{kk} + \sigma_{jk}^2$$

and

$$\sqrt{n}\left(\frac{S_{jk}}{n} - \sigma_{jk}\right) \xrightarrow{\mathcal{L}} N(0, \sigma_{jj}\sigma_{kk} + \sigma_{jk}^2). \tag{6.16}$$

If ξ is unknown, the S_{jk} given by (6.15) are independent of the X_i, and the asymptotic distribution of (6.15) is the same as that of (6.13) (Problem 6.7). ||

Example 6.5 Bivariate normal distribution. In the preceding example, it was seen that knowing the means does not affect the efficiency with which the covariances can be estimated. Let us now restrict attention to the covariances and, for the sake of simplicity, suppose that $p = 2$. With an obvious change of notation, let (X_i, Y_i), $i = 1, \dots, n$ be iid, each with density (1.4.16). Since the asymptotic distribution of $\hat{\sigma}$, $\hat{\tau}$, and $\hat{\rho}$ are not affected by whether or not ξ and η are known, let us assume $\xi = \eta = 0$. For the information matrix $I(\boldsymbol{\theta})$ [where $\boldsymbol{\theta} = (\sigma^2, \tau^2, \rho)$], we find [Problem 6.8(a)]

$$(1-\rho^2)I(\boldsymbol{\theta}) = \begin{bmatrix} \dfrac{2-\rho^2}{4\sigma^4} & \dfrac{-\rho^2}{4\sigma^2\tau^2} & \dfrac{-\rho}{2\sigma^2} \\[2ex] \dfrac{-\rho^2}{4\sigma^2\tau^2} & \dfrac{2-\rho^2}{4\tau^4} & \dfrac{-\rho}{2\tau^2} \\[2ex] \dfrac{-\rho}{2\sigma^2} & \dfrac{-\rho}{2\tau^2} & \dfrac{1+\rho^2}{1-\rho^2} \end{bmatrix}. \tag{6.17}$$

Inversion of this matrix gives the covariance matrix of the $\sqrt{n}(\hat{\theta}_j - \theta_j)$ as [Problem 6.8(b)]

$$\begin{bmatrix} 2\sigma^4 & 2\rho^2\sigma^2\tau^2 & \rho(1-\rho^2)\sigma^2 \\ 2\rho^2\sigma^2\tau^2 & 2\tau^4 & \rho(1-\rho^2)\tau^2 \\ \rho(1-\rho^2)\sigma^2 & \rho(1-\rho^2)\tau^2 & (1-\rho^2)^2 \end{bmatrix}. \tag{6.18}$$

Thus, we find that

$$\begin{aligned} &\sqrt{n}(\hat{\sigma}^2 - \sigma^2) \xrightarrow{\mathcal{L}} N(0, 2\sigma^4), \\ &\sqrt{n}(\hat{\tau}^2 - \tau^2) \xrightarrow{\mathcal{L}} N(0, 2\tau^4), \\ &\sqrt{n}(\hat{\rho} - \rho) \xrightarrow{\mathcal{L}} N[0, (1-\rho^2)^2]. \end{aligned} \tag{6.19}$$

On the other hand, if σ and τ are known to be equal to 1, the MLE $\hat{\hat{\rho}}$ of ρ satisfies (Problem 6.9)

$$\sqrt{n}\,(\hat{\hat{\rho}} - \rho) \xrightarrow{\mathcal{L}} N\left(0, \frac{(1-\rho^2)^2}{1+\rho^2}\right), \tag{6.20}$$

whereas if ρ and τ are known, the MLE $\hat{\hat{\sigma}}$ of σ satisfies (Problem 6.10)

$$\sqrt{n}(\hat{\hat{\sigma}}^2 - \sigma^2) \xrightarrow{\mathcal{L}} N\left(0, \frac{4\sigma^4(1-\rho^2)}{2-\rho^2}\right). \tag{6.21}$$

‖

A criterion for comparing $\hat{\rho}$ to $\hat{\hat{\rho}}$ is provided by the *asymptotic relative efficiency.*

Definition 6.6 If the sequence of estimators δ_n of $g(\theta)$ satisfies $\sqrt{n}[\delta_n - g(\boldsymbol{\theta})] \xrightarrow{\mathcal{L}} N(0, \tau^2)$, and the sequence of estimators $\delta'_{n'}$, where $\delta'_{n'}$ is based on $n' = n'(n)$ observations, also satisfies $\sqrt{n}\,[\delta'_{n'} - g(\boldsymbol{\theta})] \xrightarrow{\mathcal{L}} N(0, \tau^2)$, then the *asymptotic relative efficiency* (ARE) of $\{\delta_n\}$ with respect to $\{\delta'_n\}$ is

$$e_{\delta,\delta'} = \lim_{n\to\infty} \frac{n'(n)}{n},$$

provided the limit exists and is independent of the subsequences n'.

The interpretation is clear. Suppose, for example, the $e = 1/2$. Then, for large values of n, n' is approximately equal to $(1/2)n$. To obtain the same limit distribution (and limit variance), half as many observations are therefore required with δ' as with δ. It is then reasonable to say that δ' is twice as efficient as δ or that δ is half as efficient as δ'.

The following result shows that in order to obtain the ARE, it is not necessary to evaluate the limit $n'(n)/n$.

Theorem 6.7 *If* $\sqrt{n}\,[\delta_{in} - g(\theta)] \xrightarrow{\mathcal{L}} N(0, \tau_i^2)$, $i = 1, 2$, *then the ARE of* $\{\delta_{2n}\}$ *with respect to* $\{\delta_{1n}\}$ *exists and is* $e_{2,1} = \tau_1^2/\tau_2^2$.

Proof. Since

$$\sqrt{n}\,[\delta_{2n'} - g(\theta)] = \sqrt{\frac{n}{n'}}\sqrt{n'}\,[\delta_{2n'} - g(\theta)],$$

it follows from Theorem 1.8.10 that the left side has the same limit distribution $N(0, \tau_1^2)$ as $\sqrt{n}\,[\delta_{1n} - g(\theta)]$ if and only if $\lim\left[n/n'(n)\right]$ exists and

$$\tau_2^2 \lim \frac{n}{n'(n)} = \tau_1^2,$$

as was to be proved. □

Example 6.8 Continuation of Example 6.5. It follows from (6.19) and 6.20) that the efficiency of $\hat{\rho}$ to $\hat{\hat{\rho}}$ is

$$e_{\hat{\rho},\hat{\hat{\rho}}} = \frac{1}{1+\rho^2}. \tag{6.22}$$

This is 1 when $\rho = 0$ but can be close to 1/2 when $|\rho|$ is close to 1. Similarly,

$$e_{\hat{\sigma}^2,\hat{\hat{\sigma}}^2} = \frac{2(1-\rho)^2}{2-\rho^2}. \tag{6.23}$$

This efficiency is again 1 when $\rho = 0$ but tends to zero as $|\rho| \to 1$. This last result, which at first may seem surprising, actually is easy to explain. If ρ were equal to 1, and $\tau = 1$ say, we would have $X_i = \sigma Y_i$. Since both X_i and Y_i are observed, we could then determine σ without error from a single observation. ||

Example 6.9 Efficiency of nonparametric UMVU estimator. As another example of an efficiency calculation, recall Example 2.2.2. If $X_1, \ldots, X_n$ are iid according to $N(\theta, 1)$, it was found that the UMVU estimator of

$$p = P(X_1 \le a)$$

is

$$\delta_{1n} = \Phi\left[\sqrt{\frac{n}{n-1}}(a - \bar{X})\right]. \tag{6.24}$$

Suppose now that we do not trust the assumption of normality; then, we might, instead of (6.24), prefer to use the nonparametric UMVU estimator derived in Section 2.4, namely

$$\delta_{2n} = \frac{1}{n}(\text{No. of } X_i \le a). \tag{6.25}$$

What do we lose by using (6.25) instead of (6.24) if the X's are $N(0, 1)$ after all? Note that p is then given by

$$p = \Phi(a - \theta) \tag{6.26}$$

and that

$$\sqrt{n}(\delta_{1n} - p) \to N[0, \phi^2(a - \theta)].$$

On the other hand, $n\delta_{2n}$ is the number of successes in n binomial trials with success probability p, so that

$$\sqrt{n}(\delta_{2n} - p) \to N(0, p(1 - p)).$$

It thus follows from Theorem 6.7 that

$$e_{2,1} = \frac{\phi^2(a - \theta)}{\Phi(a - \theta)[1 - \Phi(a - \theta)]}. \tag{6.27}$$

At $a = \theta$ (when $p = 1/2$), $e_{2,1} = (1/2\pi)/(1/4) = 2/\pi \approx 0.637$. As $a - \theta \to \infty$, the efficiency tends to zero (Problem 6.12). It can be shown, in fact, that (6.27) is a decreasing function of $|a - \theta|$ (for a proof, see Sampford, 1953). The efficiency loss resulting from the use of δ_{2n} instead of δ_{1n} is therefore quite severe. If the underlying distribution is not normal, however, this conclusion could change (see Problem 6.13). ||

Example 6.10 Normal mixtures. Let $X_1, \ldots, X_n$ be iid, each with probability p as $N(\xi, \sigma^2)$ and probability $q = 1 - p$ as $N(\eta, \tau^2)$. (The Tukey models are examples of such distributions with $\eta = \xi$.) The joint density of the X's is then given by

$$\prod_{i=1}^{n}\left\{\frac{p}{\sqrt{2\pi}\sigma}\exp\left[-\frac{1}{2\sigma^2}(x_i - \xi)^2\right] + \frac{q}{\sqrt{2\pi}\tau}\exp\left[\frac{1}{2\tau^2}(x_i - \eta)^2\right]\right\}. \tag{6.28}$$

This is a sum of non-negative terms of which one, for example, is proportional to

$$\frac{1}{\sigma\tau^{n-1}}\exp\left[-\frac{1}{2\sigma^2}(x_1 - \xi)^2 - \frac{1}{2\tau^2}\sum_{i=2}^{n}(x_i - \eta)^2\right].$$

When $\xi = x_1$ and $\sigma \to 0$, this term tends to infinity for any fixed values of η, τ, and $x_2, \ldots, x_n$. The likelihood is therefore unbounded and the MLE does not exist. (The corresponding result holds for any other mixture with density $\Pi\{(p/\sigma)f[(x_i - \xi)/\sigma] + (q/\tau)f[(x_i - \eta)/\tau]\}$ when $f(0) \neq 0$.)

On the other hand, the conditions of Theorem 5.1 are satisfied (Problem 6.10) so that efficient solutions of the likelihood equations exist and asymptotically efficient estimators can be obtained through Theorem 5.3. One approach to the determination of the required $\sqrt{n}$-consistent estimators is the method of moments. In the present case, this means equating the first five moments of the X's with the corresponding sample moments and then solving for the five parameters. For the normal mixture problem, these estimators were proposed in their own right by K. Pearson (1894). For a discussion and possible simplifications, see Cohen 1967, and for more details on mixture problems, see Everitt and Hand 1981, Titterington et al. 1985, McLachlan and Basford 1988, and McLachlan 1997.

A study of the improvement of an asymptotically efficient estimator over that obtained by the method of moments has been carried out for the case for which it is known that $\tau = \sigma$ (Tan and Chang 1972). If $\Delta = (\eta - \xi)/\sigma$, the AREs for the estimation of all four parameters depend only on Δ and p. As an example, consider the estimation of p. Here, the ARE is < 0.01 if $\Delta < 1/2$ and $p < 0.2$; it is < 0.1 if $\Delta < 1/2$ and $0.2 < p < 0.4$, and is > 0.9 if $\Delta > 0.5$. (For an alternative starting point for the application of Theorem 5.3, see Quandt and Ramsey 1978, particularly the discussion by N. Kiefer.) ‖

Example 6.11 Multinomial experiments. Let $(X_0, X_1, \ldots, X_s)$ have the multinomial distribution (1.5.4). In the full-rank exponential representation,

$$\exp[n \log p_0 + x_1 \log(p_1/p_0) + \cdots + x_s \log(p_s/p_0)]h(x),$$

the statistics T_j can be taken to be the X_j. Using the mean-value parameterization, the likelihood equations (6.10) reduce to $np_j = X_j$ so that the MLE of p_j is $\hat{p}_j = X_j/n$ $(j = 1, \ldots, s)$. If X_j is 0 or n, the likelihood equations have no solution in the parameter space $0 < p_j < 1$, $\sum_{j=1}^{s} p_j < 1$. However, for any fixed vector $\mathbf{p}$, the probability of any X_j taking on either of these values tends to zero as $n \to \infty$. (But the convergence is not uniform, which causes trouble for asymptotic confidence intervals; see Lehmann and Loh 1990.) That the MLEs $\hat{p}_j$ are asymptotically efficient is seen by introducing the indicator variables $X_{j\nu}$, $\nu = 1, \ldots, n$, which are 1 when the νth trial results in outcome j and are 0 otherwise. Then, the vectors $(X_{o\nu}, \ldots, X_{s\nu})$ are iid and $T_j X_{j1} + \cdots + X_{jn}$, so that asymptotic efficiency follows from Example 6.3. ‖

In applications of the multinomial distribution to contingency tables, the p's are usually subject to additional restrictions. Theorem 5.1 typically continues to apply, although the computation of the estimators tend to be less obvious. This class of problems is treated comprehensively in Haberman (1973, 1974), Bishop, Fienberg, and Holland (1975), and Agresti (1990). Empty cells often present special problems.

7 Extensions

The discussion of efficient likelihood estimation so far has been restricted to the iid case. In the present section, we briefly mention extensions to some more general situations, which permit results analogous to those of Sections 6.3–6.5. Treatments not requiring the stringent (but frequently applicable) assumptions of Theorem 3.10 and 5.1 have been developed by Le Cam 1953, 1969, 1970, 1986, and others. For further work in this direction, see Pfanzagl 1970, 1994, Weiss and Wolfowitz 1974, Ibragimov and Has'minskii 1981, Blyth 1982, Strasser 1985, and Wong(1992).

The theory easily extends to the case of two or more samples. Suppose that the variables $X_{\alpha 1}, \ldots, X_{\alpha n_\alpha}$ in the αth sample are iid according to the distribution with density $f_{\alpha,\theta}$ $(\alpha = 1, \ldots, r)$ and that the r samples are independent. In applications, it will typically turn out that the vector parameter $\boldsymbol{\theta} = (\theta_1, \ldots, \theta_s)$ has some components occurring in more than one of the r distributions, whereas others may be specific to just one distribution. However, for the present discussion, we shall permit each of the distributions to depend on all the parameters.

The limit situation we shall consider supposes that each of the sample sizes n_α tends to infinity, all at the same rate, but that r remains fixed. Consider, therefore, sequences of sample sizes $n_{\alpha,k}$ ($k = 1, \ldots, \infty$) with total sample size $N_k = \Sigma_{\alpha=1}^r n_{\alpha_k}$ such that

$$n_{\alpha,k}/N_k \to \lambda_\alpha \quad \text{as } k \to \infty \tag{7.1}$$

where $\Sigma\lambda_\alpha = 1$ and the λ_α are > 0.

Theorem 7.1 *Suppose the assumptions of Theorem 5.1 hold for each of the densities $f_{\alpha,\theta}$. Let $I^{(\alpha)}(\theta)$ denote the information matrix corresponding to $f_{\alpha,\theta}$ and let*

$$I(\theta) = \Sigma\lambda_\alpha I^{(\alpha)}(\theta). \tag{7.2}$$

The log likelihood $l(\theta)$ is given by

$$l(\theta) = \sum_{\alpha=1}^{r}\sum_{j=1}^{n_\alpha} \log f_{\alpha,\theta}(x_{\alpha j})$$

and the likelihood equations by

$$\frac{\partial}{\partial\theta_j} l(\theta) = 0 \quad (j = 1, \ldots, s). \tag{7.3}$$

With these identifications, the conclusions of Theorem 5.1 remain valid.

The proof is an easy extension of that of Theorem 5.1 since $l(\theta)$, and therefore each term of its Taylor expansion, is a sum of r independent terms of the kind considered in the proof of Theorem 5.1 (Problem 7.1). (For further discussion of this situation, see Bradley and Gart 1962.)

That asymptotic efficiency continues to have the meaning it had in Theorems 3.10 and 5.1 and follows from the fact that Theorem 2.6 and its extension to the multiparameter case also extends to the present situation (see Bahadur 1964, Section 4).

Corollary 7.2 *Under the assumptions of Theorem 7.1, suppose that for each α, all off-diagonal elements in the jth row and jth column of $I^{(\alpha)}(\theta)$ are zero. Then, the asymptotic variance of $\hat{\theta}_j$ is the same when the remaining θ's are unknown as when they are known.*

Proof. If the property in question holds for each $I^{(\alpha)}(\theta)$, it also holds for $I(\theta)$ and the result thus follows from Problem 6.3. □

The following four examples illustrate some applications of Theorem 7.1.

Example 7.3 Estimation of a common mean. Let $X_1, \ldots, X_m$ and $Y_1, \ldots, Y_n$ be independently distributed according to $N(\xi, \sigma^2)$ and $N(\xi, \tau^2)$, respectively, with ξ, σ, and τ unknown. The problem of estimating ξ was considered briefly in Example 2.2.3 where it was found that a UMVU estimator for ξ does not exist. Complications also arise in the problem of asymptotically efficient estimation of ξ.

Since the MLEs of the mean and variance of a single normal distribution are asymptotically independent, Corollary 7.2 applies and shows that ξ can be estimated with the efficiency that is attainable when σ and τ are known. Now, in that case, the MLE—which is also UMVU—is

$$\hat{\xi} = \frac{(m/\sigma^2)\bar{X} + (n/\tau^2)\bar{Y}}{m/\sigma^2 + n/\tau^2}.$$

It is now tempting to claim that Theorem 1.8.10 implies that the asymptotic distribution of $\hat{\xi}$ is not changed when σ^2 and τ^2 are replaced by

$$(7.4) \qquad \hat{\sigma}^2 = \frac{1}{m-1}\Sigma(X_i - \bar{X})^2 \quad \text{and} \quad \hat{\tau}^2 = \frac{1}{n-1}\Sigma(Y_j - \bar{Y})^2$$

and the resulting estimator, say $\hat{\hat{\xi}}$, is asymptotically normal and efficient. However, this does not immediately follow. To see why, let us look at the simple case where $m = n$ and, hence, $\text{var}(\hat{\xi}) = (\sigma^2 + \tau^2)/n$. Consider the asymptotic distribution of

$$(7.5) \qquad \sqrt{n}\frac{\hat{\hat{\xi}} - \xi}{\sqrt{\sigma^2 + \tau^2}} = \sqrt{n}\frac{\hat{\hat{\xi}} - \hat{\xi}}{\sqrt{\sigma^2 + \tau^2}} + \sqrt{n}\frac{\hat{\xi} - \xi}{\sqrt{\sigma^2 + \tau^2}}.$$

Since $\hat{\xi}$ is efficient, efficiency of $\hat{\hat{\xi}}$ will follow if $\sqrt{n}(\hat{\hat{\xi}} - \hat{\xi}) \to 0$, which is not the case. But Theorem 7.1 does apply, and an asymptotically efficient estimator is given by the full MLE (see Problem 7.2). ||

Example 7.4 Balanced one-way random effects model. Consider the estimation of variance components σ_A^2 and σ^2 in model (3.5.1). In the canonical form (3.5.2), we are dealing with independent normal variables Z_{11} and $Z_{i1}, (i = 2, \ldots, s)$, and $Zij, (i = 1, \ldots, s, j = 2, \ldots, n)$. We shall restrict attention to the second and third group, as suggested by Thompson (1962), and we are then dealing with samples of sizes $s - 1$ and $(n-1)s$ from $N(0, \tau^2)$ and $N(0, \sigma^2)$, where $\tau^2 = \sigma^2 + n\sigma_A^2$. The assumptions of Theorem 7.1 are satisfied with $r = 2$, $\boldsymbol{\theta} = (\sigma^2, \tau^2)$, and the parameter space $\Omega = \{(\sigma, \tau) : 0 < \sigma^2 < \tau^2\}$. For fixed n, the sample sizes $n_1 = s - 1$ and $n_2 = s(n-1)$ tend to infinity as $s \to \infty$, with $\lambda_1 = 1/n$ and $\lambda_2 = (n-1)/n$.

The joint density of the second and third group of Z's constitutes a two-parameter exponential family; the log likelihood is given by

$$(7.6) \qquad -l(\boldsymbol{\theta}) = n_2 \log\sigma + n_1 \log\tau + \frac{S^2}{2\sigma^2} + \frac{S_A^2}{2\tau^2} + c$$

where $S^2 = \sum_{i=1}^s \sum_{j=2}^n Z_{ij}^2$ and $S_A^2 = \sum_{i=1}^s Z_{i1}^2$. By Example 7.8, the likelihood equations have at most one solution. Solving the equations yields

$$(7.7) \qquad \hat{\sigma}^2 = S^2/n_2, \quad \hat{\tau}^2 = S_A^2/n_1,$$

and these are the desired (unique, ML) solution, provided they are in Ω, that is, they satisfy

$$(7.8) \qquad \hat{\sigma}^2 < \hat{\tau}^2.$$

It follows from Theorem 5.1 that the probability of (7.8) tends to 1 as $s \to \infty$ for any $\theta \in \Omega$; this can also be seen directly from the fact that $\hat{\sigma}^2$ and $\hat{\tau}^2$ tend to σ^2 and τ^2 in probability.

What can be said when (7.8) is violated? The likelihood equations then have no root in Ω and an MLE does not exist (the likelihood attains its maximum at the boundary point $\hat{\sigma}^2 = \hat{\tau}^2 = (S_A^2 + S^2)/(n_1 + n_2)$ which is not in Ω). However, none of this matters from the present point of view since the asymptotic theory has nothing to say about a set of values whose probability tends to zero. (For small-sample computations of the mean squared error of a number of estimators of σ^2 and σ_A^2, see Klotz, Milton and Zacks 1969, Portnoy 1971, and Searle et al. 1992.)

The joint asymptotic distribution of $\hat{\sigma}^2$ and $\hat{\tau}^2$ can be obtained from Theorem 6.7 or directly from the distribution of S_A^2 and S^2 and the CLT, and a linear transformation of the limit distribution then gives the joint asymptotic distribution of $\hat{\sigma}^2$ and $\hat{\sigma}_A^2$ (Problem 7.3). ||

Example 7.5 Balanced two-way random effects model. A new issue arises as we go from the one-way to the two-way layout with the model given by (3.5.5). After elimination of Z_{111} (in the notation of Example 5.2), the data in canonical form consist of four samples Z_{i11} ($i = 2, \ldots, I$), Z_{1j1} ($j = 2, \ldots, J$), Z_{ij1} ($i = 2, \ldots, I$, $j = 2, \ldots, J$), and Z_{ijk} ($i = 1, \ldots, I$, $j = 1, \ldots, J$, $k = 2, \ldots, n$), and the parameter is $\theta = (\sigma, \tau_A, \tau_B, \tau_C)$ where

$$(7.9) \quad \tau_C^2 = \sigma^2 + n\sigma_C^2, \quad \tau_B^2 = nI\sigma_B^2 + n\sigma_C^2 + \sigma^2, \quad \tau_A^2 = nJ\sigma_A^2 + n\sigma_C^2 + \sigma^2$$

so that $\Omega = \{\theta : \sigma^2 < \tau_C^2 < \tau_A^2, \tau_B^2\}$. The joint density of these variables constitutes a four-parameter exponential family. The likelihood equations thus again have at most one root, and this is given by

$$\hat{\sigma}^2 = S^2/(n-1)IJ, \quad \hat{\tau}_C^2 = S_C^2/(I-1)(J-1),$$
$$\hat{\tau}_B^2 = S_B^2/(J-1), \quad \hat{\tau}_A^2 = S_A^2/(I-1)$$

when $\hat{\sigma}^2 < \hat{\tau}_C^2 < \hat{\tau}_A{}^2, \hat{\tau}_B{}^2$. No root exists when these inequalities fail.

In this case, asymptotic theory requires that both I and J tend to infinity, and assumption (7.1) of Theorem 7.1 then does not hold. Asymptotic efficiency of the MLEs follows, however, from Theorem 5.1 since each of the samples depends on only one of the parameters σ^2, τ_A^2, τ_B^2, and τ_C^2. The apparent linkage of these parameters through the inequalities $\sigma^2 < \tau_C^2 < \tau_A^2, \tau_B^2$ is immaterial. The true point $\theta^0 = (\sigma^0, \tau_A^0, \tau_B^0, \tau_C^0)$ is assumed to satisfy these restrictions, and each parameter can then independently vary about the true value, which is all that is needed for Theorem 5.1. It, therefore, follows as in the preceding example that the MLEs are asymptotically efficient, and that $\sqrt{(n-1)IJ}(\hat{\sigma}^2 - \sigma^2)$, and so on. have the limit distributions given by Theorem 5.1 or are directly obtainable from the definition of these estimators. ||

A general large-sample treatment both of components of variance and the more general case of mixed models, without assuming the models to be balanced was given by Miller (1977); see also Searle et al. 1992, Cressie and Lahiri 1993, and Jiang 1996, 1997.

Example 7.6 Independent binomial experiments. As in Section 3.5, let X_i ($i = 1, \ldots, s$) be independently distributed according to the binomial distributions $b(p_i, n_i)$, with the p_i being functions of a smaller number of parameters. If the n_i tend to infinity at the same rate, the situation is of the type considered in Theorem 7.1, which will, in typical cases, ensure the existence of an efficient solution of the likelihood equations with probability tending to 1.

As an illustration, suppose, as in (3.6.12), that the p's are given in terms of the logistic distribution, and more specifically that

$$p_i = \frac{e^{-(\alpha+\beta t_i)}}{1 + e^{-(\alpha+\beta t_i)}} \tag{7.10}$$

where the t's are known numbers and α and β are the parameters to be estimated. The likelihood equations

$$\sum n_i p_i = \sum x_i, \quad \sum n_i t_i p_i = \sum t_i x_i \tag{7.11}$$

have at most one solution (Problem 7.6) which will exist with probability tending to 1 (but may not exist for some particular finite values) and which can be obtained by standard iterative methods.

That the likelihood equations have at most one solution is true not only for the model (7.10) but more generally when

$$p_i = 1 - F\left(\sum \beta_j t_j\right) \tag{7.12}$$

where the ts are known, the βs are being estimated, and F is a known distribution function with $\log F(x)$ and $\log[1 - F(x)]$ strictly concave. (See Haberman 1974, Chapter 8; and Problem 7.7.) For further discussion of this and more general logistic regression models, see Pregibon 1981 or Searle et al. 1992, Chapter 10. ||

For the multinomial problem mentioned in the preceding section and those of Example 7.6, alternative methods have been developed which are asymptotically equivalent to the ELEs, and hence also asymptotically efficient. These methods are based on minimizing χ^2 or some other functions measuring the distance of the vector of probabilities from that of the observed frequencies. (See, for example, Neyman 1949, Taylor 1953, Le Cam 1956, 1990, Wijsman 1959, Berkson 1980, Amemiya 1980, and Ghosh and Sinha 1981 or Agresti 1990 for entries to the literature on choosing between these different estimators.)

The situation of Theorem 7.1 shares with that of Theorem 3.10 the crucial property that the total amount of information $T(\theta)$ asymptotically becomes arbitrarily large. In the general case of independent but not identically distributed variables, this need no longer be the case.

Example 7.7 Total information. Let X_i ($i = 1, \ldots, n$) be independent Poisson variables with $E(X_i) = \gamma_i \lambda$ where the γ's are known numbers. Consider two cases.

(a) $\sum_{i=1}^{\infty} \gamma_i < \infty$. The amount of information X_i contains about λ is γ_i/λ by (2.5.11) and Table 2.5.1 and the total amount of information $T_n(\lambda)$ that $(X_1, \ldots, X_n)$ contains about λ is therefore

$$T_n(\lambda) = \frac{1}{\lambda} \sum_{i=1}^{n} \gamma_i. \tag{7.13}$$

It is intuitively plausible that in these circumstances λ cannot be estimated consistently because only the early observations provide an appreciable amount of information. To prove this formally, note that $Y_n = \Sigma_{i=1}^n X_i$ is a sufficient statistic for λ on the basis of $(X_1, \ldots, X_n)$ and that Y_n has Poisson distribution with mean $\lambda \Sigma_{i=1}^n \gamma_i$. Thus, all the Y's are less informative than a random variable Y with distribution $P(\lambda \Sigma_{i=1}^\infty \gamma_i)$ in the sense that the distribution of any estimator based on Y_n can be duplicated by one based on Y (Problem 7.9). Since λ cannot be estimated exactly on the basis of Y, the result follows.

(b) $\Sigma_{i=1}^\infty \gamma_i = \infty$. Here, the MLE $\delta_n = \Sigma_{i=1}^n X_i / \Sigma_{i=1}^n \gamma_i$ is consistent and asymptotically normal (Problem 7.10) with

$$\left[\sum_{i=1}^n \gamma_i\right]^{1/2} (\delta_n - \lambda) \xrightarrow{\mathcal{L}} N(0, \lambda). \tag{7.14}$$

Thus, δ_n is approximately distributed as $N[\lambda, 1/T_n(\lambda)]$ and an extension of Theorem 2.6 to the present case (see Bahadur 1964) permits the conclusion that δ_n is asymptotically efficient.

Note: The norming constant required for asymptotic normality must be proportional to $\sqrt{\Sigma_{i=1}^n \gamma_i}$. Depending on the nature of the γ's, this can be any function of n tending to infinity rather than the customary $\sqrt{n}$. In general, it is the total amount of information rather than the sample size which governs the asymptotic distribution of an asymptotically efficient estimator. In the iid case, $T_n(\boldsymbol{\theta}) = nI(\boldsymbol{\theta})$, so that $\sqrt{T_n(\boldsymbol{\theta})}$ is proportional to $\sqrt{n}$. ||

A general treatment of the case of independent random variables with densities $f_j(x_j|\boldsymbol{\theta}), \boldsymbol{\theta} = (\theta_1, \ldots, \theta_r)$, along the lines of Theorems 3.10 and 5.1 has been given by Bradley and Gart (1962) and Hoadley (1971) (see also Nordberg 1980). The proof (for $r = 1$) is based on generalizations of (3.18)-(3.20) (see Problem 7.14) and hence depends on a suitable law of large numbers and central limit theorem for sums of independent nonidentical random variables. In the multiparameter case, of course, it may happen that some of the parameters can be consistently estimated and others not.

The theory for iid variables summarized by Theorems 2.6, 3.10, and 5.1 can be generalized not only to the case of independent nonidentical variables but also to dependent variables whose joint distribution depends on a fixed number of parameters $\boldsymbol{\theta} = (\theta_1, \ldots, \theta_r)$ where, for illustration, we take $r = 1$. (The generalization to $r > 1$ is straightforward.) The log likelihood $l(\theta)$ is now the sum of the logarithms of the conditional densities $f_j(x_j|\theta, x_1, \ldots, x_{j-1})$ and the total amount of information $T_n(\theta)$ is the sum of the expected conditional amounts of information $I_j(\theta)$ in X_j, given $X_1, \ldots, X_{j-1}$:

$$I_j(\theta) = E\left\{E\left[\frac{\partial}{\partial\theta} \log f_j(X_j|\theta, X_1, \ldots, X_{j-1})\right]^2\right\} = E\left[\frac{\partial}{\partial\theta} \log f_j(X_j|\theta)\right]^2.$$

Under regularity conditions on the f_j's, analogous to those of Theorems 3.10 and 5.1 together with additional conditions to ensure that the total amount of information tends to infinity as $n \to \infty$ and that the appropriate CLT for dependent variables is applicable, it can be shown that with probability tending to 1, there exists a root $\hat{\theta}_n$ of the likelihood equations such that $\sqrt{T_n(\theta)}(\hat{\theta}_n - \theta) \xrightarrow{\mathcal{L}} N(0, 1)$. This program has been carried out in a series of papers by Bar-Shalom (1971), Bhat (1974), and Crowder (1976).[5] [The required extension of Theorem 2.6 can be obtained from Bahadur (1964); see also Kabaila 1983.] The following illustrates the theory with a simple classic example.

Example 7.8 Normal autoregressive Markov series. Let

$$X_j = \beta X_{j-1} + U_j, \quad j = 2, \ldots, n, \tag{7.15}$$

where the U_j are iid as $N(0, 1)$, where β is an unknown parameter satisfying $|\beta| < 1$,[6] and where X_1 is $N(0, \sigma^2)$. The X's all have marginal normal distributions with mean zero. The variance of X_j satisfies

$$\text{var}\,(X_j) = \beta^2 \text{var}(X_{j-1}) + 1 \tag{7.16}$$

and hence $\text{var}(X_j) = \sigma^2$ for all j provided

$$\sigma^2 = 1/(1 - \beta^2). \tag{7.17}$$

This is the stationary case in which $(X_{j_1}, \ldots, X_{jk})$ has the same distribution as $(X_{j_1+r}, \ldots, X_{j_k+r})$ for all $r = 1, 2, \ldots$ (Problem 7.15).

The amount of information that each $X_j (j > 1)$ contains about β is (Problem 7.17) $I_j(\theta) = 1/(1-\beta^2)$, so that $T_n(\beta) \sim n/(1-\beta^2)$. The general theory therefore suggests the existence of a root $\hat{\beta}_n$ of the likelihood equation such that

$$\sqrt{n}(\hat{\beta}_n - \beta) \xrightarrow{\mathcal{L}} N(0, 1 - \beta^2). \tag{7.18}$$

That (7.18) does hold can also be checked directly (see, for example, Brockwell and Davis 1987, Section 8.8). ||

The conclusions of this section up to this point can be summarized by saying that the asymptotic theory developed for the iid case in Sections 6.2–6.6 continues to hold—under appropriate safeguards—even if the iid assumption is dropped, provided the number of parameters is fixed and the total amount of information goes to infinity.

We shall now briefly consider two generalizations of the earlier situation to which this conclusion does not apply. The first concerns the case in which the number of parameters tends to infinity with the total sample size.

In Theorem 7.1, the number r of samples was considered fixed, whereas the sample sizes n_α were assumed to tend to infinity. Such a model is appropriate when one is dealing with a small number of moderately large samples. A quite different asymptotic situation arises in the reverse case of a large number (considered as tending to infinity) of finite samples. Here, an important distinction arises between *structural parameters* such as ξ in Example 7.3, which are common to all the samples and which are the parameters of interest, and *incidental parameters* such

[5] A review of the literature of maximum likelihood estimation in both discrete and continuous parameter stochastic processes can be found in Basawa and Prakasa Rao(1980).

[6] For a discussion without this restriction, see Anderson (1959) and Heyde and Feigin (1975).

as σ^2 and τ^2 in Example 7.3, which occur in only one of the samples. That Theorem 5.1 does not extend to this case is illustrated by the following two examples.

Example 7.9 Estimation of a common variance. Let $X_{\alpha j}$ $(j = 1, \ldots, r)$ be independently distributed according to $N(\theta_\alpha, \sigma^2)$, $\alpha = 1, \ldots, n$. The MLEs are

$$\hat{\theta}_\alpha = X_{\alpha\cdot}, \quad \hat{\sigma}^2 = \frac{1}{rn}\Sigma\Sigma(X_{\alpha j} - X_{\alpha\cdot})^2. \tag{7.19}$$

Furthermore, these are the unique solutions of the likelihood equations.

However, in the present case, the MLE of σ^2 is not even consistent. To see this, note that the statistics

$$S_\alpha^2 = \Sigma(X_{\alpha j} - X_{\alpha\cdot})^2$$

are identically independently distributed with expectation

$$E(S_\alpha^2) = (r-1)\sigma^2,$$

so that $\Sigma S_\alpha^2/n \to (r-1)\sigma^2$ and hence

$$\hat{\sigma}^2 \to \frac{r-1}{r}\sigma^2 \quad \text{in probability.} \tag{7.20}$$

A consistent and efficient estimator sequence of σ^2 is available in the present case, namely

$$\hat{\hat{\sigma}}^2 = \frac{1}{(r-1)n}\Sigma S_\alpha^2.$$

||

The study of this class of problems (including Example 7.9) was initiated by Neyman and Scott (1948), who also considered a number of other examples including one in which an MLE is consistent but not efficient.

A reformulation of the problem of structural parameters was proposed by Kiefer and Wolfowitz (1956), who considered the case in which the incidental parameters are themselves random variables, identically independently distributed according to some distribution, but, of course, unobservable. This will often bring the situation into the area of applicability of Theorems 5.1 or 7.1.

Example 7.10 Regression with both variables subject to error. Let X_i and Y_i $(i = 1, \ldots, n)$ be independent normal with means $E(X_i) = \xi_i$ and $E(Y_i) = \eta_i$ and variances σ^2 and τ^2, where $\eta_i = \alpha + \beta\xi_i$. There is, thus, a linear relationship between ξ and η, both of which are observed with independent, normally distributed errors. We are interested in estimating β and, for the sake of simplicity, shall take α as known to be zero. Then, $\theta = (\beta, \sigma^2, \tau^2, \xi_1, \ldots, \xi_n)$, with the first three parameters being structural and the ξ's incidental. The likelihood is proportional to

$$\frac{1}{\sigma^n\tau^n}\exp\left[-\frac{1}{2\sigma^2}\Sigma(x_i - \xi_i)^2 - \frac{1}{2\tau^2}\Sigma(y_i - \beta\xi_i)^2\right]. \tag{7.21}$$

The likelihood equations have two roots, given by (Problem 7.20),

$$\hat{\beta} = \pm\sqrt{\frac{\Sigma y_i^2}{\Sigma x_i^2}}, \quad 2n\hat{\sigma}^2 = \Sigma x_j^2 - \frac{1}{\hat{\beta}}\Sigma x_j y_j, \quad 2n\hat{\tau}^2 = \Sigma y_j^2 - \hat{\beta}\Sigma x_j y_j, \tag{7.22}$$

$$2\hat{\xi}_i = x_i + \frac{1}{\hat{\beta}} y_i, \quad i = 1, \ldots, n$$

and the likelihood is larger at the root for which $\hat{\beta}\Sigma x_i y_i > 0$. If Theorem 5.1 applies, one of these roots must be consistent and, hence, tend to β in probability. Since $S_X^2 = \Sigma X_j^2$ and $S_Y^2 = \Sigma Y_j^2$ are independently distributed according to noncentral χ^2-distributions with noncentrality parameters $\lambda_n^2 = \Sigma_{j=1}^n \xi_j^2$ and $\beta^2\lambda_n^2$, their limit behavior depends on that of λ_n. (Note, incidentally, that for $\lambda_n = 0$, the parameter β becomes unidentifiable.) Suppose that $\lambda_n^2/n \to \lambda^2 > 0$. The distribution of S_X^2 and S_Y^2 is unchanged if we replace each ξ_i^2 by λ_n^2/n, and by the law of large numbers, $\Sigma X_j^2/n$ therefore has the same limit as

$$E(X_1^2) = \sigma^2 + \frac{1}{n}\lambda_n^2 \to \sigma^2 + \lambda^2.$$

Similarly, $\Sigma Y_j^2/n$ tends in probability to $\tau^2 + \beta^2\lambda^2$ and, hence, $\hat{\beta}_n^2 \xrightarrow{P} (\tau^2 + \beta^2\lambda^2)/(\sigma^2 + \lambda^2)$. Thus, neither of the roots is consistent. [It was pointed out by Solari (1969) that the likelihood in this problem is unbounded so that an MLE does not exist (Problem 7.21). The solutions (7.22) are, in fact, saddlepoints of the likelihood surface.]

If in (7.21) it is assumed that $\tau = \sigma$, it is easily seen that the MLE of β is consistent (Problem 7.18). For a discussion of this problem and some of its generalizations, see Anderson 1976, Gleser 1981, and Anderson and Sawa 1982. Another modification leading to a consistent MLE is suggested by Copas (1972a).

Instead of (7.21), it is sometimes assumed that the ξ's are themselves iid according to a normal distribution $N(\mu, \gamma^2)$. The pairs (X_i, Y_i) then constitute a sample from a bivariate normal distribution, and asymptotically efficient estimators of the parameters $\mu, \gamma, \beta, \sigma$, and τ can be obtained from the MLEs of Example 6.4. An analogous treatment is possible for Example 7.9. ||

Kiefer and Wolfowitz (1956) have considered not only this problem and that of Example 7.9, but a large class of problems of this type by postulating that the ξ's are iid according to a distribution G, but treating G as unknown, subject only to some rather general regularity assumptions. Alternative approaches to the estimation of structural parameters in the presence of a large number of incidental parameters are discussed by Andersen (1970b) and Kalbfleisch and Sprott (1970). A discussion of Example 7.10 and its extension to more general regression models can be found in Stuart and Ord (1991, Chapters 26 and 28), and of Example 7.9 in Jewell and Raab (1981).

A review of these models, also known as *measurement error models*, is given by Gleser (1991) and is the topic of the book by Carroll, Ruppert, and Stefanski (1995).

Another extension of likelihood estimation leads us along the lines of Example 4.5, in which it was seen that an estimator such as the sample median, which was not the MLE, was a desirable alternative. Such situations can lead naturally to replacing the likelihood function by another function, often with the goal of obtaining a *robust* estimator.

Such an approach was suggested by Huber (1964), resulting in a compromise between the mean and the median. The mean and the median minimize, respectively, $\sum(x_i - a)^2$ and $\sum |x_i - a|$. Huber suggested minimizing instead

$$\sum_{i=1}^{n} \rho(x_i - a) \tag{7.23}$$

where ρ is given by

$$\rho(x) = \begin{cases} \frac{1}{2}x^2 & \text{if } |x| \le k \\ k|x| - \frac{1}{2}k^2 & \text{if } |x| \ge k. \end{cases} \tag{7.24}$$

This function is proportional to x^2 for $|x| \le k$, but outside this interval, it replaces the parabolic arcs by straight lines. The pieces fit together so that ρ and its derivative ρ' are continuous (Problem 7.22). As k gets larger, ρ will agree with $\frac{1}{2}x^2$ over most of its range, so that the estimator comes close to the mean, As k gets smaller, the estimator will become close to the median. As a moderate compromise, the value $k = 1.5$ is sometimes suggested.

The *Huber estimators* minimizing (7.23) with ρ given by (7.24) are a subset of the class of *M-estimators* obtained by minimizing (7.23) for arbitrary ρ. If ρ is convex and even, as is the case for (7.24), it follows from Theorem 1.7.15 that the minimizing values of (7.23) constitute a closed interval; if ρ is strictly convex, the minimizing value is unique. If ρ has a derivative $\rho' = \psi$, the M-estimators M_n may be defined as the solutions of the equation

$$\sum_{i=1}^{n} \psi(x_i - a) = 0. \tag{7.25}$$

If $X_1, \ldots, X_n$ are iid according to $F(x - \theta)$ where F is symmetric about zero and has density f, it turns out under weak assumptions on ψ and F that

$$\sqrt{n}(M_n - \theta) \to N[0, \sigma^2(F, \psi)] \tag{7.26}$$

where

$$\sigma^2(F, \psi) = \frac{\int \psi^2(x) f(x) dx}{[\int \psi'(x) f(x) dx]^2}, \tag{7.27}$$

provided both numerator and denominator on the right side are finite and the denominator is positive.

Proofs of (7.26) can be found in Huber (1981), in which a detailed account of the theory of M-estimators is given not only for location parameters, but also in more general settings. See also Serfling 1980, Chapter 7, Hampel et al. 1986, Staudte and Sheather 1990, as well as Problems 7.24-7.26.

For

$$\rho(x) = -\log f(x), \tag{7.28}$$

minimizing (7.23) is equivalent to maximizing $\prod f(x_i - a)$, and the M-estimator then coincides with the maximum likelihood estimator. In particular, for known F, the M-estimator of θ corresponding to (7.28) satisfies (7.26) with $\sigma^2 = 1/I_f$ (see Theorem 3.10). Further generalizations are discussed in Note 10.4.

The results of this chapter have all been derived in the so-called *regular case*, that is, when the densities satisfy regularity assumptions such as those of Theorems 2.6, 3.10, and 5.1. Of particular importance for the validity of the conclusions is that the support of the distributions P_θ does not vary with θ. Varying support brings with it information that often makes it possible to estimate some of the parameters with greater accuracy than that attainable in the regular case.

Example 7.11 Uniform MLE. Let $X_1, \ldots, X_n$ be iid as $U(0, \theta)$. Then, the MLE of θ is $\hat{\theta}_n = X_{(n)}$ and satisfies (Problem 2.6)

$$n(\theta - \hat{\theta}_n) \xrightarrow{\mathcal{L}} E(0, \theta). \tag{7.29}$$

Since $\hat{\theta}_n$ always underestimates θ and has a bias of order $1/n$, the order of the error $\hat{\theta}_n - \theta$, considers as an alternative the UMVU estimator $\delta_n = [(n+1)/n]X_{(n)}$, which satisfies

$$n(\theta - \delta_n) \xrightarrow{\mathcal{L}} E(-\theta, \theta). \tag{7.30}$$

The two asymptotic distributions have the same variance, but the first has expectation θ, whereas the second is asymptotically unbiased with expectation zero and is thus much better centered.

The improvement of δ_n over $\hat{\theta}_n$ is perhaps seen more clearly by considering expected squared error. We have (Problem 2.7)

$$E[n(\hat{\theta}_n - \theta)]^2 \to 2\theta^2, \quad E[n(\delta_n - \theta)]^2 \to \theta^2. \tag{7.31}$$

Thus, the risk efficiency of $\hat{\theta}_n$ with respect to δ_n is $1/2$. ‖

The example illustrates two ways in which such situations differ from the regular iid cases. First, the appropriate normalizing factor is n rather than $\sqrt{n}$, reflecting the fact that the error of the MLE is of order $1/n$ instead of $1/\sqrt{n}$. Second, the MLE need no longer be asymptotically optimal even when it is consistent.

Example 7.12 Exponential MLE. Let $X_1, \ldots, X_n$ be iid according to the exponential distribution $E(\xi, b)$. Then, the MLEs of ξ and b are

$$\hat{\xi} = X_{(1)} \quad \text{and} \quad \hat{b} = \frac{1}{n}\Sigma[X_i - X_{(1)}]. \tag{7.32}$$

It follows from Problem 1.6.18 that $n[X_{(1)} - \xi]/b$ is exactly (and hence asymptotically) distributed as $E(0, 1)$. As was the case for $\hat{\theta}$ in the preceding example, $\hat{\xi}$ is therefore asymptotically biased. More satisfactory is the UMVU estimator δ_n given by (2.2.23), which is obtained from $\hat{\xi}$ by subtracting an estimator of the bias (Problem 7.27).

It was further seen in Problem 1.6.18 that $2n\hat{b}/b$ is distributed as χ^2_{2n-2}. Since $(\chi^2_n - n)/\sqrt{2n} \to N(0, 1)$ in law, it is seen that $\sqrt{n}(\hat{b} - b) \to N(0, b^2)$. We shall now show that $\hat{b}$ is asymptotically efficient. For this purpose, consider the case that ξ is known. The resulting one-parameter family of the X's is an exponential family and the MLE $\hat{\hat{b}}$ of b is asymptotically efficient and satisfies $\sqrt{n}(\hat{\hat{b}} - b) \to N(0, b^2)$ (Problem 7.27). Since $\hat{b}$ and $\hat{\hat{b}}$ have the same asymptotic distribution, $\hat{b}$ is a fortiori also asymptotically efficient, as was to be proved. ‖

Example 7.13 Pareto MLE. Let $X_1, \ldots, X_n$ be iid according to the Pareto distribution $P(a, c)$ with density

$$f(x) = ac^a/x^{a+1}, \quad 0 < c < x, \quad 0 < a. \tag{7.33}$$

The distribution is widely used, for example, in economics (see Johnson, Kotz, and Balakrishnan 1994, Chapter 20) and is closely connected with the exponential distribution of the preceding example through the fact that if X has density (7.33), then $Y = \log X$ has the exponential distribution $E(\xi, b)$ with (Problem 1.5.25)

$$\xi = \log c, \quad b = 1/a. \tag{7.34}$$

From this fact, it is seen that the MLEs of a and c are

$$\hat{a} = \frac{n}{\Sigma \log(X_i/X_{(1)})} \quad \text{and} \quad \hat{c} = X_{(1)} \tag{7.35}$$

and that these estimators are independently distributed, $\hat{c}$ as $P(na, c)$ and $2na/\hat{a}$ as χ^2_{2n-2} (Problem 7.29).

Since $\hat{b}$ is asymptotically efficient in the exponential case, the same is true of $1/\hat{b}$ and hence of $\hat{a}$. On the other hand, $n(X_{(1)} - c)$ has the limit distribution $E(0, c/a)$ and hence is biased. As was the case with the MLE of ξ in Example 7.12, an improvement over the MLE $\hat{c}$ of c is obtained by removing its bias and replacing $\hat{c}$ by the UMVU estimator

$$X_{(1)}\left[1 - \frac{1}{(n-1)\hat{a}}\right]. \tag{7.36}$$

For the details of these calculations, see Problems 7.29–7.31. ||

Example 7.14 Lognormal MLE. As a last situation with variable support, consider a sample $X_1, \ldots, X_n$ from a three-parameter lognormal distribution, defined by the requirement that $Z_i = \log(X_i - \xi)$ are iid as $N(\gamma, \sigma^2)$, so that

$$f(x; \xi, \gamma, \sigma^2) = \frac{1}{(x-\xi)\sqrt{2\pi}\sigma} \exp\left\{-\frac{1}{2\sigma^2}[\log(x-\xi) - \gamma]^2\right\} \tag{7.37}$$

when $x > \xi$, and $f = 0$ otherwise. When ξ is known, the problem reduces to that of estimating the mean γ and variance σ^2 from the normal sample $Z_1, \ldots, Z_n$. However, when ξ is unknown, the support varies with ξ. Although in this case the density (7.37) tends to zero very smoothly at ξ (Problem 7.34), the theory of Section 6.5 is not applicable, and the problem requires a more powerful approach such as that of Le Cam (1969). [For a discussion of the literature on this problem, see, for example, Johnson, Kotz, and Balakrishnan 1994, Chapter 14. A comprehensive treatment of the lognormal distribution is given in Crow and Shimizu (1988).]

The difficulty can be circumvented by a device used in other contexts by Kempthorne (1966), Lambert (1970), and Copas (1972a), and suggested for the present problem by Giesbrecht and Kempthorne (1976). These authors argue that observations are never recorded exactly but only to the nearest unit of measurement. This formulation leads to a multinomial model of the kind considered for one parameter in Example 4.6, and Theorem 5.1 is directly applicable.

The corresponding problem for the three-parameter Weibull distribution is reviewed in Scholz (1985). For further discussion of such irregular cases, see, for example, Polfeldt 1970 and Woodrofe 1972. ||

Although the MLE, or bias-corrected MLE, may achieve the smallest asymptotic variance, it may not minimize mean squared error when compared with all other estimators. This is illustrated by the following example in which, for the sake of simplicity, we shall consider expected squared error instead of asymptotic variance.

Example 7.15 Second-order mean squared error. Consider the estimation of σ^2 on the basis of a sample $X_1, \ldots, X_n$ from $N(0, \sigma^2)$. The MLE is then

$$\hat{\sigma}^2 = \frac{1}{n}\Sigma X_i^2,$$

which happens to be unbiased, so that no correction is needed. Let us now consider the more general class of estimators

$$\delta_n = \left(\frac{1}{n} + \frac{a}{n^2}\right)\Sigma X_i^2. \tag{7.38}$$

It can be shown (Problem 7.32) that

$$E(\delta_n - \sigma^2)^2 = \frac{2\sigma^4}{n} + \frac{(4a + a^2)\sigma^4}{n^2} + O\left(\frac{1}{n^3}\right). \tag{7.39}$$

Thus, the estimators δ_n are all asymptotically efficient, that is, $nE(\delta_n - \theta)^2 \to 1/I(\theta)$ where $\theta = \sigma^2$. However, the MLE does not minimize the error in this class since the term of order $1/n^2$ is minimized not by $a = 0$ (MLE) but by $a = -2$, so that $(1/n - 2/n^2)\Sigma X_i^2$ has higher second-order efficiency than the MLE. In fact, the normalized limiting risk difference between the MLE ($a = 0$) relative to δ_n with $a = -2$ is 2, that is, the limiting risk of the MLE is larger (Problem 7.32). ||

A uniformly best estimator (up to second-order terms) typically will not exist. The second-order situation is thus similar to that encountered in the exact (small-sample) theory. One can obtain uniform second-order optimality by imposing restrictions such as first-order unbiasedness, or must be content with weaker properties such as second-order admissibility or minimaxity. An admissibility result (somewhat similar to Theorem 5.2.14) is given by Ghosh and Sinha (1981); the minimax problem is treated by Levit (1980).

8 Asymptotic Efficiency of Bayes Estimators

Bayes estimators were defined in Section 4.1, and many of their properties were illustrated throughout Chapter 4. We shall now consider their asymptotic behavior.

Example 8.1 Limiting binomial. If X has the binomial distribution $b(p, n)$ and the loss is squared error, it was seen in Example 4.1.5 that the Bayes estimator of p corresponding to the beta prior $B(a, b)$ is

$$\delta_n(X) = (a + X)/(a + b + n).$$

Thus,

$$\sqrt{n}[\delta_n(X) - p] = \sqrt{n}\left(\frac{X}{n} - p\right) + \frac{\sqrt{n}}{a+b+n}\left[a - (a+b)\frac{X}{n}\right]$$

and it follows from Theorem 1.8.10 that $\sqrt{n}[\delta_n(X) - p]$ has the same limit distribution as $\sqrt{n}[X/n - p]$, namely the normal distribution $N[0, p(1-p)]$. So, the Bayes estimator of the success probability p, suitably normalized, has a normal limit distribution which is independent of the parameters of the prior distribution and is the same as that of the MLE X/n. Therefore, these Bayes estimators are asymptotically efficient. (See Problem 8.1 for analogous results.) ||

This example raises the question of whether the same limit distribution also obtains when the conjugate priors in this example are replaced by more general prior distributions, and whether the phenomenon persists in more general situations. The principal result of the present section (Theorem 8.3) shows that, under suitable conditions, the distribution of Bayes estimators based on n iid random variables tends to become independent of the prior distribution as $n \to \infty$ and that the Bayes estimators are asymptotically efficient.

Versions of such a theorem were given by Bickel and Yahav (1969) and by Ibragimov and Has'minskii (1972, 1981). The present proof, which combines elements from these papers, is due to Bickel. We begin by stating some assumptions.

Let $X_1, \ldots, X_n$ be iid with density $f(x_i|\theta)$ (with respect to μ), where θ is real-valued and the parameter space Ω is an open interval. The true value of θ will be denoted by θ_0.

(B1) The log likelihood function $l(\theta)$ satisfies the assumptions of Theorem 2.6.

To motivate the next assumption, note that under the assumptions of Theorem 2.6, if $\tilde{\theta} = \tilde{\theta}_n$ is any sequence for which $\tilde{\theta} \xrightarrow{P} \theta$ then

$$l(\theta) = l(\theta_0) + (\theta - \theta_0)l'(\theta_0) - \frac{1}{2}(\theta - \theta_0)^2[nI(\theta_0) + R_n(\theta)] \tag{8.1}$$

where

$$\frac{1}{n}R_n(\theta) \xrightarrow{P} 0 \quad \text{as } n \to \infty \tag{8.2}$$

(Problem 8.3). We require here the following stronger assumption.

(B2) Given any $\varepsilon > 0$, there exists $\delta > 0$ such that in the expansion (8.1), the probability of the event

$$\sup\left\{\left|\frac{1}{n}R_n(\theta)\right| : |\theta - \theta_0| \le \delta\right\} \ge \varepsilon \tag{8.3}$$

tends to zero as $n \to \infty$.

In the present case it is not enough to impose conditions on $l(\theta)$ in the neighborhood of θ_0, as is typically the case in asymptotic results. Since the Bayes estimators involve integration over the whole range of θ values, it is also necessary to control the behavior of $l(\theta)$ at a distance from θ_0.

(B3) For any $\delta > 0$, there exists $\varepsilon > 0$ such that the probability of the event

$$\sup\left\{\frac{1}{n}[l(\theta) - l(\theta_0)] : |\theta - \theta_0| \geq \delta\right\} \leq -\varepsilon \tag{8.4}$$

tends to 1 as $n \to \infty$.

(B4) The prior density π of θ is continuous and positive for all $\theta \in \Omega$.

(B5) The expectation of θ under π exists, that is,

$$\int |\theta|\pi(\theta)d\theta < \infty. \tag{8.5}$$

To establish the asymptotic efficiency of Bayes estimators under these assumptions, we shall first prove that for large values of n, the posterior distribution of θ given the X's is approximately normal with

$$\text{mean} = \theta_0 + \frac{1}{nI(\theta_0)}l'(\theta_0) \quad \text{and} \quad \text{variance} = 1/nI(\theta_0). \tag{8.6}$$

Theorem 8.2 *If $\pi^*(t|\mathbf{x})$ is the posterior density of $\sqrt{n}(\theta - T_n)$ where*

$$T_n = \theta_0 + \frac{1}{nI(\theta_0)}l'(\theta_0), \tag{8.7}$$

(i) *then if* (B1)-(B4) *hold,*

$$\int \left|\pi^*(t|\mathbf{x}) - \sqrt{I(\theta_0)}\phi\left[t\sqrt{I(\theta_0)}\right]\right| dt \xrightarrow{P} 0. \tag{8.8}$$

(ii) *If, in addition,* (B5) *holds, then*

$$\int (1 + |t|)\left|\pi^*(t|\mathbf{x}) - \sqrt{I(\theta_0)}\phi\left[t\sqrt{I(\theta_0)}\right]\right| dt \xrightarrow{P} 0. \tag{8.9}$$

Proof. (i) By the definition of T_n,

$$\begin{aligned}\pi^*(t|\mathbf{x}) &= \frac{\pi\left(T_n + \frac{t}{\sqrt{n}}\right)\exp\left[l\left(T_n + \frac{t}{\sqrt{n}}\right)\right]}{\int \pi\left(T_n + \frac{u}{\sqrt{n}}\right)\exp\left[l\left(T_n + \frac{u}{\sqrt{n}}\right)\right] du} \\ &= e^{\omega(t)}\pi\left(T_n + \frac{t}{\sqrt{n}}\right)/C_n\end{aligned} \tag{8.10}$$

where

$$\omega(t) = l\left(T_n + \frac{t}{\sqrt{n}}\right) - l(\theta_0) - \frac{1}{2nI(\theta_0)}[l'(\theta_0)]^2 \tag{8.11}$$

and

$$C_n = \int e^{\omega(u)}\pi\left(T_n + \frac{u}{\sqrt{n}}\right) du. \tag{8.12}$$

We shall prove at the end of the section that

$$J_1 = \int \left|e^{\omega(t)}\pi\left(T_n + \frac{t}{\sqrt{n}}\right) - e^{-t^2 I(\theta_0)/2}\pi(\theta_0)\right| dt \xrightarrow{P} 0, \tag{8.13}$$

so that

(8.14) $$C_n \xrightarrow{P} \int e^{-t^2 I(\theta_0)/2} \pi(\theta_0)\, dt = \pi(\theta_0)\sqrt{2\pi/I(\theta_0)}.$$

The left side of (8.8) is equal to J/C_n, where

(8.15) $$J = \int \left| e^{\omega(t)} \pi\left(T_n + \frac{t}{\sqrt{n}}\right) - C_n\sqrt{I(\theta_0)}\phi\left[t\sqrt{I(\theta_0)}\right]\right| dt$$

and, by (8.14), it is enough to show that $J \xrightarrow{P} 0$.

Now, $J \le J_1 + J_2$ where J_1 is given by (8.13) and

$$J_2 = \int \left| C_n\sqrt{I(\theta_0)}\phi\left[t\sqrt{I(\theta_0)}\right] - \exp\left[-\frac{t^2}{2}I(\theta_0)\right]\pi(\theta_0)\right| dt$$
$$= \left|\frac{C_n\sqrt{I(\theta_0)}}{\sqrt{2\pi}} - \pi(\theta_0)\right| \int \exp\left[-\frac{t^2}{2}I(\theta_0)\right] dt.$$

By (8.13) and (8.14), J_1 and J_2 tend to zero in probability, and this completes the proof of part (i).

(ii) The left side of (8.9) is equal to

$$\frac{1}{C_n}J' \le \frac{1}{C_n}(J_1' + J_2')$$

where J', J_1', and J_2' are obtained from J, J_1, and J_2, respectively, by inserting the factor $(1 + |t|)$ under the integral signs. It is therefore enough to prove that J_1' and J_2' both tend to zero in probability. The proof for J_2' is the same as that for J_2; the proof for J_1' will be given at the end of the section, together with that for J_1. □

On the basis of Theorem 8.2, we are now able to prove the principal result of this section.

Theorem 8.3 *If (B1)-(B5) hold, and if $\tilde{\theta}_n$ is the Bayes estimator when the prior density is π and the loss is squared error, then*

(8.16) $$\sqrt{n}(\tilde{\theta}_n - \theta_0) \xrightarrow{\mathcal{L}} N[0, 1/I(\theta_0)],$$

so that $\tilde{\theta}_n$ is consistent[7] *and asymptotically efficient.*

Proof. We have

$$\sqrt{n}(\tilde{\theta}_n - \theta_0) = \sqrt{n}(\tilde{\theta}_n - T_n) + \sqrt{n}(T_n - \theta_0).$$

By the CLT, the second term has the limit distribution $N[0, 1/I(\theta_0)]$, so that it only remains to show that

(8.17) $$\sqrt{n}(\tilde{\theta}_n - T_n) \xrightarrow{P} 0.$$

Note that Equation (8.10) says that $\pi^*(t|\mathbf{x}) = \frac{1}{\sqrt{n}}\pi(T_n + \frac{t}{\sqrt{n}}|\mathbf{x})$, and, hence, by a change of variable, we have

$$\tilde{\theta}_n = \int \theta\pi(\theta|\mathbf{x})\, d\theta$$

[7] A general relationship between the consistency of MLEs and Bayes estimators is discussed by Strasser (1981).

$$= \int \left(\frac{t}{\sqrt{n}} + T_n \right) \pi^*(t|\mathbf{x})\,dt$$
$$= \frac{1}{\sqrt{n}} \int t\pi^*(t|\mathbf{x})\,dt + T_n$$

and hence

$$\sqrt{n}(\tilde{\theta}_n - T_n) = \int t\pi^*(t|\mathbf{x})\,dt.$$

Now, since $\int t\sqrt{I(\theta_0)}\phi\left[t\sqrt{I(\theta_0)}\right]dt = 0$,

$$\sqrt{n}|\tilde{\theta}_n - T_n| = \left| \int t\pi^*(t|\mathbf{x})\,dt - \int t\sqrt{I(\theta_0)}\phi\left[t\sqrt{I(\theta_0)}\right]dt \right|$$
$$\leq \int |t| \left| \pi^*(t|\mathbf{x}) - \sqrt{I(\theta_0)}\phi\left[t\sqrt{I(\theta_0)}\right] \right| dt,$$

which tends to zero in probability by Theorem 8.2. □

Before discussing the implications of Theorem 8.3, we shall show that assumptions (B1)-(B5) are satisfied in exponential families.

Example 8.4 Exponential families. Let

$$f(x_i|\theta) = e^{\theta T(x_i) - A(\theta)}.$$

so that

$$A(\theta) = \log \int e^{\theta T(x)} d\mu(x).$$

Recall from Section 1.5 that A is differentiable to all orders and that

$$A'(\theta) = E_\theta[T(X)],$$
$$A''(\theta) = \text{var}_\theta[T(X)] = I(\theta).$$

Suppose $I(\theta) > 0$. Then,

$$l(\theta) - l(\theta_0) = (\theta - \theta_0)\Sigma T(X_i) - n[A(\theta) - A(\theta_0)]$$
$$(8.18) \qquad = (\theta - \theta_0)\Sigma[T(X_i) - A'(\theta_0)]$$
$$-n\{[A(\theta) - A(\theta_0)] - [(\theta - \theta_0)A'(\theta_0)]\}.$$

The first term is equal to $(\theta - \theta_0)l'(\theta_0)$. Apply Taylor's theorem to $A(\theta)$ to find

$$A(\theta) = A(\theta_0) + (\theta - \theta_0)A'(\theta_0) + \frac{1}{2}(\theta - \theta_0)^2 A''(\theta^*),$$

so that the second term in (8.18) is equal to $(-n/2)(\theta - \theta_0)^2 A''(\theta^*)$. Hence,

$$l(\theta) - l(\theta_0) = (\theta - \theta_0)l'(\theta_0) - \frac{n}{2}(\theta - \theta_0)^2 A''(\theta^*).$$

To prove (B2), we must show that

$$A''(\theta^*) = I(\theta_0) + \frac{1}{n} R_n(\theta)$$

where

$$R_n(\theta) = n[A''(\theta^*) - I(\theta_0)]$$

satisfies (8.3); that is, we must show that given ε, there exists δ such that the probability of

$$\sup\{|A''(\theta^*) - I(\theta_0)| : |\theta - \theta_0| \le \delta\} \ge \varepsilon$$

tends to zero. This follows from the facts that $I(\theta) = A''(\theta)$ is continuous and that $\theta^* \to \theta_0$ as $\theta \to \theta_0$.

To see that (B3) holds, write

$$\frac{1}{n}[l(\theta) - l(\theta_0)] = (\theta - \theta_0)\left\{\frac{1}{n}\Sigma[T(X_i) - A'(\theta_0)] - \left[\frac{A(\theta) - A(\theta_0)}{\theta - \theta_0} - A'(\theta_0)\right]\right\}$$

and suppose without loss of generality that $\theta > \theta_0$.

Since $A''(\theta) > 0$, so that $A(\theta)$ is strictly convex, it is seen that $\theta > \theta_0$ implies $[A(\theta) - A(\theta_0)]/(\theta - \theta_0) > A'(\theta_0)$. On the other hand, $\Sigma[T(X_i) - A'(\theta_0)]/n \xrightarrow{P} 0$ and hence with probability tending to 1, the factor of $(\theta - \theta_0)$ is negative. It follows that

$$\sup\left\{\frac{1}{n}[l(\theta) - l(\theta_0)] : \theta - \theta_0 \ge \delta\right\}$$
$$\le \delta\left\{\frac{\Sigma[T(X_i) - A'(\theta_0)]}{n} - \inf\left[\frac{A(\theta) - A(\theta_0)}{\theta - \theta_0} - A'(\theta_0) : \theta - \theta_0 \ge \delta\right]\right\}.$$

and hence that (B3) is satisfied. ||

Theorems 8.2 and 8.3 were stated under the assumption that π is the density of a proper distribution, so that its integral is equal to 1. There is a trivial but useful extension to the case in which $\int \pi(\theta)d\theta = \infty$ but where there exists n_0, so that the posterior density

$$\tilde{\pi}(\theta|x_1, \ldots, x_{n_0}) = \frac{\prod_{i=1}^{n_0} f(x_i|\theta)\pi(\theta)}{\int \prod_{i=1}^{n_0} f(x_i|\theta)\pi(\theta)\,d\theta}$$

of θ given $x_1, \ldots, x_{n_0}$ is, with probability 1, a proper density satisfying assumptions (B4) and (B5). The posterior density of θ given $X_1, \ldots, X_n$ $(n > n_0)$ when θ has prior density π is then the same as the posterior density of θ given $X_{n_0+1}, \ldots, X_n$ when θ has prior density $\tilde{\pi}$, and the result now follows.

Example 8.5 Location families. The Pitman estimator derived in Theorem 3.1.20 is the Bayes estimator corresponding to the improper prior density $\pi(\theta) \equiv 1$. If $X_1, \ldots, X_n$ are iid with density $f(x_1 - \theta)$ satisfying (B1)-(B3), the posterior density after one observation $X_1 = x_1$ is $f(x_i - \theta)$ and hence a proper density satisfying assumption (B5), provided $E_\theta|X_1| < \infty$ (Problem 8.4). Under these assumptions, the Pitman estimator is therefore asymptotically efficient.[8] An analogous result holds in the scale case (Problem 8.5).

Theorem 7.9 can be generalized further. Rather than requiring the posterior density $\tilde{\pi}$ to be proper with finite expectation after a fixed number n_0 of observa-

[8] For a more general treatment of this result, see Stone 1974.

tions, it is enough to assume that it satisfies these conditions for all $n \geq n_0$ when $(X_1, \ldots, X_{n_0}) \in S_{n_0}$, where $P(S_{n_0}) \to 1$ as $n_0 \to \infty$ (Problem 8.7).

Example 8.6 Binomial. Let X_i be independent, taking on the values 1 and 0 with probability p and $q = 1 - p$, respectively, and let $\pi(p) = 1/pq$. Then, the posterior distribution of p will be proper (and will then automatically have finite expectation) as soon as $0 < \Sigma X_i < n$, but not before. Since for any $0 < p < 1$ the probability of this event tends to 1 as $n \to \infty$, the asymptotic efficiency of the Bayes estimator follows. ||

Theorem 8.3 provides additional support for the suggestion, made in Section 4.1, that Bayes estimation constitutes a useful method for generating estimators. However, the theorem is unfortunately of no help in choosing among different Bayes estimators, since all prior distributions satisfying assumptions (B4) and (B5) lead to the same asymptotic behavior. In fact, if $\tilde{\theta}_n$ and $\tilde{\theta}'_n$ are Bayes estimators corresponding to two different prior distributions Λ and Λ' satisfying (B4) and (B5), (8.17) implies the even stronger statement,

$$(8.19) \qquad \sqrt{n}(\tilde{\theta}'_n - \tilde{\theta}'_n) \xrightarrow{P} 0.$$

Nevertheless, the interpretation of θ as a random variable with density $\pi(\theta)$ leads to some suggestions concerning the choice of π. Theorem 8.2 showed that the posterior distribution of θ, given the observations, eventually becomes a normal distribution which is concentrated near the true θ_0 and which is independent of π. It is intuitively plausible that a close approximation to the asymptotic result will tend to be achieved more quickly (i.e., for smaller n) if π assigns a relatively high probability to the neighborhood of θ_0 than if this probability is very small. A minimax approach thus leads to the suggestion of a uniform assignment of prior density. It is clear what this means for a location parameter but not in general, since the parameterization is arbitrary and reparametrization destroys uniformity. In addition, it seems plausible that account should also be taken of the relative informativeness of the observations corresponding to different parameter values.

As discussed in Section 4.1, proposals for prior distributions satisfying such criteria have been made (from a somewhat different point of view) by Jeffreys and others. For details, further suggestions, and references, see Box and Tiao 1973, Jaynes 1979, Berger and Bernardo 1989, 1992a, 1992b, and Robert 1994a, Section 3.4.

When the likelihood equation has a unique root $\tilde{\theta}_n$ (which with probability tending to 1 is then the MLE), this estimator has a great practical advantage over the Bayes estimators which share its asymptotic properties. It provides a unique estimating procedure, applicable to a large class of problems, which is supported (partly because of its intuitive plausibility and partly for historical reasons) by a substantial proportion of the statistical profession. This advantage is less clear in the case of multiple roots where asymptotically efficient likelihood estimators such as the one-step estimator (4.11) depend on a somewhat arbitrary initial estimator and need no longer agree with the MLE even for large n.

In the multiparameter case, calculation of Bayes estimators often require the computationally inconvenient evaluation of multiple integrals. However, this difficulty can often be overcome through Gibbs sampling or other Monte Carlo-Markov chain algorithms; see Section 4.5.

To resolve the problem raised by the profusion of asymptotically efficient estimators, it seems natural to carry the analysis one step further and to take into account terms (for example, in an asymptotic expansion of the distribution of the estimator) of order $1/n$ or $1/n^{3/2}$. Investigations along these lines have been undertaken by Rao (1961), Peers (1965), Ghosh and Subramanyam (1974), Efron (1975, 1982a), Pfanzagl and Wefelmeyer (1978-1979), Tibshirani (1989), Ghosh and Mukerjee (1991, 1992, 1993), Barndorff-Nielsen and Cox (1994), and Datta and Ghosh (1995) (see also Section 6.4). They are complicated by the fact that to this order, the estimators tend to be biased and their efficiencies can be improved by removing these biases. For an interesting discussion of these issues, see Berkson (1980). The subject still requires further study.

We conclude this section by proving that the quantities J_1 [defined by (8.13)] and J_1' tend to zero in probability. For this purpose, it is useful to obtain the following alternative expression for $\omega(t)$.

Lemma 8.7 *The quantity $\omega(t)$, defined by (8.11), is equal to*

$$\omega(t) = -I(\theta_0)\frac{t^2}{2n} - \frac{1}{2n}R_n\left(T_n + \frac{t}{\sqrt{n}}\right)\left[t + \frac{1}{I(\theta_0)\sqrt{n}}l'(\theta_0)\right]^2 \tag{8.20}$$

where R_n is the function defined in (8.1) (Problem 8.9).

Proof for J_1. To prove that the integral (8.13) tends to zero in probability, divide the range of integration into the three parts: (i) $|t| \le M$, (ii) $|t| \ge \delta\sqrt{n}$, and (iii) $M < |t| < \delta\sqrt{n}$, and show that the integral over each of the three tends to zero in probability.

(i) $|t| \le M$. To prove this result, we shall show that for every $0 < M < \infty$,

$$\sup\left|e^{\omega(t)}\pi\left(T_n + \frac{t}{\sqrt{n}}\right) - e^{-I(\theta_0)t^2/2}\pi(\theta_0)\right| \xrightarrow{P} 0, \tag{8.21}$$

where here and throughout the proof of (i), the sup is taken over $|t| \le M$. The result will follow from (8.21) since the range of integration is bounded. Substituting the expression (8.20) for $\omega(t)$, (8.21) is seen to follow from the following two facts (Problem 8.10):

$$\sup\left\{\left|\frac{1}{n}R_n\left(T_n + \frac{t}{\sqrt{n}}\right)\right|\left[t + \frac{1}{I(\theta_0)\sqrt{n}}l'(\theta_0)\right]^2\right\} \xrightarrow{P} 0 \tag{8.22}$$

and

$$\sup\left|\pi\left(T_n + \frac{t}{\sqrt{n}}\right) - \pi(\theta_0)\right| \xrightarrow{P} 0. \tag{8.23}$$

The second of these is obvious from the continuity of π and the fact that (Problem 8.11)

$$T_n \xrightarrow{P} \theta_0. \tag{8.24}$$

To prove (8.22), it is enough to show that

$$\sup\left|\frac{1}{n}R_n\left(T_n+\frac{t}{\sqrt{n}}\right)\right| \xrightarrow{P} 0 \tag{8.25}$$

and

$$\frac{1}{I(\theta_0)}\frac{1}{\sqrt{n}}l'(\theta_0) \quad \text{is bounded in probability.} \tag{8.26}$$

Of these, (8.26) is clear from (B1) and the central limit theorem. To see (8.25), note that $|t| \le M$ implies

$$T_n - \frac{M}{\sqrt{n}} \le T_n + \frac{t}{\sqrt{n}} \le T_n + \frac{M}{\sqrt{n}}$$

and hence, by (8.24), that for any $\delta > 0$, the probability of

$$\theta_0 - \delta \le T_n + \frac{t}{\sqrt{n}} \le \theta_0 + \delta$$

will be arbitrarily close to 1 for sufficiently large n. The result now follows from (B2).

(ii) $M \le |t| \le \delta\sqrt{n}$. For this part it is enough to prove that for $|t| \le \delta\sqrt{n}$, the integrand of J_1 is bounded by an integrable function with probability $\ge 1-\varepsilon$. Then, the integral can be made arbitrarily small by choosing a sufficiently large M. Since the second term of the integrand of (8.13) is integrable, it is enough to show that such an integrable bound exists for the first term. More precisely, we shall show that given $\varepsilon > 0$, there exists $\delta > 0$ and $C < \infty$ such that for sufficiently large n,

$$P\left[e^{\omega(t)}\pi\left(T_n+\frac{t}{\sqrt{n}}\right) \le Ce^{-t^2 I(\theta_0)/4} \quad \text{for all } |t| \le \delta\sqrt{n}\right] \ge 1-\varepsilon. \tag{8.27}$$

The factor $\pi(T_n + t/\sqrt{n})$ causes no difficulty by (8.24) and the continuity of π, so that it remains to establish such a bound for

$$\exp\omega(t) \le \exp\left\{-\frac{t^2}{2}I(\theta_0) + \frac{1}{n}\left|R_n\left(T_n+\frac{t}{\sqrt{n}}\right)\right|\left[t^2 + \frac{(l'(\theta_0))^2}{nI^2(\theta_0)}\right]\right\}. \tag{8.28}$$

For this purpose, note that

$$|t| \le \delta'\sqrt{n} \quad \text{implies} \quad T_n - \delta' \le T_n + \frac{t}{\sqrt{n}} + \delta'$$

and hence, by (8.24), that with probability arbitrarily close to 1, for n sufficiently large,

$$|t| \le \delta'\sqrt{n} \quad \text{implies} \quad \left|T_n + \frac{1}{2} - \theta_0\right| \le 2\delta'.$$

By (B2), there exists δ' such that the latter inequality implies

$$P\left\{\sup_{|t|\le\delta'\sqrt{n}}\left|\frac{1}{n}R_n\left(T_n+\frac{t}{\sqrt{n}}\right)\right| \le \frac{1}{4}I(\theta_0)\right\} \ge 1-\varepsilon.$$

Combining this fact with (8.26), we see that the right side of (8.28) is $\leq C'e^{-t^2 I(\theta_0)/4}$ for all t satisfying (ii), with probability arbitrarily close to 1, and this establishes (8.27).

(iii) $|t| \geq \delta\sqrt{n}$. As in (ii), the second term in the integrand of (8.13) can be neglected, and it is enough to show that for all δ,

$$\begin{aligned}
&\int_{|t|\geq\delta\sqrt{n}} \exp[\omega(t)]\pi\left(T_n + \frac{t}{\sqrt{n}}\right) dt \\
(8.29)\qquad &= \sqrt{n}\int_{|\theta - T_n|\geq\delta} \pi(\theta)\exp\left\{l(\theta) - l(\theta_0)\right. \\
&\qquad \left. -\frac{1}{2nI(\theta_0)}[l'(\theta_0)]^2\right\} d\theta \xrightarrow{P} 0.
\end{aligned}$$

From (8.24) and (B3), it is seen that given δ, there exists ε such that

$$\sup_{|\theta - T_n|\geq\delta} e^{[l(\theta)-l(\theta_0)]} \leq e^{-n\varepsilon}$$

with probability tending to 1. By (8.26), the right side of (8.29) is therefore bounded above by

$$(8.30)\qquad C\sqrt{n}\,e^{-n\varepsilon}\int \pi(\theta)\,d\theta = C\sqrt{n}\,e^{-n\varepsilon}$$

with probability tending to 1, and this completes the proof of (iii).

To prove (8.13), let us now combine (i)-(iii). Given $\varepsilon > 0$ and $\delta > 0$, choose M so large that

$$(8.31)\qquad \int_M^\infty \left[C\exp\left[-\frac{t^2}{2}I(\theta_0)\right] + \exp\left[-\frac{t^2}{2}I(\theta_0)\right]\pi(\theta_0)\right] dt \leq \frac{\varepsilon}{3},$$

and, hence, that for sufficiently large n, the integral (8.13) over (ii) is $\leq \varepsilon/3$ with probability $\geq 1-\varepsilon$. Next, choose n so large that the integrals (8.13) over (i) and over (iii) are also $\leq \varepsilon/3$ with probability $\geq 1-\varepsilon$. Then, $P[J_1 \leq \varepsilon] \geq 1 - 3\varepsilon$, and this completes the proof of (8.13).

The proof for J_1' requires only trivial changes. In part (i), the factor $[1 + |t|]$ is bounded, so that the proof continues to apply. In part (ii), multiplication of the integrand of (8.31) by $[1 + |t|]$ does not affect its integrability, and the proof goes through as before. Finally, in part (iii), the integral in (8.30) must be replaced by $Cne^{-n\varepsilon}\int|\theta|\pi(\theta)\,d\theta$, which is finite by (B5).

9 Problems

Section 1

1.1 Let $X_1, \ldots, X_n$ be iid with $E(X_i) = \xi$.

(a) If the X_is have a finite fourth moment, establish (1.3)

(b) For k a positive integer, show that $E(\bar{X} - \xi)^{2k-1}$ and $E(\bar{X} - \xi)^{2k}$, if they exist, are both $O(1/n^k)$.

[*Hint*: Without loss of generality, let $\xi = 0$ and note that $E(X_{i_1}^{r_1} X_{i_2}^{r_2}\cdots) = 0$ if any of the r's is equal to 1.]

1.2 For fixed n, describe the relative error in Example 1.3 as a function of p.

1.3 Prove Theorem 1.5.

1.4 Let $X_1, \ldots, X_n$ be iid as $N(\xi, \sigma^2)$, σ^2 known, and let $g(\xi) = \xi^r$, $r = 2, 3, 4$. Determine, up to terms of order $1/n$,

(a) the variance of the UMVU estimator of $g(\xi)$;

(b) the bias of the MLE of $g(\xi)$.

1.5 Let $X_1, \ldots, X_n$ be iid as $N(\xi, \sigma^2)$, ξ known. For even r, determine the variance of the UMVU estimator (2.2.4) of σ^r up to terms of order r.

1.6 Solve the preceding problem for the case that ξ is unknown.

1.7 For estimating p^m in Example 3.3.1, determine, up to order $1/n$,

(a) the variance of the UMVU estimator (2.3.2);

(b) the bias of the MLE.

1.8 Solve the preceding problem if p^m is replaced by the estimand of Problem 2.3.3.

1.9 Let $X_1, \ldots, X_n$ be iid as Poisson $P(\theta)$.

(a) Determine the UMVU estimator of $P(X_i = 0) = e^{-\theta}$.

(b) Calculate the variance of the estimator of (a) up to terms of order $1/n$.

[*Hint*: Write the estimator in the form (1.15) where $h(\bar{X})$ is the MLE of $e^{-\theta}$.]

1.10 Solve part (b) of the preceding problem for the estimator (2.3.22).

1.11 Under the assumptions of Problem 1.1, show that $E|\bar{X} - \xi|^{2k-1} = O(n^{-k+1/2})$. [*Hint*: Use the fact that $E|\bar{X} - \xi|^{2k-1} \le [E(\bar{X} - \xi)^{4k-2}]^{1/2}$ together with the result of Problem 1.1.]

1.12 Obtain a variant of Theorem 1.1, which requires existence and boundedness of only h''' instead of $h^{(iv)}$, but where R_n is only $O(n^{-3/2})$.

[*Hint*: Carry the expansion (1.6) only to the second instead of the third derivative, and apply Problem 1.11.]

1.13 To see that Theorem 1.1 is not necessarily valid without boundedness of the fourth (or some higher) derivative, suppose that the X's are distributed as $N(\xi, \sigma^2)$ and let $h(X) = e^{x^4}$. Then, all moments of the X's and all derivatives of h exist.

(a Show that the expectation of $h(\bar{X})$ does not exist for any n, and hence that $E\{\sqrt{n}[h(\bar{X}) - h(\xi)]\}^2 = \infty$ for all values of n.

(b On the other hand, show that $\sqrt{n}\,[h(\bar{X}) - h(\xi)]$ has a normal limit distribution with finite variance, and determine that variance.

1.14 Let $X_1, \ldots, X_n$ be iid from the exponential distribution with density $(1/\theta)e^{-x/\theta}$, $x > 0$, and $\theta > 0$.

(a) Use Theorem 1.1 to find approximations to $E(\sqrt{\bar{X}})$ and $\text{var}(\sqrt{\bar{X}})$.

(b) Verify the exact calculation

$$\text{var}(\sqrt{\bar{X}}) = \left[1 - \frac{1}{n}\left(\frac{\Gamma(n + 1/2)}{\Gamma(n)}\right)^2\right]\theta$$

and show that $\lim_{n\to\infty} n\,\text{var}(\sqrt{\bar{X}}) = \theta/4$.

(c) Reconcile the results in parts (a) and (b). Explain why, even though Theorem 1.1 did not apply, it gave the correct answer.

(d) Show that a similar conclusion holds for $h(x) = 1/x$.

[*Hint*: For part (b), use the fact that $T = \Sigma X_i$ has a gamma distribution. The limit can be evaluated with Stirling's formula. It can also be evaluated with a computer algebra program.]

1.15 Let $X_1, \ldots, X_n$ be iid according to $U(0, \theta)$. Determine the variance of the UMVU estimator of θ^k, where k is an integer, $k > -n$.

1.16 Under the assumptions of Problem 1.15, find the MLE of θ^k and compare its expected squared error with the variance of the UMVU estimator.

1.17 Let $X_1, \ldots, X_n$ be iid according to $U(0, \theta)$, let $T = \max(X_1, \ldots, X_n)$, and let h be a function satisfying the conditions of Theorem 1.1. Show that

$$E[h(T)] = h(\theta) - \frac{\theta}{n}h'(\theta) + \frac{1}{n^2}[\theta h'(\theta) + \theta^2 h''(\theta)] + O\left(\frac{1}{n^3}\right)$$

and

$$\text{var}[h(T)] = \frac{\theta^2}{n^2}[h'(\theta)]^2 + O\left(\frac{1}{n^3}\right).$$

1.18 Apply the results of Problem 1.17 to obtain approximate answers to Problems 1.15 and 1.16, and compare the answers with the exact solutions.

1.19 If the X's are as in Theorem 1.1 and if the first five derivatives of h exist and the fifth derivative is bounded, show that

$$E[h(\bar{X})] = h(\xi) + \frac{1}{2}h''\frac{\sigma^2}{n} + \frac{1}{24n^2}[4h'''\mu_3 + 3h^{(iv)}\sigma^4] + O(n^{-5/2})$$

and if the fifth derivative of h^2 is also bounded

$$\text{var}[h(\bar{X})] = (h'^2)\frac{\sigma^2}{n} + \frac{1}{n^2}[h'h''\mu_3 + (h'h''' + \frac{1}{2}h''^2)\sigma^4] + O(n^{-5/2})$$

where $\mu_3 = E(X - \xi)^3$.

[*Hint*: Use the facts that $E(\bar{X} - \xi)^3 = \mu_3/n^2$ and $E(\bar{X} - \xi)^4 = 3\sigma^4/n^2 + O(1/n^3)$.]

1.20 Under the assumptions of the preceding problem, carry the calculation of the variance (1.16) to terms of order $1/n^2$, and compare the result with that of the preceding problem.

1.21 Carry the calculation of Problem 1.4 to terms of order $1/n^2$.

1.22 For the estimands of Problem 1.4, calculate the expected squared error of the MLE to terms of order $1/n^2$, and compare it with the variance calculated in Problem 1.21.

1.23 Calculate the variance (1.18) to terms of order $1/n^2$ and compare it with the expected squared error of the MLE carried to the same order.

1.24 Find the variance of the estimator (2.3.17) up to terms of the order $1/n^3$.

1.25 For the situation of Example 1.12, show that the UMVU estimator δ_{1n} is the bias-corrected MLE, where the MLE is δ_{3n}.

1.26 For the estimators of Example 1.13:

(a) Calculate their exact variances.

(b) Use the result of part (a) to verify (1.27).

1.27 (a) Under the assumptions of Theorem 1.5, if all fourth moments of the $X_{i\nu}$ are finite, show that $E(\bar{X}_i - \xi_i)(\bar{X}_j - \xi_j) = \sigma_{ij}/n$ and that all third and fourth moments $E(\bar{X}_i - \xi_i)(\bar{X}_j - \xi_j)(\bar{X}_k - \xi_k)$, and so on are of the order $1/n^2$.

(b) If, in addition, all derivatives of h of total order ≤ 4 exist and those of order 4 are uniformly bounded, then

$$E[h(\bar{X}_1, \ldots, \bar{X}_s)] = h(\xi_1, \ldots, \xi_s) + \frac{1}{2n} \sum_{i=1}^{s} \sum_{j=1}^{s} \sigma_{ij} \frac{\partial^2 h(\xi_1, \ldots, \xi_s)}{\partial \xi_i \partial \xi_j} + R_n,$$

and if the derivatives of h^2 of order 4 are also bounded,

$$\text{var}[h(\bar{X}_1, \ldots, \bar{X}_s)] = \frac{1}{n} \Sigma \Sigma \sigma_{ij} \frac{\partial h}{\partial \xi_i} \frac{\partial h}{\partial \xi_j} + R_n$$

where the remainder R_n in both cases is $O(1/n^2)$.

1.28 On the basis of a sample from $N(\xi, \sigma^2)$, let $P_n(\xi, \sigma)$ be the probability that the UMVU estimator $\bar{X}^2 - \sigma^2/n$ of ξ^2 (σ known) is negative.

(a) Show that $P_n(\xi, \sigma)$ is a decreasing function of $\sqrt{n}\,|\xi|/\sigma$.

(b) Show that $P_n(\xi, \sigma) \to 0$ as $n \to \infty$ for any fixed $\xi \neq 0$ and σ.

(c) Determine the value of $P_n(0, \sigma)$.

[*Hint*: $P_n(\xi, \sigma) = P[-1 - \sqrt{n}\xi/\sigma < Y < 1 - \sqrt{n}\xi/\sigma]$, where $Y = \sqrt{n}(\bar{X} - \xi)/\sigma$ is distributed as $N(0, 1)$.]

1.29 Use the t-distribution to find the value of $P_n(0, \sigma)$ in the preceding problem for the UMVU estimator of ξ^2 when σ is unknown for representative values of n.

1.30 Fill in the details of the proof of Theorem 1.9. (See also Problem 1.8.8.)

1.31 In Example 8.13 with $\theta = 0$, show that δ_{2n} is not exactly distributed as $\sigma^2(\chi_1^2 - 1)/n$.

1.32 In Example 8.13, let $\delta_{4n} = \max(0, \bar{X}^2 - \sigma^2/n)$, which is an improvement over δ_{1n}.

(a) Show that $\sqrt{n}(\delta_{4n} - \theta^2)$ has the same limit distribution as $\sqrt{n}(\delta_{1n} - \theta^2)$ when $\theta \neq 0$.

(b) Describe the limit distribution of $n\delta_{4n}$ when $\theta = 0$.

[*Hint*: Write $\delta_{4n} = \delta_{1n} + R_n$ and study the behavior of R_n.]

1.33 Let X have the binomial distribution $b(p, n)$, and let $g(p) = pq$. The UMVU estimator of $g(p)$ is $\delta = X(n - X)/n(n - 1)$. Determine the limit distribution of $\sqrt{n}(\delta - pq)$ and $n(\delta - pq)$ when $g'(p) \neq 0$ and $g'(p) = 0$, respectively.
[*Hint*: Consider first the limit behavior of $\delta' = X(n - X)/n^2$.]

1.34 Let $X_1, \ldots, X_n$ be iid as $N(\xi, 1)$. Determine the limit behavior of the distribution of the UMVU estimator of $p = P[|X_i| \leq u]$.

1.35 Determine the limit behavior of the estimator (2.3.22) as $n \to \infty$.
[*Hint*: Consider first the distribution of $\log \delta(T)$.]

1.36 Let $X_1, \ldots, X_n$ be iid with distribution P_θ, and suppose δ_n is UMVU for estimating $g(\theta)$ on the basis of $X_1, \ldots, X_n$. If there exists n_0 and an unbiased estimator $\delta_0(X_1, \ldots, X_{n_0})$ which has finite variance for all θ, then δ_n is consistent for $g(\theta)$.
[*Hint*: For $n = kn_0$ (with k an integer), compare δ_n with the estimator

$$\frac{1}{k}\{\delta_0(X_1, \ldots, X_{n_0}) + \delta_0(X_{n_0+1}, \ldots, X_{2n_0}) + \ldots\}].$$

1.37 Let Y_n be distributed as $N(0, 1)$ with probability π_n and as $N(0, \tau_n^2)$ with probability $1 - \pi_n$. If $\tau_n \to \infty$ and $\pi_n \to \pi$, determine for what values of π the sequence $\{Y_n\}$ does and does not have a limit distribution.

1.38 (a) In Problem 1.37, determine to what values $\text{var}(Y_n)$ can tend as $n \to \infty$ if $\pi_n \to 1$ and $\tau_n \to \infty$ but otherwise both are arbitrary.

(b) Use (a) to show that the limit of the variance need not agree with the variance of the limit distribution.

1.39 Let $b_{m,n}$, $m, n = 1, 2, \ldots$, be a double sequence of real numbers, which for each fixed m is nondecreasing in n. Show that $\lim_{n\to\infty} \lim_{m\to\infty} b_{m,n} = \lim_{m,n\to\infty} \inf b_{m,n}$ and $\lim_{m\to\infty} \lim_{n\to\infty} b_{m,n} = \lim_{m,n\to\infty} \sup b_{m,n}$ provided the indicated limits exist (they may be infinite) and where $\lim\inf b_{m,n}$ and $\lim\sup b_{m,n}$ denote, respectively, the smallest and the largest limit points attainable by a sequence b_{m_k,n_k}, $k = 1, 2, \ldots$, with $m_k \to \infty$ and $n_k \to \infty$.

Section 2

2.1 Let $X_1, \ldots, X_n$ be iid as $N(0, 1)$. Consider the two estimators

$$T_n = \begin{cases} \bar{X}_n & \text{if } S_n \le a_n \\ n & \text{if } S_n > a_n, \end{cases}$$

where $S_n = \Sigma(X_i - \bar{X})^2$, $P(S_n > a_n) = 1/n$, and $T_n' = (X_1 + \cdots + X_{k_n})/k_n$ with k_n the largest integer $\le \sqrt{n}$.

(a) Show that the asymptotic efficiency of T_n' relative to T_n is zero.

(b) Show that for any *fixed* $\varepsilon > 0$, $P[|T_n - \theta| > \varepsilon] = \frac{1}{n} + o\left(\frac{1}{n}\right)$, but $P[|T_n' - \theta| > \varepsilon] = o\left(\frac{1}{n}\right)$.

(c) For large values of n, what can you say about the two probabilities in part (b) when ε is replaced by $a/\sqrt{n}$? (Basu 1956).

2.2 If $k_n[\delta_n - g(\theta)] \xrightarrow{\mathcal{L}} H$ for some sequence k_n, show that the same result holds if k_n is replaced by k_n', where $k_n/k_n' \to 1$.

2.3 Assume that the distribution of $Y_n = \sqrt{n}(\delta_n - g(\theta))$ converges to a distribution with mean 0 and variance $v(\theta)$. Use Fatou's lemma (Lemma 1.2.6) to establish that $\text{var}_\theta(\delta_n) \to 0$ for all θ.

2.4 If $X_1, \ldots, X_n$ are a sample from a one-parameter exponential family (1.5.2), then $\Sigma T(X_i)$ is minimal sufficient and $E[(1/n)\Sigma T(X_i)] = (\partial/\partial\eta)A(\eta) = \tau$. Show that for any function $g(\cdot)$ for which Theorem 1.8.12 holds, $g((1/n)\Sigma T(X_i))$ is asymptotically unbiased for $g(\tau)$.

2.5 If $X_1, \ldots, X_n$ are iid $n(\mu, \sigma^2)$, show that $S^r = [1/(n-1)\Sigma(x_i - \bar{x})^2]^{r/2}$ is an asymptotically unbiased estimator of σ^r.

2.6 Let $X_1, \ldots, X_n$ be iid as $U(0, \theta)$. From Example 2.1.14, $\delta_n = (n+1)X_{(n)}/n$ is the UMVU estimator of θ, whereas the MLE is $X_{(n)}$. Determine the limit distribution of (a) $n[\theta - \delta_n]$ and (b) $n[\theta - X_{(n)}]$. Comment on the asymptotic bias of these estimators. [*Hint*: $P(X_{(n)} \le y) = y^n/\theta^n$ for any $0 < y < \theta$.]

2.7 For the situation of Problem 2.6:

(a) Calculate the mean squared errors of both δ_n and $X_{(n)}$ as estimators of θ.

(b) Show

$$\lim_{n\to\infty} \frac{E(X_{(n)} - \theta)^2}{E(\delta_n - \theta)^2} = 2.$$

2.8 Verify the asymptotic distribution claimed for δ_n in Example 2.5.

2.9 Let δ_n be any estimator satisfying (2.2) with $g(\theta) = \theta$. Construct a sequence δ_n' such that $\sqrt{n}(\delta_n' - \theta) \xrightarrow{\mathcal{L}} N[0, w^2(\theta)]$ with $w(\theta) = v(\theta)$ for $\theta \neq \theta_0$ and $w(\theta_0) = 0$.

2.10 In the preceding problem, construct δ'_n such that $w(\theta) = v(\theta)$ for all $\theta \neq \theta_0$ and θ_1 and $< v(\theta)$ for $\theta = \theta_0$ and θ_1.

2.11 Construct a sequence $\{\delta_n\}$ satisfying (2.2) but for which the bias $b_n(\theta)$ does not tend to zero.

2.12 In Example 2.7 with $R_n(\theta)$ given by (2.11), show that $R_n(\theta) \to 1$ for $\theta \neq 0$ and that $R_n(0) \to a^2$.

2.13 Let $b_n(\theta) = E_\theta(\delta_n) - \theta$ be the bias of the estimator δ_n of Example 2.5.

(a) Show that

$$b_n(\theta) = \frac{-(1-a)}{\sqrt{n}} \int_{-\sqrt[4]{n}}^{\sqrt[4]{n}} x\phi(x - \sqrt{n}\theta)\, dx;$$

(b) Show that $b'_n(\theta) \to 0$ for any $\theta \neq 0$ and $b'_n(0) \to (1-a)$.

(c) Use (b) to explain how the Hodges estimator δ_n can violate (2.7) without violating the information inequality.

2.14 In Example 2.7, show that if $\theta_n = c/\sqrt{n}$, then $R_n(\theta_n) \to a^2 + c^2(1-a)^2$.

Section 3

3.1 Let X have the binomial distribution $b(p, n)$, $0 \le p \le 1$. Determine the MLE of p

(a) by the usual calculus method determining the maximum of a function;

(b) by showing that $p^x q^{n-x} \le (x/n)^x [(n-x)/n]^{n-x}$.

[*Hint*: (b) Apply the fact that the geometric mean is equal to or less than the arithmetic mean to n numbers of which x are equal to np/x and $n-x$ equal to $nq/(n-x)$.]

3.2 In the preceding problem, show that the MLE does not exist when p is restricted to $0 < p < 1$ and when $x = 0$ or $= n$.

3.3 Let $X_1, \ldots, X_n$ be iid according to $N(\xi, \sigma^2)$. Determine the MLE of (a) ξ when σ is known, (b) σ when ξ is known, and (c) (ξ, σ) when both are unknown.

3.4 Suppose $X_1, \ldots, X_n$ are iid as $N(\xi, 1)$ with $\xi > 0$. Show that the MLE is $\bar{X}$ when $\bar{X} > 0$ and does not exist when $\bar{X} \le 0$.

3.5 Let X take on the values 0 and 1 with probabilities p and q, respectively. When it is known that $1/3 \le p \le 2/3$, (a) find the MLE and (b) show that the expected squared error of the MLE is uniformly larger than that of $\delta(x) = 1/2$.

[A similar estimation problem arises in ***randomized response*** surveys. See Example 5.2.2.]

3.6 When Ω is finite, show that the MLE is consistent if and only if it satisfies (3.2).

3.7 Show that Theorem 3.2 remains valid if assumption A1 is relaxed to A1′: There is a nonempty set $\Omega_0 \in \Omega$ such that $\theta_0 \in \Omega_0$ and Ω_0 is contained in the support of each P_θ.

3.8 Prove the existence of unique $0 < a_k < a_{k-1}$, $k = 1, 2, \ldots$, satisfying (3.4).

3.9 Prove (3.9).

3.10 In Example 3.6 with $0 < c < 1/2$, determine a consistent estimator of k.

[*Hint*: (a) The smallest value K of j for which I_j contains at least as many of the X's as any other I is consistent. (b) The value of j for which I_j contains the median of the X's is consistent since the median of f_k is in I_k.]

3.11 Verify the nature of the roots in Example 3.9.

3.12 Let X be distributed as $N(\theta, 1)$. Show that conditionally given $a < X < b$, the variable X tends in probability to b as $\theta \to \infty$.

3.13 Consider a sample $X_1, \ldots, X_n$ from a Poisson distribution conditioned to be positive, so that $P(X_i = x) = \theta^x e^{-\theta}/x!(1 - e^{-\theta})$ for $x = 1, 2, \ldots$. Show that the likelihood equation has a unique root for all values of x.

3.14 Let X have the negative binomial distribution (2.3.3). Find an ELE of p.

3.15 (a) A density function is *strongly unimodal*, or equivalently *log concave*, if $\log f(x)$ is a concave function. Show that such a density function has a unique mode.

(b) Let $X_1, \ldots, X_n$ be iid with density $f(x - \theta)$. Show that the likelihood function has a unique root if $f'(x)/f(x)$ is monotone, and the root is a maximum if $f'(x)/f(x)$ is decreasing. Hence, densities that are log concave yield unique MLEs.

(c) Let $X_1, \ldots, X_n$ be positive random variables (or symmetrically distributed about zero) with joint density $a^n \Pi f(ax_i)$, $a > 0$. Show that the likelihood equation has a unique maximum if $xf'(x)/f(x)$ is strictly decreasing for $x > 0$.

(d) If $X_1, \ldots, X_n$ are iid with density $f(x_i - \theta)$ where f is unimodal and if the likelihood equation has a unique root, show that the likelihood equation also has a unique root when the density of each X_i is $af[a(x_i - \theta)]$, with a known.

3.16 For each of the following densities, $f(\cdot)$, determine if (a) it is strongly unimodal and (b) $xf'(x)/f(x)$ is strictly decreasing for $x > 0$. Hence, comment on whether the respective location and scale parameters have unique MLEs:

(a) $f(x) = \dfrac{1}{\sqrt{2\pi}} e^{-\frac{1}{2}x^2}, \quad -\infty < x < \infty$ (normal)

(b) $f(x) = \dfrac{1}{\sqrt{2\pi}} \dfrac{1}{x} e^{-\frac{1}{2}(\log x)^2}, \quad 0 \le x < \infty$ (lognormal)

(c) $f(x) = e^{-x}/(1 + e^{-x})^2, \quad -\infty < x < \infty$ (logistic)

(d) $f(x) = \dfrac{\Gamma(\nu + 1/2)}{\Gamma(\nu/2)} \dfrac{1}{\sqrt{\nu\pi}} \dfrac{1}{[1 + (x/\nu)^2]^{\frac{\nu+1}{2}}}, \quad -\infty < x < \infty$ (t with ν df)

3.17 If $X_1, \ldots, X_n$ are iid with density $f(x_i - \theta)$ or $af(ax_i)$ and f is the logistic density $L(0, 1)$, the likelihood equation has unique solutions $\hat{\theta}$ and $\hat{a}$ both in the location and the scale case. Determine the limit distribution of $\sqrt{n}(\hat{\theta} - \theta)$ and $\sqrt{n}(\hat{a} - a)$.

3.18 In Problem 3.15(b), with f the Cauchy density $C(0, a)$, the likelihood equation has a unique root $\hat{a}$ and $\sqrt{n}(\hat{a} - a) \xrightarrow{\mathcal{L}} N(0, 2a^2)$.

3.19 If $X_1, \ldots, X_n$ are iid as $C(\theta, 1)$, then for any fixed n there is positive probability (a) that the likelihood equation has $2n - 1$ roots and (b) that the likelihood equation has a unique root.

[*Hint*: (a) If the x's are sufficiently widely separated, the value of $L'(\theta)$ in the neighborhood of x_i is dominated by the term $(x_i - \theta)/[1 + (x_i - \theta)^2]$. As θ passes through x_i, this term changes signs so that the log likelihood has a local maximum near x_i. (b) Let the x's be close together.]

3.20 If $X_1, \ldots, X_n$ are iid according to the gamma distribution $\Gamma(\theta, 1)$, the likelihood equation has a unique root.

[*Hint*: Use Example 3.12. Alternatively, write down the likelihood and use the fact that $\Gamma'(\theta)/\Gamma(\theta)$ is an increasing function of θ.]

3.21 Let $X_1, \ldots, X_n$ be iid according to a Weibull distribution with density

$$f_\theta(x) = \theta x^{\theta-1} e^{-x^\theta}, \quad x > 0, \theta > 0,$$

which is not a member of the exponential, location, or scale family. Nevertheless, show that there is a unique interior maximum of the likelihood function.

3.22 Under the assumptions of Theorem 3.2, show that

$$\left[L\left(\theta_0 + \frac{1}{\sqrt{n}}\right) - L(\theta_0) + \frac{1}{2}I(\theta_0)\right] / \sqrt{I(\theta_0)}$$

tends in law to $N(0, 1)$.

3.23 Let $X_1, \ldots, X_n$ be iid according to $N(\theta, a\theta^2)$, $\theta > 0$, where a is a known positive constant.

(a) Find an explicit expression for an ELE of θ.

(b) Determine whether there exists an MRE estimator under a suitable group of transformations.

[This case was considered by Berk (1972).]

3.24 Check that the assumptions of Theorem 3.10 are satisfied in Example 3.12.

3.25 For $X_1, \ldots, X_n$ iid as $DE(\theta, 1)$, show that (a) the sample median is an MLE of θ and (b) the sample median is asymptotically normal with variance $1/n$, the information inequality bound.

3.26 In Example 3.12, show directly that $(1/n)\Sigma T(X_i)$ is an asymptotically efficient estimator of $\theta = E_\eta[T(X)]$ by considering its limit distribution.

3.27 Let $X_1, \ldots, X_n$ be iid according to $\theta g(x) + (1 - \theta)h(x)$, where (g, h) is a pair of specified probability densities with respect to μ, and where $0 < \theta < 1$.

(a) Give one example of (g, h) for which the assumptions of Theorem 3.10 are satisfied and one for which they are not.

(b) Discuss the existence and nature of the roots of the likelihood equation for $n = 1$, 2, 3.

3.28 Under the assumptions of Theorem 3.7, suppose that $\hat\theta_{1n}$ and $\hat\theta_{2n}$ are two consistent sequences of roots of the likelihood equation. Prove that $P_{\theta_0}(\hat\theta_{1n} = \hat\theta_{2n}) \to 1$ as $n \to \infty$.

[*Hint*:

(a) Let $S_n = \{\mathbf{x} : \mathbf{x} = (x_1, \ldots, x_n)$ such that $\hat\theta_{1n}(\mathbf{x}) \neq \hat\theta_{2n}(\mathbf{x})\}$. For all $\mathbf{x} \in S_n$, there exists θ_n^* between $\hat\theta_{1n}$ and $\hat\theta_{2n}$ such that $L''(\theta_n^*) = 0$. For all $\mathbf{x} \notin S_n$, let θ_n^* be the common value of $\hat\theta_{1n}$ and $\hat\theta_{2n}$. Then, θ_n^* is a consistent sequence of roots of the likelihood equation.

(b) $(1/n)L''(\theta_n^*) - (1/n)L''(\theta_0) \to 0$ in probability and therefore $(1/n)L''(\theta_n^*) \to -I(\theta_0)$ in probability.

(c) Let $0 < \varepsilon < I(\theta_0)$ and let

$$S_n' = \left\{\mathbf{x} : \frac{1}{n}L''(\theta_n^*) < -I(\theta_0) + \varepsilon\right\}.$$

Then, $P_{\theta_0}(S_n') \to 1$. On the other hand, $L''(\theta_n^*) = 0$ on S_n so that S_n is contained in the complement of S_n' (Huzurbazar 1948).]

3.29 To establish the measurability of the sequence of roots $\hat{\theta}_n^*$ of Theorem 3.7, we can follow the proof of Serfling (1980, Section 4.2.2) where the measurability of a similar sequence is proved.

(a) For definiteness, define $\hat{\theta}_n(a)$ as the value that minimizes $|\hat{\theta} - \theta_0|$ subject to

$$\theta_0 - a \le \hat{\theta} \le \theta_0 + a \quad \text{and} \quad \frac{\partial}{\partial\theta} l(\theta|\mathbf{x})|_{\theta=\hat{\theta}} = 0.$$

Show that $\tilde{\theta}_n(a)$ is measurable.

(b) Show that θ_n^*, the root closest to θ^*, is measurable.

[*Hint*: For part (a), write the set $\{\hat{\theta}_n(a) > t\}$ as countable unions and intersections of measurable sets, using the fact that $(\partial/\partial\theta)\log(\theta|\mathbf{x})$ is continuous, and hence measurable.]

Section 4

4.1 Let

$$u(t) = \begin{cases} c\int_0^t e^{-1/x(1-x)}dx & \text{for } 0 < t < 1 \\ 0 & \text{for } t \le 0 \\ 1 & \text{for } t \ge 1. \end{cases}$$

Show that for a suitable c, the function u is continuous and infinitely differentiable for $-\infty < t < \infty$.

4.2 Show that the density (4.1) with $\Omega = (0, \infty)$ satisfies all conditions of Theorem 3.10 with the exception of (d) of Theorem 2.6.

4.3 Show that the density (4.4) with $\Omega = (0, \infty)$ satisfies all conditions of Theorem 3.10.

4.4 In Example 4.5, evaluate the estimators (4.8) and (4.14) for the Cauchy case, using for $\tilde{\theta}_n$ the sample median.

4.5 In Example 4.7, show that $l(\theta)$ is concave.

4.6 In Example 4.7, if $\eta = \xi$, show how to obtain a $\sqrt{n}$-consistent estimator by equating sample and population second moments.

4.7 In Theorem 4.8, show that $\sigma_{11} = \sigma_{12}$.

4.8 Without using Theorem 4.8, in Example 4.13 show that the EM sequence converges to the MLE.

4.9 Consider the following 12 observations from a bivariate normal distribution with parameters $\mu_1 = \mu_2 = 0, \sigma_1^2, \sigma_2^2, \rho$:

x_1	1	1	-1	-1	2	2	-2	-2	*	*	*	*
x_2	1	-1	1	-1	*	*	*	*	2	2	-2	-2

where "*" represents a missing value.

(a) Show that the likelihood function has global maxima at $\rho = \pm 1/2, \sigma_1^2 = \sigma_2^2 = 8/3$, and a saddlepoint at $\rho = 0, \sigma_1^2 = \sigma_2^2 = 5/2$.

(b) Show that if an EM sequence starts with $\rho = 0$, then it remains with $\rho = 0$ for all subsequent iterations.

(c) Show that if an EM sequence starts with ρ bounded away from zero, it will converge to a maximum.

[This problem is due to Murray (1977), and is discussed by Wu (1983).]

4.10 Show that if the EM complete-data density $f(\mathbf{y}, \mathbf{z}|\theta)$ of (4.21) is in a curved exponential family, then the hypotheses of Theorem 4.12 are satisfied.

4.11 In the EM algorithm, calculation of the E-step, the expectation calculation, can be complicated. In such cases, it may be possible to replace the E-step by a Monte Carlo evaluation, creating the MCEM algorithm (Wei and Tanner 1990). Consider the following MCEM evaluation of $Q(\theta|\hat{\theta}_{(j)}, \mathbf{y})$:

Given $\hat{\theta}_{(j)}^{(k)}$

(1) Generate $Z_1, \cdots, Z_k$, iid, from $k(\mathbf{z}|\hat{\theta}_{(j)}^{(k)}, \mathbf{y})$,

(2) Let $\hat{Q}(\theta|\hat{\theta}_{(j)}^{(k)}, \mathbf{y}) = \frac{1}{k}\sum_{i=1}^{k} \log L(\theta|\mathbf{y}, \mathbf{z})$

and then calculate $\hat{\theta}_{(j+1)}^{(k)}$ as the value that maximizes $\hat{Q}(\theta|\hat{\theta}_{(j)}^{(k)}, \mathbf{y})$.

(a) Show that $\hat{Q}(\theta|\hat{\theta}_{(j)}^{(k)}, \mathbf{y}) \to \hat{Q}(\theta|\hat{\theta}_{(j)}, \mathbf{y})$ as $k \to \infty$.

(b) What conditions will ensure that $L(\hat{\theta}_{(j+1)}^{(k)}|\mathbf{y}) \geq L(\hat{\theta}_{(j)}^{(k)}|\mathbf{y})$ for sufficiently large k? Are the hypotheses of Theorem 4.12 sufficient?

4.12 For the mixture distribution of Example 4.7, that is,

$$X_i \sim \theta g(x) + (1-\theta)h(x), \quad i = 1, \ldots, n, \text{ independent}$$

where $g(\cdot)$ and $h(\cdot)$ are known, an EM algorithm can be used to find the ML estimator of θ. Let $Z_1, \cdots, Z_n$, where Z_i indicates from which distribution X_i has been drawn, so

$$X_i|Z_i = 1 \sim g(x)$$
$$X_i|Z_i = 0 \sim h(x).$$

(a) Show that the complete-data likelihood can be written

$$L(\theta|\mathbf{x}, \mathbf{z}) = \prod_{i=1}^{n} [z_i g(x_i) + (1-z_i)h(x_i)]\theta^{z_i}(1-\theta)^{1-z_i}.$$

(b) Show that $E(Z_i|\theta, x_i) = \theta g(x_i)/[\theta g(x_i) + (1-\theta)h(x_i)]$ and hence that the EM sequence is given by

$$\hat{\theta}_{(j+1)} = \frac{1}{n}\sum_{i=1}^{n} \frac{\hat{\theta}_{(j)}g(x_i)}{\hat{\theta}_{(j)}g(x_i) + (1-\hat{\theta}_{(j)})h(x_i)}.$$

(c) Show that $\hat{\theta}_{(j)} \to \hat{\theta}$, the ML estimator of θ.

4.13 For the situation of Example 4.10:

(a) Show that the M-step of the EM algorithm is given by

$$\hat{\mu} = \left(\sum_{i=1}^{4}\sum_{j=1}^{n_i} y_{ij} + z_1 + z_2\right)/12,$$

$$\hat{\alpha}_i = \left(\sum_{j=1}^{2} y_{ij} + z_i\right)/3 - \hat{\mu}, \quad i = 1, 3$$
$$= \left(\sum_{j=1}^{3} y_{ij}\right)/3 - \hat{\mu}, \quad i = 2, 4.$$

(b) Show that the E-step of the EM algorithm is given by

$$z_i = E\left[Y_{i3}|\mu = \hat{\mu}, \alpha_i = \hat{\alpha}_i\right] = \hat{\mu} + \hat{\alpha}_i \quad i = 1, 3.$$

(c) Under the restriction $\sum_i \alpha_i = 0$, show that the EM sequence converges to $\hat{\alpha}_i = \bar{y}_{i\cdot} - \hat{\mu}$, where $\hat{\mu} = \sum_i \bar{y}_{i\cdot}/4$.

(d) Under the restriction $\sum_i n_i \alpha_i = 0$, show that the EM sequence converges to $\hat{\alpha}_i = \bar{y}_{i\cdot} - \hat{\mu}$, where $\hat{\mu} = \sum_{ij} y_{ij}/10$.

(e) For a general one-way layout with a treatments and n_{ij} observations per treatment, show how to use the EM algorithm to augment the data so that each treatment has n observation. Write down the EM sequence, and show what it converges to under the restrictions of parts (c) and (d).

[The restrictions of parts (c) and (d) were encountered in Example 3.4.9, where they led, respectively, to an *unweighted* means analysis and a *weighted* means analysis.]

4.14 In the two-way layout (see Example 3.4.11), the EM algorithm can be very helpful in computing ML estimators in the unbalanced case. Suppose that we observe

$$Y_{ijk} : N(\xi_{ij}, \sigma^2), \quad i = 1, \ldots, I, \quad j = 1, \ldots, J, \quad k = 1, \ldots, n_{ij},$$

where $\xi_{ij} = \mu + \alpha_i + \beta_j + \gamma_{ij}$. The data will be augmented so that the complete data have n observations per cell.

(a) Show how to compute both the E-step and the M-step of the EM algorithm.

(b) Under the restriction $\sum_i \alpha_i = \sum_j \beta_j = \sum_i \gamma_{ij} = \sum_j \gamma_{ij} = 0$, show that the EM sequence converges to the ML estimators corresponding to an unweighted means analysis.

(c) Under the restriction $\sum_i n_{i\cdot}\alpha_i = \sum_j n_{\cdot j}\beta_j = \sum_i n_{i\cdot}\gamma_{ij} = \sum_j \cdot j\gamma_{ij} = 0$, show that the EM sequence converges to the ML estimators corresponding to a weighted means analysis.

4.15 For the one-way layout with random effects (Example 3.5.1), the EM algorithm is useful for computing ML estimates. (In fact, it is very useful in many mixed models; see Searle et al. 1992, Chapter 8.) Suppose we have the model

$$X_{ij} = \mu + A_i + U_{ij} \quad (j = 1, \ldots, n_i, \; i = 1, \ldots, s)$$

where A_i and U_{ij} are independent normal random variables with mean zero and known variance. To compute the ML estimates of μ, σ_U^2, and σ_U^2 it is typical to employ an EM algorithm using the unobservable A_i's as the augmented data. Write out both the E-step and the M-step, and show that the EM sequence converges to the ML estimators.

4.16 Maximum likelihood estimation in the *probit model* of Section 3.6 can be implemented using the EM algorithm. We observe independent Bernoulli variables $X_1, \ldots, X_n$, which depend on unobservable variables Z_i distributed independently as $N(\zeta_i, \sigma^2)$, where

$$X_i = \begin{cases} 0 & \text{if } Z_i \le u \\ 1 & \text{if } Z_i > u. \end{cases}$$

Assuming that u is known, we are interested in obtaining ML estimates of ζ and σ^2.

(a) Show that the likelihood function is $p^{\Sigma x_i}(1-p)^{n-\Sigma x_i}$, where

$$p = P(Z_i > u) = \Phi\left(\frac{\zeta - u}{\sigma}\right).$$

(b) If we consider $Z_1, \ldots, Z_n$ to be the complete data, the complete-data likelihood is

$$\prod_{i=1}^{n} \frac{1}{\sqrt{2\pi}\sigma} e^{-\frac{1}{2\sigma^2}(z_i - \zeta)^2}$$

and the expected complete-data log likelihood is

$$-\frac{n}{2}\log(2\pi\sigma^2) - \frac{1}{2\sigma^2}\sum_{i=1}^{n}\left[E(Z_i^2|X_i) - 2\zeta E(Z_i|X_i) + \zeta^2\right].$$

(c) Show that the EM sequence is given by

$$\hat{\zeta}_{(j+1)} = \frac{1}{n}\sum_{i=1}^{n} t_i(\hat{\zeta}_{(j)}, \hat{\sigma}^2_{(j)})$$

$$\hat{\sigma}^2_{(j+1)} = \frac{1}{n}\left[\sum_{i=1}^{n} v_i(\hat{\zeta}_{(j)}, \hat{\sigma}^2_{(j)}) - \frac{1}{n}\left(\sum_{i=1}^{n} t_i(\hat{\zeta}_{(j)}, \hat{\sigma}^2_{(j)})\right)^2\right]$$

where

$$t_i(\zeta, \sigma^2) = E(Z_i|X_i, \zeta, \sigma^2) \quad \text{and} \quad v_i(\zeta, \sigma^2) = E(Z_i^2|X_i, \zeta, \sigma^2).$$

(d) Show that

$$E(Z_i|X_i, \zeta, \sigma^2) = \zeta + \sigma H_i\left(\frac{u-\zeta}{\sigma}\right),$$

$$E(Z_i^2|X_i, \zeta, \sigma^2) = \zeta^2 + \sigma^2 + \sigma(u+\zeta)H_i\left(\frac{u-\zeta}{\sigma}\right)$$

where

$$H_i(t) = \begin{cases} \dfrac{\varphi(t)}{1-\Phi(t)} & \text{if } X_i = 1 \\[2ex] -\dfrac{\varphi(t)}{\Phi(t)} & \text{if } X_i = 0. \end{cases}$$

(e) Show that $\hat{\zeta}_{(j)} \to \hat{\zeta}$ and $\hat{\sigma}^2_{(j)} \to \hat{\sigma}^2$, the ML estimates of ζ and σ^2.

4.17 Verify (4.30).

4.18 The EM algorithm can also be implemented in a Bayesian hierarchical model to find a posterior mode. Recall the model (4.5.5.1),

$$X|\theta \sim f(x|\theta),$$
$$\Theta|\lambda \sim \pi(\theta|\lambda),$$
$$\Lambda \sim \gamma(\lambda),$$

where interest would be in estimating quantities from $\pi(\theta|x)$. Since

$$\pi(\theta|x) = \int \pi(\theta, \lambda|x)\,d\lambda,$$

where $\pi(\theta, \lambda|x) = \pi(\theta|\lambda, x)\pi(\lambda|x)$, the EM algorithm is a candidate method for finding the mode of $\pi(\theta|x)$, where λ would be used as the augmented data.

(a) Define $k(\lambda|\theta, x) = \pi(\theta, \lambda|x)/\pi(\theta|x)$, and show that

$$\log \pi(\theta|x) = \int \log \pi(\theta, \lambda|x) k(\lambda|\theta^*, x)\,d\lambda - \int \log k(\lambda|\theta, x) k(\lambda|\theta^*, x)\,d\lambda.$$

(b) If the sequence $\{\hat{\theta}_{(j)}\}$ satisfies

$$\max_{\theta} \int \log \pi(\theta, \lambda|x) k(\lambda|\theta_{(j)}, x)\,d\lambda = \int \log \pi(\theta_{(j+1)}, \lambda|x) k(\lambda|\theta_{(j)}, x)\,d\lambda,$$

show that $\log \pi(\theta_{(j+1)}|x) \geq \log \pi(\theta_{(j)}|x)$. Under what conditions will the sequence $\{\hat{\theta}_{(j)}\}$ converge to the mode of $\pi(\theta|x)$?

(c) For the hierarchy

$$X|\theta \sim N(\theta), 1),$$
$$\Theta|\lambda \sim N(\lambda, 1)),$$
$$\Lambda \sim \text{Uniform}(-\infty, \infty),$$

show how to use the EM algorithm to calculate the posterior mode of $\pi(\theta|x)$.

4.19 There is a connection between the EM algorithm and Gibbs sampling, in that both have their basis in Markov chain theory. One way of seeing this is to show that the incomplete-data likelihood is a solution to the integral equation of successive substitution sampling (see Problems 4.5.9-4.5.11), and that Gibbs sampling can then be used to calculate the likelihood function. If $L(\theta|\mathbf{y})$ is the incomplete-data likelihood and $L(\theta|\mathbf{y}, \mathbf{z})$ is the complete-data likelihood, define

$$L^*(\theta|\mathbf{y}) = \frac{L(\theta|\mathbf{y})}{\int L(\theta|\mathbf{y})d\theta},$$
$$L^*(\theta|\mathbf{y}, \mathbf{z}) = \frac{L(\theta|\mathbf{y}, \mathbf{z})}{\int L(\theta|\mathbf{y}, \mathbf{z})d\theta}.$$

(a) Show that $L^*(\theta|\mathbf{y})$ is the solution to

$$L^*(\theta|\mathbf{y}) = \int \left[\int L^*(\theta|\mathbf{y}, \mathbf{z})k(\mathbf{z}|\theta', \mathbf{y})d\mathbf{z}\right] L^*(\theta'|\mathbf{y})d\theta'$$

where, as usual, $k(\mathbf{z}|\theta, \mathbf{y}) = L(\theta|\mathbf{y}, \mathbf{z})/L(\theta|\mathbf{y})$.

(b) Show how the sequence $\theta_{(j)}$ from the Gibbs iteration,

$$\theta_{(j)} \sim L^*(\theta|\mathbf{y}, \mathbf{z}_{(j-1)}),$$
$$\mathbf{z}_{(j)} \sim k(\mathbf{z}|\theta_{(j)}, \mathbf{y}),$$

will converge to a random variable with density $L^*(\theta|\mathbf{y})$ as $j \to \infty$. How can this be used to compute the likelihood function $L(\theta|\mathbf{y})$?

[Using the functions $L(\theta|\mathbf{y}, \mathbf{z})$ and $k(\mathbf{z}|\theta, \mathbf{y})$, the EM algorithm will get us the ML estimator from $L(\theta|\mathbf{y})$, whereas the Gibbs sampler will get us the entire function. This likelihood implementation of the Gibbs sampler was used by Casella and Berger (1994) and is also described by Smith and Roberts (1993). A version of the EM algorithm, where the Markov chain connection is quite apparent, was given by Baum and Petrie (1966) and Baum et al. (1970).]

Section 5

5.1 (a) If a vector $\mathbf{Y}_n$ in E_s converges in probability to a constant vector $\mathbf{a}$, and if h is a continuous function defined over E_s, show that $h(\mathbf{Y}_n) \to h(\mathbf{a})$ in probability.

(b) Use (a) to show that the elements of $||A_{jkn}||^{-1}$ tend in probability to the elements of B as claimed in the proof of Lemma 5.2.

[*Hint*: (a) Apply Theorem 1.8.19 and Problem 1.8.13.]

5.2 (a) Show that (5.26) with the remainder term neglected has the same form as (5.15) and identify the A_{jkn}.

(b) Show that the resulting a_{jk} of Lemma 5.2 are the same as those of (5.23).

(c) Show that the remainder term in (5.26) can be neglected in the proof of Theorem 5.3.

5.3 Let $X_1, \ldots, X_n$ be iid according to $N(\xi, \sigma^2)$.

(a) Show that the likelihood equations have a unique root.

(b) Show directly (i.e., without recourse to Theorem 5.1) that the MLEs $\hat{\xi}$ and $\hat{\sigma}$ are asymptotically efficient.

5.4 Let $(X_0, \ldots, X_s)$ have the multinomial distribution $M(p_0, \ldots, p_s; n)$.

(a) Show that the likelihood equations have a unique root.

(b) Show directly that the MLEs $\hat{p}_i$ are asymptotically efficient.

5.5 Prove Corollary 5.4.

5.6 Show that there exists a function f of two variables for which the equations $\partial f(x, y)/\partial x = 0$ and $\partial f(x, y)/\partial y = 0$ have a unique solution, and this solution is a local but not a global maximum of f.

Section 6

6.1 In Example 6.1, show that the likelihood equations are given by (6.2) and (6.3).

6.2 In Example 6.1, verify Equation (6.4).

6.3 Verify (6.5).

6.4 If $\theta = (\theta_1, \ldots, \theta_r, \theta_{r+1}, \ldots, \theta_s)$ and if

$$\text{cov}\left[\frac{\partial}{\partial \theta_i} L(\theta), \frac{\partial}{\partial \theta_j} L(\theta)\right] = 0 \quad \text{for any } i \leq r < j,$$

then the asymptotic distribution of $(\hat{\theta}_1, \ldots, \hat{\theta}_r)$ under the assumptions of Theorem 5.1 is unaffected by whether or not $\theta_{r+1}, \ldots, \theta_s$ are known.

6.5 Let $X_1, \ldots, X_n$ be iid from a $\Gamma(\alpha, \beta)$ distribution with density $1/(\Gamma(\alpha)\beta^\alpha) \times x^{\alpha-1} e^{-x/\beta}$.

(a) Calculate the information matrix for the usual (α, β) parameterization.

(b) Write the density in terms of the parameters $(\alpha, \mu) = (\alpha, \alpha/\beta)$. Calculate the information matrix for the (α, μ) parameterization and show that it is diagonal, and, hence, the parameters are orthogonal.

(c) If the MLE's in part (a) are $(\hat{\alpha}, \hat{\beta})$, show that $(\hat{\alpha}, \hat{\mu}) = (\hat{\alpha}, \hat{\alpha}/\hat{\beta})$. Thus, either model estimates the mean equally well.

(For the theory behind, and other examples of, parameter orthogonality, see Cox and Reid 1987.)

6.6 In Example 6.4, verify the MLEs $\hat{\xi}_i$ and $\hat{\sigma}_{jk}$ when the ξ's are unknown.

6.7 In Example 6.4, show that the S_{jk} given by (6.15) are independent of $(X_1, \ldots, X_p)$ and have the same joint distribution as the statistics (6.13) with n replaced by $n - 1$.

[*Hint*: Subject each of the p vectors $(X_{i1}, \ldots, X_{in})$ to the same orthogonal transformation, where the first row of the orthogonal matrix is $(1/\sqrt{n}, \ldots, 1/\sqrt{n})$.]

6.8 Verify the matrices (a) (6.17) and (b) (6.18).

6.9 Consider the situation leading to (6.20), where (X_i, Y_i), $i = 1, \ldots, n$, are iid according to a bivariate normal distribution with $E(X_i) = E(Y_i) = 0$, $\text{var}(X_i) = \text{var}(Y_i) = 1$, and unknown correlation coefficient ρ.

(a) Show that the likelihood equation is a cubic for which the probability of a unique root tends to 1 as $n \to \infty$. [*Hint*: For a cubic equation $ax^3 + 3bx^2 + 3cx + d = 0$, let $G = a^2d - 3abc + 2b^3$ and $H = ac - b^2$. Then the condition for a unique real root is $G^2 + 4H^3 > 0$.]

(b) Show that if $\hat{\rho}_n$ is a consistent solution of the likelihood equation, then it satisfies (6.20).

(c) Show that $\delta = \Sigma X_i Y_i/n$ is a consistent estimator of ρ and that $\sqrt{n}(\delta - \rho) \xrightarrow{\mathcal{L}} N(0, 1 + \rho^2)$ and, hence, that δ is less efficient than the MLE of ρ.

6.10 Verify the limiting distribution asserted in (6.21).

6.11 Let $X, \ldots, X_n$ be iid according to the Poisson distribution $P(\lambda)$. Find the ARE of $\delta_{2n} = [\text{No. of } X_i = 0]/n$ to $\delta_{1n} = e^{-\bar{X}_n}$ as estimators of $e^{-\lambda}$.

6.12 Show that the efficiency (6.27) tends to 0 as $|a - \theta| \to \infty$.

6.13 For the situation of Example 6.9, consider as another family of distributions, the contaminated normal mixture family suggested by Tukey (1960) as a model for observations which usually follow a normal distribution but where occasionally something goes wrong with the experiment or its recording, so that the resulting observation is a gross error. Under the *Tukey model*, the distribution function takes the form

$$F_{\tau,\epsilon}(t) = (1 - \epsilon)\Phi(t) + \epsilon\Phi\left(\frac{t}{\tau}\right).$$

That is, in the gross error cases, the observations are assumed to be normally distributed with the same mean θ but a different (larger) variance τ^2.[9]

(a) Show that if the X_i's have distribution $F_{\tau,\epsilon}(x - \theta)$, the limiting distribution of δ_{2n} is unchanged.

(b) Show that the limiting distribution of δ_{1n} is normal with mean zero and variance $\frac{n}{n-1}\left\{\phi\left[\sqrt{\frac{n}{n-1}}(a - \theta)\right]\right\}^2 (1 - \epsilon + \epsilon\tau^2)$.

(c) Compare the asymptotic relative efficiency of δ_{1n} and δ_{2n}.

6.14 Let $X_1, \ldots, X_n$ be iid as $N(0, \sigma^2)$.

(a) Show that $\delta_n = k\Sigma|X_i|/n$ is a consistent estimator of σ if and only if $k = \sqrt{\pi/2}$.

(b) Determine the ARE of δ with $k = \sqrt{\pi/2}$ with respect to the MLE $\sqrt{\Sigma X_i^2/n}$.

6.15 Let $X_1, \ldots, X_n$ be iid with $E(X_i) = \theta$, $\text{var}(X_i) = 1$, and $E(X_i - \theta)^4 = \mu_4$, and consider the unbiased estimators $\delta_{1n} = (1/n)\Sigma X_i^2 - 1$ and $\delta_{2n} = \bar{X}_n^2 - 1/n$ of θ^2.

(a) Determine the ARE $e_{2,1}$ of δ_{2n} with respect to δ_{1n}.

(b) Show that $e_{2,1} \geq 1$ if the X_i are symmetric about θ.

(c) Find a distribution for the X_i for which $e_{2,1} < 1$.

6.16 The property of asymptotic relative efficiency was defined (Definition 6.6) for estimators that converged to normality at rate $\sqrt{n}$. This definition, and Theorem 6.7, can be generalized to include other distributions and rates of convergence.

[9] As has been pointed out by Stigler (1973) such models for heavy-tailed distributions had already been proposed much earlier in a forgotten work by Newcomb (1882, 1886).

Theorem 9.1 *Let $\{\delta_{in}\}$ be two sequences of estimators of $g(\theta)$ such that*

$$n^\alpha[\delta_{in} - g(\theta)] \xrightarrow{\mathcal{L}} \tau_i T,\ \alpha > 0,\ \tau_i > 0,\ i = 1, 2,$$

where the distribution H of T has support on an interval $-\infty \le A < B \le \infty$ with strictly increasing cdf on (A, B). Then, the ARE of $\{\delta_{2n}\}$ with respect to $\{\delta_{1n}\}$ exists and is

$$e_{21} = \lim_{n_2\to\infty} \frac{n_1(n_2)}{n_2} = \left[\frac{\tau_1}{\tau_2}\right]^{1/\alpha}.$$

6.17 In Example 6.10, show that the conditions of Theorem 5.1 are satisfied.

Section 7

7.1 Prove Theorem 7.1.

7.2 For the situation of Example 7.3 with $m = n$:

(a) Show that a necessary condition for (7.5) to converge to $N(0, 1)$ is that $\sqrt{n}(\hat{\lambda}-\lambda) \to 0$, where $\hat{\lambda} = \hat{\sigma}^2/\hat{\tau}^2$ and $\lambda = \sigma^2/\tau^2$, for $\hat{\sigma}^2$ and $\hat{\tau}^2$ of (7.4).

(b) Use the fact that $\hat{\lambda}/\lambda$ has an F-distribution to show that $\sqrt{n}(\hat{\lambda} - \lambda) \not\to 0$.

(c) Show that the full MLE is given by the solution to

$$\xi = \frac{(m/\sigma^2)\bar{X} + (n/\tau^2)\bar{Y}}{m/\sigma^2 + n/\tau^2}, \quad \sigma^2 = \frac{1}{m}\Sigma(X_i - \xi)^2, \quad \tau^2 = \frac{1}{n}\Sigma(Y_j - \xi)^2,$$

and deduce its asymptotic efficiency from Theorem 5.1.

7.3 In Example 7.4, determine the joint distribution of (a) $(\hat{\sigma}^2, \hat{\tau}^2)$ and (b) $(\hat{\sigma}^2, \hat{\sigma}_A^2)$.

7.4 Consider samples $(X_1, Y_1), \ldots, (X_m, Y_m)$ and $(X_1', Y_1'), \ldots, (X_n', Y_n')$ from two bivariate normal distributions with means zero and variance-covariances $(\sigma^2, \tau^2, \rho\sigma\tau)$ and $(\sigma'^2, \tau'^2, \rho'\sigma'\tau')$, respectively. Use Theorem 7.1 and Examples 6.5 and 6.8 to find the limit distribution

(a) of $\hat{\sigma}^2$ and $\hat{\tau}^2$ when it is known that $\rho' = \rho$

(b) of $\hat{\rho}$ when it is known that $\sigma' = \sigma$ and $\tau' = \tau$.

7.5 In the preceding problem, find the efficiency gain (if any)

(a) in part (a) resulting from the knowledge that $\rho' = \rho$

(b) in part (b) resulting from the knowledge that $\sigma' = \sigma$ and $\tau' = \tau$.

7.6 Show that the likelihood equations (7.11) have at most one solution.

7.7 In Example 7.6, suppose that $p_i = 1-F(\alpha+\beta t_i)$ and that both $\log F(x)$ and $\log[1-F(x)]$ are strictly concave. Then, the likelihood equations have at most one solution.

7.8 (a) If the cdf F is symmetric and if $\log F(x)$ is strictly concave, so is $\log[1 - F(x)]$.

(b) Show that $\log F(x)$ is strictly concave when F is strongly unimodal but not when F is Cauchy.

7.9 In Example 7.7, show that Y_n is less informative than Y.

[*Hint*: Let Z_n be distributed as $P(\lambda\Sigma_{i=n+1}^\infty \gamma_i)$ independently of Y_n. Then, $Y_n + Z_n$ is a sufficient statistic for λ on the basis of (Y_n, Z_n) and $Y_n + Z_n$ has the same distribution as Y.]

7.10 Show that the estimator δ_n of Example 7.7 satisfies (7.14).

7.11 Find suitable normalizing constants for δ_n of Example 7.7 when (a) $\gamma_i = i$, (b) $\gamma_i = i^2$, and (c) $\gamma_i = 1/i$.

7.12 Let X_i $(i = 1, \ldots, n)$ be independent normal with variance 1 and mean βt_i (with t_i known). Discuss the estimation of β along the lines of Example 7.7.

7.13 Generalize the preceding problem to the situation in which (a) $E(X_i) = \alpha + \beta t_i$ and $\text{var}(X_i) = 1$ and (b) $E(X_i) = \alpha + \beta t_i$ and $\text{var}(X_i) = \sigma^2$ where α, β, and σ^2 are unknown parameters to be estimated.

7.14 Let X_j $(j = 1, \ldots, n)$ be independently distributed with densities $f_j(x_j|\theta)$ (θ real-valued), let $I_j(\theta)$ be the information X_j contains about θ, and let $T_n(\theta) = \Sigma_{j=1}^n I_j(\theta)$ be the total information about θ in the sample. Suppose that $\hat{\theta}_n$ is a consistent root of the likelihood equation $L'(\theta) = 0$ and that, in generalization of (3.18)-(3.20),

$$\frac{1}{\sqrt{T_n(\theta_0)}} L'(\theta_0) \xrightarrow{\mathcal{L}} N(0, 1)$$

and

$$-\frac{L''(\theta_0)}{T_n(\theta_0)} \xrightarrow{P} 1 \quad \text{and} \quad \frac{L'''(\theta_n^*)}{T_n(\theta_0)} \quad \text{is bounded in probability.}$$

Show that

$$\sqrt{T_n(\theta_0)}(\hat{\theta}_n - \theta_0) \xrightarrow{\mathcal{L}} N(0, 1).$$

7.15 Prove that the sequence $X_1, X_2, \ldots$ of Example 7.8 is stationary provided it satisfies (7.17).

7.16 (a) In Example 7.8, show that the likelihood equation has a unique solution, that it is the MLE, and that it has the same asymptotic distribution as $\delta_n' = \sum_{i=1}^n X_i X_{i+1} / \sum_{i=1}^n X_i^2$.

(b) Show directly that δ_n' is a consistent estimator of β.

7.17 In Example 7.8:

(a) Show that for $j > 1$ the expected value of the conditional information (given X_{j-1}) that X_j contains about β is $1/(1 - \beta^2)$.

(b) Determine the information X_1 contains about β.

7.18 When $\tau = \sigma$ in (7.21), show that the MLE exists and is consistent.

7.19 Suppose that in (7.21), the ξ's are themselves random variables, which are iid as $N(\mu, \gamma^2)$.

(a) Show that the joint density of the (X_i, Y_i) is that of a sample from a bivariate normal distribution, and identify the parameters of that distribution.

(b) In the model of part (a), find asymptotically efficient estimators of the parameters $\mu, \gamma, \beta, \sigma$, and τ.

7.20 Verify the roots (7.22).

7.21 Show that the likelihood (7.21) is unbounded.

7.22 Show that if ρ s defined by (7.24), then ρ and ρ' are everywhere continuous.

7.23 Let F have a differentiable density f and let $\int \psi^2 f < \infty$.

(a) Use integration by parts to write the denominator of (7.27) as $[\int \psi(x) f'(x) dx]^2$.

(b) Show that $\sigma^2(F, \psi) \geq [\int (f'/f)^2 f]^{-1} = I_f^{-1}$ by applying the Schwarz inequality to part (a).

The following three problems will investigate the technical conditions required for the consistency and asymptotic normality of M*-estimators, as noted in (7.26).*

7.24 To have consistency of M-estimators, a sufficient condition is that the root of the estimating function be unique and isolated. Establish the following theorem.

Theorem 9.2 *Assume that conditions (A0)-(A3) hold. Let t_0 be an isolated root of the equation $E_{\theta_0}[\psi(X,t)] = 0$, where $\psi(\cdot,t)$ is monotone in t and continuous in a neighborhood of t_0. If $T_0(\mathbf{x})$ is a solution to $\sum_{i=1}^n \psi(x_i,t) = 0$, then T_0 converges to t_0 in probability.*

[*Hint:* The conditions on ψ imply that $E_{\theta_0}[\psi(X,t)]$ is monotone, so t_0 is a unique root. Adapt the proofs of Theorems 3.2 and 3.7 to complete this proof.]

7.25 Theorem 9.3 *Under the conditions of Theorem 9.2, if, in addition*

(*i*) $E_{\theta_0}\left[\frac{\partial}{\partial t}\psi(X,t)|_{t=t_0}\right]$ *is finite and nonzero,*

(*ii*) $E_{\theta_0}\left[\psi^2(X,t_0)\right] < \infty$,

then

$$\sqrt{n}(T_0 - t_0) \xrightarrow{\mathcal{L}} N(0, \sigma_{T_0}^2),$$

where $\sigma_{T_0}^2 = E_{\theta_0}\left[\psi^2(X,t_0)\right] / (E_{\theta_0}\left[\frac{\partial}{\partial t}\psi(X,t)|_{t=t_0}\right])^2$.

[Note that this is a slight generalization of (7.27).]

[*Hint:* The assumptions on ψ are enough to adapt the Taylor series argument of Theorem 3.10, where ψ takes the place of l'.]

7.26 For each of the following estimates, write out the ψ function that determines it, and show that the estimator is consistent and asymptotically normal under the conditions of Theorems 9.2 and 9.3.

(a) The *least squares estimate*, the minimizer of $\sum(x_i - t)^2$.

(b) The *least absolute value estimate*, the minimizer of $\sum |x_i - t|$.

(c) The *Huber trimmed mean*, the minimizer of (7.24).

7.27 In Example 7.12, compare (a) the asymptotic distributions of $\hat{\xi}$ and δ_n; (b) the normalized expected squared error of $\hat{\xi}$ and δ_n.

7.28 In Example 7.12, show that (a) $\sqrt{n}(\hat{\hat{b}} - b) \xrightarrow{\mathcal{L}} N(0, b^2)$ and (b) $\sqrt{n}(\hat{b} - b) \xrightarrow{\mathcal{L}} N(0, b^2)$.

7.29 In Example 7.13, show that

(a) $\hat{c}$ and $\hat{a}$ are independent and have the stated distributions;

(b) $X_{(1)}$ and $\Sigma \log[X_i/X_{(1)}]$ are complete sufficient statistics on the basis of a sample from (7.33).

7.30 In Example 7.13, determine the UMVU estimators of a and c, and the asymptotic distributions of these estimators.

7.31 In the preceding problem, compare (a) the asymptotic distribution of the MLE and the UMVU estimator of c; (b) the normalized expected squared error of these two estimators.

7.32 In Example 7.15, (a) verify equation (7.39), (b) show that the choice $a = -2$ produces the estimator with the best second-order efficiency, (c) show that the limiting risk ratio of the MLE ($a = 0$) to $\delta_n(a = -2)$ is 2, and (d) discuss the behavior of this estimator in small samples.

7.33 Let $X_1, \ldots, X_n$ be iid according to the three-parameter lognormal distribution (7.37). Show that

(a)

$$p^*(\mathbf{x}|\xi) = \sup_{\gamma,\sigma^2} p(\mathbf{x}|\xi,\gamma,\sigma^2) = c/[\hat\sigma(\xi)]^n \Pi[1/(x_i-\xi)]$$

where

$$p(\mathbf{x}|\xi,\gamma,\sigma^2) = \prod_{i=1}^n f(x_i|\xi,\sigma^2),$$

$$\hat\sigma^2(\xi) = \frac{1}{n}\Sigma[\log(x_i-\xi) - \hat\gamma(\xi)]^2 \text{ and } \hat\gamma(\xi) = \frac{1}{n}\Sigma\log(x_i-\xi).$$

(b) $p^*(\mathbf{x}|\xi) \to \infty$ as $\xi \to x_{(1)}$.

[*Hint*: (b) For ξ sufficiently near $x_{(1)}$,

$$\hat\sigma^2(\xi) \le \frac{1}{n}\Sigma[\log(x_i-\xi)]^2 \le [\log(x_{(1)}-\xi)]^2$$

and hence

$$p^*(\mathbf{x}|\xi) \ge |\log(x_{(1)}-\xi)|^{-n}\Pi(x_{(i)}-\xi)^{-1}.$$

The right side tends to infinity as $\xi \to x_{(1)}$ (Hill 1963.]

7.34 The derivatives of all orders of the density (7.37) tend to zero as $x \to \xi$.

Section 8

8.1 Determine the limit distribution of the Bayes estimator corresponding to squared error loss, and verify that it is asymptotically efficient, in each of the following cases:

(a) The observations $X_1, \ldots, X_n$ are iid $N(\theta, \sigma^2)$, with σ known, and the estimand is θ. The prior distribution for Θ is a conjugate normal distribution, say $N(\mu, b^2)$. (See Example 4.2.2.)

(b) The observations Y_i have the gamma distribution $\Gamma(\gamma, 1/\tau)$, the estimand is $1/\tau$, and τ has the conjugate prior density $\Gamma(g, \alpha)$.

(c) The observations and prior are as in Problem 4.1.9 and the estimand is λ.

(d) The observations Y_i have the negative binomial distribution (4.3), p has the prior density $B(a, b)$, and the estimand is (a) p and (b) $1/b$.

8.2 Referring to Example 8.1, consider, instead, the minimax estimator δ_n of p given by (1.11) which corresponds to the sequence of beta priors with $a = b = \sqrt{n}/2$. Then,

$$\sqrt{n}[\delta_n - p] = \sqrt{n}\left(\frac{X}{n} - p\right) + \frac{\sqrt{n}}{1+\sqrt{n}}\left(\frac{1}{2} - \frac{X}{n}\right).$$

(a) Show that the limit distribution of $\sqrt{n}[\delta_n - p]$ is $N[\frac{1}{2} - p, p(1-p)]$, so that δ_n has the same asymptotic variance as X/n, but that for $p \ne \frac{1}{2}$, it is asymptotically biased.

(b) Show that ARE of δ_n relative to X/n does not exist except in the case $p = \frac{1}{2}$ when it is 1.

8.3 The assumptions of Theorem 2.6 imply (8.1) and (8.2).

8.4 In Example 8.5, the posterior density of θ after one observation is $f(x_1 - \theta)$; it is a proper density, and it satisfies (B5) provided $E_\theta|X_1| < \infty$.

8.5 Let $X_1, \ldots, X_n$ be independent, positive variables, each with density $(1/\tau)f(x_i/\tau)$, and let τ have the improper density $\pi(\tau) = 1/\tau$ $(\tau > 0)$. The posterior density after one observation is a proper density, and it satisfies (B5), provided $E_\tau(1/X_1) < \infty$.

8.6 Give an example in which the posterior density is proper (with probability 1) after two observations but not after one.

[*Hint*: In the preceding example, let $\pi(\tau) = 1/\tau^2$.]

8.7 Prove the result stated preceding Example 8.6.

8.8 Let $X_1, \ldots, X_n$ be iid as $N(\theta, 1)$ and consider the improper density $\pi(\theta) = e^{\theta^4}$. Then, the posterior will be improper for all n.

8.9 Prove Lemma 8.7.

8.10 (a) If $\sup |Y_n(t)| \xrightarrow{P} 0$ and $\sup |X_n(t) - c| \xrightarrow{P} 0$ as $n \to \infty$, then $\sup |X_n(t) - ce^{Y_n(t)}| \xrightarrow{P} 0$, where the sup is taken over a common set $t \in T$.

(b) Use (a) to show that (8.22) and (8.23) imply (8.21).

8.11 Show that (B1) implies (a) (8.24) and (b) (8.26).

10 Notes

10.1 Origins

The origins of the concept of maximum likelihood go back to the work of Lambert, Daniel Bernoulli, and Lagrange in the second half of the eighteenth century, and of Gauss and Laplace at the beginning of the nineteenth. (For details and references, see Edwards 1974 or Stigler 1986.) The modern history begins with Edgeworth (1908, 1909) and Fisher (1922, 1925), whose contributions are discussed by Savage (1976) and Pratt (1976).

Fisher's work was followed by a euphoric belief in the universal consistency and asymptotic efficiency of maximum likelihood estimators, at least in the iid case. The true situation was sorted out only gradually. Landmarks are Cramér (1946a, 1946b), who shifted the emphasis from the global to a local maximum and defined the "regular" case in which the likelihood equation has a consistent asymptotically efficient root; Wald (1949), who provided fairly general conditions for consistency; the counterexamples of Hodges (Le Cam, 1953) and Bahadur (1958); and Le Cam's resulting theorem on superefficiency (1953).

Convergence (under suitable restrictions and appropriately normalized) of the posterior distribution of a real-valued parameter with a prior distribution to its normal limit was first discovered by Laplace (1820) and later reobtained by Bernstein (1917) and von Mises (1931). More general versions of this result are given in Le Cam (1958). The asymptotic efficiency of Bayes solutions was established by Le Cam (1958), Bickel and Yahav (1969), and Ibragimov and Has'minskii (1972). (See also Ibragimov and Has'minskii 1981.)

Computation of likelihood estimators was influenced by the development of the *EM Algorithm* (Dempster, Laird, and Rubin 1977). This algorithm grew out of work done on iterative computational methods that were developed in the 1950s and 1960s, and can be traced back at least as far as Hartley (1958). The EM algorithm has enjoyed widespread use as a computational tool for obtaining likelihood estimators in complex problems (see Little and Rubin 1987, Tanner 1996, or McLachlan and Krishnan 1997).

10.2 Alternative Conditions for Asymptotic Normality

The Cramér conditions for asymptotic normality and efficiency that are given in Theorems 3.10 and 5.1 are not the most general; for those, see Strasser 1985, Pfanzagl 1985, or LeCam 1986. They were chosen because they have fairly wide applicability, yet are relatively straightforward to verify. In particular, it is possible to relax the assumptions somewhat, and only require conditions on the second, rather than third, derivative (see Le Cam 1956, Hájek 1972, and Inagaki 1973). These conditions, however, are somewhat more involved to check than those of Theorem 3.10, which already require some effort.

The conditions have also been altered to accommodate specific features of a problem. One particular change was introduced by Daniels (1961) to overcome the nondifferentiability of the double exponential distribution (see Example 3.14). Huber (1967) notes an error in Daniels proof; however, the validity of the theorem remains. Others have taken advantage of the form of the likelihood. Berk (1972b) exploited the fact that in exponential families, the cumulant generating function is convex. This, in turn, implies that the log likelihood is concave, which then leads to simpler conditions for consistency and asymptotic normality. Other proofs of existence and consistency under slightly different assumptions are given by Foutz (1977). Consistency proofs in more general settings were given by Wald (1949), Le Cam (1953), Bahadur (1967), Huber (1967), Perlman (1972), and Ibragimov and Has'minskii (1981), among others. See also Pfanzagl 1969, 1994, Landers 1972, Pfaff 1982, Wong 1992, Bickel et al. 1993, and Note 10.4. Another condition, which also eliminates the problem of superefficiency, is that of *local asymptotic normality* (Le Cam 1986, Strasser 1985, Section 81, LeCam and Yang 1990, and Wong 1992.)

10.3 Measurability and Consistency

Theorems 3.7 and 4.3 assert the existence of a consistent sequence of roots of the likelihood equation, that is, a sequence of roots that converges in probability to the true parameter value. The proof of Theorem 3.7 is a modification of those of Cramér (1946a, 1946b) and Wald (1949), where the latter established convergence almost everywhere of the sequence. In almost all cases, we are taught, convergence almost everywhere implies convergence in probability, but that is not so here because a sequence of roots need not be measurable! Happily, the θ_n^* of Theorem 3.7 are measurable (however, those of Theorem 4.3 are not necessarily). Serfling (1980, Section 4.2.2; see also Problem 3.29), addresses this point, as does Ferguson (1996, Section 17), who also notes that nonmeasurability does not preclude consistency. (We thank Professor R. Wijsman for alerting us to these measurability issues.)

10.4 Estimating Equations

Theorems 9.2 and 9.3 use assumptions similar to the original assumptions of Huber (1964, 1981, Section 3.2). Alternate conditions for consistency and asymptotic normality, which relax some smoothness requirements on ρ, have been developed by Boos (1979) and Boos and Serfling (1980); see also Serfling 1980, Chapter 7, for a detailed development of this topic. Further results can be found in Portnoy (1977a, 1984, 1985) and the discrete case is considered by Simpson, Carroll, and Ruppert (1987).

The theory of M-estimation, in particular results such as (7.26), have been generalized in many ways. In doing so, much has been learned about the properties of the functions ρ and $\psi = \rho'$ needed for the solution $\tilde{\theta}$ to the equation $\sum_{i=1}^n \psi(x_i - \theta) = 0$ to have reasonable statistical properties.

For example, the structure of the exponential family can be exploited to yield less restrictive conditions for consistency and asymptotic efficiency of $\tilde{\theta}$. In particular, the concavity of the log likelihood plays an important role. Haberman (1989) gives a comprehensive treatment of consistency and asymptotic normality of estimators derived from maximizing concave functions (which include likelihood and M-estimators).

This approach to constructing estimators has become known as the theory of *estimating functions* [see, for example, Godambe 1991 or the review paper by Liang and Zeger (1994)]. A general estimating equation has the form $\sum_{i=1}^n h(x_i|\theta) = 0$, and consistency and asymptotic normality of the solution $\tilde{\theta}$ can be established under quite general conditions (but also see Freedman and Diaconis 1982 or Lele 1994 for situations where this can go wrong). For example, if the estimating equation is *unbiased*, so that $E_\theta\left[\sum_{i=1}^n h(X_i|\theta)\right] = 0$ for all θ, then the "usual" regularity conditions (such as those in Problems 7.24 - 7.25 or Theorem 3.10) will imply that $\tilde{\theta}$ is consistent. Asymptotic normality will also often follow, using a proof similar to that of Theorem 3.10, where the estimating function h is used instead of the log likelihood l. Carroll, Ruppert, and Stefanski (1995, Appendix A.3) provide a nice introduction to this topic.

10.5 Variants of Likelihood

A large number of variants of the likelihood function have been proposed. Many started as a means of solving a particular problem and, as their usefulness and general effectiveness was realized, they were generalized. Although we cannot list all of these variants, we shall mention a few of them.

The first modifications of the usual likelihood function are primarily aimed at dealing with nuisance parameters. These include the *marginal, conditional*, and *profile* likelihoods, and the *modified profile likelihood* of Barndorff-Nielsen (1983). In addition, many of the modifications are accompanied by higher-order distribution approximations that result in faster convergence to the asymptotic distribution. These approximations may utilize techniques of small-sample asymptotics (conditioning on ancillaries, saddlepoint expansions) or possibly Bartlett corrections (Barndorff-Nielsen and Cox 1984).

Other modifications of likelihood may entail, perhaps, a more drastic variation of the likelihood function. The *partial likelihood* of Cox (1975; see also Oakes 1991), presents an effective means of dealing with censored data, by dividing the model into parametric and nonparametric parts. Along these lines *quasi-likelihood* (Wedderburn 1974, McCulloch and Nelder 1989, McCulloch 1991) is based only on moment assumptions and *empirical likelihood* (Owen 1988, 1990, Hall and La Scala 1990) is a nonparametric approach based on a multinomial profile likelihood.

There are many other variations of likelihood, including *directed, penalized*, and *extended*, and the idea of *predictive* likelihood (Hinkley 1979, Butler 1986, 1989).

An entry to this work can be obtained through Kalbfleisch (1986), Barndorff-Nielsen and Cox (1994), or Edwards (1992), the review articles of Hinkley (1980) and Bjørnstad (1990), or the volume of review articles edited by Hinkley, Reid, and Snell (1991).

10.6 Boundary Values

A key feature throughout this chapter was the assumption that the true parameter point θ_0 occurs at an interior point of the parameter space (Section 6.3, Assumption A3; Section 6.5, Assumption A). The effect of this assumption is that, for large n, as the likelihood estimator gets close to θ_0, the likelihood estimator will, in fact, be a root of the likelihood. (Recall the proofs of Theorems 3.7 and 3.10 to see how this is used.) However, in some applications θ_0 is on the boundary of the parameter space, and the ML estimator is not a root of the likelihood. This situation is more frequently encountered in testing than estimation, where the null hypothesis $H_0 : \theta = \theta_0$ often involves a boundary point. However, boundary values can also occur in point estimation. For example, in a mixture problem (Example 6.10), the value of the mixing parameter could be the boundary value 0 or 1. Chernoff (1954) first investigated the asymptotic distribution of the maximum likelihood estimator when the parameter is on the boundary. This distribution is typically not normal, and is characterized by Self and Liang (1987), who give many examples, ranging from multivariate normal to mixtures of chi-squared distributions to even more complicated forms.

An alternate approach to establishing the limiting distribution is provided by Feng and McCulloch (1992), who use a strategy of expanding the parameter space.

10.7 Asymptotics of REML

The results of Cressie and Lahiri (1993) and Jiang (1996, 1997) show that when using restricted maximum likelihood estimation (REML; see Example 2.7 and the discussion after Example 5.3) instead of ML, efficiency need not be sacrificed, as the asymptotic covariance matrix of the REML estimates is the inverse information matrix from the reduced problem. More precisely, we can write the ***general linear mixed model*** (generalizing the linear model of Section 3.4) as

$$\mathbf{Y} = X\beta + Z\mathbf{u} + \varepsilon, \tag{10.1}$$

where $\mathbf{Y}$ is the $N \times 1$ vector of observations, X and Z are $N \times p$ design matrices, β is the $p \times 1$ vector of fixed effects, $\mathbf{u} \sim N(0, D)$ is the $p \times 1$ vector of random effects, and $\varepsilon \sim N(0, R)$, independent of $\mathbf{u}$. The variance components of D and R are usually the targets of estimation. The likelihood function $L(\beta, \mathbf{u}, D, R|\mathbf{y})$ is transformed to the REML likelihood by marginalizing out the β and $\mathbf{u}$ effects, that is,

$$L(D, R|\mathbf{y}) = \int\int L(\beta, \mathbf{u}, D, R|\mathbf{y})\, du\, d\beta.$$

Suppose now that $V = V(\theta)$, that is, the vector θ represents the variance components to be estimated. We can thus write $L(D, R|\mathbf{y}) = L(V(\theta)|\mathbf{y})$ and denote the information matrix of the marginal likelihood by $I_N(\theta)$. Cressie and Lahiri (1993, Corollary 3.1) show that under suitable regularity conditions,

$$[I_N(\theta)]^{1/2}\left(\hat{\theta} - \theta\right) \xrightarrow{\mathcal{L}} N_k(0, I),$$

where $\hat{\theta}_N$ maximizes $L(V(\theta)|\mathbf{y})$. Thus, the REML estimator is asymptotically efficient. Jiang (1996, 1997) has extended this result, and established the asymptotic normality of $\hat{\theta}$ even when the underlying distributions are not normal.

10.8 Higher-Order Asymptotics

Typically, not only is the MLE asymptotically efficient, but so also are various approximations to the MLE, to Bayes estimators, and so forth. Therefore, it becomes important to be able to distinguish between different asymptotically efficient estimator sequences. For example, it seems plausible that one would do best in any application of Theorem 4.3 by using a highly efficient $\sqrt{n}$-consistent starting sequence. It has been pointed out earlier that an efficient estimator sequence can always be modified by terms of order $1/n$ without affecting the asymptotic efficiency. Thus, to distinguish among them requires taking into account the terms of the next order.

A number of authors (among them Rao 1963, Pfanzagl 1973, Ghosh and Subramanyam 1974, Efron 1975, 1978, Akahira and Takeuchi 1981, and Bhattacharya and Denker 1990) have investigated estimators that are "second-order efficient," that is, efficient and among efficient estimators have the greatest accuracy to terms of the next order, and in particular these authors have tried to determine to what extent the MLE is second-order efficient. For example, Efron (1975, Section 10) shows that in exponential families, the MLE minimizes the coefficient of the second-order term among efficient estimators.

For the most part, however, the asymptotic theory presented here is "first-order" theory in the sense that the conclusion of Theorem 3.10 can be expressed as saying that

$$\frac{\sqrt{n}(\hat{\theta}_n - \theta)}{\sqrt{I^{-1}(\theta)}} = Z + O\left(\frac{1}{\sqrt{n}}\right),$$

where Z is a standard normal random variable, so the convergence is at rate $O(1/n^{1/2})$. It is possible to reduce the error in the approximation to $O(1/n^{3/2})$ using "higher-order" asymptotics. The book by Barndorff-Nielsen and Cox (1994) provides a detailed treatment of higher-order asymptotics. Other entries into this subject are through the review papers of Reid (1995, 1996) and a volume edited by Hinkley, Reid, and Snell (1991).

Another technique that is very useful in obtaining accurate approximations for the densities of statistics is the *saddlepoint expansion* (Daniels 1980, 1983), which can be derived through inversion of a characteristic function or through the use of Edgeworth expansions. Entries to this literature can be made through the review paper of Reid (1988), the monograph by Kolassa (1993), or the books by Field and Ronchetti (1990) or Jensen (1995).

Still another way to achieve higher-order accuracy in certain cases is through a technique known as the *bootstrap*, initiated by Efron (1979, 1982b). Some of the theoretical foundations of the bootstrap are rooted in the work of von Mises (1936, 1947) and Kiefer and Wolfowitz (1956). The bootstrap can be thought of as a "nonparametric" MLE, where the quantity $\int h(x)dF(x)$ is estimated by $\int h(x)dF_n(x)$. Using the technique of Edgeworth expansions, it was established by Singh (1981) (see also Bickel and Freedman 1981) that the bootstrap sometimes provides a more accurate approximation than the Delta Method (Theorem 1.8.12). An introduction to the asymptotic theory of the bootstrap is given by Lehmann (1999), and implementation and applications of the bootstrap are given in Efron and Tibshirani (1993). Other introductions to the bootstrap are through the volume edited by LePage and Billard (1992), the book by Shao and Tu (1995), or the review paper of Young (1994). A more theoretical treatment is given by Hall (1992).

References

Abbey, I. L. and David, H. T. (1970). The construction of uniformly minimum variance unbiased estimators for exponential distributions. *Ann. Math. Statist.* **41**, 1217-1222.

Agresti, A. (1990). *Categorical Data Analysis*. New York: Wiley.

Ahuja, J. C. (1972). Recurrence relation for minimum variance unbiased estimation of certain left-truncated Poisson distributions. *J. Roy. Statist. Soc. C* **21**, 81-86.

Aitken, A. C. and Silverstone, H. (1942). On the estimation of statistical parameters. *Proc. Roy. Soc. Edinb. A* **61**, 186-194.

Aitken, M., Anderson, D., Francis, B. and Hinde, J. (1989). *Statistical Modelling in GLIM*. Oxford: Clarendon Press.

Akahira, M. and Takeuchi, K. (1981). *Asymptotic Efficiency of Statistical Estimators*. New York: Springer-Verlag.

Alam, K. (1979). Estimation of multinomial probabilities. *Ann. Statist.* **7**, 282-283.

Albert, J. H. (1988). Computational methods using a Bayesian hierarchical generalized linear model *J. Amer. Statist. Assoc.* **83**, 1037-1044.

Albert, J. H. and Gupta, A. K. (1985). Bayesian methods for binomial data with application to a nonresponse problem. *J. Amer. Statist. Assoc.* **80**, 167-174.

Amari, S.-I., Barndorff-Nielsen, O. E., Kass, R. E., Lauritzen, S. L., and Rao, C. R. (1987). *Differential geometry and statistical inference*. Hayward, CA: Institute of Mathematical Statistics

Amemiya, T. (1980). The n-2-order mean squared errors of the maximum likelihood and the minimum logit chi-square estimator. *Ann. Statist.* **8**, 488-505.

Andersen, E. B. (1970a). Sufficiency and exponential families for discrete sample spaces. *J. Amer. Statist. Assoc.* **65**, 1248-1255.

Andersen, E. B. (1970b). Asymptotic properties of conditional maximum likelihood estimators, *J. Roy. Statist. Soc. Ser. B* **32**, 283-301.

Anderson, T. W. (1959). On asymptotic distributions of estimates of parameters of stochastic difference equations. *Ann. Math. Statist.* **30**, 676-687.

Anderson, T. W. (1962). Least squares and best unbiased estimates. *Ann. Math. Statist.* **33**, 266-272.

Anderson, T. W. (1976). Estimation of linear functional relationships: Approximate distributions and connections with simultaneous equations in economics. *J. Roy. Statist. Soc. Ser. B* **38**, 1-19.

Anderson, T. W. (1984). *Introduction to Multivariate Statistical Analysis, Second Edition*. New York: Wiley.

Anderson, T. W. and Sawa, T. (1982). Exact and approximate distributions of the maximum likelihood estimator of a slope coefficient. *J. Roy. Statist. Soc. Ser. B* **44**, 52-62.

Arnold, S. F. (1981). *The Theory of Linear Models and Multivariate Analysis.* New York: Wiley.

Arnold, S. F. (1985). Sufficiency and invariance. *Statist. Prob. Lett.* **3**, 275-279.

Ash, R. (1972). *Real Analysis and Probability.* New York: Academic Press.

Athreya, K. B., Doss, H., and Sethuraman, J. (1996). On the convergence of the Markov chain simulation method. *Ann. Statist.* **24**, 69-100.

Bahadur, R. R. (1954). Sufficiency and statistical decision functions. *Ann. Math. Statist.* **25**, 423 462.

Bahadur, R. R. (1957). On unbiased estimates of uniformly minimum variance. *Sankhya* **18**, 211-224.

Bahadur, R. R. (1958). Examples of inconsistency of maximum likelihood estimates. *Sankhya* **20**, 207-210.

Bahadur, R. R. (1964). On Fisher's bound for asymptotic variances. *Ann. Math. Statist.* **35**, 1545-1552.

Bahadur, R. R. (1967). Rates of convergence of estimates and test statistics. *Ann. Math. Statist.* **38**, 303-324.

Bahadur, R. R. (1971). *Some Limit Theorems in Statistics.* Philadelphia: SIAM.

Bai, Z. D. and Fu, J. C. (1987). On the maximum likelihood estimator for the location parameter of a Cauchy distribution. *Can. J. Statist.* **15**, 137-146.

Baker, R. J. and Nelder, J. A. (1983a). Generalized linear models. *Encycl. Statist. Sci.* **3**, 343-348.

Baker, R. J. and Nelder, J. A. (1983b). GLIM. *Encycl. Statist. Sci.* **3**, 439-442.

Bar-Lev, S. K. and Enis, P. (1986). Reproducibility and natural exponential families with power variance functions. *Ann. Statist.* **14**, 1507-1522.

Bar-Lev, S. K. and Enis, P. (1988). On the classical choice of variance stabilizing transformations and an application for a Poisson variate. *Biometrika* **75**, 803-804.

Bar-Lev, S. K. and Bshouty, D. (1989). Rational variance functions. *Ann. Statist.* **17**, 741-748.

Bar-Shalom, Y. (1971). Asymptotic properties of maximum likelihood estimates. *J. Roy. Statist. Soc. Ser. B* **33**, 72-77.

Baranchik, A. J. (1964). Multiple regression and estimation of the mean of a multivariate normal distribution. Stanford University Technical Report No. 51.

Baranchik, A. J. (1970). A family of minimax estimators of the mean of a multivariate normal distribution. *Ann. Math. Statist.* **41**, 642-645.

Barankin, E. W. (1950). Extension of a theorem of Blackwell. *Ann. Math. Statist.* **21**, 280-284.

Barankin, E. W. and Maitra, A. P. (1963). Generalization of the Fisher-Darmois Koopman-Pitman theorem on sufficient statistics. *Sankhya A* **25**, 217-244.

Barnard, G. A. (1985). Pivotal inference. *Encycl. Stat. Sci.* (N. L. Johnson, S. Kota, and C. Reade, eds.). New York: Wiley, pp. 743-747.

Barndorff-Nielsen, O. (1978). *Information and Exponential Families in Statistical Theory.* New York: Wiley.

Barndorff-Nielsen, O. (1980). Conditionality resolutions. *Biometrika* **67**, 293-310.

Barndorff-Nielsen, O. (1983). On a formula for the distribution of the maximum likelihood estimator. *Biometrika* **70**, 343-365.

Barndorff-Nielsen, O. (1988). *Parametric Statistical Models and Likelihood.* Lecture Notes in Statistics 50. New York: Springer-Verlag.

Barndorff-Nielsen, O. and Blaesild, P. (1980). Global maxima, and likelihood in linear models. Research Rept. 57. Department of Theoretical Statistics, University of Aarhus.

Barndorff-Nielsen, O. and Cox, D. R. (1979). Edgeworth and saddle-point approximations with statistical applications (with discussion). *J. Roy. Statist. Soc. Ser. B* **41**, 279-312.

Barndorff-Nielsen, O. and Cox, D. R. (1984). Bartlett adjustments to the likelihood ratio statistic and the distribution of the maximum likelihood estimator. *J. Roy. Statist. Soc. Ser. B* **46**, 483-495.

Barndorff-Nielsen, O. and Cox, D. R. (1994). *Inference and Asymptotics.* Chapman & Hall.

Barndorff-Nielsen, O. and Pedersen, K. (1968). Sufficient data reduction and exponential families. *Math. Scand.* **22**, 197-202.

Barndorff-Nielsen, O., Hoffmann-Jorgensen, J., and Pedersen, K. (1976). On the minimal sufficiency of the likelihood function. *Scand. J. Statist.* **3**, 37-38.

Barndorff-Nielsen, O., Blaesild, P., Jensen, J. L., and Jorgensen, B. (1982). Exponential transformation models. *Proc. Roy. Soc. London A* **379**, 41-65.

Barndorff-Nielsen, O. Blaesild, P., and Seshadri, V. (1992). Multivariate distributions with generalized Gaussian marginals, and associated Poisson marginals. *Can. J. Statist.* **20**, 109-120.

Barnett, V. D. (1966). Evaluation of the maximum likelihood estimator where the likelihood equation has multiple roots. *Biometrika* **53**, 151-166.

Barnett, V. (1982). *Comparitive Statistical Inference, Second Edition.* New York: Wiley

Barron, A. R. (1989). Uniformly powerful goodness of fit test. *Ann. Statist.* **17**, 107-124.

Basawa, I. V. and Prakasa Rao, B. L. S. (1980). *Statistical Inference in Stochastic Processes.* London: Academic Press.

Basu, D. (1955a). A note on the theory of unbiased estimation. *Ann. Math. Statist.* **26**, 345-348. Reprinted as Chapter XX of *Statistical Information and Likelihood: A Collection of Critical Essays.* New York: Springer-Verlag.

Basu, D. (1955b). On statistics independent of a complete sufficient statistic. *Sankhya* **15**, 377-380. Reprinted as Chapter XXII of *Statistical Information and Likelihood: A Collection of Critical Essays.* New York: Springer-Verlag.

Basu, D. (1956). The concept of asymptotic efficiency. *Sankhya* **17**, 193-196. Reprinted as Chapter XXI of *Statistical Information and Likelihood: A Collection of Critical Essays.* New York: Springer-Verlag.

Basu, D. (1958). On statistics independent of sufficient statistics. *Sankhya* **20**, 223-226. Reprinted as Chapter XXIII of *Statistical Information and Likelihood: A Collection of Critical Essays.* New York: Springer-Verlag.

Basu, D. (1969). On sufficiency and invariance. In *Essays in Probability and Statistics* (Bose, ed.). Chapel Hill: University of North Carolina Press. Reprinted as Chapter XI of *Statistical Information and Likelihood: A Collection of Critical Essays.*

Basu D. (1971). An essay on the logical foundations of survey sampling, Part I. In *Foundations of Statistical Inference* (Godambe and Sprott, eds.). Toronto: Holt, Rinehart and Winston, pp. 203-242. Reprinted as Chapters XII and XIII of *Statistical Information and Likelihood: A Collection of Critical Essays.*

Basu, D. (1988). *Statistical Information and Likelihood: A Collection of Critical Essays* (J. K. Ghosh, ed.). New York: Springer-Verlag.

Baum, L. and Petrie, T. (1966). Statistical inference for probabilistic functions of finite state Markov chains. *Ann. Math. Statist.* **37**, 1554-1563.

Baum, L., Petrie, T., Soules, G., and Weiss, N. (1970). A maximization technique occurring in the statistical analysis of probabilistic functions of Markov chains. *Ann. Math. Statist.* **41**, 164-171.

Bayes, T. (1763). An essay toward solving a problem in the doctrine of chances. Phil. Trans. Roy. Soc. **153**, 370-418. Reprinted in (1958) *Biometrika* **45**, 293-315.

Bell, C. B., Blackwell, D., and Breiman, L. (1960). On the completeness of order statistics. *Ann. Math. Statist.* **31**, 794-797.

Berger, J. (1975). Minimax estimation of location vectors for a wide class of densities. *Ann. Statist.* **3**, 1318-1328.

Berger, J. (1976a). Tail minimaxity in location vector problems and its applications. *Ann. Statist.* **4**, 33-50.

Berger, J. (1976b). Admissible minimax estimation of a multivariate normal mean with arbitrary quadratic loss. *Ann. Statist.* **4**, 223-226.

Berger, J. (1976c). Inadmissibility results for generalized Bayes estimators or coordinates of a location vector. *Ann. Statist.* **4**, 302-333.

Berger, J. (1976d). Admissibility results for generalized Bayes estimators of coordinates of a location vector. *Ann. Statist.* **4**, 334 356.

Berger, J. (1976e). Inadmissibility results for the best invariant estimator of two coordinates of a location vector. *Ann. Statist.* **4**, 1065-1076.

Berger, J. (1980a). Improving on inadmissible estimators in continuous exponential families with applications to simultaneous estimation of gamma scale parameters *Ann. Statist.* **8**, 545-571.

Berger, J. (1980b). A robust generalized Bayes estimator and confidence region for a multivariate normal mean. *Ann. Statist.* **8**, 716-761.

Berger, J. O. (1982a). Bayesian robustness and the Stein effect. *J. Amer. Statist. Assoc.* **77**, 358-368.

Berger, J. O. (1982b). Estimation in continuous exponential families: Bayesian estimation subject to risk restrictions. *Statistical Decision Theory III* (S. S. Gupta and J. O. Berger, eds.). New York: Academic Press, pp. 109-141.

Berger, J. (1982c). Selecting a minimax estimator of a multivariate normal mean. *Ann. Statist.* **10**, 81-92.

Berger, J. O. (1984). The robust Bayesian viewpoint. *Robustness of Bayesian Analysis* (J. Kadane, ed.). Amsterdam: North-Holland.

Berger, J. O. (1985). *Statistical Decision Theory and Bayesian Analysis, Second Edition.* New York: Springer-Verlag.

Berger, J. O. (1990a). On the inadmissibility of unbiased estimators. *Statist. Prob. Lett.* **9**, 381-384.

Berger, J. O. (1990b). Robust Bayesian analysis: Sensitivity to the prior. *J. Statist. Plan. Inform.* **25**, 303-328.

Berger, J. O. (1994). An overview of robust Bayesian analysis (with discussion). *Test* **3**, 5-124.

Berger, J. O. and Berliner, L.M. (1986). Robust Bayes and empirical Bayes analysis with ϵ-contaminated priors. *Ann. Statist.* **14**, 461-486.

Berger, J. O. and Bernardo, J. M. (1989). Estimating a product of means: Bayesian analysis with reference priors. *J. Amer. Statist. Assoc.* **84**, 200-207.

Berger, J. O. and Bernardo, J. M. (1992a). On the development of reference priors. *Bayesian Statist. 4* (J. O. Berger and J. M. Bernardo, eds.). London: Clarendon Press, pp. 35-49.

Berger, J. O. and Bernardo, J. M. (1992b). Ordered group reference priors with application to multinomial probabilities. *Biometrika* **79**, 25-37.

Berger, J. O. and Bock, M. E. (1976). Combining independent normal mean estimation problems with unknown variances. *Ann. Statist.* **4**, 642-648.

Berger, J. O. and Dey, D. (1983a). On truncation of shrinkage estimators in simultaneous estimation of normal means. *J. Amer. Statist. Assoc.* **78**, 865-869.

Berger, J. O. and Dey, D. (1983b). Combining coordinates in simultaneous estimation of normal means. *J. Statist. Plan. Inform.* **8**, 143-160.

Berger, J. O. and Robert, C. (1990). Subjective hierarchical Bayes estimation of a multivariate normal mean: On the frequentist interface. *Ann. Statist.* **18**, 617-651.

Berger, J. and Srinivasan, C. (1978). Generalized Bayes estimators in multivariate problems. *Ann. Statist.* **6**, 783-801.

Berger, J. O. and Strawderman, W. E. (1996). Choice of hierarchical priors: Admissibility in estimation of normal means. *Ann. Statist.* **24**, 931-951.

Berger, J. O. and Wolpert, R. W. (1988). *The Likelihood Principle, Second Edition.* Hayward, CA: Institute of Mathematical Statistics

Berger, J. O., Bock, M. E., Brown, L. D., Casella, G., and Gleser, L. J. (1977). Minimax estimation of a normal mean vector for arbitrary quadratic loss and unknown covariance matrix. *Ann. Statist.* **5**, 763-771.

Berk, R. (1967a). A special group structure and equivariant estimation *Ann. Math. Statist.* **38**, 1436-1445.

Berk, R. H. (1967b). Review of Zehna (1966). *Math. Rev.* **33**, No. 1922.

Berk, R. (1972a). A note on sufficiency and invariance. *Ann. Math. Statist.* **43**, 647-650.

Berk, R. H. (1972b). Consistency and asymptotic normality of MLE's for exponential models. *Ann. Math. Statist.* **43**, 193-204.

Berk, R. H., Brown, L. D., and Cohen, A. (1981). Bounded stopping times for a class of sequential Bayes tests. *Ann. Statist.* **9**, 834-845.

Berkson J. (1980). Minimum chi-square, not maximum likelihood! *Ann. Statist.* **8**, 457-469.

Bernardo, J. M. (1979). Reference posterior distributions for Bayesian inference. *J. Roy. Statist. Soc. Ser. B* **41**, 113-147.

Bernardo, J. M. and Smith, A. F. M. (1994). *Bayesian Theory.* New York: Wiley.

Bernstein, S. (1917). *Theory of Probability.* (Russian) Fourth Edition (1946) Gostekhizdat, Moscow-Leningrad (in Russian).

Berry, J. C. (1987). Equivariant estimation of a normal mean using a normal concomitant variable for covariance adjustment. *Can. J. Statist.* **15**, 177-183.

Berry, J. C. (1990). Minimax estimation of a bounded normal mean vector. *J. Mult. Anal.* **35**, 130-139.

Bhat, B.R. (1974). On the method of maximum likelihood for dependent observations, *J. Roy. Statist. Soc. Ser. B* **36**, 48-53.

Bhattacharyya, A. (1946, 1948). On some analogs to the amount of information and their uses in statistical estimation. *Sankhya* **8**, 1-14, 201-208, 277-280.

Bhattacharyya, G. K., Johnson, R. A., and Mehrotra, K. G. (1977). On the completeness of minimal sufficient statistics with censored observations. *Ann. Statist.* **5**, 547-553.

Bhatttachayra, R. and Denker, M. (1990). *Asymptotic Statistics*. Basel: Birkhauser-Verlag.

Bickel, P. J. (1981). Minimax estimation of the mean of a normal distribution when the parameter space is restricted. *Ann. Statist.* **9**, 1301-1309.

Bickel, P. J. (1983). Minimax estimation of the mean of a normal distribution subject to doing well at a point. *Recent Advances in Statistics: Papers in Honor of Herman Chernoff on his Sixtieth Birthday* (M. H. Rizvi, J. S. Rustagi and D. Siegmund, eds.), New York: Academic Press, pp. 511-528.

Bickel, P. J. (1984). Parametric robustness: small biases can be worthwhile. *Ann. Statist.* **12**, 864-879.

Bickel, P. J. and Blackwell, D. (1967). A note on Bayes estimates. *Ann. Math. Statist.* **38**, 1907-1911.

Bickel, P. J. and Doksum, K. A. (1981). An analysis of transformations revisited. *J. Amer. Statist. Assoc.* **76**, 296-311.

Bickel, P. J. and Freedman, D. A. (1981). Some asymptotic theory for the bootstrap. *Ann. Statist.* **9**, 1196-1217.

Bickel, P. J. and Lehmann, E. L. (1969). Unbiased estimation in convex families. *Ann. Math. Statist.* **40**, 1523-1535.

Bickel, P. J. and Lehmann, E. L. (1975-1979). Descriptive statistics for nonparametric models I-III. *Ann. Statist.* **3**, 1038-1045, 1045-1069, **4**, 1139-1159; IV *Contributions to Statistics*, Hájek Memorial Volume (J. Jureckova, ed.). Prague: Academia, pp. 33-40.

Bickel, P. J. and Lehmann, E. L. (1981). A minimax property of the sample mean in finite populations. *Ann. Statist.* **9**, 1119-1122.

Bickel, P. J. and Mallows, C. (1988). A note on unbiased Bayes estimation. *Amer. Statist.* **42**, 132-134.

Bickel, P.J. and Yahav, J.A. (1969) Some contributions to the asymptotic theory of Bayes solutions. *Z. Wahrsch. Verw. Geb.* **11**, 257-276.

Bickel, P. J., Klaassen, P., Ritov, C. A. J., and Wellner, J. (1993). *Efficient and Adaptive Estimation for Semiparametric Models*. Baltimore: Johns Hopkins University Press.

Billingsley P. (1995). *Probability and Measure, Third Edition*. New York: Wiley.

Bilodeau, M. and Srivastava, M. S. (1988). Estimation of the MSE matrix of the Stein estimator. *Can. J. Statist.* **16**, 153-159.

Bischoff, W. and Fieger, W. (1992). Minimax estimators and gamma-minimax estimators of a bounded normal mean under the loss $Ł_p(\theta, d) = |\theta - d|^p$. *Metrika* **39**, 185-197.

Bishop, Y. M. M., Fienberg, S. E, and Holland, P. W. (1975). *Discrete Multivariate Analysis*. Cambridge, MA: MIT Press.

Bjørnstad, J. F. (1990). Predictive likelihood: A review (with discussion). *Statist. Sci.* **5**, 242-265.

Blackwell, D. (1947). Conditional expectation and unbiased sequential estimation. *Ann. Math. Statist.* **18**, 105-110.

Blackwell, D. (1951). On the translation parameter problem for discrete variables. *Ann. Math. Statist.* **22**, 393-399.

Blackwell, D. and Girshick, M. A. (1947). A lower bound for the variance of some unbiased sequential estimates. *Ann. Math. Statist.* **18**, 277-280.

Blackwell, D. and Girshick, M. A. (1954). *Theory of games and Statistical Decisions*. New York: Wiley. Reissued by Dover, New York, 1979.

Blackwell, D. and Ryll-Nardzewski, C. (1963). Non-existence of everywhere proper conditional distributions. *Ann. Math. Statist.* **34**, 223-225

Blight, J. N. and Rao, P. V. (1974). The convergence of Bhattacharyya bounds. *Biometrika* **61**, 137-142.

Bloch, D. A. and Watson, G. S. (1967). A Bayesian study of the multinomial distribution. *Ann. Math. Statist.* **38**, 1423-1435.

Blyth C. R (1951). On minimax statistical decision procedures and their admissibility. *Ann. Math. Statist.* **22**, 22-42.

Blyth, C. R. (1970). On the inference and decision models of statistics (with discussion). *Ann. Math. Statist.* **41**, 1034-1058.

Blyth, C. R. (1974). Necessary and sufficient conditions for inequalities of Crámer-Rao type. *Ann. Statist.* **2**, 464-473.

Blyth, C. R. (1980). Expected absolute error of the usual estimator of the binomial parameter. *Amer. Statist.* **34**, 155-157.

Blyth, C. R. (1982). Maximum probability estimation in small samples. *Festschrift for Erich Lehmann* (P. J. Bickel. K. A. Doksum, and J. L. Hodges, Jr., eds). Pacific Grove, CA: Wadsworth and Brooks/Cole.

Bock, M. E. (1975). Minimax estimators of the mean of a multivariate normal distribution. *Ann. Statist.* **3**, 209-218.

Bock, M. E. (1982). Employing vague inequality information in the estimation of normal mean vectors (Estimators that shrink toward closed convex polyhedra). *Statistical Decision Theory III* (S. S. Gupta and J. O. Berger, eds.). New York: Academic Press, pp. 169-193.

Bock, M. E. (1988). Shrinkage estimators: Pseudo-Bayes estimators for normal mean vectors. *Statistical Decision Theory IV* (S. S. Gupta and J. O. Berger, eds.). New York: Springer-Verlag, pp. 281-298.

Bondar, J. V. (1987). How much improvement can a shrinkage estimator give? *Foundations of Statistical Inference* (I. MacNeill and G Umphreys, eds.). Dordrecht: Reidel.

Bondar, J. V. and Milnes, P. (1981). Amenability: A survey for statistical applications of Hunt-Stein and related conditions on groups. *Zeitschr. Wahrsch. Verw. Geb.* **57**, 103-128.

Bondesson, L. (1975). Uniformly minimum variance estimation in location parameter families. *Ann. Statist.* **3**, 637-66.

Boos, D. D. (1979). A differential for L-statistics. *Ann. Statist.* **7**, 955-959.

Boos, D. D. and Serfling, R. J. (1980). A note on differentials and the CLT and LIL for statistical functions, with application to M-estimators. *Ann. Statist.* **86**, 618-624.

Borges, R. and Pfanzagl, J. (1965). One-parameter exponential families generated by transformation groups. *Ann. Math. Statist.* **36**, 261-271.

Box, G. E. and Tiao, G. C. (1973). *Bayesian Inference in Statistical Analysis*. Reading, MA: Addison-Wesley.

Box, G. E. P. and Cox, D. R. (1982). An analysis of transformations revisited, rebutted. *J. Amer. Statist. Assoc.* **77**, 209-210.

Boyles, R. A. (1983). On the convergence of the EM algorithm. *J. Roy. Statist. Soc. Ser. B* **45**, 47-50.

Bradley R A. and Gart, J.J. (1962). The asymptotic properties of ML estimators when sampling for associated populations. *Biometrika* **49**, 205-214.

Brandwein, A. C. and Strawderman, W. E. (1978). Minimax estimation of location parameters for spherically symmetric unimodal distributions under quadratic loss. *Ann. Statist.* **6**, 377-416.

Brandwein, A. C. and Strawderman, W. E. (1980). Minimax estimation of location parameters for spherically symmetric distributions with concave loss. *Ann. Statist.* **8**, 279-284.

Brandwein, A. C. and Strawderman, W. E. (1990). Stein estimation: The spherically symmetric case. *Statist. Sci.* **5**, 356-369.

Bravo, G. and MacGibbon, B. (1988). Improved shrinkage estimators for the mean vector of a scale mixture of normals with unknown variance. *Can. J. Statist.* **16**, 237-245.

Brewster, J. F and Zidek, J. V. (1974). Improving on equivariant estimators *Ann. Statist.* **2**, 21-38.

Brockwell, P. J. and Davis, R. A. (1987). *Time Series: Theory and Methods* New York: Springer-Verlag

Brown, L. D. (1966). On the admissibility of invariant estimators of one or more location parameters. *Ann. Math. Statist.* **37**, 1087-1136.

Brown, L. D. (1968). Inadmissibility of the usual estimators of scale parameters in problems with unknown location and scale parameters. *Ann. Math. Statist.* **39**, 29-48.

Brown, L. D. (1971). Admissible estimators, recurrent diffusions, and insoluble boundary value problems. *Ann. Math. Statist.* **42**, 855-903. [Corr: (1973) *Ann. Statist.* **1**, 594-596.]

Brown, L. D. (1975). Estimation with incompletely specified loss functions (the case of several location parameters). *J. Amer. Statist. Assoc.* **70**, 417-427.

Brown, L. D. (1978). A contirbution to Kiefer's theory of conditional confidence procedures. *Ann. Statist.* **6**, 59-71.

Brown, L. D. (1979). A heuristic method for determining admissibility of estimators - with applications. *Ann. Statist.* **7**, 960-994.

Brown, L. D. (1980a). A necessary condition for admissibility. *Ann. Statist.* **8**, 540-544.

Brown, L. D. (1980b). Examples of Berger's phenomenon in the estimation of independent normal means. *Ann. Statist.* **8**, 572-585.

Brown, L. D. (1981). A complete class theorem for statistical problems with finite sample spaces. *Ann. Statist.* **9**, 1289-1300.

Brown, L. D. (1986a). *Fundamentals of Statistical Exponential Families*. Hayward, CA: Institute of Mathematical Statistics

Brown, L. D. (1986b). Commentary on paper [19]. *J. C. Kiefer Collected Papers, Supplementary Volume*. New York: Springer-Verlag, pp. 20-27.

Brown, L. D. (1988). Admissibility in discrete and continuous invariant nonparametric estimation problems and in their multinomial analogs. *Ann. Statist.* **16**, 1567-1593.

Brown L. D. (1990a). An ancillarity paradox which appears in multiple linear regression (with discussion). *Ann. Statist.* **18**, 471-538.

Brown L. D. (1990b). Comment on the paper by Maatta and Casella. *Statist. Sci.* **5**, 103-106.

Brown, L. D. (1994). Minimaxity, more or less. *Statistical Decision Theory and Related Topics V* (S. S. Gupta and J. O. Berger, eds.). New York: Springer-Verlag, pp. 1-18.

Brown, L. D. and Cohen, A. (1974). Point and confidence estimation of a common mean and recovery of interblock information, *Ann. Statist.* **2**, 963-976.

Brown, L. D. and Farrell, R. (1985a). All admissible estimators of a multivariate Poisson mean. *Ann. Statist.* **13**, 282-294.

Brown, L. D. and Farrell, R. (1985b). Complete class theorems for estimation of multivariate Poisson means and related problems. *Ann. Statist.* **13**, 706-726.

Brown, L. D. and Farrell, R. (1990). A lower bound for the risk in estimating the value of a probability density. *J. Amer. Statist. Assoc.* **90**, 1147-1153.

Brown, L. D. and Fox, M. (1974a). Admissibility of procedures in two-dimensional location parameter problems. *Ann. Statist.* **2**, 248-266.

Brown, L. D. and Fox, M. (1974b). Admissibility in statistical problems involving a location or scale parameter. *Ann. Statist.* **2**, 807-814.

Brown, L. D. and Gajek, L. (1990). Information inequalities for the Bayes risk. *Ann. Statist.* **18**, 1578-1594.

Brown, L. D. and Hwang, J. T. (1982). A unified admissibility proof. *Statistical Decision Theory III* (S. S. Gupta and J. O. Berger, eds.). New York: Academic Press, pp. 205-230.

Brown, L. D. and Hwang, J. T. (1989). Universal domination and stochastic domination: U-admissibility and U-inadmissibility of the least squares estimator. *Ann. Statist.* **17**, 252-267.

Brown, L. D., Johnstone, I., and MacGibbon, B. K. (1981). Variation diminishing transformations: A direst approach to total positivity and its statistical applications. *J. Amer. Statist. Assoc.* **76**, 824-832.

Brown, L. D. and Low, M. G. (1991). Information inequality bounds on the minimax risk (with an application to nonparametric regression). *Ann. Statist.* **19**, 329-337.

Brown, L. D. and Purves, R. (1973). Measurable selection of extrema *Ann. Statist.* **1**, 902-912.

Brown, P. J. and Zidek, J. V. (1980). Adaptive multivariate ridge regression. *Ann. Statist.* **8**, 64-74.

Brown, L. D., Cohen, A. and Strawderman, W. E. (1976). A complete class theorem for strict monotone likelihood ratio with applications. *Ann. Statist.* **4**, 712-722.

Brown, L. D., Casella, G., and Hwang, J. T. G. (1995). Optimal confidence sets, bioequivalence, and the limaçon of Pascal. *J. Amer. Statist. Assoc.* **90**, 880-889.

Bucklew, J. A. (1990). *Large Deviation Techniques in Decision, Simulation and Estimation.* New York: Wiley.

Buehler, R. J. (1959). Some validity criteria for statistical inference. *Ann. Math. Statist.* **30**, 845-863.

Buehler, R. J. (1982). Some ancillary statistics and their properties. *J. Amer. Statist. Assoc.* **77**, 581-589.

Burdick, R. K. and Graybill, F. A. (1992). *Confidence Intervals on Variance Components.* New York: Marcel Dekker

Butler, R. W. (1986). Predictive likelihood inference with applications (with discussion). *J. Roy. Statist. Soc. Ser. B* **48**, 1-38.

Butler, R. W. (1989). Approximate predictive pivots and densities. *Biometrika* **76**, 489-501.

Carlin, B. P. and Louis, T. A. (1996). *Bayes and Empirical Bayes Methods for Data Analysis.* London: Chapman & Hall.

Carroll, R. J. and Lombard, F. (1985). A note on N estimators for the binomial distribution. *J. Amer. Statist. Assoc.* **80**, 423-426.

Carroll, R. J. Ruppert, D., and Stefanski, L. (1995). *Measurment Error in Nonlinear Models.* London: Chapman & Hall.

Carter, R. G., Srivastava, M. S., and Srivastava, V. K. (1990). Unbiased estimation of the MSE matrix of Stein-rule estimators, confidence ellipsoids, and hypothesis testing. *Econ. Theory* **6**, 63-74.

Casella, G. (1980). Minimax ridge regression estimation. *Ann. Statist.* **8**, 1036-1056.

Casella, G. (1985a). An introduction to empirical Bayes data analysis. *Amer. Statist.* **39**, 83-87.

Casella, G. (1985b). Matrix conditioning and minimax ridge regression estimation. *J. Amer. Statist. Assoc.* **80**, 753-758.

Casella, G. (1986). Stabilizing binomial *n* estimators. *J. Amer. Statist. Assoc.* **81**, 172-175.

Casella, G. (1987). Conditionally acceptable recentered set estimators. *Ann. Statist.* **15**, 1363-1371.

Casella, G. (1988). Conditionally acceptable frequentist solutions (with discussion). *Statistical Decision Theory IV* (S. S. Gupta and J. O. Berger, eds.). New York: Springer-Verlag, pp. 73-111.

Casella, G. (1990). Estimators with nondecreasing risk: Application of a chi-squared identity. *Statist. Prob. Lett.* **10**, 107-109.

Casella, G. (1992a). Illustrating empirical Bayes methods. *Chemolab* **16**, 107-125.

Casella, G. (1992b). Conditional inference from confidence sets. In *Current Issues in Statistical Inference: Essays in Honor of D. Basu* (M. Ghosh and P. K. Pathak, eds.). Hayward, CA: Institute of Mathematical Statistics, pp. 1-12.

Casella, G. and Berger, R. L. (1990). *Statistical Inference.* Pacific Grove, CA: Wadsworth/Brooks Cole.

Casella, G. and Berger, R. L. (1992). Deriving generalized means as least squares and maximum likelihood estimates. *Amer. Statist.* **46**, 279-282.

Casella, G. and Berger, R. L. (1994). Estimation with selected binomial information. *J. Amer. Statist. Assoc.* **89**, 1080-1090.

Casella, G. and Hwang, J. T. G. (1982). Limit expressions for the risk of James-Stein estimators. *Can. J. Statist.* **10**, 305-309.

Casella, G. and Hwang, J. T. (1983). Empirical Bayes confidence sets for the mean of a multivariate normal distribution. *J. Amer. Statist. Assoc.* **78**, 688-697.

Casella, G. and Hwang, J. T. (1987). Employing vague prior information in the construction of confidence sets. *J. Mult. Anal.* **21**, 79-104.

Casella, G. and Strawderman, W. E. (1981). Estimating a bounded normal mean. *Ann. Statist.* **9**, 870-878.

Casella, G. and Strawderman, W. E. (1994). On estimating several binomial N's. *Sankhya* **56**, 115-120.

Cassel, C., Särndal, C., and Wretman, J. H. (1977). *Foundations of Inference in Survey Sampling*. New York: Wiley.

Cellier, D., Fourdrinier, D., and Robert, C. (1989). Robust shrinkage estimators of the location parameter for elliptically symmetric distributions. *J. Mult. Anal.* **29**, 39-42.

Chan, K. S. and Geyer, C. J. (1994). Discussion of the paper by Tierney. *Ann. Statist.* **22**, 1747-1758

Chan, L. K. (1967). Remark on the linearized maximum likelihood estimate. *Ann. Math. Statist.* **38**, 1876-1881.

Chapman, D. G. and Robbins, H. (1951). Minimum variance estimation without regularity assumptions. *Ann. Math. Statist.* **22**, 581-586.

Chatterji, S. D. (1982). A remark on the Cramér-Rao inequality. In *Statistics and Probability: Essays in Honor of C. R. Rao* (G. Kallianpur, P. R. Krishnaiah, and J. K. Ghosh, eds.). New York: North Holland, pp. 193-196.

Chaudhuri, A. and Mukerjee, R. (1988). *Randomized Response: Theory and Techniques*. New York: Marcel Dekker.

Chen, J. and Hwang, J. T. (1988). Improved set estimators for the coefficients of a linear model when the error distribution is spherically symmetric. *Can. J. Statist.* **16**, 293-299.

Chen, L., Eichenauer-Herrmann, J., and Lehn, J. (1990). Gamma-minimax estimation of a multivariate normal mean. *Metrika* **37**, 1-6.

Chen, S-Y. (1988). Restricted risk Bayes estimation for the mean of a multivariate normal distribution. *J. Mult. Anal.* **24**, 207-217.

Chernoff, H. (1954). On the distribution of the likelihood ratio. *Ann. Math. Statist.* **25**, 573–578.

Chernoff, H. (1956). Large-sample theory: Parametric case. *Ann. Math. Statist.* **27**, 1-22.

Chib, S. and Greenberg, E. (1955). Understanding the Metropolis-Hastings algorithm. *The American Statistician* **49**, 327–335.

Chow, M. (1990). Admissibility of MLE for simultaneous estimation in negative binomila problems. *J. Mult. Anal.* **33**, 212-219.

Christensen, R. (1987). *Plane Answers to Complex Questions: The Theory of Linear Models, Second Edition*. New York: Springer-Verlag.

Christensen, R. (1990). *Log-linear Models*. New York: Springer-Verlag.

Churchill, G. A. (1985). Stochastic models for heterogeneous DNA. *Bull. Math. Biol.* (51) **1**, 79-94.

Chung, K. L. (1974). *A Course in Probability Theory, Second Edition*. New York: Academic Press.

Clarke, B. S. and Barron, A. R. (1990). Information-theoretic asymptotics of Bayes methods. *IEEE Trans. Inform. Theory* **36**, 453-471.

Clarke, B. S. and Barron, A. R. (1994). Jeffreys prior is asymptotically least favorable under entropy loss. *J. Statist. Plan. Inform.* **41**, 37-60.

Clarke, B. S. and Wasserman, L. (1993). Noninformative priors and nuisance parameters. *J. Amer. Statist. Assoc.* **88**, 1427-1432.

Cleveland, W. S. (1985). *The Elements of Graphing Data*. Monterey, CA: Wadsworth.

Clevensen, M. L. and Zidek, J. (1975). Simultaneous estimation of the mean of independent Poisson laws. *J. Amer. Statist. Assoc.* **70**, 698-705.

Cochran, W. G. (1977). *Sampling Techniques, Third Edition*. New York: Wiley.

Cohen, A. (1965a). Estimates of linear combinations of the parameters in the mean vector of a multivariate distribution. *Ann. Math. Statist.* **36**, 78-87.

Cohen, A. C. (1965b). Maximum likelihood estimation in the Weibull distribution based on complete and on censored samples. *Technometrics* **7**, 579-588.

Cohen, A. C. (1966). All admissible linear estimators of the mean vector. *Ann. Math. Statist.* **37**, 458-463.

Cohen, A. C. (1967). Estimation in mixtures of two normal distributions. *Technometrics* **9**, 15-28.

Cohen, A. (1981). Inference for marginal means in contingency tables. *J. Amer. Statist. Assoc.* **76**, 895-902.

Cohen, A. and Sackrowitz, H. B. (1974). On estimating the common mean of two normal distributions. *Ann. Statist.* **2**, 1274-1282.

Cohen, M. P. and Kuo, L. (1985). The admissibility of the empirical distribution function. *Ann. Statist.* **13**, 262-271.

Copas, J.B. (1972a) The likelihood surface in the linear functional relationship problem. *J. Roy. Statist. Soc. Ser. B* **34**, 274-278.

Copas, J. B. (1972b). Empirical Bayes methods and the repeated use of a standard. *Biometrika* **59**, 349-360.

Copas, J. B. (1975). On the unimodality of the likelihood for the Cauchy distribution. *Biometrika* **62**, 701-704.

Copas, J. B. (1983). Regression, prediction and shrinkage. *J. Roy. Statist. Soc. Ser. B* **45**, 311-354.

Corbeil, R. R. and Searle, S. R. (1976). Restricted maximum likelihood (REML) estimation of variance components in the mixed model. *Technometrics* **18**, 31-38.

Cox, D. R. (1958). Some problems connected with statistical inference. *Ann. Math. Statist.* **29**, 357-372.

Cox, D. R. (1970). *The Analysis of Binary Data*. London: Methuen.

Cox, D. R. (1975) Partial likelihood. *Biometrika* **62**, 269-276.

Cox, D. R. and Oakes, D. O. (1984). *Analysis of Survival Data*. London: Chapman & Hall.

Cox, D. R. and Reid, N. (1987). Parameter orthogonality and approximate conditional inference (with discussion). *J. Roy. Statist. Soc. Ser. B* **49**, 1-39.

Crain, B. R. (1976). Exponential models, maximum likelihood estimation, and the Haar condition. *J. Amer. Statist. Assoc.* **71**, 737-740.

Cramér, H. (1946a). *Mathematical Methods of Statistics*. Princeton, NJ: Princeton University Press.

Cramér, H. (1946b). A contribution to the theory of statistical estimation. *Skand. Akt. Tidskr.* **29**, 85-94.

Cressie, N. and Lahiri, S. N. (1993). The asymptotic distribution of REML estimators. *J. Mult. Anal.* **45**, 217-233.

Crow, E. L. and Shimizu, K. (1988). *Lognormal Distributions: Theory and Practice*. New York: Marcel Dekker.

Crowder, M. J. (1976). Maximum likelihood estimation for dependent observations. *J. Roy. Statist. Soc. Ser. B* **38**, 45-53.

Crowder, M. J. and Sweeting, T. (1989). Bayesian inference for a bivariate binomial distribution. *Biometrika* **76**, 599-603.

Daniels, H.E. (1954). Saddlepoint approximations in statistics. *Ann. Math. Statist.* **25**, 631-650.

Daniels, H. E. (1961). The asymptotic efficiency of a maximum likelihood estimators. *Proc. Fourth Berkeley Symp. Math. Statist. Prob.* **1**, University of California Press, 151-163.

Daniels, H. E. (1980). Exact saddlepoint approximations. *Biometrika* **67**, 59-63.

Daniels, H. E. (1983). Saddlepoint approximations for estimating equations. *Biometrika* **70**, 89-96.

Darmois, G. (1935). Sur les lois de probabilite a estimation exhaustive. *C. R. Acad. Sci. Paris* **260**, 265-1266.

Darmois, G. (1945). Sur les lois limites de la dispersion de certaines estimations. *Rev. Inst. Int. Statist.* **13**, 9-15.

DasGupta, A. (1985). Bayes minimax estimation in multiparameter families when the parameter space is restricted to a bounded convex set. *Sankhya* **47**, 326-332.

DasGupta, A. (1991). Diameter and volume minimizing confidence sets in Bayes and classical problems. *Ann. Statist.* **19**, 1225-1243.

DasGupta, A. (1994). An examination of Bayesian methods and inference: In search of the truth. Technical Report, Department of Statistics, Purdue University.

DasGupta, A. and Bose, A. (1988). Gamma-minimax and restricted-risk Bayes estimation of multiple Poisson means under ϵ-contamination of the subjective prior. *Statist. Dec.* **6**, 311-341.

DasGupta, A. and Rubin, H. (1988). Bayesian estimation subject to minimaxity of the mean of a multivariate normal distribution in the case of a common unknown variance. *Statistical Decision Theory and Related Topics IV* (S. S. Gupta and J. O. Berger, eds.). New York: Springer-Verlag, pp. 325-346.

DasGupta, A. and Studden, W. J. (1989). Frequentist behavior of robust Bayes estimates of normal means. *Statist. Dec.* **7**, 333-361.

Datta, G. S. and Ghosh, J. K. (1995). On priors providing frequentist validity for Bayesian inference. *Biometrika* **82**, 37-46.

Davis, L. J. (1989). Intersection-union tests for strict collapsibility in three-dimensional contingency tables. *Ann. Statist.* **17**, 1693-1708.

Dawid, A. P. (1983). Invariant prior distributions *Encycl. Statist. Sci.* **4**, 228-235.

deFinetti, B. (1937). La prevision: Ses lois logiques, ses source subjectives. *Ann. Inst. Henri Poincare* **7**, 1-68. (Translated in *Studies in Subjective Probability* (H. Kyburg and H. Smokler, eds.). New York: Wiley.)

deFinetti, B. (1970). *Teoria Delle Probabilita*. English Translation (1974) *Theory of Probability*. New York: Wiley.

deFinetti, B. (1974). *Theory of Probability, Volumes I and II*. New York: Wiley

DeGroot, M. (1970). *Optimal Statistical Decisions*. New York: McGraw-Hill.

DeGroot, M., H. and Rao, M. M. (1963). Bayes estimation with convex loss. *Ann. Math. Statist.* **34**, 839-846.

Deeley, J. J. and Lindley, D. V. (1981). Bayes empirical Bayes. *J. Amer. Statist. Assoc.* **76**, 833-841.

Dempster, A. P. (1971). An overview of multivariate data analysis. *J. Mult. Anal.* **1**, 316-346.

Dempster, A. P., Laird, N. M., and Rubin, D. B. (1977). Maximum likelihood from incomplete data via the EM algorithm. *J. Roy. Statist. Soc. Ser. B* **39**, 1-22.

Denny, J. L. (1964). A real-valued continuous function on R almost everywhere 1-1. *Fund. Math.* **55**, 95-99.

Denny, J. L. (1969). Note on a theorem of Dynkin on the dimension of sufficient statistics. *Ann. Math. Statist.* **40**, 1474-1476.

Devroye, L. (1985). *Non-Uniform Random Variable Generation*. New York: Springer-Verlag.

Devroye, L. and Gyoerfi, L. (1985). *Nonparametric Density Estimation: The L_1 View*. New York: Wiley.

Dey, D. K and Srinivasan, C. (1985). Estimation of a covariance matrix under Stein's loss. *Ann. Statist.* **13**, 1581-1591.

Dey, D. K., Ghosh, M., and Srinivasan, C. (1987). Simultaneous estimation of parameters under entropy loss. *J. Statist. Plan. Inform.* **15**, 347-363.

Diaconis, P. (1985). Theories of data analysis, from magical thinking through classical statistics. In *Exploring Data Tables, Trends and Shapes* (D. Hoaglin, F. Mosteller, and J. W. Tukey, eds.). New York: Wiley, pp. 1-36.

Diaconis, P. (1988). *Group Representations in Probability and Statistics*. Hayward, CA: Institute of Mathematical Statistics

Diaconis, P. and Ylvisaker, D. (1979). Conjugate priors for exponential families. *Ann. Statist.* **7**, 269-281.

Dobson, A. J. (1990). *An Introduction to Generalized Linear Models*. London: Chapman & Hall.

Donoho, D. (1994). Statistical estimation and optimal recovery. *Ann. Statist.* **22**, 238-270.

Donoho, D. L., Liu, R. C., and MacGibbon, B. (1990). Minimax risk over hyperrectangles, and implications. *Ann. Statist.* **18**, 1416-1437.

Doss, H. and Sethuraman, J. (1989). The price of bias reduction when there is no unbiased estimate. *Ann. Statist.* **17**, 440-442.

Downton, F. (1973). The estimation of $P(Y < X)$ in the normal case. *Technometrics* **15**, 551-558.

Draper, N. R and Van Nostrand, R. C. (1979). Ridge regression and James-Stein estimation: Review and comments. *Technometrics* **21**, 451-466.

Dudley, R. M. (1989). *Real Analysis and Probability*. Pacific Grove, CA: Wadsworth and Brooks/Cole.

Dynkin, E. B. (1951). Necessary and sufficient statistics for a family of probability distributions. English translation in *Select. Transl. Math. Statist. Prob.* **I** (1961) 23-41.

Eaton, M. L. (1989). *Group Invariance Applications in Statistics*. Regional Conference Series in Probability and Statistics. Hayward, CA: Institute of Mathematical Statistics

Eaton, M. L. (1992). A statistical diptych: Admissible inferences-recurrence of symmetric Markov chains. *Ann. Statist.* **20**, 1147-1179.

Eaton, M. L. and Morris, C. N. (1970). The application of invariance to unbiased estimation. *Ann. Math. Statist.* **41**,1708-1716.

Eaves, D. M. (1983). On Bayesian nonlinear regression with an enzyme example. *Biometrika* **70**, 367-379.

Edgeworth, F. Y. (1883). The law of error. *Phil. Mag. (Fifth Series)* **16**, 300-309.

Edgeworth, F. Y. (1908, 1909) On the probable errors of frequency constants. *J. Roy. Statist. Soc. Ser. B* **71**, 381-397, 499-512, 651-678; **72**, 81-90.

Edwards, A. W. F. (1974). The history of likelihood, *Int. Statist. Rev.* **42**, 4-15.

Edwards, A. W. F. (1992). *Likelihood*. Baltimore: Johns Hopkins University Press.

Efron, B. (1975). Defining the curvature of a statistical problem (with applications to second order efficiency). *Ann. Statist.* **3**, 1189-1242.

Efron, B. (1978). The geometry of exponential families. *Ann. Statist.* **6**, 362-376.

Efron, B. (1979). Bootstrap methods: Another look at the jackknife. *Ann. Statist.* **7**, 1-26.

Efron, B. (1982a). Maximum likelihood and decision theory. *Ann. Statist.* **10**, 340-356.

Efron, B. (1982b). *The Jackknife, the Bootstrap, and other Resampling Plans*. Volume 38 of CBMS-NSF Regional Conference Series in Applied Mathematics. Philadelphia: SIAM.

Efron, B. (1990). Discussion of the paper by Brown. *Ann. Statist.* **18**, 502-503.

Efron, B. and Hinkley, D. (1978). Assessing the accuracy of the maximum likelihood estimator: Observed vs. expected Fisher information. *Biometrica* **65**, 457-481.

Efron, B. and Johnstone, I. (1990). Fisher's information in terms of the hazard ratio. *Ann. Statist.* **18**, 38-62.

Efron, B. and Morris, C. N. (1971). Limiting the risk of Bayes and empirical Bayes estimators–Part I: The Bayes case. *J. Amer. Statist. Assoc.* **66**, 807-815.

Efron, B. and Morris, C. N. (1972a). Limiting the risk of Bayes and empirical Bayes estimators–Part II: The empirical Bayes case. *J. Amer. Statist. Assoc.* **67**, 130-139.

Efron, B. and Morris, C. (1972b). Empirical Bayes on vector observations-An extension of Stein's method. *Biometrika* **59**, 335-347.

Efron, B. and Morris, C. N. (1973a). Stein's estimation rule and its competitors–an empirical Bayes approach. *J. Amer. Statist. Assoc.* **68**, 117-130.

Efron, B. and Morris, C. (1973b). Combining possibly related estimation problems (with discussion). *J. Roy. Statist. Soc. Ser. B* **35**, 379-421.

Efron, B. and Morris, C. (1975). Data analysis using Stein's estimator and its generalizations. *J. Amer. Statist. Assoc.* **70**, 311-319.

Efron, B. and Morris, C. N. (1976a). Families of minimax estimators of the mean of a multivariate normal distribution. *Ann. Statist.* **4**, 11-21.

Efron, B. and Morris, C. N. (1976b). Multivariate empirical Bayes and estimation of covariance matrices. *Ann. Statist.* **4**, 22-32.

Efron, B. and Tibshirani, R. J. (1993). *An Introduction to the Bootstrap*. London: Chapman & Hall.

Eichenauer-Herrmann, J. and Fieger, W. (1992). Minimax estimation under convex loss when the parameter interval is bounded. *Metrika* **39**, 27-43.

Eichenauer-Herrmann, J. and Ickstadt, K. (1992). Minimax estimators for a bounded location parameter. *Metrika* **39**, 227-237.

Eisenhart, C. (1964). The meaning of 'least' in least squares. *J. Wash. Acad. Sci.* **54**, 24-33.

Ericson, W. A. (1969). Subjective Bayesian models in sampling finite populations (with discussion). *J. Roy. Statist. Soc. Ser. B* **31**, 195-233.

Everitt, B. S. (1992). *The Analysis of Contingency Tables, Second Edition*. London: Chapman & Hall.

Everitt, B. S. and Hand, D. J. (1981). *Finite Mixture Distributions*. London: Chapman & Hall.

Fabian, V. and Hannan, J. (1977). On the Cramér-Rao inequality. *Ann. Statist.* **5**, 197-205.

Faith, R. E. (1976). Minimax Bayes point and set estimators of a multivariate normal mean. Ph.D. Thesis, Department of Statistics, University of Michigan.

Faith, R. E. (1978). Minimax Bayes point estimators of a multivariate normal mean. *J. Mult. Anal.* **8** 372-379.

Fan, J. and Gijbels, I. (1992). Minimax estimation of a bounded squared mean. *Statist. Prob. Lett.* **13**, 383-390.

Farrell, R. (1964). Estimators of a location parameter in the absolutely continuous case. *Ann. Math. Statist.* **35**, 949-998.

Farrell, R. H. (1968). On a necessary and sufficient condition for admissibility of estimators when strictly convex loss is used. *Ann. Math. Statist.* **38**, 23-28.

Farrell, R. H., Klonecki, W., and Zontek. (1989). All admissible linear estimators of the vector of gamma scale parameters with applications to random effects models. *Ann. Statist.* **17**, 268-281.

Feldman, I. (1991). Constrained minimax estimation of the mean of the normal distribution with known variance *Ann. Statist.* **19**, 2259-2265.

Feldman, D and Fox, M. (1968). Estimation of the parameter *n* in the binonial distribution. *J. Amer. Statist. Assoc.* **63** 150-158.

Feller, W. (1968). *An Introduction to Probability Theory and Its Applications, Volume 1, Third Edition*. New York: Wiley.

Fend, A. V. (1959). On the attainment of the Cramér-Rao and Bhattacharyya bounds for the variance of an estimate. *Ann. Math. Statist.* **30**, 381-388.

Feng, Z. and McCulloch, C. E. (1992). Statistical inference using maximum likelihood estimation and the generalized likelihood ratio when the true parameter is on the boundary of the parameter space. *Statist. Prob. Lett.* **13**, 325-332.

Ferguson, T. S. (1962). Location and scale parameters in exponential families of distributions. *Ann. Math. Statist.* **33**, 986-1001. (Correction **34**, 1603.)

Ferguson, T. S. (1967). *Mathematical Statistics: A Decision Theoretic Approach*. New York: Academic Press.

Ferguson, T. S. (1973). A Bayesian analysis of some nonparametric problems. *Ann. Statist.* **1**, 209-230.

Ferguson, T. S. (1978). Maximum likelihood estimation of the parameters of the Cauchy distribution for samples of size 3 and 4. *J. Amer. Statist. Assoc.* **73**, 211-213.

Ferguson, T. S. (1982). An inconsistent maximum likelihood estimate. *J. Amer. Statist. Assoc.* **77**, 831-834.

Ferguson, T. S. (1996). *A Course in Large Sample Theory*. London: Chapman & Hall.

Field, C. A. and Ronchetti, E. (1990). *Small Sample Asymptotics*. Hayward, CA: Institute of Mathematical Statistics.

Finch, S. J., Mendell, N. R. and Thode, H. C. (1989). Probabilistic measures of adequacy of a numerical search for a global maximum. *J. Amer. Statist. Assoc.* **84**, 1020-1023.

Finney, D. J. (1971). *Probit Analysis*. New York: Cambridge University Press.

Fisher, R. A. (1920). A mathematical examination of the methods of determining the accuracy of an observation by the mean error, and by the mean square error. *Monthly Notices Roy. Astron. Soc.* **80**, 758-770.

Fisher, R. A. (1922). On the mathematical foundations of theoretical statistics. *Philos. Trans. Roy. Soc. London, Ser. A* **222**, 309-368.

Fisher, R. A. (1925). Theory of statistical estimation. *Proc. Camb. Phil. Soc.* **22**, 700-725.

Fisher, R. A. (1930). Inverse probability. *Proc. Camb. Phil. Soc.* **26**, 528-535.

Fisher, R. A. (1935). The fiducial argument in statistical inference. *Ann. Eugenics* **6**, 391-398.

Fisher, R. A. (1934). Two new properties of mathematical likelihood. *Proc. Roy. Soc. A* **144**, 285-307.

Fisher, R. A. (1956). On a test of significance in Pearson's Biometrika tables (No. 11). *J. Roy. Statist. Soc. B* **18**, 56-60.

Fisher, R. A. (1959). *Statistical Methods and Scientific Inference*, Second Edition. New York: Hafner. Reprinted 1990, Oxford: Oxford University Press.

Fisher, N. I. (1982). Unbiased estimation for some new parametric families of distributions. *Ann. Statist.* **10**, 603-615.

Fleming, T. R. and Harrington, D. P. (1991). *Counting Processes and Survival Analysis*. New York: Wiley.

Fourdrinier, D. and Wells, M. T. (1995). Estimation of a loss function for spherically symmetric distributions in the general linear model. *Ann. Statist.* **23**, 571-592.

Fourdrinier, D., Strawderman, W. E., and Wells, M. T. (1998). On the construction of proper Bayes minimax estimators. *Ann. Statist.* **26**, No. 2.

Foutz, R. V. (1977). On the unique consistent solution to the likelihood equations. *J. Amer. Statist. Assoc.* **72**, 147-148.

Fox, M. (1981). An inadmissible best invariant estimator: The i.i.d. case. *Ann. Statist.* **9**, 1127-1129.

Fraser, D. A. S. (1954). Completeness of order statistics. *Can. J. Math.* **6**, 42-45.

Fraser, D. A. S. (1968). *The Structure of Inference*. New York: Wiley

Fraser, D. A. S. (1979). *Inference and Linear Models*. New York: McGraw-Hill.

Fraser, D. A. S. and Guttman, I. (1952). Bhattacharyya bounds without regularity assumptions. *Ann. Math. Statist.* **23**, 629- 632.

Fréchet, M. (1943). Sur l'extension de certaines evaluations statistiques de petits echantillons. *Rev. Int. Statist.* **11**, 182-205.

Freedman, D. and Diaconis, P. (1982). On inconsistent M-estimators. *Ann. Statist.* **10**, 454-461.

Fu, J. C. (1982). Large sample point estimation: A large deviation theory approach. *Ann. Statist.* **10**, 762-771.

Gajek, L. (1983). Sufficient conditions for admissibility. *Proceedings of the Fourth Pannonian Symposium on Mathematical Statistics* (F. Konecny, J. Mogyoródi, W. Wertz, eds)., Bad Tatzmannsdorf, Austria, pp. 107-118.

Gajek, L. (1987). An improper Cramér-Rao lower bound. *Applic. Math.* **XIX**, 241-256.

Gajek, L. (1988). On the minimax value in the scale model with truncated data. *Ann. Statist.* **16**, 669-677.

Gajek, L. and Kaluzka, M. (1995). Nonexponential applications of a global Cramér-Rao inequality. *Statistics* **26**, 111-122.

Gart, J. J. (1959). An extension of the Cramér-Rao inequality. *Ann. Math. Statist.* **30**, 367-380.

Gatsonis, C. A. (1984). Deriving posterior distributions for a location parameter: A decision-theoretic approach. *Ann. Statist.* **12**, 958-970.

Gatsonis, C., MacGibbon, B., and Strawderman, W. (1987). On the estimation of a restricted normal mean. *Statist. Prob. Lett.* **6**, 21-30.

Gauss, C. F. (1821). Theoria combinationis obsercationunt erronbus minimis obnoxiae. An English translation can be found in Gauss's work (1803-1826) on the Theory of Least Squares. Trans. H. F. Trotter. Statist. Techniques Res. Group. Tech. Rep. No. 5. Princeton University. Princeton. (Published translations of these papers are available in French and German.)

Gelfand, A. E. and Dey, D. K. (1988). Improved estimation of the disturbance variance in a linear regression model. *J. Economet.* **39**, 387-395.

Gelfand, A. E. and Smith, A. F. M. (1990). Sampling-based approaches to calculating marginal densities. *J. Amer. Statist. Assoc.* **85**, 398-409.

Gelman, A. and Rubin, D. B. (1992). Inference from iterative simulation using multiple sequences (with discussion). *Statist. Sci.* **7**, 457-511.

Gelman, A., Carlin, J., Stern, H., and Rubin, D.B. (1995). *Bayesian Data Analysis*. London: Chapman & Hall.

Geman, S. and Geman, D. (1984). Stochastic relaxation, Gibbs distributions and the Bayesian restoration of images. *IEEE Trans. Pattern Anal. Mach. Intell.* **PAMT-6**, 721-740.

George, E. I. (1986a). Minimax multiple shrinkage estimators. *Ann. Statist.* **14**, 188-205.

George, E. I. (1986b). Combining minimax shrinkage estimators. *J. Amer. Statist. Assoc.* **81**, 437-445.

George, E. I. (1991). Shrinkage domination in a multivariate common mean problem. *Ann. Statist.* **19**, 952-960.

George, E. I. and Casella, G. (1994). An empirical Bayes confidence report. *Statistica Sinica* **4**, 617-638.

George, E. I. and McCulloch, R. (1993). On obtaining invariant prior distributions. *J. Statist. Plan. Inform.* **37**, 169-179.

George, E. I. and Robert, C. P. (1992). Capture-recapture estimation via Gibbs sampling. *Biometrika* **79**, 677-683.

George, E. I., Makov, Y., and Smith, A. F. M. (1993). Conjugate likelihood distributions. *Scand. J. Statist.* **20**, 147-156.

George, E. I., Makov, Y., and Smith, A. F. M. (1994). Fully Bayesian hierarchical analysis for exponential families via Monte Carlo simulation. *Aspects of Uncertainty* (P.R. Freeman and A. F. M. Smith, eds.). New York: Wiley, pp. 181-198.

Geyer, C. (1992). Practical Markov chain Monte Carlo (with discussion). *Statist. Sci.* **7**, 473-511.

Ghosh, J. K., and Mukerjee, R. (1991). Characterization of priors under which Bayesian and frequentist Bartlett corrections are equivalent in the multiparameter case. *J. Mult. Anal.* **38**, 385-393.

Ghosh, J. K. and Mukerjee, R. (1992). Non-informative priors (with discussion). In Bayesian Statistics IV (J. M. Bernardo, J. O. Berger, A. P. Dawid, and A. F. M. Smith, eds.). Oxford: Oxford University Press, pp. 195-210.

Ghosh, J. K. and Mukerjee, R. (1993). On priors that match posterior and frequentist distribution functions. *Can. J. Statist.* **21**, 89-96.

Ghosh, J. K. and Sinha, B. K. (1981). A necessary and sufficient condition for second order admissibility with applications to Berkson's bioassay problem. *Ann. Statist.* **9**, 1334-1338.

Ghosh, J. K. and Subramanyam, K. (1974). Second order efficiency of maximum likelihood estimators. *Sankhya* A **36**, 325-358.

Ghosh, M. (1974). Admissibility and minimaxity of some maximum likelihood estimators when the parameter space is restricted to integers. *J. Roy. Statist. Soc. Ser. B* **37**, 264-271.

Ghosh, M. N. (1964). Uniform approximation of minimax point estimates. *Ann. Math. Statist.* **35**, 1031-1047.

Ghosh, M. and Meeden, G. (1977). Admissibility of linear estimators in the one parameter exponential family. *Ann. Statist.* **5**, 772-778.

Ghosh, M. and Meeden, G. (1978). Admissibility of the MLE of the normal integer mean. *Sankhya* B **40**, 1-10.

Ghosh, M., Hwang, J. T., and Tsui, K-W. (1983). Construction of improved estimators in multiparameter estimation for discrete exponential families. *Ann. Statist.* **11**, 351-367.

Ghosh, M., Hwang, J. T., and Tsui, K-W. (1987). Construction of improved estimators in multiparameter estimation for discrete exponential families. *Ann. Statist.* **11**, 368-376.

Ghurye, S. G. and Olkin, (1969), Unbiased estimation of some multivariate probability densities. *Ann. Math. Statist.* **40**, 1261-1271.

Giesbrecht, F. and Kempthorne, O. (1976). Maximum likelihood estimation in the three-parameter lognormal distribution. *J. Roy. Statist. Soc. Ser. B* **38**, 257-264.

Gilks, W.R., Richardson, S., and Spiegelhalter, D.J., eds. (1996). *Markov Chain Monte Carlo in Practice*. London: Chapman & Hall.

Girshick, M. A. and Savage, L. J. (1951). Bayes and minimax estimates for quadratic loss functions. University of California Press, pp. 53-73.

Girshick, M. A., Mosteller, F., and Savage, L. J. (1946). Unbiased estimates for certain binomial sampling problems with applications. *Ann. Math. Statist.* **17**, 13-23.

Glasser, G. J. (1962). Minimum variance unbiased estimators for Poisson probabilities. *Technometrics* **4**, 409-418.

Gleser, L. J. (1979). Minimax estimation of a normal mean vector when the covariance matrix is unknown. *Ann. Statist.* **7**, 838-846.

Gleser, L. J. (1981). Estimation in a multivariate errors-in-variables regression model: Large sample results. *Ann. Statist.* **9**, 24-44.

Gleser, L. J. (1986). Minimax estimators of a normal mean vector for arbitrary quadratic loss and unknown covariance matrix. *Ann. Statist.* **14**, 1625-1633.

Gleser, L. J. (1991). Measurement error models (with discussion). *Chem. Int. Lab. Syst.* **10**, 45-67.

Gleser, L. J. and Healy, J. (1976). Estimating the mean of a normal distribution with known coefficient of variation. *J. Amer. Statist. Assoc.* **71**, 977-981.

Godambe, V. P. (1955). A unified theory of sampling from finite populations. *J. Roy. Statist. Soc. Ser. B* **17**, 269-278.

Godambe, V. P. (1982). Estimation in survey sampling: Robustness and optimality. *J. Amer. Statist. Assoc.* **77**, 393-406.

Godambe, V. P. (1991). *Estimating Functions*. UK:Clarendon Press.

Goel, P. and DeGroot, M. (1979). Comparison of experiments and information measures. *Ann. Statist.* **7** 1066-1077.

Goel, P. and DeGroot, M. (1980). Only normal distributions have linear posterior expectations in linear regression. *J. Amer. Statist. Assoc.* **75**, 895-900.

Goel, P. and DeGroot, M. (1981). Information about hyperparameters in hierarchical models. *J. Amer. Statist. Assoc.* **76**, 140-147.

Good, I. J. (1952). Rational decisions. *J. Roy. Statist. Soc. Ser. B* **14**, 107-114.

Good, I. J. (1965). The Estimation of Probabilities: An Essay on Modern Bayesian Methods. Cambridge: M.I.T. Press.

Goodman, L. A. (1970). The multivariate analysis of qualitative data: Interactions among multiple classifications. *J. Amer. Statist. Assoc.* **65**, 226-256.

Goutis, C. and Casella, G. (1991). Improved invariant confidence intervals for a normal variance. *Ann. Statist.* **19**, 2015-2031.

Goutis, C. and Casella, G. (1992). Increasing the confidence in Student's t. *Ann. Statist.* **20**, 1501-1513.

Govindarajulu, Z. and Vincze, I. (1989). The Cramér-Fréchet-Rao inequality for sequential estimation in non-regular case. *Statist. Data. Anal. Inf* 257-268.

Graybill, F. A. and Deal, R. B. (1959). Combining unbiased estimators. *Biometrics* **15**, 543-550.

Green, E. and Strawderman, W. E. (1991). A James-Stein type estimator for combining unbiased and possibly biased estimators. *J. Amer. Statist. Assoc.* **86**, 1001-1006.

Groenebroom, P. and Oosterhoof, J. (1981). Bahadur efficiency and small sample efficiency. *Int. Statist. Rev.* **49**, 127-141.

Gupta, A. K. and Rohatgi, V. K. (1980). On the estimation of restricted mean. *J. Statist. Plan. Inform.* **4**, 369-379.

Gupta, M. K. (1966). On the admissibility of linear estimates for estimating the mean of distributions of the one parameter exponential family. *Calc. Statist. Assoc. Bull.* **15**, 14-19.

Guttman, S. (1982a). Stein's paradox is impossible in problems with finite parameter spaces. *Ann. Statist.* **10**, 1017-1020.

Guttman, S. (1982b). Stein's paradox is impossible in the nonanticipative context. *J. Amer. Statist. Assoc.* **77**, 934-935.

Haas, G., Bain, L., and Antle, C. (1970). Inferences for the Cauchy distribution based on maximum likelihood estimators. *Biometrika* **57**, 403- 408.

Haberman, S. J. (1973). Loglinear models for frequency data: Sufficient statistics and likelihood equations. *Ann. Statist.* **1**, 617-632.

Haberman, S. J. (1974). *The Analysis of Frequency Data*. Chicago: University of Chicago Press.

Haberman, S. (1989). Concavity and estimation. *Ann. Statist.* **17**, 1631-1661.

Haff, L. R. (1979). Estimation of the inverse covariance matrix: Random mixtures of the inverse Wishart matrix and the identity. *Ann. Statist.* **7** 1264-1276.

Haff, L. R. and Johnson, R. W. (1986). The superharmonic condition for simultaneous estimation of means in exponential families. *Can. J. Statist.* **14**, 43-54.

Hájek, J. (1972). Local asymptotic minimax and admissibility in estimation. *Proc. Sixth Berkeley Symp. Math. Statist. Prob.* **1**, University of California Press175-194.

Hall, P. (1990). Pseudo-likelihood theory for empirical likelihood. *Ann. Statist.* **18**, 121-140.

Hall, P. (1992). *The Bootstrap and Edgeworth Expansion*. New York: Springer-Verlag.

Hall, P. and La Scala, B. (1990). Methodology and algorithms of empirical likelihood. *Int. Statist. Rev.* **58**, 109-127.

Hall, W. J., Wijsman, R. A., and Ghosh, J. R. (1965). The relationship between sufficiency and invariance with applications in sequential analysis. *Ann. Math. Statist.* **36**, 575-614.

Halmos, P. R. (1946). The theory of unbiased estimation. *Ann. Math. Statist.* **17**, 34-43.

Halmos, P. R (1950). *Measure Theory*. New York: Van Nostrand.

Halmos, P. R. and Savage, L. J. (1949). Application of the Radon-Nikodym theorem to the theory of sufficient statistics. *Ann. Math. Statist.* **20**, 225-241.

Hammersley, J. M. (1950). On estimating restricted parameters. *J. Roy. Statist. Soc. Ser. B* **12**, 192-240.

Hampel, F. R., Ronchetti, E. M., Rousseeuw, P. J., and Stahel, W. A. (1986). *Robust Statistics: The Approach Based on Influence Functions*. New York: Wiley.

Hardy, G. H., Littlewood, J. E., and Polya, G. (1934). *Inequalities*. Cambridge: Cambridge University Press.

Harter, H. L. (1974-1976). The method of least squares and some alternatives. *Int. Statist. Rev.* **42**, 147-174, 235-268, 282; **43**, 1-44, 125-190, 269-278; **44**, 113-159.

Hartley, H. O. (1958). Maximum likelihood estimation from incomplete data. *Biometrics* **14**, 174-194.

Hartung, J. (1981). Nonnegative minimum biased invariant estimation in variance component models. *Ann. Statist.* **9**, 278-292.

Harville, D. (1976). Extensions of the Gauss-Markov theorem to include the estimation of random effects. *Ann. Statist.* **4**, 384-395.

Harville, D. N. (1977). Maximum likelihood approaches to variance component estimation and to related problems. *J. Amer. Statist. Assoc.* **72**, 320-340.

Harville, D. (1981). Unbiased and minimum-variance unbiased estimation of estimable functions for fixed linear models with arbitrary covariance structure. *Ann. Statist.* **9**, 633-637.

Hastings, W. K. (1970). Monte Carlo sampling methods using Markov chains and their application. *Biometrika* **57**, 97-109.

Heath, D. and Sudderth, W. (1989). Coherent inference from improper priors and from finitely additive priors. *Ann. Statist.* **17**, 907-919.

Hedayat, A. S. and Sinha, B. K. (1991). *Design and Inference in Finite Population Sampling*. New York: Wiley.

Helms, L. (1969). *Introduction to Potential Theory*. New York: Wiley

Heyde, C. C. and Feigin, P. D. (1975). On efficiency and exponential families in stochastic process estimation. In *Statistical Distributions in Scientific Work* **1** (G. P. Patil, S. Kotz, and K. Ord, eds.). Dordrecht: Reidel, pp. 227-240.

Hill, B. M. (1963). The three-parameter lognormal distribution and Bayesian analysis of a point-source epidemic. *J. Amer. Statist. Assoc.* **58**, 72-84.

Hill, B. M. (1965). Inference about variance components in a one-way model. *J. Amer. Statist. Assoc.* **60**, 806-825.

Hinkley, D. V. (1979). Predictive likelihood. *Ann. Statist.* **7**, 718-728. (Corr: **8**, 694.)

Hinkley, D. V. (1980). Likelihood. *Can. J. Statist.* **8**, 151-163.

Hinkley, D. V. and Runger, G. (1984). The analysis of transformed data (with discussion). *J. Amer. Statist. Assoc.* **79**, 302-320.

Hinkley, D. V., Reid, N., and Snell, L. (1991). *Statistical Theory and Modelling. In honor of Sir David Cox*. London: Chapman & Hall.

Hipp, C. (1974). Sufficient statistics and exponential families. *Ann. Statist.* **2**, 1283-1292.

Hjort, N. L. (1976). Applications of the Dirichlet process to some nonparametric problems. (Norwegian). Univ. of Tromsf, Inst. for Math. and Phys. Sciences.

Hoadley, B. (1971). Asymptotic properties of maximum likelihood estimators for the independent not identically distributed case. *Ann. Math. Statist.* **42**, 1977-1991.

Hoaglin, D. C. (1975). The small-sample variance of the Pitman location estimators. *J. Amer. Statist. Assoc.* **70**, 880-888.

Hoaglin, D. C., Mosteller, F., and Tukey, J. W. (1985). *Exploring Data Tables, Trends and Shapes*. New York: Wiley, pp. 1-36.

Hobert, J. (1994). Occurrences and consequences of nonpositive Markov chains in Gibbs sampling. Ph.D. Thesis, Biometrics Unit, Cornell University.

Hobert, J. and Casella, G. (1996). The effect of improper priors on Gibbs sampling in hierarchical linear mixed models. *J. Amer. Statist. Assoc.* **91** 1461-1473.

Hodges, J. L., Jr., and Lehmann, E. L. (1950). Some problems in minimax point estimation. *Ann. Math. Statist.* **21**, 182-197.

Hodges, J. L., Jr., and Lehmann, E. L. (1951). Some applications of the Cramér-Rao inequality. *Proc. Second Berkeley Symp. Math. Statist. Prob.* **1**, University of California Press, 13-22.

Hodges, J. L., Jr., and Lehmann, E. L. (1952). The use of previous experience in reaching statistical decisions. *Ann. Math. Statist.* **23**, 396-407.

Hodges, J. L., Jr., and Lehmann, E. L. (1981). Minimax estimation in simple random sampling. In *Essays in Statistics in honor of C. R. Rao* (P. Krishnaiah, ed.). Amsterdam: North-Holland. pp. 323-327.

Hoeffding, W. (1948). A class of statistics with asymptotically normal distribution. *Ann. Math. Statist.* **19**, 293-325.

Hoeffding, W. (1977). Some incomplete and boundedly complete families of distributions. *Ann. Statist.* **5**, 278-291.

Hoeffding, W. (1982). Unbiased range-preserving estimators. *Festschrift for Erich Lehmann* (P. J. Bickel, K. A. Doksum, and J. L. Hodges, Jr., eds.). Pacific Grove, CA: Wadsworth and Brooks/Cole, pp. 249-260

Hoerl, A. E. and Kennard, R. W. (1971a). Ridge regression: Biased estimation for nonorthogonal problems. *Technometrics* **12**, 55-67.

Hoerl, A. E. and Kennard, R. W. (1971b). Ridge regression: Applications to nonorthogonal problems. *Technometrics* **12**, 69-82. (Corr: **12**, 723.)

Hoffmann-Jorgensen, J. (1994). *Probability with a View Toward Statistics, Volumes I and II*. London: Chapman & Hall.

Hogg, R. V. (1960). On conditionai expectations of location statistics. *J. Amer. Statist. Assoc.* **55**, 714 717.

Honda, T. (1991). Minimax estimators in the MANOVA model for arbitrary quadratic loss and unknown covariance matrix. *J. Mult. Anal.* **36**, 113-120.

Hora, R. B. and Buehler, R. J. (1966). Fiducial theory and invariant estimation. *Ann. Math. Statist.* **37**, 643-656.

Hsu, J. C. (1982). Simultaneous inference with respect to the best treatment in block designs. *J. Amer. Statist. Assoc.* **77**, 461-467.

Hsu, P. L. (1938). On the best unbiased quadratic estimate of the variance. *Statist. Research Mem.* **2**, 91-104.

Hsuan, F. C. (1979). A stepwise Bayes procedure. *Ann. Statist.* **7**, 860-868.

Huber, P. J. (1964). Robust estimation of a location parameter. *Ann. Math. Statist.* **35**, 73-101.

Huber, P. J. (1966). Strict efficiency excludes superefficiency (Abstract). *Ann. Math. Statist.* **37**, 1425.

Huber, P. J. (1967). The behavior of the maximum likelihood estimator under nonstandard conditions. *Proc. Fifth Berkeley Symp. Math. Statist. Prob.* **1**, University of California Press, 221-233.

Huber, P. J. (1973). Robust regression: Asymptotics, conjectures, and Monte Carlo. *Ann. Statist.* **1**, 799 -821.

Huber, P. J. (1981). *Robust Statistics*. New York: Wiley.

Hudson, H. M. (1974). Empirical Bayes estimation. Technical Report No. 58, Department of Statistics, Stanford University.

Hudson, H. M. (1978). A natural identity for exponential families with applications in multiparameter estimation. *Ann. Statist.* **6**, 473-484.

Hudson, H. M. (1985). Adaptive estimation for simultaneous estimation of Poisson means. *Ann. Statist.* **13**, 246-261.

Hudson, H. M. and Tsui, K.-W. (1981). Simultaneous Poisson estimators for a priori hypotheses about means. *J. Amer. Statist. Assoc.* **76**, 182-187.

Huzurbazar, V. S. (1948). The likelihood equation, consistency and the maxima of the likelihood function. *Ann. Eugenics* **14**, 185-200.

Hwang, J. T. (1982a). Improving upon standard estimators in discrete exponential families with applications to Poisson and negative binomial cases. *Ann. Statist.* **10**, 857-867.

Hwang, J. T. (1982b). Semi-tail upper bounds on the class of admissible estimators in discrete exponential families with applications to Poisson and binomial cases. *Ann. Statist.* **10**, 1137-1147.

Hwang, J. T. (1985). Universal domination and stochastic domination: Estimation simultaneously under a broad class of loss f .nctions. *Ann. Statist.* **13**, 295-314.

Hwang, J. T. and Brown, L. D. (1991). Estimated confidence under the validity constraint. *Ann. Statist.* **19**, 1964-1977.

Hwang, J. T. and Casella, G. (1982). Minimax confidence sets for the mean of a multivariate normal distribution. *Ann. Statist.* **10**, 868-881.

Hwang, J. T. and Casella, G. (1984). Improved set estimators for a multivariate normal mean. *Statistics and Decisions*, Supplement Issue No. 1, pp. 3-16.

Hwang, J. T and Chen, J. (1986). Improved confidence sets for the coefficients of a linear model with spherically symmetric errors. *Ann. Statist.* **14**, 444-460.

Hwang, J. T. G. and Ullah, A. (1994). Confidence sets centered at James-Stein estimators. A surprise concerning the unknown variance case. *J. Economet.* **60**, 145-156.

Hwang, J. T., Casella, G., Robert, C., Wells, M. T., and Farrell, R. H. (1992). Estimation of accuracy in testing. *Ann. Statist.* **20**, 490-509.

Ibragimov, I. A. and Has'minskii, R Z. (1972). Asymptotic behavior of statistical estimators. II. Limit theorems for the a posteriori density and Bayes' estimators. *Theory Prob. Applic.* **18**, 76-91.

Ibragimov, I. A. and Has'minskii, R. Z. (1981). *Statistical Estimation: Asymptotic Theory*. New York: Springer-Verlag.

Ibrahim, J. G. and Laud, P. W. (1991). On Bayesian analysis of generalized linear models using Jeffreys's prior. *J. Amer. Statist. Assoc.* **86**, 981-986.

Iwase,K. (1983). Uniform minimum variance unbiased estimation for the inverse Gaussian distribution. *J. Amer. Statist. Assoc.* **78**, 660-663.

Izenman, A. J. (1991). Recent developments in nonparametric density estimation. *J. Amer. Statist. Assoc.* **86**, 205-224.

Jackson, D. A., O'Donovan, T. M., Zimmer, W. J., and Deely, J. J. (1970). G_2 minimax estimators in the exponential family. *Biometrika* **70**, 439-443.

James, I. R. (1986). On estimating equations with censored data. *Biometrika* **73**, 35-42.

James, W. and Stein, C. (1961). Estimation with quadratic loss. *Proc. Fourth Berkeley Symp. Math. Statist. Prob.* **1**, University of California Press, 311-319.

Jaynes, E. T. (1979). Where do we stand on maximum entropy? *The Maximum Entropy Formalism* (R. D. Levine and M. Tribus, eds.). Cambridge, MA: M.I.T. Press, pp. 15-118.

Jeffreys, H. (1939,1948,1961). *The Theory of Probability*. Oxford: Oxford University Press.

Jennrich, R I. and Oman, S. (1986). How much does Stein estimation help in multiple linear regression? *Technometrics* **28**, 113-121.

Jensen, J. L. (1995). *Saddlepoint Approximations*. Oxford: Clarendon Press.

Jewell, N. P. and Raab, G. M. (1981). Difficulties in obtaining consistent estimators of variance parameters. *Biometrika* **68**, 221-226.

Jiang, J. (1996). REML estimation: Asymptotic behavior and related topics. *Ann. Statist.* **24**, 255-286.

Jiang, J. (1997). Wald consistency and the method of sieves in REML estimation. *Ann. Statist.* **25**, 1781-1802.

Johansen, S. (1979). *Introduction to the Theory of Regular Exponential Families*. Institute of Mathematical Statistics Lecture Notes, Vol. 3. Copenhagen: University of Copenhagen.

Johnson, B. McK. (1971). On the admissible estimators for certain fixed sample binomial problems. *Ann. Math. Statist.* **42**, 1579-1587.

Johnson, N. L. and Kotz, S. (1969-1972). *Distributions in Statistics* (4 vols.). New York: Wiley.

Johnson, N. L., Kotz. S., and Balakrishnan, N. (1994). *Continuous Univariate Distributions, Volume 1, Second Edition*. New York: Wiley.

Johnson, N. L., Kotz. S., and Balakrishnan, N. (1995). *Continuous Univariate Distributions, Volume 2, Second Edition*. New York: Wiley.

Johnson, N. L., Kotz. S., and Kemp, A. W. (1992). *Univariate Discrete Distributions, Second Edition*. New York: Wiley.

Johnson, R. A., Ladalla, J., and Liu, S. T. (1979). Differential relations in the original parameters, which determine the first two moments of the multi-parameter exponential family. *Ann. Statist.* **7**, 232-235.

Johnson, R. W. (1987). Simultaneous estimation of binomial *N*'s. *Sankhya Series A* **49**, 264-266.

Johnstone, I. (1984). Admissibility, difference equations, and recurrence in estimating a Poisson mean. *Ann. Statist.* **12**, 1173-1198.

Johnstone, I. (1988). On inadmissibility of some unbiased estimates of loss. *Statistical Decision Theory IV* (S. S. Gupta and J. O. Berger, eds.). New York: Springer-Verlag, pp. 361-380.

Johnstone, I. (1994). On minimax estimation of a sparse normal mean vector. *Ann. Statist.* **22**, 271-289.

Johnstone, I. and MacGibbon, K. B. (1992). Minimax estimation of a constrained Poisson vector. *Ann. Statist.* **20**, 807-831.

Johnstone, I. and MacGibbon, K. B. (1993). Asymptotically minimax estimation of a constrained Poisson vector via polydisc transforms. *Ann. Inst. Henri Poincaré*, **29**, 289-319.

Joshi, V. M. (1967). Inadmissibility of the usual confidence sets for the mean of a multivariate normal population. *Ann. Math. Statist.* **38**, 1180-1207.

Joshi, V. M. (1969a). On a theorem of Karlin regarding admissible estimates for exponential populations. *Ann. Math. Statist.* **40**, 216-223.

Joshi, V. M. (1969b). Admissibility of the usual confidence sets for the mean of a univariate or bivariate normal population. *Ann. Math. Statist.* **40**, 1042-1067.

Joshi, V. M. (1976). On the attainment of the Cramér-Rao lower bound. *Ann. Statist.* **4**, 998-1002.

Kabaila, P. V. (1983). On the asymptotic efficiency of estimators of the parameters of an ARMA process. *J. Time Ser. Anal.* **4**, 37-47.

Kagan, A. M. and Palamadov, V. P. (1968). New results in the theory of estimation and testing hypotheses for problems with nuisance parameters. Supplement to Y. V. Linnik, *Statistical Problems with Nuisance Parameters*. Amer. Math. Soc. Transl. of Math. Monographs **20**.

Kagan, A. M., Linnik, Yu V., and Rao, C. R. (1965). On a characterization of the normal law based on a property of the sample average. *Sankhya* A **27**, 405-406.

Kagan, A. M., Linnik, Yu V., and Rao, C. R. (1973). *Characterization Problems in Mathematical Statistics*. New York: Wiley.

Kalbfleisch, J. D. (1986). Pseudo-likelihood. *Encycl. Statist. Sci.* **7**, 324-327.

Kalbfleisch, J. D. and Prentice, R. L. (1980). *The Statistical Analysis of Failure Time Data*. New York: Wiley.

Kalbfleisch, J. D. and Sprott, D. A. (1970). Application of likelihood methods to models involving large numbers of parameters. *J. Roy. Statist. Soc. Ser. B* **32**, 175-208

Kariya, T. (1985). A nonlinear version of the Gauss-Markov theorem. *J. Amer. Statist. Assoc.* **80**, 476-477.

Kariya, T. (1989). Equivariant estimation in a model with an ancillary statistic. *Ann. Statist.* **17**, 920-928.

Karlin, S. (1958). Admissibility for estimation with quadratic loss. *Ann. Math. Statist.* **29**, 406-436.

Karlin, S. (1968). *Total Positivity*. Stanford, CA: Stanford University Press.

Karlin, S. and Rubin, H. (1956). Distributions possessing a monotone likelihood ratio. *J. Amer. Statist. Assoc.* **51**, 637-643.

Kass, R. E. and Steffey, D. (1989). Approximate Bayesian inference in conditionally independent hierarchical models (parametric empirical Bayes models). *J. Amer. Statist. Assoc.* **84**, 717-726.

Katz, M. W. (1961). Admissible and minimax estimates of parameters in truncated spaces. *Ann. Math. Statist.* **32**, 136-142.

Kemeny, J. G. and Snell, J. L. (1976). *Finite Markov Chains*. New York: Springer-Verlag

Kempthorne, O (1966) Some aspects of experimental inference. *J. Amer. Statist. Assoc.* **61**, 11-34.

Kempthorne, P. (1988a). Dominating inadmissible procedures using compromise decision theory. *Statistical Decision Theory IV* (S. S. Gupta and J. O. Berger, eds.). New York: Springer-Verlag, pp. 381-396.

Kempthorne, P. (1988b). Controlling risks under different loss functions: The compromise decision problem. *Ann. Statist.* **16**, 1594-1608.

Kester, A. D. M. (1985). *Some Large Deviation Results in Statistics*. Amsterdam: Centrum voor Wiskunde en Information.

Kester, A. D. M. and Kallenberg, W. C. M. (1986). Large deviations of estimators *Ann. Statist.* **14**, 648-664.

Khan, R. A. (1973). On some properties of Hammersley's estimator of an integer mean. *Ann. Statist.* **1**, 838-850.

Ki, F. and Tsui, K.-W. (1985). Omproved confidence set estimators of a multivariate normal mean and generalizations. *Ann. Inst. Statist. Math.* **37**, 487-498.

Ki, F. and Tsui, K.-W. (1990). Multiple shrinkage estimators of means in exponential families. *Can. J. Statist.* **18**, 31-46.

Kiefer, J. (1952). On minimum variance estimators. *Ann. Math. Statist.* **23**, 627-629.

Kiefer, J. (1957). Invariance, minimax sequential estimation, and continuous time processes. *Ann. Math. Statist.* **28**, 573-601.

Kiefer, J. (1966). Multivariate optimality results. *Multivariate Analysis*. (P. Krishnaiah, ed.). New York: Academic Press, pp. 255-274.

Kiefer, J. (1976). Admissibility of conditional confidence procedures. *Ann. Statist.* **4**, 836-865.

Kiefer, J. (1977). Conditional confidence statements and confidence estimators (with discussion). *J. Amer. Statist. Assoc.* **72**, 789-827.

Kiefer, J. and Wolfowitz, J. (1956). Consistency of the maximum likelihood estimator in the presence of infinitely many incidental parameters. *Ann. Math. Statist.* **27**, 887-906.

Kish, L. (1965). *Survey Sampling*. New York: Wiley.

Klaassen, C. A. J. (1984). Location estimators and spread. *Ann. Statist.* **12**, 311-321.

Klaassen, C. A. J. (1985). On an inequality of Chernoff. *Ann. Prob.* **13**, 966-974.

Kleffe, J. (1977). Optimal estimation of variance components-A survey. *Sankhya* B **39**, 211-244.

Klotz, J. (1970). The geometric density with unknown location parameter. *Ann. Math. Statist.* **41**, 1078-1082.

Klotz, J. H., Milton, R. C., and Zacks, S. (1969). Mean square efficiency of estimators of variance components. *J. Amer. Statist. Assoc.* **64**, 1383-1402.

Kojima, Y., Morimoto, H., and Taxeuchi, K. (1982). Two best unbiased estimators of normal integral mean. In *Statistics and Probability: Essays in Honor of C. R. Rao*, (G. Kallianpur, P. R. Krishnaiah, and J. K. Ghosh, eds.). New York: North-Holland, pp. 429-441.

Kolassa, J. E. (1993). *Series Approximation Methods in Statistics*. New York: Springer-Verlag

Kolmogorov, A. N. (1950). Unbiased estimates. *Izvestia Akad. Nauk SSSR, Ser. Math.* **14**, 303-326.

Konno, Y. (1991). On estimation of a matrix of normal means. *J. Mult. Anal.* **36**, 44-55.

Koopman, B. O. (1936). On distributions admitting a sufficient statistic. *Trans. Amer. Math. Soc.* **39**, 399-409.

Koroljuk, V. S. and Borovskich, Yu. V. (1994). *Theory of U-Statistics*. Boston: Kluwer Academic Publishers.

Kozek, A. (1976). Efficiency and Cramér-Rao type inequalities for convex loss functions. *Inst. Math., Polish Acad. Sci.* Preprint No. 90.

Kraft, C. and LeCam, L. (1956). A remark on the roots of the likelihood equation. *Ann. Math. Statist.* **27**, 1174-1177.

Kremers, W. (1986). Completeness and Unbiased Estimation for Sum-Quota Sampling *J. Amer. Statist. Assoc.* **81**, 1070-1073.

Kruskal, W. (1968). When are Gauss-Markov and least squares estimators identical? A coordinate-free approach. *Ann. Math. Statist.* **39**, 70-75.

Kubokawa, T. (1987). Admissible minimax estimation of the common mean of two normal populations. *Ann. Statist.* **15**, 1245-1256.

Kudo, H. (1955). On minimax invariant estimates of the translation parameter. *Natural Sci. Report Ochanomizu Univ.* **6**, 31-73.

Kullback, S. (1968). *Information Theory and Statistics*, Second Edition. New York: Dover. Reprinted in 1978, Gloucester, MA: Peter Smith.

Laird, N., Lange, N., and Stram, D. (1987). Maximum likelihood computation with repeated measures: Application of the EM algorithm. *J. Amer. Statist. Assoc.* **82**, 97-105.

Lambert, J. A. (1970). Estimation of parameters in the four-parameter lognormal. *Austr. J. Statist.* **12**, 33-44.

LaMotte, L. R. (1982). Admissibility in linear estimation. *Ann. Statist.* **10**, 245-255.

Landers, D. (1972). Existence and consistency of modified minimum contrast estimates. *Ann. Math. Statist.* **43**, 74 83.

Landers, D. and Rogge, L. (1972). Minimal sufficient σ-fields and minimal sufficient statistics. Two counterexamples. *Ann. Math. Statist.* **43**, 2045-2049.

Landers, D. and Rogge, L. (1973). On sufficiency and invariance, *Ann. Statist.* **1**, 543-544.

Lane, D. A. (1980). Fisher, Jeffreys, and the nature of probability. In *R. A. Fisher: An Appreciation.* (S. E. Fienberg and D. V. Hinkley, eds.). Lecture Notes in Statistics 1. New York: Springer-Verlag.

Laplace, P. S. (1774). Mémoire sur la probabilité des causes par évènements. *Mem. Acad. Sci. Sav. Etranger* **6**, 621-656.

Laplace, P. S. de (1820). *Théorie analytique des probabilités, Third Edition* Paris: Courcier.

Lavine, M. (1991a). Sensitivity in Bayesian statistics: The prior and the likelihood. *J. Amer. Statist. Assoc.* **86**, 396-399.

Lavine, M. (1991b). An approach to robust Bayesian analysis for multidimensional parameter spaces. *J. Amer. Statist. Assoc.* **86**, 400-403.

Le Cam, L. (1953). On some asymptotic properties of maximum likelihood estimates and related Bayes' estimates. Univ. of Calif. Publ. in Statist. **1**, 277-330.

Le Cam, L. (1955). An extension of Wald's theory of statistical decision functions. *Ann. Math. Statist.* **26**, 69-81.

Le Cam, L. (1956). On the asymptotic theory of estimation and testing hypotheses. *Proc. Third Berkeley Symp. Math. Statist. Prob.* **1**, University of California Press, 129-156.

Le Cam, L. (1958). Les propriétés asymptotiques des solutions de Bayes. *Publ. Inst. Statist. l'Univ. Paris VII*, Fasc. 3-4, 17-35.

Le Cam, L. (1969). *Théorie Asymptotique de la Décision Statistique.* Montréal: Les Presses de l'Université de Montréal.

Le Cam, L. (1970). On the assumptions used to prove asymptotic normality of maximum likelihood estimates. *Ann. Math. Statist.* **41**, 802-828.

Le Cam, L. (1979a). On a theorem of J. Hajek. *Contributions to Statistics, J. Hajek Memorial Volume* (J. Jureckova, ed.). Prague: Academia.

Le Cam, L. (1979b). Maximum Likelihood: An Introduction. Lecture Notes in Statistics No. 18. University of Maryland, College Park, Md.

Le Cam, L. (1986). *Asymptotic Methods in Statistical Decision Theory.* New York: Springer-Verlag.

Le Cam, L. (1990). Maximum likelihood: An introduction. *Int. Statist. Rev.* **58**, 153-171.

Le Cam, L. and Yang, G. L. (1990). *Asymptotics in Statistics: Some Basic Concepts.* New York: Springer-Verlag.

Lee, A. J. (1990). *U-Statistics, Theory and Practice.* New York: Marcel Dekker.

Lehmann, E. L (1951). A general concept of unbiasedness. *Ann. Math. Statist.* **22**, 587-592.

Lehmann, E. L. (1981). An interpretation of completeness and Basu's theorem. *J. Amer. Statist. Assoc.* **76**, 335-340.

Lehmann, E. L. (1986). *Testing Statistical Hypotheses, Second Edition* (TSH2). New York: Springer-Verlag

Lehmann, E. L. (1983). Estimation with inadequate information. *J. Amer. Statist. Assoc.* **78**, 624-627.

Lehmann, E. L. (1999). *Elements of Large-Sample Theory*. New York: Springer-Verlag.

Lehmann, E. L. and Loh, W-L. (1990). Pointwise versus uniform robustness of some large-sample tests and confidence intervals *Scan. J. Statist.* **17**, 177-187.

Lehmann, E. L. and Scheffé, H. (1950, 1955, 1956). Completeness, similar regions and unbiased estimation. *Sankhya* **10**, 305-340; **15**, 219-236. (Corr: **17**, 250.)

Lehmann, E. L. and Scholz, F. W. (1992). Ancillarity. *Current Issues in Statistical Inference: Essays in Honor of D. Basu* (M. Ghosh and P. K. Pathak, eds.). Institute of Mathematical Statistics, Hayward, CA: 32-51.

Lehmann, E. L. and Stein, C. (1950). Completeness in the sequential case. *Ann. Math. Statist.* **21**, 376-385.

Lele, S. (1993). Euclidean distance matrix analysis (EDMA): Estimation of the mean form and mean form difference. *Math. Geol.* **5**, 573-602.

Lele, S. (1994). Estimating functions in chaotic systems. *J. Amer. Statist. Assoc.* **89**, 512-516.

Leonard, T. (1972). Bayesian methods for binomial data. *Biometrika* **59**, 581-589.

LePage, R. and Billard, L. (1992). *Exploring the Limits of Bootstrap*. New York: Wiley.

Letac, G. and Mora, M. (1990). Natural real exponential families with cubic variance functions. *Ann. Statist.* **18**, 1-37.

Levit, B. (1980). On asymptotic minimax estimators of the second order. *Theor. Prob. Applic.* **25**, 552-568.

Liang, K.-Y. and Waclawiw, M. A. (1990). Extension of the Stein estimating procedure through the use of estimating functions. *J. Amer. Statist. Assoc.* **85**, 435-440.

Liang, K. Y. and Zeger, S. L. (1994). Inference based on estimating functions in the presence of nuisance parameters (with discussion). *Statist. Sci.* **10**, 158-166.

Lindley, D. V. (1962). Discussion of th paper by Stein. *J. Roy. Statist. Soc. Ser. B* **24** 265-296.

Lindley, D. V. (1964). The Bayesian analysis of contingency tables. *Ann. Math. Statist.* **35**, 1622-1643.

Lindley, D. V. (1965). *Introduction to Probability and Statistics*. Cambridge: Cambridge University Press.

Lindley, D. V. (1965). *Introduction to Probability and Statistics from a Bayesian Viewpoint. Part 2. Inference* . Cambridge: Cambridge University Press.

Lindley, D.V. and Phillips, L. D. (1976). Inference for a Bernoulli process (a Bayesian view). *Amer. Statist.* **30**, 112-119.

Lindley, D. V. and Smith, A. F. M.. (1972). Bayes estimates for the linear model (with discussion). *J. Roy. Statist. Soc. Ser. B* **34**, 1-41.

Lindsay, B and Yi, B. (1996). On second-order optimality of the observed Fisher information. Technical Report No. 95-2, Center for Likelihood Studies. Pennsylvania State University.

Linnik, Yu V. and Rukhin, A. L. (1971). Convex loss functions in the theory of unbiased estimation. *Soviet Math. Dokl.* **12**, 839-842.

Little and Rubin, D. B. (1987). *Statistical Analysis with Missing Data*. New York: Wiley.

Liu, C. and Rubin, D. B. (1994). The ECME algorithm: A simple extension of EM and ECM with faster monotone convergence. *Biometrika* **81**, 633-648.

Liu, R. and Brown, L. D. (1993). Nonexistence of informative unbiased estimators in singular problems. *Ann. Statist.* **21**, 1-13.

Loh, W-L. (1991). Estimating the common mean of two multivariate distributions. *Ann. Statist.* **19**, 297-313.

Louis, T. A. (1982). Finding the observed information matrix when using the EM algorithm. *J. Roy. Statist. Soc. Ser. B* **44**, 226-233.

Lu, K. L and Berger, J. O. (1989a). Estimation of normal means: frequentist estimation of loss. *Ann. Statist.* **17**, 890-906.

Lu, K. L. and Berger, J. O. (1989b). Estimated confidence procedures for multivariate normal means. *J. Statist. Plan. Inform.* **23**, 1-19.

Luenberger, D. G. (1969). *Optimization by Vector Space Methods*. New York: Wiley.

Maatta, J. and Casella, G. (1987). Conditional properties of interval estimators of the normal variance. *Ann. Statist.* **15**, 1372-1388.

Maatta, J. and Casella, G. (1990). Developments in decision-theoretic variance estimation (with discussion). *Statist. Sci.* **5**, 90-120.

MacEachern, S. N. (1993). A characterization of some conjugate prior distributions for exponential families. *Scan. J. Statist.* **20**, 77-82.

Madansky, A. (1962). More on length of confidence intervals. *J. Amer. Statist. Assoc.* **57**, 586-589.

Makani, S. M. (1972). Admissibility of linear functions for estimating sums and differences of exponential parameters. Ph.D. Thesis. University of California, Berkeley.

Makani, S. M. (1977). A paradox in admissibility. *Ann. Statist.* **5**, 544-546.

Makelainen, T., Schmidt, K., and Styan, G. (1981). On the existence and uniqueness of the maximum likelihood estimate of a vector-valued parameter in fixed-size samples. *Ann. Statist.* **9**, 758-767.

Mandelbaum, A. and Rüschendorf, L. (1987). Complete and symmetrically complete families of distributions. *Ann. Statist.* **15**, 1229-1244.

Marazzi, A. (1980). Robust Bayesian estimation for the linear model. Res. Repon No. 27, Fachgruppe f. Stat., Eidg. Tech. Hochsch., Zurich.

Maritz, J. S. and Lwin, T. (1989). *Empirical Bayes Methods, Second Edition*. London: Chapman & Hall.

Marshall, A. and Olkin, I. (1979). *Inequalities — Theory of Majorization and its Applications*. New York: Academic Press.

Mathew, T. (1984). On nonnegative quadratic unbiased estimability of variance components. *Ann. Statist.* **12**, 1566-1569.

Mattner, L. (1992). Completeness of location families, translated moments, and uniqueness of charges. *Prob. Theory Relat. Fields* **92**, 137-149.

Mattner, L. (1993). Some incomplete but boundedly complete location families. *Ann. Statist.* **21**, 2158-2162.

Mattner, L. (1994). Complete order statistics in parametric models. *Ann. Statist.* **24**, 1265-1282.

McCullagh, P. (1991). Quasi-likelihood and estimating functions. In *Statistical Theory and Modelling: In Honor of Sir David Cox* (D. Hinkley, N. Reid, and L. Snell, eds.). London: Chapman & Hill, pp. 265-286.

McCullagh, P. (1992). Conditional inference and Cauchy models. *Biometrika* **79**, 247-259.

McCullagh, P. and Nelder, J. A. (1989). *Generalized Linear Models, Second Edition*. London: Chapman & Hall.

McLachlan, G. (1997). *Recent Advances in Finite Mixture Models*. New York: Wiley

McLachlan, G. and Basford, K. (1988). *Mixture Models: Inference and Applications to Clustering*. New York: Marcel Dekker.

McLachlan, G. and Krishnan, T. (1997). *The EM Algotithm and Extensions*. New York: Wiley

Meeden, G. and Ghosh, M. (1983). Choosing between experiments: Applications to finite population sampling. *Ann. Statist.* **11**, 296-305.

Meeden, G., Ghosh, M., and Vardeman, S. (1985). Some admissible nonparametric and related finite population sampling estimators. *Ann. Statist.* **13**, 811-817.

Meng, X-L. and Rubin, D. B. (1993). Maximum likelihood estimation via the ECM algorithm: A general framework. *Biometrika* **80**, 267-278.

Messig, M. A. and Strawderman, W. E. (1993). Minimal sufficiency and completeness for dichotomous quantal response models. *Ann. Statist.* **21**, 2141-2157.

Metropolis, N., Rosenbluth, A. W., Rosenbluth, M. N., Teller, A. H., and Teller, E. (1953). Equations of state calculations by fast computing machines. *J. Chem. Phys.* **21**, 1087–1092.

Meyn, S. and Tweedie, R. (1993). *Markov Chains and Stochastic Stability*. New York: Springer-Verlag.

Mikulski, P. W. and Monsour, M. (1988). On attainable Cramér-Rao type lower bounds for weighted loss functions. *Statist. Prob. Lett.* **7**, 1-2.

Miller, J. J. (1977). Asymptotic properties of maximum likelihood estimates in the mixed model of the analysis of variance. *Ann. Statist.* **5**, 746-762.

Moore, T. and Brook, R. J. (1978). Risk estimate optimality of James-Stein estimators. *Ann. Statist.* **6**, 917-919.

Moors, J. J. A. (1981). Inadmissibility of linearly invariant estimators in truncated parameter spaces. *J. Amer. Statist. Assoc.* **76**, 910-915.

Morris, C. N. (1977). Interval estimation for empirical Bayes generalizations of Stein's estimator. *Proc. Twenty-Second Conf. Design Exp. in Army Res. Devol. Test.*, ARO Report 77-2.

Morris, C. N. (1982). Natural exponential families with quadratic variance functions. *Ann. Statist.* **10**, 65-80.

Morris, C. N. (1983a). Parametric empirical Bayes inference: Theory and applications (with discussion). *J. Amer. Statist. Assoc.* **78**, 47-65.

Morris, C. N. (1983b). Natural exponential families with quadratic variance functions: Statistical theory. *Ann. Statist.* **11**, 515-529.

Morton, R. and Raghavachari, M. (1966). On a theorem of Karlin regarding admissibility of linear estimates in exponential populations. *Ann. Math. Statist.* **37**, 1809-1813.

Mosteller, F. (1946). On some useful inefficient statistics. *Ann. Math. Statist.* **17**, 377-408.

Müller-Funk, U., Pukelsheim, F. and Witting, H. (1989). On the attainment of the Cramér-Rao bound in $\mathcal{L}_r$-differentiable families of distributions. *Ann. Statist.* **17**, 1742-1748.

Murray, G. D. (1977). Discussion of the paper by Dempster, Laird and Rubin. *J. Roy. Statist. Soc. Ser. B* **39**, 27-28.

Murray, M. K. and Rice, J. W. (1993). *Differential Geometry and Statistics*. London: Chapman & Hall.

Natarajan, J. and Strawderman, W. E. (1985). Two-stage sequential estimation of a multivariate normal mean under quadratic loss. *Ann. Statist.* **13**, 1509-1522.

Natarajan, R. and McCulloch, C. E. (1995). A note on the existence of the posterior distribution for a class of mixed models for binomial responses. *Biometrika* **82**, 639-643.

Nelder, J. A. and Wedderburn, R. W M. (1972). Generalized linear models. *J. Roy . Statist. Soc. A* **135**, 370-384.

Newcomb, S. (1882). Discussion and results of observations on transits of Mercury from 1677 to 1881. *Astronomical Papers*, Vol. 1, U.S. Nautical Almanac Office, 363-487.

Newcomb, S. (1886). A generalized theory of the combination of observations so as to obtain the best result. *Amer. J. Math.* **8**, 343-366.

Neyman, J. (1934). On the two different aspects of the representative method: The method of stratified sampling and the method of purposive selection. *J Roy. Statist. Soc.* **97**, 558-625.

Neyman, J. (1935). Sur un teorema concernente le cosidette statistiche sufficienti. *Giorn. Ist. Ital. Att.* **6**, 320-334.

Neyman, J. (1949). Contributions to the theory of the χ^2 test. *Proc. First Berkeley Symp. Math. Statist. Prob.* **1**, University of California Press, 239-273.

Neyman, J. and Pearson, E. S. (1933a). The testing of statistical hypotheses in relation to probabilities a priori. *Proc Camb. Phil. Soc.* **24**, 492-510.

Neyman, J. and Pearson, E. S. (1933b). On the problem of the most efficient tests of statistical hypotheses. *Phil. Trans. Roy. Soc. A.* **231**, 289-337.

Neyman, J. and Scott, E. L. (1948). Consistent estimates based on partially consistent observations. *Econometrica* **16**, 1-32.

Noorbaloochi, S. and Meeden, G. (1983). Unbiasedness as the dual of being Bayes. *J. Amer. Statist. Assoc.* **78**, 619-623.

Nordberg, L. (1980). Asymptotic normality of maximum likelihood estimators based on independent, unequally distributed observations in exponential family models. *Scand. J. Statist.* **7**, 27-32.

Norden, R. H. (1972-73). A survey of maximum likelihood estimation. *Int. Statist. Rev.* **40**, 329-354; **41**, 39-58.

Novick, M. R. and Jackson, P. H. (1974). *Statistical Methods for Educational and Psychological Research.* New York: McGraw-Hill.

Oakes, D. (1991). Life-table analysis. *Statistical Theory and Modelling, in Honor of Sir David Cox, FRS.* London: Chapman & Hall, pp. 107-128.

Obenchain, R. L. (1981). Good and optimal ridge estimators. *Ann. Statist.* **6**, 1111-1121.

Olkin, I. and Pratt, J. W (1958). Unbiased estimation of certain correlation coefficients. *Ann. Math. Statist.* **29**, 201-211.

Olkin, I. and Selliah, J. B. (1977). Estimating covariances in a multivariate distribution. In *Statistical Decision Theory and Related Topics II* (S. S. Gupta and D. S. Moore, eds.). New York: Academic Press, pp. 313-326.

Olkin, I. and Sobel, M. (1979). Admissible and minimax estimation for the multinomial distribution and for independent binomial distributions. *Ann. Statist.* **7**, 284-290.

Olkin, I., Petkau, A, and Zidek, J. V. (1981). A comparison of n estimators for the binomial distribution. *J. Amer. Statist. Assoc.* **76**, 637-642.

Olshen, R. A. (1977). Comments on "A note on a reformulation of the S-method of multiple comparisons." *J. Amer. Statist. Assoc.* **72**, 144-146.

Oman, S. (1982a). Contracting towards subspaces when estimating the mean of a multivariate distribution. *J. Mult. Anal.* **12**, 270-270.

Oman, S. (1982b). Shrinking towards subspaces in multiple linear regression. *Technometrics* **24**, 307-311.

Oman, S. (1985). Specifying a prior distribution in structured regression problems. *J. Amer. Statist. Assoc.* **80**, 190-195.

Oman, S. (1991). Random calibration with many measurements: An application of Stein estimation. *Technometrics* **33**, 187-195.

Owen, A. (1988). Empirical likelihood ratio confidence intervals for a single functional. *Biometrika* **75**, 237-249.

Owen, A. (1990). Empirical likelihood ratio confidence regions. *Ann. Statist.* **18**, 90-120.

Padmanabhan, A. R. (1970). Some results on minimum variance unbiased estimation. *Sankhya* **32**, 107-114.

Pathak, P. K. (1976). Unbiased sampling in fixed cost sequential sampling schemes. *Ann. Statist.* **4**, 1012-1017.

Pearson, K (1894). Contributions to the mathematical theory of evolution. *Phil. Trans. Royal Soc Ser. A.* **185**, 71-110.

Peers, H. W. (1965). On confidence points and Bayesian probability points in the case of several parameters. *J. Roy. Statist. Soc. Ser. B* **27**, 16-27.

Peisakoff, M. P. (1950). *Transformation Parameters*. Ph. D. Thesis, Princeton University, Princeton, NJ.

Perlman, M. (1972). On the strong consistency of approximate maximum likelihood estimators. *Proc. Sixth Berkeley Symp. Math. Statist. Prob.* **1**, University of California Press, 263-281.

Perlman, M. (1983). The limiting behavior of multiple roots of the likelihood equation. *Recent Advances in Statistics: Papers in Honor of Herman Chernoff on his Sixtieth Birthday* (M. H. Rizvi, J. S. Rustagi and D. Siegmund, eds.). New York: Academic Press.

Perng, S. K. (1970). Inadmissibility of various 'good' statistical procedures which are translation invariant. *Ann. Math. Statist.* **41**, 1311-1321.

Pfaff, Th. (1982). Quick consistency of quasi maximum likelihood estimators. *Ann. Statist.* **10**, 990-1005.

Pfanzagl, J. (1969). On the measurability and consistency of minimum contrast estimators. *Metrika* **14**, 249-272.

Pfanzagl, J. (1970) On the asymptotic efficiency of median unbiased estimates. *Ann. Math. Statist.* **41**, 1500-1509.

Pfanzagl, J. (1972). Transformation groups and sufficient statistics. *Ann. Math. Statist.* **43**, 553-568.

Pfanzagl, J. (1973). Asymptotic expansions related to minimum contrast estimators. *Ann. Statist.* **1**, 993-1026.

Pfanzagl, J. (1979). On optimal median unbiased estimators in the presence of nuisance parameters. *Ann. Statist.* **7**, 187-193.

Pfanzagl, J. (1985). *Asymptotic Expansions for General Statistical Models.* New York: Springer-Verlag

Pfanzagl, J. (1990). Large deviation probabilities for certain nonparametric maximum likelihood estimators. *Ann. Statist.* **18**, 1868-1877.

Pfanzagl, J. (1994). *Parametric Statistical Theory.* New York: DeGruyter.

Pfanzagl, J. and Wefelmeyer, W. (1978-1979). A third order optimum property of the maximum likelihood estimator. *J. Mult. Anal.* **8**, 1-29; **9**, 179-182.

Piegorsch, W. W. and Casella, G. (1996). Empirical Bayes estimation for Logistic regression and extended parametric regression models. *J. Ag. Bio. Env. Statist.* **1**, 231-249.

Ping, C. (1964). Minimax estimates of parameters of distributions belonging to the exponential family. *Chinese Math.* **5**, 277-299.

Pitcher, T. S. (1957). Sets of measures not admitting necessary and sufficient statistics or subfields. *Ann. Math. Statist.* **28**, 267-268.

Pitman, E. J. G. (1936). Sufficient statistics and intrinsic accuracy. *Proc. Camb. Phil. Soc.* **32**, 567-579.

Pitman, E. J. G. (1939). The estimation of the location and scale parameters of a continuous population of any given form. *Biometrika* **30**, 391-421.

Pitman, E. J. G. (1979). *Some Basic Theory for Statistical Inference.* London: Chapman & Hall.

Plackett, R. L. (1958). The principle of the arithmetic mean. *Biometrika* **45**, 130-135.

Plackett, R. L. (1972). The discovery of the method of least squares. *Biometrika* **59**, 239-251.

Polfeldt, T. (1970). Asymptotic results in non-regular estimation. *Skand. Akt. Tidskr. Suppl. 1-2.*

Polson, N. and Wasserman, L. (1990). Prior distributions for the bivariate binomial. *Biometrika* **77**, 901-904.

Portnoy, S. (1971). Formal Bayes estimation with application to a random effects models. *Ann. Math. Statist.* **42**, 1379-1402.

Portnoy, S. (1977a). Robust estimation in dependent situations. *Ann. Statist.* **5**, 22- 43.

Portnoy, S. (1977b). Asymptotic efficiency of minimum variance unbiased estimators. *Ann. Statist.* **5**, 522-529.

Portnoy, S. (1984). Asymptotic behavior of M-estimators of p regression parameters when p^2/n is large. I. Consistency. *Ann. Statist.* **12**, 1298-1309.

Portnoy, S. (1985). Asymptotic behavior of M estimators of p regression parameters when p^2/n is large: II. Normal approximation. *Ann. Statist.* **13**, 1403-1417. (Corr: **19**, 2282.)

Pratt, J. W. (1976). F. Y. Edgeworth and R. A. Fisher on the efficiency of maximum likelihood estimation. *Ann. Statist.* **4**, 501-514.

Pregibon, D. (1980). Goodness of link tests for generalized linear models. *Appl. Statist.* **29**, 15-24.

Pregibon, D. (1981). Logistic regression diagnostics. *Ann. Statist.* **9**, 705-724.

Pugh, E. L. (1963). The best estimate of reliability in the exponential case. *Oper. Res.* **11**, 57-61.

Pukelsheim, F. (1981). On the existence of unbiased nonnegative estimates of variance covariance components. *Ann. Statist.* **9**, 293-299.

Quandt, R. E and Ramsey, J. B. (1978). Estimating mixtures of normal distributions and switching regressions. *J. Amer. Statist. Assoc.* **73**, 730-752.

Quenouille, M. H. (1949). Approximate tests of correlation in time series. *J. Roy. Statist. Soc. Ser. B* **11**, 18-44.

Quenouille, M. H. (1956). Notes on bias in estimation. *Biometrika* **43**, 353-360.

Raiffa, H. and Schlaifer, R. (1961). *Applied Statistical Decision Theory*. Cambridge, MA: Harvard University Press.

Ralescu, S., Brandwein, A. C. and Strawderman, W. E. (1992). Stein estimation for non-normal spherically symmetric location families in three dimensions. *J. Mult. Anal.* **42**, 35-50.

Ramamoorthi, R. V. (1990). Sufficiency, invariance, and independence in invariant models. *J. Statist. Plan. Inform.* **26**, 59-63.

Rao, B. L. S. Prakasa (1992). Cramer-Rao type integral inequalities for estimators of functions of multidimensional parameter. *Sankhya* **54**, 53-73.

Rao, C. R. (1945). Information and the accuracy attainable in the estimation of statistical parameters. *Bull. Calc. Math. Soc.* **37**, 81-91.

Rao, C. R. (1947). Minimum variance and the estimation of several parameters. *Camb. Phil. Soc.* **43**, 280-283.

Rao, C. R. (1949). Sufficient statistics and minimum variance estimates. *Proc. Camb. Phil. Soc.* **45**, 213-218.

Rao, C. R. (1961). Asymptotic efficiency and limiting information. *Proc. Fourth Berkeley Symp. Math. Statist. Prob.* **1**, University of California Press, 531-546.

Rao, C. R. (1963). Criteria of estimation in large samples. *Sankhya* **25**, 189-206.

Rao, C. R. (1970). Estimation of heteroscedastic variances in linear models. *J. Amer. Statist. Assoc.* **65**, 161-172.

Rao, C. R. (1976). Estimation of parameters in a linear model. *Ann. Statist.* **4**, 1023-1037.

Rao, C. R. (1977). Simultaneous estimation of parameters—A compound decision problem. In *Decision Theory and Related Topics* (S. S. Gupta and D. S. Moore, eds.). New York: Academic Press, pp. 327-350.

Rao, C. R. and Kleffe, J. (1988). *Estimation of variance components and applications*. Amsterdam: North Holland/Elsevier.

Rao, J. N. K. (1980). Estimating the common mean of possibly different normal populations: A simulation study. *J. Amer. Statist. Assoc.* **75**, 447-453.

Redner, R. (1981). Note on the consistency of the maximum likelihood estimate for non-identifiable distributions. *Ann. Statist.* **9**, 225-228.

Reeds, J. (1985). Asymptotic number of roots of Cauchy location likelihood equations. *Ann. Statist.* **13**, 775-784.

Reid, N. (1988). Saddlepoint methods and statistical inference (with discussion). *Statist. Sci.* **3**, 213-238.

Reid, N. (1995). The role of conditioning in inference (with discussion). *Statist. Sci.* **10**, 138-166.

Reid, N. (1996). Likelihood and Bayesian approximation methods. *Bayesian Statistics 5*, J. M. Bernardo, ed., 351-368.

Rényi, A. (1961). On measures of entropy and information. *Proc. Fourth Berkeley Symp. Math. Statist. Prob.* **1**, University of California Press, pp. 547-561.

Resnick, S. I. (1992). *Adventures in Stochastic Processes*. Basel: Birkhauser.

Ripley, B. (1987). *Stochastic Simulation*. New York: Wiley.

Robbins, H. (1951). Asymptotically subminimax solutions of compound statistical decision problems. In *Proc. Second Berkeley Symp. Math. Statist. Probab. 1*. Berkeley: University of California Press.

Robbins, H. (1964). The empirical Bayes approach to statistical decision problems. *Ann. Math. Statist.* **35**, 1-20.

Robbins, H. (1983). Some thoughts on empirical Bayes estimation. *Ann. Statist.* **11**, 713-723.

Robert, C. (1991). Generalized inverse normal distributions, *Statist. Prob. Lett.* **11**, 37-41.

Robert, C. P. (1994a). *The Bayesian Choice: A Decision-Theoretic Motivation*. New York: Springer-Verlag.

Robert, C.P. (1994b). Discussion of the paper by Tierney. *Ann. Statist.* **22**, 1742-1747.

Robert, C. (1995). Convergence control methods for Markov chain Monte Carlo algorithms. *Statist. Sci.* **10**, 231-253.

Robert, C. and Casella, G. (1990). Improved confidence sets in spherically symmetric distributions. *J. Mult. Anal.* **32**, 84-94.

Robert, C. and Casella, G. (1994). Improved confidence statements for the usual multivariate normal confidence set. *Statistical Decision Theory V* (S. S. Gupta and J. O. Berger, eds.). New York: Springer-Verlag, pp. 351-368.

Robert, C. and Casella, G. (1998). *Monte Carlo Statistical Methods*. New York: Springer-Verlag.

Robert, C., Hwang, J. T., and Strawderman, W. E. (1993). Is Pitman closeness a reasonable criterion? (with discussion). *J. Amer. Statist. Assoc.* **88**, 57-76.

Roberts, A. W. and Varberg, D. E. (1973). *Complex Functions*. New York: Academic Press.

Robinson, G. K. (1979a). Conditional properties of statistical procedures. *Ann. Statist.* **7**, 742-755.

Robinson, G. K. (1979b). Conditional properties of statistical procedures for location and scale parameters. *Ann. Statist.* **7**, 756-771.

Robson, D. S. and Whitlock, J. H. (1964). Estimation of a truncation point. *Biometrika* **51**, 33.

Romano, J. P. and Siegel, A. F. (1986). *Counterexamples in Probability and Statistics*. Monterey, CA: Wadsworth and Brooks/Cole.

Rosenblatt, M. (1956). Remark on some nonparametric estimates of a density function. *Ann. Math. Statist.* **27**, 832-837.

Rosenblatt, M. (1971). Curve estimates. *Ann. Math. Statist.* **42**, 1815-1842.

Rosenblatt, M. (1971). *Markov Processes. Structure and Asymptotic Behavior*. New York: Springer-Verlag

Ross, S. (1985). *Introduction to Probability Models, Third Edition*. New York: Academic Press.

Rothenberg, T. J. (1977). The Bayesian approach and alternatives in econometrics. In *Studies in Bayesian Econometrics and Statistics. Vol. 1* (S. Fienberg and A. Zellner, eds.). Amsterdam: North-Holland, pp. 55-75.

Roy, J. and Mitra, S. K. (1957). Unbiased minimum variance estimation in a class of discrete distributions. *Sankhya* **8**, 371-378.

Royall, R. M. (1968). An old approach to finite population sampling theory. *J. Amer. Statist. Assoc.* **63**, 1269-1279.

Rubin, D. B. and Weisberg, S. (1975). The variance of a linear combination of independent estimators using estimated weights. *Biometrika* **62**, 708-709.

Rubin, D. B. (1976). Inference and missing data. *Biometrika* **63**, 581-590.

Rubin, D. B. (1987). *Multiple Imputation for Nonresponse in Surveys*. New York: Wiley.

Rudin, W. (1966). *Real and Complex Analysis*. New York: McGraw-Hill.

Rukhin, A. L. (1978). Universal Bayes estimators. *Ann. Statist.* **6**, 1345-1351.

Rukhin, A. L. (1987). How much better are better estimators of the normal variance? *J. Amer. Statist. Assoc.* **82**, 925-928.

Rukhin, A. L. (1988a). Estimated loss and admissible loss estimators. *Statistical Decision Theory IV* (S. S. Gupta and J. O. Berger, Eds.). New York: Springer-Verlag, pp. 409-420.

Rukhin, A. L. (1988b). Loss functions for loss estimation. *Ann. Statist.* **16**, 1262-1269.

Rukhin, A. L. (1988c). Improved estimation in lognormal regression models. *J. Statist. Plan. Inform.* **18**, 291-297.

Rukhin, A. (1995). Admissibility: Survey of a concept in progress. *Int. Statist. Rev.* **63**, 95-115.

Rukhin, A. and Strawderman, W. E. (1982). Estimating a quantile of an exponential distribution. *J. Amer. Statist. Assoc.* **77**, 159-162.

Rutkowska, M. (1977). Minimax estimation of the parameters of the multivariate hypergeometric and multinomial distributions. *Zastos. Mat.* **16**, 9-21.

Sacks, J. (1963). Generalized Bayes solutions in estimation problems. *Ann. Math. Statist.* **34**, 751-768.

Sampford, M. R. (1953). Some inequalities on Mill's ratio and related functions. *Ann. Math. Statist.* **10**, 643-645.

Santner, T. J. and Duffy, D. E. (1990). *The Statistical Analysis of Discrete Data*. New York: Springer-Verlag.

Särndal, C-E., Swenson, B., and Wretman, J. (1992). *Model Assisted Survey Sampling*. New York: Springer-Verlag.

Savage, L. J. (1954, 1972). *The Foundations of Statistics*. New York: Wiley. Rev. ed., Dover Publications.

Savage, L. J. (1976). On rereading R. A. Fisher (with discussion). *Ann. Statist.* **4**, 441-500.

Schervish, M. (1995). *Theory of Statistics*. New York: Springer-Verlag.

Scheffé, H. (1959). *The Analysis of Variance*. New York: Wiley.

Scholz, F. W. (1980). Towards a unified definition of maximum likelihood. *Can. J. Statist.* **8**, 193-203.

Scholz, F. W. (1985). Maximum likelihood estimation. In *Encyclopedia of Statistical Sciences* **5**, (S. Kotz, N. L. Johnson, and C. B. Read, eds.). New York: Wiley.

Sclove, S. L., Morris, C., and Radhakrishnan, R. (1972). Non optimality of preliminary-test estimators for the mean of a multivariate normal distribution. *Ann. Math. Statist.* **43**, 1481-1490.

Seal, H. L. (1967). The historical development of the Gauss linear model. *Biometrika* **54**, 1-24.

Searle, S. R. (1971a). *Linear Models*. New York: Wiley.

Searle, S. R. (1971b). Topics in variance component estimation *Biometrics* **27**, 1-76.

Searle, S. R. (1987). *Linear Models for Unbalanced Data*. New York: Wiley.

Searle, S.R., Casella, G., and McCulloch, C. E. (1992). *Variance Components*. New York: Wiley.

Seber, G. A. F. (1977). *Linear Regression Analysis*. New York: Wiley.

Self, S. G. and Liang, K-Y. (1987). Asymptotic properties of maximum likelihood estimators and likelihood ratio tests under nonstandard conditions. *J. Amer. Statist. Assoc.* **82**, 605-610.

Sen, P. K. and Ghosh, B. K. (1976). Comparison of some bounds in estimation theory. *Ann. Statist.* **4**, 755-765.

Sen, P. K. and Saleh, A. K. Md. (1985). On some shrinkage estimators of multivariate location. *Ann. Statist.* **13**, 272-281.

Sen, P. K. and Saleh, A. K. Md. (1987). On preliminary test and shrinkage M-estimation in linear models. *Ann. Statist.* **15**, 1580-1592.

Serfling, R. J. (1980). *Approximation Theorems of Mathematical Statistics*. New York: Wiley.

Seshadri, V. (1963). Constructing uniformly better estimators. *J. Amer. Statist. Assoc.* **58**, 172-175.

Seth, G. R. (1949). On the variance of estimates. *Ann. Math. Statist.* **20**, 1-27.

Shaffer, J. P. (1991). The Gauss-Markov theorem and random regressors. *Amer. Statist.* **45**, 269-273.

Shao, J. and Tu, D. (1995). *The Jackknife and the Bootstrap*. New York: Springer-Verlag

Shao, P. Y-S. and Strawderman, W. E. (1994). Improving on the James-Stein positive-part estimator. *Ann. Statist.* **22**, 1517-1538.

Shemyakin, A. E. (1987). Rao-Cramer type integral inequalities for estimates of a vector parameter. *Theoret. Prob. Applic.* **32**, 426-434.

Shinozaki, N. (1980). Estimation of a multivariate normal mean with a class of quadratic loss functions. *J. Amer. Statist. Assoc.* **75**, 973-976.

Shinozaki, N. (1984). Simultaneous estimation of location parameters under quadratic loss. *Ann. Statist.* **12**, 233-335.

Shinozaki, N. (1989). Improved confidence sets for the mean of a multivariate distribution. *Ann. Inst. Statist. Math.* **41**, 331-346.

Shorrock, G. (1990). Improved confidence intervals for a normal variance. *Ann. Statist.* **18**, 972-980.

Sieders, A. and Dzhaparidze, K. (1987). A large deviation result for parameter estimators and its application to nonlinear regression analysis. *Ann. Statist.* **15**, 1031-1049.

Silverman, B. W. (1986) *Density Estimation for Statistic and Data Analysis*. London: Chapman & Hall.

Simons, G. (1980). Sequential estimators and the Cramér-Rao lower bound. *J. Statist. Plan. Inform.* **4**, 67-74.

Simpson, D. G., Carroll, R. J., and Ruppert, D. (1987). M-estimation for discrete data: Asymptotic distribution theory and implications. *Ann. Statist.* **15**, 657-669.

Simpson, T. (1755). A letter to the Rignt Honorable George Earl of Macclesfield, President of the Royal Society. on the advantage of taking the mean of a number of observations, in practical astronomy. *Phil. Trans. R. Soc. London* **49** (Pt. 1), 82-93.

Singh, K. (1981). On the asymptotic accuracy of Efron's bootstrap. *Ann. Statist.* **9**, 1187-1995.

Sivagenesan, S. and Berger, J. O. (1989). Ranges of posterior measures for priors with unimodal contaminations. *Ann. Statist.* **17**, 868-889.

Smith, A. F. M. and Roberts, G. O. (1993). Bayesian computation via the GIbbs sampler and related Markov chain methods (with discussion). *J. Roy. Statist. Soc. Ser. B* **55**, 3-23.

Smith, W. L. (1957). A note on truncation and sufficient statistics. *Ann. Math. Statist.* **28**, 247-252.

Snedecor, G. W. and Cochran, W. G. (1989) *Statistical Methods, Eighth Edition*. Ames, IA: Iowa State University Press.

Solari, M. E. (1969). The 'maximum likelihood solution' of the problem of estimating a linear functional relationship. *J. Roy. Statist. Soc. Ser. B* **31**, 372-375.

Solomon, D. L. (1972a). Γ-minimax estimation of a multivariate location parameter. *J. Amer. Statist. Assoc.* **67**, 641-646.

Solomon, D. L. (1972b). Γ-minimax estimation of a scalemeter. *J. Amer. Statist. Assoc.* **67**, 647-649.

Spruill, M. C. (1986). Some approximate restricted Bayes estimators of a normal mean. *Statist. Dec.* **4**, 337-351.

Sriram, T. N. and Bose, A. (1988). Sequential shrinkage estimation in the general linear model. *Seq. Anal.* **7**, 149-163.

Srivastava, M. S. and Bilodeau, M. (1989). Stein estimation under elliptical distributions. *J. Mult. Anal.* **28**, 247-259.

Staudte, R. G., Jr. (1971). A characterization of invariant loss functions. *Ann. Math. Statist.* **42**, 1322-1327.

Staudte, R. G. and Sheather, S. J. (1990). *Robust Estimation and Testing*. New York: Wiley.

Stefanov, V. T. (1990). A note on the attainment of the Cramér-Rao bound in the sequential case. *Seq. Anal.* **9**, 327-334.

Stein, C. (1950). Unbiased estimates of minimum variance. *Ann. Math. Statist.* **21**, 406-415.

Stein, C. (1955). A necessary and sufficient condition for admissibility. *Ann. Math. Statist.* **26**, 518-522.

Stein, C. (1956a). Efficient nonparametric testing and estimation. *Proc. Third Berkeley Symp. Math. Statist. Prob.* **1**, University of California Press, 187-195.

Stein, C. (1956b). Inadmissibility of the usual estimator for the mean of a multivariate distribution. *Proc. Third Berkeley Symp. Math. Statist. Prob.* **1**, University of California Press, 197-206.

Stein, C. (1959). The admissibility of Pitman's estimator for a single location parameter. *Ann. Math. Statist.* **30**, 970-979.

Stein, C. (1962). Confidence sets for the mean of a multivariate normal distribution. *J. Roy. Statist. Soc. Ser. B* **24**, 265-296.

Stein, C. (1964). Inadmissibility of the usual estimator for the variance of a normal distribution with unknown mean. *Ann. Inst. Statist. Math.* **16**, 155-160.

Stein, C. (1965). Approximation of improper prior measures by prior probability measures. In *Bernoulli, Bayes, Laplace Anniversary Volume*. New York: Springer-Verlag.

Stein, C. (1973). Estimation of the mean of a multivariate distribution. *Proc. Prague Symp. on Asymptotic Statistics*, pp. 345-381.

Stein, C. (1981). Estimation of the mean of a multivariate normal distribution. *Ann. Statist.* **9**, 1135-1151.

Steinhaus, H. (1957). The problem of estimation. *Ann. Math. Statist.* **28**, 633-648.

Stigler, S. M. (1973). Laplace, Fisher, and the discovery of the concept of sufficiency. *Biometrika* **60**, 439-445.

Stigler, S. M. (1980). An Edgeworth Curiosum. *Ann. Statist.* **8**, 931-934.

Stigler, S. M. (1981). Gauss and the invention of least squares. *Ann. Statist.* **9**, 465-474.

Stigler, S. (1983). Who discovered Bayes's theorem? *Amer. Statist.* **37**, 290-296.

Stigler, S. (1986). *The History of Statistics: The Measurement of Uncertainty before 1900.* Cambridge, MA: Harvard University Press.

Stigler, S. (1990). A Galtonian perspective on shrinkage estimators. *Statist. Sci.* **5**, 147-155.

Stone, C. J. (1974). Asymptotic properties of estimators of a location parameter. *Ann. Statist.* **2**, 1127-1137.

Stone, M. (1965). Right Haar measure for convergence in probability to quasi posterior distributions. *Ann. Math. Statist.* **36**, 440-453.

Stone, M. (1967). Generalized Bayes decision functions, admissibility and the exponential family. *Ann. Math. Statist.* **38**, 818-822.

Stone, M. (1970). Necessary and sufficient conditions for convergence in probability to invariant posterior distributions. *Ann. Math. Statist.* **41**, 1349-1353.

Stone, M. (1976). Strong inconsistency from uniform priors (with discussion). *J. Amer. Statist. Assoc.* **71**, 114-125.

Stone, M. and Springer, B. G. F. (1965). A paradox involving quasi prior distributions. *Biometrika* **59**, 623-627.

Stone, M. and von Randow, R. (1968). Statistically inspired conditions on the group structure of invariant experiments. *Zeitschr. Wahrsch. Verw. Geb.* **10**, 70-80.

Strasser, H. (1981). Consistency of maximum likelihood and Bayes estimates. *Ann. Statist.* **9**, 1107-1113.

Strasser, H. (1985). *Mathematical Theory of Statistics.* New York: DeGruyter.

Strawderman, W. E. (1971). Proper Bayes minimax estimators of the multivariate normal mean. *Ann. Math. Statist.* **42**, 385-388.

Strawderman, W. E. (1973). Proper Bayes minimax estimators of the multivariate normal mean vector for the case of common unknown variances. *Ann. Statist.* **1**, 1189-1194.

Strawderman, W. E. (1974). Minimax estimation of location parameters for certain spherically symmetric distributions. *J. Mult. Anal.* **4**, 255-264.

Strawderman, W. E. (1992). The James-Stein estimator as an empirical Bayes estimator for an arbitrary location family. *Bayesian Statist.* **4**, 821-824.

Strawderman, W. E and Cohen, A. (1971). Admissibility of estimators of the mean vector of a multivariate normal distribution with quadratic loss. *Ann. Math. Statist.* **42**, 270-296.

Stuart, A. (1958). Note 129: Iterative solutions of likelihood equations. *Biometrics* **14**, 128-130.

Stuart, A. and Ord, J. K. (1987). *Kendall' s Advanced Theory of Statistics, Volume I, Fifth Edition.* New York: Oxford University Press.

Stuart, A. and Ord, J. K. (1991). *Kendall' s Advanced Theory of Statistics, Volume II, Fifth Edition.* New York: Oxford University Press.

Sundberg, R. (1974). Maximum likelihood theory for incomplete data from an exponential family. *Scand. J. Statist.* **2**, 49-58.

Sundberg, R. (1976). An iterative method for solution of the likelihood equations for incomplete data from exponential families. *Comm. Statist. B* **5**, 55-64.

Susarla, V. (1982). Empirical Bayes theory. In *Encyclopedia of Statistical Sciences* **2** (S. Kotz, N. L. Johnson, and C. B. Read, eds.). New York: Wiley.

Tan, W. Y. and Chang, W. C. (1972). Comparisons of method of moments and method of maximum likelihood in estimating parameters of a mixture of two normal densities. *J. Amer. Statist. Assoc.* **67**, 702-708.

Tan, M. and Gleser, L. J. (1992). Minimax estimators for location vectors in elliptical distributions with unknown scale parameter and its application to variance reduction in simulation. *Ann. Inst. Statist. Math.* **44**, 537-550.

Tanner, M. A. (1996). *Tools for Statistical Inference, Third edition.* New York: Springer-Verlag.

Tanner, M. A. and Wong, W. (1987). The calculation of posterior distributions by data augmentation (with discussion). *J. Amer. Statist. Assoc.* **82**, 528-550.

Tate, R. F. and Goen, R. L. (1958). Minimum variance unbiased estimation for a truncated Poisson distribution. *Ann. Math. Statist.* **29**, 755-765.

Taylor, W. F. (1953). Distance functions and regular best asymptotically normal estimates. *Ann. Math. Statist.* **24**, 85-92.

Thompson, J. R. (1968a). Some shrinkage techniques for estimating the mean. *J. Amer. Statist. Assoc.* **63**, 113-122.

Thompson, J. R. (1968b). Accuracy borrowing in the estimation of the mean by shrinking to an interval. *J. Amer. Statist. Assoc.* **63**, 953-963.

Thompson, W. A., Jr. (1962). The problem of negative estimates of variance components. *Ann. Math. Statist.* **33**, 273-289.

Thorburn, D. (1976). Some asymptotic properties of jackknife statistics. *Biometrika* **63** 305-313.

Tiao, G. C. and Tan, W. Y. (1965). Bayesian analysis of random effects models in analysis of variance, I. Posterior distribution of variance components. *Biometrika* **52**, 37-53.

Tibshirani, R. (1989). Noninformative priors for one parameter of many. *Biometrika* **76**, 604-608.

Tierney, L. (1994). Markov chains for exploring posterior distributions (with discussion). *Ann. Statist.* **22**, 1701-1762.

Tierney, L. and Kadane, J. B. (1986). Accurate approximations for posterior moments and marginal densities. *J. Amer. Statist. Assoc.* **81**, 82-86.

Tierney, L., Kass, R. E., and Kadane, J. B. (1989). Fully exponential Laplace approximations to expectations and variances of nonpositive functions. *J. Amer. Statist. Assoc.* **84**, 710-716.

Titterington, D. M., Smith, A. F. M., and Makov, U. E. (1985). *Statistical Analysis of Finite Mixture Distributions*. New York: Wiley.

Trybula, S. (1958). Some problems of simultaneous minimax estimation. *Ann. Math. Statist.* **29**, 245-253.

Tseng, Y. and Brown, L. D. (1997). Good exact confidence sets and minimax estimators for the mean vector of a multivariate normal distribution. *Ann. Statist.* **25**, 2228-2258.

Tsui, K-W. (1979a). Multiparameter estimation of discrete exponential distributions. *Can. J. Statist.* **7**, 193-200.

Tsui, K-W. (1979b). Estimation of Poisson means under weighted squared error loss. *Can. J. Statist.* **7**, 201-204.

Tsui, K-W. (1984). Robustness of Clevenson-Zidek estimators. *J. Amer. Statist. Assoc.* **79**, 152-157.

Tsui, K-W. (1986). Further developments on the robustness of Clevenson-Zidek estimators. *J. Amer. Statist. Assoc.* **81**, 176-180.

Tukey, J. W. (1958). Bias and confidence in not quite large samples. *Ann. Math. Statist.* **29**, 614.

Tukey, J. W. (1960). A survey of sampling from contaminated distributions. In *Contributions to Probability and Statistics* (I. Olkin, ed.). Stanford, CA: Stanford University Press.

Tweedie, M. C. K. (1947). Functions of a statistical variate with given means, with special reference to Laplacian distributions. *Proc. Camb. Phil. Soc.* **43**, 41-49.

Tweedie, M. C. K. (1957). Statistical properties of the inverse Gaussian distribution. *Ann. Math. Statist.* **28**, 362.

Unni, K. (1978). The theory of estimation in algebraic and analytical exponential families with applications to variance components models. PhD. Thesis, Indian Statistical Institute, Calcutta, India.

Unni, K (1981). A note on a theorem of A. Kagan. *Sankhya* **43**, 366-370.

Van Rysin, J. and Susarla, V. (1977). On the empirical Bayes approach to multiple decision problems. *Ann. Statist.* **5**, 172-181.

Varde, S. D. and Sathe. Y. S. (1969). Minimum variance unbiased estimation of reliability for the truncated exponential distribution. *Technometrics* **11**, 609-612.

Verhagen, A. M. W. (1961). The estimation of regression and error-scale parameters when the joint distribution of the errors is of any continuous form and known apart from a scale parameter. *Biometrika* **48**, 125-132.

Vidakovic, B. and DasGupta, A. (1994). Efficiency of linear rules for estimating a bounded normal mean. *Sankhya Series A* **58**, 81-100.

Villegas, C. (1990). Bayesian inference in models with Euclidean structures. *J. Amer. Statist. Assoc.* **85**, 1159-1164.

Vincze, I. (1992). On nonparametric Cramér-Rao inequalities. *Order Statistics and Nonparametrics* (P. K. Sen and I. A. Salaman, eds.). Elsevier: North Holland, 439-454.

von Mises, R. (1931). *Wahrscheinlichkeitsrecheung*. Franz Deutiche: Leipzig.

von Mises, R. (1936) Les lois de probabilitité pour les functions statistiques. *Ann. Inst. Henri Poincaré* **6**, 185-212.

von Mises, R. (1947). On the asymptotic distribution of differentiable statistical functions. *Ann. Math. Statist.* **18**, 309-348.

Wald, A. (1939). Contributions to the theory of statistical estimation and hypothesis testing. *Ann. Math. Statist.* **10**, 299-326.

Wald, A. (1949). Note on the consistency of the maximum likelihood estimate. *Ann. Math. Statist.* **20**, 595-601.

Wald, A. (1950). *Statistical Decision Functions.* New York: Wiley.

Wand, M.P. and Jones, M.C. (1995). *Kernel Smoothing.* London: Chapman & Hall.

Wasserman, L. (1989). A robust Bayesian interpretation of the likelihood region. *Ann. Statist.* **17**, 1387-1393.

Wasserman, L. (1990). Recent methodological advances in robust Bayesian inference (with discussion). In *Bayesian Statistics 4* (J. M. Bernardo, J. O. Berger, A. P. David, and A. F. M. Smith, eds.). Oxford: Oxford University Press, pp. 483-502.

Watson, G. S. (1964). Estimation in finite populations. Unpublished report.

Watson, G. S. (1967). Linear least squares regression. *Ann. Math. Statist.* **38**, 1679-1699.

Wedderburn, R. W. M. (1974). Quasi-likelihood functions, generalized linear models, and the Gauss-Newton method. *Biometrika* **61**, 439-447.

Wedderburn, R. W. M. (1976). On the existence and uniqueness of the maximum likelihood estimates for certain generalized linear models. *Biometrika* **63**, 27-32.

Wei, G. C. G. and Tanner, M. A. (1990). A Monte Carlo implementation of the EM algorithm and the poor man's data augmentation algorithm. *J. Amer. Statist. Assoc.* **85**, 699-704.

Weiss, L. and Wolfowitz, J. (1974). *Maximum Probability Estimators and Related Topics.* New York: Springer-Verlag.

Wijsman, R. (1959). On the theory of BAN estimates. *Ann. Math. Statist.* **30**, 185-191, 1268-1270.

Wijsman, R. A. (1973). On the attainment of the Cramér-Rao lower bound. *Ann. Statist.* **1**, 538-542.

Wijsman, R. A. (1990). *Invariant Measures on Groups and Their Use in Statistics.* Hayward, CA: Institute of Mathematical Statistics

Withers, C. S. (1991). A class of multiple shrinkage estimators. *Ann. Inst. Statist. Math.* **43**, 147-156.

Wolfowitz, J. (1946). On sequential binomial estimation. *Ann. Math. Statist.* **17**, 489-493.

Wolfowitz, J. (1947). The efficiency of sequential estimates and Wald's equation for sequential processes. *Ann. Math. Statist.* **18**, 215-230.

Wolfowitz, J. (1965). Asymptotic efficiency of the maximum likelihood estimator. *Theory Prob. Applic.* **10**, 247-260.

Wong, W. (1992). On asymptotic efficiency in estimation theory. *Statistica Sinica* **2**, 47-68.

Woodrofe, M. (1972). Maximum likelihood estimation of a translation parameter of a truncated distribution. *Ann. Math. Statist.* **43**, 113-122.

Woodward, W. A. and Kelley, G. D. (1977). Minimum variance unbiased estimation of $P(Y < X)$ in the normal case. *Technometrics* **19**, 95-98.

Wu, C. F. J. (1983). On the convergence of the EM algorithm. *Ann. Statist.* **11**, 95-103.

Yamada, S. and Morimoto, H. (1992). Sufficiency. In *Current Issues in Statistical Inference: Essays in Honor of D. Basu* (M. Ghosh and P. K. Pathak, eds.). Hayward, CA: Institute of Mathematical Statistics, pp. 86-98.

Young, G. A. (1994). Bootstrap: More than a stab in the dark (with discussion). *Statist. Sci.* **9**, 382-415.

Zacks, S. (1966). Unbiased estimation of the common mean of two normal distributions based on small samples of equal. size. *J. Amer. Statist. Assoc.* **61**, 467-476.

Zacks, S. and Even, M. (1966). The efficiencies in small samples of the maximum likelihood and best unbiased estimators of reliability functions. *J. Amer. Statist. Assoc.* **61**, 1033-1051.

Zehna, P. W. (1966). Invariance of maximum likelihood estimation. *Ann. Math. Statist.* **37**, 744.

Zellner, A. (1971). *An Introduction to Bayesian Inference in Econometrics*. New York: Wiley.

Zidek, J. V. (1970). Sufficient conditions for the admissibility under squared error loss of formal Bayes estimators. *Ann. Math. Statist.* **41**, 446-456.

Zidek, J. V. (1973). Estimating the scale parameter of the exponential distribution with unknown location. *Ann. Statist.* **1**, 264-278.

Zidek, J. V. (1978). Deriving unbiased risk estimators of multinormal mean and regression coefficient estimators using zonal polynomials. *Ann. Statist.* **6**, 769-782.

Zinzius, E. (1981). Minimaxschätzer für den Mittelwert Θ einer normalverteilten Zufallsgröße mit bekannter Varianz bei vorgegebener oberer und unterer Schranke für Θ. *Math Operationsforsch. Statist., Ser. Statistics* **12**, 551-557.

Zinzius, E. (1982). Least favorable distributions for single parameter point estimation (German). *Metrika* **29**, 115-128.

Author Index

Subject Index

Springer Texts in Statistics *(continued from page ii)*

Madansky: Prescriptions for Working Statisticians
McPherson: Applying and Interpreting Statistics: A Comprehensive Guide, Second Edition
Mueller: Basic Principles of Structural Equation Modeling: An Introduction to LISREL and EQS
Nguyen and Rogers: Fundamentals of Mathematical Statistics: Volume I: Probability for Statistics
Nguyen and Rogers: Fundamentals of Mathematical Statistics: Volume II: Statistical Inference
Noether: Introduction to Statistics: The Nonparametric Way
Nolan and Speed: Stat Labs: Mathematical Statistics Through Applications
Peters: Counting for Something: Statistical Principles and Personalities
Pfeiffer: Probability for Applications
Pitman: Probability
Rawlings, Pantula and Dickey: Applied Regression Analysis
Robert: The Bayesian Choice: From Decision-Theoretic Foundations to Computational Implementation, Second Edition
Robert and Casella: Monte Carlo Statistical Methods
Rose and Smith: Mathematical Statistics with *Mathematica*
Santner and Duffy: The Statistical Analysis of Discrete Data
Saville and Wood: Statistical Methods: The Geometric Approach
Sen and Srivastava: Regression Analysis: Theory, Methods, and Applications
Shao: Mathematical Statistics, Second Edition
Shorack: Probability for Statisticians
Shumway and Stoffer: Time Series Analysis and Its Applications
Simonoff: Analyzing Categorical Data
Terrell: Mathematical Statistics: A Unified Introduction
Timm: Applied Multivariate Analysis
Toutenburg: Statistical Analysis of Designed Experiments, Second Edition
Whittle: Probability via Expectation, Fourth Edition
Zacks: Introduction to Reliability Analysis: Probability Models and Statistical Methods

Printed in the United States
By Bookmasters